Integrated Pest and Disease Management

Edited by :
RAJEEV K. UPADHYAY
K.G. MUKERJI
B.P. CHAMOLA
O.P. DUBEY

A.P.H. PUBLISHING CORPORATION
Ansari Road, Darya Ganj
New Delhi-110 002

Published by
S.B. Nangia
A P H Publishing Corporation
4435-36/7, Ansari Road, Daryaganj
New Delhi 110002
Ph.: 23274050
E-mail : aphbooks@gmail.com

2026

Rs. 5995/-

Printed at
Balaji Offset
Navin Shahdara, Delhi 110032

Contents

1 Integrated Pest Management : Concepts and Prospects

Rajeev K. Upadhyay, K.G. Mukerji, B.P. Chamola and O.P. Dubey

CONTENTS

Integrated Pest and Disease Management
Rajeev K. Upadhyay, K.G. Mukerji, B.P. Chamola and O.P. Dubey (eds.)
APH Publishing Corporation, New Delhi-110002, India

ABSTRACT

A relook has been made to define and describe Integrated Pest Management (IPM) and their components including cultural, biological, host resistance, regulatory and chemical controls. Future prospects and challenges of IPM have been discussed in view of economic, social, environmental and biotechnological perspectives.

1. INTRODUCTION

Despite large increases in agricultural production over the past 50 years, our search for food remains important as escalating demands are placed on the earth's resources. It has been estimated that the world's population will increase from the present level of 5.5 billion to between 9 and 15 billion by the middle of the next century. The demand for food over the hext few decades will rise two to four times of anticipated increases in population and living standards occur. There is very limited scope left for increasing food production through horizontal enlargement in the agricultural land but through vertical expansion only primarily due to populations increase and land requirement for other than agriculture. Food Production per unit land is limited by many factors including organisms (pests) that feed on or compete with food plants. Food plants of the world are damaged by 100,000 microbial pathogens (like fungi, viruses, bacteria, and other microorganism), 10,000 species of insects, 1000 species of nematodes, and 30,000 species of weeds. It has been estimated that, with current pest control technologies, yields are 30 to 35% less (more than US$ 300 billions annually) on average than they would be in the absence of pests. Additional losses of 10 to 20% occur after harvest (14). Losses each year in grain crops alone are estimated to exceed the gross and net food grain deficits of the developing world. Pests not only cause quantitative losses, but also effect the quality of agricultural produce which may become noticeable through a deterioration of taste, flavour, fibre strength, processing characteristics, seed germination etc. It is essential for humans to take action to limit the destruction to food crops by pests so as to protect the security of the world food supply. Infect, any progress in agricultural production can be made only if measures aiming at an expansion and intensification of agricultural production are simultaneously supported by an adequate system of integrated crop protection and pesticide management (9).

Since ages, tillage, hand picking, mechanical barriers, crop rotation, removal of affected plants and crop residue, burning or burying crop residue, and eliminating plants harbouring pests provided some control of insects and plant diseases which remained unsatisfactory until the introduction of commercial chemical pesticides and scientific plant breeding. Tools now used to control plant pests include genetic resistance to pests in plant, safe and judicious use of pesticides, quarantines and regulatory controls, biological control biotechnological approaches and a wide diversity of cultural practices (17).

Pesticides are basically poisons and help or harm us depending on whether they are used properly or improperly. With the introduction of first commercial pesticide, arsenical compound (in 1867) followed by Bordeaux mixture (in 1885), lime sulfur (in 1880), lead arsenate (in 1892), thiram (in 1931), DDT (in 1939), and 2, 4-D (in 1942), heavy reliance was placed on chemicals to control pests in 1940s, 50s and 60s which replaced most of the other pest control practices. This led to the overuse and misuse of pesticides and the emergence of several social, environmental, health. economic and new pest problems.

2. DEFINITION OF IPM

In view of environmental and health concerns, a determined effort was made to reduce and rationalize the use of pesticides and to manage pests more effectively which led to the emergence of a new discipline called IPM (1, 2, 4,). In the classical version of IPM, a pesticide is applied only when insect populations reach a level that would lead to economic losses in the crop. In other words classical IPM concentrates on reducing the destructiveness of pests primarily by reducing their numbers with pesticides. However, IPM has been defined by the Panel of Experts on Integrated Pest Management, Rome (17), as : "A pest management system that, in the context of the associated environment and the population dynamics of the pest species, utilises all suitable techniques and methods in as a compatible manner as possible and maintains the pest population at levels below those causing economic injury". IPM in the broad sense has been defined as "the optimization of pest control in an economically and ecologically sound manner, accomplished by the coordinated use of multiple tactics to assure stable crop production and to maintain pest damage below the economic injury level while minimizing hazards to humans, animals, plants and the environment". Moreover, a comprehensive IPM places emphasis on the plant rather than the pest or a particular method of pest control and recognize that the health of the plant depends on the interaction of the plant with pests, with a multitude of other organisms, and with the physical and chemical environment in which the plant is growing (9). Thus IPM moves beyond a one plant-one pest-one control view of pest control to an integrated view of plant health as the resultant of complex interactions between the plant, other organisms, and the physical and chemical environment. Moreover, the basic concern of IPM is with designing and implementing pest management practices that meet the goals of farmers, consumers and government in reducing pest losses while at the same time, safeguarding against the longer term risks of environmental pollution, hazard to human health and reduced agricultural sustainability (17).

3. COMPONENTS OF IPM

Some pest management practices have a preventive character; other, a curative effect. Therefore, one need to learn the advantages and disadvantages of all components which might be applied in an IPM system in order to make an appropriate choice (13).

3.1. Cultural control

Historically, cultural practices have been the most important method for controlling pests and preventing crop losses. Cultural and crop management practices provide means to manipulate the environment to make it less favourable for pests, whilst maintaining the best possible conditions for high productivity, thereby achieving economic control of pests or reducing their rate of increase and damage. Although cultural control is often associated only with mechanical operations, such as tillage or burning, it involves many aspects of crop and soil management, inclduing crop rotation, time of planting and harvesting, trap cropping system diversification (9,11). Cultural control practices may inflict direct mortality on pest species, but these practices more often are oriented at prevention of pest outbreaks rather than control of an existing pest infestation. When cultural control practices are integrated easily with other cultural practices, they are usually readily adopted by growers.

3.1.1. Crop rotation and intercropping

Crop rotation and fallowing systems offer numerous advantages in soil structure, fertility and erosion management, as well as aiding in control of various pest species. The principle of rotating cropping pattern is to interrupt the specific relationship between pests and host plants which favours or limits the development of these damaging organisms (13). Crop rotation for pest management consists of a planting pattern alternating susceptible and nonsusceptible crops. The interval necessary between susceptible crops depends upon the length of the pest's life cycle, reproductive potential, degree of specificity and dispersal characteristics of the target pest. This approach is most useful for fairly immobile, soil dwelling pest species such as rootworms, and also those pests with a restricted host range or a life cycle of one year or more such as wireworms. In cotton, rotation sequences of crops such as cotton/sorghum/cotton, cotton/grain/cotton, and cotton/sorghum/grain aid in reducing propagules of soil pathogens. Most pathogens of cotton do not attack sorghum and other grain crops.

The growing of two or more crops on the same plot or field is a common practice in subsistence farming of many tropical countries. The mixed planting of food crops, vegetables and fruit trees often provides relatively high yields owing to the optimum use of soil, nutrients, water, space and other factors (13). Interplanting of trap crops induce on a large scale the development of parasites without giving them a chance to complete their life cycle because of resistance, or because the trap crop concerned is destroyed by man to save the main crop.

3.1.2. Planting and harvesting dates

One of the oldest method used to avoid excessive pest damage is manipulation of planting and harvesting dates. Planting at the optimum time for rapid seed germination and seedling growth is important to avoid conditions that will tend to delay crop development at any stage (9). Planting before or after certain dates can disrupt a synchronised crop-pest association and enable the plant to escape damage from pests during a susceptible growth stage. Careful selection of planting date makes it possible to avoid the egg-laying period of a particular insect, and allow young plants to establish a tolerant stage before the attack occurs. This practice allows a shorter period of susceptibility during which the insect may attack; the crop may reach a mature stage before a certain pest becomes abundant. Many cotton insect pests, especially those that attack fruit, cause greater damage during later parts of the growing season. Therefore, the tactic of delaying the planting date on an area-wise basis has been quite successful in minimising damage. A classical example is the delayed planting of wheat as a means of controlling the Hessian fly. In Virginia soybean production, early-planted corn and soybeans are far less susceptible to corn earworm damage than are the late-planted crops. Early harvest of alfalfa can often be used to control both alfalfa weevil and the potato leafhopper (8). The harvesting process, particularly where the hay is chopped for haylage, destroys most weevil larvae and is an effective control. Early harvest is also used to minimize potato leaf hopper damage.

3.1.3. Solarization and phytosanitation

Covering the soils with plastic sheets for the purpose of raising temperature and thereby killing noxious soil organisms is called solarization (15). Mulching with vegetative matter or polythene film (plastic) plays a very important role in reducing evaporation of soil, humidity, furthering humification processes and beneficial soil organisms and in suppressing weed growth (13).

Phytosanitation includes measures aimed at the removal and destruction of infected plant material as the potential source from which the pest or disease could spread collecting or buring crop debris and stubble from the field removes many diapausing larvale, eggs and pathogens. Eradication of weeds deprives many noxious organisms of their intermediate hosts, and rodents of food and shelter. Further more, cleaning and disinfecting planting tools and machines, containers and soil, are necessary measures to prevent the spreading pathogens, nematodes and weeds.

3.2. Quarantine and regulatory control

The introduction of new pests and diseases from sources of infestation within or outside a country of often caused by the movement to people, commodities and equipment carrying the contaminant. Quarantine and containment law or regulations and other legislations have been enacted and forced in many countries to help overcome this problem. Plants and plant products to be imported into a country must be free from pests and diseases; and plants with the potential of becoming weeds must be strictly prohibited. Other examples of regulatory control are the prohibition to grow certain trees, plants or crops which are intermediate hosts to important pests or diseases of the economically most important crops of an area or the compulsory arrangement of crop rotation sequences. Governments are responsible for overseeing such regulatory control programmes which include eradication, containment and suppression.

The plant quarantine regulations the world over aim at enforcing legal provisions on the movement of plants/plant materials between countries (Foreign quarantine) and between states within the country (Domestic quarantine) to check the infiltration/ spread of exotic pests/ diseases. The International Conference on Phytopathology held in Rome in 1914 recommended several quarantine measures to preclude the spread of pests and diseases across national boundaries. In consonance with those recommendations the Government of India enacted "Destructive Insects and Pests Act, 1914 (DIP Act, 1914). The plants/plant materials are imported into the country as per regulations enshrined under "The Plants, Fruits, and Seeds" (Regulation of Import into India) order, 1989 (PFS order) issued under DIP Act aiming at preventing the introduction of exotic pests/diseases and weeds into the country (17).

3.3. Biological control

One of the most successful, non-chemical approaches to pest management is that of biological control which indicates the use of control agents in which the active principle is living organism or virus for regulating the incidence of pests or pathogens (13, 18). Biological control envisages use of parasites (specific natural enemies), predators (non-specific natural enemies), the sterile male method, micro-organisms and sex attractants. Numerous organisms exist in nature that feed upon, or infect insect pests, pathogens, and weeds. Collectively these organisms provide a significant level of 'natural control', in many cases preventing many inset species from ever reaching the status of pests. The importance of natural control is demonstrated when natural enemy populations are destroyed by insecticides and a previously unimportant insect suddenly escapes from natural control and becomes a major pest. The cabbage looper is an example of this phenomenon. Thus natural control is defined as suppression, maintenance or regulation of organisms by natural enemies sufficient to maintain pests below the action level (economically damaging levels) is the definition of an inaction level (3). The establishment of inaction levels for natural enemies is suggested as one of the first steps in optimising pest management decision so that the impact of mortality factors that impinge on pests can be considered. Management decisions based solely on the abundance of pests can result in unneeded expenses for chemical insecticides and harm to the environment.

Biological control is defined as applied natural control wherein man intervenes to improve the efficacy of natural enemies including parasites, predators, and pathogens of pest species by introductions, conservation, or augmentation to maintain pest populations below economically injurious levels (18). Some natural enemies become permanent established *viz.*, cassava mealybug is controlled for a long term by inoculative introductions of the few wasps *E. lopez*. The inoculative method of biological control is most successful in perennial crops, orchards and forests or those crops which are practically always available in the field such as cassava and ratoon sugarcane. However, some natural enemies need inundative introductions in large numbers due to their short time survival and effectiveness. Inundative biological control aims at directly increasing mortality in a pest population that has already built up in a crop, whereby the released natural enemy is used as a biological insecticide. For instance, for the control of caterpillars on crops such as cabbage, sometimes a release of more than 100,000 *Trichogramma* wasps per hectare is required in split dosages (7).

Augmentations of natural enemies through rearing and mass release have been used extensively for pest control, including release of predatory phytoseid mites for control of spider mites in strawberries (12), mass release of coccinellid predator for control of avocado brown mite (10), and control of bollworm and budworm in cotton through release of green lace wing larvae (16).

Recent researches using entomogenous nematodes in the genera *Steinernema (Neoplectana)* and *Heterorhabditis* have shown promise for control of several insect pests, particularly those living in the soil (6). *Romanomermis* spp. are effective against insects of the order Diptera (flies, gall midges, leaf miners) and *Neoplectana* spp. against caterpillars of the order Lepidoptera and termites.

Biological control by means of entomopathogens and other microbial pest control agents involves the application of micro-organisms on to the crop for ingestion by insect pests or directly on to the noxious insects, fungus or weed with the objective of destroying them. These biocontrol agents include bacteria, protozoa, fungi, viruses and nematodes. Although these microbials probably play a role in pest control, their density is not generally monitored to assist in making pest management decisions. Microbials such as *Bacillus thuringiensis* and the nuclear polyhedrosis virus, applied to the field as insecticides, have been used in the management of *Heliothis* spp. in many areas of the world. The major advantage of using microbial pathogens to control pests is that they inflict little or no harm to other natural enemies. Fungi have long been recognized as having the potential of killing insects. *Bauveria bassiana* is reportedly used in the USSR for the control of Colorado beetles in potato fields.

Pheromones are substances emitted by insects and mites for communication between individuals of the same species. In biological control, sexpheromones are used for trapping male adults and for communication disruption. Juvenile hormones are substances that regulate development in insects and that may prevent insects from reaching maturity. These substances can also be used as biochemical pest control agents. Growth regulators are natural chemical substances in plants and animals that regulate process of growth and any serious disturbance may obviously interfere with process of hatching, molting and pupation in insects, thereby allowing their use for the control of caterpillars, bugs, mosquitoes, flies and ants (13).

Since ancient times, man has practiced biological control of plant pathogens through cultural practices which provided for the biological destruction and/or suppression of disease organisms. Biological control aimed directly at the pathogen or mediated through adjustments in the host offers unlimited opportunities to reduce losses caused by biotic and abiotic stresses. Biological control of plant pathogens seeks a solution in terms of restoring and maintaining the biological balance within the ecosystem and must be considered part of modern agriculture. Biological control offers a powerful means to improve the health and hence the productivity of plants by suppression or destruction of pathogen inoculum, protection of plants against infection, or increasing the ability of plants to resist pathogens.

3.4. Host-plant resistance and biotechnology

Crop breeding and selection has been an important element in the evolution of Agriculture since pre-historic times. The choice of cultivar probably is the single most important management decision the grower makes in an integrated crop management system. The cultivar sets the framework for the level of susceptibility to pests, the tactics applied to manage the crop, and production costs. Resistant cultivars provide the cornerstone for a successful IPM system. Resistant cultivars, even those with low to moderate levels of resistance, are highly compatible with all other control tactics; they contribute stability and offer advantage to IPM. Resistance may either be a contributing factor or the primary means for controlling pests. Genetic resistance is most likely to be used in relation to other pest control measures, which include cultural, biological, and chemical approaches. Resistant cultivars may not require as many treatments or as high rates of pesticide application to achieve adequate pest control. This results in reduced production cost and risk, and increased profits. Genetic resistance is recognised as the most effective, economical, and reliable means of maintaining healthy plants and reducing crop losses.

Development and use of pest resistant varieties has had a tremendous impact on global agricultural production. Recent advances in genetic engineering (Biotechnology) have raised hopes of greatly accelerating classical crop breeding efforts, as well as incorporating new resistance mechanisms. Particular significance to pest management is the ability of genetic engineering to overcome the narrow taxonomic restrictions on the sources of genes that can be transferred to a given crop. This kind of genetic manipulation can be achieved by means of recombinant-DNA technology or somatic cell fusion. r-DNA technology offers, in principle, the possibility of overcoming incompatibilities between species and to pave the way for the development of genetic manipulation of micro-organisms, cells and eukaryotic tissues (13). The successful examples of transfer of a toxin producing gene from a bacterium (*Bacillus thuringiensis*) to tobacco (5) using r-DNA technology has opened a new era in pest control technologies. Genetically engineered pest resistance in crops could offer powerful economic and environmental incentives for large-scale adoption by growers.

In vitro regeneration techniques offer the possibility to grow complete plants from plant-cells, calluses and plant tissue fragments; this technique is used for the vegetative production of disease-free plants from plants infected by viruses, fungi or bacteria (13).

3.5. Pesticides

These control agents have been, are now and will continue to be, in the foreseeable future basic tools in pest management. They provide a dependable, rapid, effective and economical means of controlling whole complexes of crop pests. Pesticides are applied either as a preventive treatment to protect crops or stored products against infection but diseases or infestation by animal pests, or as a curative treatment to destroy or limit population development of noxious organisms. It is, however, essential that the reasons for using pesticides and the consequences of misusing them be carefully analysed in order to obtain maximum benefit from their application, while at the same time preventing and remedying their possible hazardous effects on non-target organisms and the environment. The proper, restrained and further refined use of pesticides must be a major objective of any integrated pest management strategy in order to delay resistance and to avoid harmful interactions with biological control agents and the environment. Insecticides which have a specific or systemic action and a short residual activity, or which are stomach instead of contact poisons, are less harmful to natural enemies. Also some formulations such as granules and baits, or methods of application such as spot treatments, adding pheromones or trunk injection, are relatively safe for natural enemies and thus fit better in an integrated control approach than, for instance, blanket spraying.

4. STATUS AND IMPACT OF IPM IN INDIA

IPM programmes are being led by the Directorate of Plant Protection, Quarantine & Storage, Government of India, Ministry of Agriculture for the developmental activities, Indian Council of Agricultural Research and State Agricultural Universities for research, education and training activities and State Departments of Agriculture for extension work. Based on the experiences gained through biocontrol programme during Vth and VIth plans, survey and surveillance of insect pests, diseases weeds and natural enemies during VIth and VIIth Plan, and IPM demonstrations at farmer's field during VIth Plan, IPM was adopted as core of plant protection activities during VIIth plan. During VIIIth plan (1992-93 to 1996-97), IPM gained a momentum and attained the status of social movement under the ambit of total crop production programme, particularly with the help of externally aided projects sponsored by FAO, ADB and UNDP and also by huge domestic plan allocation to the tune of 45 crore.

Since 1993, much emphasis is being given to human resource development in IPM technology through rigorous field oriented training of state extension functionaries and farmers. Training programmes envisage their tier approach comprising of (i) residential training to master trainers through season long training (SLT), (ii) training of agriculture extension officers (AEOs) and farmers through establishment of farmers' field schools (FFSs) and (iii) popularisation of IPM practices among the farmers through conduct of IPM demonstrations. During 1994-95 and 1995-96, seventeen SLTs were conducted wherein 567 master trainers were trained in IPM of cotton, rice, vegetables, oilseeds and pulses all over the country. Apart from this, so far 2592 FFSs were established wherein 12,245 AEOs and 77,556 farmers were trained in IPM technology including agro-ecosystem analysis. Over and above, demonstrations of field tested IPM package of practices are being undertaken at the farmers' field to train farmers. During 1996-97, 8 SLTs have been proposed to train 320 master trainer, and 2100 FFSs to train 10,500 AEOs and 63,000 farmers. Working group for agriculture (inputs-pesticides) for IXth plan in principle has recommended for giving major emphasis to IPM under crop production programme for the safety of human and environmental health as well as to restore biodiversity and to reduce cost of cultivation.

Introduction of IPM has brought following changes in the approach and strategy of scientists, technocrats, farmers, and policy makers with regards to crop protection.

1. **Policy change :** Govt, has withdrawn all kinds of subsidy and encouragement aimed for the enhanced use of pesticides. Instead, Government is promoting exploitation of natural biocontrol potential, use of resistant/tolerant varieties and cultural operations so as to minimise pesticide consumption. During VIIIth plan an amount of Rs. 45 crores have been earmarked for IPM which includes 15 crores for the establishment of new state biocontrol laboratories. Besides, a lot of money is being funded by the Department of Biotechnology and ICAR for accelerating the research activities on different components of IPM, particularly other than application of pesticides.

2. **Awareness towards health of environment and man :** Recent upsurge insecticide related social environmental and economic problems forced society at large and farmers and pest managers specifically, to relook and to rethink about their decision to use pesticide. Now, advanced farmers in villages are aware of food chain getting contaminated with pesticides and their impact on human health. As a result, farmers have started taking utmost care while making pest management decision and go for pesticide use only when other options are already exhausted.

3. **Decline in pesticide consumption :** There is substantial reduction in pesticide consumption which declined from 75033 MT (1990-91) to 61357 MT (1994-95). During 1995-96, the pesticides consumption is anticipated to be around 60,000 MT. In Punjab, AP and TN, farmers used to apply 10-20 sprays in cotton which has come down to 5-7 sprays due to adoption of IPM strategy.

4. **Banning of hazardous pesticides :** Contamination of ecosystem with pesticides and associated hazards to human health, environment and deleterious effects on natural pest biocontrol agents inspired Government to ban already registered but most hazardous pesticides. Consequently 20 pesticides have already been banned, 18 pesticides have been refused registration, 25 are under review and 11 are allowed for restricted use only.

5. **Promotion of biopesticides :** Microbial pesticides and pesticides of plant origin are compatible with the IPM programme and are being popularised by the Government and NGOs amongst farmers. Government has liberalised registration process while considering grant of registration to such pesticides. As a result, neem products, *Bacillus thuringiensis* and *B. sphaericus* have been granted registration.

5. NEW CHALLENGES AND FUTURE PROSPECTS

Crop losses vary on an average between 30 to 45% and range to 100%. The value of pesticides in protection of crops can be questioned when it is noted that estimated percentage losses in food production attributed to pests are as great today as they were 50 years ago when organic pesticides began to be widely used. Further, it has been estimated that pesticide use could be reduced by 35 to 50% without lowering crop yields or causing an appreciable increase in the price of food (14). Thus basic questions are : Do we actually need pesticides ? and Do we need to control plant pests? It seems inevitable that less pesticide will be used in the future and that greater reliance will be placed on biologically based technologies. Nevertheless, without pesticides we could not produce certain crops economically. Fundamental changes in our mind set, in definitions of pest and of damage, and in our tolerance of damage would be needed to accept immediate cessation of all pesticides use in food production. The pesticides debate will continue and pesticides will contribute to pest control for some time. The future prospect and challenges of IPM are :

1. Emphasis should be placed on the development of pesticides that are active at lower doses, more specific for the target organism, less toxic to the user, consumer, wildlife, biocontrol agents, and less persistent in the environment. Continued improvements are needed in pesticide application technology and in methods to manage or prevent pesticide resistance in pests. Thresholds need relook and further exploration on the basis of biologic, economic, or psychologic considerations.

2. Research must continue on genetic approaches to pest resistance in plants which is facilitated by preserving land races having better resistance gene pool as source of resistance and biotechnology comprising of recombinant DNA technology. These techniques result in transgenic plants with resistance to viruses, insects and herbicides. Several fungal diseases, like wheat rust, barley powdery mildew, and late blight of potato have been controlled by introducing resistant varieties bred through traditional breeding programme. However, it is important that varietal resistance must be polygenic in nature.

3. Biological control is an emerging technology to control pests, diseases and weeds. Genetic engineering will play a vital role in production of transgenic biocontrol agents having high biocontrol potential and ecological acceptability.

4. In implementing classical IPM or comprehensive IPM, improved methods for integrating information will be required. Pesticides must be integrated with other pest control technologies including genetic resistance in plants, cultural practices, biological control, and biotechnology. Technologies for the control of one group of pest must be integrate pest management with crop production. Integration of IPM must be done in the context of agroecosystem management.

5. IPM and other forms of pest management were built on the pre-existing research base. The challenge before us is to develop new science, new technology, new management skills, and new concepts of integration in order to control plant pests, protect our environment, and provide a continous supply of safe and nutritious food in abundance for a rapidly expending world population.

6. REFERENCES

1. Apple, J. L. and Smith, R. F. (eds.) 1976. Integrated Pest Management. Plenum Press, New York.

2. Flint, M. L. and van den Bosch, R. 1981. Introduction to Integrated Pest Management, Plenum Press, New York.

3. Fillman, D. A. and Sterling, W. L. 1985. Inaction levels for the red imported fire ant, *Solenopsis invicta* (Hyms Formicidal) : a predator of the boll weevil, *Anthonomus grandis* (Col : Curculionidae) Agric. Ecpsust & Environ.**13** : 93-102.

4. Glass, E. H. 1975. Integrated Pest Management : Rational, Potential, Needs and Improvement. Science **236** : 48-54.

5. Goodman, R. M., Hanptli, H., Crossway, A. and Knanf, V. C. 1987. Gene transfer in crop improvement. Science **236** : 48-54.

6. Kard, B. M., Hain, F. P. and Brooks, W. M. 1988. Field suppression of three whitegrub species (coleopetra : scarabacidae) by the entomogenous nematodes *Steinerma feltae* and *Heterorhabitis heliothis*. J. Econ. Ent. **81** : 1033-1039.

7. Lingren, P. D. 1970. Biological control-can it be effectively used in cotton production today? In : Proceedings, second Annual Texas Conference on insect, Plant diseases, Weed and Bush Control. Texas A&M University, College Station.

8. Luna, J. M. 1987. Forage crop pest management. In : Pest management guide for forages and pastures. Virginia Cooperative Extension Service, Blackburg.

9. Luna, J. M. and House, G. J. 1990. Pest management in sustainable agriculture system. In : Sustainable Agricultural system (eds. C. A. Edwards, Rattan Lal, P. Maden, R. H. Miller and G. House) St. Lucie Press, Florida, pp. 157-173.

10. McMurthy, J. A., Johnson, H. G. and Seriven G. T. 1969. Experiments to determine effects of mass releases *Stethoorus picipes* on the level of infestation of the avocado brown mite. J. Ecpm. Ent. **62** : 1216-1221.

11. Metcalf, R. L. and Luckman, W. H. 1975. Introduction to insect pest management. John Wiley and sons, New York.

12. Oatman, E. R., McMurthy, J. A. and Voth, V. 1968. Suppression of the two-spotted spider made on strawberry with mass releases of *Phytoseiulus persimilis*. J. of Economic Entomology **61** : 1517-1521.

13. Oudejans, J. H. 1991. Agro-pesticides : . roperties and Functions in Integrated crop protection, United National, Bangkok.

14. Pimental, D. (eds). 1991. CRC Handbook of Pest Management in Agriculture, 2nd edition, IIIrd volume, CRC Press, Boca Raton, FL.

15. Porter, I. J. 1992, Soil solarisation and biological control. In : Recent developments in biocontrol of plant diseases (eds. K. G. Mukerji, J. P. Tewari, D. K. Arora and Geeta Saxena) Aditya Book Pvt. Ltd., New Delhi, pp. 164-186.

16. Ridgeway, R. L. and Jones, S. L. 1969. Inundative release of *Chrysopa carnea* for control of *Heliothis* on cotton. J. Eco. Ent. **62** : 177-180.

17. Upadhyay, R. K., Rajak, R.L. and Mukerji, K. G. 1996. Integrated Pest Management. In : IPM System in Agriculture, Vol. I. Principles and Perspectives (eds. R. K. Upadhyay, K. G. Mukerji and R. L. Rajak) Aditya Books Private Ltd., New Delhi, pp. 1-23.

18. Upadhyay, R. K., Sharma O.P., Mukerji, K. G. and Dubey O. P. 1996. Biological control in IPM : Needs and opportunities. In : IPM System in agriculture, Vol. I. Principles and Perspectives (eds. R. K. Upadhyay, K. G. Mukerji and R. L. Rajak) Aditya Books Pvt. Ltd., New Delhi, pp. 400-412.

2 Major Diseases of Potato and their Management

S.M. Paul Khurana, I.D. Garg, B.P. Singh and A.V. Gadewar

CONTENTS

Integrated Pest and Disease Management
Rajeev K. Upadhyay, K.G. Mukerji, B.P. Chamola and O.P. Dubey (eds.)
APH Publishing Corporation, New Delhi-110002, India

ABSTRACT

The potato, a nutritive food crop, now grown in many parts of the world and next only to cereals in importance, was moved out of its native home in the Andes (South America) only during 15th Century. Though the list of pathogens infecting potato has remained almost unchanged over many decades, there has been steady expansion in the area under potato and improvement in tuber yields/ productivity. Obviously it has been made possible mainly by new cultivars and better disease management practices based on research in areas like breeding for disease and pest resistance, virus-detection technology for selecting virus-free stocks, monitoring of vectors responsible for virus spread, *in vitro* multiplication of virus-free stocks, breeding of lines suitable for production of true potato seeds, elimination of virus(es) through meristem tip culture and chemo-and thermotherapy, chemical control of insect vectors and fungal diseases, control of bacterial wilt through cultural practices, etc. During the last decade efforts have been made to incorporate virus resistance by incorporation of a part of the viral genome into the plant genome. Long term impact of this approach is being assessed. Areas which need attention for research include (i) interactions amongst weather, potato viruses and the vectors under field conditions to develop models for predicting the levels of virus infection(s) in the produce and for advising farmers to adopt suitable remedial measures and (ii) interactions amongst weather, potato cultivars and late blight pathogens at various locations for developing the disease forecasting systems.

1. INTRODUCTION

Commercial potato propagation is generally through the 'seed tubers'. Due to the vegetative mode of potato propagation, virus(es) infection, having once taken place, goes on increasing in intensity and incidence with a successive propagations ultimately leading to a situation where tuber yields are usually not economical. Intracellular occurrence and dependence on host cell machinery for multiplication makes the viruses very difficult agents to be controlled chemically. Therefore, they are controlled by employing various means, collectively called as integrated virus disease management. These include virus-free stocks production, exclusion, vector-avoidance during crop growth, vector control, cultural practices to minimise virus contamination/spread, elimination of the virus(es) through tissue culture and chemotherapy from valuable clones/cultivars, resistance breeding, and incorporation of virus-genome derived resistance into desired varieties through genetic engineering (203).

2. POTATO VIRUSES

Potatoes are infected by over two dozen plant viruses belonging to different virus groups (Table 1). Their important characteristics like shape, size, genome, group, variability, transmission mode, symptoms, etc. are listed in Tables 1-2. Some of these viruses are mainly associated with and are dependent on potatoes for perpetuation and spread while others can perpetuate and spread through other hosts as well (Table 3). The major types of viral diseases, caused by these viruses, may be categorised as mosaics (202, 203, 276), leafroll, necrotic spots/necrosis, etc. Mosaics and leafroll are important due to their ubiquitous distribution and tuber yield reduction. Potato viruses causing mosaic symptoms include potato viruses X, S, A, Y, V, M. Andean potato latent virus (APLV), Andean potato mottle virus (APMV), Potato aucuba mosaic virus (PAMV), Tobacco/rattle virus (TRV), Tobacco streak virus (TSV), Potato mop top virus (PMTV), Potato yellow dwarf virus (PYDV), while leafroll is caused by Potato leafroll virus (PLRV) and necrotic leaf spots & stem/petiole necrosis by a tomato spotted wilt group virus (Tospovirus) (204). Besides these important viruses, potatoes are also readily infected by Potato spindle tuber viroid (PSTVd). Mosaics under natural conditions are caused generally by a combined infection of more than one of the following described viruses depending on the seed source, region, variety, etc.

Table 1 : Characteristics of important viruses infecting potatoes

Virus (Group)	Disease caused	Particle Size/nm (RNA stands) shape	Seroaffinity/ strains	Mode of Transmi-ssion*	Host Range
PVA (Poty-virus)	Supermild mosaic	730x13 (1) flexuous	PVY/few	Aphids^{NP}	Solanaceae
PLRV (Luteo-virus)	Leafroll	24-26, isom (1)	BWYV	Aphids^{PI}	Restricted
PVM (Carla-virus)	'Leaf rolling mosaic (Para crinckle)	650x12 (1) flexuous	PVS/few	Contact	Restricted
PVS (Carla-virus)	Latent	650x12 (1) flexuous	PVM/few Car LMV	Contact	Restricted
PVX (Potex-virus)	Latent/ Mild mosaic	515x13 (1)	White clover mosaic virus	Contact	Wide
PVY (Poty-virus)	Severe/ Rugose mosaic	730x11(1)	PVA/Many	Aphids^{NP}	-do-
PVT^{Q} (Capill-virus)	Mild/ Transient mosaic	640x12 (1)	Apple stem grooving	Pollen/ seed	-do-
APLV^{Q} (Tymo-virus)	Severe mosaic	28, isom (1)	Some Cocumo-viruses/few/some	Contact/beetle seed	-do-
AMV (Alfalfa mosaic virus)	Calico/ yellow mosaic	58, 204 x 18 (3)	-	Aphids^{NP}	-do-
PAMV (Potex virus)	Aucuba mosaic	580x12 (1) flexuous	Potexvirus/ some	Contact/ aphids	Restricted
PMTV^{Q} (Furo virus) (Tobam-ovirus)	(i) Mop top (ii) TMV -PMTV	300x18 (1) Rods	TMV/some -do-	Injury/ fungi	Wide
PYDV (Rhabdovirus)	Yellow dwarf mosaic	380x75 (1) bacilliform	?/2	Sap/leaf hopper	Restricted
TSV^{Q} (Ilarvirus)	Faint Mosaic	28, isom (4)	?/some		Wide

TNV^{Q} (Necrovirus)	Necrotic mosaic	26, isom (1)	?/Few	Fungi	Wide
TRV^{Q} (Tobravirus)	Rings-pots/Leaf-reduction & "spraing"	45 x 90 $180\text{-}210^{18\text{-}25(2)}$	?/Few	Seed/ nematodes	Extensive
TSWV (Bunyavirus)	Stem-/ leaf necrosis	80 (70-90) (1) (Spheroidal)	Many	Thrips	Extensive
SALCV (?)	Apical leafcurl	17 (Triplets)**	?	(White flies)	Restricted
PSTVd (Viroid)	Spindle tuber	Small, circular free naked RNA (i)	?	Contact/ seed	Wide

NP/P = Non-persistently/Persistently aphid-borne.
HC = Helper component required for aphid transmission.
* Tuber Transmision is common for all potato viruses (except TSWV)
** The Only DNA virus infecting potatoes.

PVX : Some of the important characteristics of the viruses are listed in Table 1-2. The virus being highly contagious is readily sap and contact transmitted, but not through insect vectors. It is most widely distributed wherever potatoes are grown and often causes latent infection but also faint distinct mosaic depending on the virus strain, potato cultivar and climatic conditions. It possesses a lot of variability and four strain groups are recognised based on their reaction to resistance genes Nx and Nb (61). In addition, another strain designated PVX-HB has been reported (235) which is able to overcome all the resistance genes against PVX, and infects PVX-HB also systemically but latently infects *Gomphrena globosa* which is used as a local lesion indicator for PVX. Serologically PVX strains can be distinguished only through cross absorption of antisera and not by serological differentiation indices which differ by not more than two or three units (108, 369). A broad spectrum of potato virus X isolates collected from America, Europe, Asia and Africa could be divided into two serotypes viz PVX^{o} and PVX^{a} by using a panel of monoclonal antibodies. Strains of serotype PVX^{o} are worldwide in distribution while those of PVX^{a} are confined to Peru and Bolivia (100).

PAMV: It is another member of the potexvirus group infecting potatoes. Its important characters are given in Tables 1-2. It is readily sap-transmitted but also by the aphid (*Myzus persicae*) from plants co-infected either with PVA or PVY, but not from plants containing PAMV alone (194). Further, aphids having fed previously on a PVY infected plant can acquire PAMV from a second plant and transmit the virus (195). Furthermore, aphids carrying the PVY helper component can also acquire and transmit PAMV from sap extracts (139, 196). The ability of aphids to acquire PAMV under above said conditions has been attributed to the occurrence of a triplet of amino acids 'DAG' near the N-terminal of PAMV coat protein (21).

PVS : It is another most contagious virus readily contact and sap-transmitted. Some isolates of the virus also get transmitted by *Myzus persicae* and *Aphis nasturtii* in a non-persistent manner (365). It is widely distributed and causes latent infection in most of the potato cultivars (14). In combination with PVX, it may cause appreciable reduction in tuber yield (370). Two strains of the virus, namely ordinary (PVS^{o}) and Andean (PVS^{A}) are recognised on the basis of reaction in *Ghenopodium* species. Isolation of PVS^{o} from North America and Europe cause local lesions whereas those of the PVS^{A} cause only systemic symptoms in *C. quinea, C. amaranticolor* and *C. rubrum.* The Andean isolates sometimes may also cause chlorosis and bronzing of leaves in European potato cultivars (268). Important characteristics of the virus are given in Tables 1-2.

Table 2 : Genomic properties of important potato viruses

Sl. No.	Virus	No. of genomic RNA species, and sense	Terminal structures at 51	31	Translation strategy (ies)	No. of open reading frames per genomic nucleic acid	Known/rutative function of ORF products
1.	PVX	1,+	m^7Gpp cap	poly (A) tail	subgenomic m-RNA	5	1 = Replicase; 2-4 MP[a]; 5 = CP[b]
2.	PAW	1,+	m^7Gppp cap*	poly (A) tail*	subgenomic m-RNA	5	1 = Replicase; 2-4 = MP; 5 = CP
3.	PVS	1,+	?	Poly (A) tail	subgenomic m-RNA(?)	6	1 = Replicase; 2-4 = MP; 5 = CP; 6 = metal binding and nucleic acid binding 'finger'
4.	PVM	1,+	?	Poly (A) tail	-do-	6	1 = Replicase; 2-4 = MP, 5 = CP; 6=metal binding nucleic acid binding 'finger'
5.	PVA	1,+	VPg	Poly (A) tail	Processing of a polyprotein	9*	2 = HC[c]; 5 = CI[d] ; 7 = NI[e]; 8 = NIb[1]; 9 = CP
6.	PVY	1,+	VPg	Poly (A) tail	-do-	9	2 = HC[c]; 5 = CI[d]; 7 = NIa[e] ; 8 = NIb[f]; 9 = CP
7.	PVT	1,+	?	Poly (A) tail	?	3 determined in the 3 terminal part	2 = MP
8.	TRV	2,+	m^7 Gppp	t=RNA like structure	subgenomic m-RNA and read-through of a leaky in -itiation codon	43 in RNA1, and 1 to 3 in RNA2	RNA1 : 1 & 2 = Replicase, 3 = MP RNA 2 : 1' = CP
9.	PMIV	3,+	?	?	subgenomic mRNA; and read-through	4 in RNA-2, and 1 in RNA-3	RNA-2:1-3=MP(?) RNA-3:CP

10.	PLRV 1, +	VP_g	?	Subgenomic mRNA; frame-shift, by-pass, etc.	6	2b = Replicase ; 4 = CP
11.	APLV 1,+	m^7 Gppp	tRNA like	Subgenomic m-RNA*	3*	1-MP; 2=Replicase; 3 = CP
12.	APMV 2,+	VP_g	Poly(A) tail	Polyprotein processing	B* RNA=4M* RNA=3	?
13.	ALMV 3,+	m^7 Gppp	Hair pin like structures	Subgenomic mRNA	RNA1=1, RNA2-1; RNA3=2	RNA1 & RNA2 Productes = Replicase; RNA3 : 1 = MP ; 2 = CP
14.	PYDV 1,—	*comple— mentary i.e. + sense RNA has m Gppp	*complement ary+sense RNA has poly (A) tail	?	?	?
15.	PSTVd 1	-	-	-	-	-
16.	TSWV 3,	Inverted complementary repeats of 65-70 nts at the ends of all the RNAs		Ambisense soding in S-RNA	S-RNA=2; M-RNA=2?; L-RNA=1	S-RNA:N protein and Ns protein; M-RNA; Glycoproteins G1 & G2 & NSm (?); L-RNA; polymerase

a = Movement protein ; b = coat protein ; c = Helper component ; d = cylindrical inclusion protein ; e = Nuclear inclusion protein a; f = Nuclear inclusion protein b; * = based on information from other members of the group.

Table 3 : Distribution of viruses by their association on the potato (276)

Viruses dependent on potatoes for survival and spread	Virus not dependent on potatoes for survival and spread	Undetermined
Potato virus X (PVX)	Tobacco rattle virus (TRV)	Potato black ringspot virus (PBRV)
Potato virus Y (PVY)	Tobacco mosaic virus (TMV)	Potato spindle tuber viroid (PSTVd)
Potatovirus S (PVS)	Tomato spotted wilt virus (TSWV)	Potato yellow vein virus[a] (PYVY)
Potato virus M (PVM)	Cucumber mosaic virus (CMV)	Potato deforming mosaic virus (PDMV)[a]
Potato mop-top virus (PMTV)	Tomato blacking virus (TBRV)	
Potato aucuba mosaic virus (PAMV)	Tobacco ringspot virus (TRSV)	
Wild aucuba mosaic virus (WAMV)	Solanum apical leaf curling virus (SALCV)	
Potato virus T (PVT)	Potato yellow dwarf virus (PYDV)	
Potato leafroll virus (PLRV)	Beet curly top virus (BCTV)	
Andean potato latent virus (APLV)	Alfalfa mosaic virus (AMV)	
Andean potato mottle virus (APMV)	Tobacco necrosis virus (TNV)	
Potato virus V (PVV)	Tobacco streak virus (TSV)	
Potato stunt virus (PSV)		

[a]Their virus origin has not been yet determined.

PVY : It is probably the most important potato virus so far as its wide distribution, high incidence and economic losses are concerned (191). Combined infections with PVA, PVX and/or PVS may cause severe diseases resulting in heavy tuber yield losses. It is sap and non-persistently aphid-transmitted. The most efficient vector in *Myzus persicae.* Isolates of PVY show variation in their serological properties, host range and the symptoms incited by them. Most of the isolates have been grouped into three clusters of strains based on symptoms produced in potato and tobacco. The ordinary strains (PVY^o) cause mosaic without veinal necrosis in tobacco, while tobacco veinal necrosis (PVY^c) also called 'Stipple streak' due to production of star-like necrotic lesions in certain potato cultivars, differs from PVY^o and PVY^N in being unable to overcome the dominant hypersensitivity gene Nc present in many potato cultivars (61, 71, 180).

Molecular biology of aphid-transmission of PVY has been studied. First evidence of the involvement of a non-structural virus-coded protein called helper component (HC) in the aphid-transmission of PVY was given by Govier *et al.* (140) and its association with the cytoplasmic rod-like structures called amorphous inclusions in virus-infected cells was shown by Baunoch *et al.* (22). Besides the HC, aphid transmission is also determined by a triplet of amino acids 'DAG' present near the N-terminal of the virus coat protein. A single amino acid substitution in the 'DAG' of the coat protein or in the HC renders the virus aphid non-transmissible (124, 212).

PVM: It is another member of the carlavirus-group and related to PVS. Many of its characteristics are given in Table 1-2. It is sap and non-persistently aphid-transmitted, most efficiently by *Myzus persicae*. PVM causes different types of symptoms such as faint mottle, mosaic, crinkling and also rolling of leaflets and stunting of infected plants. It has a narrow host range. Several biological strains can be recognised. Mild strains produce a mosaic between the veins of the leaflet tips and some deformation of the leaves. This deformation is in the form of twisting of leaflet tips, sometimes severe, and some even cause rolling of the top leaves.

PVA : It belongs to potyvirus group and like PVY is readily sap and aphid transmitted. *Myzus persicae* is the efficient vector of PVA with acquisition period of 20 seconds and retention period of 20 minutes (226). Isolates of the virus differing in the severity of symptoms produced on potato have been encountered and classified into four groups, viz. very mild, mild, moderately severe and severe (226). The virus is widespread in some cultivars. In combination with PVX it usually causes a severe disease known as potato crinkle. Similarly, its coinfection with PVY results in a severe mosaic disease. Many potato cultivars are hypersensitive to most strains of PVA, viz. Bintje, Duke of York, Katahdin, Sebago, etc.

PVV: It is also a potyvirus, serologically but distantly related to PVY^C. It causes severe necrotic reactions when inoculated to potato cultivars carrying hypersensitivity gene Nc against PVY. Due to this very reason, originally it was classified as PVY^C (61) who called it PVY^CG1 isolate. Subsequently, PVY-AB from Ireland (52) and PVY-UF from Central America (109) were shown to be serologically closely related to one another but distantly related to PVY and consequently given a separate name PVV (110). PVV is more closely serologically related to Peruvian tomato virus (PTV) and wild potato mosaic virus (WPMV). Out of the eight monoclonal antibodies raised against PVV, only four reacted strongly with isolates of WPMV from Peru while 4 to 7 monoclonal antibodies reacted with different isolates of PTV. Besides, PTV was the only other virus to share with PVV more than one of the four conformation-sensitive epitopes that were detectable (96).

PLRV : It probably occurs worldwide and is next only to PVY in importance among potato viruses responsible for fast degeneration of potato seed stocks and heavy yield depressions. Under natural conditions, it is mainly spread by aphid vectors and through infected tubers. It can be artificially transmitted by stem grafting but with difficulty also by tuber-core grafting. It cannot be sap transmitted. Severity of PLRV symptoms depends on the cultivar and environmental conditions. Current season infection expresses as upright habit with pale yellowish or purplish pigmentation and rolling of the young top leaves only if the infection occurs quite early and environmental conditions are favourable. Late season infection does not express while progeny tubers may show partial infection. Secondary (infection) symptoms are generally severe. The whole plant often looks erect but stunted than the healthy ones; older leaves are rolled and younger ones are pale. Old rolled leaves are stiff, leathery and brittle. Besides, older leaves depending on the cultivar are either tinged purple or show severe necrosis, especially at the margins. Tubers of certain cultivars like those of Green Mountain, Nordgold and Russet Burbank react to PLRV infection in the form of net necrosis. PLRV infection normally causes phloem necrosis in stems and petioles.

Three to four strains, differing in severity of symptom on potato cultivars and other hosts, *Physalis floridana* in particular may be distinguished (179, 341). The genome of PLRV is unique in employing several strategies for its expression. The first of the six **open reading frames** (ORF) are ORF_1 is translated conventionally, second ORF (ORF 2a) is apparently expressed through by-pass of the ORF 1 initiation codon 1, 3rd ORF (ORF 2b) is thought to be expressed through frame-shift during translation of ORF 2a; 4th ORF (ORF 3) is expressed through a subgenomic m-RNA (231), 5th ORF (ORF 4) is the second ORF expressed through by-pass of an upstream initiation codon while the 6th

ORF (ORF 5) is expressed by read through of the ORF 3 stop codon to give a fusion protein containing domains corresponding to both ORF 3 and ORF 5 (16, 223).

Aphid (*M. persicae*) is the most efficient vector which transmits PLRV persistently. Different clones of the vector differ in efficiency (367). Besides, vector efficiency also depends on the plant species of potato cultivar and PLRV isolate (323). PLRV transmission efficiency from *Physalis floridana* is greater than that from *Solanum tuberosum* (253). Both larval and adult stages and morphs of *M. persicae transmit* PLRV through larval stages generally perform better than the adult morphs (262). Other aphid species which transmit PLRV less efficiently include *Aphis naxturtis, Aulocorthum solani* and *Macrosiphum euphorbiae* (326). Clones of *Aphis gossypii* also have been reported to vary in their efficiency to transmit PLRV (323).

APLV: It is a member of the tymovirus group and occurs mainly in the Andeas in South America. It is serologically related to *Ononis* yellow mosaic virus and *Dulcamara* virus (133). The virus is readily transmitted through contact between plants and comparatively poorly through true seed and the potato flea beetle *(Epitrix* sp.). Infection may be latent but occasionally with severe symptoms or mottling or chlorotic netting of minor leaf veins. Secondary infection often causes mild mosaic. However, rugosity and severe mosaic are caused in some potato cultivars. Two serologically distinct strains have been recognised (133).

APMV : It belongs to comoviruses and is highly contagious being readily transmitted through contact between plants, contaminated farm implements and not through true seed. Like APLV, it occurs in the Andean region at elevations of 2000-4000 m amsl. It may be suspected present in germplasm material originating from Andens. Infection induces mild mottle in primary infections but more severe symptoms in secondarily infected plants.

PVT: It is a member of the newly recognised group 'Capilloviruses' and occurs in some South American regions. It is easily sap transmitted, and is most readily seed transmitted in *Nicandra physalodies, Datura stramonium* and *Solanum damissum* (275, 276). Infection is latent in many Solanaceous hosts. Its infection causes slight veinal necrosis and chlorotic spotting in potato cv. King Edward. Graft inoculation of potato cv. Cara results in top necrosis. Many isolates of the virus are known to occur and it is serologically related to apple stem grooving virus.

AIMV: Though the virus has worldwide distribution, it is of only a little significance in potatoes. It is transmitted through sap, stem, and tuber-core grafting, infected tubers and also non-persistently through aphid vectors, viz. *Aphis nasturtii, Aulcorthum solani, Macrosiphum euphorbiae* and *Myzus persicae* (73, 173, 263). The virus spreads from lucerne to potatoes when aphids migrate from the former to the latter (25). Two main groups of strains, namely potato calico, and potato tuber necrosis have been recognised. It is a multipartite virus with split genome, i.e. virus with different sized particles, wherein RNA-1 and RNA-2 code for the virus replicase while RNA-3 codes for capsid protein which is essential for the replication to occur (343). RNA-3 also determines the types of symptoms on tobacco (83).

PYDV: It belongs to rhabdoviruses group and is the only leafhopper-borne virus that is also sap transmissible. It has been reported from Canada and U.S.A.. Virions are very unstable as they even cannot withstand drying over $CaCl_2$ or storage at-20°C. Two forms of PYDV, viz. one transmitted by *Aceratagallia* spp. and the other by *Agallia* spp. are recognised. Both forms are transmitted persistently. It is not true-seed transmitted and is of little economic significance in most of the potato growing areas.

TRV : It belongs to tobraviruses group and is transmitted under natural conditions through nematodes. It can be transmitted by stem grafting, through true seeds and also by sap inoculation. It persist in the trichodorid nematode vectors for a long time. It occurs in many potato growing areas, only on slight and peaty soil, and causes a serious disease called stem mottle or 'corky ringspot'. It generally occurs in localized patches in a field. Two groups of strains, namely potato stem mottle (infecting one or only few stems/hill) and potato spraing (causing necrosis in tuber flesh) are recognised and they cause characteristic symptoms in *Nicotiana tabacum.* The antisera prepared against some strains react specifically only to a limited number of other strains indicating that many serotypes (at least eight) exist. Different isolates of the virus differ widely in their symptoms in test plants, but this is not correlated with the serological variation. Another kind of variability, unique to TRV, is the existence of so-called NM-type isolates (which comprise long particles carrying only RNA-1). These isolates can cause infection but without the assembly of virions.

Nematode transmission involves attaching of virions to the inner surface of the nematode oesophagus from where they are thought to be ejected with saliva to infect root cells that are punctured by the stylet. Each nematode vector species is able to transmit only some specific virus isolates and it is suggested that the specificity is governed by the virus isolate (266).

PMTV : It belongs to furoviruses group, and it occurs in many European countries. It is transmitted through rot-infecting fungus *Spongospora subterranea.* The virus produces symptoms in the tubers of sensitive potato cultivars. Tuber symptoms comprise brown arcs and lines in the tuber flesh, and lines may also be visible on the tuber surface. Yellow 'V' markings are produced on leaves of plants grown from infected tubers and the haulms of some cultivars become stunted, giving the 'mop-top' appearance to the plants. PMTV genome consists of three distinct RNA molecules, but size of the RNA vary in different species (347).

TSWV: It is a member of tospoviruses group and occurs worldwide (27). It is less common in Europe. It is also quite unstable *in vitro* and is transmitted by thrips, viz. *Thrips tabaci* and *Frankliniella* spp. Mechanical transmission requires special precautions esp. while extracting the sap for inoculation. It causes 'local' necrotic spots of foliage but mainly necrotic streaks on stems hence the name 'stem-necrosis' (204). Due to severe mass infection the entire crop develops foliar and stem necrosis, gets stunted badly due to which the field gives 'blighted' appearance despite the dry weather. Depending on the time of infection and disease index, it reduces tuber yields considerably (255). It is normally not transmitted through tubers and infection in the crop occurs through viruliferous thrips which enter potato fields from outside.

PSTVd : It is a small, single-stranded, circular RNA with 359 nucleotides and with a high degree of intramolecular complementarity (150). Cleavage of circular viroid RNA molecules results in linear forms which are also infectious (152). Three main strains, viz. severe, intermediate and mild, are recognised on the basis of severity of symptoms produced in potato and tomato (278). These strains differ by 6 to 10 nucleotide exchange (289).

Sensitive potato cultivars respond to infection by stunted and upright growth, tubers become spindly and reduced in size and may show growth cracks. Disease is severe under warm climatic conditions. Depending on the potato cultivars, PSTVd strain and the climatic conditions, tuber yield reduction varies from negligible to 97% (254, 327).

The viroid is highly contagious and readily transmitted through mechanical contact, sap inoculation, contaminated farm machinery (229) and mainly through true seeds (183). PSTVd occurs in many potato growing countries, but strangely is absent in Andean potato cultivars (5) and only recently recorded in some wild *Solanum* spp. (246).

2.1 Yield losses

Potato tuber yield reduction due to virus infections is influenced by a variety of factors like the virus strain, host cultivars, current or secondary infection, time of current infection, how many seasons old is the secondary infection, proportion of plants infected, growth behaviour of the cultivar, climatic and edaphic conditions. Due to too many factors affecting tuber yields of virus-infected potato plants, precise estimation of tuber yield loss due to a particular virus is quite difficult under field conditions. It is further complicated by the occurrence of mixed virus infections which are so common under natural conditions. Generally, tuber yield losses are 5-15% when all the plants are secondarily infected with PVX and PVS; 15-30% for 100% secondary infection of PVY^N, and 40 to 70% for infection with PLRV and PVY^o(10, 37, 38). Further, these figures only indicate the range of losses often experienced in the temperate zone under normal climatic conditions. Adverse or extreme climatic conditions like high temperatures and drought may result in higher losses.

Blodgett (34) was first to develop a mathematical model to asses the yield loss due to PLRV infection. DeWit (80), proved that estimates of potato yields were very close to the actual yield of crops, infected with viruses. This formula (given below) consisted of two components, one for the yield of the healthy proportion of the plants and one for the yield of the diseased proportion of the plants.

$$Y = \frac{K \times Zh}{K \times Z_h \times Z_l} \times \frac{M_h Z_l}{K \times Z_h + Z_l} \times M_l$$

Where -

Y = yield of a potato crop

K = Crowding coefficient

Zh = Proportion of healthy plants

Z1 = Proportion of diseased plants.

Mh = Yield of the crop with 100% healthy plants

M1 = Yield of the crop with 100% diseased plants.

Van der Zaag (357) calculated the yield losses for three crowding coefficients, viz, 1, 5, 2 and 2.5 for three kinds of viruses differing in their effect on tuber yield.

Epidemiology : Potato viruses spread from plant to plant within a field and between fields through mechanical means as well as vectors. PVX, PVS and PSTVd are contagious and spread through cutting of tubers into small pieces at the time of planting and/or mechanical injury to tubers and sprouts and contact of infected foliage with the healthy, workers' contaminated clothes, implements, farm machinery, etc. No insect vector is known for PVX and PSTVd. However, PSTVd may be disseminated through the females of root knot nematode, *Meloidogyne incognita* passively (7). Besides, PSTVd is also most readily true seed transmitted.

PVX, PVS and PSTVd and highly stable and retain their infectivity on contaminated objects for varying lengths of time. Consequently, the main source of primary inoculum of these agents are the infected tubers used as seed, and even weeds (131, 325).

Amongst the aphid vector transmitted viruses, PVY and PLRV are more important. Many apterous forms of aphids have been shown to transmit PVY under experimental conditions. But apterous forms are important in plant to plant spread of the virus(es). About 30 species or species groups of alatae aphids may transmit PVY (74,154). The main attributes which determine vector efficiency are vector behaviour and vector mobility (331). Efforts have been made to devise models/ programmes to predict the relationship between vector activity and the incidence of vector PVY, viz, VanHarten (360) calculated vector pressure from relative efficiency factors (REF) of related aphid species prevalent in the area. Gabriel (116, 117) calculated contamination index 'W' from the number of aphids trapped or observed on 100 potato leaves over 80 days and the efficiency with which tubers got infected with PVY. Vector pressure has been used to examine annual variations in the incidence of PVY in the Netherlands (360), Sweden (312) and Scotland (353). REF values assigned to different aphid species vary and depend on vector efficiency, relative abundance, time of migration, age of potato crop, etc. Thus, *Brachycaudas helichrysi* had a higher REF than *M. persicae* at Harpenden during 1984 due to higher efficiency and larger number (155); *Rhopalosiphum padi* was more important in Sweden (318) and the Netherlands (74) due to early migration in larger numbers at a time when potato crop is young or more susceptible to PVY.

Like PVY, attempts have also been made to predict the incidence of PLRV from factors like temperature during winter, and/or during the crop growth and the vector activity. There was a link between the number of aphids trapped in suction trap, and the winter temperature, spring migration of aphids and the spread of PLRV in the field (331). Winter temperature during February and dates of first catches of anholocyclic specimens of *M. persicae* showed a positive correlation (351) and this information has been employed to forecast the spring migration of *M. persicae* (352) and to advise farmers whether or not to use the pesticide at the time of planting for aphid control.

2.2 Management

Potato viruses adopt different strategies for their spread and perpetuation. Consequently, they have to be contained or managed through a package of strategies, viz. exclusion, therapy, healthy seed production, vector control, resistance breeding and incorporation of virus resistance through genetic engineering.

Exclusion : Some of the potato viruses occur only in certain specified areas and not the others. Thus, APMV, APLV, PVT and PVV normally occur in the Andean region. Elsewhere they have reached only through germplasm which originated from the Andes. Besides, these viruses (except, APMV, PVV) are also true seed transmitted. Ascertaining freedom of the material from these viruses under Quarantine before release will check their spread.

Therapy : Valuable germplasm and potato cultivars, if found infected with viruses, are rescued by freeing them from the virus infection generally through meristem top culture with or without heat and chemo-therapy. Survival of meristem tips, less than 0.5 mm (required for virus elimination) is generally very limited. This problem is overcome by first growing the infected plant for several weeks at 36°C before taking meristem tips which could be much larger in this case (334). Chemo-therapy using antimetabolites and antiviral chemicals, viz. Virazole (Ribavirin) (81, 363) has also been reported to improve the recovery of virus-free plantlets from larger meristem sections. Though multiplication of many of the potato viruses is inhibited at elevated temperatures, that of PSTVd is accelerated. Elimination of PSTVd needs cold treatment below 10°C (214). Efficiency of virus elimination depends on the virus, the potato cultivar and method of elimination.

Elimination of potato viruses through chemo- and thermo-therapy without the meristem tip culture has been tried (149). *In vitro* potato plantlets were exposed to an alternating' 4h cycle of 35 and 31°C and ribavirin (20mg/litre) therapy for 4 weeks. They were further propagated via nodal cuttings at room temperature without exposure to ribavirin. Ribavirin alone or in combination with heat not only reduced PVM, PVS and PVX titres (ten and sixty folds) but also freed many plants from other viruses including PVY and PLRV (64).

Vector control : Many insecticides are effective against the spread of PLRV whereas virtually none is effective against the non-persistently aphid-transmitted viruses like PVY and PVA (317, 336, 361). This is due to very short periods required for acquisition and transmission of PVY by aphids. Only pyrethroids with quick knock down effect on the vector have been reported to show some promise in the control of PVY (252). Besides pyrethroids, mineral oils have been reported to reduce the spread of non-persistently aphid transmitted viruses (44). Fifty to 90% prevention of PVY spread with mineral oil sprays has been reported (286). Nonetheless, use of mineral oils for PVY control has not become popular probably due to high material and labour costs (285).

Insect repellents, like alarm pheromones and their derivatives, and antifeedants have been tested for the control of potato virus(es) spread. (E)-B-fernesene (sequiterpene hydrocarbon), a derivative of the common alarm pheromone, has been found to inhibit acquisition of PVY by aphids (72). The same pheromone occurs in the leaf hairs of the wild potato *S. berthaulti* and is responsible for the inhibition of aphid colonization (134). Inhibition of aphid colonization and PVY acquisition has also been reported to occur with the sequiterpene dialdehyde polygodial, an antifeedant derived from *Polygonum hydropiper* (135, 239).

2.3 Healthy Seed Production

Virus free or healthy planting stocks are the first requirement for any viable potato cultivation. Consequently many of the potato producing countries have established organisation which are responsible for a regular supply of the healthy seed-potatoes. Improvements in the virus detection technology coupled with the *in vitro* methods of rapid multiplication of the virus free stocks have resulted in better health standards of potato crops with increased tuber yields (84).

Generally, virus free stocks are multiplied in high hills exposed to heavy wind currents which minimise the vector pressure. Alternatively, they are multiplied in regions having climate suitable for potato crop production. Such regions should either lack the aphid vector or have vector population below a certain critical level during the crop growth. In India, virus-free potato stocks, referred to as breeders seed, are multiplied in high hills (above 2000 metres) in summer, and in the Indo-gangetic plains during autumn (September/October to December/January) when vector activity is very low.

Another method of multiplication of virus-free stocks is through *in vitro* multiplication, and is gaining importance in many countries with tropical climate (84). The pool of virus-free stocks after specified multiplications serve as the basic or nuclear seed. Certified seed is raised from the basic seed by further multiplication on either government-owned farms or on the farms of registered growers under direct supervision of the qualified persons. Different health grades (categories) of seed potatoes conform to different tolerance limits of virus infection. Tolerance limits for different categories vary from country to country (Table-4).

Only a few potato viruses are transmitted through true potato seed (TPS) and use of hybrid TPS, capable of giving uniform/homogeneous crop stands, holds promise at least for the areas having high cost and short supply of seed-potatoes. The first commercial potato production using TPS was tried in China in 1972 (330). Use of TPS for potato production in India was attempted during 1950s but had to be abandoned due to heterogeneity in crop maturity and the produce. Efforts have been again initiated in South Asia to exploit the use of TPS for potato production (148). Many of the previously experienced bottlenecks are being overcome through further research and as a result many improved and promising TPS families have since been selected and their successful performance at specified locations demonstrated (Table-5).

Table 4 : Field tolerances (no. of affected plants in%) for basic and certified seed, as applied in four countries

		Canada[1]	France[1,3]	The Netherlands[1,3]	U.K.[2]
Severe mosaic, including leaf-roll, stipple streak & crinkle	Basic seed	0.25	0.33	0.09	0.1
	Certified	1.0	1.0	0.25	2.0
Mild mosaic	Basic seed	0.25	0.33	0.09	0.5
	Certified seed	1.0	1.0	2.0	5.0
Erwinia sp.	Basic seed	0.1	0.5	nil	0.5
	Certified seed	2.0	1.0	5 plants/ha	2.0
Varietal Impurity	Basic seed	0	0.1	0	0.05
	Certified seed	0.1	0.2	0	0.5

1. Three official inspections for basic seed and two for certified seed.
2. Two official inspections for basic seed and one for certified seed.
3. The data presented for basic and certified seed concern those for the grades E and A respectively (grades E and A are generally marketed).

Table 5 : Tuber yields (t/ha) in three successive clonal generations of 8 TPS progeny families and the cultivar Kufri Bahar (148)

TPS progeny	Clonal generation		
	C_1 (1985-86)	C_2 (1986-87)	C_3 (1987-88)
TPS-2 (OP)	27.9	26.0	28.6
TPS-3 (OP))	22.0	25.1	28.8
TPS-7 (OP)	33.8	28.7	34.6
HPS 4/3	27.2	29.1	24.1
HPS 7/III	34.3	33.2	33.6
HPS 1/13	33.4	32.6	37.3
HPS 1/III	35.8	35.7	32.9
HPS 2/III	36.8	36.7	30.5
Kufri Bahar	36.1	32.2	34.7

Virus-Resistance Breeding : Resistance breeding is one of the most economic, effective and eco-friendly method of virus disease management. Resistance genes to different potato viruses and conferring different levels of protection against these viruses have been identified in many wild potato specis (180). Many of these genes have been incorporated in many of the recently released potato cultivars.

PVX : The genes conferring resistance against PVX have been designated as *Nx, Nb, Rxadg* and *Rxacl* according to the type of resistance response and the source of the resistance gene. Thus, *Nx, Nb* and PVX strain specific whereas *Rx* confers in extreme type of resistance which is durable and virus strain non-specific. *Rxadg* has been derived from *S. tuberosum* subsp. *andigena* while *Rxacl* has been derived from *S. acaule* (61, 270). These resistance genes have been and are being used for incorporating resistance against PVX in potato cultivars (225). Barring the stray occurrence of resistance breaking strain of PVX in greenhouse experiments, resistance breaking strains have seldom been found occurring naturally in field (102).

Major-gene-resistance genes against PVX occur as simplex. Clones with duplex, triplex and quadruplex forms of these genes are now available for breeding (225). Use of these clones makes the incorporation of resistance easier and quicker (270, 271).

Immunity to PVX has been incorporated into many potato cultivars namely Saco, Kufri Lalima, SanJuan, Annett, Natta, Santanlalla, etc. (6).

PVY : Virus strain specific resistance is controlled by the resistance genes *Ny* while the extreme resistance in controlled by *Ry*. *Ny* genes are found in a large number of cultivars including Pentland Krown, Pentland Ivory, King Edward and Cara, and in hybrids derived from wild species like *S. chacoense, S. domissum* and *S. microdontum* (272). The source of *Ry* genes is *S. stoloniferum* (Rystol) and *S. tuberosum* subsp. *andigena* (Ryndg) (336). *Ry* is inherited as a single dominant gene and is easy to breed for. The use of clones duplex for *Ry* gene raises the percentage of immune progeny of a cross to 82% (47). Currently released potato cultivars carry the *Ry* gene either singly or in combination with the resistance genes for PVX and PLRV.

PVA : Some potato cultivars carrying the resistance gene *Na* are hypersensitive to PVA. *S. stoloniferum* carries extreme resistance to PVY and PVA and the resistance is controlled by six genes at three different loci. The genes are differentiated from the phenotypic expression of plants after inoculation with PVY or PVA (61). Potatoes possessing extreme resistance to PVA are also bred by employing *Ry (Sto)* and *Ry*(adg) genes (75, 99).

PVS: Hypersensitive type resistance imparted by the gene *Ns* is found in *S. tuberosum* subsp. *andigena* and *S. megistacrolobum* (15). Potato cv. Saco is immune to PVS and the gene conferring immunity is inherited as a recessive. It is also present in potato cv. Jemseg (15).

PERV: Resistance to PLRV has been found to be oligogenic and it has been detected in *Solanum brevidens* (184) and *S. tuberosum* (189). Transfer of resistance genes from *S. brevidens* has been through protoplast fusion (136). Whereas transfer of resistance from *S. tuberosum* has been through bridge crosses (170).

Among the hexaploid somatic hybrids derived from *S. tuberosum* and *S. brevidens* protoplast fusion, some hybrids with high PLRV resistance were detected (250, 251, 355). In the case of resistance from *S. etuberosum, S. tuberosum* is first crossed to *S. pinnatisectum* and chromosome number of the hybrids is doubled to get allotetraploids (4x - EP) which are fertile (171). Many of these 4x - EP resist the build-up of PLRV titre upon inoculation (55, 56). Their further crossing to *S. acuale* yielded the hexaploids 4 x-EP=*S. acaule* (2n=72; 6x-AEP) which were then used to obtain fertile hybrids by crossing to *S. phureja* pollen parent. Resistance segregation in the progeny of the last cross suggested simple inheritance of resistance (55).

The oligogenic resistance to PLRV appears to comprise different components like resistance to virus multiplication, resistance to virus infection and the spread with the plant, etc. (13). Evaluation of PLRV resistance in some clones has revealed that restriction of virus multiplication may be controlled by a single gene (19, 20, 173, 223). This type of resistance has been found in *S. tuberosum, S. brevidens* and *S. etuberosum* (189, 261).

Cultural-Practices: Cultural practices greatly influence the spread of potato viruses. Some practices bring together large communities of crop plants and weeds associated with their cultivation. Some of the weeds serve as reservoirs of viruses and their vectors. Factors like date of planting, mode of weeding, isolation of the crop, removal of volunteer plants and virus-infected plants at appropriate time, disinfection of tools and farm equipments, killing of vines at appropriate time and prevention of resprounting of dehaulmed stems, play an important role in determining the spread and dissemination of potato viruses. Thus, reinfection of healthy potato stocks during their multiplication can be minimized by following measures (i) selecting a field free from soil-borne disease and pests; (ii) selecting a planting date which assures a minimum of 75 days of vector-free period suitable for crop growth and tuberization; (iii) soil application of systemic granular insecticides at the time of planting to control vectors; (iv) blind (or full) earthing up at planting and use of herbicides for weed control, minimize the chances of spread of contagious viruses and PSTVd; (v) crop inspection to remove the off-types and virus-infected plants; (vi) isolation of the crop from ware-potato crop; (vii) dehaulming at right time; (viii) proper handling during harvest and storage (18, 203).

2.4 Resistance Incorporation Through Genetic Engineering

Successful virus infection of a plant requires uncoating of the nucleic acid immediately after its entry into the cell, production of virus-genome encoded products required for genome and coat protein replication as well as for virus spread within the plant. Blocking of any step will check virus infection or restrict it to the cells initially entered into. Incorporation of the viral coat protein gene into the genome of the host plant through genetic engineering has shown that the transformed plant becomes resistant to that virus (23). This type of virus resistance called virus coat-protein mediated resistance or protection (VCPMP) has been incorporated into potato cvs. Bintje and Escort with respect to PVX (176), and Russet Burbank with respect to both PVX and PVY (211). Generally transformed plants expressing the coat protein (CP) gene accumulate CP in their cells and there seems to be a relation between the CP accumulation and the degree of virus resistance. However, some transformed plants, viz. transformed potato cv. Russet Burbank expressing the CP gene does not accumulate CP in their cells and there seems to be a relation between the CP accumulation and the degree of virus resistance. However, some transformed plants, viz. transformed potato cv. Russet Burbank, expressing the CP gene, does not accumulate CP (211). In a field trial, one clone (-3030) of Russet Burbank expressing CP for PVX and PVY showed higher tuber yield than non-transformed plants when all the plants were inoculated with PVY and PVX (192). The mechanism underlying the CP-mediated resistance is still not clear and may operate in one of more ways, viz. (i) prevention of uncoating of the invading virion, (ii) inhibition of replication of released viral RNA, if uncoating occurred; (iii) hinderance in the cell-to-cell movement of the virus (23).

Transformation of plants with the viral RNA replicase gene or part of the replicase gene has been shown to impart virus resistance to the plant. Thus, transgenic tobacco lines expressing the putative replicase gene of PVX were resistant to PVX (45).

Other techniques of incorporating virus-resistance in plants through genetic engineering, viz. satellite RNA's defective interferring nucleic acids, antisense RNAs ribozymes have so far not been exploited in potato.

3. FUNGAL DISEASES

Fungal diseases of economic importance to potato can be broadly categorised into two groups, viz. foliar diseases and soil and tuber borne diseases (209). Each group contains a large number of diseases (Table-6) but only a few of them matter, for example, amongst foliar diseases, late/early blights and Phoma blight are the most important whereas dry rot, common scab, black scurf, wart, *Verticillium* wilt and *Fusarium* wilt are important amongst the soil and tuber borne diseases.

Table 6 : Important Potato Diseases (Fungal)

Diseases		Pathogen
Folliar		
1.	Late Blight	*Phytophthora infestans*
2.	Early Blight	*Alternaria solani*
3.	Grey Mold	*Botrytis cineria*
4.	Phoma Blight	*Phoma* spp.
5.	Leaf Blotch	*Cercospora concors (Mycovellosiella concors)*
Soil & Tuber Borne		
1.	Black Leg & Bacterial Soft rot	*Erwinia* spp.
2.	Brown rot	*Pseudomonas solanacearum*
3.	Ring rot (not recorded in India)	*Corynebacterium sepedonicum*
4.	Common scab	*Streptomyces* spp.
5.	Powdery scab	*Spongospora subterranea*
6.	Wart	*Synchytrium endobioticum*
7.	Skin spot	*Polyscytalcum pustulans*
8.	Leak	*Pythium* spp.
9.	Pink rot	*Phytophthora erythroseptica*
10.	White mold	*Sclerotinia sclerotiorum*
11.	*Sclerotium* wilt	*Sclerotium rolfsii*
12.	*Rhizoctonia* Canker (Black scurf)	*Rhizoctonia solani*
13.	Silver scurf	*Helminthosporium solani*
14.	Black dot	*Colletotrichum* atramentarium
15.	Charcoal rot	*Macrophomina phaseolina*
16.	Gangrene	*Phoma exigua* var. *foveata*
17.	*Fusarium* wilt & dry rot	*Fusarium* spp.
18.	*Verticillium* wilt	*Verticillium* spp.
19.	Smut (not recorded in India)*	*Angiosorus solani**
20.	Rust (not recorded in India)* of quarantine	*Puccinia pittieriana**

* Significance for India.

3.1 Disease Estimation

The blight was first observed in Europe of 1845 at Courtrcu in Belgium in late June to early July. Subsequently, it spread to most of the European countries, including U. K. and Ireland, causing the worst ever famine known as 'Irish Potato Famine' Since then it has spread far and wide. Today, it occurs wherever the potato crop is grown. In India, it reached through Europe in 1870 and 1880 and was recorded for the first time in Nilgiris hill (51). Later, it spread to other parts of the country climate is cool and humid, typical of Europe, parts of U.S.A. and all hilly regions of the world including India. Intensity of the disease also varies from variety of variety. For example, susceptible varieties like Kufri

Chandramukhi and Kufri Bahar, are more prone to blight. Their crop is killed within a period of 8-10 days under blight favourable period. Whereas, in case of resistant varieties, it takes little longer time.

The losses due to this disease are of two kinds; losses caused by foliage infection which leads to premature death of the plant and consequently a reduction in tuber yield; and those caused by tuber infection and loss through rottage of infected tubers in the field and stores (265). In the hills, the losses may be as high as 90% in a susceptible variety. In the plains (irrigated crop) losses mainly depend on the time of disease appearance. In blight year, losses range from 10-40% (30).

Several methods have been devised to estimate the severity of the disease in laboratory, glasshouse, and fields (78). For rating the disease level in the laboratory on detached leaves, Simmonds and Malcolmson (319) developed a disease rating scale with susceptibility grade 1 (resistant) to 5 (highly susceptible). Disease rating scales for measuring the level of disease in whole plants/seedlings have been developed (32, 95, 218). Based on type of lessions and the estimated area infected, Black (32) devised a 0-5 scale for seedling screening where 0-represents the immune reaction and 5-the highly susceptible. Malcolmson's scale and Estrada *et. al.* scales' are similar except that grade 1 of Malcolmson's scale was equivalent to grade 9 of Estrada *et. al.* scale and vice-a-versa.

Various methods have been used to qualtify the disease levels in the field. BMS scale (4) is the oldest of all which is still used extensively for the disease estimation in field. Vander Plank (356) was the pioneer in devising a method for comparing the build up of the disease under different set of conditions. He developed the concept of apparent infection rate (r) for comparing the disease epidemics. However, apparent infection rate does not take into account the variation in the disease build up which often occur due to changes in weather during the course of epidemic. Wilcoxon *et. al.* (366) developed a concept of area under disease progress curve (AUDPC) which takes care of such variations. Shaner and Finney (295) improved this method by incorporating time component in it. McRae and Platt (228) further improved the concept of AUDPC (ARDPC) by expressing the cultivar resistance relative to a standard cultivar.

Appearance of late blight and its subsequent buildup and spread depends on several factors such as : source of inoculum (193), host resistance per se (88), plant protection through fungicides (112, 113), deployment of host resistance in the field (30) and weather conditions (158, 185). Epidemiological dimensions of these factors are discussed below.

3.2 Sources of Inoculum

Infected tubers, vines, refuge piles, perennial hosts, etc. are some of the source through which the late blight pathogen *P. infestans* might survive. Over summering of this disease through cold stored tubers and over wintering through seed tubers/refuge piles are well documented (36, 258). The seed tubers and rejuge piles in the temperate regions and cold stored tubers in subtropical regions serve as primary source of the disease. The fungus does not survive in soil except in Central Mexico where both mating types (A_1 and A_2) occurring equal proportion and form oospores in nature (327, 329). Oospores are thick walled resting bodies which survive in soil for many months and serve as source of the disease in the next crop season. Fortunately, A_2 type was restricted to Central Mexico till early eighties. However, since 1984, A_2 mating type has been recorded from most of the European countries Egypt (302, 303, 304), Japan (236) and India (322). It has opened the floodgates for the assumption that sexual reproduction might be occurring in nature in these regions also, leading to formation of oospores. If it is true (there are no valid reasons to think otherwise), it would have a far reaching impact on the disease epidemiology. Formation of oospores in potato tissues by inoculating it with equal proportion of A_1 and A_2 types was demonstrated (237) but occurrence of oospores in nature outside Mexico is yet to be recorded.

Role of potato/tomato stalks in the parennation of the disease is not certain (60). Tomato seeds get *P. infestans* infection in the field and act as primary source of the disease (362). Potato seeds also carry late blight infection but they are yet to be established as the primary source of the disease (129).

3.3 Disease Appearance

Appearance of late blight depends primarily on three factors, viz. inoculum load, host resistance and weather conditions. Inoculum load affects the disease in two ways: (i) if overall inoculum load (per cent tuber infection) is more, it would lead to formation of more infection foci and, (ii) if the inoculum load is more on tubers (per cent tuber area infected), seed tubers would rot. Tubers having <10% infection germinate and serve as primary source of the disease (321).

3.4 Forecasting

Harrison (158) reviewed features of most of the late blight forecasting systems evolved so far, Van Everdingen (359) for the first time set criteria, popularly known as 'Dutch Rules' for forecating late blight appearance in Holland. Since then various workers from different countries, mostly from U. K., U.S.A., Germany, Norway and Eire either improved upon the original concepts of Van Everdingen, by bringing in more weather parameters or by discarding the less important ones. Most of the earlier forecasting systems have been based mostly on impirical results from field observations or from experiments in which the environment was not controlled precisely (158). However, over the years, more precise methods of disease forecasting have been evolved. Nonetheless, there is every little difference between the sensitivity of these methods. There are broadly two ways through which disease forecasting is accomplished; i) the weather parameters known to influence disease, viz. temperature, rainfall, R.H., dew, sunshine, etc. are recorded and used to estimate the extent of the disease buildup and then apply fungicides to contain the disease. This approach has been used successfully by several workers (24, 93, 185, 206, 328, 359), to manage this disease. In the second approach, the future blight progress is forecast based on the assessment of future weather conditions, the extrapolation of these conditions for development of suitable microclimate within the potato crop, and estimation of likely disease progress (158). This approach is promising but difficult to apply as it is mainly based on the accurate measurement of future weather conditions (43, 274, 364).

It has been seen that the most primitive and simpler methods devised in the fifties and sixties had worked very well in predicting the disease in areas/season conductive for the disease but failed to predict the disease as the weather was not so conducive. For such situations studies are required. Host resistance, age and previous environment, pathogen race etc. affect the disease (114, 175). These factors ought to be incorporated in the future predicting systems.

3.5 Buildup and Spread

Buildup and spread of late blight has mainly been studied in relation to a point source. Pattern of blight development around the infection focus primarily depends on the dispersal of *P. infestans* spores which is mainly restricted to a short distance (232, 321), although long distance dispersal has also been recorded (174, 267). Blight favourable weather favours long distance dispersal. Primary infection focus is limited to an area of 10 meters of the source. Secondary infection, it begin with, starts at confluence of the primary infection focus which flatterns the disease gradiant with time. Mixed planting of susceptible and resistant cultivars in 4:1 ratio reduces the disease buildup; though only to a limited period.

Buildup of late blight has mainly been studied with respect to time variable using Van der Planks's model (356). Its spread from a point source has also been studied in respect of space (101, 210). Nonetheless, methods used by these workers considered time and space independent of each other. Minoque and Fry (232, 233) proposed a model to study the relationship between the two parameters but they did not integrate the two parameters in one model. Jager (187) for the first time proposed eight models in which time and space were considered together. Singh *et. al.* (322) estimated the goodness of fit of these models by using the empirical data. They observed that two dimensional models improved the disease predictability significantly (19-31%) over one dimensional models. Out of two parameters, distance contributed more (16.3-43%) than time (16.3-27.9%).

3.6 Environmental Factors

The role of weather conditions in the development of the late blight is a subject of text books. In fact early associations of late blight epidemics with certain weather conditions in Ireland and England were so strong that the disease was first attributed to effects of bad weather on potato. Later, it became evident that *P. infestans* and late blight epidemics are highly responsive to change in weather.

Effect of environmental factors in disease development reflects the interaction of conditions favourable for survival, germination, penetration, sporulation and reinfection of the host. Interaction of temperature and moisture affect all the stages of disease development but expansion of late blight lesions is governed mainly by temperature (273, 373). Soil temperature during and after rainfall need to be below 18°C for successful tuber infection (279). Similarly the moisture requirements for *P. infestans* to infect foliage and growing tubers also differs greatly a few hours of rain or dew at suitable temperature suffice for foliage infection (181), whereas a much greater accumulation of water in the soil is required for tuber infection (208, 279). In Israel cool temperature at night with heavy dew enable *P. infestans* to grow on irrigated potato crop in spite of very warm and dry days (273).

3.7 Host-Pathogen Interaction

P. infestans forms an intermediary link in the evolution of fungal pathogens. Presence of cell wall degrading enzymes and consequently the tissue necrosis that they produce in the potato tissue is the indication of the fact that at one time the fungus might have existed as a necrotroph. Production of intracellular haustoria in the host tissue on the other hand suggest a more balanced type of parasitism which is a measure of biotrophy. Although it is not certain what role fungal enzymes play in the overall pathogenesis, still there is every likelihood that endo-cellulase and gluconase may help in locally dissolving the host cell wall fabric and to degrade wall oppositions (178).

Under optimum disease conditions, the fungus zoosporangium germinates on the host surface either by germ tube or indirectly through the release of zoospores. The fungus mostly penetrates through the epidermal, cells just adjacent to stomatal guard cells (368). In side the host cell the fugus produces primary hypha/infection vesicle (98). They produce secondary hyphae which are both intracellular and transcellular.

As already mentioned the role of fungal enzymes in the host penetration is not well defined. *P. infestans* produces galactanase and polygalacturonase in culture filtrates. Galactnase has a specific role in pathogenesis through the removal of galactose from cell walls, but role of polygalacturonase is not known. Once the fungus establishes in the cells, it multiplies mostly through intercellular spaces and throughout haustoria in the healthy host cells. Extrahaustorial matrix gets deposited around the houstorium. Penetration of the fungus into the host triggers off many biochemical and cytochemical reactions which vary with genotypes. In susceptible genotypes, there is no cell necrosis in early stages of infection. Cytoplasm of infected cell shows no perceptible changes in organelles (178). After 24-48 hours post-infection, the cells get necrosed. This however, does not affect the development of the fungus in the adjoining healthy cells/tissue.

Race-specific and race-non-specific types of resistance are similar as far as cytological events are concerned except that in case of race-specific resistance the fungus gets confined to the site of initial penetration. Host cells respond more rapidly to the invading fungus by triggering a hypersensitive response (62).

Following infection and tissue colonization by *P. infestans,* the biochemical processes undergo a drastic change in the infected tissue. In incompatible interactions, chlorogenic acid increased more pronouncedly than compatible ones (85, 86, 87). Lignin deposition is higher in incompatible interactions. It is now believed that increase in phenylalanine ammonia lyase (PAL) activity and lignin deposition in early phase of infection are indications of a race-specific reaction of potato tuber tissue to *P. infestans.* Deposition of insoluble phenolic or lignin-like compounds is mostly associated with non-race specific resistance (3). Whereas scopolin accumulation in infected tubers is mostly likely an indication of compatible interaction.

P. infestans also has a profound effect on accumulation of sesquiterpenoid compounds in the necrotic tissue. Rishitin was the first such compound (phytoalexin) discovered in infected tissue. Since then many more phytoalexins have been isolated but only rishitin, lubimin and phytuberin are main phytoalexins which play role in disease resistance (111).

It is now an established fact that hypersensitive reaction coupled with accumulation of phytoalexins are involved in the host defens mechanism (111). But how these phenomena gets triggered on and off is a matter of discussion and/or investigation.

3.8 Genetics of Pathogen

Three main period of genetic studies on *P. infestans* can be recognized (301). This was first reviewed by Gallegly (129) following the discovery in Mexico in 1956 of the two mating types (126, 127, 128, 141, 241). Crosses between A and A isolates were made and the progeny of a single oospores was studied (125). The second period of genetic analysis of *P. infestans* sought to explain the mechanisms generating variations in asexual populations of the fungus. Antibiotic resistant and dependent strains were selected and used as markers, but the underlying mechanism(s) generating these somatic recombinants could not be ascertained (166). The most recent period stems from the collection and isozyme characterization of A_1 and A_2 Mexican isolates of *P. infestans* (332, 346) and accidental discovery of A_2 isolates in Egype and Western Europe (177, 219, 342). Shattock *et al.* (297, 298, 299, 300) found that crosses between parent isolates which were homozygous for different isozymes alleles yielded F_1 progeny which was heterozygous. Crosses between heterozygous parents yielded expected 1:2:1 ratios of homogygous: heterozygous : homogygous genotypes. Further results (301), indicated that in some cases segregation ratios did not follow Mendelian expectation. The cause of these disturbed segregations are not known.

Sweigard *et al.* (340) studied genetics of virulence in *P. infestans* by using isozymes as independent markers. Segregation in 1:1 ratios was observed for virulence/avirulence to both host genes (R_2 & R_4) and the two virulences appeared to be linked (132). These results indicated a simple genetic control of the factors virulence but dominance relationships could not be ascertained.

3.9 Management

Late blight is primarily a foliar disease but also badly infects the tubers. Therefore, its effective management entails reduction of both foliar and tuber infections. This cannot be achieved by employing any single method. Instead, the three way approach i.e. use of cultural and chemical methods and host resistance have to be used for an integrated control of the disease (Table 7).

Table 7 : Integrated Management of Late blight

Cultural Practices

1. use disease free and/or treated seed.
2. Remove/Rouge out all ground keepers.
3. Adjustment in planting time is helpful-in the plains, crop should be planted by 20th October.
4. High ridges and proper earthing up must be adopted to avoid exposure and infection of seed-tubers. This prevents further carrying over and dissemination of *P. infestans* spores.
5. Water management-Allow only light/or withold irrigation when the blight is likely to appear.
6. Killing of haulms of 75% crop damage.

Chemical

1. Application of a prophylactic spray of contact fungicide and subsequently only need based fungicide application is recommended. In years of prolonged winter-rains, in the plains one (two)spray(s) of systemic fungicide is desirable and helpful.
2. In the hills, one or third sprays of systemic fungicides need to be applied alternating with contact fungicides.

Host Resistance

1. Cultivation of late blight resistant varieties-cv. Kufri Jyoti in N-W hills; Kufri Megha in Meghalaya hills, Kufri Kanchan in Darjeeling hills and Kufri Badshah, JH-222 and JI-5857 (hybrid) in the plains.

Cultural Practices

These methods mainly aim at eliminating/reducing the initial inoculum load in seed tubers and other sources and to check the spread of the disease. Infected seed tubers are the main source of the disease. Therefore, care should be taken to avoid the seed infection by cutting the infected haulms. The tubers should be harvested after proper skin curing. If possible no seed should be taken from the infected crop. Alternatively through checking for sorting out the diseased tubers should be carried out before storage. In hills, refuse piles can serve as primary source of inoculum. Therefore, infected tubers and vines should be collected and buried, deep in the soil. Besides, ground keepers and volunteer plants harbour the pathogens, and should, therefore, be removed immediately after emergence. Higher ridges covering the tubers thoroughly with soil prevent growing out of fungus from mother tuber and consequent infection of progeny tubers. In the plains, the disease normally appears with the onset of winter rains. Therefore, irrigation should be completely stopped during the rainy period (30).

Host Resistance

During the early period of potato breeding, general resistance/field resistance was the only type of resistance available to breeders (354). Although wild species like *S. demissum* were put to use for exploiting their late blight resistance potential immediately after the Irish Potato Famine, still it was only in the beginning of this century that *S. demissum* was crossed with *S. tuberosum* and their progeny used in the resistance breeding programme in U. K. and Germany (33). Later on several wild species were exploited for their late blight resistance potential but *S. demissum* has been so far the most important source of resistance. In the beginning, *S. demissum* derived resistance brought success as it provided immunity (complete freedom) to the disease. However, the success was short lived as new compatible physiologic races of *P. infestans* appeared almost simultaneously (282) making the use of race-specific resistance infrustuous yet breeding for race-specific resistance continued until the end of the 1960. Slowly, the attention was shifted to breeding for general field resistance (35). This kind of resistance does not exert any directional selection pressure on the fungus hence the worked very well so far. There is hardly any increase in aggressiveness of *P. infestans* in association with the cultivars having a general resistance (188).

Field resistance is polygenically controlled and therefore difficult to handle in a back crossing programme as most of the genes flow only to a low proportion of progeny and that too disappear during the back crossing process (272). Nevertheless, field resistance had become the back bone of blight resistance breeding programme in most countries. Yet another problem faced in the exploitation of field resistance was its identification in absence of complex races of the fungus throughout the world except Toluca valley. Consequently, most of the countries had to get their breeding material evaluated there is Mexico. CPRI also entered into such a collaboration with Rock Feller Foundation Programme in Mexico in 1962 for screening the advanced hybrids, which led to selection of a large number of promising hybrids and were released for commercial cultivation viz. cv. Kufri Jyoti, Kufri Jeevan, Kufri Khasigaro and Kufri Muthu.

Besides, several other *Solanum* species, viz. *S. microdontum, S. bulbocastanum, S. chacoense* and *S. verrucosum* also possess high degree of field resistance. Some of these species like *S. verrucosum* and *S. bulbocastanum* have already become part of the breeding strategies in some selected laboratories (207, 296). The current level of Field resistance in *S. tuberosum* can be strengthened by combining this resistance with partial resistance from related *Solanum* species (26, 169, 170). Alternatively, non-host resistance from related species can be transferred to *S. tuberosum* (63).

Concept of 'neo-tuberosum' and its utility in late blight resistance breeding is a subject of discussion. Native *andigena* does not possess any resistance to late blight (319). However, *andigena* had the potential to respond to selection for field resistance. The main aim of the programme was to re-create a new subspecies of *Tuberosum* from South American primitive cultivars of the *andigena* type using a much larger genetic base (138). This material is now being used for breeding commercial cultivars and a reasonable degree of resistance has been buildup in Scotland and elsewhere but its nature is yet to be ascertained.

Chemicals

In the beginning, inorganic molecules were evaluated and those found promising against late blight were used for its management. Copper salts were found highly effective and, therefore, used extensively during 1885-1934 (288) against late blight. Despite their limitation of being strictly protective and requiring repeated applications, they were able to reduce the infection pressure resulting in an overall increase in crop yield. In 1934, dithiocarbamates and 1, 2-bisdithiocarbamates were discovered. They proved highly effective against this disease and were immediately put to wide use. They had since been used extensively for managing this disease. In between other groups, of fungicides were also discovered but all of them (excepting triphenyltin compounds) were contact/protective fungicides. A new group of fungicide, viz. cyanoacetamide-oxime (Cymoxanil) was discovered in 1976 which had better fungicidal activity and strong acropetal movement thus exhibiting both residual as well as curative action against *P. infestans* (288). Since then a host of other systemic fungicides have been discovered with better promise. Metalaxyl (Phenylamides) has by far proved the most potent fungicide ever evolved against late blight and within a few years of its discovery, its use increased rapidly and by 1980 it became the major fungicide used for the control of late blight in Ireland (335). Simultaneously, it was also used extensively in most of other European countries. However, within a very short period of its use, it led to the development of metalaxyl resistance strains in *P. infestans* (89, 90). Since then efforts are on to minimise the development of fungicide resistant strains in *P. infestans* (287, 335). Use of systemic fungicides alone over a long period leads to the formation of fungicide-resistant strains. Therefore, systemic fungicides need to be used in combination with contact fungicides which helps both to reduce the disease pressure and checks build up of the resistance strains. In some cases, mixing of contact fungicides with systemic fungicide alone may not help reduce the risk of development of the fungicide resistant strains. In such cases, other systemic fungicide alongwith contact fungicides need to be mixed. Samoucha and Gohen (277) reported synergistic interactions of cymoxanil mixtures in the control of metalaxyl resistant isolates of *P. infestans.* Alternation of fungicides also helps to prevent development of resistant strains but fungicide mixtures are superior (190).

Recently, systemic fungicides have also been found to eradicate the tuber borne inocula (320) but this may not be commercially exploited as it may pose some other problems. Whatever be the merits and demerits of systemic fungicides, they continue to hold the key to the successful management of late blight. However, it is advisble to use them only under compelling circumstances.

4. BACTERIAL WILT

Bacterial wilt of potato caused by *Pseudomonas solanacearum* was probably first recorded by Burril (197) in southern USA. While this is the first bacterial disease observed in India (53, 77).

Internationally, the disease is christened to express the symptoms caused and is known as bacterial wilt or brown rot. Locally in different countries, it is named differently (197) and in India it is known as rassa (high moisture disease), bangle blight or banqdi, parrya, ghera and ukta.

The disease is prevalent in all the six continents of globe but is more severe in tropical, subtropical and warm climats and devastating nature of the disease has changed cropping patterns in some parts of countriees like India, Indonesia and Peru (106, 151, 216). Before 1920, the disease was considered to be a limiting factor in potato cultivation in many parts of India (51, 220) but later surveys revealed negligible level of wilt incidence (31, 249). Latest surveys (1, 2, 123, 307), revealed prevalence of the disease in sixteen states of India. In some countries like Indonesia, China and others, the disease continues to be a problem (164, 216).

Assessment of economic losses are difficult as the crop plants wilt before tuberization and produce from infested fields continue to rot in storage. Reported damages are often based on the wilting in the fields. In India losses recorded are upto 55% in Kumaon hills (153, 247), 0.33-40% in Maharashtra (247), 20-25% at Hyderabad in Andhra Pradesh (240), and over 75% in some localities of Karnataka (123).

4.1 Symptoms

Infected plants shows sudden wilting but some times may show temporary recovery, followed by wilting in some branches or complete collapse of the plants. Potato tubers from diseased plants may or may not show external symptoms. Cross sections of the infected tubers indicate brown discolouration, but not essentially, in the vascular region and a slight pressure makes bacterial mass come out of the vascular region. In advanced stages of the disease, bacterial masses even ooze out from eyes of the tuber.

In North Western hills of India, lenticel infection of potato tubers due to *P. solanacearum* is observed. Water soaked lesions develop at the site of infected lenticels. The lesion enlarges from 0.5-1 cm in diameter and form 5 mm deep pits. The tuber skin remains intact but becomes shrivelled. These symptoms are often confused with *Fusarium* dry rot and are mostly associated with non wilted plants. However, when such tubers are cut bacterial mass is detected below the skin (314).

4.2 Pathogen, Distribution and Biology

P. infestans is an aerobic, Gram negative, rod bacterium measuring approximately 0.5-0.7 x 1.5 x 2.5 mm. Virulent forms are mainly non flagellated. Avirulent forms usually bear 1-4 flagella and are highly motile (199). Polar fimbriae present with which twitching motility and spreading growth on solid media are associated (168). Cells contain inclusion of poly-beta-hydroxybutyrate which are sudanophilic and refractile under phase contrast microscope (159) and commonly show biopolar staining.

It is chemoorganotroph with aerobic respiratory metabolism; catalase and Kovac's oxidase positive; optimum temperature for growth varies from 27-37°C depending on the strain (249, 374); nitrate is reduced to nitrite (160). *P. solanacearum* usually show low level of salt tolerance.

The bacterium lack fluorescence, phenazine and carotenoid pigments. A brown to black, diffussive pigment is often produced on variety of agar media containing tyrosine.

Two systems have been adopted to further classify *P. solanacearum* one placing emphasis on host affinity and establishment of 5 races (50, 165), and the other making use of selected biochemical properties as the basis for delineating it into 5 biovars (8, 49, 50, 160, 165).

Strains of race 1 and 3 affect potato crop and have generally been placed in biovar I, III and IV (race 1) and II (race 3). Biovar I & II are nutritionally less versatile than biovar III & IV (162, 248). Genetic diversities in races and biovars are being indicated by several recent studies carried at DNA level. Studies of 1991 reveals that biovar 2 have specific 2 kb DNA (the same DNA was also present in only 5 strains of race 1) and another large fragment was also associated with biovar 2 isolates. Biovar I and II are distinct from biovars III, IV & V on the basis of DNA probes and RFLP analysis (68). Within one RFLP group there are several clonal lines which can further be divided on the basis of whole genome analysis. Gilling (137) studied RFLP group by while genome analysis and found that RFLP group strain can further be divided on the basis of host range. There are also marked differences in geographical distribution of biovars. The race 1/biovar I is less frequent than race 3 biovar II in South America where the potatoes have originated (103). The only other parts of the world where the biovar I has been observed are East and West Africa (157, 242) and Philippines but these supposedly got distributed through infected seed material.

Race 1/biovar IIII is most predominant is Asia. Race 1 and biovar III and IV occur in Australia (256); China (164, 165), India (307), Indonesia (216), Papua New Guinea (345) and Sri Lanka (290). Biovar III is predominant in low land areas. Biovar III and IV are rare on potato in the center of origin of the crop (103). In India a large number of isolates (from plains and plateau) belonged to race 1/biovar III. Biovar IV has been recorded only from two locations in eastern plains and two locations in plateau. Although race 1 primarily occurs in low lands, in some cases it has been recorded in high lands of Peru (104) and India (339).

Race 3 biovar II often referred to as potato or low temperature race is wide spread in higher altitude of globe; Colombia (146), Peru to Southern Brazil, Uruguay (130), Argentina and Mexico (115), Sweden (244) and higher altitude in Costa Rica (107), Sri Lanka (290, 344), India (307), Papua New Guinea, China and Australia (160).

In many countries of southern Europe and Mediterranean area (160), and Argentina, Chile, Uruguay biovar II is the sole biovar (221). Strains of biovar II from potato in Australia formed a uniform phenotype whereas those from America showed some heterogeneity and were divisable in several distinct phenotypes (163, 222). Stray instances of this biovar have been reported from Indian lowlands but they are believed to have come through the seed. Martin *et. al.* (221) reported biovar II strains in low land tropical jungle of Perù which were more versatile than high land biovar II. These strains fall in RLFP division 2 but differ from RFLP group of high land biovar II.

Recent advances have shown that RFLP analysis of genome can provide objective basis for revaluations of taxonomy of *P. solanacearum* and also can determine races and biovars (65, 67).

P. Solanacearum in host plants or *in vitro* culture undergoes spontaneous shift from wild virulent types to avirulent or weakly virulent types at varying rates and associated changes are acquisition of flagella, ability to cause hypersensitive raction (HR) and others. These changes can easily be identified by growing *P. solanacearum* on nutrient rich Kelman's agar medium (198) on the basis of colony morphology, microscopic and plant inoculation tests.

Lipopolysaccharide (LPS) analysis of HR inducing and non inducing spontaneous mutants revealed that LPS of HR inducing mutant was rough (incomplete LPS, lack part or complete 0 antigen) whereas that of non inducing was smooth (complete LPS). Acridine orange resistant avirulent mutants also had changed LPS and EPS (exopolysaccharides) (91, 230).

LPS of *P. solanacearum* plays an important role in the virulence (17, 167). In *P. solanacearum* the wild type EPS molecule apparently is required to prevent agglutination by certain plant lectins (292). Most, but not all of the mutations which are affected in LPS synthesis are also affected in virulent (167).

Most of the known or suspected virulence factors in *P. solanacearum* are extracellular products including cell wall degrading enzymes, extracellular polysaccharides, plant hormones, auxins, cytokinis and ethylene. Efforts are being made to identify gene(s) responsible for expression and production of these products by creating artificially well defined mutants. Other genes whose gene products are not yet known are also identified and have been found to affect pathogenicity in *P. solanacearum.*

Two polygalacturonase (endo Pg and exo Pg) have been identified in AW1 strain of *P. solanacearum*. The major one corresponds to endo Pg and its importance was assessed by creating Pg1A deficient strain that had normal level of EPS. Such mutant caused delayed wilt in tomato plants (281). Endogluconase (EGL) is excreted at high level by several virulent strains isolated from various geographical region and EGL is produced at low level by spontaneous pleiotropic avirulent mutants. EGL (B-1, 4 endogluconase) cleave the soluble cellulose apparently releasing cellobios (280). Eg1 deficient mutants had delayed onset of wilt and took longer time to kill the plants. Restoration of virulence was achieved by complementation of gene Eg1A. These indicated that eg1 plays role in pathogenesis but it is not absolutely essential for wilt (264).

Mechanism of wilt induction is not clearly understood. The extracellular polysaccharide produced by *P. solanacearum* in plant is believed to contribute to plugging of xylem vessels, interfering with the water transport and appears to be most formidable arsenal of virulence. It has been studied most extensively as the virulence factor. Both EPS and LPS which are the component of outer membrane are required to control the exchange of charged molecules with the environment and can be important for virulence. Complex molecule like EPS require many enzymes for biosynthesis.

Staschawitcz *et al.* (333), Denny *et al.* (76) and Xu *et al.* (372) obtained EPS lacking, Tn5 induced avirulent mutants. In their studies loss in virulence was associated with loss in EPS. However, several mutants have been obtained which and fluidal colony types but lacked virulence. Quantitative and qualitative changes in the EPS have been found to affect virulence in *P. solanacearum.* Some fluidal mutants induced by Tn5 insertions lose virulence and have been shown to be due to interruptions in virulence gene whereas when the mutants were avirulent and lacked EPS, the gene affected have shown interruption in gene which control both EPS synthesis and virulence. In virulent mutant which lacked EPS, the interruption has been observed in the EPS synthesis gene (371, 372).

A gene cluster of 6.5 required for afluidal phenotype and virulence was identified in avirulent mutant of K 60 (66). Only insertions in 6.5 kb region affected colony morphology or virulence or both. Five complementation units were identified. Of these epsA-epsD were required for fluidal morphology and virulence and rvrA (reduced virulence) was required for virulence only. All the mutants affecting colony morphology affected the virulence. Furthermore all EPS-mutants were less virulent than EPS i (impaired) mutants.

Studies on *P. solanacearum* strain GMI 1000 revealed that large proportions of Tn5 mutants produced colonies which were intermediate in fluidity and also afluidal and retained pathogenic characters but varied in aggressiveness to cause wilt (39). Denny *et al.* (76) demonstrated the role of EPS in aggressiveness by isolating Tn5 induced mutants varying in their EPS production. Supernatants from EPS affected colonies were less viscous than wild types. These mutants fell into three classes according to aggressiveness and additional traits. Class I mutants were impaired in EPS production (EPS i), produced limited amount of EPS both in rich and minimal medium and were weak in their ability to induce wilting in tomato plants. Class II mutants did not produce EPS when grown in rich media but produced normal amount when grown in minimal medium or in plants and retained wild type aggressiveness towards potato plants. Class III mimmicked the avirulent spontaneous mutants in all respect in their ability to cause wilt. These studies indicate that EPS production affects aggressiveness of the pathogen. Atleast three cluster of genes involved in the biosynthesis of EPS have been mapped by transposon mutagenesis. Class I and II mutants were maped in two adjacent region of the chromosome spanning over a minimum of 9 and 2.6 kb respectively. Mutations in EPS I affected virulence because of affected EPS production in plants but was not affected by mutation in EPS II. A third cluster of genes spanning more than 6.5 kb was identified in K 60 controlling EPS production. A mutation in any of the ive complementation units affected aggressiveness.

A key question has been that how and what can be responsible for spontaneous shift from fluidal types to the afluidal avirulent types and in a process change in associated traits. It has been shown by random and site specific (Tn5, Tn7, Tn3 and Tn10) mutagenesis that interruptions in biosynthetic and virulence genes change the colony types and virulence.

Plasmids have been recorded in most of the strains of *P. solanacearum*, irrespective of geographical origin or host plants. Their number varied from 1-2 (69, 234, 269). As the spontaneous avirulent mutants in *P. solanacearum* were not reverting it was though that loss of plasmid is responsible for the change of virulence. In some strains plasmids have not been detected. It may be due to harsh methods used in isolation or extremely small size of the plasmid not forming an apparent band. Currier and Morgan (69) and Morales and Sequeira (234) compared the plasmid contents of wild types and their spontaneous avirulent mutants. They found that avirulent mutants harboured the same plasmids and of same size as of their parents. Ofuya and Wood (243) indicated lack of plasmid in avirulent strains however.

Afluidal variants of two Indian field isolates (GSC 26 and CSG 39) collected from two agroclimatic zones showed tendency to throw out fluidal revertants in the range of 3-8% (122, 315) on TZC medium without an inducer or deliberate genetic manipulation. The frequency of reversions has been found to increase in presence of antibiotics; oxytetracycline and tetracycline, but not by chloremphenicol, streptomycin, vancomycin. Afluidal variants caused slight wilt in tomato plants and was attributed to the presence of fluidal cells in inoculum, as all the afluidal colonies used for inoculations contained 3-8% fluidal cells and such infected plants often recovered. Spontaneous and antibiotic induced fluidal revertants were similar in potential to cause wilt in tomato plants (57). It was therefore inferred that these isolates produce instable afluidal variants. The reasons for such instability are not still understood (54, 122).

A mutant AW 83 obtained from wild type AW1 by Tn5 mutation could be induced to produce EPS on cultivation close to wild type. Such induction was not possible in spontaneously produced avirulent mutants (PC type) or other Tn5 induced mutants (PC like). Tn5 insertions were recorded in 4 kb Eco R I fragment in case of PC like mutant whereas it was in 10 kb Eco R I fragment in inducible in 5 mutants. Transformation of 2.2 kb DNA fragment from wild type to PC type or PC like mutants resorted wild type character but not in inducible mutants. The gene on this fragment is named pheA. Introduction of this gene in many PC type from other isolates restored to wild types but not all PC type like B1 (48). This indicate that there are other genes which are also responsible for causing PC type mutations. This gene pheA is known to code 38 kd protein. The loss of expression or functional product of this gene has shown to be the reason for the production of avirulent afluidal types and appears to have regulatory function responsible for set of PC associated traits.

A locus for causing phenotypic shift has also been identified in wild type K 60 and its spontaneous avirulent mutant B1 and multicopy of single gene espR was found responsible. This gene encode protein of about 25 kd. phenotypic shift in K 60 is probably caused by rearrangements that result in high level of gene expression (181).

Avirulent mutant of *P. solanacearum* GM 1000 and GM1 1229 were obtained using Tn5 random mutagenesis (39). Broadly these mutants could be divided into two classes : a) Mutant lost ability to cause hypersensitivity on tobacco and wilt in tomato plants (GMI 1301, 1304, 1305, 1331). Similar mutants were also obtained by Macol and Denny (217) in another strain AW1. b) Mutants lost ability to wilt tomato plants but retained ability to cause hypersensitive reaction on tobacco plants (GM1 1229, 1313, 1314, 1315, 1330).

All these mutants infected tomato seedlings to varying degree except GM1 1313 (Hr+). One mutant GM1 1305 (Hr-1) was most invasive and multiplied at a higher rate than any other mutant but was not upto the extent of wild type (348). These findings indicate that hrp genes are not necessary for infection of plant roots and multiplication. These mutants could be isolated from vermiculite around tomato plants and grow easily *in vitro*. Reduced growth in plants could be limited by availability of nutrient of the plant. If so hrp genes could play role in diverting certain plant metabolites from plant to bacteria.

Acridine resistant avirulent mutants of GM1 1000 were deficient in atleast 85 kb region from megaplasmid. Most of the hrp genes are clustered on a particular region of megaplasmid and span more than 23 kb and organised in six transcriptional units (9, 40, 41). Homologous sequences have been detected in more than 90 *P. solanacearum* strains representing most of the biovars, host and geographical area (42, 68). Mutants affected in this locus are non aggressive on potato and are leaky in their ability to induce HR on to tobacco. This locus have been organised into two transcriptional units of 0.7 and 1.3 kb. Another hrp gene cluster has been identified in strain K 60 (182) which does not share homology with the above cluster.

4.3 Epidemiology

Latently infected potato tubers are the single most important carrier of *P. solanacearum* from one continent to another. In crop grown under low soil inoculum level, wilt is rarely noticed but latent infection is observed to an extent of 60% (119, 120), and pathogen is carried in vascular tissues, on surface and in lenticels (324, 338).

Once the pathogen is introduced in the field soils, it causes infections in the susceptible plants. Bacteria spread from plant to plant by root contact or through the water film or even flowing water. Bacterial slime from the plants may form small pillets that enhance the bacterial survival (314). Population of *P. solanacearum* in field soils may increase or decrease but rarely become extinct, depending upon the strain present, cropping pattern and native vegetation, Field soils formed the major inoculum source in potato crops cultivated in Belgaum; India (123, 314). It has been observed that the inoculum reaches to the threshold level of 10^7 cfu before the infection starts (79).

Soil inoculum of bacteria is affected by several factors. *P. solanacearum* is non spore forming and therefore low soil moisture and high temperatures are expected to affect soil survivals. Reports regarding soil survival of this organism are not consistant. The bacterium survived for 15 months in pot soils and less than a year in soil (70, 260).

Shekhawat and Perombelon (316) studied survival of strains of biovar III (race1) and biovar II (race 3) in artificially infested clay and sand mixed in different proportions. At 30°C both the races reached nondetectable level in 3 weeks in dry sand and 4-6 weeks in dry clay. In wet soils (60% WHC) the pathogen perished in 8-10 weeks at 35°C and 12 weeks at 5°C. The race 1 isolate in most congenial conditions i.e. clay soil at 60% WHC and 15°C declined at the rate of Log no. 0.33. This rate of decline would bring down population to nil in 23 weeks.

Under field conditions, soil temperature at diferent locations may affect the survival. Shekhwat *et al.* (306, 307) observed that the pathogen does not survive in northern plains of India where mean monthly maximum temperature remains above 40°C accompanied with water deficit in soil continuously for 45 days during summer and also in hills where mean monthly minimum temperature during Dec.-Feb. remain below 0°C.

Strains of race 1 are versatile and have greater adaptability than the race 3 strains. These strains are mostly prevalent in South East Asia and Australian region. Survival is greatest under wet but well drained conditions and is adversely affected by soil dessication. Although *P. solanacearum* is regarded anaerobic organism, field studies in Sri Lanka demonstrated that *P. solanacearum* is not eradicated by flooding soils for several weeks (291). In India too the experience has been very much similar (123, 310, 313). In wilt endemic areas (Balgaum region of Karnataka), it is a common practice to alternate potato crop with paddy. Under such situations, usually a higher potato wilt incidence has been observed. Submerged soils of paddy fields are not uniformly devoid of oxygen. In puddled soil culture, soil is a complex mosaic consisting aerobic and anaerobic microecosystems (257). The rice rhizosphere and rhizoplane provide aerobic environment due to diffusion of atmospheric oxygen through the root (11). Therefore, even in flooded fields, soils are not truly anaerobic.

Remnants of plant parts, latently infected left over tubers and self sown infected but germinated tubers contribute inoculum for next seasons crop (92, 142, 238).

In spite of short survival in free soils and host debris recurrence of bacterial wilt has been reported when planting of host crop plants were resumed after keeping the fields fallow or were under rotations with non host crops for two to seven years (120) indicating long term survival of the pathogens in field conditions. Pathogen is carried in sheltered sites inside the plants with or without wilt symptoms or sustain them in rhizosphere. Significant inoculum is maintained in such ways and inoculum continues to remain in the field for long durations. The magnitude of damage to the crop may depend upon strains prevalent in the area and host crop planted. The extensive host range includes several species. Alone in Indian different workers have reported 114 kinds of plants representing 37 families as host for *P. solanacearum* (314), many of them are symptomless. A perennial symptomless plant *Solanum dulcamara* and *S. cinereum* carries biovar II in Sweden and Australia respectively providing inoculum to potato crops (144, 245). In India *P. solanacearum* has been isolated from surface sterilised roots of *Ageratum conyzoides* and *Ranunculus scelaratus* showing no symptoms (338). *P. solanacearum* has also been reported to multiply in the rhizosphere of *Portulaca oleracea* in Philippines (259).

Extended survival is also achieved by infecting non host crop plants (147). Shekhawat *et al.* (313) found infections in the roots of maize, wheat, paddy and beans. Gadewar and Shekhawat (121) found infections in roots of wheat, sorghum, maize, cabbage, cauliflower but not in roots of finger millet. Survival in deep soil layers have also been reported (145, 201, 224, 338).

Environmental factors affecting plant pathogen interactions

In general increase in the temperature enhanced wilt both in tomato and potato (58, 79, 172). Sequeira and Rowe (293) reported that any fluctuations in temperature may upset the balance between host resistance and multiplication of the pathogen and shift host reaction either towards higher resistance or greater susceptibility. Tremendous variations in strain of *P. solanacearum* in relation to the ecological conditions has been described (106). Tung *et al.* (350) showed that strains of biovar III and IV has increased virulence when the temperature is increased. These data support the concept that high temperature exerts its effect mainly by favouring rapid disease development in plants.

High soil moisture favours survival of the bacterium in the soil, infection, disease development after infection and exit of the bacterium from the host and spread in the soil (197).

Combined effect of temperature and moisture is very much glaring in North Western High hills of India where the potatoes are planted in March and in subsequent months congenial temperature (25-30°C) prevails during May and first fortnight of June but the disease appears only in mid July after onset of monsoon. Singh (324) observed that soil population of *P. solanacearum* as well as wilt incidence were maximum during July-August when the soil moisture was comparatively higher.

4.4 Control

The bacterial wilt pathogens forms a heterogenous group on the basis of its host range, its soil borne nature and host parasite interactions which are greatly affected by environmental factors. Hence its control practices ought to be directed towards reducing the inoculum in the field and avoiding introduction of fresh inoculum, growing the crop when the environmental conditions are not favourable for the pathogens and developing resistant varieties or by biological means such as use of the avirulent strains of the pathogens, and/or rhizobacteria.

Use of disease free seed

Use of healthy seed in infested fields reduced wilt incidence up to 81.5% (311). Farmers' fields in Deccan plateau (Karnataka, India) revealed a high wilt incidence where the seed used was from the wilt endemic areas than the seed from the wilt free-areas (123).

In the past, areas in Deccan plateau (India) like Pune (Maharashtra), Hassan, Chikbalapur, Dharwar (Karnataka) were considered to be 'hot spots' for bacterial wilt but now the disease is either absent or has low incidence. This has happened because, in the past potato seed was mainly obtained from the wilt endemic areas but now, the seed used is either from the wilt-free areas like higher hills of the Himachal Pradesh or from plains of Punjab (123, 314). A similar situations occurs in Australia.

Soil disinfection

Under Indian conditions, Shekhawat *et al.* (310, 311) observed that application of bleaching power @ 12 kg/ha mixed with fertilizer or soil drenching before or after first earthing up reduced the bacterial wilt incidence by 80% and also increased the potato yield but more than one application of bleaching powder suppressed the yield.

The pathogen is susceptible to higher temperature. Bacterial wilt incidence is negligible where the soil burning (Jhumming) is practiced. These observations were further substantiated by conducting experiments in eastern and western hills and plateau regions of India where experimental plots were covered with straw and burnt (308).

Soil amendments

An application of urea @ 100 kg/ha has been found to be effective in reducing the wilt incidence (197, 305, 309). Lee *et al.* (213) also observed marked reduction of tobacco bacterial wilt in plots receiving 2 and 4 kg borax and boric acid/ha. In Java development of the potato wilt was reduced by application of copra and peanut meals though the final wilt incidence was similar to that in untreated plots (358).

Biological Control

Biological control involving rhizobacteria and avirulent forms of the pathogen can be a potential way to manage the disease. The mechanism involves exclusion of *P. solanacearum* by colonisation, antagonism and may also induce resistance by avirulent forms of *P. solanacearum* and release of bacteriophages by lysogenic strains of *P. solanacearum* (57). Certain soil and rhizosphere bacteria like *P. fluorescens, Bacillus* spp., *B. polymyxa* and actinomycetes have been found to reduce the wilt development and incidence (12, 120).

In our trials *Bacillus* spp. reduced potato bacterial wilt incidence upto 79% both under glass house and field conditions. Similarly *P. fluorescens* isolates reduced incidence by 43 to 75% and actinomycetes upto 79% (314).

Spontaneous avirulent mutants of *P. solanacearum* had been tried to protect potato plants from virulent strains (200, 227). It was observed that the wilt incidence was reduced in some trials and was influenced by change in the temperature. Protection effect was lost if temperature were higher favouring the multiplication of the virulent types.

Transposon (Tn5) induced avirulent form invading unwounded tomato root system were effective in preventing subsequent colonization by wild type strains but the degree of protection depended upon maintaining ratio of virulent to avirulent cells in the inoculum at equal or lower than 0.1% and the timing of inoculation (349).

Resistant varieties

Solanum phureja based resistant cultivars viz. Serrana, Caxamarca and Molinera are reported to remain effective in South American conditions but the Indian isolates caused wilt in them within 5-10 days of inoculation (309).

The specificity of pathogen-host-environment interaction has prevented the breeding of the genotypes with universal durable wilt resistance fit for locations different from those under which they have been selected (337). New sources of resistance have now been identified in diploid *Solanum* species viz. *S. raphanifolium, S. sparsipilum, S. chacoense* and *S. microdontum* which are currently being used (59) for resistance breeding (283, 284). Recent evidence obtained with wild species suggest that contrary to the major genes inheritance observed in *S. phureja,* there may also be polygenic inheritance in some of the resistant wild species.

Hot and cold weather cultivation

The pathogen persists in debris, glued in soil and deep remnant of infected roots. Pathogens can be exposed to adverse either cold or hot climate conditions by deep ploughing in off season (244). Trial in plateau and hills of India revealed that summer and winter ploughing alone reduced potato wilt incidence by 66 and 50%, respectively (309) and support the findings of Llyod (215) that soil disturbance has deleterious effect on survival of the pathogen.

Crop rotations and/or intercropping

Crop rotation as a means to reduce wilt incidence has been successful in both hills and in Indian plains (123, 309, 314). Crop rotations with wheat, maize, oat, barley, sunhemp, finger millet and vegetables like cabbage, onion, garlic and others reduced the wilt incidence to the extent of 94%. Recent studies (313) in north western hills further reveals the effectiveness of crop rotations. Rotations with crop viz. bean, paddy, wheat, maize and finger millet reduced the wilt incidence in the range of 25-81% (Table-8). Roots of finger millet were free of *P. solanacearum* infection whereas paddy, maize, wheat and beans carried the pathogen. Such infection also have been reported in maize, wheat, sorghum, cabbage and knol-khol but not in finger millets in field studies in Indian plains in Bangalore. However, rhizosphere population of the pathogen was found to be reduced (123, 313). The pathogen infects the roots and produce small necrotic lesions in parenchyma in which buildup of bacterial population remains low, not adequate for release in the soil and therefore wilt is reduced.

Table 8 : Effect of crop rotation on wilt incidence and yield loss of potato at Bhowali North-western hills (313)

	Agerave wilt incidence		Yield loss
	1987	1990	during 1990
Potato-potato-potato-potato	26.9	51.7 (+2)	56.9
Potato-bean-bean-potato	26.3	19.7 (-25)	18.3
Potato-paddy-paddy-potato	25.5	12.0 (-53)	29.7
Potato-wheat-wheat-potato	25.3	9.2 (-63)	18.8
Potato-maize maize-potato	25.7	8.7 (-66)	13.1
Potato-finger millet-finger millet-potato	24.8	44.7 (-81)	7.1

Note : Figures in parentheses indicate per cent increase (+) or decrease (-) of wilt incidence over 1987).

Control of the bacterial wilt race 3 strains is easier because of its limited host range. In New South Wales, Australia *S. nigrum* (161) and *S. cinereum* (143) are known hosts for this strain but they do not survive vegetatively in cool winters in northern table lands. Shamsuddin *et al.* (294) examined survival of the pathogen in the soils in NSW and showed that in bare fallow and weed fallow soils the bacterium survived for two years but the population declined remarkably. An effective control is obtained by following rotation with the pasture for two years coupled with use of clean seed.

It is difficult to control bacterial wilt in a region where active *P. solanacearum* is maintained in plants. This is possible in a situation where soil moisture is always present to perpetuate flora which always propogates *P. solanacearum*. Jackson and Gonzalez (186) found that in of controlling weeds and crop rotations with maize, beans and sweet potatoes. In some parts of Deccan plateau (Belgaum area) the soil moisture is always present due to intensive cropping and weeds are present or the symptomless plants like *Amaranthus virdis* (118) either grown as an intercrop vegetable or occurring as weed, harbour *P. solanacearum*. Under such a situation, the control of the bacterial wilt is difficult.

Intercropping has been used as a means of reducing pathogen population in soil and root to root spread of *P. solanacearum* race 3 biovar II. Combination of potato with maize or cowpea were the most effective among the several crops tried (205).

In Mauritus, bacterial wilt is endemic in sugarcane fields and it has been suspected that rhizosphere of this plant is good for this pathogen (97). In Belgaum, where sugarcane is grown either as an inter or rotation crop, incidence of potato bacterial wilt has always been high (123).

Integrated control

It has been well established that the seed borne inoculum and inoculum in the fields (short term in infected left overs, self sown infected tubers, roots of non host crops, soil inocula or plant debris and long term inocula; in deep soils, weed hosts and rhizosphere of weed and crop host plants) are the main sources of primary inocula. In addition, the strains differ in their ability to infect one or several plants and aggressiveness. Host interactions with strains is positively influenced by rise in the temperature (104). Therefore, evaluation of contribution of each source has to be made before undertaking a management schedule.

Use of certified seed has been successful in coastal areas of Peru, where the soil is antagonist to the pathogen (104, 105). In warmer climate of Rio Negro in Colombia (94), a six months fallow reduced the wilt. In Kenya two year rotations with pasture was useful (156). In New South Wales, rotation with pastures has been successful. However any single method may not be useful to reduce the wilt to an economically affordable level at all the locations and a combination of control measures has to be worked out for any particular location.

Under Indian conditions the major source of the pathogen seems be the infected seed material moving from one region to another. Other inoculum sources are:

1. Crop plants which are host of the pathogen and are under regular cultivation in a region.
2. Weed hosts or symptomless carriers.
3. Isolates in India (presently reported only in a few) have tendency to revert from avirulent types to virulent types may have a role in survival. Their contribution in invoking the disease has not been so far investigated.

Based on inoculum source and pattern of occurrence of the disease, three zones nave been identified in India (312) viz. i) non endemic areas (Gujarat, Maharashtra, North western and North central plains and North western high hills), ii) North western mid hills (upto 2200 MSL), North Eastern hills and Nilgiris and iii) Eastern plains and Deccan plateau. Trials were carried at places represent in the above zones (Table-9). In non endemic areas, use of disease free seed alone is adequate to control the disease. In mid hills, use of disease free seeds, cold weather cultivation and application of stable bleaching powder @ 12 kg/ha are adequate. Application of bleaching powder may be substituted with two years crop rotation with crop like wheat, barley, finger millet, cabbage, cauliflower, knol khol, carrot, onion, garlic etc. Eastern plains and Deccan plateau are likely to have many symptomless hosts of the pathogen and therefore control of the disease is most difficult. However, disease can be reduced to a considerable extent by following cultural practices, like i) use of disease free seeds, ii) adjusting the date of planting, iii) application of bleaching powder, iv) straw burning in infested soils, v) blind earthing up. vi) hot weather cultivation and vii) crop rotations (85, 311, 312, 314).

Table 9 : Control of Bacterial wilt and Brown rot (1)

	Percent Reduction		Percent yield increase
	Wilt	B. rot	
T1 = Healthy seed	50.76	40-46	18-27
T1 + SPP 12 Kg/ha + Blind Earthing	7-94	50-85	22-28
T1 + Summer Ploughing in Plains	50-83	50-70	5-25
T1 + Winter Ploughing in Hills	60-75	62-78	35-40
T1 + Early Planting in hills	77-80	75-85	44-68

At Deesa and Chindwara, 100% disease control by disease free seed alone.

Based on 3 years data from seven locations viz. Deesa. Chhindwara, Ranchi, Kalyani, Bhubaneswar, Shillong & Bhowali.

5. CONCLUSIONS

During the recent years there has been increase in potato tuber yields in many parts of the world, and also expansion of potato cultivation in newer areas mainly due to increased supply of healthy seed-potatoes locally made possible through better virus-detection technology, *in vitro* rapid multiplication of healthy stocks and also incorporation of resistance to some of the important potato viruses in new cultivars. Incorporation of virus resistance against common potato viruses through transformation by genetic engineering has been attempted and is being tested for its relevance at field scale. The areas needing more attention include understanding the mechanism of resistance incorporated through plant transformation, development of models for predicting virus and late blight incidence and yield losses, and use of TPS as an alternative for potato production in traditionally non-seed areas.

6. REFERENCES

1. AICPIP, 1989. 17th Progress Report. All India Coordinated Potato Improvement Project. CPRI, Shimla pp. 122-132.

2. AICPIP. 1990. 18th Progress Report. All India Coordinated Potato Improvement Project. CPRI, Shimla. India.

3. Ampomah, Y. A. and Friend, J. 1988. Insoluble phenolic compounds and resistance of potato tuber disc to *Phytophthora* and *Phoma.* Phytochemistry, **27** : 2533-2541.

4. Anonymous. 1947. The measurement of potato blight. Trans. Br. Mycol. Soc. **31** : 140–141.

5. Anonymous. 1981. Annual Report, International Potato Center (CIP) for 1980. CIP, Lima Peru, 198 pp.

6. Anonymous. 1991. Annual Report, International Potato Center, Lima-Peru for1990, pp. 53-58.

7. Anonymous. 1992. Annual Report, International Potato Center, CIP. Lima-Peru.

8. Aragaki, M. and Guinon, V. L. 1965. Bacterial wilt of ornamental gingers *(Hedychium* spp.) caused by *Pseudomonas solanaçearum* Plant Dis. Reptr. **49** : 378-379.

9. Arlat, M., Gough, C. L., Zischeck, C., Barberis, P. A. and Trigalet, A. 1992. Transcriptional organization and expression of the large *hrp* gene cluster of *Pseudomonas solanacearum.*Mol. Plant-Microb. Interact. **5** : 187-193.

10. Arenz. B. and Hunnius, W. 1959. Der Einflusz verschiedener Virusarten auf die Ertrajsbildung hei der Kartoffel. Bayerisches landwirts chaftliche Jaharbush, **36** : 163-173.

11. Armstrong, W. 1967.The oxidizing activity of roots in water logged soil. Physiol. Plant, **20** : 920-926.

12. Aspiras, R. B. and de la Cruz, A. R. 1986. Potential biological control of bacterial wilt in tomato and potato with *Bacillus* Polymyxa FU 6 and *Pseudomonas fluorescens*. In "Bacterial Wilt Disease in Asia and South pacific". (ed. Persley, G. J) ACIAR Proc. No. 13, pp. 89-92.

13 Austin, S., Baer, M. A. and Halojeson, J. P. 1986. Transfer of resistance to potato leafroll virus from *Solanum brevidens* into *S. tuberosum* by somatic fusion. Plant Sci. **39** : 75-82.

14. Bagnall, R. H. 1981. Potato virus X. . Compendium of Potato Diseases, Ed. W. J. Hooker, pp. 75-47. Am. Phytopathol. Soc., St. Paul, Minn. 125pp.

15. Bagnall, R.H. 1988. Role of plant resistance in control of the potato viruses. In : Potato Pest Management in Canada Ed. G. Boiteau, *et al,* pp 70-76, Canada Agric., N.B.

16. Bahner, I., Lamb, J., Mayo, M. A. and Hay, R. T. 1990. Expression of the genome of potato leafroll virus: readthrough of the coat protein termination codon *in vivo.* J. Gen. Virol. **71** : 2251-2256.

17. Baker, J. M., Neilson, J., Sequeira, L. and Keegstra, K. G. 1984. Chemical characterization of the lipid of *Pseudomonas solanacearum.* Appl. Environ. Microbio. **47** : 1096-1100

18. Banttari, E. E., Ellis, P. J. and Khurana, S. M. Paul. 1993. Management of diseases caused by viruses and virus like pathogens. In : Potato Health Management, Ed. R. C. Rowe, Potato Assoc. of Ameerica & Amer, Phytopath Soc. St. Paul, MN (USA), pp. 127-134.

19. Barker, H. and Harrison, B. D. 1985. Restricted multiplication of potato leafroll virus in resistant potato genotypes. Ann. Appl. Biol. **107** : 205-212.

20. Barker, H. and Harrison, B. D. 1986. Restricted distribution of potato leafroll virus antigen in resistant potato genotypes and its effect on transmission of the virus by aphids. Ann. Appl. Biol. **109** : 595-604.

21. Baulcombe, D. C., Lloyd, J., Manoussopoulos, N., Roberts, V. M. and Harrison, B. D. 1993. Signal for potyvirus dependent aphid transmission of potato ancuba mosaic virus and effect of its transfer to potato virus x. J. Gen. Virol. **74** : 1245-1253.

22. Baunoch, D. A., Das, P. and Hari, V. 1990. Potato virus Y helper component protein is associated with amorphous inclusions. J. Gen. Viron. **71** : 2479-2482.

23. Beachy, R. N., Loesch-Fries, I. and Tumer, N. E. 1990. Coat protein-mediated resistance against virus infection. Ann. Reg. Phytopathol. **28** : 451-474.

24. Beaumont, A. 1947. The dependence on the weather of the dates of out-break of potato blight epidemics. Trans. Br. Mycol. Soc. **31** : 45-53.

25. Beemster, A.B.R., and de Bokx, J. A. 1987. Surfey of properties and symptoms. In : Viruses of Potatoes and Seed-Potato Production, 2nd. Edn., Eds. J. A. de Bokx and J. P. H. van der Want, pp. 94 Pudoc, Wagenigen.

26. Behnke, M. 1980. General resistance to late blight of *Solanum tuberosum* plants from callus resistant to culture filtrates of *Phytophthora infestans.* Theoret. Appl. Genet. **56** : 151-152.

27. Best, R. J. 1968. Tomato spotted wilt virus. Adv. Virus Res. **13** : 65-146.

28. Bhattacharya, S. K., Phadtare, S. G. and Khanna, R. N. 1982a. Races of *Phytophthora infestans* recorded in India. pp. 148-453. In : Workship on *Phytophthora* diseases of Tropical cultivated plants Ed. K. K Nambiar, *et al.* CPCRI, Kasargod, Kerala, India.

29. Bhattacharyya, S. K. Raj, S., Singh, D. S., Khanna, R. N. and Shiv Ram. 1982b. Forecasting late blight of potato in Indian hills. In : Potato in developing countries, Eds: B. B. Nagaich, *et al.*, IPA, Shimla, pp. 414-424.

30. Bhattacharyya, S. K., Shekhawat, G. S., and Singh, B. P. 1990. Potato late blight. Tech. Bull. No. 27. CPRI, Shimla., p. 40.

31. Bhide, V. P. 1959. Bacterial wilt of Solanaceous vegetables. Proc. Seminar on Diseases of Horticultùral Plants. ICAR, New Delhi. pp. 172-175.

32. Black, W. 1960. Races of *Phytophthora infestans* and resistance problems in potatoes. Ann. Rept. pp. 29-38. Scot. Plant Breeding Sta., Edinburgh (U. K.)

33. Black, W. 1971. Research on potatoes. pp. 52-60. ann. Rept. 1970-71. Scottish Plant Breeding Station, Edinburgh (U.K.)

34. Blodgett, F. M. 1941. A method for the determination of losses due to diseased or missing plants. Amer. Potato J. **18** : 132-135.

35. Bonde, R. 1932. A promising blight resistant potato. Amer. Potato J. **9** : 49-54.

36. Bonde, R. and Schultz, E. S. 1943. Potato refuse piles as a factor in the dissemination of late blight. Maine. Agric. Exp. Stn. BNull. No. 416, 229 p.

37. Bondfe, R., Schultz, E. S. and Raleigh, W. P. 1943. Rate of spread and effect on yield of potato virus diseases. Maine Agri. Expt. Stn. Bull . 421, 28 pp.

38. Borchardt, G., Bonde, O., Bartels, R. and Holz, W. 1964. Utersu-chungen uber die Minderung des Ertrages von Kartoffel-pflanzen durch Virusinfektionen. J. Plant Pathol. **81** : 168-475.

39. Boucher, C. Barberis, P. A., Trigalet, A. P. and Demery, D. A. 1985. Transposon mutagenesis of *Pseudomonas solanacearum :* isolation of Tn5 induced avirulent mutant. J. Gen. Microbiol. **131** : 2449-2457.

40. Boucher, C., Martinel, A., Barberis, P., Alloing, G. and Zischeck, C. 1986. Virulence genes are carried by a megaplasmid of the plant pathogen *Pseudomonas solancearum.* Mol. Gen. Genet. **205** : 270-275.

41. Boucher, C. A., Van Gijsegens, F., Barberis, P. A., Arlat, M., and Zischek, C. 1987. *Pseudomonas solanacearum* genes controlling both pathogenicity and hypersensitivity on tobacco are clustered. J. Bacteriol. **169** : 5626-5632.

42. Boucher, C. A., Barberris, P. A. and Arlat, M. 1988. Acridine orange selection for deletion of hrp genes in all races of *Pseudomonas solanacearum* Mol. Plant-Microb. Interact. **1** : 282-288.

43. Bourke, P.M.A. 1955. Working Group on Weather and Plant Pathological Problems (Commision for Agric. Meteorol., World Meteor, Organization, pp. 49, Irish Meteorological Service, Dublin, Ireland). Report Pl. Path. and Lab. Ann. Epiphyt. **18** : 54-96.

44. Bradley, R. H. E., Moore, C. E. and Pond, D. D. 1966. Spread of Potato virus Y curtailed by oil. Nature (London), **209** : 1370-1371.

45. Braun, C. J. and Hemenway, C. L. 1992. Expression of aminoterminal portions or full-length viral replicase genes in transgenic plants confers resistance to potato virus X infection. Plant Cell, **4** : 735-744.

46. Brown. C. R. 1980. Incorporation of virus resistance and future plants. Report of Plant Conf. Strat. Virus Manag. In Potatoes. CIP. Lima, Peru.

47. Brown, C. R., Salazar, L. F., Ocha, C. and Chuquillanqui, C. 1984. Strain-specific immunity to PVX. HB is controlled by a single dominant gene, Proc. 9th Triennial Conf. Eur. Assoc. Pot. Res., Interlaken, 1984, pp. 249-250. (abstr.)

48. Brumbley S. M. and Denny, T. P. 1990. Cloning of *Pseudomonas solanacearum* phcA, a gene that alter expression of multiple traits that controbute to virulence. J. Bacteriol. **172**: 5677-5685.

49. Buddenhangen, I. W. 1986. Bacterial wilt revisted. In : "Bacterial Wilt Disease in Asia and South Pacific". Ed. G. J. Persley. ACIAR Proc. No. 13 pp. 126-143.

50. Buddenhangen, I. W. and Kelman, A. 1964. Biological and Physiological aspects of bacterial wilt caused by *Pseudomonas solanacearum.* Ann. Rev. Phytopathol. **2**: 203-230.

51. Butler, E. J. 1918. Fungi and Disease in Plants. Thacker Spink and Co., Calcutta, 574. pp.

52. Calvert, E. L., Cooper, P. and McClure, J. 1980. An aphidtransmitted strain of P?VYc recorded in potatoes in Northern Ireland. Record Agri. Res., Dept. Agri. for Northern Ireland, **28**: 63-74.

53. Cappel, E. L. 1892. A note on potato diseases prevalent in Poona district and elsewhere. Rep. Dept. Ld. Rec. Agric. Bombay, 16 pp.

54. Chakrabarti, S. K. Shekhawat, G. S. and Gadewar, A. V. 1993. Phenotypic reversion in *Pseudomonas solanacearum* Smith from weakly virulent, afluidal colony types to highly virulent fluidal stable type. IPS, PKV, Akola, Jan. 1993. pp. 1. (Abstr.).

55. Chavez, R., Brown, C. R. and Iwanaga M. 1988a. Transfer of resistance to potato leafroll virus titre buildup from *Solanum etuberosum* to a tuber-bearing *Solanum* gene pool. Theor. Appl. Genetics, **76** : 129-135.

56. Chavez, R., Brown, C. R. and Iwanga, M. 1988b. Application of interspecific requiploidy to introgression of potato leafroll virus resistance from non-tuber bearing *Solanum etuberosum* to cultivated potato germplasm. Theor. Appl. Genetics. **76** : 47-500.

57. Chen, W. Y. and Echandi, E. 1984. Effects of avirulent bacteriocin producing strains of *Pseudomonas solanacearum* on the control of bacterial wilt of tobacco. Plant Pathol. **33** : 245-253.

58. Ciampi, L. and Sequeira, L. 1980. Influence of temperature on virulence of race 3 strains of *Pseudomonas solanacearum* . Amer. Potato J. **57** : 207-317.

59. CIP. 1987. Annual Report 1986-87. International Potato Center (CIP) Peru.

60. Clayton, A. M. and Robertson, N. F. 1956. Survival of *Phytophthora infestans* in potato stem lesions. Plant Pathology, **5** : 30.

61. Cockerham, G. 1970. Genetical studies on resistance to potato viruses X and Y. Heredity, **25**: 309-348.

62. Coffey, M. D. and Wilson, U. E. 1983. An ultrastructural study of the late blight fungus *Phytophthora infestans* and its interaction with the foliage of two potato cultivars possessing different levels of general resistance. Can. J. Bot. **61** : 2669-2685.

63. Colon, L. T. and Budding, D.J.K. 1988. Resistance to late blight *(Phytophthora infestans)* in ten wild *Solanum* species. Euphytica Supplement, 77-86.

64. Conrad, P. L. 1991. Potato virus S-free plants obtained using an-tiviral compounds and nodal segment culture of potato. Amer. Pot. J. **68** : 507-513.

65. Cook, D. R. and Sequeira, L. 1988. The use of restriction fragment length polymorphism (RFLP) analysis in taxonomyand diagnosis. ACIAR Bacterial Wilt Newsl. **4** : 4.

66. Cook, D. and Sequeira, L. 1991. Genetic and biochemical characterization of gene cluster from *Pseudomonas solancearum* required or exopolysacccharide production and virulence. J. Bacteriol. **173**: 1654-1662.

67. Cook, D. and Sequiera, L. 1993. Strain differrentiation of *Pseudomonas solancearum* by molecular genetic methods. In "Bacterial Wilt". Eds. G. L. Hartman and A. C. Hayward, AICAR Proc. No. 45. pp. 96.

68. Cook D., Barlow, E. and Sequiera, L. 1989. Genetic diversity of *Pseudomonas solancearum* : detection of restriction fragment length polymorphism with DNA probes that specify virulence and the hypersensitive response. Mol. Plant Microb. Interact. **2** : 113-121.

69. Currier, T. C. and Morgan, M. K. 1981. Plasmids are not associated with the formation of noncapsulated variants *Pseudomonas solancearum* Proc. 5th Int. Conf. on Plant Pathogenic Bact. CIAT, Cali, col. pp. 420-425.

70. Dass, C. R. and Chattopathyay, S. B. 1955. Bacterial wilt of eggplant. Indian Phytopath. **8** : 130-135.

71. Davidson, T. M. W. 1980. Breeding for resistance to virus diseases of the potato *(Solanum tuberosum)* at the Scottish Plant Breeding Stastion. Report Scottish Plant Breeding Stn., Edinburgh (UK).

72. Dawson, G. W., Gibson, R. W., Griffithis, D. C., Pickett, J. A., Rice, A. D. and Woodcock, C. M. 1982. Aphid alarm pheromone derivatives affecting settling and transmission of plant viruses. J. Chem. Ecology, **8** : 1377-1388.

73. de Bokx, J.A. 1987. Biological properties. In : Virusees of Potatoes and Seed-Potato Production, Eds. J. A. de Bokx and J. P. H. vander Want, Pudoc, Wageningen, pp. 58-82.

74. de Bokx, J. A. and P. G. M. Prione. 1990. Relative efficiency of a number of aphid species in the transmission of potato virus Y^N in the Netherlands. Nath. J. Plant Pathol. **96** : 237-246.

75. Delhey, R. 1975. Zur Natur der extremen Virusresistenz beider kartoffel. II. Das Y-Virus. Phytopath. Z. **82** : 163-168.

76. Denny, T. P., Makini, F. W. and Brumbley, S. M. 1988. Characterization of *Pseudomonas solancearum* Tn5 mutants deficient in extracellular polysaccharide. Mol. Plant-Microb. Interact. **1** : 215-223.

77. Devi, L. R., Menon, M. R. and Aiyer, R. S. 1982. Population thresh hold of *Pseudomonas solancearum* at the onset of first symptoms of wilt in tomato. Indian J. Microbiol. **22** : 41-43.

78. De Weille, G. A. 1963. Laboratory result regarding potato blight and their significance in the epidemiology of blight. Europ. Potato J. **6** : 121-129.

79. De Weille, G. A. 1964. Forecasting crop infection by the potato blight fungus. Mededelingen en Verhandelingen, Koninklijk Nederlandsch Mteorologisch Instituut (Utrecht). No. 82, pp. 144.

80. DeWit, C. T. 1960. On competition. Agricultural Research Reports 66.8 (end edn., 1964), 82 pp.

81. Dhingra, M. K., Khurana, S. M. Paul, Lakhanpal, T. N. 1987. Effect of virazole and 2, 4,-D on the growth and potato virus X content of potato leaf callus. J. Indian Potato Assoc. **14** : 100-104.

82. Dhingra, M. K., Khurana, S. M. Paul, Lakhanpal, T. N. and Chandra, R. 1988. Thermotherapy in potato tissue culture. J. Indian Potato Assoc. **15** : 53-59.

83. Dingjan-Versteegh, A. M., van Vloten Doting, L. and Jaspars, E. M. J. 1972. Alfalfa mosaic virus hybrids constructed by exchanging nucleoprotein components. Virology, **49** : 716-722.

84. Dodds, J. H. and Horton, D. 1990. Impact of biotechnology on potato production in developing countries. Agbiotech. News and Information, **2** : 397-400.

85. Doke, N. 1985. NADPH-dependent I-generation in membrane fractions isolated from wounded potato tubers inoculated with *Phytophthora infestans*. Physiological Plant Pathology, **27** : 311-322.

86. Doke, N. and Chai, H. B. 1985. Activation of superoxide generation and enhancement of resistance against compatible races of *Phytophthora infestans* in potato plants treated with digitonin. Physiological Plant Pathology, **27** : 323-334.

87. Doke, N., Tomiyama, K. and Furuichi, N. 1982. In : Plant Infection : The physiological and biochemical basis. Eds. Y. Asoda, W. R. Bushnell, S. Ouchi and C. P. Vance pp. 79-86. Japan Sci. Soc. Press, Tokyo. (Quoted from Advances in Pl. Pathology, Vol. 7., pp. 110, Eds. D. S. Ingram and P. H. Williams. Academics Press, 1991).

88. Doster, M. A., Sweigard, J. A. and Fry, W. E. 1989. The influence of host resistance and climate on the initial appearance of foliar blight of potato from infected seed tubers. Amer. Poatato J. **66** : 227-233.

89. Dowley, L. J. and E. O. 'Sullivan. 1981. Metalaxyl resistant strains of *Pseudomonas solancearum* (Mont) de Bary in Ireland. Potato Research, **24** : 417-421.

90. Dowley, L. J. and O'Sullivan, E. 1981. Metalaxyl resistant strains of *Phytophthora infestans* in Ireland. Potato Res. **24** : 417-421.

91. Drigues, P., Demery-Lafforgue, D., Trigalet, A., Dupin, P., Samaian, D. and Asserinean, J. 1985. Comparative studies of lipopolysaccharide and expopolysaccharide from a virulent strain of *Pseudomonas solancearum* and from three avirulent mutants. J. Bacteriol. **162** : 504-509.

92. Dukes, P. D., Jenks, S. F., Jaworski, C. A. and Morton, D. J. 1965. The identification and persistance of an indigeneous race of *Pseudomonas solancearum* in soils in Georgia. Plant Dis. Reptr. **49** : 586-590.

93. Duvauchelle, S. 1991. Lutte contre le mildiou de la pomme de terre en France : moddes de prevision pour leg specialites contentant des mattierres actives systemiques face a la resistance. EPPO Bulletin. **21** : 49-55.

94. Eddins, A. H. 1936. Brown rot of Irish potatoes and control. Univ. Florida Agric. Expt. Stn. Tech. Bull. 299, 44 pp.

95. Estrada-Ramos, N., Perez-Alvarez, O., Henfling, J., and Malamud, O., 1983. Testing CIP and ICA potato selections for late blight resistance and yield at Rionegro. (Laselva) station in Colombia. pp. 78-79. In : Research for the Potato in the yiear 2000", Ed. W. J. Hooker. CIP Lima, Peru.

96. Farmer, M-J., Milne, E. W., Roberts, I. M. and Harrison, B. D. 1989. Monoclonal antibodies to pottao V potyvirus (PVV). Scottish Crops Research Institute, Dundee (UK), Annual Report, 1988, pp. 172-173.

97. Felix, S. and Ricaud, C. 1979. A practical methode for assessment of the bacterial wilt potential of soil in Mauritius. Proc. CIP Workshop on bacterial wilt of potato. Los Banos and Baguio, Philippines. Feb 5-19, 1978. 63-65 pp.

98. Ferris, V. R. 1955. Histological study of pathogen-suscept relationship between *Phytophthora infestans* and derivation of *Solanum demissum*. Phytopathology, **45**: 546-552.

99. Fernandez-Northcote, E. N. 1983. Prospects for stability of resistance to potato virus Y. Research for the Potato in the year 2000. Proc. of the International Congress 10th Anniversary of the International Potato Center, Lima-Peru, 1982, pp. 82.

100. Fernandez-Northcote, E. N. and Lizarraga, C. L. 1991. Geographical distribution of potato virus X serotypes. Fitopatologia, **26** : 13-18.

101. Fitt, B.D.L., Gregory, P. H., Todd, A. D., Mc Cartney, H. A. and Macdonald, O.C. 1987. Spore dispersal and plant disease gradients, a comparison between two empirical models. J. Phytopathology, **118** : 227-242.

102. Foxe, M. J. 1992. Breeding for viral resistance : conventional methods. Netherlands. J. Plant Pathol. **98** (supplement 2) : 13-20.

103. French., E. R. 1979a. Classification, distribution and origin of *Pseudomonas solancearum*. Developments in Control of Potato Bacterial Wilt Disease. Rept. Planning Conf. CIP. 1979. pp. 28-35.

104. French, E. R. 1979b. Progress in the integrated control of bacterial wilt. Developments in control of potato bacterial wilt disease. Rept. Planing Conf. CIP. Lima, Peru, June 12-15, 1979. pp. 72-81.

105. French, E. R. 1986a. Persistence of bacterial wilt of potato in different regions of Peru. ACIAR bact. Wilt Newsl. **1** : 5-6.

106. French, E. R. 1986b. Interactions between strains of *Pseudomonas solancearum* its host and environment. In : Bacterial Wilt Disease in Asia and South Pacific, Ed. G. J. Persley. ACIAR Proc. No. 13 pp. 99-104.

107. French, E. R. and Martin, C. 1977. American biotypes and races of *Pseudomonas solancearum* of the potato. Fitopatologia, **12** : 40 (Abstr).

108. Fribourg, C. E. 1975. Studies on potato virus X strains isolated from Peruvian potatoes. Potato Research, **18** : 216-226.

109. Friboiurg, C. E. and Nakashima, J. 1983. A PVY strain that does not induce local lesions in clone A6. In: Research for the Potato in the year 2000, pp. 86-87. International potato Center, Peru.

110. Fribourg, C. E. and Nakashima, J. 1983. Characterization of a new potyvirus from potato. Phytopathology, **74** : 1363-1369.

111. Friend, J. 1991. The biochemistry and cell biology of interaction. pp. 85-130. In Advances in Plant Pathology, Eds. D. S. Ingram and H.P.H. Williams, vol. 7. *Phytophthora infestans* the cause of late blight of potato. Academic Press, New York.

112. Fry, W. E. 1975. Integrated affects of polygenic resistance and a protective fungicide on development of potato late blight. Phytopathology, **65**: 908-911.

113. Fry, W. E. 1977. Integrated control of potato late bight : Effects of polygenic resistance and techniques of timing fungicide applications. Phytopathology, **67** : 415-420.

114. Fry, W. E. and Apple, A.E. 1986. Disease management implications of age-related changes in susceptibility of potato foliage of *Phytophthora infestans*. Amer. Potato J. **63** : 47-56.

115 Fucikovsky, L. 1978. Distribution of *Pseudomonas solancearum* in Mexico and its early detection in potato tubers. Proc. 4th Intn. Conf. on Pl. Pathogenic bacteria; Angers, France, Aug. 27-Sept. 2., 1978, pp. 863-867.

116. Gabreil, W. 1961. L'importance de certainers especes de pucerons pour la propagation des maladies a virus de la pomme de terre en Pologne. Proc. 4th Confr. Potato Virus Diseases, Braunschweig 1960, p. 126-137.

117. Gabreil, W. 1981. Essais d' amelioration de la previsin de l'infection des tubercules des pommes de terre par la virus Y. Potato Research, **24** : 301-308.

118. Gadewar, A. V. 1986. A note on natural infection of *Amaranthus viridis* by *Pseudomonas solancearum* Smith. J. Indian Potato Assoc. **13** : 83-84.

119. Gadewar, A. V. and Shekhawat, G. S. 1988a. *Pseudomonas fluorescens* a potentional biocontrol agent for *Pseudomonas solancearum* wilt in potato. J. Indian Potato Assoc. **15** : 194 (Abstr).

120. Gadewar, A. V. and Shekhawat, G. S. 1988b. Field persistence of bacterial wilt of potato in the state of Karnataka. In : Advances in Research on Plant Pathogenic Bacteria. Eds. A. Mahadevan and S. S. Gnanamanickam. Today and Tommorrow Printers and Publishers, New Delhi, pp. 83-86.

121. Gadewar, A. V. and Shekhawat, G. S. 1990. Survival and population fluctuation of *Pseudomonas solanacearum* in rhizosphere of potato Strategies for Potato. Production, Marketing, Storage and Processing (Abstr.) IPA & CPRI, Shimla, pp. 72.

122. Gadewar, A. V. Shekhawat, G. S. and Chakrabarti, S. K. 1993. Antibiotic-Induced virulence and changes in colony morphology of *Pseudomonas solanacearum.* In : Bacterial Wilt Ed. G. L. Hartman, and A. C. Hayward. ACIAR Proc. No. 45 pp. 261-268.

123. Gadewar, A. V., Trivedi, T. P. and Shekhawat, G. S. 1991. Potato in Karnataka. Tech. Bull. No. 17. CPRI, Shimla pp. 33.

124. Gal-On, A., Antignus, Y., Rosner, A. and Raccah, B. 1992. A zucchini yellow mosaic virus coat protein gene mutation restores aphid transmissibility but has no effect on multiplication. J. Gen. Virol. **73** : 2183-2187.

125. Gallegly, M. E. 1968. Genetics of pathogenicity of *Phytophthora infestans.* Ann. Rev. Phytopathol, **6** : 375-396.

126. Gallegly, M. E. and Galindo, J. 1957. The sexual stage of *Phytophthora infestans* in Mexico. Phytopathology, **47** : 13. (Abstr).

127. Gallegly, M. E. and Galindo, J. 1958. Mating types and oospores of *Phytophthora infestans* in nature in Mexico. Phytopathology, **48** : 274 - 227.

128. Gallegly, M. E. and Neiderhauser, J. S. 1959. Genetic controls of host parasite interactions in the *Phytophthora* late blight disease. Plant Pathol. Probl. Progr. 1908-1958. pp. 168-182.

129. Gallegly, M. E. and Walker, J. C. 1949. Plant nutrition in relation to disease development. V. Bacterial wilt of tomato. Am. J. Bot. **36** : 613-623.

130. Garcia, S. 1976. Status of potato brown rot in Uruguay. Proc. Ist Intn. Planning Conf. and Workshop on the ecology and control of bacterial wilt, Eds. L. Sequeira & A. Kelman). Raleigh, NC, 1976. pp. 78-83.

131. Garg, I. D., Khurana, S. M. Paul and Singh, M. N. 1990. Common weed reservoirs of potato viruses x, S and Y in Shimla hills. Plant Dis. Res. **5**: 84-86.

132. Gees, R. and Hohl, H. R. 1988. Cytological comparision of specific (R3) and general resistance to late blight in potato leaf tissue. Phytopathology, **78** : 350-357.

133. Gibbs. A. J., Hecht-Pionar, E., Woods, R. D. and Mckee, R. K. 1966. Some properties of three related viruses : Andean potato alatent, *Dulcmara* mottle *Onions* yellow mosaic. J. Gen. Microbiol. **44** : 177-193.

134. Gibson, R. W. and Pkckett, J. A. 1983. Wild potato repels aphids by release of aphid alarm pheromone. Nature, **302** : 608-609.

135. Gibson, R. W., Rice A. D. and Sawicki, R. M. 1982. Effects of the pyrethroid deltamethrin on the acquisition and inoculation of viruses by *Myzus persicae.* Ann. Appl. Biol. **100** : 49-54.

136. Gibson, R. W., Jones, M. G. K. and Fish, N. 1988. Resistance to potato leafroll viruses and potato viruses Y in somatic hybrids between dihaploid *Solanum tuberosum* and *S. brevidens.* Theor. Appl. Genetics, **676**: 113-117.

137. Gillings, M. and Fahy, P. 1993. Genomic fingerprinting and PCR analysis : Rapid, sensitive and inexpensive means of differntiating strains of *Pseudomonas solanacearum*. In : Bacterial Wilt Eds. G. L. Hartman and A. C. Hayward. ACIAR Proc. No. 45 pp. 85-92.

138. Glendnning, D.R. 1981. Annual Report, Scottish Crop Research Institute, Dundee (U. K.), pp. 183-184.

139. Govier, D. A. and Kassanis, B. 1974. A virus induced component of plant rap needed when aphids acquire potato viruses from purified preparations. Verology, **61** : 420-426.

140. Govier, D. A., Kassanis, B. and Pirone, T. P. 1977. Partial purification and characterization of the potato virus Y helper component. Virology, **78** : 306-314.

141. Gobinda, A. J., Gallegly, M. E. 196. The nature of sexuality in *Phytophthora infestans*. Phytopathology, **50** : 1123-128.

142. Graham, J., Jones, D. A. and Lloyd, A. B. 1979. Survival of *Pseudomonas solanacearum* race 3 in plant debris and latently infected potato tubers. Phytopathology, **69** : 1100-1102.

143. Graham, J. and Lloyd, A. B. 1978a. An improved indicator plant method for the detection of *Pseudomonas solanacearum* race 3 in soil. Plant Dis. Reptr. **62** : 35-37.

144. Graham, J. and Lloyd, A. B. 1978b. *Solanum cinereum* R. Br. Wild host of *Pseudomonas solanacearum* biotype II. J. Ajustr. Inst. Agric. Sci. **44** : 124-126.

145. Graham, J. and Lloyd, A. B. 1979. Survival of potato strain (race 3) of *Pseudomonas solanacearum* in deeper soil layers. Aust. J. Agric. Res. **30** : 489-496.

146. Granada, G. A. 1976. Current status of potato bacterial wilt research in Colombia. Proc. Ist Int. Planing Conf. and Workshop on ecology and control of bacterial wilt, Eds. L. Sequeira and A. Kelman, Railegh, N.C, 1976, pp. 91.

147. Granada, G. A. and Sequeira, L. 1983. Survival of *Pseudomonas solanacearum* in soil, rhizosphere and plant roots. Can J. Microbiol. **29** : 433-440.

148. Grewal, J. S. (Ed.) 1992. True potato seed for potato production in India Revised Edn. CPRI, Shimla Tech. Bull. **24** : 16 pp.

149. Griffiths, H. M., Slack, S. A. and Dodds, J. H. 1990. Effect of chemical and heat therapy on virus concentration in *in vitro* potato plantlets. Can. J. Bot. **68** : 1515-1521.

150. Gross, H. N., Domdey, H., Lossow, C., Jank, P., Raba, M. 1978. Nucleotide sequence and secondary structure of potato spindle tuber viroid. Nature, **273** : 203-208.

151. Gufran, S. M. and Chakrabarti, B. P. 1960. Potato wilt in Ranchi. Proc. Bihar Acad. Agri. Sci. **8 - 9** : 126-127.

152. Hadidi, A. and Diener, T. O. 1978. *In vitro* synthesis of potato spindle tuber viroid : kinetic relationship between the circular and linear forms. Virology, **86** : 57-59.

153. Hari Kishore and Pushkarnath. 1963. Brown rot of potatoes in India and its control. Indian Potato J. **6** : 3-10.

154. Harrington, R. and Gilson, R. W. 1989. Transmission of potato virus Y by aphids trapped in potato crops in Southern England. Potato Res. **32** : 167-174.

155. Harrington, R., Katis, N. and Gilson, R. W. 1986. Field asessment of the relative importance of different aphid species in the transmission of potato virus Y^N. Potato Res. **29** : 67-76.

156. Harris, D. 1972. The significance of bacterial wilt in development of potatoes in Kenya. In : Prospect for the potato in developing World, Ed. E. R. French, CIP, Lima pp. 220-205.

157. Harris, D. C. 1976. Bacterial wilt in Kenya with particular reference to potato Proc. Ist Int. Planning Conf. and Workshop in the ecology and control of bacterial wilt, Ed. L. Sequiera and A. Kelman, Raleigh, NC 1976. pp. 84-85.

158. Harrison, J. G. 1992. Effect of the aerial environment of late blight of potato foliage-a review. Plant Pathology, **41** : 384-416.

159. Hayward, A. C. 1960. A method for chaeracterising *Pseudomanas solancearum.* Nature, **186**: 265-277.

160. Hayard, A. C. 1974. Characteristic of *Pseudomanas solanacearum.* J. Appl. Bacteriol, **27** : 265-277.

161. Hayard, A. C. 1975. Biotypes of *Pseudomanas solanacearum* in Australia. Australian Pl. Path. Soc. Newsl, **4** : 9-11.

162. Hayward, A. C. 1988. Classification and identificatin of *Pseudomanas solanacearum.* Rept. of Planning Conference on bacterial diseases of Potato, CIP Lima, 1987 pp. 113-121.

163. Hayward, A. C., El-Nashaar, H. M., Nydegger, U. and de Lindo, L. 1990. Variation in nitrate metabolism in biovars of *Pseudomanas solancearum.* J. Appl. Bacteriol. **69** : 269-280.

164. He, L. Y. 1986. Bacterial wilt in Peoples Republic of China in Bacterial Wilt Disease in Asia & South Pacific, Ed. G. J. Persley, ACIAR Proc. No. 13 pp. 40-48.

165. He, H. Y., Sequeira, L. and Kelman, A. 1983. Characteristics of strain of *Pseudomanas solancearum* in China. Pl. Dis. **67** : 1357-1361.

166. Helgeson, J. P., Hunt, G. J., Haberlach, G. T. and Austin, S. 1986. Somatic hybrids between *Solanum brevidens* and *S. tuberosum :* Expression of a late blight resistance gene and potato leafroll virus resistance. Plant Cell Reporter, **3** : 212-214.

167. Hendrick, C. and Sequeira, L. 1984. Lipopolysaccharide defective mutants of the wilt pathogen *Pseudomanas solanacearum.* Appl. Environ. Microbio, **48** : 94-101.

168. Henrichsen, J. 1975. The occurrence of twitching motility among Gram negative bacteria. Acta Path. Microbiol. Scand. Sect. B. **83** : 171-178.

169. Hermsen, J. G. T. 1980. Recent progress and future plans for utilizing Mexican wild species for pest and disease resistance. In : Report of Plant Conf. Utilization of Genetic Resources of Potato III, CIP, Lima, Peru.

170. Hermsen, J. G. T. and Taylor, L. M. 1979. Successful hybridization of non-tuberous *Solanum etuberosum* Lindl. and tujber-bearing *S. pinnatisectum* Dum. Euphytica, **28** : 1-7.

171. Hermsen, J. R. T., Ramanna, M. S. and Sawor, Z. 1981. The effect of chromosome doubling on fertility, meotic behaviour and crossability of *Solanum etubrosum* and *S. pinnatisectum.* Euphytica, **30** : 33-39.

172. Hingorai, M. K., Mehta, P. P. and Singh N. J. 1956. Brown rot of potato in India. Indian Phytopath, **9** : 67-71.

173. Hinz, B. 1966. Versuche zur ubetragung des Blattrollvirus der Kartoffel durch die Grunstreifige Kartoffalblottlaus *Macrosiphum euphorbiae* (Thomas). Wissenschaftlicha Zeitschrift Universitat Rostock, **15** : 289-293.

174. Hirst, J. M. and Stedman, O. J. 1960. The epidemiology of *Phytophthora infestans* II. The source of inoculum. Ann. Appl. Biol. **48** : 489-517.

175. Hodgson, W. A. 1961. Laboratory testing of the potato for partial resistance of *Phytophthora infestans.* Amer. Potato J. **38** : 259-264.

176. Hoekema, A., Huisman, M. J., Molendijk, L., van den Flkezen, P. J. M. and Cornelissen, B. J. C. 1989. The genetic engineering of two commercial potato cultivars for resistance to potato virus X. Biotechnology, **7** : 273-278.

177. Hohl, H. R. and Iselin, K. 1984. Strains of *Phytophthora infestans* from Switzerland with A2 mating type behaviour. Trans. Br. Mycol. Soc. **83** : 529-530.

178. Hohl, H. R. and Stossel, P. 1976. Host parasite interfaces in a resistant and a susceptible cultivar of *Solanum tuberosum* inoculated with *Phytophthora infestans.* Tuber tissue. Can. J. Bot, **54** : 900-912.

179. Horvath, J. 185. A check list of new host plants for identification and separation of twelve potato viruses. Potato Res. **28** : 7189.

180. Horvath, J. 1988.Potato gene centres, wild *Solanum* species, viruses and aphid vectors. Acta Phytopath. Entamol. Hungarica, **23** : 423-448.

181. Huang, Y., Xu, P. and Sequeira, L. 1990. A second cluster of genes that specify pathogenicity and host response in *Pseudomanas solanacearum*.. Mol. Plant - Microb. Interact. **3** : 48-53.

182. Huang, Y. and Sequeira, L. 1990. Identification of locus that regulate multiple functions in *Pseudomanas solanacearum*.. J. Bacteriol. **172** : 4728-4731.

183. Hunter, D. E., Darling, D. H. and Baele, W. L. 1969. Seed transmission of potato spindle tuber virus. Am. Potato J. **46** : 247-249.

184. Hyre, R. A. 1954. Progress of forcasting late blight of potato and tomato. Plant Dis. Reptr. **38** : 245-53.

185. Hyre, R. A. and Bonde, R. 1955. Forecasting late blight of potato in Maine. Am. Potato J. **32** : 119-125.

186. Jackson, M. T. and Gonzalez, L. C. 1979. Persistance of *Pseudomanas solanacearum* in an inceptisol in Costa Rica : Rept. Planning Conf. on the develpments in the control of bacterial disease of potato. CIP, Lima, Peru, pp. 66-81.

187. Jager, M. J. 1983. Analysis of epidemics in time and space. Plant Pathology, **32** : 5-11.

188. James, R. V. and Fry, W. E. 1983. Potential of *Phytophthora infestans* population to adapt to potato cultivars with rate reducing resistance.

189. Jones, R. A. C. 1979. Resistance to potato leafrol virus in *Solaum brevidens.* Potato Res. **22** : 149-152.

190. Kable, P. F. and Jaffery, H. J. 1980. Selection for tolerance in organisms exposed to sprays of biocide mixtures : AS theoretical appreoach. Phytopatholoty, **70** : 8-12.

191. Kalra, A., Grover, R. K., Rishi, N. and Khurana, S. M. Paul. 1989. Influence of different factors on potato virus Y induced resistance to *Phytophthora infestans.* J. Plant Diseases and Prot. **96** : 470-477.

192. Kaniwski, W. K., Lawson, C., Sammons, B., Haley, L., and J. 1990. Field resistance of transgenic Russet Burbank Potato to effects of infection by potato virus X and potato virus Y. Biotechnology, **8** : 750-754.

193. Karolev. A. A. 1978. Slightly infected tubers as the source of *Phytophthora.* Zasbechita Rastenii. **8** : 49.

194. Kassanis, B. 1961. The transmission of potato aucuba mosaic virus by aphids from plants also infected by potato virus A or Y. Virology, **13** : 93-97.

195. Kassanis, B. and Govier, D. A. 1971a. New evidence on the mechanism of aphid transmission of potato C and potato aucuba mosaic viruses. J. Gen. Virol. **10** : 99-101.

196. Kasanis, B. and Govier, D. A. 1971b. The role of the helper virus in aphid transmission of potato aucuba mosaic virus and potato virus. C. J. Gen. Virol. **13** : 221-228.

197. Kelman, A. 1953. The bacterial wilt caused by *Pseudomanas solanacearum*.. NC Agric. Exp. Sta. Tech. Bull. 99, pp. 194.

198. Kelman, A. 1954. The relationship of pathogenicity of *Pseudomanas solanacearum*. to colony appearance in a tetrazolium medium. Phytopathology, **44** : 693-695.

199. Kelman, A. and Hruschka, J. 1973. The role of motility and aerotaxis in the selective increase of avirulent bacteria in still broth culture of *Pseudomanas solanacearum*. J. Gen. Microbiol. **76** : 177-188.

200. Kempe, J. and Sequeira, L. 1983. Biological control of bacterial wilt of potatoes: attempt to induce resistance by treating tubers with bacterial. Pl. Dis. **67** : 499-503.

201. Keshwal, R. L.1980. Studies on bacterial wilt of Solanaceous crops caused by *Pseudomanas solanacearum*. EF Smith with special reference to its control. Ph. D. Thesis, Univ. Jabalpur. 201 pp.

202. Khurana, S. M. Paul, 1992. Potato viruses and viral disease. Tech. Bull. 35 : 23 pp. CPRI, Shimla.

203. Khurana, S. M. Paul and Garg. I. D. 1992. Potato mosaics, In : Plant Diseases of International Importance Vol II, Eds. H. H. Chaube *et al*, Prentice Hall, N. J. (USA) pp. 148-164.

204. Khurana, S. M. Paul, Phadtare, S. G. Garg, I. D., Singh, M. N. and Bhardwaj, V. P. 1989. Potato Stem necrosis epidemic due to tomato spotted wilt virus in India. Proc. IVth Int. Pl. Virus epidem. Workshop, Montpellier (France), Sept. 3-8, 1989, pp. 301.

205. Kloos, J. P., Tulong, B. and Tumapon, A. S. 1987. Effect of inter-cropping on potato bacterial wilt. Philipp. Agir. **70** : 83-90.

206. Krause, R. A., Massie, L. B. and Hyre, R. A. 1975. Blitecast : a computerized forecast of potato late blight. Plant Dis. Reptr. **59** : 95-98.

207. Landeo, J. 1989. Late blight breeding strategy at CIP. pp. 57-75. In : Fugal diseases of the potato. Report Planning Conference on Fungal Diseases of the Potato, CIP, Lima, Sept.l 21-25, 1987.

208. Lapwood, D. H. 1977. Factors affecting the field infection of potato tubers of different cultivars by blignt (*Phytophthora infestans)*. An. Appl. Biol. **85** : 23-42.

209. Large, E. C. 1940. The advances of fungi. The Hollen Street Press Ltd. London, W. I. 477 p.

210. Latin, R. X. 1980. Factors in potato that reduce the rate of epidemic increase of potato late blight. Ph.D. Thesis. The Pennsylvania State Univ., U.S.A. 48 pp.

211. Lawson, C., Kaniewski, W., Haley, L., Rojman, R., and Newell, C. 1990. Engineering resistance to mixed virus infection in a commercial potato cultivar. Resistance to potato virus X and potato virus Y in transgenic Russet Burbank. Biotechnology, **8** : 127-134.

212. Lecoq, H. 1993. Transmission of potyviruses by their aphid vectors : Molecular and structural interactions. Indo-French Symp. Mol. Plant Pathol. & Plant -Pathogen Interactions, March 10-12, 1993, New Delhi (Abstr.) pp. 29.

213. Lee. J. D., Ban, Y. S. and Yu, I. S. 1982. Effect of boron application on the outbreak of Granville wilt in flue cured tobacco. Korean J. Plant Prot. **21** : 73-77.

214. Lizarraga, R. E., Salazar, L. F., Roca, W. M. and Schilde-Rentschler, L. 1980. Elimination of potato spindle tuber viroid by low temperature and meristem culture. Phytopathology, **70** : 754-755.

215. Lloyd, A. B. 1976. Bacterial wilt in cold temperature climate of Australia. In : Planning Conf. and Workshop on the ecology and control of bacterial wilt Raleigh, N. C. pp. 134-135.

216. Machmud, M. 1986. Bacterial wilt in Indonesia. In : Bacterial wilt disease in Asia and the South Pacific Ed. G. J. Persley, ACIAR Proc. No. 13 pp. 30-34.

217. Macol, L. A. and Denny, T. P. 1989. Cloning of DNA region of *Pseudomanas solanacearum* strain AW that affects both hypersensitive response on tobacco and pathogenicity on tomato and eggplant. Phytopathology, **79** : 1211.

218. Malcolmson, J. F. 1976. Assessment of field resistance to blight *(Phytophthora infestans)* in potatoes. Trans. Br. Mycol. Soc. **67** : 321-341.

219. Malcolmson, J. F. 1985. *Phytophthora infestans :* A2 compatibiity type recorded in Great Britain. Trans. Brit. Mycol. Soc. **85** : 531.

220. Mann, H. H. and Nagpurkar, S. D. 1920. The ring disease of potato. Dept. Agric. Bombay Bull. No. **102**: 37-57.

221. Martin, C., French, E. R. and Nydegger, U. 1981. Bacterial wilt in the Amazon Basin. Dis, **65**: 246-248.

222. Martin, C., French, E. R. and Naydeggr, U. 1982. Strains of *Pseudomanas solanacearum* affecting solanaceae in America. Pl. Dis. **66** : 458-460.

223. Mayo, M. A. and Robinson, D. J. 1988. The genome of potato leafroll luteovirus. Ann. Rept. Scottish Crops Res. Inst. Dundee (UK), pp. 160-161.

224. McCarter, S. M., Dukes, P.D. and Jaworski, C. A. 1969. Vertical distribution *Pseudomanas solanacearum* in several sils. Phytopathology, **59** : 1675-1677.

225. McKay, G. R. 1984. Parental line breeding at the tetraploid level. In : Parental Line Breedikng and Selection in Potato Breeding, Eds. K. M. Louwes, H. A. J. M. Toussaint and L.M.W. Dellaert, pp. 31-135. Pudoc, Wageningen.

226. McLachlan, D. S., Larson, R. H. and Walker, J. C. 1954. Potato virus A. Amer. Potato J. **131** : 67-69.

227. McLaughlin, R. J. and Sequeira, L. 1988. Evaluation of an avirulent strain of *Pseudomanas solanacearum* for biological control of bacterial wilt of potato. Am. Potato J. **65** : 255-267.

228. Mc Rae, K. B. and Platt, H. W. 1987. An index for cultivar resistance based on disease progress curves. Phytopathology, **77** : 1181-1186.

229. Merriam, D. and Bonde, R. 1954. Dissemination of potato spindle tuber by contaminated tractor wheels and by foliage contact with diseased plants. Phytopathology, **44** : 111 (abstr).

230. Message, B., Biostard, P., Pitrat, M., Schmit, J. and Boucher, C. 1978. A new class of fluidal avirulent mutants of *Pseudomanas solanacearum* unable to induce a hypersensitive reaction. Proc. 4th Intn. Conf. Pl. Pathogenic Bacteria, **2** : 823-833.

231. Miller, J. S. and Mayo, M.A. 1991. The location of the 5" end lof the potato leafroll luteovirus subgenomic coat protein in RNA. J. Gen. Virol. **72** : 2633-2638.

232. Minoque, K. P. and Fry, W. E. 1983a. Models for the spread of disease : Some experimental results. Phytopathology, **73** : 1168-1173.

233. Minoque, K. P. and Fry, W. E. 1983b. Models for the spread of disease. Phytopathology, **73** : 1173-1176.

234. Morales, V. M. and Sequeira, L. 1985. Indigenous plasmids in *Pseudomanas solanacearum*. Phytopathology, **75** : 767-770.

235. Moreira, A., Jones, R. A. C. and Fribourg, C. E. 1980. Properties of a resistance-breaking strain of potato virus X. Ann. Appl. Biol. **95** : 93-103.

236. Mosa, A. A., Kato, M., Sato, N., Kobayashi, K. and Ogoshi. A. 1989. Occurrence of the A2 mating type of *Phytophthora infestans* on potato in Japan. Ann. Phytopath. Soc. Japan, **55** : 615-620.

237. Mosa, A. A., Kobayashi, K., Ogoshi, A., Sato, M. and Sato, N. 1991. Formatilon of oospores by *Phytophthora infestans* in inoculated potato tissues. Ann. Phytopath. Soc. Japan. **57** : 334-338.

238. Nagaich, B. B., Shekhawat, G. S., Khurana, S. M. Paul and Bhattacharya, S. K. 1974. Pathological problems of the potato cultivation in India. J. Indian Potato Assoc. **1** : 32-44.

239. Nakanishi, K. and Kubo, I. 1977. Studies on warburganal, mujigadial and related compounds. Israel J. Chem. **16** : 28-31.

240. Nath, V. R., Vaheeduddin, S. and Reddy, D. P. 1958. Bacterial Wilt (ring disease) of potato in Hyderabad. Sci. and Cult. **23** : 605-606.

241. Niederhauser, J. S. 1956.. The blight, the blighter and the blighted. Trans. New York Acad, Sci. **19** : 55-63.

242. Ofuya, C. O. and Wood, R.K.S. 1978. Pathogenmicity and biochemical characteristics of isolates of *Pseudomanas solanacearum* in Nigeria. Proc. Ist Intn. Conf. Plant Pathogenic Bacteria. Angers, France. 1978. pp. 835-840.

243. Ofuya, C. D. and Wood, R.K.S. 1981. Cell wall degrading polysaccharidases and extrachromosomal DNA in wild caused by *Pseudomanas solanacearum*. Proc. 5th Intn. Conf. on Plant Pathogenic Bacteria, Colombia, Ed. J. C. Lozano pp. 270-279.

244. Olosson, K. 1976a. Overwintering of *Pseudomanas solanacearum* in Sweden. In : Proc. Ist Intn. Conf. and Workshop Ecology and Control of Bacterial Wilt, Eds. L. Squeira, A. Kelman and N. C. Raleigh, pp. 166.

245. Olosson, K. 1976b. Experience of brown rot caused by *Pseudomanas solanacearum* (Smith) Smith in Sweden EPPO Bull. **6** : 199-207.

246. Owens, R. A., Khurana, S. M. Paul, Smith, D. R., Singh, M. N. and Garg, I. D. 1992. A new mild strain of potato spindle tuber viroid isolated from wild *Solanum* spp. in India. Plant Disease, **76** : 527-529.

247. Paharia, K. D. 1963. Brown rot of potato. Proc. 2nd All India Potato Workers Conf., Shimla, pp. 198-202.

248. Palleroni, N. J. and Doudoroff, M. 1971. Phenotypic characterisation and deoxyribonucleic acid homologies of *Pseudomanas solanacearum*. J. Bacteriol. **107** : 690-696.

249. Patel, M. K., Kulkarni, N. B. and Kulkarni, Y. S. 1952. Ring Disease of potato in India. Curr. Sci. **21** : 47-48.

250. Pehu, E., R. W. Gibson, Jones, M. G. K. and Karp, A. 1990a. Studies on genetic basis of resistance to potato leagroll virus, potato virus Y and potato virus X in *Solanum brevidens* using somatic hybrids of *S. brevidens* and *S. tuberosum*. Plant Sci. **69** : 95-101.

251. Pehu, E., Thomas, M., Poutala, T., Karp, A. and Jones, M. G. K. 1990b. Studies on genetic basis of resistance to potato leagroll virus, potato virus Y and potato virus X in *Solanum brevidens* using somatic hybrids of *S. brevidens* and *S. tuberosum*.

252. Perrin, R. M. and Gibson, R. W. 1985. Control of some insect-borne plant viruses with the pyrethroid, pp. 321 (Karate). International Pest Control, **27** : 142-145.

253. Peters, D. 1987. Spread of viruses in potato crops. In : Viruses of Potatoes and Seed-Potato Production. Eds. J. A. de Bokx and J. P. H. van der Want. PUDOC, Wageingen, pp. 126-145.

254. Pfannenstiel, M. A. and Slack, S. A. 1980. Response of potato cultivars to infection by potao spindle tuber viroid. Phytopathology, **70** : 1015-1017.

255. Phadtare, S. G. Khurana, S. M. Paul, Gar, I. D. and V. P. Bhardwaj. 1989. Stem necrosis-a viral disease of potatoes in Central India. J. Indian Potato Assoc. **16** : 164-165 (Abstr).

256. Pitkethley, R. N. 1981. Host range and biotypes of *Pseudomanas solanacearum* in the northern territory. Austr. Pl. Pathol. **10** : 46-47.

257. Ponnamperuma, F. N. 1972. The chemistry of submerged soils. Adv. in Agron, **24** : 29-96.

258. Pushkarnath and Paharia, K. D. 1963. Survivarl of *Phytophthora infestans* on infected tubers in cold storage in the plains of India. Indian Potato J. **5** : 48-51.

259. Quimio, A. J. and Chan, H. H. 1979. Survival of *Pseudomanas solanacearum* E.F.S. in rhizosphere of some weeds and economic plant species. Philippines Phytopathology, **15** : 108-121.

260. Rangaswami, G. and Trunavukarasu, V. 1964. Studies on the survival of plant pathogens added to the soil III. On four phytopathogenic bacterial species. Indian Phytopath. **17** : 202-207.

261. Rizvi, S.A.H. 1983. Extreme resistance to potato leafroll virus (PLRV) in seedlings in *Solanum tuberosum* x *S. pinnatisectum* (EP) with 4 x chromosomes. In : Proc. Int. Confr. Res. Potato for the year 2000. CIP, Lima, Peru.

262. Robert, Y. 1971. Epidemiologier de l'enroulment de la pomme de terre : cacacite" vectrice de stades etde formes des pucerons *Aulacorthum solani* Kltb., *Macsoriphum euphorbiae* Thom. et *Myzus persicae* Sulz. Potato Res. **145** : 130-139.

263. Robert, Y. and Maury, Y. 1971. Capacite's vectrices comparees de plusieurs souches de *Myzus persicae* Sulz., *Aulacorthum solani*Kltb. et *Macrosiphum euphorbiae* Thom. Dans l'etude de la transmission de l'enroulement de la omme de terre. Potato Res. **13** : 199-209.

264. Robert, D. P., Denny, T. P. and Schell, M. A. 1988.Cloning of Egl gene of *Pseudomanas solanacearum* and analysis of its role in phytopathogenicity. J. Bacteriol. **171** : 1445-1451.

265. Robertson, N. F. 1994. The challenge of *Phytophthora infestans.* pp. 1-25. In : Advances in Plant Pathology. Vol. 7. *Phytophthora infestans,* the cause of late blight of potato, Eds. D. S. Ingram and P. H. Wiliams, Academic Press, NY.

266. Robinson, D. J., Ploeg, A. T., Brown, D. J. F. and Legorburu, F. J. 1992. Tobraviruses and transmissibility by nematodes. Annual report Scottish Crops Res. Inst., Dundee (UK) 1991. pp. 77-80.

267. Rogoshin, A. N. and Filipov, A. V. 1983. Distribution and conidial viability of *Phytophtora infestans* in the air above pootato fields. Mikologiyei Fitopatologia, **17** : 225-227.

268. Rose, D. G. 1983. Some properties of an unusual isolate of potato virus S. Potato Res. **126** : 49-62.

269. Rosenberg, C., Casse-Delbart F., Dusha, I., David, M. and Boucher, C. 1992. Megaplasmid in the plant associated *Rhizobium meliloti* and *Pseudomanas solanacearum* J. Bacteriol. **150** : 402-406.

270. Ross, H. 1954. Die Vererbung der 'Immunitat' gegen das X-Virus in tetraploidem *Solanum acaule* Proc. 9th Intern. Confr. Genetics. Bellagio, Italy (1953) Caryologia, **6** (supplement) : 1128-1132.

271. Ross, H. 1966. Use of wild *Solanum* species in German potato of the past and today. Am. Potato J. **43** : 63-80.

272. Ross, H. 1986. Potato breeding-problems and perspectives. In : Advances in Plant Breeding, No. 13. Pal Parey, Berlin, Hamburg.

273. Rotem, J. and Cohen, Y. 1974. Epidemiological patterns of *Phytophthora infestans* under semi arid conditions. Phytopathology, **74** : 711-714.

274. Royer A. H., Russo, J. M. and Kelley, G.W. 1989. Plant disease prediction using a mesoscale weather forecasting technique. Plant Disease, **73** : 618-624.

275. Salazar, L. F. and Harrison, B.D. 1978a. Host range, purification and properties of potato virus T. Ann. Appl. Biol. **89** : 223-235.

276. Salazar, L. F. and Harrison, B. D. 1978b. Potato virus T. CMI/AAB Description of Plant Viruses No. 187, 4pp.

277. Samoucha, Y. and Gohen, G. 1988. Synergistic interaction of cymoxanil mixtures in the control of metalaxyl resistant *Phytophthora infestans* of potato. Phytopathology, **73** : 636-640.

278. Sanger, H. L. 1982. Biology, structure, functions and possible origin of vivoids. In : Encyclopedia of Plant Physiology, New Series, Vol. 14B. Nucleic Acids and Proteins in Plants, Eds. B. Parthier and D. Boulter, Springer-Verlag, Berlin, pp. 368-454.

279. Sato, W. 1980. Sources of inoculum and sites of infection of potato trubers by *Phytophthora infestans* in soil. Ann. Phytopath. Soc. Japan. **46** : 231-240.

280. Schell, M. A. 1987. Purification and characterization of endoglucanse from *Pseudomanas solanacearum.* Appl. environ. Microbiol. **53** : 2231-2241.

281. Schell, M. A., Robert, D. P. and Denny, T. P. 1988. Analysis of polygalacturonase encoded by pg1A and its involvement in pathogenicity. J. Bacteriol. **170** : 4507-4508.

282. Schick, R. 1932. Uber das verhalten von *Sonanum tuberosum,* Und ihren Bastaeden gegenumber verschiedenen Herkunften von *Phytophthora infestans,* Zuchter **4** : 233-237.

283. Schmiediche, P. 1985. Breeding bacterial wilt *Pseudomanas solanacearum* resistant germplasm. In : Present and future strategies for potato breeding and improvement. CIP, Lima, Peru. pp. 45-55.

284. Schmiediche, P. 1985. Breeding potatoes for resistance to bacterial wilt caused by *Pseudomanas solanacearum.* . In : Bacterial wilt disease in Asia and South Pacific. Ed. G. J. Persley. ACIAR Proc. No. 13 pp. 105-111.

285. Schepers, A., Bus, C. B. and Styszko, L. 1978. Effects of application of mineral oil on seed potatoes. Proc. 7th Trien. Conf. Eur. Assoc. Pot. Res., Warsaw, 1978, pp. 269-270. (Abstr.)

286. Schepers, A., Harrewijn, P. and Bus, C. B. 1984. Non-insecticidal treatments to reduce aphid development and infection with stylet-borne virus diseases in seed potatoes. Proc. 9th Trien. Conf. Eur. Assoc. Pot Res., Interlaken, pp. 204-206. (Abstr.)

287. Schwinn, F. J. 1979. Control of Phycomycetes, a changing scene. Proc. 10th Br. Crop Prot. Conf. **3** : 791-802.

288. Schwinn, F. J. and P. Margot. 1991. Control with chemical, pp 225-266. In : Advances in Plant Pathology, Eds. D. S. Ingram & P. H. Williams Vol. 7. *Phytophthora infestans,* the cause of late blight of potato. Academic Press, NY.

289. Semancik, J. S. 1979. Small pathogenic RNA in plants-the viroids. Ann. Rev. Phytopathol. **17** : 461.

290. Seneviratne, S. N. de S. 1969. On the occurrence of *Pseudomanas solanacearum* in hill country of Ceylon. J. Hort. Sci. **44** : 393-402.

291. Senevirantne, S. N. de S. 1976. Bacterial wilt in solanaceious crops grown in rice fields. In : Proc. Ist. Int. Planning Conf. and Workshop on the Ecology and Control of Bacterial wilt, Eds. L. Sequeira and A. Kelman and Raleighs, N. C. pp. 95-101.

292. Sequeira, L. and Graham, T. L. 1977. Agglutination of avirulent strain of *Pseudomanas solanacearum* by potato lectin. Physiol. Pl. Path. **11** : 43-54.

293. Sequeira, L. and Rowe, P. R. 1969. Selection and utilization of *Solanum phureja* clones with high resistance to different strains of *Pseudomonas solanacearum*. . Am. Potato. J. **46** : 451-462.

294. Shamsuddin, N., Llyod, A. B. and Grahm, J. 1978. Survival of potato strain of *Pseudomanas solanacearum.* in soil. Australian Inst. Agric. Sci. **44** : 212-215.

295. Shaner, G. and Finney, R. F. 1977. The effect of nitrogen fertilization on the expression of slow mildewing resistance in Knox wheat. Phytopathology, **67** : 1051-1056.

296. Sharma, K. P., Pandey, S. K. and Tyagi, B. R. 1982. Breeding for field resistance to late blight of potato. pp. 92-95. In : Potato in Developing Countries, Eds. B. B. Nagaich, *et al.* Indian Potato Assc., CPRI, Shimla.

297. Shattock, R. C. 1976. Variation in *Phytophthora infestans* on potatoes grownin walk-in polythene tunnels. Ann. Appl. Biol. **32** : 227-232.

298. Shattock, R. C., Shaw, D. S., Fyfe, A. M., Dunn., J. R., Loney, K. H. and Shaltock, J. A. 1990. Phenotypes of *Phytophthora infestans* response to metalaxyl and isozyme analysis. Plant Pathology,.**39** : 242-248.

299. Shattock, R. C., Tooley, P. W. and Fry, W. E. 1986 a. Genetics of *Phytophthora infestans :* Characterization of single oospores cultures from A1 isolats induced to selt by intraspecific stimulation. Phytopathology, **76** : 407-410.

300. Shattock, R. C., Tooley, P. W. and Fry, W. E. 1986 b. Genetic of *Phytophthora infestans :* determization of recombination, segregation and selfing by isozyme analysis. Phytopathology, **76** : 410-413.

301. Shattock, R. C., Tooley, P. W. Sweigard, J. and Fry, W. E. 1987. Genetic of *Phytophthora infestans :* pp. 175-185. In : Genetics and Plant Pathogenesis, Eds. P. R. Day and G. J. Jellis. Brown rot. Ann. Sci. Rept., CPRI, Shilackwell Sci. Publ., Oxford (UK).

302. Shaw, D. S. 1983. The cytogenetics and genetics of *Phytophthora.* In : *Phytophthora* its Biology, Taxonomy, Ecology and Pathology pp. 81-94, Eds. D. C. Erwin, Brtnicki-Garcia, S. and Tsao, P. H. Am. Phytopath. Soc. Press, St. Paul.

303. Shaw, D. S. 1991. Genetics. In : Advances in Plant Pathology Vol. 7, *Phytophthora infestans,* the cause of late blight of potato. Eds. D. S. Ingram and P.H. Williams, pp. 131-167, Academic Press, 267 pp.

304. Shaw, D. S., Fyfe, A. M. Hibberb, P. G. and Abedl Sattar, M. A. 1985. Occurrence of rate A2 mating type of *Phytophthora infestans* on imported Egyptian potato and production of sexual progeny with A1 mating types with the U.K. isolates. Plant Pathology, **34** : 552-556.

305. Shekhawat, G. S., Bahal, V. K., Garg, I. D., Sharma, G. D., Verma, R. K. and Ansari, M. M. 1981. Brown rot. Ann. Sci. Rept., CPRI, Shimla, pp. 72-73.

306. Shekhawat, G. S., Kishore, V., Singh, D. S., Khanna, R. N., Singh, R. and Bahal, V. K. 1979. Survival of *Pseudomonas solanacearum* under diverse agroclimates in India. Indian J. Agric. Sci. **49** : 335-338.

307. Shekhawat, G. S., Singh, R, and Kishore, V. 1978. Distribution of bacterial wilt and races and biotypes of the pathogen in India. J. Indian Potato Assoc. **5** : 155-165.

308. Shekhawat, G. S., Sharma, G. D., Bahal, V. K., Garg, I. D., Kishore, V. and Verma, R. K. 1982a. Studies on brown rot. Ann. Sci. Rept. CPRI, Shimla, pp. 110-111.

309. Shekhawat, G. S., Kishore, V., Singh, D. S., Khanna, R. N., Singh, R. and Bahal, V. K. 1982b. Perpetuation of *Pseudomonas solanacearum* in India. In : Potato in Developing Countries, Eds. B. B. Nagaich, *et al.* Indian Potato Associac. CPRI, Shimla, pp. 279-283.

310. Shekhawat, G. S., Kishore, V., Sunaina, V., Bahal, V. K., Gadewar, A. V., Verma, R. K. and Chakrabarti, S. K. 1988a. Latency and management of *Pseudomonas solanacearum* . Ann. Sci. Rept. CPRI, Shimla. pp. 117-121.

311. Shekhawat, G. S., Gadewar, A. V., Bahal, V. K. and Verma, R. K. 1988b. Cultural practices for managing bacterial wilt of potato. In : Bacterial diseases of Potato : report of the planning conference on bacterial diseases of potato, 1987. CIP, Lima, Peru. pp. 65-84.

312. Shekhawat, G. S., Bahal, V. K., Kishore, V., Sanger, R. B. S., Patel, R. L., Dey, B. K., Sinhal, S. K. and Pani, A. K. 1990. Control of bacterial wilt of potatoes by agronomic practices. J. Indian Potato Assoc. **17**: 52-60.

313. Shekhawat, G. S., Kishore, V., Gadewar, A. V., Sunaina, V., Verma, R. K., Bahal, V. K. and Chakrabarti, S. K. 1991. Ecology, detection and magement of *Pseudomonas solanacearum* Ann. Sci. Rept. CPRI, Shimla, pp. 76-82.

314. Shekhawat, G. S., Chakrabarti, S. K. and Gadewar, A. V. 1992a. Potato Bacterial Wilt in India. Tech. Bull. No. 38. CPRI, Shimla, 52 pp.

315. Shekhawat, G. S., Gadewar, A. V. and Chakrabarti, S. K. 1992b. Spontaneous phenotypic reversion from afluidal to fludial state in strains of *Pseudomonas solanacearum.* ACIAR Bacterial Wilt Newsl. **8** : 5-6.

316. Shekhawat, G. S. and Perombelon, M. C. M. 1991. Factors affecting survival in soil and virulence of *Pseudomonas solanacearum.* J. Plant Dis. Prot. **98** : 258-267.

317. Sigvald, R. 1984. The relative efficiency of some aphid species as vectors of potato virus Y^o (PVY^o). Potato Res. **27** : 285-290.

318. Sigvald, R. 1987. Aphid migration and the importance of some aphid species as vectors of potato virus Y^o (PVY^o) in Sweden. Potato Res. **30** : 267-283.

319. Simmonds, N. W. and Malcolmson, J. F. 1967. Resistance to late blight in Andigena potatoes. Eur. Potato J. **10** : 161-166.

320. Singh, B. P. and S. K. Bhattacharyya. 1990a. Control of late blight tuber infection by systemic fungicides. Indian J. Plant Prot. **18** : 119-123.

321. Singh, B. P. and Bhatacharyya, S. K. 1990b. Appearance, buildup and spread of late blight in relation to source of inoculum. Indian Phytopathology, **43** : 393-400.

322. Singh, B. P., Sharma, H. C. and Bhattacharyya, S. K. 1991. Spread of late blight in time and space : Comparison of some descriptive models. Asian Potato J. **2** : 47-49.

323. Singh, M. N., Khurana, S. M. Paul, Nagaich, B. B. and Agrawal. H. O. 1988. Environmental factors influencing aphid transmissionof potato virus Y and Potato leafroll virus. Potato Res. **31** : 501-509.

324. Singh, Rajpal. 1980. Studies on *Pseudomonas solanacearum* the Incidant of bacterial wilt of solanaceous crops. Ph.D. Thesis, Univ. Gorkhpur, India.

325. Singh, R. P. 1988. Role of weeds in potato virus spread. In : Potato Pest management in Canada, Eds. G. Boiteau *et al.*, Canada Agric. NB, pp 355-362.

326. Singh, R. P. and Bioteau, G. 1986. Re-evaluation of potato aphid *Macrosiphum euphorbiae* (Thomas) as vector of potato virus Y and potato leafroll virus. Amer. Potato. J. **63** : 335-340.

327. Singh, R. P., Finnie, R. F., and Bangnall, R. E. 1971. Losses due to the potato spindle tuber virus. Am. Potato J. **48** : 262-264.

328. Smith, L. P. 1956. Potato blight forecasting by 90 per cent humidity criteria. Plant Path. **5** : 83-87.

329. Smoot, J. J., Gough, F. J., Lamey, H. A., Eichenmuller, J. J. and Gallegly, M. E. 1958. Production and germination of oospores of *Phytophthora infestans*. Phytopathology, **48** : 165-171.

330. Song Bo-fu. 1984. Use of true potato seed in China. CIP Cirucular, **12 (2)** : 6-7.

331. Sparrow, L.A.D. 1976. Recent trends in the activity of aphids infesting potatoes in south-east Scotland in relation to virus incidence in the crop. Scotish Hort. Res. Inst. Assoc. Bull. **11** : 8-14.

332. Spielman, L. J., Sweigard, J. A. Shattock, R. C. and Fry, W. E. 1986. Use of isozymes for the analysis of unexpected segregation ratios in F2 and back cross progeny of *Phytophthora infestans*. Phytopathology **76** : 1102 (Abstr.).

333. Staskawitsz, B. J., Dahlbeck, D., Miller, J. and Damm, D. 1983. Molecular analysis of virulence gene in *Pseudomonas solanacearum*. In : Molecular genetics of the bacteria-plant interaction, Ed. A. Pulher, Springer Verlag, Berlin, pp. 345-352.

334. Stace-Smith, R. 1985. Virus-free clones through plant tissue culture. In : Comprehensive Biotechnology, Vol. 4, Eds. C. W. Robinson and J. A. Howell. Pergamon, Oxford, pp. 169-179 (U.K.)

335. Staub, T. and Schwin, F. J. 1980. Fungicidal properties of metalaxyl and its strategies. pp. 154-157. Proc. Congr. Mediaterr. Phytopathol. Soc., Athens, Greece.

336. Stelzner, G. 1950. Virusresistenz der Wildkartoffel. Archiv fur Pflanzenzuchtung, *29* : 135-158.

337. Stuart, W. 1905. Disease-resistant potatoes. pp. 135-140. In Vt. Agric. Exp. Stn. Bull 115.

338. Sunaina, V., Kishore, V. and Shekhwat, G. S. 1989. Latent survival of *Pseudomonas solanacearum* race 1 in potato tubers and weeds. Z. Pfanzenkr Pfanzenschutz, **96** : 361-364.

339. Sunaina, V., Kishore, V. and Shekhawat, G. S. 1991. Occurrence of *Pseudomonas solanacearum* race 1 in high bills, a new report from India. J. Indian Potato Assoc. **18** : 106-107.

340. Sweigard, J. A., Spielman, L. J., Tooley, P. W., Shattock, R. C. and Fry. W. E. 1987. The genetic control of virulence o in *Phytophthora infestans*. Phytopathology, **77** : 122 (Abstr).

341. Tamada, T., Harrison, B. D. and Robeats, J. M. 1984. Variation among British isolates of potato leafroll virus. Ann. Appl. Biol, **104** : 107-116.

342. Tantius, P. H., Fyfe, A. M., Shaw, D. S., and Shatlock, R. C. 1986. Occurrence of the A2 mating type and self fertile isolates of *Phytophthora infestans* in England and Wales. Plant Pathology, **35** : 578-581.

343. Taschmer, P. E.M., van der Kuyl. A. C., Neeleman, L. and Bol, J. F. 1991. Replication of an incomplete alfalfa mosaic virus genome in plants transformed with viral replicase genes. Virology, **181** : 445-450.

344. Thurston, 1963. Bacterial wilt of potato in Colombia. Am. Potato J. **40** : 381-390.

345. Tomilinson, D. L. and Gunther, M. T. 1986. Bacterial wilt in Papua New Guinea. In : Bacterial Wilt in Asia and South pacific. Ed. G. J. Persley, ACIAR Proc. No. 13 pp.

346. Tooley, P. W., Fry. W. E. and Villarreal Gonzalez, M. N. 1985. Isozyme characterization of sexual and a sexual *Phytophthora infestans* populations. J. Heredity, **76** : 431-455.

347. Torrance, L., Gowan, G. H., Scott, K. P., Pereiva, L. G., Roberts, I. M., Reavy, B. and Harrison, B. D. 1992. Potato mop top : detection and diagnosis. Ann. Rept. 1991. Scottish Crops Res. Inst. Dundee (U.K.)

348. Trigalet, A. and Demergy, D. 1986. Invasiveness in tomato plants of Tn 5 induced *Pseudomonas solanacearum*. Physiol. Mol. Plant Pathol. **28** : 423-430.

349. Trigalet, A. and Demery, D. 1990. Use of avirulent mutants of *Pseudomonas solanacearum* for the biological control of bacterial wilt of tomato plants. Physiol. Mol. Plant Pathol. **36** : 27-38.

350. Tung, P. S., Rasco, E. T., Van der Zaag, P. and Schmiediche, P. 1990. Resistance to *Pseudomonas solanacearum* in the potato : II Aspect of host-pathogen-environment interaction. Euphytica, **45** : 211-215.

351. Turl, L.A.D. 1980. An approach to forecasting the incidence of potato and cereal aphids in Scotland. EPPO Bull. **10** : 135-141.

352. Turl, L. A. D. 1983. The effect of winter weather on the survival of aphid populations on weeds in Scotland. EPPO Bull. **13** :139-143.

353. Turl, L. A. D. and MacDonald, D. M. 1987. The relationship between suction trap catches and the spread of potato virus Y° in south-east Scotland. In : Proc. EC Expert Meet. Aphid Migration and Forecasting 'euraphid' Systems in European Countries', Ed. R. Cavalloro, pp. 177-185.

354. Umaerus, V., Umayus, M., Erijefalt, L. and Nilsson, B. A. 1983. Control of *Phytophthora* by host resistance: Problems and Progress. pp. 315-326. In : *Phytophthora*, its biology, taxonomy, ecology and Pathology. Ed. D. C. Erwin, Bartricki-Garcia and P. H. Tsao, The American Phytopathological Soc., St. Paul, Minnesota, USA.

355. Valkonan, J. P. T., Pehu, E., Jones, M. G. K., and Gilson, R. W. 1991. Resistance in *Solanum brevidens* to both PVY and PVX may be acciated with slow cell to cell spread. J. Gen. Virol. **72 (2)** : 231-236.

356. Van der Plank, J. E. 1963. Pant disease : Epidemics and Control Acad. Press, New York & London, 349 pp.

357. Van der Zaag, D. E. 1977. De betekenis van infectie van poters met virus op de opbrengst bij aardappelen. Landbouukundig Tijdschrijt, **89** : 154-157.

358. Van Eek, T. H. and Thung, T. H. 1950. Resultanten van onderzoekingen omtrent aaradappelziekten op. Java. Landouw (Bogor), **22** : 303-346.

359. Van Everdingen, E. 1926. The relation between weather conditions and potato blight. *Phytophthora infestans*. Tijdschr. Plantenziekten, **32** : 129-140.

360. Van Harten, A. 1983. The relationship between aphid fights and the spread of potato virus $Y^{N (PVYN)}$ in the Netherlands. Potato Res. **26** : 1-15.

361. Van Hoof, H. A. 1980. Aphid vectors of potato virus Y^n . Netherlands. J. Plant Pathol. **86** : 159-162.

362. Vertanian, V. G. and Endo. R. M. 1982. Survival of *Phytophthora infestans* in seeds extracted from infected tomato fruit. Phytopathology, **75** : 375-378.

363. Walkey D. G. A. 1980. Production of virus-free plants tissue culture. In : Tissue Culture Methods for Plant Pathologists, Eds. D. S. Imgram and J. P. Helgeson. Blackwell Oxford. pp. 109-117.

364. Wallin, J. R. and Riley, J. A. 1960. Weather map analysis - an aid in forecasting potato late blight . Plant Dis. Reptr. **44** : 227-234.

365. Weardrop, E. A., Gray, A. B., Singh, R. P. and Peterson, J. F. 1989. Aphid transmission of potato virus S. Am. Potato J. **66** : 449-459.

366. Wilcoxon, R. D., Komana, B. S. and Atif., A. H. 1975. Evaluation of wheat cultivars for their ability to retard development of stem rust. Ann. Appl. Biol. **80**: 275-281

367. Williams, W. L. and Ross, A. F. 1957. Aphid transmission of potato leafroll virus as affected by the feeding of non-viruliferous aphids in the test plants and by vectors variability. Phytopathology, **47** : 538.

368. Wilson, U. E. and Coffey, M. D. 1980. Cytological evaluation of general resistance to *Phytophthora infestans* in potato foliage. Annals of Botany, **45** : 81-90.

369. Wright, N. S. 1963. Detection of strain-specific serological activity in antisera to tobacco mosaic virus, Clover yellow mosaic virus, and potato virus X, by complement fixation. Virology, **20** : 131-136.

370. Writht, N. S. 1977. The effect of separate infections by potato virus X and S on Netted Gem potato. Am. Potato J. **54** : 147-149.

371. Xu, P., Iwata, M., Leong, S. and Sequeira, L. 1990. Highly virulent strains of *Pseudomonas solanacearum* that are defective in extracellular polysacccharide production. J. Bacteriol. **172** : 3946-3951.

372. Xy, P., Leong, S. and Sequeira, L. 1988. Molecular cloning of genes that specify virulence in *Pseudomonas solanacearum.* J. Bacteriol, **170** : 617-622.

373. Zann. K. 1962. Activity of *Phytophthora infestans* in soil in relation to tuber infection. Trans. Br. Mycol. Soc. **45** : 205-221.

374. Zehr, E. I. 1971. Cultural, physiological and biochemical properties of isolates of Philippine *Pseudomonas solanacearum.* Philipine Phytopathology, **6** : 29-43.

3 Management of Fungal Diseases of Rice and Scope of Biotechnological Approaches

R. Jayarajan, S. Mohan, P. Balasubramanian and V. Mariappan

CONTENTS

Integrated Pest and Disease Management
Rajeev K. Upadhyay, K.G. Mukerji, B.P. Chamola and O.P. Dubey (ed.)
APH Publishing Corporation, New Delhi-110002, India

ABSTRACT

Recent advances in the management strategies of fungal diseases affecting rice are discussed in this paper. Epidemiological studies which have been useful in managing rice blast in the recent past could be applied in case of other fungal diseases also. Forecasting an epidemic is now possible in case of blast alone. Importance of more work is emphasized for development of such models also for sheath blight, brown spot and sheath rot diseases which pose a serious problem in rice cultivation. Use of topical fungicides as a part of IPM strategies has been highlighted with recent reports on the economically important fungal diseases of rice. More emphasis needs to be given in IPM strategies for biocontrol systems developed so far again rice fungal pathogens. There is immense potential in using chitinase gene form *Trichoderma* or elsewhere against *Pyricularia oryzae, Helminthosporium oryzae, Rhizoctonia solani* and *Sarocladium oryzae*, using modern biotechnological tools like *Agrobacterium*-mediated gene transfer systems developed for rice transformation, Biolistic process, use of ribosome - inactivating proteins use of detoxifying genes in pathogens, manipulation of susceptibility of pathogens to phytoalexins etc. have also been indicated.

1. INTRODUCTION

The over-use and over-dependence of chemical pesticides following world war II have resulted in undesirable ecological changes, such as development of pesticide-resistant strains of pathogens, minor pathogens becoming major ones, resurgence, destruction of antagonistic organisms besides contamination of food grains and environment. In rice, IPM for fungal diseases assumes added importance in view of the fact that the crop is usually grown in seasons which are very favourable for pathogens also. In many places, the crop is grown continuously all round the year ensuring susceptible stage of the crop to particular diseases throughout the year which helps in survival of the pathogen. Many weed hosts of rice pathogens grow commonly in rice field bunds. During boom phase of a variety it occupies vast stretches in a tract making it ideal for outbreak of epiphytotic. The resource-poor rice farmers in developing countries can ill-afford timely application of several rounds of costly pesticides. The high sporulation potential of the fungal pathogens help them in fast spread of disease in a very short time. Recently Tamil Nadu Agricultural University, Coimbatore conducted several experiments in farmers' fields to demonstrate the efficacy and economy of adopting IPM. The results clearly indicated that this is a cost-effective technology.

Recently, biotechnology has come as a handy tool to solve many plant disease problems. The principle strategies in using biotechnological methods in rice disease management include identification of the genes encoding resistance for diseases and incorporating them into cultivated rice using tissue culture, diagnosis of diseases well before they become highly destructive, and containing them using various methods involving recombinant DNA technology. The progress made in solving rice fungal diseases problems and future scope are indicated.

Table 1

Blast	*Magnaporthe grisea (= Pyricularia oryzae;* Syn. *P. grisea)*
Brown leaf spot	*Cochliobolus miyabeanus (= Helminthosporium oryzae)*
Sheath blight	*Thanetophorus cucumeris (= Rhizoctonia solani)*
Sheath rot	*Sarocladium oryzae* (Syn. *Acrocylindrium oryzae)*

Irregular stem rot	*Magnaporthe salvinii* (= *Helminthosporium sigmoideum* var. *irregulare; Sclerotium oryzae)*
Stem rot	*Leptosphaeria salvinii* (= *H. sigmoideum)*
Foot rot	*Gibberella fujikuroi* (= *Fusarium moniliforme)*
Seedling blight	*Magnaporthe salvinii* (= *Sclerotium rolfsii)*
Narrow brown leaf spot	*Sphaerulina oryzina* , *Cercospora oryzae*
Leaf smut	*Entyloma oryzae*
Kernel smut	*Tilletia barciayana*
Bunt	*Neovossia horrida*
False smut	*Ustilaginoidea virens*
Downy mildew	*Sclerophthora macrospora*
White leaf streak	*Ramularia oryzae*
Stackburn	*Alternaria padiwickii* (Syn. *Trichoconis padwickii)*
Leaf scald	*Rhynchosporium oryzae*
Collar rot	*Ascochyta oryzae*
Rust	*Puccinia graminis* f. sp. *oryzae, Uromyces coronatus*
Sheath rot blotch	*Cylindrocladium scoparium*
Crown sheath rot	*Ophilobolus oryzinus; O. oryzae*
Seath blotch	*Pyrenochaeta oryzae*
Seedling damping off	*Fusarium* spp., *Pythium* spp., *Achyla* spp., *Pythiomorpha* spp.
Udbatta disease	*Ephelis oryzae*
Black kernel	*Curvularia* spp.
Minute leaf and grain spot	*Nigrospora* spp.
Glume blight	*Phyllosticta glumarum*
Scab	*Gibberella zeae*
Red blotch of grains	*Epicoccum purpurascens*
Speckled blotch	*Septoria* spp.

2. MAJOR DISEASES

2.1. Blast : *Pyricularia oryzae*

It is the most serious one in almost all countries and appeared in an epiphytotic from in India, Korea, Japan and Philippines. The yield loss is 50.2% in Korea and 90% in Philippines.

2.1.1 Inoculum source

The pathogen overwinters in both infected straw and seed as mycelia and spores. The spores and mycelia on the surface of the straw died before the next spring but those within the straw pile overwintered. The pathogen attacks a number of weeds like *Echinochloa colona, Leersia hexandra, Brachiaria distachya, Leptochloa chinensis* and *Panicum repens.*

2.1.2 Disease development

Rice plants at seedling, tillering and heading stages are highly susceptible. In the dry nursery, it occurs severely after three weeks of growth. In the planted crop, it is severe at the time of vigorous tillering. Close planting increases the severity. Application of nitrogenous fertilizer always induces heavy incidence of blast regardless of application of other fertilizers. Farmers usually apply greater part of nitrogen fertilizers before planting and top dress the rest before the panicle formation stage. As a result severe leaf and panicle blast occurs under weather conditions favourable for an out break of epiphytotics.

Disease incidence is great in zinc-deficient soil. Furnace slag containing 36% silicon oxide significantly reduces the infection. Phosphatic fertilizers reduced lesions/leaf.

2.1.3 Epidemiology

With a view to elucidating the role of weather factors in the outbreak of epiphytotics, laboratory and field studies have been conducted and the part played by each factor analyzed in many countries.

Temperature has a profound influence on all stages of infection. Germination of conidia and formation of appressoria occur between 10°C and 33°C, the optimum being 25°-28°C (61). A minimum dew period of 6-8 hours at 25°C was sufficient to cause infection but in higher temperatures, longer period was taken (22). Invasion of host cell took 10 hr at 32°C, 8 hr at 28°C and 6 hr at 24°C. Lesions were observed in 13-18 days at 9-10°C, 7-9 day at 17-18°C days at 24-25°C and 4-5 days at 26-28°C (14). Maximum sporulation occurred at 12 to 34°C, optimum being 28°C. Water molecules act as agents for the release of conidia. The conidia are produced on the lower surface of the leaf to the extent of 2-4 times greater then on the upper surface (49). From this, it was estimated that one acre of rice crop with 25% leaf area infected might produce 4000 x 10^9 conidia per day. From aerobiological studies, it was found that a high spore load in air (1562/m^3) was recorded in September. In the air 98% of the spores were viable. This can account for the rapid spread of the disease over long distance (16, 17, 49).

Sun light inhibited appressorial formation. A leaf wetness period of 12 hr was required for infection at 20-25°C but 16 hours are required at a sub-optimum temperature of 18°C (21). Penetration rate and the lesion number were higher in darkness than in light.

Minimum temperature of 18-26°C and relative humidity 90% were conducive for the out break of epiphytotics (46). Continuous cloud with a temperature 22°C and more than 93% RH were considered optimum for penetration (25,53). Dew played an equally important role as temperature in the disease development. The important role of weather in the development of blast epiphytotics can be highlighted by the performance of the blast susceptible a variety IR 50 in Tamil Nadu (India) where this variety recorded little blast incidence when grown in July-September (temperature 28°-32°C) but it succumbed to blast epiphytotics in December, 1983 when the weather conditions were conducive for the disease.

Sporulation occurred twice in 24 hr period around 6 to 7 pm and from 4 to 5 am (30). Trapping of spores in India confirmed the occurrence of double peaks in the quantity of spores in the atmosphere. Release of conidia is minimum between 8 am and 4 pm but maximum between 12 pm and 4 am (29). Alternate changes of light and darkness under saturated humidity are essential for release of conidia. From 6 to 8 hours of darkness following a period of light are required for the release of conidia. Strong wind cause mechanical release of conidia. Water drops induce release of conidia by touching the juncture between conidia and conidiophore.

2.1.4 Forecasting

The management of this disease depends upon the ability to anticipate epidemic outbreak and modify spray schedule, approximate action for disease control. The association of specific weather conditions with blast have been observed and the forecasting systems were based on determining when the weather conditions are favourable for disease development. The computerized system for rice blast forecasting was developed by the International Rice Research Institute (IRRI), Philippines, the International Centre for Tropical Agriculture, Colombia, the Rural Development Administration, Korea and the Pennsylvania State University in the United States. The model was synthesized and used a combination of mean air temperature, the hours of leaf wetness and hours of relative humidity above 90% to derive daily mean values of blast units of severity. The appearance of the disease was recorded on the crop only after a minimum period of 10 days following the spore catch (30). Therefore many attempts have been made to forecast the outbreak of epiphytotics based on the number of spores caught in air. In India, bund grasses often grow luxuriantly and infection can be seen almost any time of the year. Heavy sporulation could be seen on grasses like *Panicum repens* (1).

The number of days with maximum temperature of 20°C was suggested to be the reason for blast development. The occurrence of a minimum temperature ranging from 20 to 25°C along with humidity of 95% and above lasting for a week or more were found to be associated with the blast epidemic (47). A minimum period of 5 days or more in a month with night temperature of about 24°C along with humidity of 9o% and above were found to initiate the development of blast. Low night temperature leads to exaggerated nitrogen metabolism and this may well account for the disease proneness of rice plants.

Management of blast depends upon anticipating epidemic outbreaks so that appropriate control measures can be scheduled in time. Earlier systems of bast forecasting were based on weather conditions and/or number of spores in the air which were correlated by simple or multiple regression or mathematical analysis. With the advent of computers, simulation techniques have been used to forecast epidemics of blast (24). Manibhusanrao and Krishnan (40) devised a computerized simulation named EPIBLA based on FORTRAN 77 that uses field data on dew, humidity, temperature, number of spores, disease incidence and predicted number of spores to forecast rice blast epidemics. Two models were developed to forecast blast outbreaks. The first one, called "the spore prediction model" uses daily weather parameters as independent variables and the number of spores as the dependent variable. The second model, "the disease progression model" uses the proportion of disease incidence as the dependent variable and the sum of predicted spores and average weather parameters for the seven days preceding disease onset as the independent variables. The general multiple regression equation,

$$Y = a + b_1X_1 + b_2X_2 + \ldots\ldots\ldots\ldots\ldots\ldots b_nX_n$$

where, a = intercept, b_n = slope of line (the partial regression coefficient value), and X_n = independent variable.

EPIBLA requires monitoring only a few variables. It reads the data from the data files and predicts the disease for each seven day period from the 15th day after transplanting. A forewarning of at least one week prior to the onset of the epidemic is possible so that the farmer can take up crop protection measures within the period. However, the model has been tested only on IR 50 (a highly susceptible variety) and IR 20 (a tolerant variety). More studies are needed to extend the use of this computer programme to other varieties and situations and to improve its prediction accuracy.

2.1.5 Management

Chemical : In developed countries like Japan and United States major emphasis is given to fungicidal sprays. Antibiotics like blastomycin and blasticidin-A showed high promise in Japan. The organophosphorus fungicides edifenphos and iprobenphos were found to be very effective (56). Among systemic fungicides, carbendazim, benomyl and thiophanate methyl lowered the incidence. One important point to be considered is the correct timing of application. Experiments conducted world over have indicated that the spray should be given as soon as the lesion reaches grade two because increased sporulation due to favourable climate will cause heavy damage. Continuous use of edifenphos, pyrozophos and kasugamycin created resistance in *Pyricularia oryzae* (66).

Recently several granular formulations like isoprothiolane have been reported to be effective. In view of the fact that appearance of the disease in the nursery wipes out the seedlings, the need for a systemic seed treatment fungicide was keenly felt. Pyroquilon an antipenetrant was able to afford protection to the seedlings for 35 days. Neem oil (3%) or neem seed kernal extract (5%) gave appreciable reduction of the disease spread.

Several classes of compounds have the potential to act as inducers of natural resistance mechanisms, though they themselves are not antimicrobial. Those chemicals with such indirect mode of action offer an attractive alternative to existing contact fungicides in integrated disease management. Probenazole evokes an antimicrobial alteration in rice physiology which results in resistance to blast (55). Another synthetic compound, N-phenylsulfonyl-2-chloroisonicotinamide was found to induce an activity in rice which significantly protects against blast. Similar effect has also been observed consequent to application of salicyclic acid (51).

Cultural methods:

Upland nursery seedlings are highly susceptible due to lower silicon content of the epidermal cell. The from of nitrogenous fertilizer also influenced disease severity. In plots receiving urea it was more severs than plots supplied with ammonium sulfate and calcium ammonium nitrate. Plants given nitrogen from nitrate were more susceptible than those provided with NH_4^+ nitrogen. Application of lesser amounts of nitrogenous fertilizer and in split doses reduce disease severity. The soluble nitrogen that accumulates in plants may serve as a nutrient for the pathogen. Plants receiving larger amounts of nitrogen have fewer silicated epidermal cells and thus lower resistance. Application of nitrogen also reduce hemicellulose and lignin in the cell wall and weakens the plant's resistance. Thus selection of suitable form of nitrogenous fertilizer and its application in split doses offers a good strategy. Neem blended urea and neem coated urea gave slow release of nitrogen and reduced blast disease. Reduction in disease intensity was obtained by application of calcium silicate. It is believed that Si association with cell wall constituents makes them less accessible to the enzymatic degradation. In sandy soil supplemented with hill side clay P_2O_5, SiO_2 and $Ca(OH)_2$ the incidence of blast was reduced.

Weed hosts act as initial foci of infection and often the disease first appeared in rice plants adjacent to the weed hosts in which profuse sporulation occurs. Therefore removal of these grasses from rice field bunds will go a long way in reducing the initial foci of infection. Coating rice seeds with spore suspension of *Chaetomium globosum* prevented the early infection. *Nigrospora oryzae* was identified as hyperparasite in blast lesions and reduced lesion size (63). Pre-inoculation of rice seedling with some incompatible *P. oryzae* pathogen suppressed disease severity. However the biocontrol of foliar pathogens in general has not yet attained the take off stage in view of the fact that the biocontrol agents should be able to grow and sporulate in the infection court using the leaf exudates as nutrients.

Varietal Resistance:

(a) Resistance through classical breeding programme

Blast is a disease for which disease resistance in germplasm has been exploited from the beginning of this century. In India, Madras state took the lead and released blast resistant variety Co_4 in 1926. Durable blast resistance has been identified in rice cultivars IR20, IR36 and IR42 under tropical conditions.

(b) Resistance through tissue culture

Selection of reliable source of blast resistance requires exacting conditions of relative humidity, temperature etc. which not be possible always. This is a constraint in large scale screening and evaluation of germplasm. Somaclonal variation refers to the appearance of new traits that regenerate from a callus in tissue culture and is due to single gone mutations. Thus, culture of plant tissues is the simplest form of genetic engineering. By subjecting large populations of cultured cells to the action of a selective agent in the Petri dish, it is possible to select new varieties which retain all the desired qualities of the existing variety while adding one or more characters such as disease resistance. Toxins produced by pathogen are used as molecular sieves to select calli from which resistant plants could be regenerated in tissue cultures.

Several toxins viz., picolinic acid, pyricularin, pyriculol, and a glycopeptide have been isolated from liquid culture of *P. oryzae*. Toxin tolerant plants show long term stability and exhibit horizontal resistance. Crude toxin extracts prepared from mycelia or spores of *P. oryzae* produced typical blast symptoms on the treated parts and hence could be used as substitutes for the pathogen in screening for blast resistance.

(c) Resistance through cloning of avirulent genes

Classical genetic studies with several pathogens and their hosts have shown that race specific reactions are controlled by avirulence genes in parasite and resistance genes in host. Avirulence genes of *P. oryzae* towards the rice cultivar CO_{39} seggregated in the cross between two isolates of the fungus. Backcross and sib-cross studies suggested that avirulence to this cultivar is controlled by a single gene. The gene was mapped to a small linkage group corresponding to part of chromosome 1. In addition to walking towards this locus, approach of site-specific cleavage was used to determine the relation between genetic and physical distance in this region of genome. By using these approaches coupled with RFLP studies it might become possible to clone the avirulent gene of the blast fungus into rice(7).

(d) Mainpulating genes encoding sporulation

Any interference with the sporulation should provide potentially a useful strategy for controlling blast disease which has multiple conidial/infection cycles. A detailed understanding of the genetic regulation of sporulation in the rice blast fungus would help design rice genotypes that can withstand blast epidemics. The following single gene mutations control the sporulation in the blast fungus: *Con1* which causes abnormal morphology as well as a 97% reduction in conidial production and *Con2* which produces conidiophores with swollen hyphal tips but fails to develop conidia. Two complementary genes, namely *aleu1* and *aleu2* cause the production of thick-walled pigmented conidia called aleuriospores.

Based on the rationale that knowledge of the products of genes controlling sporulation may help identify target sites for intervention, Leung and Shi (33) focussed their efforts on the isolation of the *Con1* and *Con2* genes. A mapping technique based on random amplification of polymorphic DNA (RAPD) and bulked segregant analysis was used to rapidly generate DNA markers flanking the target gene. A linked marker was also obtained. They also proposed the use of closely linked marker genes for cloning *Con1* and *Con2* genes by chromosome walking.

Considering the fact that one blast lesion produces 2000 - 6000 conidia per day which amounts to 28000 - 84000 in 15 days (44), this approach would help in curtailing the spore load in atmosphere.

2.2. Brown leaf spot (*Helminthosporium oryzae*)

The disease was a major factor behind the failure of rice crops leading to the Great Bengal famine of 1942-43.

2.2.1 Inoculum source

The fungus can remain on the husk and in the kernel for more than three years. It is both internally and externally seed-borne as mycelia. This pathogen is a facultative saprophyte on organic matter under shade or in soil. Stubbles or infected leaves in soil may also serve as primary sources of infection. Several weeds like *Leersia hexandia*, *Cynodon dactylon*, *Echinochloa colona*, *Arunello donux* and crops like *Avena sativa*, *Zeamays*, *Sorghum vulgare*, *Eleusine coracana*, *Saccharium officinarum*, *Setaria italica*, *Pennisetum americanum* and *Panicum miliare* act as collateral hosts (10).

The pathogen overwinter on stubbles and seeds and its survival period varied under different conditions. However, the survival of the pathogen in unsterilized soil was low due to its poor competitive saprophytic ability and high susceptibility to antagonism of *Trichoderma, Streptomyces* and *Bacillus* (28). In irrigated soil, it survived longer in straw (70 days) than on leaves (40 days), longer at soil depth of 30 cm than at 15 cm. (27).

2.2.2 Epidemiology

Primary infection from seed-brone inoculum was severe when the soil temperature was below 26°C. The incidence was found to be higher in dry soil than in wet soil. It is more severe in rainfed than in irrigated/flooded fields.

A relative humidity of over 89% and 25°C favoured disease establishment. The disease control occurred in a severe form during cold weather viz., November to March in South India. Sunshine hours below 60 and above 110 in a fortnight reduced the intensity (9). Crop planted before the middle of August developed less infection than that planted afterward (59). Incidence was 40-50% in November-December. Susceptibility was more during boot leaf and flowering stage.

Conidia are present in the air throughout the year (28). The conidial content of air over rice filed was correlated with disease incidence. The highest number of conidia were caught during November when wind velocity was 7.0 -7.5 km/hr, RH 90%, maximum and minimum temperatures were 27.0-27.4°C and 15.3 -15.8°C respectively.

2.2.3 Disease Development

The disease development increased with deficiency as well as excess of nitrogen and leaching of nutrients from soil (28). The presence of trace elements like Mg, Mn, I, Zn and ca decreased susceptibility where as Cd and Co increased it (23). Low phosphorus and potassium level pre-disposed the plant to the disease. Higher concentration of non-reducing sugars and acid-hydrolyzable polysaccharide sugar and lower concentration of reducing sugars in the tissue make the rice leaves less susceptible.

2.2.4 Management

Adding silicate slag, manganese, potassium, neem cake with urea, sulphur-coated urea, and urea super granules significantly reduced the incidence (23). Foliar spray with $FeSo_4$ or $CuSo_4$ recorded less incidence (42).

Chemical culture

Since the pathogen is primarily seed-brone, presoaking of seeds in cold water for 8 hr followed by hot water treatment at 52°C for 15-20 min eliminated the pathogene. Organ mercurial seed treatment came into the during 1950s. Seed treatment with Furmecyclox, mancozeb and carboxin was found to be effective in reducing seed-borne inoculum (31). Compete control was observed by seed treatment with new fungicides like quazatine, fenapanil, parasanlone, imazalil and filoxazol. Seed treatment with sodium salenite and thioglycalic acid induced phytoalexin in rice plant up to eight weeks after sowing (58).

Foliar spray with mancozeb, iprobenphos and edifenphos were found to be effective in managing this disease (12). A combination of fentin acetate and mancozeb in the proportion of 1:5 gave best control of the disease. Spraying Epigallo catechin - 3 gauata 10% emulsion reduced the number of lesions. Seed treatment with *Bacillus subtilis* followed by two sprays on 30th and 60th day reduced, secondary spread. Spraying trace elements like boron, copper, manganese and zinc reduced the severity of the disease. Foliar spray with ferric chloride, nickel chloride, nickel nitrate, sodium molybdate, DL-methionine, sodium selenite, thioglycolic acid, sodium malonate and calcium chloride gave appreciable protection against the pathogen by inducing phytoalexin production (8). However these experiments have been conducted only under green house conditions and their efficacy under field conditions is yet to be studied.

Varietal resistance

Anatomical characters like thicker cuticle and silicated bulliform cells were associated with resistance. A minimum of three major genes controlled lesion size and at least one gene controlled lesion number. Three genes conditioned differences in disease rating which included factors for both lesion size and number.

Ling *et al.* (34) produced regenerates of IR8 and IR54 that were resistant to toxin produced by *H. oryzae*. Resistant was maintained in the subsequent generations also. Of the 202 isolates of *H. oryzae* tested for their pathogenicity only isolates 134, 136, 147 and 1202 showed differential reaction on rice cultivars IR8, TN1, Co45 and Co20. Compatible isolates grew faster than incompatible isolates in the 4 cultivars. The difference in reaction of calli to different isolates was more discernible on the third day after inoculation. Since calli and whole plants behaved in response to *H. oryzae* inoculation, selection of resistant calli *in vitro* to produce resistant plant is a possibility (64).

Four toxin resistant calli were selected and plants regenerated from 2 of these resistant calli showed resistance to the pathogen. This resistance was heritable and stability of resistance was observed in the R1, R2 and R3 generations (65).

2. 3. Sheath rot (*Sarocladium oryzae*)

In recent years, sheath rot has become a serious threat to rice production in several countries. The per cent yields loss is 57.4 in Tamil Nadu, 26 in Orissa and 85 in Andhra Pradesh states. Ratna, Annapoorna, IET 2234, Kannagi, Soorya, Cauvery, ADT 30 and Jaya were most susceptible.

2.3.1 Inoculum source

The pathogen is seed-borne both internally and externally. The weeds *Cyperus difformis* and *Echinochloa crusgalli, E. colona* were the collateral hosts for the pathogen.

2.3.2 Disease development

Nutrition of the host plant influenced the incidence of the disease. Application of heavy doses of nitrogenous fertilizers is conducive to the development of the disease. Nitrogen increased the susceptibility of the host by decreasing the phenolic content. The reduction in sheath rot incidence due to potassium application was attributed to increase in phenolic content.

Dense planting increased the disease incidence. Infestation with bug *(Heterococus rehi)* and stemborer *(Shirpophaga incertulas)* increased the incidence of this disease.

2.3.3 Epidemiology

The disease intensity is influenced by environmental factors like temperature and relative humidity. It is maximum when the minimum temperature is 17-20°C and relative humidity is high at flowering stage. The period of susceptibility to sheath rot is from December to May.

2.3.4 Management

Chemical

Benlate and panoctine improved germination of sheath rot infection even 6 months after seed treatment. Among several chemicals tested, benomyl 0.2% spray reduced the disease incidence to 6.3% from 41.4 % in unsprayed control plots. Three sprays of carbendazim (0.05%) or brestanol (0.01%) gave good disease control in the field. Tridemorph at 100 ml + Carbendazim at 100g/ha or tridemorph at 200 ml/ ha at 40 and 60 days after transplanting reduced the disease incidence and increased the yield (57).

Cultural

The weed hosts should be destroyed around the fields to avoid survival of the pathogen. Significant reductions in disease spread and increase in grain and straw yields were observed with wide spacing (15 X 15 cm) compared to closer spacing (10 X 10 cm). The disease incidence increased with increased N but decreased with increasing K (54). In a field experiment, ammonium chloride significantly reduced the disease. Foliar application of N and K together with carboxin reduced panicle damage by the pathogen (43). Combination of urea and farm yard manure recorded lesser disease incidence (8.5%) and higher yields (5.3 tons/ha) compared to 31.1% disease and 5.0 tons in application of urea alone. The disease incidence was reduced by the application of manganese, ferrous sulfate and borax. Spraying with calcium sulphate 2% or zinc sulphate 1% respectively. The integrated management of the disease was achieved by nitrogen application in three splits (50% basal, 25% on 25 days after planting and 25% on 40 days after planting) and potassium in two splits (50% basal and 50% on 25 days after planting) with foliar spray of calcium sulphate 2% and carbendazim 0.1%. Neem cake-coated urea can be applied in two splits (50% basal and 50% on 25 days after planting) instead of N fertilizer.

Spraying neem seed kernal extract (2%) thrice reduced the disease incidence to 8.6% from 14.7% in control. Application of coal tar-coated urea or need-coated urea (5:1 w/w/ratio) to soil and spraying neem seed extract 2% or neem cake extract 5% to the plant reduced sheath rot intensity to 3.3, 4.3, 5.7 and 5.7 respectively from 7.7 in control. Neem coated urea increased the grain yield and straw yield.

Recently the role of soil solarization on the propagule density of pathogen in soil has been investigated. It was found that soil mulching with transparent polythene sheets for seven days reduced the population of propagule (63).

Biocontrol

Seed bacterization with *Pseudomonas fluorescens* reduced the sheath rot incidence significantly over control (52). Soil amendments encourage growth and multiplication of soil microflora and thus kill the pathogen in soil by competition or antibiosis. The application of different organic amendments like saw dust, mahua cake, rice chaff, neem cake, neem leaf and *Ipomoea crasaicalus* reduced the saprophytic activity of the pathogen and shortened the survival period of the pathogen. Infection was reduced by their application (19). Amending the soil with non-edible oil cakes will be a good practice in rice cultivation.

Resistant varieties

BJ.1, TKM 6 and Sigadis were completely free from the disease at Tamil Nadu Rice Research Institute, Aduthurai. However, there is no confirmed source of resistance to the disease.

2.4. Sheath blight (*Rhizoctonia solani*)

In India, USA and Philippines yield loss due to this disease ranges from 5.2 to 50.0%. In Japan the loss due to the disease has been estimated as 24-33x10^3 tonnes annually.

2.4.1 Inoculum source

The pathogen overwinters in seeds and causes seedling infection. But predominant infection occurs only through soil-borne sclerotia. Dormant mycelium in crop refuse may remain active in soil for almost a year. Survival period of sclerotia varies based on microenvironment, soil texture, soil temperature, organic matter, green manuring, nutrient status and bacterial activity etc. Collateral hosts include crops like *Zea mays, Eleusine coracana, Sorghum vulgare* and weeds like *Imperata cylindrica, Saccharum spontaneum, Urocloa panicoides* and *Echinochloa colona, Cynodan dactylon* and *Cyperus pilosus.* They help in annual replenishment of sclerotial population in soil (18).

2.4.2 Disease development

Late maturing varieties are severely affected. The old (5-6 weeks) leaf sheaths or blades are highly susceptible. High nitrogen favours lesion development. Rain water and flood irrigation act as good dispersal agents of sclerotia (11). Source and size of sclerotia and their fragments, melanin status and quality, membrane permeability endogenous respiration etc., of the sclerotia H-ion concentration of the clay increase disease severity. The rice root nematode, *Hirschmanniella oryzae* was shown to favour infection.

2.4.3 Epidemiology

Temperature range of 28-32°C and the relative humidity of 81-97% are conducive for the disease development (38, 45). Lesion development was rapid in light. Disease severity was maximum at 50% light intensity, medium at normal sunshine and least in complete shading.

2.4.4 Management

Chemical

The chemical management of the disease has been attempted both by as seed treatment as well as foliar spray. Seed treatment with carbendazim followed by dipping seedlings in 0.05% solution for 10 min was better than seed treatment alone. Seed treatment with quintozene + two foliar sprays with 0.1% edifenphos checked the spread of the disease (4). Three sprays with edifenphos, iprobenphos or zineb or carbendazim at 10 days intervals starting from tillering stage provided good protection against sheath blight (15). Flutolanil a systemic benzanilide with acropetal translocation afforded protective and curative effects when applied as foliar spray at 25 ppm or seed treatment at 3 mg ai/kg. It reduced disease by 80% compared to control. Pencyuron and propiconazole spray effectively reduced disease and increased the yield (6). Under upland and lowland conditions soil drenching with iprodione and benomyl singly or in combination effectively reduced blight. Soil application of thiram and quintozene before transplanting followed by one spray with edifenphos at maximum tillering stage proved effective.

Another recent approach is the efficacy of phytoalexin inducing chemicals like sodium selonite, ferric chloride or cycloheximide. When sprayed, they reduced the sheath blight by 45% (53). This finding has the potential of field use considering the resistance imparted to the plant for which further experiments are necessary.

Cultural

Destruction of collateral hosts during off season will be helpful in reducing the survival of the pathogen. Balanced levels of NPK, application of N in split doses, as slow release forms like grilled urea or large urea granules decreased the disease severity (3). The phenol content was increased in plants receiving high doses of K. The 'P' at higher levels partly reversed the ill effects of high doses of N. Two sprays of borax and sulphate of Zn, Cu and Fe each at 0.05% at 10 days intervals reduced disease severity and increased yield. Wide spacing reduced the disease incidence and countered the effects of high N.

Green manuring with legumes and non-legumes reduced the survival of sclerotia and saprophytic activity of *R. solani* and enhanced mycostasis (3,39). *Glyricidia* leaves inhibited the pathogen considerably. The extracts of neem, *Calotropis procera*, and *Datura stromonium* were found to be toxic (39). These findings are to be exploited under field condition for the disease management.

Biocontrol

Application of oil cakes, farm yard manure, agricultural and industrial wastes to soil reduced sclerotial survival and sheath blight incidence (20, 32, 39, 45). Cotton, neem, groundnut, gingally and mahua cakes stimulated the soil microflora and brought about biocontrol.

Antagonistic fungi like *Gliocladium virens, Trichoderma viride, T. harzianum, T. pseudokoningii, Aspergillus niger, A. terreus, Neurospora crassa, Pencillium* spp. and *Laetiseria arvalis* inhibited mycelial growth, suppressed formation and germination of sclerotia, caused hyperparasitism and lysis and exhausted substrates by early and rapid competitive saprophytic colonization (38, 45). Fluorescent and non-fluorescent pseudomonads inhibited mycelial growth and sclerotial viability of the pathogen (41). *Trichoderma* population is widespread in tropical rice soils, higher in rainfed than in irrigated soils. *T. harzianum* releases extracellular chitinase, 1,3-ß-D-glucanase, etc. and degrade the *R. solani* cell wall. *T. harzianum* and *T. viride* coil tightly around *R. solani* hyphae penetrate and erode them in natural field soil. Foliar spray with spore suspension of *Trichoderma* spp. or application in soil (37) or standing water before planting gave effective biological control. In field trials, seed treatment with *Pseudomonas fluorescens* reduced sheath blight by 65-72% (5). Besides, presence of plasmids and dsRNA mycoviruses of *Rhizoctonia solani* (as of AG-2) make them slow-growing and weakly pathogenic (13,62).

Varietal resistance

Resistance might be governed either by oligogenic or polygenic character. Moderate resistance was reported in Katuktara DA-2 (IRRI) 12814/SI 26. In IRRI 7,614 lines were screened and 72 have been identified as resistant. Among them, 48 belonged to upland, 17 wild rice and only three among elite ones. In field, 37 genotypes showed resistance to *R. solani* (50). But only 11 remained resistant under high relative humidity. In Tamil Nadu, 20 genotypes were identified as resistant. So for no cultivated variety has been found to exhibit desired level of resistance to this disease.

Trichoderma spp. produce chitinases and glucanases and cause lysis of fungal pathogens and consequently biocontrol. If the genes encoding chitinases / glucanases from *Trichoderma* or similar organism could be cloned into rice, then the plant could defend by itself from the attack of this pathogen. This can be achieved by *Agrobacterium* mediated transformation systems and/or Biolistic systems. An alternative approach is that the genes encoding chitinases in *Trichoderma* could be cloned into fluorescent pseudomonads which are capable of surviving the rice rhizosphere. The transgenic pseudomonad offer great scope as a biocontrol agent.

3. FUTURE BIOTECHNOLOGICAL PERSPECTIVES

3.1 Tissue culture

Tissue culture techniques have a great potential in evolving disease resistant crop varieties, through their usefulness has not been tapped fully. The tissue culture offers great scope especially in situations where traditional plant breeding strategies are difficult to apply in evolving disease resistant varieties. Continued efforts in this directions would lead to better understanding of the host-pathogen interaction. Such efforts are indispensable in combating the devastating diseases like the sheath blight and blast.

3.2 Genetic engineering

(a) *Agrobacterium*-mediated gene transfer system

Agrobacterium is regarded as the Nature's Genetic Engineer. Several *Agrobacterium*-based vector system are being used extensively now for the transfer of various traits to crop plants as well as for the study of gene function in plants. Several laboratories are attempting to effect such gene transfer in rice. Embryo-derived cultures of two rice cultivars have been transformed using two *Agrobacterium* mediated gene transfer systems. Raineri *et al.*, (48) inoculated mature embryos in the variety Nipponbure with a wide host range (WHR) supervirulent strain of *Agrobacterium tumefaciens* and found tumorigenic callus tissue that grew on hormone-free medium. The transformation of this tissue was confirmed by DNA hybridization analysis that showed T-DNA present in the rice genome. Embryos of another variety Fujisaka 5, gave a hypersensitive response when inoculated with *Agrobacterium* strain A281. Extensive root proliferation was observed following inoculation with the limited host range (LHR) strain A856. These roots grew on hormone-free medium and produced octopine. Fujisaka 5 embryos, subsequently inoculated with a disarmed WHR strain conferring both kanamycin resistance and GUS activity, produced callus that grew on high levels of kanamycin and synthesized ß-glucuronidase. The results suggested in rice cells. Recently Li *et al*, (35) have reported the possibility of transforming an *indica* rice plants using *Agrobacterium* systems. However, till date normal rice plants have not been regenerated from the transgenic tissues. Upon further experimentation and verification, the *Agrobacterium* system may become a useful alternative to currently employed procedures for rice transformation.

(b) Biolistic method for introducing DNA into intact rice cells

Biolistic process (better know as the 'gene gun' or the particle gun bombardment method) is an effective method for the delivery of foreign DNA into intact plant cells (26). To date there are different types of biolistic devices, which differs mainly in their method of accelerating the microprojectile (tungsten or gold particles). Its potential in evolving newer transgenic rices is being investigated in many laboratories the world over. This technique will be of value in transforming material from rice cultivars that may not be readily amenable to plant regeneration from protoplasts.

(c) Chitinases and ß-1, 3-glucanases

Chitinases and ß-1, 3-glucanases are active *in vitro* against various fungal pathogens. Since these are individually encoded by single genes, these defence systems should be highly amenable for manipulation by gene transfer. Significant reduction in growth of *Rhizoctonia solani* and subsequent delay in disease development were observed using these systems (2). Host pathogen combinations involving rice and blast/sheath blight disease offer a good scope of exploiting this approach. Excellent promoter systems like 35S CaMV can be of value in such an attempt.

(d) Ribosome-inactivating proteins (RIP)

Ribosome-inactivating proteins such as those found in barley seed or its analogues are unique in their antimicrobial activity that they do not affect 'self' ribosome but show an immense activity against 'alien' ribosomes from unrelated species including fungi. Expression of a barley RIP cDNA under the control of a wound-inducible promoter in transgenic *Rhizoctonia solani* (36). Any attempt in getting such RIPs expressed in rice would be of great value in managing dreaded disease like blast and sheath blight.

(e) Manipulation of inhibitors of fungal pectic enzymes

Plant inhibitors of fungal α-1, 4-D polygalacturonases offer a novel venue for formulating strategies against fungal disease of rice. Several plant species contain specific inhibitors of fungal polygalacturonases which have no effect on plant or other microbial endopolygalacturonases. Cloning of a gene encoding such inhibitor from bean provides the basis for manipulations of these inhibitors for managing rice fungal diseases.

(f) Use of detoxifying genes in pathogens

Since pathogens inducing pathotoxins for invasion are not affected by the toxins they produce, it should be worthwhile studying the genes in *Pyricularia oryzae* and *Helminthosporium oryzae* encoding such detoxification and cloning them into rice. This approach has been demonstrated in the case of tabtoxin produced by *Pseudomonas syringae* var. *tabaci* causing tobacco wildfire disease. The transgenic plants effectively prevented the wildfire symptom production.

(g) Susceptibility of pathogens to phytoalexins

Pathogens are less tolerant to phytoalexins produced by non-hosts and virulence of fungal pathogens is conditioned by their ability to detoxify those phytoalexins. Therefore it may be worthwhile to examine the possibilities of using such pathogen non-host combinations and subsequently cloning those genes from non-hosts into rice with a view to imparting resistance to fungal diseases.

(h) Use of novel antimicrobial substances

A bovine lysozyme produced by *Pichea pastoralis* which can be used for topical applications, as it can withstand a rigorous environment such as of rice ecosystem was found to be useful in lying microbes. An important feature of this sort of approach is that it saves efforts in producing transgenic plants to express such lytic activity constitutively, as the compound is highly stable. This method is attractive for rice that is difficult to be transformed.

4. REFERENCES

1. Barksdale, T. K. 1969. Spread of rice blast in small fields. Plant Dis. Reptr. **51**: 243-247.

2. Broglie, K., Chet, I., Holliday, M., Cressman, R., Biddle, P., Knowlton, C., Mavais, C. J. and Broglie, R. 1991. Transgenic plants with enhanced resistance to fungal pathogen *Rhizoctonia solani*. Science, **254**: 1194-1197.

3. Chang, K. K., Dong, S. R. and Hong, S. M. 1985. Ecological studies of rice sheath blight caused by *Rhizoctonia solani*. III cultural methods and disease development. Korea J. Plant Prot. **24**: 7-12.

4. Das, S. R. and Mishra, B. 1990. Field evaluation of fungicides for control of sheath blight of rice. Indian Phytopath. **43**: 94-96.

5. Devi, T. V., Vizhi, R. M., Sakthivel, N. and Ganamanickam, S. S. 1989. Biological control of sheath blight of rice in India with antagonist bacteria. Plant & Soil, **119**: 325-330.

6. Eckhout, E. Van, Rush, M. C. and Blackwell, M. 1991. Effects of rate and timing of fungicide applications on incidence and severity of sheath blight and grain yield of rice. Plant Dis. **75**: 1254-1264.

7. Farman, M., Skinner, D., Budde, A., Smith, R., Leung, H., Cantone, F., and Leong, S. 1993. Towards map-based cloning of avirulence genes in the rice blast fungus, Manaporthe on rice biotechnology, Feb. 1-5, 1993, Chiaing Mai Thailand p. 105.

8. Giri, D. N. and Sinha, A. K. 1979. Effect of non-toxic chemicals on brown spot disease in rice seedlings, Int. Rice Res. Newsl. **4**: 10-11.

9. Gangopadhyay, S. 1978. Sunshine hours and incidence of brown spot diseases incidence in upland rice. Oryzae. **18**: 176-182.

10. Gangophdhyay, S., and Chakrabarti, N. K. 1987. Epidemiology of brown spot disease of rice. In: Advances in rice pathology, Ed. S. Kannaiyan, pp.138-148.

11. Gangophadhyay, S. and Chakarbarti, N. 1982. Sheath blights of rice. Rev. Plant Pathol. **61**: 451-453.

12. Gowda, S. S. and Gowda, K. T. F. 1985. *In vitro* studies on the efficacy of certain fungicides against *Drechslera oryzae*. Pesticides, **19**: 41-42.

13. Hashiba, T., Homma, Y., Hykumadu, M. and Mastuda, I. 1984. Insolation of a DNA plasmid in the fungus *Rhizoctonia solani*. J. gen. Microbiol. **130**: 2067-2072.

14. Hemmi, T., Abe, T., Ikeya, J., and Inoue, Y. 1936. Studies on the rice blast disease. IV. Relation of the environment to the development of blast disease and physiologic specialization in the rice blast fungus. Mater. Rural Improv. Dept. for Japan, **105**: 145-148.

15. Jaganathan, R. and Kanniyan, S. 1978. Studies in the chemical control of sheath blight disease of rice. Indian J. Pl. Prot. **6** :30-32.

16. Jayarajan, R. 1989. Importance and scope of aerobiology in plant disease forecasting. In: Recent researches in ecology, environment and pollution. **3**: 1-10, Ed. S. T. Tilak, Today and Tomorrow's Printers and Publishers, New Delhi.

17. Jeyarajan, R. and Palaniswami, A. 1993. Epidemiology and forecasting of rice blast disease. In: Advances in mycology and aerobiology, Today and Tomorrows Printers and Publishers, New Delhi (In press).

18. Kannaiyan, S. 1987. Sheath blight disease of rice. In: Advances in rice pathology. S. Kannaiyan (ed.) Coimbatore, Tamil Nadu Agric. Univ. pp. 210-233.

19. Kannaiyan, S. and Prasad, N. N. 1981. Effect of certain green manures on the saprophytic survival of *Rhizoctona solani* in soil. National Acad. Sci. Letter, **4**: 5-7.

20. Kannaiyan, S. and Prasad, N. N. 1980. Dicot weed hosts of *Rhizoctonia solani* causing sheath blight of rice. Indian J. Microbiol. **20**: 49-50.

21. Kapoor, A. S. and Singh, B.N. 1977. Influence of some environment factors on spore germination and infection of rice by *Pyricularia oryzae*. Indian Phytopath. **30**: 369-373.

22. Kato, H. 1974. Epidemiology of rice blast disease. Rev. Plant Prot. Res. **7**: 1-6.

23. Kaur, P., and Padmanabhan, S. Y. 1974. Control of *Helminthosporium* disease of rice with soil amendments. Curr. Sci.**43**: 78-81.

24. Kim, C. H., Mackenzie, D. R. and Rush, M. C. 1988. Field testing a computerized forecasting system for rice blast disease. Phytopathology, **78**: 931-934.

25. Kin, C. K., Min, H. S. and Yoshino, R. 1990. Epidemiological studies of rice blast disease caused by *Pyricularia oryzae* cavara III. Dinurnal pattern of conidial release and dispersal under the natural conditions. Ann. Phytopath. Soc. Japan, **56**: 315-321.

26. Klein, T. M., Wolf, E. D., Wu, R., Sanford, J. C. 1987. High-velocity microprojectiles for delivering nucleic acids into living cells. Nature, **327**: 70-71.

27. Kulkarni, S., Ramakrishnan, K. and Hegde, R. H. 1982. Epidemiology and control of brown leaf spot of rice in Karnataka IX Atmoshperic Variation in spore load of *Drechslera oryzae*. Indian Phytopath. **35**: 80-84.

28. Kulkarni, S., Ramakrishnan, K. and Hegde, R. H. 1981. Epidemiology and control of brown leaf spot of rice in Karnataka. II survival of *Drechslera oryzae* in nature. Curr. Res. **10**: 67-69.

29. Kuribayashi, K. 1928. Studies on overwintering, primary injection control of rice blast fungus. *Pyricularia oryzae*. Ann. Phytopath. Soc. Japan, **2**: 99-117.

30. Kuribayashi, K. and Ichikawa, H. 1952. Studies on the forecasting of rice blast. Specl. Rep. Nagano agric. Exp. Stn. **13**: pp. 299.

31. Lakshmanan, P. and Mohan, S. 1988. Effect of seed treatments on brown spot disease and their influence on rice seedlings. Madras agric. J. **75**: 57-58.

32. Lakshmanan, P. and Nair, M. C. 1984. Effects of soil amendments on the viability of sclerotia of *Rhizoctonia solani* in soil. Madras agric. J. **71**: 526-529.

33. Leung, H. and Shi, Z. 1993. Genetic dissection of sporulation in the rice blast fungus. Sixth Annual Meeting of the International Program on rice biotechnology, Feb. 1-5, 1993. Chiang Mai, Thailand. p.106.

34. Ling, D., Vidhyasekaran, P., Borromeo, E. S., Zapata, F. J. and Donghua, M. 1986. *In vitro* screening of rice germplasm for resistance of brown spot disease using phytotoxin. Acta Genet. Sin. **13**: 9-14.

35. Li, X-Q, Liu, C. N., Ritchile, S. W., Peng, J. M., Geluin, S. B. and Hodges, T. K. 1992. High efficiency transient expression of ß-glucuronidase in rice mediated by *Agrobacterium tumefaciens*. Plant Molec. Biol. **20**: 1037-1048.

36. Logemann, J., Jach, G., Tommerup, H., Mundy, J. and Schell, K. 1992. Expression of a barely ribosome inactivating protein leads to increased fungal protection in transgenic tobacco plants. Biotechnology, **10**: 305-308.

37. Manian, S. and Paulsamy, S. 1980. Biological control of sheath blight disease of rice. J. Biol. control **1**: 57-61.

38. Manibhusanrao, K., Baby, U. I. and Joe, Y. 1990. Basic research on rice sheath blight disease. In: Basic Research for Crop Disease Management. Daya Publishing House, Ed. P. Vidhyasekaran, Delhi. pp. 376-381.

39. Manibushanrao, K., Baby, U. I. and Joe, Y. 1988. Influence of various amendments of soil microflora in relation to sheath blights of rice (Abstr.) in 5th International Congress of Plant Pathology, 1988 Kyoto, Japan.

40. Manibushanrao, K. and Krishnan, P. 1991. Epidemiology of blast (EPIBLA): a simulation model and forecasting system for tropical rice in India. Rice blast modelling and forecasting, IRRI, Manila, pp.31-38.

41. Mew, T. W. and Rosales, A. N. 1984. Relationship of soil microorganisms to rice sheath blight development in irrigated and dryland rice cultures. Tech. Bulletin Asia.

42. Muthusamy, S., Mariappan, V., Narasimhan, M., Muthusamy, M. and Eswaramoorthy, S. 1988. Effect of micronutrients on rice brown spot incidence. Int. Rice Res. Newsl. **13**: 32-33.

43. Nair, B. K., Balakrishnan, B. and Nair, M. C. 1988. Managing rice sheath rot disease in Kerala, India. Int. Rice. Res. Newsl. **13**: 20-21.

44. Ono, K. 1965. Principles methods and organization of blast forecasting. The rice blast disease. IRRI, John Hopkins Press, Maryland, pp. 173-194.

45. Ou, S. H. (ed.) 1985. Rice Disease. Commonwealth Agricultural Bureaux, 380 pp.

46. Padhi, B. and Chakrabarti, N. K. 1982. Variation in the occurrence of blast disease incidence at monthly intervals in relation to meterological factors. Z. Pl Krankh. **89**: 144-150.

47. Padmanabhan, S. Y., Chakrabarti, N. K. and Row, K. V. S. R. K. 1971. Forecasting and control of rice disease. Indian Acid. Sci. **37B**: 423-429.

48. Raineri, D. M., Bottino, P., Gordon, M. P. and Nester, E. W. 1990. *Agrobacterium*-mediated transformation of rice (*Oryza sativa* L.), Biotechnology, **8**: 33-38.

49. Ramalingam, A. 1969. Further studies on the effective dispersal of *Pyricularia oryzae* Cav. J. Mysore Univ. **B22**: 39-46.

50. Rani, N. S. and Satyanarayana, K. 1992. Testing for sheath blight resistant in rice. Int. Rice Res. Newsl. **7(5)**: 7.

51. Raskin, I. 1992. Salicylate, a new phytohormone, Plant Physiol. **99**: 799-803.

52. Sakthivel, N. and Gnanamanickam, S. S. 1986. Bacterization of rice with Psueudomonas fluorescens reduces sheath rot infection. Int. Rice. Res. Newsl. **11**: 17-18.

53. Sanker, M. L. and Sinha, A. K. 1989. Use of phytoalexin inducing chemicals to control rice sheath blight. Int. Rice. Res. Newsl. **14**: 23.

54. Sekhar, M. J. and Prasad, N. N. 1989. Effect of different levels of potassium at each of the two nitrogen level on the introduction of sheath rot growth and yield of rice. Madras agric. J. **76**: 1-4.

55. Sekizama, Y., Ilaga, M., Iwata, M., Hamamto, A., Chihara, C. and Takino, Y. 1985. Probenazole and bust of respiration in rice leaf tissue infected with blast fungus. J. Pesticide sci. **10**: 225-231.

56. Sharma, S. K., Maheswari, S. K. and Deo, P. P. 1979. Effectiveness of fungicides in the control of blast (neck rot). Int. Rice Res Newsl. **6(2)**: 11.

57. Singh, R. A. and Raju, C. A. 1981. Studies on sheath rot of rice. Int. Rice. Res. Newsl. **4**: 10-11.

58. Sinha, A. K. and Hait, G. N. 1982. Host sensitization as a factor in induction of resistance in rice against *Drechslera* by seed treatment with phytoalexin inducers. Trans. Brist. Mycol. Soc. **70**: 213.

59. Sooriachandraselvan, N. and Ranganathan, T. B. 1989. Effect of planting season on the incidence of rice diseases. Madras agric. J. **76**: 297-300.

60. Suzuki, H. 1974. Behaviour of rice blast fungus spores and application to outbreak forecast of rice blast disease. Japan Agric. Res. Qtly. **8**: 78-83.

61. Suzuki, H. 1969. Diurnal periodicity in spore discharge of rice blast fungus. Rev. Plant. Prot. Res. **2**: 64-66.

62. Tavantzis, S. M. and Bandy, B. P. 1988. Properties of mycovirus from *Rhizoctonia solani* and its virion-associated RNA polymerase. J. gen. Virol. **69**: 1465-1469.

63. Venkata Rao, A. and Thangamani, N. 1988. Role of *Nigrospora oryzae* (Berks & BR Petch) in blast lesions of rice caused by *Pyricularia oryzae* Cav. Curr. Sci. **57**: 272-273.

64. Vidhyasekaran, P., Borromeo, E. S., Ling, D. H. and Mew, T. W. 1991. Relationship between growth rate of *Helminthosporium oryzae* isolates on calluses of rice cultivars and their disease reaction on rice plants. Plant Cell, Tissue and Organ Culture, **24**: 237-241.

65. Vidhyasekaran, P., Ling, D. H., Borromeo, E. S., Zapata, F. J. and Mew, T. W. 1990. Selection of brown spot resistant plants from the *Helminthosporium oryzae* toxin-testing calluses. Ann. Appl. Biol. **117**: 515-523.

66. Viswanathan, R. and Narayanasamy, P. 1991. Occurrence of resistance in rice pathogen to fungicides. Indian J. Mycol. Pl. Pathol. **21(1)**: 63-65.

67. Yoshida, H., Konishi, K. Nakagawa, I., Sekido, S., Yamaguchi, I. 1990. Characteristics of N-phenysulfonyl-2-chlorosoni continamide as an anti-rice blast agent. J. Pesticide Sci. **15**: 199-203.

68. Yashino, R. and Yamaguchi, T. 1974. Influence of sunshine and shade conditions on the occurrence of rice blast caused by *Pyricularia oryzae* Cav. Bull. Hokurku Nat. agric. Sta. **16**: 61-119.

4 Biological Control of Sheath Blight and Blast of Rice

A. Premlatha Dath, K.G. Mukerji, Rajeev K. Upadhyay, B.P. Chamola and Rajee N. Kumar

CONTENTS

Integrated Pest and Disease Management
Rajeev K. Upadhyay, K.G. Mukerji, B.P. Chamola and O.P. Dubey (eds.)
APH Publishing Corporation, New Delhi-110002, India

ABSTRACT

Sheath blight *(Rhizoctonia solani* Kuhn.) and blast disease of rice caused by *Pyricularia oryzae* Cav. has emerged as important diseases in modern rice production in both the temperate and tropical production rice growing countries causing serious crop losses. In efforts to develop alternate strategies for rice disease management, the idea of biological control of rice diseases has became a particularly attractive alternative for plant pathologists. Many microorganisms both fungi and bacteria have been attempted as potential biocontrol agents of sheath blight and blast disease of rice. Several fungi especially *Trichoderma* spp. were found to be antagonistic to *R. solani in vitro. T. harzianum* when introduced under rainfed condition, decomposed rice straw and reduced the survival of *R. solani.* More than 400 strains of bacteria isolated from IRRI rice fields were screened in the laboratory for antagonism towards *Pyricularia oryzae* and *Rhizoctonia solani.* 3 strains of *Pseudomonas fluorescens,* 5 strains of *Bacillus* spp. and a strain of *Enterobacter* were identified which suppressed sheath blight (Sh-B) *in vitro.* In field studies, bacterial treatments-afforded significant Sh-B control and performed better than fungicides. Integrated control using *Trichoderma longibrachiatum* and *Gliocladium virens* and two organic amendments (neem cake and *Gliricidia* leaves) significantly controlled sheath blight. Although biocontrol of sheath blight and blast diseases of rice are still in its infancy but it is gaining momentum and it can replace fungicides to some extent.

1. INTRODUCTION

Rice is the staple food for around 60% of the world's population, with a world production of 500 million tonnes from 150 million hectares of land. Since in India, our stocks amount to only two month's supply, many people are at risk from famine. People have not forgotten the devastating Benal famine in 1942 due to brownsport disease in rice which has gone into history. Moreover, it has been estimated that the world requirement in 2020 will be about 760 million tonnes, an increase of 50%. This pressure of increase in population on food makes efficient disease and pest management vital and is the reason for awareness among plant pathologists to develop alternate strategies for rice disease management. But the intensive cultivation practices unfortunately favour the build-up of pests and diseases. Continuous cropping also leads to a high population of soilborne diseases.

Rice sheath blight caused by *Rhizoctonia solani* (Kuhn), *Thanetophorus cucumeris* (Frank Donk), once a minor disease, of late has emerged as one of the most important diseases in modern rice production in both the temperate and tropical rice growing countries causing serious crop losses (36).

Likewise, blast disease caused by *Pyricularia oryzae* Cav. is the second most important disease of rice in the rice growing areas in our country viz. W. Bengal, Assam and other North Eastern states.

This disease causes severe production constraint in dryland and unbended upland rice cultures where rice farmers have no access to nor can they afford costly fungicides (56). Because of the instability of the rice blast fungus and the marked variability in pathogenicity found in different strains of this organism, control through breeding for resistance has met with only partial success. In recent years total crop failures have occurred in some parts of our country where farmers use very high seeding rates, use of resistant cultivars is not successful because adequate level of host resistance has not been found. Therefore, alternate disease management strategies are sought that are ecologically sound and cost effective.

The causal fungus is a soilborne organism which survives in soil mainly as sclerotia and also as mycelia. The saprophytic ability of the pathogen and its wide host range make the disease control very difficult. Thus, control of this pathogen through biological means by employing antagonistic microoganisms seems to be a viable alternative.

The idea of biocontrol of plant pathogens has become a particularly attractive alternative as it is economical, self perpetuating and usually free from residual side effects. Successful strategies for biocontrol of certain insect and weed pests and economic considerations of disease control costs also encourage the use of microoganisms to control diseases. Recent studies conducted in Australia, Canada, China, India, Netherlands, Phillippines and the United States reveal the potential of root colonizing microorganisms to inhibit or displace soilborne pathogens at the root soil interface and thereby protect the root sheath of perennial and annual plants (39) such as cotton, potato, tobacco, flax, radish, cucumber, wheat and rice (11, 19, 24, 25, 29, 51, 53, 55). These microorganisms inhibit pathogens by producing antibiotics, siderophores (compounds that chelate biologically available iron) and possibly substances that stimulate plant growth (54, 58, 60, 61). This chapter will focus more on the biological control of rice diseases mainly sheath blight and blast using microorganism viz., fungi and bacteria.

2. BIOCONTROL AGENTS

2.1 Role of saprophytic microorganisms in the survival of the pathogen

Sclerotia of *R. solani* has poor competitive saprophytic ability in the presence of other root-inhabiting microorganisms. However, if the sclerotia were contaminated with either arable or virgin (newly reclaimed) soil after the sclerotia had initiated germination, the antagonistic effect of the soilborne microbes was reduced. Sclerotial activity was also influenced by the number of microorganisms present in their surrounding or adhering on their surfaces-lesser number of microorganisms was ineffective to arrest germination (45). *Aspergillus terreus* parasitized sclerotia of *R. solani* causing loss of their viability when incubated in dual culture. However, treating the sclerotia with spore suspension of *A. terreus* was ineffective to reduce their viability (47).

Soil amendment with green manures such as Dhaincha *(Sesbania aculeata)* increased the population of microorganisms of which many were found to be antagonistic to *R. solani* (Table 1, 2). This amendment greatly reduced the survival period of sclerotia in soil (37).

Table 1 : Fungal and bacterial population in different green manure amended soils and their antagonistic effect against sheath blight pathogen of rice (37)

Green manure	Total population of microorganisms/g of soil		No. of micro-organisms isolated	No. of anta-gonists	Average inhibition zone (mm) of each antagonist
Azolla	Fungi	5.0 x 10^7	4	2	4.0, 6.0
	Bacteria	1.9 x 10^7	5	1	5.0
Dhaincha	Fungi	4.4 x 10^7	7	3	10.2, 4.0, 5.4
	Bacterial	3.6 x 10^7	8	4	3.4, 2, 4, 5.3, 8.2
Ipomoea corrnea	Fungi	5.4 x 10^7	8	-	
	Bacteria	1.0 x 10^8	3	-	
Mung	Fungi	1.1 x 10^8	7	3	3.4, 5.0, 6.2
	Bacteria	1.0 x 10^8	8	1	6.0

Neem	Fungi	4.4×10^7	8	2	2.2, 3.4
	Bacteria	3.3×10^7	4	-	
Sunhemp	Fungi	1.3×10^8	5	1	4.2
	Bacteria	5.4×10^8	9	-	
Control (No	Fungi	1.8×10^7	4	-	
green manure)	Bacteria	1.1×10^7	8	2	6.2, 4.0

— = Not antagonistic

Table 2 : Incidence of sheath blight and growth characters of rice plants in pots with different inoculations of Rhizoctonia solani and Aspergillus terreus (47)

Treatment (inoculation per pot)		Plant height (cm)	Dry weight of shoot (g)	Dry weight of root (g)	% sheath infected (in angular values)
R. solani	30g	92.0	6.9	1.57	68.02a
R. solani	15g+	90.0	6.3	1.19	45.43b
A. terreus	15g				
R. solani	30g+	94.2	7.3	1.19	58.23a
A. terreus	15g				
A. terreus	15g	95.0	6.6	1.16	
		NS	NS	NS	SED = 4.9
					CD (5%) = 11.1

2.2 Fungal antagonists

Several fungi have been isolated from rhizosphere, rhizoplane, phylloplane and soil of paddy fields (45). Some of the fungi viz., *Aspergillus niger* (16); *A. terreus* (33, 44, 47), *Fusarium solani, Penicillium ehrlichi, Pseudodeurotium multisporum* (45); *Trichoderma aureoviride, T. glaucum, T. hamatum, T. harzianum, T. koningii* (34); *T. lignorum* (26); *T. longibrachiatum* (27, 34); *T. piluliferum, T. polysporum, T. pseudokoningii, T. reesii* (37) and *T. viride* (16, 34, 42, 43) are routinely isolated from the rhizosphere of various paddy fields which have been implicated in biocontrol. At IRRI (International Rice Research Institute, Phillippines) a lot of work has been done on biocontrol of rice diseases (34). *Trichoderma harzianum* and *T. pseudokoningii* were found to be most suitable potential biocontrol fungi in upland and lowland rice fields respectively, while *T. aureoviride* was more abundant in both upland and lowland fields in Phillippines.

Trichoderma has been an exceptionally good model with which to study biocontrol because it is ubiquitous, easy to isolate and culture, grows rapidly on many substrates, affects a wide range of plant pathogens, is rarely pathogenic on the host crop, acts as a mycoparasite, competes well for food and site, produces antibiotics and has an enzyme system capable of attaching a wide range of plant blight pathogents (62).

Table 3 : Effect of fungal antagonists and organic amendments on sheath blight severity and yield in rice (28)

Treatment	Disease incidence (%)	Protection (% over contol)	Straw yield (t/ha)	Grain yield (t/ha)
Control	15.11 a	-	5.3bc	3.0 f
Gliricidia	8.71 b	42.4	5.5abc	3.6 ef
Neem	8.99 b	40.5	5.4bc	4.6 bcd
Gliocladium virens (Gv)	9.27 b	38.6	5.2c	3.9 de
Trichoderma longibrachiatum (Tl)	9.47 b	37.3	5.0c	4.4 cde
Gliricidia + Gv	7.40 b	49.4	6.2a	5.9 a
Gliricidia + Tl	6.66 b	55.9	5.8ab	5.7 a
Neem + Gv	8.15 b	46.1	5.6abc	5.4 ab
Neem + Tl	7.29 b	51.8	5.5abc	5.3 abc

Values followed by the same letters are not significantly different (P=0.01) by Duncan's multiple range test

2.2.1 Sheath blight disease

Ogura and Akai (35) and Hashioka and Fukita (17) were the first to study the antagonistic action of *Trichoderma, Gliocladium* and *Acremonium* in attempts to combat the disease. The occurrence of *R. solani* was remarkably reduced in the presence of *Neurospora crassa* in the soil and *N. crassa* did not have any effect on the seedling growth and this may be used as a biocontrol agent of sheath blight. Similarly, presence of *Aspergillus terreus* in soil reduced the infection of *R. solani* in rice seedlings (44). Later, Roy (47) found that the growth of sclerotia inoculated on potato dextrose agar 24 hour after seeding with spores of *A. terreus* was suppressed but not when it was inoculated simultaneously. However, infection of sheath blight on rice plants grown in pots containing sterilized soil with inocula of *R. solani* and *A. terreus* was reduced. *Trichoderma harzianum* was readily isolated from pieces of rice straw buried in the soil of dryland and rainfed wet land fields but not from irrigated fields with standing water (21, 29). In India, sclerotia from 5-15 cm soil depths were also found to be colonised by different organisms including *T. viride,* but those retrieved from below 5 cm were colonized most and those from below 15 cm least (43). When the culture of *T. viride* was incorporated into sterilized soil along with *R. solani* (both cultures grown in 4% maize meal sand medium) sheath blight infection on rice was slightly reduced, but when *T. viride* spores were sprayed on aerial parts of the plants before inoculation with *R. solani,* the disease could not be checked (42). The pathogen mycelium stopped growing when it comes in contact with *Trichoderma.* Microscopic examination of the hyphae showed that *Trichoderma* frequently coiled around the aerial hyphae of *Rhizoctonia solani.* The cytoplasm vacuolated and coagulated followed by hyphal burst of the pathogen (40, 48). Earlier, Hashioka and Fukita (17) also recorded mycoparasitic activity of *T. brachiatum* and *T. polysporum* on *R. solani.*

The microbial decomposition of rice straw greatly affected the survival of the rice sheath blight pathogen. *T. harzianum* isolated from a dryland rice field was found to decompose rice straw. When *Trichoderma* and *R. solani* were mixed together with the rice straw, recovery of rice sheath blight pathogen was reduced to 20% at 2 weeks after incubation (20). When *T. harzianum* was introduced under rainfed condition it decomposed rice straw and by deplenishing the substrates reduced the survival of *R. solani* (29).

In addition to *Trichoderma* spp., *Gliocladium* spp. is another potential biocontrol agent of sheath blight pathogen. Mycoparasitic and antibiotic potential of *Gliocladium virens* and *T.*

longibrachiatum to *R. solani* has been described (27). No specific correlation was observed between antibiotic production and mycoparasitic potential of either of these antagonists. Both were able to utilize various components of fungal cell walls as carbon sources.

The microbial decomposition of rice straw greatly affected the survival of the rice sheath blight pathogen. *T. harzianum* isolated from a dryland rice field was found to decompose rice straw. When *Trichoderma* and *R. solani* were mixed together with the rice straw, recovery of rice sheath blight pathogen was reduced to 20% at 2 weeks after incubating (20). When *T. harzianum* was introduced under rainfed condition, it decomposed rice straw and by deplenishing the substrates reduced the survival of *R. solani* (29). Thus the ability of *T. harzianum* to decompose rice straw, thereby affecting the survival of *R. solani* has a potential use in disease management for dryland crops grown after rice. Recently, biocontrol of sheath blight (Sh.B) of rice using *Aspergillus terreus* has been reported by Gogoi and Roy (15). By foliar spraying of the antagonist the incidence of the disease was reduced and under controlled conditions. Cartwright *et al.* (4) also reported rice sheath blight using indigenous rice field microorganisms which includes bacteria, *Sclerotium* sp., *Trichoderma* spp.

A paddy field whether it is rainfed or irrigated is subject to flooding for several months during a crop season, the fungal saprophytes potentially antagonistic to *R. solani* may be reduced to a minimum (29). *T. harzianum* was more effective in inhibiting *R. solani* in dry season which limits the wide adaptation of such potential fungal saprophytes in the biocontrol of sheath blight under diverse ecosystems.

2.2.2 Rice blast disease

Biocontrol of rice blast disease using fungal antagonists has not been reported by many workers. The mycoparasite *Verticillium biguttatum* causes some reduction in the pathogen population (49).

2.3 Bacterial antagonists

2.3.1 Sheath blight disease

Two groups of bacteria the fluorescent *Pseudomonas* and the non-fluorescent strains were isolated from rice field soil, water, rice tissues and rhizosphere of healthy and infected plants (29, 40): these strains which were available in IRRI's culture collection showed strong antagonistic activity against the pathogen. These bacterial strains inhibited the mycelial growth of the pathogen and they were mostly isolated from sclerotia of the pathogen (29). Rosales *et al.* (41) identified 23 bacteria from rice fields in the tropics antagonistic to several rice fungal pathogens and all the strains identified inhibited the mycelial growth of *R. solani in vitro.* Seed bacterization with these bacterial strains provided a sheath blight protection of 4.5 to 75% in the glasshouse trials. Sarker *et al.* (52) and Cartwright et al. (4) also isolated several bacterial strains for biocontrol of rice sheath blight (Sh-B) disease. Results obtained in glasshouse experiments suggest that soil pH that is optimal for lowland rice production (pH 5.5 to 6.5) and acid pH (5.0) are suitable for obtaining significant results on sheath blight reduction by bacterial treatments (14). The soils used in these experiments had active Fe-contents of 1.3 to 2.8%. It is known that fluorescent siderophores are not produced in the presence of moderate amounts of Fe (25, 31). Most bacterial treatments performed better in a direct-seeded crop than in a transplanted crop. In *Pseudomonas fluorescens* strain 33, 1-14-1 and 7-14 treatments had lower Sh-B severities of 6.4, 0.2 and 2.1% respectively in direct-seeded and had higher sheath blight severities of 9.3, 6.0 and 5.0% respectively in transplanted IR58 crops. This may be due to the greater crop canopy in a direct seeded crop and the ideal ecological niche it provides for the epiphytic survival and colonization by bacteria. Bacteria introduced into the rice seed migrate to aerial parts of the rice plant (29). This has proved to be useful to farmers as they have resorted to direct-seeding as the culture practice in intensive rice production (Tables 7 to 9).

Sclerotia in irrigated fields are more vulnerable to colonization by bacteria (Table 4). Most of these bacteria inhibited the mycelial growth of *R. solani in vitro.* But the ability of different bacterial isolates to inhibit the fungal growth varied (Table 5, 6).

Use of such potential antagonistic microorganisms as biological control agents in reducing sclerotial survival and infection in the field seems to offer good scope.

Table 4 : Viability of Rhizoctonia solani sclerotia collected after the rice harvest from the soil of dryland and irrigated rice fields with a moderate level of sheath blight infection (30)

Field Type	No. sclerotia[1]	No. viable sclerotia[2]	No. non-viable sclerotia	
			With bacteria	Without bacteria
Irrigated	16.4	5.5 (33.5)	9.7 (59.1)	1.2 (7.3)
Dryland	36.0	12.2 (33.9)	12.6 (35.0)	11.2 (31.1)

[1] Per 100 g soil (oven dry wt.)

[2] Figure in parenthesis denotes percentage

Table 5 : Inhibition of mycelial growth of Rhizoctonia solani by saprophytic bacteria isolated from the sclerotia (30)

Fluorescent type of Bacteria	**Inhibition Zone (mm)***
In-b-4	12.5
In-b-24	12.0
In-b-20	10.8
In-b-23	6.5
In-b-22	4.8
In-b-27	4.0
Nonfluorescent type of Bacteria	**Inhibition Zone (mm)***
In-b-3	21.0
In-b-17	12.5
In-b-19	10.8
In-b-16	8.8
In-b-31	8.5
In-b-26	6.8
In-b-29	6.5
In-b-30	6.0
In-b-28	5.5
In-b-5	5.5
In-b-25	5.0
In-b-21	4.2
In-b-9	4.2

* Determined by measuring the clearance between the tip of mycelial growth and the margin of the bacterial colony (mean of 3 replications)

Table 6 : Effect of antagonistic bacteria on the incidence and severity of sheath blight (30, 40)

Treatment[a]	Sheath blight[b]	
	Incidence (%)	Severity
Pathogen + In-b-17	30	0.6
Pathogen + In-b-24	75	1.3
Pathogen alone	100	2.0

[a] Bacterial suspension (1×10^6 cells/ml) was sprayed at the basal portion of rice plants 1 day before inoculation of the pathogen (sclerotia)

[b]
$$\text{Incidence} = \frac{\text{No. of tillers infected}}{\text{Total No. of tillers (3 hills)}}$$

Severity = Sheath blight lesion (cm)

Table 7 : Effect of bacterization and two sprays of stems and leaves of rice plants on sheath blight severity in direct seeded and transplanted IR 58 rice (WS 1988) (14)

Treatment	Sheath blight severity	
	Direct seeded	Transplanted
Validamycin	5.0 bc[a]	2.8 c
In-b-33	6.4 bc	9.3 ab
In-b-1-14-1	0.2 c	6.0 bc
In-b-784	5.1 bc	4.4 bc
In-b-33 + In-b-1-14-1	7.7 b	5.0 bc
Check	13.4 a	12.8 a

[a] Each value is an average of 4 replications. Means followed by a common letter are not significantly different from each other at the 5% level by Duncan's multiple range test

Table 8 : Effect of Pseudomonas fluorescens (rice strain Pfr 11 and citrus strain Pfcp) on sheath blight development in rice seedlings (59)

Rice cultivar and treatment	Sheath blight severity	Reduction in sheath blight severity (%)
IR 20		
Pfr 11 - treated	0.92	31.34**
Pfcp - treated	0.78	41.79**
Not treated (check)	1.34	-
TKM 9		
Pfr 11 - treated	0.40	74.36**
Pfcp - treated	0.65	58.33**
Not treated (check)	1.56	-

Each value is based on observation from 100 seedlings (2 weeks old) and is a mean of 3 replications

Calculated 't' value significant at P = 0.01 (**)

Table 9 : Effect of seed bacterization on sheath blight lesion size in field experiment (59)

Cultivar	Seed treatment	Sheath blight lesions length (mm)	No. of tillers per hill	Grain yield per plot (kg)
IR 50	Pfcp - treated	30.0**	17	2.5*
	NF 2 - treated	31.1**	17	2.5*
	Not treated (check)	109.0	11	2.2
TKM 9	Pfcp - treated	32.4**	20	2.7*
	NF 2 - treated	31.0**	15	2.4
	Not treated (check)	94.0	10	2.2

Each value is a mean of 100 measurements
Calculated 't' value significant at P = 0.01 (**)
P = 0.05 (*) and not significant (ns)

2.3.2 Rice blast disease

Two bacterial strains *Pseudomonas* spp. and *Bacillus* spp. has been tested in Phillippines to cause significant leaf blast reduction with rice variety UPL Ri-5. The crude extract prepared from the strain 7-14 of *Pseudomonas fluorescens* provided 70-100% inhibition of conidial germination at 1.0 ppm. The antiblast extract also protected IR50 rice seedlings from infections of *P. oryzae*. Evidences suggested that siderophore was unlikely to be involved in the mechanism of strain 7-14 in its antagonism against the rice blast fungus. *In vitro* test showed that inhibition of *P. oryzae* by strain 7-14 was not reversed by Fe amendments (13). Seed treatment and spray application with bacteria led to reductions in leaf blast severity in both IR50 and C22 rice crops. These strains have been selected on the basis of *in vitro* tests for antibiosis for reduction of leaf and neck blast severities. In *Bacillus* strains 33 and 4-03, and in a particular strain 33, caused maximum leaf blast reductions (80%) in variety C22 and IR50 rice crops (12). Another bacteria *Pseudomonas fluorescens* strains (7-14 and 4-15) in spite of their lower population were more effective in reducing leaf blast and neck blast then the *Bacillus* strains in upland sites. Evidences suggested that siderophore production, one of the important mechanisms known to mediate bacterial antagonism in fungal inhibition is unlikely to be involved in highly acidic soils (31) such as soil pH 4.0 and with an Fe-content that ranges from 3.7 to 4.6% (10).

In India, biological control of rice diseases especially sheath blight has been attempted in Central Rice Research Institute. Strains of fluorescent and nonflourescent strains of bacteria isolated from rice rhizosphere showed antagonism towards *R. solani*. In the field, plants raised from bacterized seeds of two varieties of paddy (IR20 and TKM9) showed 65-72% control of sheath blight disease (27). The possibility of utilizing chitinase production as a criterion for selecting bacterial antagonists for biological control of sheath blight was studied by Thara and Gnanamanickam (57). Of the total 1000 bacterial strains tested, 43% were negative for chitinase production 33.6% were weak producers and 23.4% were strong producers. But chitinase production and sheath blight suppression did not show a positive correlation in the greenhouse and seedbed evaluations and thus they were of the opinion that chitinase production test cannot be considered the sole indicator in selecting promising bacterial strains for biological control of sheath blight.

3. SEED TREATMENT

When used in seed bacterization (seeds were soaked in bacterial suspension of about 10^9 cells/ml for 24 hr before sowing), the fluorescent and nonfluorescent antagonists suppressed the disease and promoted the plant growth. Subsequent planting (after the first crop in which the seeds were treated with bacteria) on the same soil also showed reduced disease severity. The bacteria appeared to establish in the soil after sowing of seeds treated with antagonistic bacteria. Though both nonfluorescent and fluorescent bacteria reduced disease incidence and severity, compared to plants without bacteria (Table 10), nonfluorescent bacteria appeared to be more effective than fluorescent bacteria (29, 30). The bacteria could also establish on rice plants after seed treatment and they were capable of moving from seeds to upper parts of plants enhancing their efficacy as biocontrol agent (22). Seed bacterization not only suppressed sheath blight but also increased grain yield (Table 11). Though incidence of sheath blight was lower when plants were grown either from seeds treated with or sprayed with bacteria, grain yield was higher in seed treatment than in spray application of bacteria (29).

Table 10 : Effect of seed treatment with fluorescent and nonfluorescent types of bacteria on the incidence and severity of sheath blight, and on rice plant growth (30)

Treatment[1]	Incidence (%)[2]	Severity (lesion length, cm)	Plant height (cm)[3]
Pathogen alone	93.7	9.7	28.1
Pathogen + Fluorescent type of bacteria			
Pathogen + In-b-4	81.4	5.3	29.9
Pathogen + In-b-20	84.7	4.1	48.4
Pathogen + In-b-22	86.5	2.6	44.7
Pathogen + In-b-27	87.0	4.8	41.3
Pathogen + In-b-23	89.2	3.6	38.9
Pathogen + In-b-24	91.0	2.8	43.1
Pathogen + Nonfluorescent type of bacteria			
Pathogen + In-b-17	60.2	1.7	40.0
Pathogen + In-b-9	80.9	3.7	33.0
Pathogen + In-b-3	86.3	4.0	48.2

1 Pathogen and autoclaved soil were mixed together, then sown with IR 36 seeds treated with antagonistic bacteria for 10 minutes (1 x 10^9 cells/ml)

2 Data taken 26 days after sowing

3 Data taken 60 days after sowing in seed-box planting

Table 11 : Effect of fluorescent (F) and nonfluorescent (NF) bacteria and spray application 45 days after transplanting on plant height, tiller number and grain yield of rice culture IR 36, and on sheath blight incidence in soil in a screen house (29)

Bacterial isolate		Plant height* (cm)		Tiller[x] (No.)		Grain yield[y] (g)		Incidence (%)	
		ST[z]	S	ST	S	ST	S	ST	S
NF	In-b-17	52.9ab	49.9b	10.0a	7.6abcd	107.5a	87.0bc	14.3bcd	9.3d
F	In-b-24	55.7ab	51.0b	9.0abc	6.7cd	105.7ab	84.5c	13.0cd	13.3cd
F	In-b-150	52.0b	54.0ab	8.0abcd	9.3ab	108.7a	86.5bc	12.3d	23.3abcd
NF	In-b-590	59.5a	53.5ab	9.3ab	9.0abc	97.0abc	93.0ab	19.3abcd	18.0abcd
NF	In-b-591	59.8a	54.5ab	9.0abc	8.7abcd	93.0cd	95.0a	32.3a	28.7ab
Check		51.0b	49.4b	6.3d	7.3bcd	88.0cd	87.0bc	27.6abc	22.3abcd

* Mean of 6 tillers measured 1 month after transplanting

x Mean of 3 replications, 6 hills per replicate measured at maximum tillering

y Grain weight taken from 6 hills. Means followed by a common letter are not significantly different (P = 0.05) by Duncan's multiple range test.

z ST = Soil treatment; S = Spray application to plants at about 10^9 cfu/ml.

4. SOIL AMENDMENT

Soil amendment with green manures such as Dhaincha (*Sesbania aculeata*) increased the population of microorganisms of which many were found to be antagonistic to *R. solani.* This amendment greatly reduced the survival period of sclerotia in soil (Premlatha Dath, 1982).

5. INTEGRATED CONTROL

Manibhushan Rao and Baby (28) and Baby and Manibhushan Rao (1) have evaluated the efficacy of two antagonistic fungi viz., *T. longibrachiatum* and *G. virens* and two organic manures (*Gliricidia maculata* leaves) and *Azadirachta indica* (neem) cake, both individually and combined to control sheath blight. They found that both the antagonists and amendments alone and in combination significantly decreased sheath blight intensity and increased grain yield. Integration of these two systems significantly enhanced the efficacy over their individual effects. In addition to disease suppression, the growth of rice plants as well as grain yield were increased to varying degrees in different treatments. The antagonists applied to the soil as wheat bran dust preparation survived well in the soil and reduced the pathogen population. When antagonists were supplemented with organic substrates an increase in the colony forming units of the antagonists and a marked reduction in the pathogen survival were noticed. In most of the integrated studies with amendments and antagonists, it showed better control of the pathogen than with antagonists alone. A combination of *Gliricidia* and *Gliocladium virens* systems showed the best results with the maximum yield of the paddy crop.

Integrated management of the rice blast disease was tried by Correa and Zeigler (9) using several disease management practices. Both water management and chemical control were combined to effectively control the disease. There is a lot of environmental influence on rice blast disease (3). A self defensive substance was synthesized from D-mannose for protection against the rice blast disease (63). Chin and Bhandhufalck (8) have discussed the importance of crop growth stages for determining the application timing of disease control agents on rice.

5.1 Effect of plant extracts

The rice blast pathogen *P. oryzae* was effectively controlled by using plant extracts of *Adhatoda vasica, Azadirachta indica, Polyalthia longifolia, Prosopis juliflora, Vitex negundo in vitro* in India (38). Soybean lecithin inhibited 96% appresorial penetration of *P. oryzae* at spores 5 ppm and 500 ppm concentration (18).

6. BIOTECHNOLOGICAL ADVANCES IN DISEASES CONTROL

Antifungus compounds were isolated in the resistant cultivars of rice. Five epoxides and five allyl alcohols which are oxygenated fatty acids which inhibited the rice blast fungus activity in rice plants (23).

Recently, Chen and Chen (6) and Chen *et al.* (5) have screened many rice blast resistant mutants using phytotoxins. The application of α-picolinic acid for the screening of rice blast resistant varients has been studied (6). Saifulla (50) also screened various rice genotypes for blast disease.

Genetic diversity of rice blast fungus has been studied in detail using DNA finger printing (32). By DNA finger printing using *Pyricularia grisea,* 151 isolates of the rice blast fungus were collected from 15 rice cultivars in a blast disease breeding nursery in Colombia. This study is useful in selecting germ plasm sources for more durable blast resistant breeding. Genetic analysis of blast resistant in somaclonal mutants of rice has been studied by Chen *et al.* (7). Bernardo *et al.* (2) has reported a rapid method for DNA finger printing using rice blast fungus *P. grisea.*

Although biological control of rice sheath blight and blast diseases is still in its infancy but it is gaining momentum and promising results are being reported all over the world. These results can thus be exploited to improve rice disease management. Further, if genes for antiblast and antisheath blight diseases could be identified, cloned and incorporated into rice genotypes by using latest genetic engineering techniques biological control can be further strengthened. This will help the resource poor farmers of the tropics particularly India where we cannot afford expensive chemicals, in their efforts to sustain rice yields.

7. REFERENCES

1. Baby, U.I. and Manibhushan Rao, K. 1993. Control of rice sheath blight through the integration of fungal antagonists and organic amendments. Trop. Agric. **70(3)** : 240-244.

2. Bernardo, M.A., Nagui, N., Leung, H., Zeigler, R.S. and Nelson, R.J. 1993. A rapid method for DNA finger printing of the rice blast fungus *P. grisea.* Int. Rice Res. Notes, **18(1)** : 48-50.

3. Bonman, J.M. 1992. Durable resistance to rice blast disease : environmental influences. Euphytica, **63(1-2)** : 115-123.

4. Cartwright, R.D., Crippen, D.L. and Templeton, G.E. 1993. Biocontrol of rice sheath blight using indigenous rice-field microorganisms. Phytopathology, **83(12)** : 1347-1349.

5. Chen, Q., Chen, Z. and Wang, J. 1993. Studies of the screening of rice blast-resistant mutant using phytotoxin. Acta Genetica Sinica, **20(4)** : 340-347.

6. Chen, Z. and Chen, Q. 1993. The application of α-picolinic acid on the screening of rice blast resistant variants. Acta Botanica Sinica, **35(3)** : 165-170.

7. Chen, Z., Chen, Q. and Hou, Q. 1993. Genctic analysis of blast resistant somaclonal mutants of rice (*Oryza sativa* L.). Acta Genetica Sinica, **20(4)** : 334-339.

8. Chin, K.M. and Bhandhufalck, A. 1990. The importance of crop growth stages for determining the application timing of diseased control agents on rice. In Pest management in rice. Ed. Grayson, B.T., Green, M.B. and Copping, L.G., Elsevier Applied Science Publi., London and New York. pp. 131-142.

9. Correa, F. and Zeigler, R. 1991. Integrated management of the rice blast disease *(P. oryzae).* Fitopatologia Colombiana, **15(1)** : 35-37.

10. Garrote, B.P., Mercado, A. and Garrity, D.P. 1986. Soil fertility management in acid upland environment. Phillipp. J. Crop. Sci., **11** : 113-123.

11. Gnanamanickam, S.S. and Mew, T.W. 1989. Biological control of rice blast with antagonistic bacteria. International Rice Res. Newsl. **14(2)** : 34-35.

12. Gnanamanickam, S.S. and Mew, T.W. 1990. Biological control of rice diseases (Blast and Sheath blight) with bacterial antagonists : An alternate strategy for disease management. In Pest Management in rice Ed. Grayson, B.T., Green, M.B. and Copping, L.G., Elsevier Applied Science Publ., London and New York. pp. 87-110.

13. Gnanamanickam, S.S. and Mew, T.W. 1992. Biological control of blast disease of rice *(Oryza sativa* L.) with antagonistic bacteria and its mediation by a *Pseudomonas* antibiotic. Ann. Phytopathol. Soc. Jpn. **58(3)** : 380-385.

14. Gnanamanickam, S.S., Candole, B.L. and Mew, T.W. 1992. Influence of soil factors and culture practice on biological control of sheath blight of rice with antagonistic bacteria. Plant and Soil, **144** : 67-75.

15. Gogoi, R. and Roy, A.K. 1993. International Rice Research Notes, **18(3)** : 31-32.

16. Gokulapalan, C. and Nair, M.C. 1984. Antagonism of a few fungi and bacteria against *Rhizoctonia solani* Kuhn. Indian J. Microbiol. **24** : 57-58.

17. Hashioka, Y. and Fukita, T. 1969. Ultrastructural observations on mycoparasitism of *Trichoderma, Glicladium* and *Acremonium* to phytopathogenic fungi. Rep. Tottori Mycol. Inst. **7** : 8-18.

18. Homma, Y., Takahashi, H. and Imoto, Y.A. 1992. Studies on the mode of action of soybean lecithin : Part 3. Effect on the infection process of rice blast fungus, *P. oryzae.* Ann. Phytopathol. Soc. Jpn. **58(4)** : 514-521.

19. Howell, C.R. and Stipanovic, R.D. 1979. Control of *Rhizoctonia solani* on cotton seedlings with *Pseudomonas fluorescens* and with an antibiotic produced by the bacterium. Phytopath. **69** : 480-482.

20. IRRI, 1981. International Rice Research Institute, Annual Report for 1980. Los Banos, Philippines. 165-167.

21. IRRI, 1983. International Rice Research Institute, Annual Report for 1982. Los Banos, Philippines. 170-172.

22. IRRI, 1985. International Rice Research Institute, Annual Report for 1984. Los Banos, Phillipines, **46-49** : 144-146.

23. Kato, T., Yamaguchi, Y., Namai, T. and Hirukawa, T. (1993). Oxygenated fatty acids with anti-rice blast fungus activity in rice plants. Biosci. Biotechnol Biochem. **57(2)** : 283-287.

24. Kloepper, J.W. 1983. Effect of seed piece inoculation with plant growth-promoting rhizobacteria on populations of *Erwinia carotovora* on potato roots and daughter tubers. Phytopath. **73** : 217-219.

25. Kloepper, J.W., Leong, J., Teintz, M. and Schroth, M.N. 1980. Enhanced plant growth by siderophores produced by plant growth promoting rhizobacteria. Nature, **286** : 885-886.

26. Kohli, C.K. 1967. Some cultural studies on *Rhizoctonia solani* the causal organism of sheath blight of paddy. J. Res. Ludhiana, **4** : 401-405.

27. Manibhushanrao, K., Sreenivasa Prasad, S., Baby, U.I. and Joe, Y. 1989. Susceptibility of rice sheath blight pathogen to mycoparasites. Curr. Sci. **58** : 515-518.

28. Manibhushanrao, K. and Baby, U.I. 1991. Managing rice sheath blight (Sh. B) using fungal antagonists and organic amendments. Int. Rice Res. Newsl. **16(6)** : 19-20.

29. Mew, T.W. and Rosales, A.M. 1986. Bacterization of rice plants for control of sheath blight caused by *Rhizoctonia solani.* Phytopathology, **76** : 1260-1264.

30. Mew, T.W. and Rosales, A.M. 1984. Relationship of microorganisms to rice sheath blight development in irrigated and dryland rice cultures. In soil borne crop diseases in Asia. FFTC, Book series No. 26. Food and Fertilier Technology Centre for the Asian and Paliyic Region, Taiwan, pp. 147-158.

31. Misaghi, I.J., Olsen, M.W., Cotty, P.J. and Dondellinger, C.R. 1988. Fluorescent siderophore-mediated iron deprivation-acontingent biological control mechanism. Soil Biol. Biochem. **20** : 573-574.

32. Morris, L., Correa-Victoria, F.J., Zeigler, R.S., Xu, S. and Hamer, J.E. 1993. Genetic diversity of rice blast fungus in a disease nursery in Colombia. Phytopathologia, **83(12)** : 1427-1433.

33. Mukherjee, B. 1987. Interaction of *Aspergillus* with two rice pathogens. Environment and Ecology, **5** : 395-401.

34. Nagamani, A. and Mew, T.W. 1987. *Trichoderma* in Phillipines rice field soils. Int. Rice Res. Newsl. **12(4)**: 25-27.

35. Ogura, H. and Akai, S. 1965. Studies on *Rhizoctonia solani* Kuhn. *(Pellicularia filamentosa* Pat. Rogers). IV. The activity of antagonists to *Rhizoctonia solani* Kuhn. Ann. Phytopath. Soc. Japan, **30** : 219-224.

36. Ou, S.H. 1985. Rice diseases. 2nd Ed. Commonwealth Mycological Institute, Kew, Surrey, England, 380 pp.

37. Premalatha Dath, A. 1982. Effect of soil amendment with some green manures on the survival of sclerotia of *Corticium sasakii.* Indian Phytopath. **35** : 523-525.

38. Rajeswari. E. and Mariappan, V. 1992. Effect of plant extracts on *in vitro* growth of rice blast (BI) pathogen, *P. oryzae.* Int. Rice Res. Newsl. **17(6)** : 24-26.

39. Research Briefings, 1987. Report of the Research Briefing panel on Biological control in managed ecosystems, National Academy Press, Washington. D.C.

40. Rosales, A.M. and Mew, T.W. 1982. Antagonistic effect of soil microorganisms on rice sheath blight pathogen. Int. Rice Res. Newsl. **7(6)** : 12-13.

41. Rosales, A.M., Vantomone, R., Swings, J., Dc Ley, J. and Mew, T.W. 1993. Identification of some bacteria from paddy antagonistic to several rice fungal pathogens. J. Phytopathol (Berl.), **138(3)** : 189-208.

42. Roy, A.K. 1977. Parasitic activity of *Trichoderma viride* on the sheath blight fungus on rice *(Corticium sasakii)* Pflkrankh. Journal of Plant diseases and protection, **84** : 675-683.

43. Roy, A.K. 1980. Survival of sclerortia of *Corticium sasakii* at different soil depths. Int. Rice Res. Newsl. **5(6)** : 12-14.

44. Roy, A.K. 1984. Inhibitory effect of *aspergillus terreus* Thom. against *Rhizoctonia solani* f. sp. *sasakii.* Int. Rice. Res. Newsl. **9(3)** : 12-14.

45. Roy, A.K. 1985. Growth of sclerotia of *Rhizoctonia solani* f. sp. *sasakii* in presence of soil-inhabiting microorganisms. Indian Phytopath. **38** : 45-49.

46. Roy, A.K. 1986. Survival of sclerotia of *Rhizoctonia solani* f. sp. *sasakii* in relation to moisture regime of soil. Indian Phytopath. **39** : 259-263.

47. Roy, A.K. 1991. Inhibitory effect of *Aspergillus terreus* on *Rhizoctonia solani* causing sheath blight of rice. J. Indian Bot. Soc. **70** : 95-98.

48. Roy, A.K. and Sayre, R.M. 1984. Electron microscopical studies of *Trichoderma harzianum* and *T. viride* and mycoparasitic activity of the former on *Rhizoctonia solani* f. sp. *vasakii.* Indian Phytopath. **37** : 719-712.

49. Sadanov, A.K., Achio, S., Kurmanbaev, A.A. and Tulemisooa, K.A. 1992. Antagonism of a *Tricoderma* culture to rice seed pathogens. Izv. Akad. Nauk. Resp. Kaz. Ser. Biol. **0(4)** : 68-71.

50. Saifulla, M. 1992. Screening of rice genotypes for blast disease. Agric. Sci. Dig. **12(4)** : 178-180.

51. Sakthivel, N. and Gnanamanickam, S.S. 1987. Evaluation of *Pseudomonas fluorescens* for suppression of sheath rot diseases and for enhancement of grain yields in rice, *Oryza sativa.* L. Appl. Environ. Microbiol. **53** : 2056-2059.

52. Sarker, D.K., Sharma, N.R. and Shahjahan, A.K. 1992. Antagonistic soil bacteria for biological control of rice sheath blight (Sh.B) disease. Int. Rice Res. Newsl. **17(6)** : 22-23.

53. Scher, F.M. and Baker, R. 1982. Effect of *Pseudomonas putida* and a synthetic iron chelator on induction of soil supressiveness to *Fusarium* wilt pathogens. Phytopathology, **72** : 1573-1577.

54. Schippers, B., Lugtenberg, B. and Weisbeek, P.J. 1987. Plant growth control by fluorescent *Pseudomonas*. In Innovative approaches to Plant disease control Ed. I. Chet. John Wiley and Sons, New York, pp. 19-39.

55. Stutz, E.W., Defago, G. and Kern, H. 1986. Naturally occuring *Pseudomonas fluorescens* involved in suppression of black root rot of tabacco. Phytopathology, **76** : 184-185.

56. Sugiyama, M. 1988. Rice sheath blight and chemical control in Japan. Japan Pesticide Information, **52** : 9-12.

57. Thara, K.V. and Gnanamanickam, S.S. 1992. Evaluation of chitinase production as a criterion for selecting bacterial antagonists for biological control of rice sheath blight (Sh.B). Int. Rice Res. Newsl. **12(1)** : 25-26.

58. Thomashow, L.S. and Weller, D.M. 1988. Role of phenazine antibiotic from *Pseudomonas fluorescens* in biological control *Gaeumannomyces graminis* var. *tritici*. J. Bacteriol. **170** : 3499-3508.

59. Vasantha Devi, T.V., Malar Vizhi, R., Sakthirel, N. and Gnanamanickam, S.S. 1989. Biological control of sheath blight of rice in India with antagonistic bacteria. Plant and Soil, **199** : 325-330.

60. Weller, D.M. 1988. Biological control of soilborne plant pathogens in the rhizosphere with bacteria. Annu. Rev. Phytopathol. **26** : 319-407.

61. Weller, D.M. and Cook, R.J. 1985. In Iron. Siderophores and Plant disease, Ed. T.R. Swineburne, Plenum Publ. Corp. New York. pp. 71-82.

62. Wells, H.D. 1988. *Trichoderma* as a biocontrol agent. In : Biocontrol of plant diseases vol. I. (Eds. K.G. Mukerji and K.L. Garg), CRC Press, Inc. Florida pp. 71-82.

63. Wu, W. and Wu, Y.L. 1993. Stereo selective synthesis of methyl (115, 125, 135, 92, 15Z) 11-hydroxy-12, 13-epoxyoctode cadienoate from D-mannose. Tetrahedron, **49(21)** : 4665-4670.

5 Nematode Problems of Wheat and Barley and their Management

D.S. Bhatti and I.J. Paruthi

CONTENTS

Integrated Pest and Disease Management
Rajeev K. Upadhyay, K.G. Mukerji, B.P. Chamola and O.P. Dubey (eds.)
APH Publishing Corporation, New Delhi-110002, India

ABSTRACT

Wheat and barley are important cereal crops in India and are vulnerable to a large number of fungal, bacterial and nematode pathogens. A large number of plant parasitic nematodes are commonly encountered around the rhizosphere of both these crops, but only two species *Anguina tritici* and *Heterodera avenae* are considered of economic importance in all wheat and barley growing areas of the country. Earcockle and 'tundu' or yellow ear rot diseases caused by *Anguina tritici* alone and *A. tritici* + *Clavibacter tritici,* respectively, are initially diagnosed by basal swelling of the stem, crinkling, curling and twisting of leaves, increased tillering and stunted growth of the diseased plants. But during humid atmosphere (low temperature and high humidity) yellow bacterial mass can be seen on the leaves and earheads in case of 'tundu' disease. Generally, no grain is formed in diseased plants due to the destruction of floral primordia by the bacterial pathogen. However, few healthy grains of poor quality are obtained in partially diseased earheads. In case of molya disease, the diseased plants remain stunted with yellowish leaves and less tillering. However, this disease is confirmed by seeing root system (bushy roots alongwith white glistening females during the months of December and January). These three diseases are of common occurrence in Northern region of the country and cause enormous losses to both the crops under field conditions. A detailed literature on history, distribution, host spectrum, life history, symptomatology, association with other organisms and various management practices of these nematodes has been given alongwith some information available on other nematodes of minor importance.

1. INTRODUCTION

Wheat and barley are important cereal crops. Wheat occupies a unique position and provides sustenance to over one third of the world's population. Barley is considered as a poor man's diet. Both the crops are vulnerable to a large number of insect pests and pathogens which cause a significant loss in their yield. Amongst various disease causing agents, the nematodes play an important role. A large number of nematodes reside in the rhizosphere of wheat (Table 1) and barley (Table 2) of these. Wheat seed gall nematode *(Anguina tritici)* and cereal cyst nematode *(Heterodera avenae)* are considered as major nematodes because of their common occurrence and more damaging nature. This chapter deals with various aspects of these nematodes alongwith brief information on other nematodes of minor importance.

Table 1 : Plant parasitic nematodes found associated with wheat rhizosphere in India

Nematode species	Distribution	Reference
Anguina tritici	Bihar, Delhi, Haryana, Himachal Pradesh, Punjab, Rajasthan and U.P.	94, 168
Basiria nasikensis	Maharashtra	47
Heterodera avenae	Bihar, Delhi, Haryana, Himachal Pradesh, Punjab, J. & K., Rajasthan, and U.P.	21, 36, 81, 90 104, 189 203
Hirschmanniella mucronata	U.P.	162
H. oryzae	U.P.	162
Hoplolaimus indicus	U.P.	162

Longidorus brevicaudatus	U.P., Rajasthan	92
Meloidogyne javanica	North India	158
M. incognita	Rajasthan	156
Pratylenchus brachyurus	North West India	177
P. coffeae	North West India	177
P. delatteri	Rajasthan	213
P. penetrans	North West India	177
P. thornei	North West India	177
Psilenchus graminis	Haryana	10
Rotylenchulus reniformis	North India	91, 162
Tylenchorhynchus brevidens	North West India	176
T. ewingi	North West India	176
T. mashoodi	North West India	176
T. vulgaris	Delhi	205
Xiphinema americanum	U.P.	92

Table 2 : Plant parasitic nematodes found associated with barley rhizosphere in India

Nematode species	Distribution	Reference
Anguina tritici	Haryana, Delhi	20, 53
Heterodera avenae	Haryana, Punjab, Rajasthan & U.P.	94
H. zeae	Delhi	194
Hoplolaimus indicus	U.P.	162
Merlinius brevidens	Delhi	206
Pratylenchus neglectus	Delhi	177
P. scribneri	Delhi	177
Pseudhalenchus indicus	North India	175
Tylenchorhynchus brevidens	North India	176
T. capitatus	North India	176
T. ewingi	North India	176
T. mashoodi	North India	176
T. vulgaris	Gujarat	155, 205
Meloidogyne javanica	Gujarat	153
M. incognita	Rajasthan	156

2. MAJOR NEMATODES

2.1 Wheat seed gall nematode-*Anguina tritici*

2.1.1 History

This is the first plant-parasitic nematode reported on wheat (12). Instead of forming normal seeds infected plants develop dark hypertrophied seeds filled with white fibrous nematodes (165). Initially it was named as *Vibrio tritici* (196) but later it was given the present accepted name, *Anguina tritici* (30). Lot of work has been done on life-cycle of the nematode and development of the disease (30, 36, 78).

2.1.2 Distribution and host range

Earlier this nematode was reported from all wheat growing areas of the world but now more or less it has been controlled by using modern techniques. However, it is still a troublesome problem in certain developing and underdeveloped countries. In India, the nematode is widely distributed in many states (75, 100).

The host spectrum of this nematode is restricted. Wheat is considered as the main host but some other crops like emmer, spelt, rye, barley, triticale, maize and some other graminaceous weeds in wheat fields can also be parasitised by this nematode (4, 20, 44, 51, 132, 137, 140, 143, 146, 151, 199).

2.1.3 Biology

(a) Life history

Life history of this nematode has been studied in detail (58, 153). The nematode completes one generation in a year. The life cycle of the nematode synchronises with the development of the plant and takes about 4-5 months depending upon the plant cultivar and other edaphic factors. By sowing infested wheat seed (containing nematode galls), on contact with the soil moisture, the second stage larvae of *A. tritici* emerge from the softened walls of the gall and find their way to the growing point of wheat seedling (Fig. 1). The time taken by larval egress from the galls and reaching the growing point is influenced by the temperature and soil type (133, 145). Normally higher temperature and heavy soils facilitate the early release of the larvae (133). These larvae remain as ectoparasites and are carried upward mechanically alongwith the growth of culms. According to Leukel (97) larval feeding during this phase is negligible whereas reverse is reported by Midha *et al.* (119) since they observed comparatively high metabolic activity of the growing point of nematode infested seedlings. At the time of floral initiation, these larvae become endoparasitic and get enclosed in the cavities formed by the overgrowth of initials which turn into galls. The larvae moult within 3-5 days (201) and develop into males and females usually in the ratio of 1:2 (121). The females start laying eggs within 6-10 days. One female lays about 200 eggs within the freshly formed galls (193). The adults die soon after oviposition and the eggs hatch quickly to first stage and moult to second stage. These larvae are resistant to desiccation and remain viable inside the gall in anhydrobiotic form for several years (30, 100).

(b) Physiological strains/races

Existence of physiological races in *Anguina tritici* has been repeated from India (65, 70) and Iraq (2, 150).

2.1.4 Diseases

This nematode alone produces earcockle disease but in association with a bacterium, *Clavibacter tritici* produces another disease known as yellow ear rot or 'tundu'. Presence of nematode is must for production of both the diseases which are common in wheat fields in northern region of India and can be diagnosed easily by seeing above ground symptoms.

(a) Earcockle

The disease is initially characterised by basal swelling of stem (71), crinkling, curling and twisting of leaves (Plate 1A) with increased tillering and stunted growth (71, 132, 149). Such symptoms on leaves are more pronounced in early stage of plant growth which start disappearing at the later stage. Expression of symptoms, however, depends upon the level of nematode infestation (120, 131).

Typical diseased earheads are smaller or broader than the healthy ones with or without awns on them (Plate 1B) but under field conditions some normal looking diseased earheads also appear (141). Normally the galls are formed in place of grains (Plate 1D) due to nematode attack on the floral primordia but these have also been reported to be formed on other organs like leaves, awns, glumes, lemma and palea (71, 139). The galls on such organs are considered of minor importance because of their rare occurrence and less gall content (40). In addition to these symptoms two new abnormalities, i.e., sterility in wheat earheads and elongation of distance between first and second spikelet of the earhead (148) are also observed.

(b) 'Tundu'

It is also known as yellow ear rot disease (Plate 1C) which is caused by *Clavibacter tritici* in association with *A. tritici* (nematode). Since the disease can be caused by the bacterium only in the presence of nematode hence the initial symptoms of this disease are similar to those of earcockle (149). At the later stage, the disease is characterised by the production of yellow bacterial slime or gum on the leaf surface of young plants and also on the abortive ears (71). This yellow slime can be seen trickling down the tissues under humid conditions (Plate 1C), which becomes hard and brittle on drying. Culms which are infected die in young stages or grow until headings. In later case the emerging spike becomes narrow with the grains partially or completely replaced by the bacterial mass (71, 152).

2.1.5 Epidemiology

Development of earcockle and 'Tundu' diseases is dependent on larval and bacterial concentrations, environmental conditions like temperature and humidity, age of the seedling and source of galls (120). Under experimental conditions, minimal nematode population to induce the disease is 10^4 larvae/kg soil. Generally, the expression of 'Tundu' symptoms is more under low temperature and high humidity conditions. Such conditions favour the rapid bacterial multiplication and profuse production of bacterial mass suppressing nematode activity and gall formation.

2.1.6 Crop losses

This nematode causes both quantitative and qualitative losses in wheat either by reducing its yield or its acceptability for human consumption due to flour colour change (142). Unlike other species of *Anguina*, particularly *A. agrostis*, this nematode does not produce any toxin (121). However, as high as 30-70% quantitative loss due to this nematode (Table 3) is not unusual (163).

Table 3 : Losses due to Anguina tritici in wheat and barley under field conditions

Crop	% loss due to disease incidence	Country	Reference
Wheat	30-70	U.S.A.	163
	10-30	China	180
	22.9-45	Iraq	197
	50.0	Tigre Province	80
	1-3	Delhi, Haryana	136, 138
	30.0	U.P. (India)	182
Barley	40.0	Iraq	197

2.1.7 Interaction with other microorganisms

(a) Essential

The nematode is known to interact essentially with a fungus, *Dilophospora alopecuri* (9) and a bacterium, *Clavibacter tritici* (72, 208) causing a plumed spore and yellow ear rot or 'tundu' diseases, respectively. Presence of nematode is considered necessary in causation of both these diseases. The nematode, *A. tritici* has been reported to act as a vector of the pathogen *C. tritici.* The nematode carries the bacterium to the vicinity of apical meristem and confers the beneficial role for the bacterium by changing the physiology of the host plant (199), whereas the latter is antagonistic to the nematode (72, 134). The bacterium is located inside the galls and on the surface of nematode body (72).

The bacterium does not remain viable for more than seven months (34) but inside the galls the longevity is reported upto 5 yeas (109). Longer longevity inside the galls is attributed to the presence of food material (proteins and vitamins) in the galls for the bacterium (118).

(b) Non-essential

Non-essential simultaneous associations of *A. tritici* with several fungi in wheat have also been reported by several workers as given below.

Anquina tritici + *Neovossia indica* (135).

A. tritici + *Tilletia foetida* (110).

A. tritici + *Ustilago tritici* (13, 150).

A. tritici + *C. tritici* + *U. tritici* (14).

A. tritici + *C. tritici* + *Sclarophthora macrospora* (147).

2.1.8 Management strategies

The diseases, being seed-borne in nature, can very effectively be controlled by use of clean or healthy gall-free seed for sowing purposes (86, 97).

(a) Physical and Mechanical

(i) Fanning and Winnowing: Winnowing the infested seed material in open air or in front of fan separates the cockles from the grains. These practices though commonly used by the farmers in India and China are not effective because some large sized heavy galls fall alongwith the seed while cleaning. Chu (41) from China reported that about 45% of the galls remained in wheat seed even if the fan revolved at 850 rpm. Similarly, Jat *et al.* (86) reported that none of the samples cleaned with winnowing basket was free from cockles.

(ii) Sieving: The method is based on the difference in size between gall and seed. Complete removal of galls is not achieved by this method since some large sized bold galls are retained on the sieve alongwith the seed. Paruthi and Bhatti (144) reported that sieving the infested seed material of different wheat varieties resulted in 77.5 to 91.8% removal of the galls.

(iii) Hot water treatment : Putting the infested wheat seed in hot water at 54-56°C for 10-12 minutes, 122°F for 30 minutes (99) and 30°C for 4 hrs has been found effective in destroying the larvae of *A. tritici.* Chu (41) advocated presoaking of galls in cold water before putting them in the hot water.

(iv) Water flotation/salt sedimentation : Putting contaminated seed in a drum containing water or salt solution is very effective practice of removing cockles from the grains. The galls, being lighter in weight, float on the surface of water, can be removed with a sieve and destroyed. According to Paruthi and Bhatti (144), more than 90% seed galls could be removed by this method. However, some large sized compound galls start sinking down to the bottom after absorbing water. Chu (41) reported that the galls remain on the surface of water for about 5 minutes. Paruthi and Bhatti (144) observed complete removal of cockles at and above 10% concentration of salt. Different workers have used various concentrations of salt solution ranging from 10-30% (83, 208). Salt sedimentation is though more effective than water floatation yet it is comparatively more cumbersome because seeds treated with salty water are required to be washed with water several times before sowing to avoid the injurious effect of salt on germination of seed (83).

(b) Cultural

(i) Crop rotation : Sowing of non-host crop for about 1-2 years in the infested area has been reported to reduce the nematode infestation to a great extent (54, 135). In tropical and subtropical countries like India and China where the temperature in the summer goes very high, the nematodes in galls left in the soil, do not play any role in disease causation in the next cropping season (54, 175).

(ii) Roguing : Infected plants should be removed and destroyed.

(c) Plant resistance

Several workers have screened a lot of wheat germplasm for resistance against this nematode without much success. However, some resistant lines (Table 4) have been reported from various countries.

Table 4 : Resistant response of wheat and barley genotypes/varieties against *Anguina tritici*

Crop	Varieties/lines	Country	Reference
A. Wheat	Kanred	U.S.A.	-
	Novokrymka and Co-operatorka	U.S.S.R.	159
	PC 876, P 63 and Hsuchow Resistant No. 2	China	54
	Concho, Funone, Bison, Dica, Marquis and Mequrele	Rumania	215
	Acodde	Yugoslavia	263
	Nainari 60, Pictic 62, V 5009 and V 5040	Pakistan	7, 109
	India 66 and Saberbeg, 342, 346, 347, 348 (genotypes derived from Saberbeg x Mexicap by irradiation)	Iraq	220
	V 509, Raj 1774, CPANI 676 and HD 2177, E 650, HD 1738, HD 2021, HP 4530, Karachia 65, Moti, NI 747-9 and NP 888	India	3, 272
B. Barley	BR 70, DL, 70, C 164, BP 264, BP 344, BP 380, BP 393, BP 409, BP 416, BP 618, BP 628, BP 624, BP 627, BP 630, BP 639, BP 644, BP 652, BP 670, BP 676, BP 688, BP 693, BP 698, BP 708, BP 716 and T 5	India	181

(d) Chemical

A host of workers have tried a few substances for the control of *A. tritici.*

(i) In vitro screening : Two hrs. in 0.5% sulfuric acid six hours in 25% alcohal, one month in glycerine, five hrs. in formaldehyde (1:240), four and a half hours in mercuric chloride (1:1000) are required to kill free living larvae of *A. tritici* (205).

(ii) Seed Treatment : Seed soaking : Treating contaminated seeds with chemicals has been recommended by various workers. Sarauer (247) recommended soaking of infested seeds in dilute sulfuric acid. Chu (54) tried some common disinfectants and obtained good results with corrosive sublimate. The resistance of galls to chemicals varied according to their state of dryness. Dry galls are comparatively less resistant to chemicals than the pre-soaked galls. According to some workers seeds cannot be thoroughly disinfected without injuring the wheat grains (205, 246).

Fumigation : Effect of some fumigants against *A. tritici* on wheat has also been reported by some workers (191, 216). Romascu (216) tested some fumigants against *A. tritici* on spread wheat seeds and found that methyl bromide @ 40-80 g/m^3 seed for 24 h caused higher mortality of *A. tritici* than the other treatments. Paruthi (175) reported that fumigation of wheat seed galls with celphos @ 3g/25 l pitcher for 7 days showed 35.83% larval mortality as compared to 3.0% in untreated galls.

(iii) Soil treatment : Application of aldicarb (Temik) @ 2.5 kg a.i./ha has been found effective against *A. tritici* (160).

(iv) Foliar spray : Srivastava and Katiyar (252) reported that three applications of diazinon @ 0.15% reduced the infection of *A. tritici* from 30.6 to 2.48%. However, it showed some phytotoxicity to the plants. Foliar application of ethyl parathion completely eliminated the galls of *A. tritici* on wheat (240).

(e) Phytotherapeutic/organic amendments

Aqueous extracts of *Anagalis arvensis* (170), *Erigeron linifolius* (169) and *Azadirachta indica* (209) have been reported to be effective against *A. tritici*. Nematicidal activity of oils of 3 cymbopogon grasses against *A. tritici* has been reported by Sangwan *et al.* (223). Kumari *et al.* (132) reported 75-100% mortality of *A. tritici* in methanolic extracts of leaves, stems and buds of *Datura stramonium, Ipomea carnea, Tagetes patula* and *Lawsonia altre*. Malik *et al.* (141) also reported nematicidal value of essential oils, extracts of root and stem of *Xanthium strumarium* against this nematode. Gokte and Swarup (83) observed a significant reduction in larval penetration of *A. tritici* by treating wheat seeds with neem powder @ 600-800 mg/100 g seed. Cultivation of mustard and rapeseed with wheat reduced the infection of *Anguina tritici* (271).

(f) Biological

Very little information is available regarding biocontrol of *A. tritici* in wheat (65, 79, 85, 86). Dhawan *et al.* (65) found a fungus, *Arthrobotrys oligospora,* parasitising *A. tritici.* Gokte and Swarup (85) reported ten bacterial isolates from the cockles out of which five species of *(Bacillus subtilis, B. pumilus, B. cereus)* and two species of *Pseudomonas* were found to be larvicidal. They further evaluated the efficacy of these isolates under pot conditions as seed treatment against *A. tritici* and found best response by *B. subtilis* and *B. pumilus* (86).

2.2 Cereal cyst nematode - *Heterodera avenae*

2.2.1 History

Heterodera avenae was first seen on the roots of wheat, oat and barley in Germany (131) and was originally named as the oat race of *H. schachtii* and later as *H. schachtii* var. *avenae* Schmidt. Franklin (72) renamed this as *H. major* and later as *H. avenae* Wollenweber.

2.2.2 Distribution and Host range

Earlier it was thought that *H. avenae* occurred only in temperate areas of the world but now it is believed that it is cosmopolitan in distribution and found in several countries (154). In India, Prasad *et al.* (208) reported this nematode for the first time from Rajasthan state. Later on, several workers reported its occurrence from other states like Bihar, Haryana, Punjab, Himachal Pradesh, Jammu and Kashmir and Uttar Pradesh (15, 26, 48, 107, 261).

It is the main parasite of several cultivated crops of family gramineae however wheat, barley, oat and rye are well established hosts of this nematode. Gair (73) reported that oat and barley are better hosts than wheat and rye. In India, maize and sorghum have also been reported as hosts of this nematode (76, 276). In addition to these crops several weeds and grasses, mainly of gramineae family have also been reported to be parasitised (23).

2.2.3 Biology

(a) Life history

H. avenae exhibits sexual dimorphism and life cycle is similar to that of other cyst nematodes. During optimum favourable environmental conditions in the crop seasons, the second-state juveniles after emerging from the cysts invade the developing roots and undergo morphological developments causing the formation of syncytia (specialised cells upon which the nematode feeds) which interfere with the ability of the plant to obtain adequate nutrients and water from the soil (Fig. 2). The time of penetration and development is dependent on several edaphic factors and plant cultivars (146, 147). The development of the nematode inside the root undergoes three moults. The larvae that become female start swelling to lemon shape with the enlargement of body and protrude out of root tissues. The larvae destined to become males remain vermiform and move out of the root tissues into the soil and fertilise the females. Under Indian conditions, white coloured females of *H. avenae* can be seen on the roots with naked eyes during January and February. Such females start forming eggs which fill up the whole body. After some days, the females die and body cuticle becomes hard and changes in colour from white to brown and get detached from the host roots into the soil. The period of cyst formation has been reported to vary from one geographic region to other depending upon the environmental conditions and availability of host plant. These cysts in the soil remain in dormant stage and the dormancy seems to be facultative since it can be induced by high temperature.

Although different periods of development are observed from different geographical areas but the life cycle is similar. There is only one generation in a year.

Unlike several other *Heterodera* species, emergence of larvae is not affected by root leachates (146). Several edaphic factors, however, affect the juvenile hatching. Gair (73) found cysts in all soil types. Koshy and Swarup (130), and Mathur *et al.* (147) from India reported occurrence of more cysts in light soils. Temperature also influences the nematode survival (34), storage at high temperature reduce the cyst viability. The eggs in cyst have been reported to survive in air dried soil from 6-24 months (68, 146).

(b) Biotypes

Regarding existence of biotypes of this nematode, there are variable reports from various countries (8, 9, 38, 56, 70, 171, 213). Till today one biotype has been reported from Australia (38); two from Sweden (8), Denmark (9) and Britain (56, 70), four from France (213), five from India (144) and six from Germany (171).

2.2.4 Nature of damage

(a) Symptomatology

The disease caused by *Heterodera avenae* is known as 'Molya' in India and occurs in patches (Plate 2B). The diseased plants show yellowing of leaves, stunted growth, reduced tillering, delayed emergence of ears with reduced number of spikelets and grains (145). Such attributes can be ascribed to moisture stress and nutritional deficiencies. On uprooting such plants, nematode symptoms can readily be detected by a matted or bushy root (Plate 2A) resulting in a root growth proliferation. Presence of white females (attached on the roots) during January and February is diagnostic of the disease. Exhibition of symptoms and reduction in yield depends upon the amount of nematode inoculum present in the soil (63, 156). The damage threshold level of *H. avenae* differs in plant cultivars and geographical areas. Decker (63) found that 1-3 juvenile per cubic centimeter of soil is a moderate population, 3-5 juveniles per c.c. is a severe and over 5 juveniles is very severe.

(b) Histopathology

Second-state juveniles penetrate the roots usually in the meristematic region behind the root cap and other parts of the roots also. The larvae move through the cortex both intercellularly and intracellularly and establish for feeding in stelar region and lie parallel to the stele. The cells surrounding the head begin to enlarge ad their protoplasm becomes dense. Cell wall dissolution and merging of protoplasm results in the formation of elongate multinucleate syncytia and with thickened out boundaries (274). Formation of syncytia in susceptible and resistant cultivars of wheat and barley differed (11, 57). Initiation of giant cells in resistant barley cv. sabrilis occurred but differed in size and appearance from those in susceptible cultivar (11). Cotten (57) also reported the initiation of syncytia in resistant barley cv. KLV 91, but subsequently these cells become vacuolated with less dense granular cytoplasm than susceptible roots.

2.2.5 Crop losses

In India, Van Berkam and Seshadri (268) estimated losses worth 60 million rupees in wheat and barley in Rajasthan State alone. Seshadri (224) reviewed the work done in India and observed losses in wheat upto 100%. Several other workers (31, 100, 102, 149, 155) also reported losses due to *H. avenae* from various parts of the country (Table 5).

Table 5 : Crop losses estimated due to Heterodera avenae in wheat and barley

Crop	Avoidable loss (per cent)	State	Reference
Wheat	50-100		224
	100.0		224, 268
	41.6-66.6	Rajasthan	90
	4.1-32.1	Punjab	224
	56.0-91.5	Rajasthan	238
	8.2-48.8	Haryana	31
	32.4-66.6	Rajasthan	149
	29.8	Rajasthan	102
Barley			
Cv. Rc-59	54.88-87.2	Rajasthan	100
Husked barley	38.7	Rajasthan	102
Huskless barley	30.8	Rajasthan	102

2.2.6 Interaction with other micro-organisms

Very limited information is available regarding association of *H. avenae* with other organisms. Gill and Swarup (77) reported *Helminthosporium gramineum* and *Fusarium moniliforme* on barley plants infested with *H. avenae*. Both the fungi singly suppressed the nematode population.

2.2.7 Management strategies

(a) Cultural

(i) Crop rotation and fallowing : The nematode being host-specific can very effectively be managed by sowing non-host crops for 2-3 years in the nematode infested areas (91, 95, 96, 98, 129, 158). Kort (129) has suggested avoiding wheat and barley for atleast two years in heavily infested areas through rotation with non-hosts like, beet and alfalfa. Meagher and Rooney (158) found improved wheat yield by following wheat-legume rotation. Crop rotation with non-hosts and keeping the field fallow during rabi reduced disease incidence by 50% and 75% in one and two years, respectively (95, 96, 98). Singh (236) demonstrated that non-host crops reduced the population significantly. The resistant barley when included in rotation, reduced population by 57-58% and mustard reduced it by 50%. Under field conditions, resistant barley cv. Rajkiran reduced by nematode population by 84.8% (239). Growing of 'sarson', 'toria', 'raya' and 'taramirah' for one season resulted in 87-100% reduction in cyst population (241). Handa (91) in his trial with non-host and fallowing found 70% reduction in nematode population with continuous rotation with non-host crops. He also found that yield of barley was increased by 56% with two year rotation of non-host crops.

(ii) Deep summer ploughing : Under Indian conditions, 2-3 ploughings at an interval of 10-15 days during May and June (summer), reduce the nematode population resulting in a substantial increase in yields (21, 150). Deep summer ploughings under Indian conditions have their own advantages. Soil temperatures during May and June remain more than 40°C. By ploughing the soil is loosened and freshly formed cysts are disturbed and come in the upper zone of the soil (0-20 cm). The nematode is sensitive to desiccation, hence its contents start disintegrating.

(iii) Trap crop : Use of wheat (cv. C 306) as a trap crop has been suggested by Bhatti and Dabur (22). According to their observation, the final population of *H. avenae* with aldicarb and trap crop decreased by 7.5 and 9.4%, respectively and corresponding yield increases were 9.0 and 13.5%.

(iv) Mixed cropping : Mixed cropping of wheat and barley as 'Gajra' is common practice in Rajasthan. Handa *et al.* (94) reported the beneficial effect of mixed cropping of wheat (cv. Kalyan Sona) with resistant barley (cv. Rajkiran) than that of wheat alone.

(v) Sowing time : Early sown crop suffers comparatively less damage than the late sown crop (155).

(vi) Irrigation : Irrigation is one of the most important input in agriculture especially in India where water is scanty. Williams (274) did not observe any change in post-harvest population of *H. avenae* with increase in soil moisture. Mathur *et al.* (147) reported higher multiplication of the nematode in well irrigated fields in wheat and barley. Bhatti *et al.* (30) reported 36.2% increase in wheat yield by 2 additional post harvest irrigations.

(b) Plant Resistance

Use of resistant lines is a vital component in the management of any pest. For the control of *H. avenae* source of resistance in barley has been met with more success (39, 40, 66, 102, 153, 234, 277) than in wheat (Table 6).

Table 6 : Resistant response of wheat and barley genotypes/varieties against Heterodera avenae

Crop	Varieties/lines	Country	Reference
A. Wheat	Loros	Denmark	173
	Katyil, Aus. 18558 and Gerek	Australia	275
	Pane	Spain	222
	PI 183868, PI 185205, PI 185207 (USDA Collections), HD 2257, HD 2265, WH 250, WL 922, WL 2199, G 736, Raj 1409, Raj 1470, Raj 1487, PV 18, Sonara 64, Lerma Rojo and Agra local	India	111, 114, 234, 257
B. Barley	Morocco (CI 3902), Marocaine (CI 18334), Martin 403-2, Marocaine 079, Sabarlis, Nile, C 18147 and anthenais	Australia	39, 40, 69, 153
	Primus, Swanhals, Chevalier, Ansgar	Sweden	172
	Drost, Fero, Kron, PL 191, No. 14, No. 291, Drost, 746534 and 747475	Denmark	9, 167
	Morocco and Martin	Indonesia	245
	AQ 253, AQ 853, AQ 1127, BP 262, 264, C 164, BH 75, DL 349, DL 375, DL 376, DL 379, C 164, Rajkiran	India	24, 66, 277

(c) Chemical

Several chemicals have been found effective in bringing down he population of this nematode in the field but their use on large scale by a common farmer is avoided because of cost factor. A large number of chemicals (fumigants and non-fumigants) has been tested by various workers (27, 29, 59,92, 93, 112, 118, 165, 242, 260) as soil/seed treatment (Table 7). Soil application with aldicarb and carbofuran was identified as the most effective. However, with the non-availability of aldicarb in the market carbofuran is the only chemical which can be used effectively against this nematode.

(d) Phytotherapeutic/Organic Amendments

Bhatti and Dhawan (32) observed reduction in cyst populations with the addition of crushed seeds of carrot (1, 2, 4% W/W) and coriander (4%). Sharma *et al.* (231) demonstrated that mustard cake amendments (1 to 5% W/W) significantly reduced the cyst population and increased number of tillers and shoot weight. Dahiya *et al.* (60) reports 52.85 and 48.66% reduction in cyst population with addition of 'subabool' and sunhemp leaves, respectively. Bhatti and Verma (35) reported significant increase in wheat yield with application of castor *(Ricinus communis)* and 'aak' *(Calotropis procera)* in *H. avenae* infested field. Mathur (143) reported that oil cakes, farmyard manure, compost and saw dust applications resulted in improved plant growth and subdued multiplication of the nematode. Sakhuja *et al.* (219) reported increase in wheat yield with the addition of nitrogenous fertiliser. Barry *et al.* (16) found that applied nitrogen did not have much influence on grain yield. They also reported that yield increase was independent of source of nitrogen. Mukhopadhyaya *et al.* (166) reported increase in cyst population with the use of nitrogenous fertiliser.

Table 7 : Efficacy of various pesticides on Heterodera avenae in wheat and barley

A. As soil treatment			Yield	Cost bene-	
1. Fumigants	Dose	Crop	increase	fit ratio	Reference
Chemical	1 kg/ha				
D.D.	300 lt	Wheat & Barley	-	-	260
	400 lt	Wheat	83.9%	-	29
	400 kg	Wheat	3 fold	-	225
	400 kg	Barley	13 fold	-	225
	450 lt	Wheat	-	-	112
	500 lt	Wheat	253%	-	165
DBCP	30 lt	Wheat	125.8%	1:1.45	27
	45 lt	Wheat	95%	-	237
	45 lt	Wheat	12.5-95.3%	-	29
2. Non-Fumigants					
Aldicarb	1.5-2.0 kg	Wheat	-	1:4 to 1:11	29, 59, 92
Carbofuran	1.5-4.0 kg	Wheat	-	-	29, 59, 92
B. As seed treatment					
Aldicarb	2.5-5.0	Wheat	-	-	118
Aldicarb sulfone	2.5-5.0	Wheat	-	-	118
Carbofuran	5.0-10.0	Wheat	-	-	118
Fensulfothion	1.0-2.0	Wheat	-	-	118
Isofenphos	1.0-5.0	Wheat	-	-	118

(e) Biological

Recently attention has been directed towards exploitation of natural enemies in bringing down the nematode population in soil. Not much information on control of *H. avenae* through biocontrol agents is available except for few preliminary reports of endosparasitic fungi attacking eggs, juveniles and females (51, 52, 84, 121, 123). Chowdhary and Dhawan (51) reported infection of *Catenaria vermicola* in juveniles and eggs of *H. avenae.* The infection in the second stage juveniles occurs through zoospores. The fungus completes its life cycle within the nematode body in 2-3 days (64). Chowdhary and Kaushal (52) found, *Verticillium uniseptum* infecting eggs of *H. avenae.* several other fungi like *Catenaria auxiliaris, Nematophthora gynophila, Verticillium chlamydosporium, Tarichium auxiliare* and *Cylindrocarpan destructants* are also known to parasitise *H. avenae* in Britain (121, 123). Sharma (230) isolated 17 organisms including 15 fungi and a bacterium, out of which *Bacillus penetrans, Glomus fasciculatum, Catenaria auxiliaris, C. vermicola, Allomyces anomalus* and a species of *Olpidium* appeared to possess potential for their exploitation in the management of cyst populations. *Glomus fasciculatum* and *Pasteuria penetrans* were found to be the most destructive bioagents against *H. avenae.* Bhattarcharya and Swarup (20) reported the effect of *P. penetrans* on the biology and control of *H. avenae.* The bacterium complete its life cycle in 49 days at 17°C. Direct mixing of bacterial spores with infested soil was found to be the most effective.

(f) Integrated approach

Amendments in cultural practices like summer ploughing, crop rotation, irrigation, use of organic amendments and treatment of soil with chemicals have been tested individually by various workers but integration of two or more control practices for *H. avenae* has not received much attention (28, 98, 101, 102, 219, 239). Handa *et al.* (98) recorded an increase in yield from 18.29 to 126.74 and 35.01 to 105.23% in wheat and barley, respectively when nitrogen was combined with carbofuran and aldicarb. Sakhuja *et al.* (219) used DBCP @ 45 l/ha alongwith 150 kg nitrogen per hectare, resulting in grain yield of 49.33 q/ha as compared to 27.5 l/ha in control. Application of aldicarb 10G @ 2 kg a.i./ha in early November sowing gave better yield and 81.6% reduction in cyst population (28). Singh and Yadav (239) recorded a greater reduction in nematode population over control when organic amendments were combined with nitrogen fertilizer.

3. MINOR NEMATODES

In addition to the above mentioned nematodes, some other species of phytonematodes are commonly found around the rhizosphere of these crops. Important among these are given here.

3.1 Stunt nematode-*Tylenchorhynchus* spp.

These nematodes are of common occurrence in wheat and barley fields under Indian conditions (202, 203). Amongst various species *T. vulgaris* is the most common. Patel *et al.* (202) reported outbreaks of *T. vulgaris* in Anand area of Gujarat.

Symptoms

The symptoms of *T. vulgaris* are stunted growth, burning and yellowing of leaves (203, 266).

Hosts

This is a polyphagous pest. Wheat and barley are reported as very good hosts of this nematode (266).

Interaction with other organism

This nematode has been commonly found in concomitant occurrence with *Merlinius brevidens.* Upadhyaya and Swarup (267) reported decline in the population of *T. vulgaris* in the presence of *M. brevidens.*

Control

This nematode can easily be controlled by the application of some granular pesticides like phenamephos, phorate, carbofuran and ethoprop @ 1 kg a.i./ha (203).

3.2 Root-knot nematode - *Meloidogyne* spp.

Wheat and barley were considered to be the non-host crops for root knot nematodes. But some cases of parasitization of these crops by *Meloidegyne* spp. have been reported (74, 201, 202, 204).

Symptoms

The affected plants show poor growth, yellowing of leaves with very small galls on the roots (202).

Hosts

This is a polyphagous pest. Wheat is considered as the poor host of this nematode.

Recently Gaur and Khan (74) reported a new species of root knot nematode which is a parasite of wheat and rice. This nematode completes one generation in 31 days. Gaur and Singh (75) compared the effect of various types of puddling practices alone and in combination with nitrogenous fertilizers.

Minimum soil population was recorded in tractor disc puddling treatment. However, the differences due to puddling were evident during September under rice and February-March in wheat crop.

3.3 Other nematodes

A high population of several other phytonematodes like *Pratylenchus, Helicotylenchus, Hoplolaimus* etc. was reported by Chawla and Prasad (47) in maize-wheat rotation. Prasad and Gaur (207) reported a negative correlation between wheat yield and population of some nematodes.

4. CONCLUSION

Anguina tritici and *Heterodera avenae* are economically two major nematode pests of wheat and barley crops. *A. tritici* alone causes earcockle disease but in association with a bacterium, *Clavibacter tritici,* produces another disease called 'tundu' or yellow ear rot. *Heterodera avenae* produces 'molya' disease in both wheat and barley crops. Besides these two key nematode pests, a few other plant parasitic nematodes are associated with the rhizospheres of wheat and barley as minor pests.

Information available on various economic aspects of *A. tritici* and *H. avenae* such as host range, distribution, biology, nature and extent of damage, disease syndromes, histopathology, interaction with other microorganisms and different control tactics facilitating integrated management approach systems has been presented in the text.

The necessity to work on some pivotal aspects of these two nematodes, likely to yield potential information for integrated nematode management, such as host-parasite relationships, population dynamics study in relation to crop losses, cultural, biological, genetic, phytotherapeutic and chemical control has been projected.

5. FUTURE THRUSTS

Anguina tritici and *Heterodera avenae* are the two major nematode pests of wheat and barley in the leading wheat growing states of the country. Both the nematodes are responsible for causing tangible crop losses depending upon initial level of field infestations.

A. tritici is the oldest plant parasitic nematode. The mechanical control of this nematode pest (making seed gall-free by sieving, water floatation and salt sedimentation) is the cheapest and most effective. Even then if the crop losses are severe due to this nematode the fault lies with ignorance of the growers or poor extension system. However, future thrust is required to work on management of the earcockle and 'tundu' diseases when they appear in the field following wheat sowing. This disease can be tackled by developing suitable chemical pesticides, besides, conducting serious research on biocontrol agents, phytotherapeutic substances, organic amendments of soil and re-organisation of extension system.

H. avenae is another major nematode pest of wheat and barley, causing 'molya' disease. This nematode, being host-specific, is not difficult to control. But future research emphasis is needed on host-parasite relations, population dynamics in relation to crop losses, cultural, biological, genetic, phytotherapeutic and chemical control tactics for developing integrated nematode management.

Although the other nematodes are still not posing any economic problem, recently a new species of root knot nematode *(Meloidogyne* sp.) described by Gaur and his associates is a potential pest of both wheat and rice. Wheat and rice, being the most economic rotation and practised extensively in the northern India, the new *Meloidogyne* sp. is likely to become the most devastating pest in future. Therefore, all the economic aspects of this nematode need to be studied very thoroughly. Similarly, the root knot nematode *(M. javanica)* reported from Gujarat occurs extensively where tobacco-barley is the common rotation. This nematode is likely to become serious in areas where tobacco may be rotated with vegetable crops. Therefore, population dynamics of this nematode needs to be studied in different cropping systems.

ACKNOWLEDGEMENTS

The authors are thankful to Dr. R.S. Kanwar, Department of Nematology, CCS Haryana Agricultural University, Hisar for his constructive help in preparation of this chapter.

6. REFERENCES

1. Akhtar, C.M. and Aqil, T. 1986. Biological control of earcockle of wheat. J. Agric. Res., **24** : 327.

2. Alam, M. 1983. Reaction of certain varieties of wheat to the seed gall nematode, *Anguina tritici.* Indian J. Plant Path., **1** : 210-211.

3. Alsabie, R.F. and Evans, A.A.F. 1980. Intraspecific variation, host range and infection process of *Anguina tritici* (abstracts). 15th International Nematology Symposium of the European Soc. of Nematologists, Bari, Italy, 24-30 Aug. 1980.

4. Al-Talib, N.Y., Al-Taae, A.K.M., Nimer, S.M., Stephan, S.A. and Al-Baldawi, A.S. 1986. New record of *Anguina tritici* on barley from Iraq. Int. Nematol. Network Newsl., **3** : 25-27.

5. Amin, M. 1970. Relative resistance of 25 varieties of wheat against earcockle of wheat *(Anguina tritici)* (Abstracts). In Pakistan Science Conf. (21-22nd) Rajshahi Part III, Proceedings, Lahore, Pakistan Association for Advancement of Science, A 80 - A 82.

6. Andersen, K. and Andersen, S. 1970. Changes in population of cereal eelworm under grass species and resistant barley varieties. Tijdschr Planteavl., **74** : 559-565.

7. Andersen, S. 1959. Resistance of barley to various populations of the cereal root eelworm *(Heterodera major).* Nematologica, **4** : 91-98.

8. Anon 1972. Britain Welsh Plant Breeding Station. Aberystowyth, Wales : University College 1972, 129 p. (In : Plant Nematology, pp. 45-47.

9. Atanasoff, d. 1925. The Dilophospora disease of cereals. Phytopathology, **15** : 11-40.

10. Bajaj, H.K., Kaushik, H.D. and Bhatti, D.S. 1981. *Psilenchus graminus* sp. n. and *Deladenus durus* (Cobb, 1922) Throne, 1941, a new record from India. Indian J. Nematol., **11** : 137-140.

11. Bajaj, H.K. and Walia, R.K. 1985. A new report on the occurrence of cereal cyst nematode, *Heterodera avenae* Woll. 1924 in Ambala district (Haryana). Geobios, **4** : 169-170.

12. Barry, E.R., Brown, R.H. and Erloit, B.R. 1974. Cereal cyst nematode *(Heterodera avenae)* in Victoria: Influence of cultural practices on grain yields and nematode population. Aust. J. Expt. Agric. Anim. Husb., **14** : 566-571.

13. Bedi, K.S., Chohan, J.S. and Chahal, D.S. 1959. Simultaneous occurrence of *Ustilago tritici* (Pers.) Rost and *Anguina tritici.* Haryana Agric. Univ. J. Res., **6** : 162-163.

14. Beniwal, M.S., Gupta, A. and Karwasra, S.S. 1991. Simultaneous occurrence of loose smut and 'tundu' in a single earhead of wheat. Research and Development Reporter, **8** : 76-77.

15. Bhattacharya, D. and Swarup, G. 1988. *Pasteuria penetrans,* a pathogen of the genus *Heterodera,* its effect on nematode biology and control. Indian J. Nematol., **18** : 61-70.

16. Bhatti, D.S. and Dabur, K.R. 1982. Effect of summer ploughings on *Heterodera avenae* population and wheat yield. J. Nematol., **14** : 429.

17. Bhatti, D.S. and Dabur, K.R. 1982. Use of Wheat as a trap crop for the control of *Heterodera avenae*. J. Nematol., **14** : 429.

18. Bhatti, D.S., and Dahiya, R.S. 1992. Nematode pests of wheat and barley - *Heterodera avenae*. In : Nematode pests of crops (eds. D.S. Bhatti and R.K. Walia) CBS Publishers and Distributors, Delhi, India, pp. 27-42.

19. Bhatti, D.S., Dahiya, R.S., Dalal, M.R. and Dhawan, S.C. 1976. Resistant barley varieties for the control of *Heterodera avenae*. curr. Sci., **45** : 678.

20. Bhatti, D.S., Dahiya, R.S. and Dhawan, S.C. 1978. New record of tundu and earcockle in barley. Nematologica, **24** : 331-332.

21. Bhatti, D.S., Dahiya, R.S., Gupta, D.C. and Malhan, I. 1980. Plant Parasitic nematodes associated with various crops in Haryana. Haryana agric. Univ. J. Res., **10** : 413-414.

22. Bhatti, D.S. and Dalal, M.R. 1972. Yield responses in wheat to nematicidal treatments in a soil heavily infested with wheat cyst nematode. Haryana Agric. Univ. J. Res., **2** : 231-232.

23. Bhatti, D.S., Dalal, M.R., Dahiya, R.S. and Dhawan, S.C. 1980. Integrated control of *Heterodera avenae* in wheat. J. Nematol., **12** : 214-215.

24. Bhatti, D.S., Dalal, M.R., Dahiya, R.S. and Malhan, I, 1981. Chemical control of *Heterodera avenae* in wheat. J. Nematol., **13** : 433.

25. Bhatti, D.S., Dalal, M.R. and Malhan, I. 1980. Effect of additional irrigation on population of *Heterodera avenae* and wheat yield. Haryana Agric. Univ. J. Res., **10** : 571-572.

26. Bhatti, D.S., Dalal, M.R. and Malhan, I. 1981. Estimation of loss in wheat yield due to cereal cyst nematode, *Heterodera avenae*. Tropical Pest Management, **27** : 375-378.

27. Bhatti, D.S., and Dhawan, S.C. 1980. Effect of crushed seeds of carrot and coriander on wheat plant growth and multiplication of *Heterodera avenae*. Haryana agric. Univ. J. Res., **10** : 419-20.

28. Bhatti, D.S. and Malhan, I. 1982. Effect of temperature on larval emergence from cysts of *Heterodera avenae*. Indian J. Nematol., **12** : 73.

29. Bhatti, D.S. and Verma, K.K. 1991. Phytotherapy of *Heterodera avenae* in wheat. Pl. Dis. Res., **6** : 28-34.

30. Bird, A.F. and Buttrose, M.S. 1974. Ultra structural changes in the nematode, *Anguina tritici* associated with anhydrobiosis. J. Ultra Structure Research., **48** : 177-189.

31. Brown, R.H. 1969. The occurrence of biotypes of the cereal cyst nematode *(Heterodera avenae)* in Victoria, Aust. J. Expt. Agric. Anim. Husb., **9** : 453.

32. Brown. R.H. 1974. Further studies of the victorian biotype of cereal cyst nematode *(H. avenae)*. Aust. J. Expt. Agric. Anim. Husb., **14** : 394-398.

33. Brown. R.H. 1985. The selection of management strategies for controlling nematodies in cereals. Agric. Ecosystems Env., **12** : 371-388.

34. Chand, J.N. 1967. Longevity of *Corynebacterium tritici* (Hutchinson) Burk. Causing tundu disease of wheat in Háryana. Sci. Cult., **33** : 539.

35. Chawla, M.L. and Prasad, S.K. 1972. Multiple cropping and nematodes II. The effect of intensity of cropping on nematode populations. Indian J. Nematol., **2** : 75-77.

36. Chhabra, H.K. 1973. Distribution of *Heterodera avenae* Wollenweber, cereal cyst nematode in Punjab. Burr. Sci., **42** : 441.

37. Chitwood, B.G. 1935. Nomenclatorial notes. Proc. Helminth. Soc. Wash., **II (I)** : 51-54.

38. Chowdhry, P.N. and Dhawan, S.C. 1984. *Catenaria vermicola* in *Heterodera avenae* nematode-a new record. Curr. Sci., **53** : 96.

39. Chowdhry, P.N. and Kaushal, K.K. 1984. *Cylindro carpon uniseptatum* sp nov. a new fungus from India. Curr. Sci., **53** : 205-207.

40. Christie, J.R. 1959. Plant Nematodes, their bionomics and control. Phytopathology Florida Agricultural Experiment Station, 256.

41. Chu, V.M. 1945. The prevalence of wheat nematode in China and its control. Phytopathology, **35** : 288-295.

42. Cotten, J. 1963. Resistance in barley and oats to the cereal root eelworm, *Heterodera avenae.* Wollen Weber. Nematologica, **9** : 81-94.

43. Cotten, J. 1970. Some aspects of breeding for cereal root eelworm *(H. avenae)* resistance in barley. Proc. IX Int. Nematol. Symp., pp. 231-237.

44. Dahiya, R.S. and Bhatti, D.S. 1980. Incidence of 'tundu' on weed plants. Haryana Agric. Univ. J. Res., **10** : 413-414.

45. Dahiya, R.S. and Bhatti, D.S. 1983. Field evaluation of aldicarb for the control of *Heterodera avenae.* III Nematol. Symp. held at HPKVV Solan, India, pp. 39-40.

46. Dahiya, R.S., Dutt, R. and Gupta, D.C. 1986. Effect of subabool and sunnhemp leaves on cereal cyst nematode, *Heterodera avenae* (Woll. 1924) in wheat. Haryana Agric. Univ. J. Res., **16** : 389-91.

47. Darekar, K.S. and Khan, E. 1978. Soil and plant parasitic nematodes from Maharashtra, India. VI. Three new species of *Helicotylenchus* Steiner, 1945. (Tylenchida : Nematoda). Indian J. Nematol., **8** : 132-139.

48. Davaine, C. 1857. Pecherches sur. I 'anguillule, double' nielle, Consideree au point de vuede I'histoire naturelle et de I'agriculture, Compt, Rend. Soc. Biol; Paris (1856) 2 Ser, **3** : 201-207.

49. Decker, H. 1972. In : plant nematodes and their control (ed. N.M. Sveshnikova) Translated from Russia, USDA and NSF, Washington, D.C. Amerind Publ. Co. Pvt. Ltd., New Delhi.

50. Dhawan, S.C. 1985. Mode of infection and life cycle of *Catenaria vermicola* infecting juveniles *Heterodera avenae.* Indian Phytopath., **38** : 380-381.

51. Dhawan, S.C., Jain, K.K. and Swarup, G. 1979. A nematode trapping fungus of *Anguina tritici* larvae. Indian J. Nematol., **9** : 180-183.

52. Dhawan, S.C. and Sethi, C.L. 1983. Resistance in barley to the cereal cyst nematode *(Heterodera avenae* Woll). Indian J. Nematol., **13** : 235-237.

53. Dhawan, S.C. and Swarup, G. 1979. *Anguina tritici* infestation in barley. Indian J. Nematol., **9** : 51-52.

54. Duggan, J.J. 1960. Effect of soil drying on the viability of *Heterodera major* cyst. Nature, **185** : 534-555.

55. Eilis, S.E. and Brown J.A.M. 1976. The reactions of resistant cultivars of barley to some Australian populations of *Heterodera avenae.* Nematologica, **22** : 87-93.

56. Fiddian, W.E.H. and Kimber, D.S. 1965. A study of biotypes of the cereal cyst nematode *(Heterodera avenae* Woll) in England and Wales. Nematologica, **10** : 631.

57. Franklin, M.T., Throne, G. and Oostenbaink, M. 1959. Proposal to stabilise the scientific name of the cereal-root eelworm (class Nematoda) : Z.N. (S.) 375. Bull. Zool. Nomencl., **17** : 76-85.

58. Gair, R. 1965. Cereal root eelworm. In : Plant Nematology (ed. J.F. Southey) Tech. Bull (2nd Ed.) Minist. Agric. Fish. HMSO. London, pp. 199-221.

59. Gaur, H.S. and Khan, E. 1992. The life cycle and pathogenicity of a new cereal root knot nematode, *Melordogyne* spp. on wheat and rice (Abstract). First Afro Asian Nematology Symp. Afro Asian Soc. of Nematologists, Nov. 29-Dec. 3, 1992, Deptt. of Botany, Aligarh Muslim University, Aligarh, pp. 9.

60. Gaur, H.S. and Singh, A.K. 1993. Effect of puddling practices and nitrogen levels on the root knot nematode infecting wheat and rice. National Symp. on Recent Approaches in integrated Nematode Management of Agricultural Crops (August 6-7, 1993) at CCS HAU Hisar, Haryana, India (Abstract).

61. Gill, J.S. and Swarup, G. 1971. On the host range of the cereal cyst nematode, *Heterodera avenae* Woll 1924, the causal organism of molya disease of wheat. Indian J. Nematol., **1** : 63-67.

62. Gill, J.S. and Swarup, G. 1977. Effect of interaction between *Heterodera avenae* Woll. 1924, *Fusarium moniliforme* and *Helminthosporium gramineum* on barley plants and nematode reproduction. Indian J. Nematol., **7** : 42-45.

63. Gokte, N. and Swarup, G. 1985. Bacteria as potential biocides against *Anguina tritici.* Indian J. Nematol., **15** : 266-267.

64. Gokte, N. and Swarup, G. 1987. Studies on morphology and biology of *Anguina tritici.* Indian J. Nematol., **17** : 306-307.

65. Gokte, N. and Swarup, G. 1988. On the identification of physiological specialisation in *Anguina tritici* populations. Indian J. Nematol., **18** : 78-83.

66. Gokte, N. and Swarup, G. 1988. Effect of neem (Azadirachta indica) seed kernel powder treatment on penetration and development of *Anguina tritici* on wheat. Indian J. Nematol, **18** : 149.

67. Gokte, N. and Swarup, G. 1988. On the potential of some bacterial biocides against root knot and cyst nematodes. Indian J. Nematol., **18** : 152-153.

68. Gokte, N. and Swarup, G. 1988. On the association of bacteria with larvae and galls of *Anguina tritici.* Indian J. Nematol., **18** : 313-318.

69. Gokte, N. and Swarup, G. 1988. Efficacy of some bacterial isolates as seed treatment on penetration of *Anguina tritici* larvae to the growing point of wheat seedlings and viability of the larvae in the galls formed. Indian J. Nematol., **18** : 368-369.

70. Gupta, P. 1966. Studies on earcockle and 'tundu' diseases of wheat. Ph.D. thesis, I.A.R.I., New Delhi, India.

71. Gupta, P. and Swarup, G. 1968. The earcockle and yellow ear rot diseases of wheat : I. Symptoms and histopathology. Indian Phytopath., **21** : 318-323.

72. Gupta, P. and Swarup, G. 1972. Earcockle and yellow ear rot disease of wheat : II. Nematode bacterial association. Nematologica, **18** : 320-324.

73. Handa, D.K. 1983. Studies on molya disease *(heterodera avenae* Woll.) of wheat and barley with special emphasis on its control. Ph.D. thesis, Univ. of Rajasthan, 340 p.

74. Handa, D.K. and Mathur, B.N. 1982. Control of cereal cyst nematode *(Heterodera avenae)* of barley through systemic nematicides. Nematologica, **28** : 150.

75. Handa, D.K. Mathur, B.N., Sharma, G.L. and Yadav, B.D. 1980. Effect of mixed cropping of susceptible wheat and resistant barley on the control of molya disease. Indian Jour. of Mycol. & Pl. Path., **10** : 7-8.

76. Handa, D.K. Mathur, B.N., Yadav, B.D. and Sharma, G.L. 1989. Cultural practices and their effects on *Heterodera avenae* and yields of cereals in Rajasthan, India. Indian J. Nematol., **19** : 250-253.

77. Handa, D.K. Mathur, R.L. and Mathur, B.N. 1975. the effect of crop rotation and nematicides on the cereal cyst Nematode *Heterodera avenae* and yield of wheat and barley. Indian J. of Mycol. & Pl. Path., **5** : 20-21.

78. Handa, D.K. Mathur, R.L. and Mathur, B.N. 1978. Studies on the integrated control of molya disease of wheat and barley caused by *Heterodera avenae* in Rajasthan. Indian J. Mycol. & Pl. Path., **8** : 70.

79. Handa, D.K. Mathur, R.L., Mathur, B.N. and Yadav, B.D. 1985. Estimation of looses in barley due to cereal cyst nematode in sandy and sandy loam soils. Indian J. Nematol., **15** : 163.

80. Hingorani, M.K. and Beckele, G. 1969. Outbreaks and new records. Pl. Prot. Bull. FAO, **17** : 41-43.

81. Husain, S.I., Siddiqui, Z.A. and Siddiqui, M.R. 1989. Pmrevalence and geographical distribution of cyst forming nematodes in Uttar Pradesh, India, Indian J. Nematol., **19** : 108-114.

82. Ilyas, M. and Hussain, M. 1985. Screening of genetic stocks of *Triticum aestivum* L. em. Thell. for resistance to earcockle *(Anguina tritici).* J. of Agri. Res. Pakistan.

83. Jain, J.S, Sinha, R.P. and Mathur, R.S. 1956. Controlling earcockle of wheat by floating seed in salt solution. Agric. Anim. Husb. India, **5**:3.

84. Jain, R.K. and Sehgal, S.P. 1975. Testing of different varieties, strains and species of wheat for resistance to the cereal cyst nematode *Heterodera avenae.* Indian J. Mycol. & Pl. Path., **5** : 18.

85. Jain, R.K. and Sehgal, S.P. 1975. Efficacy of nematicides in controlling the cereal cyst nematode, *Heterodera avenae* in wheat in light sandy soils in Rajasthan. Indian J. Mycol. & Pl. Path., **5** : 19.

86. Jat, R.R., Pathak, V.N. and Verma, O.P. 1986. Removal of earcockles from contaminated wheat seeds. Seed Sci. and Technol., **14** : 237-243.

87. Kaushal, K.K. and Seshadri, A.R. 1986. Evaluation of chemicals as seed dressers against *Heterodera avenae* on wheat. Indian J. Nematol., **16** : 140.

88. Kerry, B.R. 1975. Fungi and the decrease of cereal cyst nematode population in cereal monoculture. European Plant Protection Organisation Bull., **5** : 353-361.

89. Kerry, B.R. and Crump, D.H. 1979. Two fungi parasitic on females of cyst nematodes *(Heterodera* spp.). Trans. Br. Mycol. Soc., **74** : 119-125.

90. Khan, E., Gaur, H.S. and Lalitha, Y. 1990. Occurrence of cereal cyst nematode, *Heterodera avenae* at high altitude in Leh, India. Int. Nematol. Network Newsl. **9** : 3.

91. Khan, F.A. and Khan, A.M. 1969. Reaction of certain economically important plants to *Rotylenchulus reniformis.* All India Nematology Symp. New Delhi, pp. 8-9.

92. Khan, F.A. and Khan, A.M. 1972. Studies on distribution and population of *Longidorus brevicaudatus, Xiphinema basiri* and *X. americanum* in Uttar Pradesh and Rajasthan with description of *Longidorus psidii* n. sp. (Nematoda : Dorylaimoidea). Indian Phytopath., **25** : 269-274.

93. Kort, J.J. 1959. Some observations on soil populations of cereal root eelworm, *Heterodera avenae (H. major)* under the influence of various crops in sandy soils. T. Pl. Ziekten, **65** : 1-4.

94. Koshy, P.K. and Swarup, G. 1971. Distribution of *Heterodera avenae, H. zeae, H. cajani* and *Anguina tritici* in India. Indian J. Nematol., **1** : 106-111.

95. Kuhn, J. 1874. Uber das vor Kommen Von Rubennematoden an den wurzehdea Halmfruchte. Landw., Jbr. **3** : 47-50.

96. Kumari, R., Verma, K.K., Dhindsa, K.S. and Bhatti, D.S. 1986. Datura Ipomea, Teqestrs and Lawsonia as control of *Tylenchulus sempenetrans* and *Anguina tritici.* Indian J. Nematol., **16** : 236-240.

97. Leukel, R.W. 1924. Investigations on the nematode disease of cereals caused by *Tylenchus tritici* J. Agric. Res., **27** : 925-955.

98. Leukel, R.W. 1957. Nematode disease of wheat and rye. U.S. Dept. Agric. Bull. 1607 (Revised). 16 p.

99. Limber, D.P. 1938. Notes on hot water treatment of *Anguina tritici* galls on wheat and comparison of Indian and Chinese collection by use of weight criteria. Proc. Helminth Soc. Wash., **5** : 20-23.

100. Limber, D.P. 1973. Notes on the longevity of *Anguina tritici* (Steinbuch, 1799) Filipjev, 1936 and its ability to linvade wheat seedlings after 32 years of dormancy. Proc. Helminth. Soc. Wash., **40** : 272-274.

101. Malik, M.S., Sangwan, N.K., Dhindsa, K.S. and Bhatti, D.S. Nematicidal activity of extracts of *Xanthium strumarium* Pesticides **21** : 19-20.

102. Mathur, B.N. 1969. Studies on the cereal root eelworm *(Heterodera avenae* Woll.) with special reference to 'Molya' disease of wheat and barley in Rajasthan. Ph.D. Thesis, Univ. Rajasthan, Jaipur, 233 p.

103. Mathur, B.N., Arya, H.C., Mathur, R.L. and Handa, D.K. 1974. The occurrence of biotypes of the cereal cyst nematode *(Heterodera avenae)* in the light soil of Rajasthan and Haryana, India, Nematologica, **20**: 19-26.

104. Mathur, B.N., Arya, H.C., Handa, D.K. and Mathur, R.L. 1975. On the symptoms and distribution of 'molya' disease of wheat and barley caused by *Heterodera avenae* in the light soils of Rajasthan. Indian J. Mycol. & Pl. Path., **5** : 91-94.

105. Mathur, B.N., Arya, H.C., Handa, D.K. and Mathur, R.L. 1975. The biology of cereal cyst nematode, Heterodera avenae in India. II. Factors affecting the emergence of larvae from cysts. Indian J. Mycol. Pl. Path., **5** : 149-155.

106. Mathur, B.N., Arya, H.C., Handa, D.K. and Mathur, R.L. 1981. The biology of cereal cyst nematode, Heterodera avenae in India. III. Factors affecting population level and damage to host crops. Indian J. Mycol. & Pl. Path., **11** : 5-13.

107. Mathur, B.N., Handa, D.K., Swarup, G., Sethi, C.L., Sharma, G.L. and Yadav, B.D. 1986. On the loss estimation and chemical control of molya disease of wheat caused by *Heterodera avenae* in India. Indian J. Nematol., **16** : 152.

108. Mathur, B.N., Swarup, G., Sharma, G.L. and Handa, D.K. 1992. Effect of summer ploughings and nitrogenous fertilisers on the cereal cysts nematodes, *Heterodera avenae* and yield of wheat. Afro Asian J. Nematol., **1** : 108-111.

109. Mathur, R.S. and Ahmad, Z.U. 1964. Longevity of *Corynebacterium tritici* causing tundu disease of wheat. Proc. Nat. Acad. Sci. India, **34B** : 335-336.

110. Mathur, R.S. and Misra, M.P. 1961. Simultaneous occurrence of *Tilletia foetida* (Wallr.) Liro and *Anguillulina tritici* in the ear and grains of wheat in Pauri Garhwal, Uttar Pradesh. Curr. Sci., **30** : 307.

111. McLeod, R.W., Wong, P.T.W. and Southwell, R.J. 1986. Biology and control of cereal cyst nematode in Northern New South Walls. Aust. J. Expt. Agric., **26** : 375-381.

112. Meagher, J.W. 1977. World dissemination of the cereal cyst nematode *(Heterodera avenae)* and its potential as a pathogen of wheat. J. Nematol., **9** : 9-15.

113. Meagher, J.W. 1982. Yield losses caused by *Heterodera avenae* in cereal crops grown in a Mediterranean climate Bull. EPPO, **12** : 325-331.

114. Meagher, J.W., Brown, R.H. and Rovira, A.O. 1978. The effects of cereal cyst nematode *(Heterodera avenae)* and *Rhizoctonia solani* on the growth and yield of wheat. Aust. J. Agrl. Res., **29** : 1127-1137.

115. Meagher, J.W. and Rooney, D.R. 1966. The effect of crop rotations in the Victorian Wimmora on the cereal cyst nematode *(Heterodera avenae),* nitrogen fertility and wheat yield. Aust. J. Exp. Agric. Anim. Husb., **6** : 425-431.

116. Meyer, A.A. 1939. The wheat nematode, *Anguillulina tritici* in the Crimea. Leningrad Academy of Agricultural Sciences, Institute for Plant Protection, Collected works on nematode of Agricultural crops Ex. by E.S. Kiryanova Moscow and Leningrad, pp. 160-167.

117. Midha, S.K. 1969. Studies on the earcockle nematode and the development of earcockle and tundu diseases of wheat. Ph.D. Thesis, IARI, New Delhi, India.

118. Midha, S.K. 1971. Investigations on the nature of association between *Anguina tritici* and *Corynebacterium tritici.* Proc. 2nd Int. Symp. Pl. Path. IARI, New Delhi. pp. 8.

119. Midha, S.K., Chatrath, M.S. and Swarup, G. 1971. On the feeding of *Anguina tritici* on growing point of wheat seedlings, Indian J. Nematol., **1** : 93-94.

120. Midha, S.K. and Swarup, G. 1972. Factors affecting development of earcockle and tundu diseases of wheat. Indian J. Nematol., **2** : 97-104.

121. Midha, S.K. and Swarup, G. 1974. Studies on the wheat see gall nematode caused by *Anguina tritici.* Indian J. Nematol., **4** : 53-63.

122. Mukhopadhyaya, M.C., Dalal, M.R. and Kharub, S.S. 1971. Effects of nematicides on the cyst nematode *(Heterodera avenae)* and wheat yield. Haryana agric. Univ. J. Res., **1** : 36-38.

123. Mukhopadhyaya, M.C., Dalal, M.R., Saran, S. and Kharub, S.S. 1972. Studies on the malya disease of wheat and barley. Indian J. Nematol., **2** : 11-20.

124. Munk, A., Clausen, K. and Hjortscholm, K. 1980. Report No. 34 on the work of Agril. Cereal Breeding Station, Horsens, Denmark, 35 p.

125. Needham, T. 1743. A letter concerning certain chalky tubulous concretions called malm; with some microosopical observations on the farina of red lily and of worms discovered in smutty corn. Philos. Trans. Roy. Soc. Lond., **42** : 173, 174, 634-641.

126. Nene, Y.L. and Kumar, K. 1967. Nematostatic Properties of *erigeron linifolius* extracts. Indian Phytopath., **20** : 172-174.

127. Nene, Y.L. and Thapliyal, P.L. 1966. Nematicidal properties of *Anagalis arvensis* L. extracts. Indian Phytopath., **19** : 26-29.

128. Neubert, E. 1967. Uber das Vorkommen Von biotypen des Haferzestenaichens *(Heterodera avenae* Wollenweber, 1924) in Norden der DDR Nbl. Dtsch Pflanzenschutzd (Berlin), **21** : 66-68.

129. Nilsson-Ehle, H. 1920. Uber Resostenz gagen *Heterodera schachtii* be gewssen Gersten sortenihre verez. bangs weise und Bedeutung furdie Praxis. Hereditas, **1** : 1-34.

130. Nilsen, C.H. 1966. Onter suchwagen uber die vorerbung der Resistenz gengan die Getreidenematiden *(H. avenae)* beim weizen. Nematologica, **12** : 575-578.

131. Paruthi, I.J. 1982. Studies on *Anguina tritici* in wheat and barley. Ph.D. thesis, Haryana Agricultural University, Hisar, India.

132. Paruthi, I.J. 1992. Nematode pests of wheat and barley *Anguina tritici.* In : Nematode Pests of Crop (ed. D.S. Bhatti, and R.K. Walia) CBS Publishers & Distributors, Delhi, India, pp. 14-26.

133. Paruthi, I.J., Bajaj, H.K. and Bhatti, D.S. 1988. Some factors affecting the emergence of *Anguina tritici* larvae from wheat seed galls and their ascendence on the seedlings. Indian J. Ecol., **15** : 82-86.

134. Paruthi, I.J., Bajaj, H.K. and Bhatti, D.S. 1989. Further observations on *Anguina tritici* and *Corynebacterium michiganense* pv. *tritici* causing earcockle and yellow ear rot in wheat. Nematologica, **35** : 491-493.

135. Paruthi, I.J. and Bhatti, D.S. 1980. Concomitant occurrence of *Anguina tritici* and Karnal bunt fungus *(Neovossia indica)* in wheat seed gall. Indian J. Nematol., **10** : 256-257.

136. Paruthi, I.J. and Bhatti, D.S. 1981. Estimation of losses in wheat yield due to *Anguina tritici* and seed gall ingestation in market grains. Indian J. Nematol., **11** : 15-18.

137. Paruthi, I.J. and Bhatti, D.S. 1981. Studies on *Anguina tritici* in barley. Nematologica, **27** : 463-466.

138. Paruthi, I.J. and Bhatti, D.S. 1985. Estimation of losses in yield and incidence of *Anguina tritici* on wheat in Haryana. Int. Nematol. Network. Newsl., **2** : 13-16.

139. Paruthi, I.J. and Bhatti, D.S. 1986. Comparative morphology and contents of galls caused by *Anguina tritici :* Nematoda (Anguinidae) on various parts of wheat plants. Nematologica., **33** : 243-245.

140. Paruthi, I.J. and Bhatti, D.S. 1987. Host differential responses of some cereals to *Anguina tritici.* Int. Nematol. Network Newsl., **4** : 8-10.

141. Paruthi, I.J. and Bhatti, D.S. 1987. Variation in earcockle symptoms on wheat ears caused by *Anguina tritici.* Nematologica, **33** : 360-363.

142. Paruthi, I.J. and Bhatti, D.S. 1988. Effect on consumption and marketing of wheat contaminated with seed galls of *Anguina tritici.* Haryana Agric. Univ. J. Res., **18** : 173-176.

143. Paruthi, I.J. and Bhatti, D.S. 1990. Wheat seed gall nematode, *Anguina tritici,* a review, Curr. Nematol., **1** : 119-136.

144. Paruthi, I.J. and Bhatti, D.S. 1992. Efficacy of different practices for the control of earcockle in wheat caused by *Anguina triticia.* Curr. Nematol., **3** : 1-4.

145. Paruthi, I.J. Bhatti, D.S. and Bajaj, H.K. 1990. Effect of some abiotic factors on the host parasite relationship of Angu*ina tritici* in wheat. National Conf. on Plant Res. and Dev. Feb. 26-March 1, 1990 (Plant Science Colloquim) Dept. Plant Breeding, Haryana Agricultural University, Hisar, India, pp. 50.

146. Paruthi, I.J. Bhatti, D.S., Bajaj, H.K. and Dalal, M.R. 1992. Earcockle-a new disease of lesser canary grass *(Phalaris minor).* Nematologia Medit. **20** : 197-198.

147. Paruthi, I.J. Bhatti, D.S. and Singh, S. 1992. A new record of simultaneous occurrence of yellow ear rot and downy mildew diseases on wheat and wild canary grass. Curr. Nematol., **3** : 95-96.

148. Paruthi, I.J. and Gupta, D.C. 1985. Two unrecorded abnormalities in wheat ears infested with *Anguina tritici.* Nematologica, **32** : 357-360.

149. Paruthi, I.J. and Gupta, D.C. 1986. Studies on combined occurrence of *Ustilago tritici* and *Angina tritici* in wheat. Haryana Agric. Univ. J. Res., **16** : 392-394.

150. Paruthi, I.J. and Gupta, D.C. 1986. Control of earcockle and yellow ear rot diseases of wheat. Indian Fmg., **36** : 32-35.

151. Paruthi, I.J. and Gupta, D.C. 1987. Incidence of 'tundu' in barley and kanki in wheat field infested with *Anguina tritici.* Haryana Agric. Univ. J. Res., **17** : 78-79.

152. Paruthi, I.J., Singh, M. and Gupta, D.C. 1987. Quantitative and qualitative losses in wheat *(Triticum aestivum)* grains due to earcockle and tundu. Seed Res., **15** : 83-86.

153. Patel, G.J., Patel, D.J., Jogani, D.K. and Patel, S.T. 1986. Wheat-a host of *Melodogyne javanica.* Indian J. Nematol., **16** : 134.

154. Patel, D.J., Thakar, N.A. and Patel, C.C. 1987. Outbreak of stunt nematode on wheat in Gujarat. Curr. Sci., **56** : 14.

155. Patel, B.N. and Thakar, N.A. 1989. Chemical control of the stunt nematode, *Tylenchorphnchus vulgaris* on wheat. Indian J. Nematol., **19** : 277-278.

156. Pathak, A.K. and Yadav, B.S. 1978. Host range of root knot nematode, *Meloidogyne incognita* on selected *graminaceous* field crops. Indian J. Myçol. & Pl. Path., **8** : 74.

157. Pennetier, G. 1886. Limitede La resistance violate des anguillules de la nielle. Compt. Rend. Acad. Sci. (Paris)., **103** : 284-286.

158. Prasad, S.K., Das Gupta, D.R. and Mukhopadhaya, M.C. 1964. Nematodes associated with commercial crops in North India and host range of *M. javanica.* Indian J. Ent., **26** : 235-238.

159. Prasad, S.K. and Gaur, H.S. 1974. Relation between nematode population and yield of wheat. Indian J. Nematol., **4** : 152-159.

160. Prasad, N., Mathur, R.L. and Sehgal, S.P. 1959. 'Molya' disease of wheat and barley in Rajasthan. Curr. Sci., **28** : 453.

161. Qamar, F., Saeed, M., Khan, A. and Seema, N. 1989. Control of Anguinosis of wheat using plant extracts. Pakistan J. of Scientific and Industrial Research, **32** : 260-261.

162. Rashid, A., Khan, F.A. and Khan, A.M. 1973. Plant parasitic nematodes associated with vegetables, fruits, cereals and other crops in North India. Indian J. Nematol., **3** : 8-23.

163. Reitz, I.P. 1976. Wheat in the United States. Agriculture information Bull., Agriculture Re. Service, U.S.D.A. No. **386** : 57.

164. Rivoal, R. 1975. Cereal cyst nematode, *Heterodera avenae* Woll in France, damage, biological characteristics and prospects of control. Bull. EPPO, **12** : 517-523.

165. Rofferdi, D.M. 1775. Memori sur l'origine des petits vers on anguilles due beld rachitique. Observet Mem. Phys. Hist. Nat., **5** : 1-19.

166. Romascu, E. 1969. Resistance of some autumn and spring varieties of wheat *Anguina (Tylenchus) tritici* Steinb. Communicri Zhool, Bucuresti, pp. 91-98.

167. Romascu, E. 1972. Contribution to the study of chemical control of *Anguina tritici* Steinb. Anable Institertuli de Cercelari Pentru Protectia Plantelor, **9** : 253-259.

168. Sakhuja, P.K., Singh, I., Kang, L.S. and Kang, M.S. 1990. Prevalence and losses caused by *Anguina tritici* to wheat in Punjab. Plant Dis. Res., **5** : 182-185.

169. Sakhuja, P.K., Singh, I. and Sharma, S.K. 1978. Effect of nitrogen alone and in combination with some nematicides on the population of cereal cyst nematode, *Heterodera avenea* and yield of wheat. Indian J. Nematol., **8** : 159-162.

170. Saleh, H.M. and Fattah, F.A. 1988. Reaction of wheat genotypes to infection by *Anguina tritici.* Revue de Nematolo, **12** : 429-431.

171. Sanchez, A., Scristan, J.C. and Bellò, A. 1985. Investigation of cereal varieties resistant to the nematode *Heterodera avenae* woll. Anales de Edofologia y Agrobiologia, **44** : 169-173.

172. Sangwan, N.K., Verma, K.K., Verma, B.S., Malik, M.S. and Dhindsa, K.S. 1985. Nematicidal activity of essential oils from Cymbopogone grasses. Nematologica, **31** : 93-99.

173. Seshadri, A.R. 1970. Nematology. In : Agricultural Year Book : new vistas in crop yields. Indian Council of Agril. Research, New Delhi, pp. 370-411.

174. Seshadri, A.R. 1971. Recent research on control of plant parasitic nematodes in India. Proc. Symp. Epidemiology, Forecasting & control of plant diseases. Indian Nat. Sci. Acad, pp. 15.

175. Sethi, C.L. and Swarup, G. 1967. *Pseudhalenchus indicus,* a new nematode species from India (Tylenchinae : Nematoda). Indian Phytopath., **20** : 26-28.

176. Sethi, C.L. and Swarup, G. 1968. Plant-Parasitic nematodes of North Western India I. The genus *Tylenchorphynchus.* Nematologica, **14** : 77-88.

177. Sethi, C.L. and Swarup, G. 1971. Plant parasitic nematodes of North Western India III. The genus *Pratylenchus.* Indian Phytopath., **24** : 410-412.

178. Sharma, R. and Swarup, G. 1988. Pathology of cyst nematodes. Malhotra Publishing House, New Delhi, pp. 87.

179. Sharma, S.K., Sakhuja, P.K. and Singh, I. 1981. Effect of mustard cake on the hatching and cyst population of *Heterodera avenae* on wheat. Indian J. Nematol., **11** : 11-14.

180. Shen, T.H., Tai, S.E. and Chia, W.I. 1934. A preliminary report on the inheritance of nematode resistance and length of beak in a certain wheat cross. Bull. Coll. Agric. & Forest. Univ. Nanking., **19** : 1-17.

181. Sikora, R.A., Koshy, P.K. and Malek, R.B. 1972. Evaluation of wheat selections for resistance to the cereal cyst nematode. Indian J. Nematol., **2** : 81-82.

182. Singh, H. 1985. Effect of crop rotation on the population of cereal cyst nematode. Indian J. Nematol., **15**: 263-264.

183. Singh, H. and Yadav, B.S. 1978. Control of 'molya' disease of wheat caused by *Heterodera avenae.* Indian J. Mycol. & Pl. Path., **8** : 73.

184. Singh, H. and Yadav, B.S. 1980. Yield losses in wheat due to cereal cyst nematode *Heterodera avenae.* Indian J. Mycol. & Pl. Path., **10** : 15.

185. Singh, H. and Yadav, B.S. 1986. Effect of integrated control on wheat yield and population of *Heterodera avenae.* Nat. Conf. Plant Parasitic nematodes of India-Problems and progress, pp. 97.

186. Singh, I. and Prasad, S.K. 1972. Effects of some pesticides and growth regulators on *Anguina tritici* infesting wheat. Indian J. Ent., **34** : 77-79.

187. Singh, I., Sakhuja, P.K. and Sharma, S.S. 1987. Influence of some crucifers on *Heterodera avenae* - the cyst nematode. Plant Dist. Res., **2** : 93-94.

188. Singh, I., Sakhuja, P.K. and Sharma. S.K. 1988. Studies on the time of application of nematicides against cyst nematode, *Heterodera avenae.* Plant Dis. Res., **3** : 144-145.

189. Singh, I., Sharma, N.K. and Chohan, J.S. 1976. Incidence of cereal cyst nematode, *Heterodera avenae* Woll. in Jammu Kashmir state. Food Fmg. & Agric., **8** : 20.

190. Soetopo, L. 1985. Genetic analysis of the resistance in barley *(Hordeum vulgare* L.) to the cereal cyst *(Heterodera avenae* woll.). Part I. The cultivar reaction Australian Pathotype. Agrivita, **185-186** : 8-9.

191. Sommerville, W. 1919. Earcockle in wheat. Jour. Bo, Agr. (London), **26** : 907-909.

192. Soraver, P. 1913. Handbuch der. Pflanzenkran kheeton Aufl., **3** : Bd. 3.

193. Southey, J.F. 1972. *Anguina tritici* CIH Descriptions of Plant-Parasitic nematodes. Set I No. 13. Commonwealth Agricultural Bureaux publication.

194. Srivastava, A.N. and Swarup, G. 1975. Preliminary studies on some graminaceous plants for their susceptibility to maize cyst nematode, *Heterodera zeae* Koshy, *et al.* 1970. Indian J. Nematol., **5** : 257-259.

195. Srivastava, A.S. and Katiyar, K.P. 1957. Field scale trials on effect of diazinon, salicylic acid and benzoic acid on wheat nematode. Proc. 44th Indian Sci. Congr. Part-III : pp. 301.

196. Steinbuch, J.G. 1799. Das Grasa chen V*ibrio agrostic.* Naturforscher, **28** : 233-259.

197. Stephen, Z.A. 1988. Plant parasitic nematodes on cereals and legumes in Iraq. In nematodes parasitic to cereals and legumes in temperate and semi arid regions. ICARDA Proceedings Workshop held in Larnace Cyprus, pp. 155-159.

198. Stephen, Z.A. and Antoon, B.G. 1990. Biotypes of earcockle nematode, *Anguina tritici* in Iraq. Curr. Nematol., **1** : 85-88.

199. Suryanarayana, D. and Mukhopadhayaya, M.C. 1971. Earcockle and 'tundu' diseases of wheat. Indian J. Agric. Sci., **41** : 407-413.

200. Swarup, G. and Gill, J.S. 1969. Varietal response of wheat and barley to infestation of *Heterodera avenae.* All India Nematology Symposium, pp. 64-65.

201. Swarup, G. and Gupta, P. 1971. On the earcockle and 'tundu' disease of wheat II. Studies on *Anguina tritici* (Steinbuch) Filpjev, and *Corynebacterium tritici* Burkh. Indian Phytopath., **24** : 359-365.

202. Swarup, G., Mathur, R.I., Seshadri, A.R., Raski, D.J. and Mathur, B.N. 1976. Response of wheat and barley to soil fumigation by DD and DBCP against 'molya' disease caused by *Heterodera avenae.* Indian J. Nematol., **6** : 150-155.

203. Swarup, G., Sethi, C.L., Kaushal, K.K. and Siyanand 1982. Distribution of *Heterodera avenae,* the causal organism of molya disease of wheat and barley in India. Curr. Sci., **51** : 896-897.

204. Tesic, T. 1969. A contribution to the investigation of the resistance of wheat to wheat nematode. Savremen Polijofr Nosi Sad., **17** : 541-543.

205. Upadhyaya, K.D. and Swarup, G. 972. Culturing host range and factors affecting multiplication of *Tylenchorhynchus vulgaris* on maize. Indian J. Nematol., **2** : 139-145.

206. Upadhyaya, K.D. and Swarup, G. 1981. Growth of wheat in the presence of *Merlinius brevidens* singly and in combination with *Tylenchrohynchus vulgaris.* Indian J. Nematol., **11** : 42.

207. Van Berkam, J.B. and Seshadri, A.R. 1970. Some important nematode problems in India. Xth International Nematology Symp. Pescara, Italy, pp. 136-137 (abstr.).

208. Vasudeva, R.S. and Hingorani, M.K. 1952. Bacterial disease of whet caused by *Corynebacterium tritici* (Hutchinson) Bergey et al., Phytopathology, **42** : 291-293.

209. Verma, M.K. 1991. Intercultivation of mustard and rapeseed on the management of earcockle disease of wheat. International J. of Tropical Plant Disease, **19** : 91-94.

210. Verma, M.K. and Yadav, B.S. 1986. Studies on *Anguina tritici* III Screening of aestivum wheat cultivars to earcockle disease caused by *Anguina tritici.* Nat. Conf. Pl. Parasitic Nematodes of India : Problems and Progress, December 17-20, 1986, pp. 106.

211. Williams, T.D. 1969. The effects of formalin, nabam, irrigation and nitrogen on *Heterodera avenae* Woll., *Ophiobolus graminis* Sacc. and growth of spring wheat. Ann. appl. Biol., **64** : 325-334.

212. Wilson, R.E., Holiamby, G.J. and Bayarktar, A. 1983. Proc. Aust. Pl. Br. Conf. Adelaide, South Australia.

213. Yadav, B.S. and Verma. A.C. 1971. Cereal cyst eelworm and other nematodes associated with maize in Rajasthan. Indian J. Nematol., **1** : 97-98.

214. Yadav, R., Behl, R.K., Singh, D. and Singh, A. 1987. Inheritance of resistance against *Heterodera avenae* in barley. Crop Improv., **14** : 188-190.

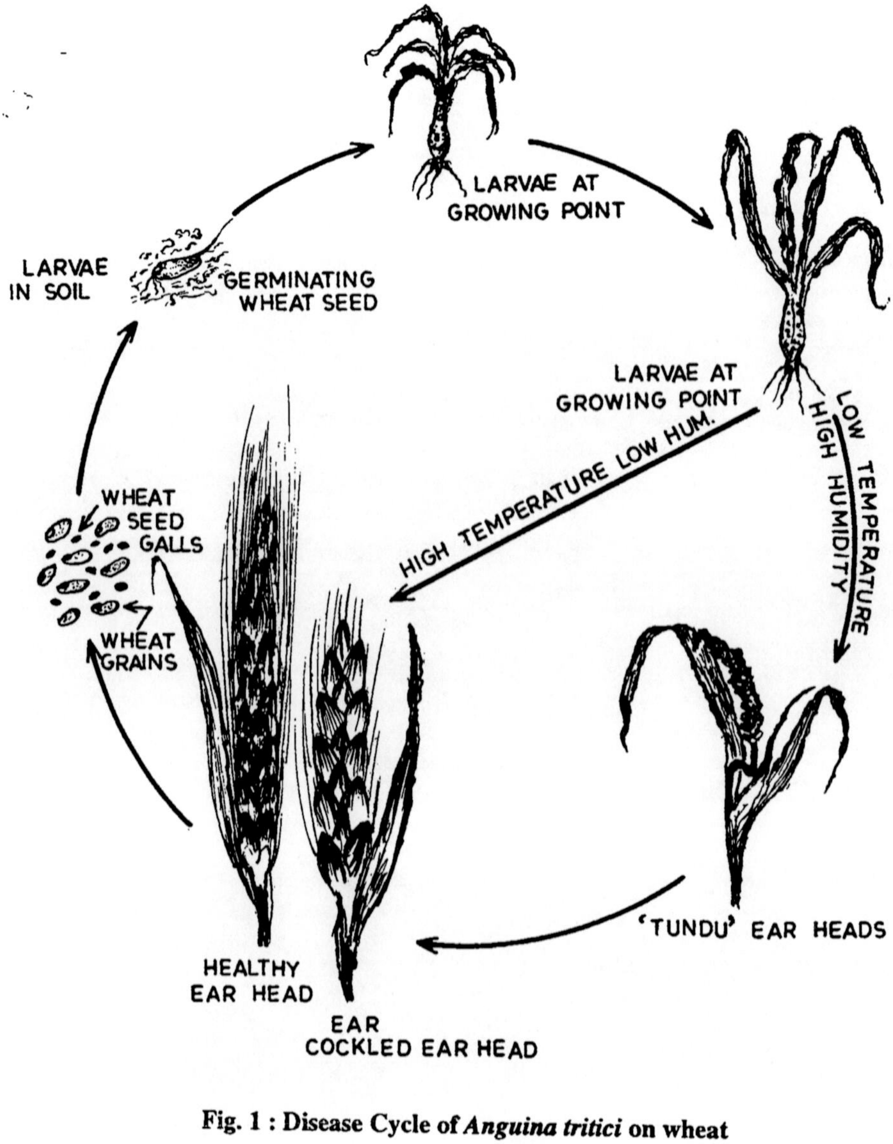

Fig. 1 : Disease Cycle of *Anguina tritici* on wheat

Plate 1A

Plate 1B

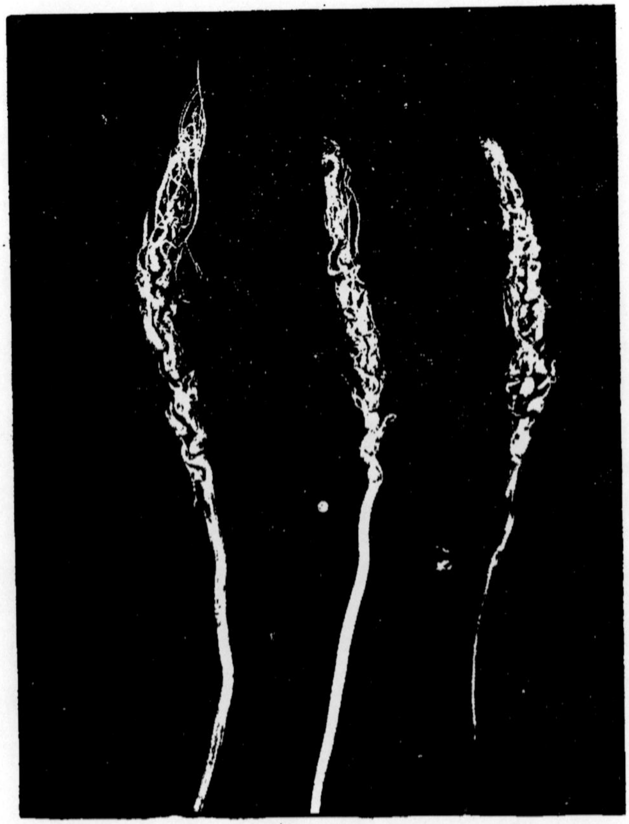

Plate 1C

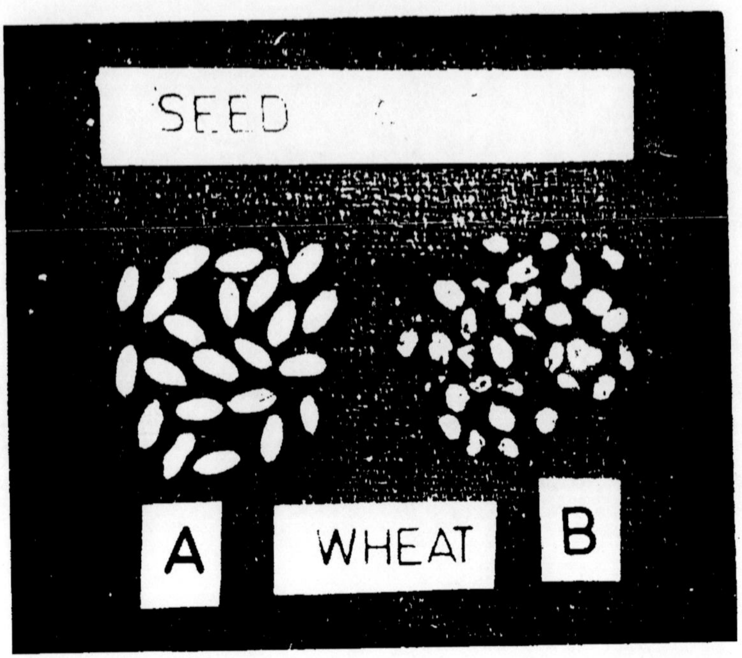

Plate 1D

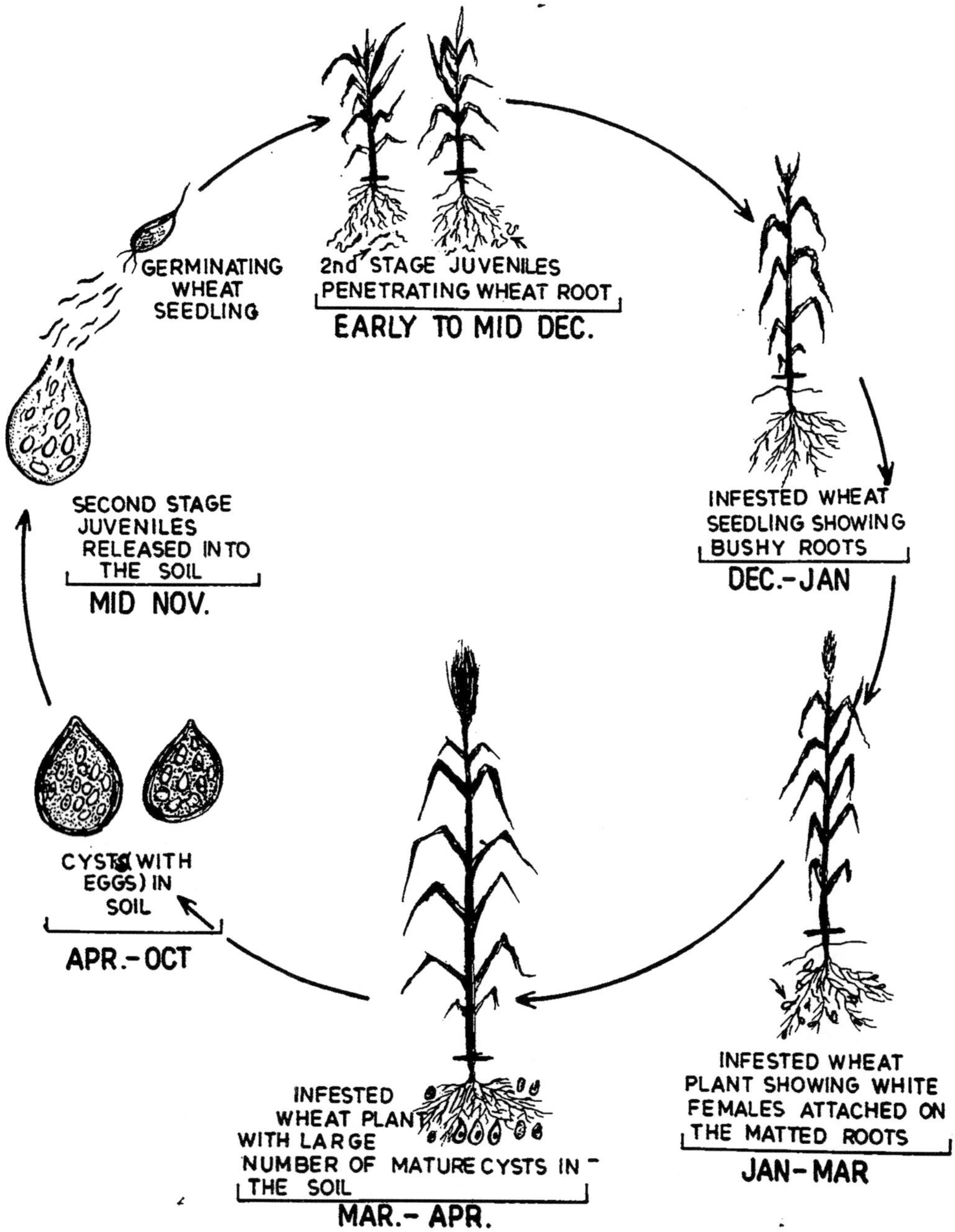

Fig. 2 : Disease Cycle of *Heterodera avenae* on Wheat

Plate 2B

6 Research, Development and Management for Production of Wheat

J. P. Tondon

CONTENTS

Integrated Pest and Disease Management
Rajeev K. Upadhyay, K.G. Mukerji, B.P. Chamola and O.P. Dubey (eds.)
APH Publishing Corporation, New Delhi-110002, India

ABSTRACT

This review article deals research and developmental aspects of wheat cultivation as onfarm technology. Both insect pests and diseases creating problems in kwheat cultivation from sowing till harvest have been identified and defined to draw the attention of the scientists, so as to develop an appropriate technology. Host plant resistance has been identified as thrust area for the suppression of pests and diseases.

1. INTRODUCTION

Wheat is world's most important food source and its production during 1989 was around 548 million tonnes which is about one fifth of the total food stuffs (dry weight basis) produced. The production of the two others important food grains namely, rice and maize was 509 and 449 million tonnes respectively during the same period. Wheat also provides for a major share of proteins in human diet and accounts for nearly 18% of total food protein sources of world including animal products which contribute about 25%. The major wheat producers during eighties have been China (87.88 mt) erstwhile USSR (84.56mt), USA (60.07 mt) and India (45.57mt).

In India wheat is the second largest food source after rice which is grown on more than 42 million hectares with production of around 75 million tonnes. Wheat is grown on nearly 24 million hectares with production of about 57 million tonnes. Three species of wheat are under cultivation in India. These are bread wheat *(Triticum aestivum)* macaroni wheat *(T. durum)* and emmer wheat *(T. dicoccum)*. Another species *T. sphaerococcum* is also reported to have been under cultivation in some parts earlier, but presently it is completely out of cultivation.

The country has recorded a rapid expansion in area, production and productivity of wheat which have gone up by 2.5, 8.9 and 3.6 times respectively during the period 1949-50 to 1991-92. The increase has been specifically spectacular since 1967-68 i.e. immediately after the introduction of semidwarf types in 1965 which spread very rapidly and changed the wheat scenario in the country. Consequently the country became self sufficient in food grain production. The per capita availability of wheat grains has markedly increased and has gone up from 65.7 gm/head/day in 1951 to 163.6 gm/head/day in 1991. Now considerable reserve stock of food grains (major part of which is contributed by wheat) are maintained for reasons of food security against climatic adversities and limited exports are also taking place.

The achievements recorded in wheat have been made possible due to a well organised and systematic wheat research programme; active extension support, governmental policies to ensure input supplies, credit facilities and remunerative support price for grains and enthusiasm of the farmers.

2. STATUS OF WHEAT RESEARCH

Wheat research carried out in India has been reviewed (8, 11) and the status with regard to specific aspects is as follows :

2.1 Varietal Improvement

2.1.1 Early Indian wheat varieties

Systematic research on wheat was initiated in the start of this century with the establishment of Indian (earlier Imperial) Agricultural Research Institute in 1905. Howard and Howard (4) made a collection of wheat types grown in India and systematically classified them into distinct forms on the basis of their various morphological and physiological characters. The Indian wheat types at that time were grouped as sorts on the basis of their place/region of cultivation/some very distinct grain quality feature/morphological plant characteristic and were named as **Sharbati, Safed Pissi, Lal Pissi, Chandausi, Darra, Mundia, Karachi choice. Hard red Calcutta** etc. in case of bread wheat; and **Bansi, Kathia, Khandwa, Malvi** etc. in case of macaroni type; and **Khapli, Poptia, Sambha** etc. in case of emer wheat type.

2.1.2 Improvement of local types

Several varieties have been introduced from Canada, America, England, Australia, Egypt and Italy during the end of 19th century aid beginning of this century (8). The exotic types were found unadapted to the Indian conditions and not suitable for direct introduction. Among these Ridley, Federation and Bena were released for cultivation in Hills. Ridley became fairly popular in hills because of its bold amber grain. However, several of exotic types possessed a high degree of disease resistance for which these were used as donors. Howard and Howard (4) were the first to make systematic attempts on varietal improvement in India. They made several pure line selections from the local 'sorts' and some of the important isolates were NP4, NP6 and NP12. Among these NP4 (earlier IP4) isolated from the parental stock 'Mundia' was found outstanding and became extensively popular, not only in India but also in several other countries such as Australia, South Africa, Rhodesia and Hungary (4, 5). Even now it is reported to be widely grown in Burma (Myanmar). Grains of NP4 were extensively used as parental stock as source for grain quality improvement in many countries. Similar research at the regional level lead to the development of type 8A, types 9D and type 11 in Punjab; K13 (earlier C13) of bread wheat and K53 and K54 of macaroni type of Uttar Pradesh and A0 13, A0 90 of *aestivum* type and Motia (Bansi 168), Gulab (Bansi 224) of *durum* type in Maharashtra and A 206 in Gujarat.

Fairly soon it was realised that hybridization work must be undertaken to combine desirable features from different pure lines as also exotic sources particularly for rust resistance. Some of the most notable tall wheat varieties developed and released till date are NP52, NP80-5, NP165, NP710, NP718, NP761, NP770, NP792, NP809, NP824, NP846, NP401 *durum* and NP 404 *(durum)* by Indian Agricultural Research Institute; PbC 518, PbC 591, PbC 273, PbC 281, C 306 in Pubjab; Niphad 4, Kenphad 28, Kenphad 32 and N59 (durum) in Maharashtra; Bijaga yellow *(durum)* Bijaga red *(durum)* and Amrut (*durum*) in Karnataka; K68 in UP; RS 31-1 in Rajasthan and Hyb 65 in Madhya Pradesh. Several of these such as C 306, N59, Hyb 65, K68 and A 206 are still under cultivation on large scale in rainfed areas of Central and peninsular India. Among these NP 809 is considered a landmark since for the first time a wheat variety combined resistance to all the three rusts prevalent in India and was derived from a complex cross involving five parents including two Indians and three exotics. C 306 is another outstanding achievement among the traditional tall types which is still grown on very large area under rainfed conditions as also under restricted irrigation water availability. All the NP wheats in the 700 and 800 series listed above combine resistance to rusts with resistance to loose smut (1, 8).

2.1.3 Development of semi-dwarf types

Need for developing varieties responsive to high doses of synthetic fertilizers was realized by the start of the fifties in view of the country's need to increase food production through intensive farming practices. The tall indigenous types have limited capacity to respond to the use of synthetic fertilizers and suffer severe damage due to lodging if attempts are made to use more than about 60 kg N/ha. Their yield potential is restricted to a maximum of around 3.5 tonnes/ha under most favourable condition under high risk of lodging. Attempts to exploit characters like straw breaking strength, weight per unit length of straw, number of vascular bundles, lignin percentage, silica content, length of lower internodes, length of sclerenchymatous fibres etc. did not prove very useful because of their highly inconsistent association with lodging and cumbersome measuring techniques (8). However, the awareness did help to develop varieties like NP 824 and Pbc 518 which have relatively lower lodging tendency. Similarly, efforts to breed for reduced plant height using sources such as *T. compactum, T., sphaerococcum* and some erectoid mutants did not prove useful since the height reducing genes in all these sources were associated with compact short clubbed heads and low yields (10).

Attention of the wheat workers was attracted by semi dwarf entries, observed at flowering stage during 1962, in the International Rust Screening Nursery received from USDA for evaluation in the 1961-62 crop season (9). These new types did not suffer from the disadvantages recorded earlier in dwarf types as these had perfectly normal lax long ears. Attempts to identify the source of origin of these materials lead to the Ford Foundation Programme on wheat improvement in Mexico headed by Dr. N. E. Borlaug. Hea had developed these materials using semi-dwarf winter wheat derivatives bred by Dr. C. A. Vogel of Washington state University, Agricultural Research Station, USA. Dr. Vogel had crossed a Japanese semi-dwarf variety Norin-10 with a local winter type variety Brewor and developed a series of semi-dwarf winter types, including the first semidwarf high yielding variety 'Gains'. Norin-10 in turn owed its origin to a Japanese semi-dwarfing source named 'Daruma' through its crosses with US winter wheat variety 'Fulz' followed by crossing of the derivatives of this cross with US variety Turkey red. Dr. Borlaug was requested to visit India, which he did in March 1963 and on return supplied bulk seed of four of his latest spring wheat varieties namely, Lerma Rojo 64A, Sanora 63, Sanora 64 and Mayo 64A along with 630 most promising improved breeding lines in advance generations (1). These materials formed the basis of development of semi-dwarf varieties programme in India.

Extensive evaluation of the four semi-dwarf varieties supplied by Dr. Borlaug showed that these could yield 30 to 50% more than the traditional tall types under well managed irrigated timely sown conditions. According to Dr. Borlaug the varieties supplied by him had a yield potential of around six tonnes per hectare and Lerma Rojo 64A did produce upto 6.8 tonnes/ha in a large scale demonstration plot (10). The yield levels of the semi-dwarf trials grown in 1963-64 were low since adequate agronomy for their cultivation had not been worked out. The best realisable yield of the traditional tall types is about 3.5 tonnes under adequate input use situations but under high risk of lodging. Lerma Rojo 64A and Sonora 64 were released for general cultivation in 1965. To popularises these varieties on an extensive scale, large qualities of seeds i.e. 250 metric tonnes in 1965 and 18,000 tonnes in 1966, were introduced directly from Mexico. The seeds available locally from trails/ multiplication were also rapidly multiplied and consequently, area under the semi-dwarf types increased rapidly. Wheat production got a boost when it jumped up from the earlier best of 12.26 million tonnes in 1964-65 to 16.54 million tonnes in 1967-68. This was taken as the point when the country entered the "Wheat revolution" phase which enabled it to become self sufficient in food grain production.

2.1.4 Improvement of semi-dwarf types

The first set of semi-dwarf varieties released namely, Lerma Rojo 64A and Sonora 64 were red grained and the processing and product qualities were not as per Indian consumer needs. It was realised that unless varieties with acceptable grain colour and processing quality were developed the wheat revolution may fizzle out. White grained forms of these two types i.e., Sharbati Sonora as mutant of Sonora 64 and Safed Lerma as backcross derivative of Lerma Rojo 64 A were developed and also two new semi-dwarf amber grain types namely, Kalyan Sona and Sonalika were released in 1967 (10). The later caught the attention of wheat farmers due to their very high grain yield potential, acceptable grain quality and wide adaptability. These varieties rapidly covered large areas all across the country and strenghtened the wheat revolution.

Since then an extensive wheat varietal development programme has been evolved using well adapted semidwarf types in crosses with local as well as exotic varieties. It was soon realised that in view of the extensive large area under wheat, cultivation of few genotypes all over the country can be disastrous and by taking this as well as several other factors in view the wheat growing regions of the country were grouped into zones, the constitution of which has been changed from time to time. At present five major wheat zones and a small sixth are recognised. Varieties are developed separately for each zone for different cultural conditions. A large number of wheat varieties have been released to meet varied needs of specific cultural situations. As a result the country is covered by a mosaic pattern of varietal distribution thereby helping to contain rust diseases and help improve average productivity continuously (Table 1).

2.2 Development of agronomical practices

Agronomic research has always formed an important component of wheat programme. Methods of seed bed preparation, sowing time, seed rate, row spacing, irrigation scheduling, weed management, fertilizer doses and application schedule to achieve high yield had been worked out for the traditional types (8). However, several of these technologies were found unsuitable for semi-dwarf types. It was realised very early that the seeding of semi-dwarf types has to be shallower than the traditional types to get good germination since these have shorter coleoptile length. Similarly, the new types could respond favourably to fertiliser doses of upto 120 kg/ha of N as against less than 60 kg/ha in case of the tall types. The new types also have a distinctly different developmental phenology needing specific adjustments in management practices such as irrigation and fertiliser scheduling. Being short in stature certain types of weeds pose more serious threat to a crop of semi-dwarf varieties. Under highly intensive cropping sequences, made feasible by low photosensitivity of semidwarf varieties, the soil nutritional imbalances pose threat to increasing productivity or even to maintain it at high levels. It was found that not only a more balanced use of major nutrients (NPK) is required but also secondary and micronutrients need to be taken care of and used wherever necessary. Work on all these aspects has been reviewed(11).

2.3 Crop protection

Wheat crop in India has always remained under threat from rust diseases of wheat. All the three rusts namely, leaf/brown rust *(Puccinia recondita)*, stem/black rust *(P. graminis tritici)* and stripe/yellow rust *(P. striiformis)* are important singly or in combination of two or all the three in different parts of the country. These have been responsible for severe damage to crop and even famines from time to time. The pioneering work of Dr. K. C. Mehta (6,7) in first half of this century had established the mode of perpetuation and existence of varied forms (races/pathotypes) of these rusts in India. This has helped to make rust resistance breeding an integral part of wheat improvement work from very early phases as a strategy for their control. Since extensive information on their racial pattern and mode of perpetuation is available and is being constantly updated (11), the containment strategy involves varietal diversification within as well as over regions to maintain a broad genetic base for disease resistance and continuous replacement of older types with new types carrying resistances of newer pathotypes which gain importance over time. This has helped to keep rusts under control and now for nearly two decades there has not been any major damage due to these diseases in any part of the country.

Table 1 : Centrally released popular/latest wheat varieties likely to continue to contribute to yield gains till the end of this century under varied irrigated situations in different parts of India

	States	Irrigated Timely sown condeion	Irrigated Late sown conditions	Rainfed Timely sown condition	Special Situations
1*	Punjab, Haryana, Western UP, Rajasthan, Plains Regions of Himachal and J & K.	HD 2329, HD 2009, HGD 2428 PBW 154, Raj 2077, WH 416 WH 283 CPAN 3004, WH 542 UP 2338, PDW 34(d) PDW 215 (d), PDW 233 (d) WH 896 (d)	HD 2285, PWB 226 HD 2270, WH 291 UP 2338, Raj 3077	PBW 175, PBW 299, PBW 65, DL 153-B	Salt affected soils KRL 1-4, Raj 3077 PBW 65. January sowing : HD 2285, HD 2402
2.	Eastern UP, Bihar, West Bengal, Assam, Orissa	JUW 206, HP 1102, UP 2+62 HD 2402, K. 8604, K 9004 DL 784-3	HUW 234, UP 1209 HD 2307, HJP 1633	C 305, K 8027	Salt effected soils same as above Rainfed late sown : HDR 77, K8962 Jan. Sowning : HD 2235, HD 202
3.	Madhya Pradesh, Gujarat Kota, Region of Rajasthan, Jhansi Div. of UP	HD 2236, Hl 1077, GW 190 DL 803-3, Raj 155 (d) H1 8381 (d)	Swata, J 405 GW 173	C 306, Sujata A 9-30-1 (d) A 206 (d)	Salt effected area : Same as above
4.	Maharashtra, Karnataka, Andhra Pradesh	HJD 2189, HD 2380, DWR 39 MACS 2496, DWR 162	H1 977, HD 2501 QHD 2610, DWR 195	Nl 5439, N 59 (d) MACS 1967 (d) Bijge Yellow (d)	
5.	Himachal Pradesh, UP Hills, J & K.	HS 240, HD 2380, UP 1109 HB 208		HS 240, HD 2380, UP 1109, HB 208	Very high hills : HPW 42 Rainfed early sown : VL 606, HS 277, Rainfed late sown : HS 207, HS 295.

*1. North Western plains zone; 2. North Eastern plains zones; 3. Central zone; 4. Peninsular zone; 5. Northern Hills zone
**(d) -durum wheat variety.

Other important diseases of wheat in India are loose smut *(Ustilago nuda* var. *tritici)*, Karnal bunt *(Neovosia indica)*, hill bunt *(Tilletia foetida)*, powdery mildew *(Erysiphe graminis tritici)* foot rot *(Helminthosporium, Sclerotium, Fusarium* and *Pythium* spp,). and flag smut *(Urocystis agropyri)*. Among these loose smut, Karnal bunt and foliar blights do considerable damage over large areas. Foliar blights have been found to be gaining importance over a wider range of situations. Previously their occurrence was severe in the north eastern regions only. Now these are occuring more vigorously even in other zones specifically the north western. Rice-wheat rotation appears to have some impact on the more severe occurrence of this disease but the exact relationship still requires to be worked out. Similarly powdery mildew is gaining importance. For control of diseases of sporadic and restricted importance emphasis has been given to use fungicides except in the case of loose smut and Karnal bunt where work on resistance breeding has been carried out. The varietal improvement strategy has been carried out. The varietal improvement strategy has been very helpful in the containment of Karnal bunt also but in case of loose smut work is restricted due to tedius screening procedures, although genetically the diseases is well understood. Recently an increased emphasis on resistance breeding has been given to foliar blights and some reliable sources of resistance has been identified for use.

Other biotic problems of wheat crop include ear cockle *(Anquina tritici)* and Molya/root knot nematode *(Heterodera avenae)* among nematodes and aphids *(Macrosiphum miscanthi)*, shoot fly *(Atherigona* sp.), brown wheat mite *(Petrobia latens)* and several polyphagoup insects among insect pests. However, their occurrence is extremely sporadic and restricted to small regions. Recently aphids have started posing more serious threat over wider area. Work on these aspects has been reviewed (11). All these problems are controlled through the use of chemicals wherever the incidence occurs since resistant varieties are not available. Recently some sources of resistance to Molya have been identified and are being exploited in breeding. Similarly attempts are in progress to identify sources of resistance/ tolerance to the insect pests listed above.

2.4 Research on other aspects

Research in other disciplines such as genetics and cytogenetics, physiology, soils, cropping system, economics etc. is directed to specific issues related to understanding the fundamentals of functioning of the wheat plant in relation to its production potential and with the objective of manipulating the production and productivity aspects (8, 11). The geneticists have worked out procedures for selection of parents to be involved in crossing programmes to maximise the probability of occurance of superior yielding segregates combining several desirable features. Inheritance of genes governing resistance to rusts and other diseases has been worked out. Genes controlling various other features of the plant have been studied and location on different chromosomes have been mapped. Several other cytogenetical and cytological issues have also been investigated. Physiologists have paid specific attention to heat and drought tolerance and phenological development of the plant. Recently research on problems associated with the rice-wheat sequence has been intensified since it is feared that the system is not sustainable in the long run and it accounts for a fairly large area in our most productive region of the north western plains. Work on biotechnological aspects is also being intensified.

3. ORGANISATION OF COORDINATED WHEAT RESEARCH

Till the beginning of sixties most wheat research was carried out by individual scientists or groups of scientists in each disciplines working within an organisation. Interaction among scientists in different research disciplines, as also different organisations, was very limited. In 1961 a group of wheat breeders, working with different organisations decided to cooperate with pathologists and agronomists on a voluntary basis in the conduct of commonly constituted varietal yield evaluation trials. Later All India Coordinated Wheat Improvement Programme (AICWIP) was formally, established in 1965 when financial support to several cooperating centres was approved by the Indian Council of Agricultural Research (ICAR) as part of its activities during the third five year plan (1961-65) and a full time coordinator was appointed. The project adopts an inter disciplinary approach in resolving various production problems of the wheat crop.

The AICWIP organise more than 700 trials every year to identify superior varieties as also production and protection technologies. These include more than 500 varietal evaluation trials involving about 850 strains. These trials are organised in co-operation with about 600 test sites located all across the five major wheat zones based on variations in climatic conditions, diseases, soil types and cropping systems (Table 1) Attempts are made to identify region/location specific varieties adapted to specified cultural situations of each zone such as irrigated, rainfed, timely sown, late sown etc. Special trials are organised to identify specific varieties for special problems through separate trial series for situations like salt effected soils diara lands, very late sown (January) irrigated conditions etc. Major emphasis is on resolving, as many as possible, various crop production and crop protection issues through varietal development for which the various research disciplines co-operate as inter-disciplinary teams. Every year more than 10,000 fresh cross combinations are made and 100,000 progeny rows in various generations are studied and nearly ten thousand new bulks are made in advance generations jointly by cooperating breeders located at different research centres. All new bulks are critically evaluated in station trials and screening nurseries before entry into coordinated testing. Every year nearly 450 fresh entries are inducted into testing programme. The existing set up has proved extremely useful and helped to find answers to most wheat production problems. Since the establishment of AICWIP in 1965 a total of 150 improved wheat varieties have been developed and released for general cultivation in different parts of the country. Presently almost all irrigated wheat area is under varieties developed by this project.

The research system has responded very well to resolve new threats to wheat production which have emerged from time to time. These include changing racial patterns of rusts which have been kept under check through regular replacement of susceptible types with new resistant varieties. Now for nearly two and half decade there has not been any serious epidemic capable of causing major damage. Similarly, the Karnal bunt disease, which posed a major new threat in the seventies after the release of semi-dwarf types, was controlled through development of tolerant varieties and proper crop management practices. *Phalaris minor,* a grassy weed, which became very important after the introduction of semi-dwarf types was brought under control through appropriate crop management practices and development of chemical control measures.

4. WHEAT PRODUCTION TECHNOLOGY

The latest wheat production technology recommended for increasing wheat productivity under varied cultural situation include the following important aspects.

4.1 Varieties

Since 1965 when the first set of high yielding cultivars was released 150 varieties have been approved by the central varietal release committee. Some additional varieties have also been released by state committees. These are suitable for cultivation under varied agroclimatic conditions. A list of varieties currently relevant for cultivation in different zones and under varied situations are given in table 1.

4.2 Cultural practices

A well prepared seed bed with adequate moisture in the seeding zone is an essential pre-requisite for obtaining good plant stand. Sowing of irrigated crop is recommended when average daily temperature falls below 22-23°C which is normally sometime in November all across the country. The rainfed crop is required to be sown at the end of October or even in middle of October in central and peninsular India to take advantage of the residual moisture from monsoon rains to get proper crop stand. The varieties recommended for rainfed sowing are specifically adapted to withstand relatively higher temperature at seeding time and under these situations only such varieties should be used. Sowing of irrigated wheat can be continued even as late as end of December or beginning of January. However, irrespective of the varieties used the potential yield is always lower under later sowing as compared to timely sowing. It is estimated that roughly there is a loss of one percent in yield potential per day of delay in sowing beyond 15th of November and the reduction becomes more drastic after middle of December. The choice of proper varieties can only help reduce the rate of loss in yield potential under later sowing. A seed rate of 100 kg/ha under timely sowing and 125 kg/ha under late sowing is recommended in the major wheat area in the north western parts of the country but 125 kg/ha under normal sowing and 150 kg/ha under late sowing in the north eastern and peninsular zone. Seed rate should be increased for bold seeded varieties. A row spacing of 22.5 cms for timely sowing and 15-18 cm for late sowing is recommended. Under late sown conditions cross sowing is also recommended.

A fertilizer application @ 120 kg N and 60 kg P_2O_5 is recommended as a blanket dose and patash to be used only if soil tests show its deficiency. For rainfed situations 40 kg of N and 20 kg P_2O_5 per hectare is considered optimum. Other elements such as Zn, S and Mn may be required to be applied under intensively cropped situations-particularly in light soils under rice-wheat rotation. Recommendations for number of irrigations varies from four to six and are required to be scheduled according to crop growth stage and adjusted according to seasonal rainfall. The first irrigation is to be applied at the crown root initiation (CRI) stage followed by irrigations at tiller completion, late jointing, flowering, milk and dough stages. In case of limited water availability the irrigation at CRI stage must be applied which could be followed by watering at the boot stage. If third irrigation can be given it should be applied at milk stage.

When weed problem is too severe to be controlled by mechanical methods, use of weedicides is recommended. For the control of broad leaved weeds use of 2,4-D @ 0.4 kg a.i. as post emergence spray is recommended. For *Phalaris minor,* Isoproturon @ 0.75 kg/ha after 30-35 days of sowing or pendimethalin @ 1 kg a.i./ha 2-3 days after sowing is recommended.

4.3 Crops protection

4.3.1 Diseases

The main thrust is to control diseases like rusts through breeding for resistance and varietal diversification. Replacement of older varieties, as soon as these becomes susceptible, with the latest improved varieties, which are always resistant to prevalent rust pathotypes at the time of identification, is recommended. Resistant/tolerant varieties are also available in case of loose smut and Karnal bunt where efforts have been made to breed such types.

For the control of loose smut use caboxin @ 2.5 gms/kg of infected seed before sowing, special care of seed plots is recommended. A rigorous careful rouging alone can ensure disease free seed for the next crop. If the seed is ensured to be free from infection there is no danger of loose smut damage to the crop raised from it. Karnal bunt can be kept under check through the use of tolerant varieties and seed sanitation. Appropriate crop management practices such as timely sowing, avoiding irrigation at anthesis and crop rotation help to contain this diseases. Hill bunt and flag smut are also controlled by the use of carboxin, while sulfur dust or karathane spray provide effective control of powdery mildew. Dithane M 45 is affective in the control of leaf blight damage whereas MEMC and Brassicol use can help in controlling foot rot.

4.3.2 Insect pests

Aldrin for the control of termites, cypermethrin for shoot fly, formithion, oxydemeton methyl and phosphomidon for brown wheat mite, and dimethoate or phosphamidon for aphids are recommended.

4.3.3 Nematodes

In case of nematodes, ear cockle can be easily controlled through seed sanitation by removal of galls by physical means since this is the only mode of its perpetuation. In the case of root cyst nematode summer ploughing, rotation with non-host crops and use of some nematocides is recommended.

4.4 Integrated wheat crop health management

As given earlier the wheat plant from seed to seed is under threat of damage from rots, blights, rusts, mildews, bunts, smuts, etc. caused by fungi, insect pests, nematodes, rodents and birds and interference from weeds. Soil physical, chemical and biological properties effect favourably or adversely the wheat plant health (2). The latter can also influence the prevalence of the problems caused by disease pathogens and insect pests by making plant more prone or tolerant to these organisms. Management of the health of the wheat plant for getting higher yield from the crop can be done through an integrated management of cultural practices, soil chemical and biological properties, genetic resistance to diseases and pests, use of well adopted varieties, use of disease free quality seeds, exploitation of beneficial micro-organisms and insect pests to minimise the use of fungicides, insecticides, nematocides etc.

For raising a healthy crop use of well adopted appropriate variety, and healthy seed, timely sowing, use of balanced doses of fertilisers including micronutrients wherever needed and proper irrigation scheduling are essential. In addition there may be a need for use of chemicals to control diseases and pests. Wheat is probably one of the examples where the principles of IPM are being exploited fairly well. Except for weedicides the use of fungicides and insecticides is minimum. This is primarily due to the fact that a major effort is being devoted to exploit genetic resistance to control its most important disease problems. Rusts are being kept under check primarily through genetic means and use of fungicides is not recommended for their control. Growers are encouraged to replace the older varieties, which become susceptible in the course of their large scale cultivation through the evolution/multiplication of new pathotypes with the latest improved types which are continuously being developed through an active and vigorous breeding effort and critical racial pattern monitoring. The strategy has helped to keep the rusts under control over the last more than two decades. Other diseases for which efforts are in progress to control through genetic means are loose smut, powdery mildews and Karnal bunt for which strong sources of resistance have been identified and are being used in development of resistant varieties.

A healthy and vigorous crop which is essential for high yields and raised with the adoption of all recommended practices, is also subjected to more severe attack by rusts, powdery mildew, Karnal bunt, foliar blights and aphids since a dense growth leads to more humid inner crop atmosphere which is congenial for fungal growth. Such a crop needs more critical monitoring and specific control measures. It is, therefore, absolutely essential to grow resistant varieties under the high yielding environments. Fortunately highly resistant varieties are available for rusts, powdery mildews and Karnal bunt. Other practices which can be helpful in reducing or controling various diseases are as follows :

4.4.1 Crop rotation

A long crop rotation, where wheat is grown once in few years, helps to control soil borne diseases such as foot rot, Karnal bunt, foliar blights and flag smut. Root cyst nematode can also be controlled by growing non-host crops during the intervening period. It is also recommended for the control of obnoxious weeds like *Phalaris minor*. However, the land constraint and need for growing wheat every year in the same fields makes the practice almost ineffective and is adopted only under very serious situations.

4.4.2 Clean field cultivation

Inoculum of several diseases such as foliar blight, foot rot, Karnal bunt, flag smut, and some insect pests survive in plant residues. Removal of straw or its burning in the field can help reduce spore load but can have adverse effects on field nutritional balance. However, straw removal is a regular practice in India and few farmers incorporate crop residue into the soil.

4.4.3 Summer ploughing

Ploughing dry fields during the hot dry months of May and June can help to reduce the incidence of diseases like Karnal bunt, flag smut and root rot. It is a recommended practice for the control of root cyst nematode. It also helps to reduce the incidence of weeds and some insect.

4.4.4 Use of clean and healthy seeds

Pathogen free help in reducing the incidence of seed borne diseases. This is recommended for the control of loose smut, hill bunt and are cockle nematode.

4.4.5 Timely sowing

Sowing during the most appropriate time which for irrigated crops is around middle of November in most parts of the country, helps to reduce the incidence of root diseases, Karnal bunt and shootfly. The late sown crop is more liable to the attack of Karnal bunt and shootfly whereas early sown crop may suffer more from root diseases. It is also helpful in keeping weeds under check.

4.4.6 Irrigation management

Irrigation during flowering tends to increase the incidence of Karnal bunt by creating congenial conditions for spore germination and infection so its avoidance at this stage helps control this disease. Excessive irrigation and temporary water logging increases the incidence of downy mildews so its avoidance helps to contain these.

4.4.7 Use of biological agents

Several diseases, insect pests and weeds can be controlled with the use of biological agents like parasitic fungi, bacteria, viruses and insects. For example *Trichoderma* and *Gliocladium* spp. can prove effective for the control of Karnal bunt, *Penicillium* and *Trichothecium* spp. can parasitise rusts postules. Lady bird beetle can help control aphids. Similarly some organisms can be used for the control of weeds. However the efficacy of their use is highly dependent on climatic factors and their population built up. These factors have not yet been fully investigated and research is needed on these aspects to take full advantage of their beneficial role.

An integrated use of above factors can help to save wheat crop from various diseases, pests and weeds. However, when their population level is beyond a particular level the use of chemical control measures become essential. There are also situations where the use of chemicals may be much easier than breeding for disease resistance. Loose smut is one example where use of fungicides in preliminary seed production stages such as breeders/nucleus seed stage where the quantities involved are small and avoidance of infection thereafter is effective in keeping this disease under control. As such a balanced use of various methods including the use of fungicides would be essential for maintaining the wheat crop in good health.

5. STRATEGY FOR SUSTAINING/INCREASING PRODUCTION TRENDS

The rapidly increasing Indian population, which is expected to reach one billion by the turn of the century necessitates the maintenance of continuous high increasing trend in wheat production. It is estimated that by the year 2000 India would need 225 million tonnes of food grains to maintain the existing food availability level of 540 gms/head/day. One third of this quality i.e. 75 million tonnes is expected to be contributed by wheat. The easy options of increasing wheat production by bringing more area under the crop, replacement of traditional low yielding varieties of wheat with high yielding semi-dwarf types and increasing proportion of area under irrigated wheat do not exist since these have been already exploited to a fairly large extent. All future increases in wheat production have to come through productivity enhancement involving increased and efficient use of inputs and more precise technological innovations.

The productivity potential of the latest improved semi-dwarf varieties under well managed adequately fertilized timely sown irrigated conditions is estimated to vary between 6.5 tonnes per hectare in the cool and relatively longer winter season is the north western plains to 4.5 tonnes per hectare in the peninsular zone where winters are warmer and crop duration is much shorter. The potential is considerably lower under late sown conditions in all zones. Other restrictive environments include poor quality soils such as shallow, water logged, salt affected with inferior physical and biological properties etc. However, to realise the full production potential of the high yielding types the following issues need to be kept in view.

5.1 Maintaining the existing genetic potential of varieties

The existing varieties are constantly under threat of loosing their yield potential through attack from (i) newer forms of pathotypes/virulence of important diseases namely, rusts; (ii) possibility of lesser important diseases gaining importance due to changed field ecology and/or inadequate attention in screening programmes resulting in release of varieties with extremely high susceptibility to these, (iii) development of more aggressive forms of insect pests which have already started attacking the wheat crop and (iv) degeneration of seed stock through admixtures, (v) genetic slipage/drift due to inadequate yield evaluation tests of the new improved varieties. These issues necessitate that an aggressive and viable varietal improvement programme is continuously maintained through interdisciplinary co-operation.

5.2 Enhancing the genetic productivity potential

As stated earlier there is still a large gap between the realisable potential and actually realised yields in all regions and bringing this gap will help to sustain increases in wheat production for a fairly long time. However, a situation is arising in some areas, particularly in the north western parts, where a large number of farmers are harvesting virtually the full exploitable yield potential such as Punjab and Haryana. Since such farmers play a major role in sustaining high production levels, it is important that varieties with enhanced productivity potential are produced. Recently CIMMYT has reported aa 10% gain in productivity of latest wheat varieties due 15 the induction of 1B/1R translocation. Similar gains can be expected in the future also.

5.3 Emerging cultural constraints

In the cropping systems involving the growing of more than one crop (some times 3 to 4 crops) within a period of one year and also the use of imbalanced fertilizers results in rapid depletion of some of the nutrients which are not included or are used in less than required quantities i.e. P and K. The crop grain harvest and other biomass removed from the fields takes away large quantities of these elements resulting deficiencies of these elements and reduced yield levels. It is estimated that every tonne of grain produced has 1.5 to 2 tonnes of straw. Removal of one tonne of biomass takes with it about 6-7 kg nitrogen, 1.5 to 2 kg of phosphorus, 12 kg potash, 2.5 kg sulphur, 4 kg calcium, 2.5 kg magnesium and certain amounts of trace elements such as zinc, copper, iron, magnese, boron and molybdenum (2). In case of high yield levels in intensive agriculture where more than 4.5 tonnes of grains are being harvested from each of the crop within a year such as rice and wheat grown in sequence in north western parts of the country, these effects can be very large and a careful watch is, therefore, required to be maintained on these issues. Application of all required mineral nutrients in appropriate forms is essential to ensure long term sustainability of the systems.

5.4 Other issues of concern

5.4.1 Economics

Farmers grow wheat with the objective of earning returns. Thus the extent of use of inputs is highly influenced by their cost as compared to wheat prices. All farm operations, energy sources and inputs cost money which is required to be earned from the produce in addition to profits and therefore, the wheat prices must go up correspondingly to keep wheat cultivation remunerative (3).

5.4.2 Environmental issues

Use of fertilizers, weedicides, fungicides etc. poses issues of environmental concern. Some of these not only built up in soils over time but also leach out and pollute other natural resources such as underground and above ground water reservoirs resulting in human health problems. All these issues need to be watched carefully.

5.4.3 Governmental policies

The Governmental policies have a great impact on wheat production. Subsidising inputs and making them available in time for proper use, and fixing remunerative prices of wheat grains are issues of governmental concern and involve policy decision as to weather a country should attain self sufficiency or import food grains. The implementation of General Agreement on Trade and Tariffs (GATT) and establishment of World Trade Organisation (WTO) are likely to result in liberalized import/export policies which may effect production systems drastically. The wheat production is likely to be affected by world market prices to a larger extent. Cheap international wheat prices may result in larger imports and lesser production and *vice versa.*

6. CONCLUSIONS

Wheat research in the country is well set to take care of the production needs of the current century as also early parts of the next century. Technology is available to maintain a steady and rapid growth rate of increased production provided adequate supply of inputs is ensured and wheat grain prices are kept remunerative. The future yield increases would require more skilful management of the inputs used as also the entire production system. Maintaining a rapid rate of replacement of older varieties with the latest resistant types, increased use of balanced fertilizers, adoption of appropriate methods of application to enhance their efficiency, use of micronutrient wherever needed, more efficient crop husbandry practices to ensure timely sowing, proper field bed preparation, weed control, proper management of available water resources as also to enhance its availability inadequately irrigated areas etc. are some of the components which can ensure maintenance of high increase in productivity trends in the future.

The issues related to various economic aspects such as cost of wheat production and marketing of wheat grains would require to be monitored carefully. With the establishment of World Trade Organisation (WTO) and the expected more liberal world trade policies, wheat production may get a positive or negative thrust depending upon various features. Relatively lower production cost compared to world market could lead to larger exports and increased wheat production.

Germplasm collection and maintenance would need very special attention since with the introduction of breeders rights/intelectual property rights and variety/gene patenting, easy flow of germplasm may decline. So far exchange of germplasm has been extremely free and has been one of the most important factors for rapid progress in wheat improvement programmes. Restrictions on movement of germplasm can have adverse effect on varietal development programmes. For the same teasons activities of the international agricultural research organisations, which have been providing extremely valuable support in this respect to National Research Programmes of developing countries, would get restricted and more reliance would have to be placed on local resources and know how.

7. REFERENCES

1. Aggarwal, R. K. 1986. Development of improved varieties. Twenty-five years of Coordinated Wheat Research 1961-86 WPD, IARI- pp. 34-93.

2. Cook R. J. and Veseth, R. J. 1991. Wheat health management, APS Oris, The Am. Phyto. Society, USA.

3. Derek, Byerlee 1991. Economic issues for wheat research in South Asia, 29th Annual Workshop AICWIP.

4. Howard, G.L.V. 1910. Wheat in India, its production, varieties and improvement. Thacher Spink & Co. Calcutta, 288 p.

5. Howard, G. L. V. 1928. Improvement of Indian Wheat, . A brief summary of investigations carried out at Pusa from 1905 to 1924 including an account of the Pusa Hybrids. Bull. Pusa Agric. Res. Inst. **171**:26.

6. Mehta, K. C. 1931. Cereal rust problem in India. Ind. J. Agri. Sci., **1** : 702-705.

7. Mehta, K. C. and Pal, B. P. 1940. Rust resistant wheats for India. Nature Land, **146** : 98.

8. Pal, B. P. 1966. Wheat, Ind. Coun. Agric. Res. Pub.

9. Swaminathan, M. S. 1965. Impact of dwarfing genes on wheat production. Journal of Post Graduate School, IARI, **3 (1)**: 57-62.

10. Swaminathan, M.S. 1993. Wheat Revolution-A Dialogue. Macmillan India Ltd. Pub.

11. Tandon, J. P., and Sethi, A. P. 1986. Twenty Five Years of Coordinated Wheat Research, WPD, IARI, 287 p.

7 Maize Disease Management in India

R. C. Sharma

CONTENTS

Integrated Pest and Disease Management
Rajeev K. Upadhyay, K.G. Mukerji, B.P. Chamola and O.P. Dubey (eds.)
APH Publishing Corporation, New Delhi-110002, India

ABSTRACT

In India, maize is an important crop. About 65% of the total produce is consumed as human food while the balance is utilized in industry and poultry feed. The average productivity of the crop in India is low and about 1.7 t/ha. Among the factors adversely affecting productivity, ubiquitous incidence of diseases in the pre-harvest stage is prominent. Of the 61 diseases of maize recorded so far, 16 play major role in limiting production. The major diseases are : foliar diseases, stalk rots, downy mildews and sheath diseases. Information on ear, cob and kernel rots, smut and viruses is given. An account of post harvest microbial problems is presented. Appropriate disease management strategies have been outlined under respective disease groups. The most important approach is considered to be the host resistance followed by chemical control. Information on adoption of certain agronomic/cultural practices in the management of diseases is also provided.

1. INTRODUCTION

In the Indian agriculture, maize occupies an important place. It is not only utilized as a staple food by the lower strata of the society, but is also a crop par excellence for industrial use. Human food accounts for about 65% of the produce while the balance finds use in industry and poultry feed. During the last few years, there has been a progressive escalation in demand for maize grain for the value-added products, like glucose, sorbitol and dextrose, besides live stock, poultry and animal feed and manufacture of starch and starch based products. During the crop years 1992-93, maize was sown in 5.8 million hectare giving total production of approximately ten million tonnes of grain. As compared to the other maize growing countries of Asian region-China, Korea, Thailand etc., the average productivity of maize in India is quite low (1.7 t/ha) (5).

Among the factors adversely affecting productivity, ubiquitous incidence of diseases in the pre-harvest stage is prominent. Since there is practically no possibility of increasing maize area, the productivity can only be raised significantly by providing seed of improved cultivars, better agronomic practices and protection against diseases and pests.

Based on research efforts for the last three decades or so under the aegis of All India Coordinated Maize Improvement Project, 16 out of 61 diseases adversely affecting this crop have been identified as major ones (31).

The potential crop loss has been estimated individually for 10 diseases (1, 12, 29, 41, 45). *Pythium* stalk rot, bacterial stalk rot and *Peronosclerospora*-incited downy mildews possess the highest potential to reduce maize yield. The total loss in economic product of the crop due to diseases has been estimated to be of the order of 13.2% (31). Based on 1992-93 production figures the loss due to diseases in grain yield comes to 1340000 tonnes per annum

2. MAJOR DISEASES AND THEIR DISTRIBUTION

A listing of economically important diseases alongwith their distribution pathogen involved, plant parts affected, extent of yield loss and mode of survival is given in Table 1. These include diseases which occur in both tropical and temperate environments as delineated by Renfro and Ullstrup (35).

2.1 Seed and seedling blights

These diseases pose a problem in reducing plant stands in temperate areas where soil temperatures are low during planting time. This group of diseases does not constitute a serious threat in the major tropical environments of India because of rapid emergence of seedlings. However, fungicide seed treatment has demonstrated a definite improvement in plant stands and seedling vigour, which was reflected in yield increases of 8-10% (25).

Although a variety of pathogens are associated with seed rots and seedling blights including *Pythium, Fusarium, Acremonium, Sclerotium*, etc., no effort has been made towards the development of resistant cultivars because fungicide treatment is a simple and cheap method of protection against this group of diseases.

2.2 Foliar diseases

About 20 foliar diseases are known to occur on maize in India (30). However, Turcicum leaf blight, Maydis leaf blight and Common Rust are considered to be important in terms of geographical distribution and potential to cause significant yield reductions.

Turcicum leaf blight

This disease is prevalent in the areas where maize is planted in the highlands, where temperate conditions prevail and in winter plantings in the plains as the cool/moderate conditions favour disease development.

Both quantitative and qualitative types of resistance operate against this disease (16). The quantitative or polygenic resistance controls the number but not the size of the lesions on plants, the qualitative or specific in expressed as chlorotic lesions (16). Four single dominant to completely dominant genes (Ht_1, Ht_2, Ht_3 and HtN) have been identified to confer chlorotic lesion type of resistance. Based on their virulence on these genes four races designated as Race 1, Race 2, Race 2-3 and Race 2-3 N have been identified (26). In India only Race 2 has been recorded (33). Role of phytoalexins and cyclic hydroxamic aids has also been demonstrated in the expression of resistance. The gene designated as *B*x appears of influence the expression of the Ht_1 gene regarding production of cyclic hydroxamic acid. DIMBOA (12).

Maydis leaf blight

This disease is distributed all over the world in warm temperate to tropical (20-30°C) maize producing areas. Three races of the pathogen, Race C, Race 0 and Race T have been reported so far. Race 0 and T differ in symptoms produced, cytoplasmic specificity, production of toxins, optimum growth temperature, reproductive rates and plant parts attacked (14). Three types of resistance are known to occur in maize to this pathogen. These are qualitative and quantitative nuclear genes and the other influenced by the cytoplasm. The chlorotic lesion type of resistance reported from Nigeria (6) is expressed as small, circular chlorotic lesions supporting little sporulation. This specific reaction was subsequently reported to be controlled by two-linked recessive genes. A single dominant gene was identified in some lines in Illinois, USA and was designated as rhm, which is located on short arm of chromosome 6.

Resistance to race T is conditioned by both cytoplasmic factors and nuclear genes, cytoplasmic component being the most important (15). Race T produces a phytotoxin (HM-T toxin) that is specific to plants having texas male sterile type of cytoplasm (CMS-T). Race C found in China has a specific virulence on cms-C cytoplasm. Larger lesions are produced on leaves of plants with cms-T, cms-S or N-cytoplasm in both seedlings and adult plants.

Table 1 : Important diseases of maize

	Disease	Pathogen	Plant parts affected	Extent of yield loss	Distribution	Mode of survival
1.	Turcicum leaf blight	*Exserohilum turcicum*	Leaf	66%	Jammu and Kashmir, Himachal Pradesh, Sikkim, West Bengal, Meghalaya, Tripura, Assam, Rajasthan, Uttar Pradesh, Bihar, Madhya Pradesh, Gujarat, Maharashtra, Andhra Pradesh, Karnataka, Tamil Nadu.	Soil, Plant debris
2.	Maydis leaf Blight	*Drechslera maydis*	Leaf, leaf sheath, Ear dusk	30%	J&K, HP, Sikkim, Meghalaya, Soil, Plant Punjab, Haryana, Rajasthan, Delhi, UP, Bihar, MP, Gujarat, Maharashtra, AP, Karnataka, T.N.	Soil, Plant
3.	Common rust	*Puccinia sorghi*	Leaf, Leaf sheath, Ear husk, Tassel	32%	J&K, HP, Sikkim, West bengal, Punjab (winter), Haryana, Rajsthan, Delhi, UP, Bihar, MP, Gujarat, Maharashtra, AP, Karnataka, T.N.	Uredospores
4.	Phaeosphaeria leaf Spot	*Phaeosphjaeria maydis*	Leaf	-	Sikkim, WB, UP, HP, Maharashtra	Not known
5.	Chrysanthemi stalk rot	*Erwinia Chrysanthemi* pv. *zeae*	Stalk	100%	HP, Sikkim, WB, Punjab, Haryana, Rajasthan, Delhi, UP, Bihar, MP, AP.	Soil
6.	Pythium stalk rot	*Pythium aphanidermatum*	Stalk	100%	Sikkim, HP, WB, Punjab, Haryana, Rajasthan, Delhi UP, Bihar	Soil
7.	Fusarium Wilt and stalk rot	*Fusarium moniliform*	Stalk, Ear	-	Rajasthan, UP, Bihar, AP	Soil
8.	Late wilt	*Cephalosporium maydis*	Stalk, Ear	50.9%	Rajasthan, HP, Bihar, AP	Soil, Seed
9.	Charcoal rot	*Macrophomina phaseolina*	Stalk, Ear	39.5%	JK, WB, Pubjab, Haryana, Delhi, UP, MP, AP, Karnataka, TN.	Soil, Oospores
10.	Brown stripe downy mildew	*Sclerophtora raysiae* varzeae.	Leaf	63%	HP, Sikkim, WB, Meghalaya, Punjab, Haryana, Rajasthan, Delhi, UP, Bihar, MP, Gujarat,	Soil, Oospore
11.	Philippine downy mildew	*Peronosclerospora philippinensis*	Systemic	100%	Punjab, Haryana, Delhi, UP, Bihar, MP.	Kans
12.	Sorghum downy mildew	*P. sorghi*	Whole plant	100%	Rajasthan, Gujarat, Maharashtra, AP, Karnataka, Tamil Nadu	Soil, Oospores
13.	Sugarcane downy mildew	*P. sacchari*	While plant	100%	UP, Bihar	-
14.	Brown spot	*Physoderma maydis*	Leaf and sheath	27%	JK, HP, Sikkim, West Bengal, Punjab, Rajasthan, MP	Soil, Resting
15.	Banded leaf and sheath blight	*Rhizoctonia solani* f. sp. *sasakii*	Leaf and sheath	40.5%	JK, HP, Sikkim, Punjab, Haryana, Rajasthan, Delhi, UP, Bihar	Soil
16.	Cyst nematode	*Heterodera zeae*	Root	-	Rajasthan, MP, Bihar, UP.	Soil

Both race T and 0 of the pathogen occur in India, however the most prevalent one remains to be race 0. Moreover, in absence of utilization of male sterility in Indian maize programme, race T does not pose any threat to maize cultivations.

Common rust

This disease is prevalent all over the world in temperate to semi-temperate environments. In Asian continent the disease is present almost in all the countries. Aecial infection occurs infrequently on *Oxalis* spp. in Nepal. *O. corniculata* has been found to be susceptible under artificial inoculations with teleospores of *P. sorghi* in India. However, aecial infection so far has not been observed in nature.

Both specific and nonspecific or generalized types of resistance occur in maize to *P. sorghi.* Five major loci have been identified with the designations Rp_1, Rp_3, Rp_4, Rp_5 and Rp_6. Generalised or mature plant resistance has been studied in detail (20, 38) and it was established to be polygenic but the number of genes governing resistance may not be large.

The major approach of management of these diseases has been the utilization of host resistance. Several hybrids and other materials possessing resistance have been released for cultivation. Some of these are Ganga 11 and Deccan 103.

Atleast two inbred lines CM104 and CM105 have been demonstrated to possess durable resistance to Turcicum and Maydis leaf blights (40).

Carbamate group of fungicides especially zinc ethylene bisdithiocarbamate has proved to be suitable for controlling this group of diseases. Complete control was obtained by a schedule of four sprays applications at a concentration of 0.2% a.i., on a susceptible cultivar Basi. For large scale high volume sprays 1.2-2.6 kg of fungicide in 900-1200 litres of water/ha, is recommended if the plant population is around 56000/ha (Table 2).

Table 2 : Chemical control of maize diseases

	Disease	Chemical	Methods and time of application	Dosage
1.	Seed and Seedling blights	Thiram/ Captan	Seed treatment (before sowing)	2 g/kg of seed
2.	Pythium Stalk Rot	Captan	Soil drench (5-7 week growth stage)	11 g a.i./100 litres of water
3.	Bacterial Stalk Rot	Bleaching Powder	Soil drench (5-7 week growth stage)	3 g/10 litres of water
4.	Foliare Diseases (Turcium Leaf Blight, Maydis Leaf Blight and Common Rust)	Zineb	Foliar spray (4 week to flowering)	1.8-2.8 kg/900-1200 litres of water/hectare
5.	Peronosclerospora Downy mildews	Metalaxyl	Seed treatment (before sowing)	2.45 g a.i./kg

2.3 Stalk rots

This group of diseases is broadly divisible into two categories-pre-flowering and post-flowering types. In the former category fall Pythium stalk rot and bacterial stalk to while others such as Late wilt, Acremonium stalk rot. Fusarium wilt and Charcoal rot fall in the latter category.

Pythium stalk rot

In this case the rot is confined to a single basal internode leading to destruction of pith parenchyma and subsequent weakening of the stalk. The affected plants topple but do not die for upto two weeks after attack. The incidence of this disease is significantly influenced by both environmental and host factors. In the former case are included time of planting, soil moisture, organic/inorganic nutrient application, temperature, and relative humidity. In respect to host factors plant maturity and plant population were found decisive in determining disease incidence.

Accordingly, the recommendations that have emerged for managing this disease are :

1. regulate planting time between 10 and 20 July, in North India.
2. plant population not to exceed 50000/ha
3. good field drainage and
4. removal of previous crop debris/wheat straw.

Some highly resistant materials have been identified. Planting of hybrid Ganga Safed 2, Hi-starch and composite Suwan 1 is recommended for the disease prone areas. Application of 75% Captan @ 12 g a.i./100 l of water as soil drench at the base of the plants when crop is 5 to 7 week old, can effectively check this disease.

Bacterial stalk rot

The high incidence of this disease is favoured by high temperature (28°C and above) and high ambient moisture which commonly prevail in most of the maize growing areas 3 to 4 week after planting.

Resistance to this disease has been identified in hybrid Ganga Safed 2 and inbred lines CM 104 and CM 600 (45). This pathogen is highly sensitive to chlorine. Efficient control of this disease by using bleaching powder ($CaOCl_2H_2O$, containing 33% chlorine) could be obtained by drenching of the basal stalk region when plants are at knee-high stage (44).

Avoidance of water logging and proper drainage also helps in reducing disease incidence. Planting of the crop on ridges rather than in flat soil is also advocated.

Post flowering stalk rots

Fusarium stalk rot

Two species of Fusarium, *F. moniliforme* and *F. graminearum* are associated with the disease. The incidence of the former is more severe in low land tropical areas while the latter is present in cooler high lands. The symptoms become conspicuous when the crop enter senescence phase that is when kernels are in 'blister' stage to the phase when dry matter is near maximum in them. The pathogen commonly affects the roots, crown region and lower internodes, when split open, the stalks show a pink-purple discolouration. In plants affected by *F. graminearum,* blue purple mycelial mat of *Gibberela zeae* develop on the rind.

Charcoal rot

This disease is prevalent is comparatively drier maize growing areas. The symptoms of the disease becomes apparent as the plants approach maturity. Affected plants dry prematurly, the effected internodes become straw coloured, pith becomes disintegrated and the presence of small pin-head like black sclerotia on the rind of the stalks is a distinguishing character. Water stress at or after flowering has been found to predispose the plants to infection.

Promising germplasm both in terms of resistance and yield has been the single cross CM202 x CM111. The 'stay green' character in which plants remains green after their physiological maturity has been associated with resistance to certain post-flowering stalk rots. Charcoal rot and late wilt incidence is reduced by avoiding water stress at flowering.

Accumulation of aluminium and iron in maize plants leads plants to susceptibility to invasion by stalk rot pathogens especially if the crop was growing in acidic soils. In such cases soil amendments, therefore, with phosphate and lime reduced stalk rot severity.

Dodd (10,11) has postulated that "resistance to maturity related stalk rots has complex inheritance patterns linked to environmental and physiological interactions in plants". A photosynthetic stress-translocation balance concept of predisposition to root and stalk rots has been developed by Dodd (9). Predisposition of stalk and root rot is initiated with the beginning of senescence of root tissues because of inadequate carbohydrate supply. The aging cells become more Vulnerable to attack by microorganisms that are weakly pathogenic in vigorous cells.

Generally, all maize cultivars are resistant to stalk rot in the pre-silking period and as the crop proceeds towards maturity so does it susceptibility. A compound called DIMBOA, C2-4-Dihydroxy-7-glucoside) has been implicated in stalk rot resistance.

2.4 Downy mildews

This group of diseases constituted the single most important limiting factor to maize production in Asia. The important downy mildews present in India are the Philippine DM, Sorghum DM, Sugarcane DM and Brown stripe DM. Coordinated research to study the etiology, epidemiology of the pathogen and the breeding downy mildew resistant cultivars in Asia, was initiated by the Inter Asian corn programme (IACP) an informal association created and funded by the Rockeffeller foundation in 1960's. Considerable information of a basic and applied nature was generated and useful exchanges took place through annual workshops.

Epidemiologically, it has been determined that the crop is most vulnerable to DM infection during the seedling stage 15 to 20 days after planting (17). Systemic fungicides mainly based on acylalamines such as, metalaxyl (Ridomil 25 wp, Apron 35 SD, Apron 35 FN) have been very effective in controlling both *Peronosclerospora* and *Sclerophthora* incited downy mildews (23, 46) seed treatment with this group of fungicides provides a sort of protection upto 30 days of growth from DM infection; secondary infection if any may be checked by foliar spray of ridomil.

Philippine DM in particular may be kept under check by removing the Kans *(Sacharum spontaneum)*, growing around the maize fields either as fencing or in random fashion.

Sun drying of the seeds has been found to be very effective for inactivation of mycelium present in it and also reduces moisture levels. Seed with less than 20% moisture content produces healthy plants. Thus use of solar energy in post-harvest drying is a good measure for preventing outbreaks of these diseases through seed.

Cultivation of hybrids Ganga 11, Deccan, Deccan 103 and composite suwan 1 is recommended for Peronosclerospara incited DM endemic areas, whereas planting of hybrid Ganga 5, Sartaj, Deccan 1 or Deccan 103 is suggested for Brown Stripe DM prone areas.

2.5 Sheath blights

As the name indicates this group of diseases mainly affects sheaths although they do infect leaves, husks and other parts. Three diseases in this group have been recorded in India, but only two i.e. brown spot and banded leaf sheath blight cause significant damage. The other disease recorded is vertical banded blight caused by *Marasmiellus paspali* (34).

Brown spot

This disease mainly occurs in sub-tropical areas. Resistance have been identified on the basis of natural incidence only. Lal and Chakravarti (22) estimated that the yield loss in hybrid Ganga 5 is of the order of 27% while in the local open pollinated variety Malan it was 24.5%.

Banded leaf and sheath blight

This disease was considered insignificant until 1960s. However, in the subsequent years, it has become one of the major diseases n the countries of tropical Asia. In India, it is prevalent in hot humid foot-hill regions in Himalayas and in plains.

The pathogen has a predilection of spreading from the basal sheaths to the developing ears under favourable environmental conditions. The developing ear is completely damaged and dries up prematurely with cacking of the husk leaves. Sharma and Hembram (37) have shown that stripping of lower 2-3 leaves alongwith their sheaths considerably lowers disease incidence and also does not affect grain yield. This method has been found to be effective in checking the disease spread significantly not only in India but in other countries of Asia also, where this disease causes considerable loss to maize crop.

This group of diseases can be managed through adoption of host resistance. Chemical control, is also possible but so far the work has been of a preliminary nature. Fungicides Rhizolex and Thiophenate M, effectively inhibited fungal growth *in vitro*. Antibiotic Validamycin proved very effective in checking the disease spread in rice. The fungus causing rice sheath blight also belongs to same group.

2.6 Smuts

Two smuts occur on maize, namely the common smut *(Ustilago maydis)* and head smut *(Sporisorium reilianum)*. Head smut attacks maize, sorghum and other grasses and has been found moderately destructive disease in the sub-temperate Himalayas and in the hilly areas of Rajasthan. There appear to be two races/strains of the pathogen, one attacks maize and the other sorghum (19). In case of head smut sanitation, rotation and balanced fertility are the measures to be adopted for management.

The common smut is primarily soil borne though occasionally seed borne infection may occur. It may be controlled by avoiding injury to the plants during field operations, by maintaining balanced fertility and by sanitation (rogueing and burning of smut galls before they rupture) (4).

2.7 Ear, cob and kernel rots

This group of diseases can be classified into three types-those which affect cob, which affect kernel and those which affect both cob and kernel. As many as 10 ear rot diseases occur. Among the most prevalent are Fusarium ear rot *(F. moniliforme)*, Aspergillus kernel rot *(A. flavus, A. niger)* and Botryodiplodia cob rot *(B. theobromae)*. Occasionally, Nigrospora and Macrophomina cob rots also have been observed. Field observations suggest that their incidence is high in humid moist environments. Insect injuries also, contribute to increase in incidence. In the breeding programme emphasis should be given on developing cultivars with tight husk covers on the ears.

2.8 Virus diseases

Three viruses occur on maize in India. These are : maize mosaic virus I, maize mosaic and vein enation (39). Maize mosaic has been found to be prevalent in several states and the incidence ranges from 2.2 to 10.6%. Resistance of a high order has been identified in inbred lines CM103 and CM104. The released cultivars have not been reported to suffer much from viruses.

2.9 Nematodes

Several nematode pathogens have been found associated with maize; cyst nematode *(H. zeae; Pratylenchus zeae* Graham; *Tylenchorhynchus vulgare* Upadhyay, Swarup and Sethi; *T. masoodii* Siddiqui; *Helicotylenchus dihysteria* (Cobb) Sher; *Hoplolaimus indicus* Sher and *Meloidogyne incognita* (Kofoid & White) Chit (21). The cyst nematode is a serious problem in the states of Bihar, MP, Rajasthan and UP (21).

Carbofuran (3G) @ 2 kg a.i./ha; Sebuphos (10G) !@ 1 kg a.i./ha and seed treatment with carbosulfan @ 2-3% of seed have been found to be effective in reducing nematode damage under field conditions. Crop rotation mainly planting of a non-susceptible host/crop helps in reducing nematode population. Against cyst nematode sorghum, pearl millet, pulses, vegetables, wheat and oat have been suggested as alternative crops in crop rotation. Apart from the practice of rotation, *H. zeae* can also be managed by eradication of *Setaria*- grass on which this organism thrives. Cultural practices like sanitation, tillage practices etc. have also been found effective in reducing nematode populations.

2.10 Post harvest problems

The presence of as many as 69 fungal species has been found on or in maize seed (34). Three types of microbial problems during post harvest period need to be recognized:

1. Those of pathological significance, e.g. *Cephlosporium acremonium, Fusarium moniliforme, Peronosclerospora philippinensis* etc.
2. Those having toxicological importance, e.g. *Aspergillus flavus, A. niger, Macrophomina phaseolina, Penicillium funiculosum;*
3. Those having storage significance, these include 11 species of *Aspergillus* and seven species of *Penicillium.*

Aspergillus flaus Link ex Fries, the fungus responsible for aflatoxin contamination occurs both in pre-and post-harvest stages. Bilgrami et. al (2, 34) have studied the incidence of *A. flavus,* in pre-harvest maize. In the *rabi* season, 17% of the samples showed *A. flavus* contamination, whereas in *kharif* season the incidence was as high as 72%. In stored grain samples incidence was 35%.

Propionic acid has been determined to be effective in reducing microbial spoilage in stored maize (7, 18). Apart from propionic acid, boric acid and sorbic acid also can effectively reduce incidence of several storage fungi (24).

Certain physical measures such as proper drying and reducing seed moisture to safer levels, also helps in elimination of the storage fungi. Genetic differences do occur in maize cultivars for the production of aflatoxin (27). This field must be investigated more vigorously.

3. REFERENCES

1. Anonymous 1972. Maize Research at Pantnagar, 1966-72; Govind Ballabh Pant University of Agriculture and Technology, Pantnagar, India.

2. Bilgrami, K. S., Prasad, T., Mishra R. S. and Sinha, K. K. 1981. Aflatoxin contamination in maize under field conditions. Indian Phytopathology, **34** : 67-68.

3. Change, S. C. and Chen, C. K. 1969. Studies on seed transmission of corn downy mildew. Report of Corn Research Center, No. 7.

4. Christensen, J. J. 1963. Corn smut caused by *Ustilago maydis* American Phytopathological Society Monograph, 2, pp. 41.

5. CIMMYT 1929. 1991-1992 CIMMYT World maize facts and trends. Maize Research Investment and Impact in Developing Countries, CIMMYT, El Batan, Mexico.

6. Craig, J. and Daniel, Kalio, L. A. 1968. Chlorotic lesion resistance to *Helminthosporium maydis* in maize. Plant Dis. Rep. **52** : 134-136.

7. Dhanraj, K. S., Chohan, J. S., Sunar, M. S. and Lal, S. 1973. Preliminary studies in the use of Luprosil in the prevention of microbial spoilage of wet maize during storage. Indian Phytopathology, **26** : 63-67.

8. Diwakar, M. C. and Payak, M. M. 1980. Influence of some environmental and host factors on phythium stalk rot of maize. Indian Phytopathology, **33** : 10-15.

9. Dodd, J. L. 1977. A photosynthetic stress translocation balance concept of corn stalk rot. Proc. 32 Annu. Corn. Sorghum Res. Conf. **32** : 122-136.

10. Dodd, J. L. 1978. The photosynthetic stress translocation balance concept of sorghum stalk rots. pp. 300-305. Proc. Intern. Workshop Sorghum Diseases, ICRISAT, Patancheru, Hyderabad, India.

11. Dodd, J. L. 1980. The role of plant stresses in development of corn stalk rots. Plant Disease **64** : 533-537.

12. Dunn, G. M. and Routley, D.F. 1982. Role of *Ht* and *Bx* genes in resistance of corn to *Helminthosporium turcicum* in Proc. 27th Corn Res. Conf. Chicago, Ill. Am. Seed Trade Assoc. Washington D. C. 168.

13. Handoo, M. I. 1969. Genetical analysis of resistance to *Helminthosporium maydis* Nishikago in maize *(Zea mays L.)*. Ph.D. Thesis, Indian Agricultural Research Institute, New Delhi. India.

14. Hooker, A. L. 1972. Southern leaf blight of corn, present status and future prospects. J. Environ. Qual., 7: 244-249.

15. Hooker, A. L. 1978. Genetics of disease resistance in maize in Maize Breeding and Genetics (Ed. Welden, D. B.) John. Wiley & Sons, New York, Chap. 21.

16. Jha, T. D. and Dhawan, N. L. 1970. Genetic analysis of resistance to *Helminthosporium turcicum* in maize inbred lines. In Plant Disease Problems (Eds. S. P. Raychoudhry et al.), Indian Phytopathological Society, New Delhi, India.

17. Kajiwara, T. 1975. Some experiments on downy mildew of maize. Tropical Agriculture Research Series, **8** : 121-123.

18. Kandhari, J., Payak, M. M., Mehta, S. L. and Sharma, R. C. 1980. Investigations on microbial toxins in stored maize samples from Rajasthan. In: Proceedings of the Symposium on Environmental Pollution and Toxicology (Ed. S. P. Raychaudhury), Today and Tomorrow Press, New Delhi, pp. 319-328.

19. Khanna, A. and Payak, M. M. 1972. Teliospore morphology of some fungi V. *Shaecelotheca reiliana.* Nova Hedwigia, **23** : 907-903.

20. Kim, S. K. and Brewbaker, J. L. 1976. Sources of general resistance of *P. sorghi* of maize in Hawaii. Plant Dis. Reptr. **60** : 551.

21. Koshy, P. K. and Swarup, G. 1971. Distribution of *Heterodera avenae, H. zeae, H. cajani* and *Anguina tritici* in India. Indian Journal of Nematology, **1** : 106-111.

22. Lal, B. B. and Chakravarti, B. P. 1976. Assessment of loss due to brown spot of maize caused by *Physoderma maydis*. Indian Phytopathology, **29** : 449-450.

23. Lal S., Bhargawa, S. K. and Saxena, S. C. 1980. Management of sugarcane and brown stripe downy mildews in maize with metalaxyl. Newsletter Working Group on Graminicolous Downy Mildews, 2.

24. Lal, S. P. and Kapoor, J. N. 1980. Efficacy of certain food preservatives in the control of species of *Aspergillus* in stored maize. Indian Phytopathology, **33**: 336-338.

25. Laxminarayan, C., Singh, S. K., Aujla, S. S., Consul, S. K., Singh, N. N., Payak, M. M., Nene, Y. L. and Renfro, B. L. 1967. Fungicide seed treatment of maize in India. Bulletin of the Indian Phytopathological Society, **3** : 18-26.

26. Leonard, K. J., Levy, Y. and Smith, D. R. 1989. Proposed nomenclature for pathogenic races of *Exserohilum tureicum* on corn. Plant Dis. **73** : 776-777.

27. Nagerajan, V. and Bhat, R. V. 1972. Factors responsible for varietal differences in aflatoxin production in maize. Journal of Agriculture and Food Chemistry, **20** : 911-914.

28. Payak, M. M. and Renfro, B. L. 1974. A decade of research of maize diseases : impact on production and its International cooperative outreach. In: Current Trends in Plant Pathology (Eds. S. P. Raychaudhury and J. P. Verma) p. 160-170.

29. Payak, M. M. and Sharma, R. C. 1978. Research on diseases of maize. Pl. 480 Project Final Technical Report (April 1969-March 1975), Indian Council of Agricultural Research, New Delhi, pp. 228.

30. Payak, M. M. and Sharma, R. C. 1980. An Inventory and Bibiliography of Maize Diseases in India. Division of Mycology and Plant Pathology, Indian Agricultural Research Institute, New Delhi, India.

31. Payak, N. M. and Sharma, R. C. 1982. Maize diseases and approaches to their management in Indian. Tropical Pest Management, **31** : 302-310.

32. Payak, M. M. and Sharma, R. C. 1986. Vertical banded blight an unusual Marasmiellus disease of maize, Current Science, **55** : 1135-1137.

33. Payak, M. M., Sharma, R. C., Singh, B. M. and Lilaramani, J. 1974. Performance of maize lines carrying single gene resistance of *Helminthosporium turcicum* and *Puccinia sorghi* in India. Indian Journal of Genetics and Plant Breeding, **34** : 31-35.

34. Payak, M. M., Sharma, R. C. and Rao, S. R. 1987. Aflatoxin in India. In Aflatoxin in Maize (Eds. Zuber, M. S., Lillehoj, E. B. and Renfro, B. L.). A Proceeding of the workshop. CIMMT, Mexico, DF.

35. Renfro, B. L. and Ullstrup, A. J. 1976. A comparison of maize diseases in temperate and tropical environments. Pest Articles and News Summaries, **22** : 491-498.

36. Shankarlingam, S., Balasubramaniam and Reddy, R. R. 1989. Nature and gene action governing resistance to *Helminthosporium* leaf blight in maize *(Zea mays* L.). Genetika, **79** : 121-127.

37. Sharma, R. C. and Hembram, D. 1990. Leaf stripping a method of control banded leaf and sheath blight of maize. Current Science, **59** : 745-746.

38. Sharma, R. C. and Payak, M. M. 1979. Resistance to common rust of maize in India. Maize Genetics Cooperation Newsletter, **53** : 665-66.

39. Sharma, R. C. and Payak, M. M. 1983. An overview of virus and virus like diseases of maize in India. In: Proceedings of the Second International Maize Virus Disease Colloquium and Workship (Eds. D. T. Goordon, L. R. Nault and R. M. Ritter), Ohio Agriculture Research and Development Centre, Wooster, Ohio, USA.

40. Sharma, R. C. and Payak, M. M. 1990. Durable resistance to two leaf blights in two maize inbred lines. Theoretical and Applied Genetics, **80** : 542-544.

41. Sharma, R. C., Payak, M. M., Shankarlingam, S. and Laxminarayan, C. 1982. A comparison of two methods of estimating yield loss in maize caused by common rust. Indian Phytopathology, **35** : 18-20.

42. Singh, B. M. and Sharma, Y. R. 1976. Evaluation of maize germplasm to banded leaf and sheath blight and assessment of yield loss. Indian Phytopathology. **29** : 129-132.

43. Sinha, K. K. 1983. Aflatoxin problem in storage and standing maize crops. In: Proceedings of the Symposium of Mycotoxins in Food and Feed (Ed. K.S. Bilgrami), Allied Press, Bhagalpur, India.

44. Thind, B.S. and Payak, M. M. 1972. Antibiotics and bleaching powder in the control of bacterial stalk rot of maize. Hindustan Antibiotics Bulletin, **15** : 9-13.

45. Thind, B. S. and Payak, M. M. 1978. Evaluation of maize germplasam and estimation of losses to Erwinia stalk rot. Plant Disease Reporter, **62** : 319-323.

46. Venugopal, M. N. and Safeeulla, K. M. 1978. Chemical control of the downy mildews of pearl millet and maize. Indian Journal of Agricultural Sciences, **48** : 537-539.

8 Biological Control of Arthropod Pests of Pulse Crops

A. Ganeswara Rao and T. Ramesh Babu

CONTENTS

Integrated Pest and Disease Management
Rajeev K. Upadhyay, K.G. Mukerji, B.P. Chamola and O.P. Dubey (eds.)
APH Publishing Corporation, New Delhi-110002, India

ABSTRACT

Although many insects feed upon pulse crops from the seedling stage, most of the economic damage is caused by pests that feed upon flowers and pods. Losses caused by insect pests to these crops are variable nut substantial, particularly for pigeonpea. Efforts have been made to suppres the major pests through biological means, particularly to *Helicoverpa armigera.* Egg parasitism is always preferable as the larva is killed before it emerges to damage the crop. However, egg parasitoids are not effective in these crops. Among the larval parasitoids *Campoletis chloridae* Uchida is a potential candidate because it parasitises 1-3 instar larvae and is active. Though hyperparasitism of *C. chloridae* at pupal stage is a constraint, hyperparasitism is low in pulse crops (pigeonpea and chick-pea) compared to cereal crops. Among the pathogens, nuclear polyhedrosis virus (NPV) appear to be promising. There are reports to the effect that they are very effective in some parts of the country even at farmer's level. Largely it has been reported that these viruses are effective when used in combination with the insecticides like endosulfan. The scope for conservation and augmentation of the natural enemies to suppress these pests on pulses is discussed.

1. INTRODUCTION

Over 200 species of insects having been recorded in India (28) few inflict economic damage at pod formation stage that greatly reduce the crop yields. Thus the pod damaging insects are the most important pests on pulse crops. Various insect pests that attack pulse crops from seedling stage to harvest are listed in Table 1. The attention bestowed towards the plant protection in pulse crops is rather poor, since the returns are not assured under rainfed conditions. Although a number of plant protection measures based on insecticides have been recommended, over a time, excessive exploitation of synthetic insecticides has resulted in serious problems.

Table 1 : Insect pests of pulses at different crop growth stages

A. Pests attacking roots	
(1) Nodule damaging fly	*Rivellia angulata* Hendel (Diptera: Platystomatidae) *Metopina ciceri* Disney (Diptera: Phoridae)
(2) Termites	*Odontotermis* spp. *Microtermes* spp. (Isoptera: Termitidae)
(3) Sitona Weevils	*Sitona macularius* (Marsham) (Coleoptera: Curculionidae)
B. Pests of Stem	
(1) Jewel Beetles	*Sphenoptera indica* Laporte & Gory (Coleoptera:Curculionidae)
(2) Stem Weevils	*Alcidodes fabrici* (Fabricius) *Alcidodes collaris* (Pascoe) (Coleoptera: Curculionidae)

(3) Stem Flies	*Ophiomyia phaseoli* (Tryo.) *Ophiomyia centrosematis* (Maijere) (Diptera: Agromyzidae)
(4) Cow Bugs	*Otinotus oueratus* W. *Oxyrachis tarandus* F. (Hemiptera: Membracidae)
(5) Scale Insects	*Ceroplastodes cajani* Maskell *Icerya purchasi* Maskell (Hemiptera: Coccidae)
(6) Cut Worms	*Agrotis ipsilon* (Hufnagel) (Lepidoptera: Noctuidae)

C. Pests of Foliage

(1) Jassids	*Empoasca fabae* (Harris) *Empoasca kerri* Pruthi *Jacobiasca lybica* (deBeryeven) (Hemiptera:Cicadellidae)
(2) Aphids	*Aphis craccivora* Koch *Aphis fabae* Scopoli *Myzus persicae* (Sulzer) *Macrosiphum* spp. (Hemiptear: Aphididae)
(3) Eriophid Mites	*Aceria cajani* Channabasavanna (Acarina: Eriophyidae)
(4) Red Spider Mites	*Schizotetranychus cajani* Gupta *Tetranychus* spp. (Acarina:Tetranychidae)
(5) Whiteflies	*Bemisia tabaci* (Genn.) (Hemiptera: Aleyrodidae)
(6) Bud-Sucking Bugs	*Campylomma* spp. *Creontiades pallidua* (Rambir) *Eurystylus* spp. *Taylorilygus vosseleri* (Poppius) (Hemiptera: Miridae)
(7) Leaf Cutter Bees	*Megachile* spp. (Hymenoptera: Megachilidae)
(8) Grasshoppers and Locusts	*Catantops erubesceus* Walk. *Colemania spheneroides* (Boliver) *Cyrtacanthacris tatarica* (L.) *Patanga succinata* (L.) *Schistocerca gregaria* Forsk. (Orthoptera:Acrididae)
(9) Leaf Damaging weevils	*Myllocerus undecimpustulatus* Faust *Nematocerus* spp. *Phyllobius* spp. *Systates* spp. (Coleoptera:curCulionidae)

(10) Leaf Webbers	*Grapholita (Cydia) critica* Meyr. (Lepidoptera:Tortricidae)
(11) Leaf Miners	*Liriomyza cicerina* (Rondani) (Diptera:Agromyzidae)
(12) Army Worm	*Spodoptera litura* (Fb.) *Spodoptera exigua* (Fb.) (Lepidoptera:Noctuidae)

D. Pests of Flowers and Pods

(1) Thrips	*Megalurothrips usitatus* (Bagnall) (Thysanoptera: Thripidae)
(2) Blister Beetles	*Mylabris pustulata* Thunberg *Coryna* spp. (Coleoptera:Meloidae)
(3) Pod-Sucking Bugs	*Clavigralla gibbosa* Spinola (Hemiptera: Coreidae) *Nezara viridula* (Hemiptera: Pentatomidae)
(4) Pod Borers	*Helicoverpa armigera* (Hubner) (Lepidoptera:Noctuidae)
(5) Lable Pod Borer	*Adisura atkinsoni* (Moore) (Lepidoptera:Lycaenidae)
(6) Blue Butterflies	*Lampides boeticus* (L.) (Lepidoptera: Pyralidae)
(7) Cowpea Pod Borer	*Maruca testulalis* (Geyer) (Lepidoptera: Pyralidae)
(8) Plume Moth	*Exelastis atomosa* (Walsingham) (Lepidoptera: Pterophoridae)
(9) Lima Bean Pod Borer	*Etiella zinckenella* Treitschke (Lepidoptera: Pyralidae)
(10) Podflies	*Melanagromyza obtusa* Mallach (Diptera:Agromyzidae)
(11) Bruchids	*Callosobruchus chinensis* (L.) *Callosobruchus maculatus* (F.) (Coleoptera: Bruchidae)

WHY BIOLOGICAL CONTROL

With the advent of DDT and other chlorinated hydrocarbon pesticides, emphasis in the pest control swung from natural and biological control to almost total reliance on chemical control. The rapid development of resistance led to more frequent application of increased dosages of even more potent pesticides which inturn lead to the degradation of the environment, the occurrence of pesticide residues in our food and water resources and eventually to the development of "SUPER PESTS" which could not be controlled by any of the chemical insecticides registered for use. These problems have indicated the need for effective, environmentally safe and low cost pest control strategy based on indigenous renewable resources.

Although biological control will never totally replace chemical insecticides, its expanded use would significantly improve the quality of the ecosystem (29). Organisms used in biological control are environmentally safe, do not cause pollution and usually affect the target organism. Biological control is compatible with conventional as well as low-input and sustainable agriculture. It is cost-effective and self-sustaining. Besides insect parasitoids and predators several pathogens including viruses such as nuclear polyhedrosis virus (NPV) and granulosis virus (GV), bacteria like *Bacillus thuringiensis* (BT), fungi like *Metarhizium*, *Beauveria* and *Verticillium* are also known in nature which cause diseases in insects and destroy them.

The purpose of this chapter is to draw attention to the status of biological control at present in the management of pests of pulse crops and to future possibilities and to encourage investment in this environmentally sound method of pest control.

2. BIOLOGICAL SUPPRESSION OF KEY PESTS OF PULSES

2.1 Parasitoids

The key pest on pules is gram pod borer, *Helicoverpa armigera* (Hb.). There has been considerable effort on the use of egg parasitoids, particularly *Trichogramma* spp. for the control of *Helicoverpa* on pigeonpea and chickpea. These tiny insects can be multiplied cheaply in large numbers and inundative releases have given good control of *Helicoverpa* in extensive field tests. However, several morphological attributes of these plants influenced the successful parasitization by the egg parasitoid, *T. chilonis* and *T. brasiliensis* . It was reported that *Trichogramma chilonis* (Ishii) was deterred from parasitizing *Helicoverpa* and other eggs on pigeonpea, particularly those on the pods (37).

Further, attempts were made to select *Trichogramma* species that would more readily accept pigeonpea plant and varieties. Influence of host plant on the parasitism of *Helicoverpa* eggs by *Trichogramma* spp. was also reported (19, 41). Although substantial differences were established between the species and varieties it was concluded that *Trichogramma* has little potential on pigeonpea. On chickpea also in India egg parasitism appears to be rare and often non-existent. It is probable that the acid exudate deters the tiny parasite adults (27). However, it was documented that replease of *Trichogramma* spp. on chickpea led to a reduction of both *Helicoverpa* spp. and *Spodoptera* spp. in Mexico (9). Interestingly, in contrast it was reported that parasitization of the eggs of *H. armigera* by *T. chilonis* was as high as 85% in cowpea (6).

Parasitism levels in larvae of *H. armigera* collected from pigeonpea were generally greater than those collected on chickpea. While dipteran parasitoids predominant on pigeonpea, the hymenopterans are abundant on chickpea particularly early in the season. These native dipterans did not appreciably reduce larval population or damage on pigeonpea were of lesser importance since the mortality of *Heliothis* mostly occurred at prepupal phase after the larvae had already caused pod damage (7). However, the early larval parasitism was more on chickpea (31.9%) whereas late larval parasitism was more on pigeonpea (16.4%)(22). The hymenopteran, *Campoletis chloridae* Uchida was an early larval parasitoid whereas the dipteran *Corcelia illiota* Curron was most common on late larval population from legumes. The per cent parasitism by *C. chloridae* on chickpea ranged from 7.5 to 10.1 in different varieties in Chattishghar region of Madhya Pradesh (5). The larval population at Ranchi, Bihar remains fairly low during December because of low temperature and high parasitism (50-53%) by *C. chloridae* (24). Similarly parasitism in Kanpur, U.P. peaks in December (40). Average parasitism was more at flowering (11.89%) followed by at podding stage (11.26%) and less at harvesting stage. The economical mass breeding for augmentative release of *C. chloridae* and *C. illota* could not be standardized so far.

As most of the parasitoids found in *Helicoverpa* on pigeonpea were dipteran it was considered worthwhile to establish an exotic dipteran, *Eucelatoria bryani* Sabrosky and this species was imported from Commonwealth Institute of Biological Control, USA to India by ICRISAT and successfully bred on *Helicoverpa* larvae in the laboratory for many generations, but attempts to establish in the filed were not successful (29). Similarly, the authors attempted multiplication and field acclimatization of another exotic parasitoid, *Hyposoter didymator* Thumb an ichneumonid against *Helicoverpa* on pigeonpea. The parasitoids be multiplied successfully in the laboratory both on *Helicoverpa* and *Spodoptera* early larval instars however, the parasitoids could not be recovered from field released population (2).

Six genera of parasitoids were found parasitising the pigeonpea pod fly, *Melanagromyza obtusa* Mallach among which two genera viz.,*Omytes* and *Euderus* are the potential candidates for the biological suppression of this pest in Andhra Pradesh (38). It was also suggested that a parasitoid, *Bracon* spp. prevalent on a related species of pod fly in Africa may be a candidate for introduction. An egg parasitoid, *Gryon antestiae* (Scelionidae) on *Clavigralla gibbosa,* the pigeonpea pod bug, was active throughout the oviposition period of the bug and attacked to an extent of 62% (35). *Cotesia orientalis* and *C. diurnii* (10) and *Diadegma trichoptilus* (32) have been recorded parasitizing plume moth, *Exelastis atomosa* Walsingham on pigeonpea in Maharashtra.

Four parasitoid species, *Elasmus anticles, Apanteles sanros, Eurytoma* spp. and *Bracon* spp. have been reported parasitizing the leaf tier, *Eucozoma (Cydia) critica* (Meyr.) from Gujarat (16). Of eight species of insect parasites recorded from the leaf tier, which is common pest of pigeonpea through the vegetative stages, none transferred to *Helicoverpa* which mainly attacks the crop at flowering and podding stage. The only parasites found in both leaf tier and *Helicoverpa* larvae were the nematodes *(Overmermis* and *Hexamermis).* An ichnenmonid, *Eriborus argenteopilosus* (Cameron) and a tachinid, *Palexorista* sp. were recorded from both *Helicoverpa* and *Adisura* sp. a relatively rare lepidopteran pod borer. The survival and shifts of the natural enemies across each season, if considered, the larval parasite *Campoletis chloridae* could be recovered could be recovered from *Helicoverpa* larvae from 20 different host plants and can be found, on one host or another throughout the year. This clearly indicates the complexity of the natural situation, for even on a single pest there is considerable variation from crop to crop and from season to season. An additional complication exists in the presence of hyperparasitoids. For example, ten species of hyperparasites were reported from *Campoletis chloridae*. Pteromalid, *Sphegigastor brunneicornis* parasitises agromyzid, *Ophiomyia phaseoli* on cowpea and *Chlorocytus* sp. has been recorded in *O. phaseoli* on soybean (17). The population of leaf rollers, *(Hedylepta indica* and *Lamprosema diemensalis)* on soybean was kept under check by several natural enemies viz., spiders, *Brachymeria* sp. and *Xanthopimpla punctata* (8).

2.2 Predators

There has been relatively little research on the potentiality of the predator on pulses. However, 19 species of predatory insects were reported any many species of spiders and birds preying in the eggs and larvae of *Helicoverpa.* Most of the life stages of the pod fly, *Melanagromyza obtusa* Malloch are protected from predators; but many of the adults have been reported to be trapped in spider's webs. Mud wasps such as *Delta* spp. carry off many large *Helicoverpa* larvae to feed their young. *Chrysopa,* the green lace wing, a potential and promising candidate against the eggs and early larval instars of *Helicoverpa* is still to be tried in field on pulse crops. The arthropod predators of *Helicoverpa* spp. including neuropterans and arachinids that are common on other crops are seldom found on chickpea. However, substantial predation of larvae on this crop by insectivorous birds reported (1).

Investigations on the provision of perching for birds in India have shown some promise (33). At Gujarat (Anand) five species of birds cattle egret, rosy paster, common mynah, drongo and house sparrow were recorded predating in large larvae of *H. armigera* on chickpea. The birds recorded the toll to an extent of 84 per cent (34). House sparrow and bank mynah are important biotic agents of *Helicoverpa* in Punjab (36). The cattle egret and river stern are observed feeding on *Helicoverpa* in Orissa. The birds are particularly effective towards the end of January month (13). Large number of adults of the predatory wasps *Polistis habraeus* and *Vespa orientalis* are attracted to pests on greengram in Delhi region (31).

2.3. Microbial Agents

Some of the microbial organisms which are amenable for mass multiplication have proved quite effective against several pests of pulses. Virus diseases have been reported in may insect species, in particular, members of Lepidoptera. Viruses can either promote lethal infections and induce disease epizootics or produce chronic disease and remain at low levels with in the population. Baculoviruses have been extensively studied as biological control agents and their main advantage lies in a narrow host range. Although baculoviruses infection can be experimentally induced by injecting virions into the haemocoel of susceptible hosts, the natural route of infection is by Occlusion Bodies (OB's) or virions.

On chickpea the efficacy of HaNPV *(Helicoverpa armigera* NPV*)* has been demonstrated in Gujarat and Hyderabad where high mortality of *Helicoverpa* larvae is obtained. Around 47% increase in grain yield has been indicated over untreated controls. Birds have been found quite helpful in dispersing HaPNV (3). In Tamilnadu HaPNV at the rate of 1.5 10^{12} polyhedral occlusion bodies (POB)/Ha. sprayed three to four times in the evening hours at an interval of 7-10 days when the pest is in the very early stages, effectively controlled *Helicoverpa* on pigeonpea, (30) chickpea, (25) lab lab bean (15) and blackgram (18). Application of HaNPV alongwith 0.035% endosulfan has been found to give better control of *Helicoverpa* on chickpea (25) and pigeonpea (30). The authors also obtained effective control with 250 LE of HaNPV/Ha. or 125 LE HaNPV/Ha. + endosulfan 0.035% against *Helicoverpa* on chickpea in Hyderbad (4). Two spraying of endosulfan 0.07% or HaNPV @ 500 LE/Ha. alternated with endosulfan significantly reduced the pod damage in pigeonpea and increased grain yield (39). HaPNV did not effect the development of the parasitoid, *Eucelatoria bryani* maggot development, adult emergence, sex ratio, longevity, larviposition and behaviour of adult flies emerging from diseased larvae was normal (21). The development of *Campoletis chloridae* is however suppressed by HaNPV infection in *H. armigera* larvae (34). The interesting feature is *C. chloridae* transmitted the virus to *H. armigera* larva directly through oviposition causing 100% mortality and *E. bryani* transmits HaNPV during larviposition (21). Perhaps, the greatest potential for biological suppression of the key pest *Helicoverpa armigera* in pulse crops lies in the use of Nuclear Polyhedrosis Virus. Such a virus is already being used commercially for *Helicoverpa* spp. management in the USA and elsewhere, but with limited success.

AaNPV first isolated from its host *Adisura atkinsoni* has proved effective at 250 LE/Ha. applied in combination with 0.035% endosulfan twice at fifteen days interval and increased the yields of field beans (21).

Crude preparation of NPV can be made economically by crushing larvae that have been diseased and killed by the virus and mixing with water. Although, it can be suggested, there is some concern that crude prepared mixtures i. e. unpurified may be dangerous since there may be other pathogens in deed insects which may threaten the health of other animals.

One of the most critical factors in the cost effective use of baculoviruses as insecticides is their relatively short persistence on crop leaf surfaces. "Half lives" of less than one day appear common for unprotected virus. Virus deposits are often surprisingly reinfest. Of the environmental factors which modify virus persistence on a plant it is generally considered that the ultra violet (u.v.) Component of sunlight is the key factor in virus inactivation (14), although leaf exudates on some plants, particularly cotton may also produce substantial inactivation (11). To protect baculoviruses from solar u.v. radiation two main groups of additives have been tested including substances which are opaque and reflect u. v., and substances which absorb u.v. energy. Many of the results are inconclusive. Recent laboratory and field tests with u.v. absorbing benzophenone compounds have been more encouraging. The host range of baculoviruses is another most important factor influencing their use as microbial insecticides. A high degree of host specificity amongst many baculoviruses has often been considered as an advantage because of the reduced ecological hazard resulting from their use.

Another microbial pesticide that has been most and commercially available is the spore forming bacterium, *Bacillus thuringiensis* Berliner (BT). This bacterium has many subspecies, most of which can be grown in large quantities on a variety of inexpensive media. These are stomach poisons, highly specific for insects and since there is no contact activity, are ineffective against aphids, scales, whiteflies and leafhoppers. Over 30 Bt subspecies have been discovered. *B.t. Kurstaki* (B. t. k.) only produce proteins significantly active against lepidopterous insects. Similarly B.t. *tenebrionis* (B. t. t.) active against coleopterous insects (12). These commercial formulations of Btk have been widely tested for the control of *Helicoverpa* spp. in pulses with occasional success and are being used on a limited scale in some countries. Our efforts with some of the commercially available Bt formulations against *H. armigera* on pigeonpea have not given very encouraging results (4). However, new and more potent strains are becoming available and these may have promise.

Mermithid nematodes are very common insect parasitic nematodes, particularly on *Helicoverpa* larva preferably from weed hosts in uncultivated areas. In one case 93% of *Helicoverpa* larvae collected during August were paralised by mermithids (20). This can not be exploited since the artificial multiplication is almost impossible.

3. FUTURE POSSIBILITIES

Scope for conservation

Efforts to preserve and increase the effectiveness of natural enemies via environmental manipulation, particularly the judicious use of insecticides, is biological control by conservation. It might be possible to improve upon the conservation of the natural enemies by adjusting cropping systems and plant protection practices. Ideally a low population of the pests should be maintained throughout the year as a food source for the natural enemies to survive. Greater parasitism could be obtained from pesticide free plots, however, the total pesticide use can not be dispensed with. Therefore judicious application on need base with relatively non persistent and safer insecticides is to be suggested against these crop pests so as to facilitate the conservation of the existing natural enemies in an ecosystem.

Scope for augmentation

The immediate means for augmentation of the natural enemies is to develop and standardise economical and very easy methods of mass production of the indigenous natural enemies. "Know Your Insect" philosophy is applicable heer and hence, behavioural studies of density dependent natural enemies are to be contemplated for the most potential candidates of biocontrol for each of the pest as well as under each of the agroclimatic cropping system. It is important to point out that behavioural aspects established for a given natural enemy in a particular cropping system and region are not readily exportable and implementable under different systems. Too many of intricacies involved are different in different crops, regions, host insects and natural enemy it self. The mass production and dissemination of microbial agents, particularly of NPV, may be relatively cheap and simple.

4. FUTURE PROSPECTS

The release of parasitoids and predators against eggs and early larval instars and use of microbial pathogens will result in discouraging indiscriminate insecticidal sprays which inturn will encourage the development of natural enemies on other development stages thereby creating a sequence of parasitoids and predators on different stages of the pest.

Suitable strains of *Trichogramma* are to be determined for use in pigeonpea and chickpea ecosystems. Lot of work needs to be done on the methods of use of parasitoids/ predators/pathogens. The rate number and time of release of natural enemies differ for different agroclimatic regions and hence needs to be established.

To achieve the twin objectives of pest management and environment protection the other possible environment friendly inputs of pest management which encourage biological suppression like pheromones for monitoring, use of resistant/tolerant varieties, easily biodegradable safer insecticides of selective action and finally use of synomones to attract natural enemies to plants and kairomones for effective parasitization of the hosts are to be advocated.

There is a need to establish a net work of large scale natural enemy multiplication units throughout the country so that, parasitoids, predators and microbial preparations are made available to the farmers.

5. REFERENCES

1. Anonymous, 1987. Annual Report All India Coordinated Research Project on Biological Control of Crop Pests and Weeds I. I. H. R. Bangalore, India.

2. Anonymous, 1989. Annual Report All India Coordinated Research Project on Biological Control of Crop Pests and Weeds Biological Control Centre, Bangalore, India.

3. Anonymous, 1992. Annual Report All India Coordinated Research project on Biological Control of Crop Pests and Weeds. Biological Control Centre, Bangalore, India.

4. Anonymous, 1993. Annual Report All India Coordinated Research Project on Biological Control of Crop Pests and Weeds. Biological Control Centre, Bangalore, India.

5. Bharadwaj, D., Kaushik, U. K. and Pawar. A. D. 1987. Incidence of *Heliothis armigera* (Hubner) and parasitism by *Campoletis chloridae* (Uchida) in Chattishgarh, Madhy Pradesh. Journal of Biological Control **1** (**2**): 79-82.

6. Bhatnagar, V. S. and Davies, J. C. 1979. Pest management in intercrop subsistence farming In: Proceedings of International Workshop in Intercropping. ICRISAT, Patancheru, A. P., India. 249-257.

7. Bhatnagar, V.S., Lateef, S.S., Sithanantham, S., Pawar, C.S., and Reed, W. 1982. Research on *Heliothis* at ICRISAT. Proceedings of International Workshop on *Heliothis* Management, ICRISAT, A.P., India. 385-396.

8. Bhattacharjee, N. S. 1977. Record of the parasites and predators of soybean leaf rollers in India. Indian Journal of Entomology **38** (**4**): 383-384.

9. Castanos Lugo, C. 1979. Control Integrado de plages eu el cultivo de garbanzo. In: Reunion nacional de control biologica. pp.164-167. Vercruz, Mexico, Comite Organizador de la VII Reunion Nacional de control biologica.

10. Chalikwar, M. R., Rao, S. N. and Nikam, P. K. 1984. Two new species of *Cotesia* from India (Hymenoptera: Braconidae) Oriental Insects. **18**:17-23.

11. Entwistle, P. F. and Evans, H. F. 1985. Viral control (p347-412) In: Comprehensive Insect Physiology Biochemistry and Pharmacology. Vol. 12: Eds. L.I Gilbert, and A.G. Krekut, Pergamon, Oxford.

12. Federici, B. A. 1991. Biological Control : Microbial Insecticides. Pesticide Outlook. **2(3)**: 22-28.

13. Ghode, M. K., Nayak, U. K., Ghosh, P. K. and Pawar, A. D. 1988. Vian predation of gram pod borer *(Heliothis armigera)* in Orissa. Journal of Advanced Zoology. **9(2)**: 148.

14. Ignoffo, C. M. and Hostetter, D. L. 1977. (Eds) Environmental stability of microbial insecticides. Misc. Publ. Entomol. Soc. Am. **10(3)**: 1-80.

15. Jayaraj, S., Rabindra, R. J. and Santharam, G. 1987. Control of *Heliothis armigera* (Hubner) on chickpea and lab bean by nuclear polyhedrosis virus. Indian Journal of Agriculture Sciences **57(10)**: 738-741.

16. Kabaria, B., Goyal, S. N., Jose, V. T. and Shah, A. H. 1988. New parasitoids recorded on leaf weber *(Cydia critica* Mevric) of pigeonpea in Gujarat, India. International pigeonpea Newsletter No.**7**: 34.

17. Kundu, G. G. 1985. *Melanagromyza phaseoli* (Tryon) (Agromyzidae: Daptera) - a new host of *Chlorocytus* spp. (Pteromalidae : Chalcidoidea) from India. Journal of Entomological Research **9(1)**: 110-111.

18. Mahadevan, N. R. and Kumarswamy, T. 1980. A note on the effectiveness of nuclear polyhedrosis virus against *Spodoptera litura* (F.) on blackgram *Phaseolus mungo* L. Madras Agricultural Journal **67**: 138-140.

19. Manjunath, T. H., Phalak, V. R. and Subramanian, S. 1970. First records of egg parasites of *Heliothis armigera* (Huber) (Lepidoptera : Noctuidae) in India. Technical Bulletin Commonwealth Institute of Biological Control. **13**: 111-115.

20. Murthy, M. S. 1987. Biological suppression of pulse and oilseed pests. Lead paper presented at National Workshop of All India Coordinated Research Project on Biological Control of Crop Pests and Weeds, Lucknow, India, October, 1987.

21. Narayanan, K. 1980. Studies on the interaction between host, *Heliothis armigera* (Hubaer), pathogen (NPV) and parasite *(Eucelatoria* sp. nr. *armigera* Coq.): In: Proceedings of Third Workshop of AICRP on Biological Control of Crop Pests and weeds. 27-30. October 1980, Ludhiana, Punjab, India.

22. Pawar, C. S., Bhatnagar, V. S. and Jadhav, D. R. 1986. *Heliothis* species and their natural enemies with their potential for biological control. Proceedings of India Academy of Sciences (Animal Science) **95(6)**: 695-703.

23. Peter, C. and Balasubramaniam, R. 1984. Record of a new parasitoid of the beanfly. Current Research. **13(7/9)**: 60-61.

24. Prasad, D., Chand, P., Deka, H. K. and Prasad, R. 1989. Population dyanamics of *Heliothis armigera* (Hb.) on chickpea. Giornafe Italiaco di Entomologica. **4**: 223-228.

25. Rabindra, R. J. and Jayaraj, S. 1987. Evaluation of some adjuvants for NPV of *Heliothis armigera* (Hb.) International *Heliothis* Biological Control Workshop Newsletter **6**:12.

26. Rabindra, R. J. and Jayaraj, S. 1988. Evaluation of certain adjuvents for NPV of *Heliothis armigera* (Hb.) on chickpea. Indian Journal of Experimental Biology. **26(1)**: 60-62.

27. Reed, W., Cardona, C., Sithanantham , S. and Lateef, S. S. 1987. Chickpea insect pests and their control. In: The chickpea, Eds. M.C. Saxena, and K.B., Singh, C. A. B. International and ICARDA, U. K. 409 pp.

28. Reed, W. and Lateef, S. S. 1990. Pigeonpea : Pest Management. Chapter 14 in the Pigeonpea, Eds. Nene, Y. L. Susan D. Hall., and V. K. C. Sheila, A. B. International and ICRISAT, India.

29. Sachan. J. N. 1990. Present status of *Heliothis* on pules and strategies for its management. Paper presented at First National Workshop on *Heliothis* management-Current Status and Future Strategies, held at Directorate of Pulses Research, Kanpur, 208 024, India.

30. Santharam, G. and Balsubramanian, M. 1981. Effect of nuclear polyhedrosis virus (NPV) used alone and in combination with insecticides in controlling *Heliothis armigera* (Hb.) on Bengalgram. Journal of Entomological Research. **6(2)**: 179-181.

31. Sardana, H.R. and Verma, S. 1987. A note on the association of predatory wasps *Plistes herbaeus* Fb. and *Vespa orientalis* Fb. with the complex of greengram *Vigna radiata*. Plant Protection Bulletin, India **39 (4)**: 36.

32. Sathe, T. V. and Nikam, P.K. 1985. Studies on the host age selection by *Diadegma trichoptilus* (Cameron), a larval parasitoid of *Exelastis atomosa* (Walsingham). Current Science. **54 (15)** : 762-763.

33. Saxena, H.P. 1980. Report of research work on pests of rabi pulses 1979-1980. All India Coordinated Project on Improvement of pulses, IARI, New Delhi, India. 65pp.

34. Singh, S.P. 1993. potential for biological control of pests of Oilseeds and pulses. In: Compendium of Lectures of Summer Institute on Recent Advances in Entomology with Special Reference to Oilseeds and Pulses, (Ed.) D.D.R. Reddy, APAU, Hyderabad. 120-129.

35. Singh, K.J. and Singh, O.P. 1987. Natural enemies of Tur pod bug, *Clavigralla gibbosa* Spinols infesting pigeonpea in Madhya Pradesh. Journal of Biological Control. **1 (2)** : 83-85.

36. Singh, J. S. and Singh, M. L. 1990. Ecology of *Helicoverpa armigera* (Hb.) on chickpea in Punjab. Journal of Insect Science, **3(1)**: 47-52.

37. Sithanantham, S., Bhatnagar, V.S. Jadhav, D. R. and Reed, W. 1982. Some aspects of *Trichogramma* spp. in eggs of *Heliothis armigera* (Hb.) (Lepidoptera: Noctuidae). Presented at the International Symposium on *Trichogramma*, 20-23 April, 1986, Antibes, France.

38. Sithanantham, S., Rao, V. R. and Reed, W. 1987. Parasites of the pigeonpea pod fly, *Melanagromyza obtusa* (Malloch). Indian Journal of Biological Control. **1(1)**: 10-16.

39. Snap, M. M. and Deshmukh, R. B. 1988. Efficacy of nuclear polyhedrosis virus against *Heliothis armigera* (Hb.) infesting pigeonpea. Indian Journal of Pulse Research. **11**: 80-82.

40. Yadav, C. P., Lal, S. S. and Dias, C. A. R. 1985. *Campoletis chloridae* Uchida, a larval parasitoid of *Heliothis armigera* (Hb.) infesting chickpea. Bulletin of Entomological Research. **26(1)**: 99-100.

41. Yadav, D. N. and Patel, R. C. 1981. Egg parasitism of *Heliothis armigera* (Hb.) (Lepidoptera: Noctuidae) in Gujarat. Gujarat Agricultural University Research Journal. **7**:19-22.

9 Diseases of Cowpea and their Management

T. B. Anil Kumar and Shoba Cherian

CONTENTS

Integrated Pest and Disease Management
Rajeev K. Upadhyay, K.G. Mukerji, B.P. Chamola and O.P. Dubey (eds.)
APH Publishing Corporation, New Delhi-110002, India

ABSTRACT

Cowpea crop is affected by diseases caused by fungi (rust, septoria leaf spot, powdery mildew), viruses (cucumo virus) and bacteria (blights). The losses may be as high as 75%, particularly due to seed borne and other aforsaid diseases. The etiology, disease cycle, epidemiology and management practices have been described. Host plant resistance has been recommended as a key component of the management technology.

1. INTRODUCTION

Though the plant genus *Vigna* comprises of over 150 species, only a few of them are being cultivated. Cowpea (*Vigna unguiculata* (L) Walp) is mainly grown in West Africa, Brazil and India. Cowpea is relatively a minor legume in India, mainly grown as vegetable and fodder in crop in Rajasthan, Karnataka, Tamil Nadu, and Orissa (183). While Africa is the origin of cowpea, a great genetic diversity exists in the germplasm collection of Africa and India. Cowpea is known by various names such as southern pea, protopea, yardlongbean, catjungbean etc.

As in the most other cultivated plants, cowpea too is subjected to attack by various pathogens during seed germination to seed production and maturity. While over 50 fungal pathogens, 20 viruses and a few bacterial pathogens are known to attack cowpea, only rust, powdery mildew, *Septoria* & *Cercospora* leaf spots, bacterial blight and cowpea banding mosaic occur in throughout India in severe form and result in substantial yield losses. In addition, root-knot nematode could appear in a very severe form as well. Many seed-borne pathogens are known to cause extensive damage by way of seed decay and seedling mortality, and in certain instances it could be as high as 75%. The losses in many instances such as anthracnose, *Septoria*, rust, bacterial blight and certain mosaics are around 50% or more (26, 148, 178, 193). Many of the pathogens which attack pods and peduncles are known to result in shrivelled and or infected seeds which become unfit for either sowing or consumption. In the following pages a brief information on most of the pathogens attacking cowpea is given while certain economically important diseases have been chosen for detailed review.

2. FUNGAL PATHOGENS

Most of the important fungal pathogens causing seed and seedling diseases, collar, root and stem diseases, foliar and pod diseases have been listed in Tables 1-4. Few diseases namely rust, septoria leaf spot, and powdery mildew causing epidemics have been discribed in detail.

2.1 Rust

Two different rusts are known to occur on cowpea. These are brown rust caused by *Uromyces phaseoli* var. *vignae* and pink rust caused by *Phakopsora pachyrhizi*. The former is more destructive and widespread, being reported from most of the cowpea growing regions of the world. On the other hand, *P. pachyrhizi* is locally important in Nigeria though it is known to occur in Asia and Puerto Rico as well.

2.1.1 Brown rust

Brown rust is one of the most destructive diseases of cowpea. The disease was first reported by Barclay in 1891 (15) and was considered identical to the rust of beans caused by *Uromyces appendicualatus*. Fromme (41) was the first to describe cowpea rust as a clearly defined species and

Table 1 : Seed rot and seedling diseases of cowpea

Sl. No.	Pathogen	Symptoms	Management	References
1.	*Fusarium equiseti*	The seedlings show sudden and outright wilting. Roots of affected plants exhibit the symptoms of rot-rot. In later stages blackening of the stem base and appearance of black streaks on the roots are common.	Seed treatment with ceresan dry, difolatan, captan, thiram or benomyl	161, 170.
2.	*Fusarium solani*	The hypocotyl region exhibits a water soaked appearance and the leaves drop. Followed by rotting at the collar region.	Seed treatment with mancozeb of captafol	97
3.	*Macrophomina phaseolina*	Characterised by brown to greyish discolouration of the hypocotyl region which later turns black and the infected seedlings die. If the soil inoculum to high or seed is infected, seed or germling death is common.	Seed treatment with ceresan dry difolatan or captan	42, 170
4.	*Pythium aphanidermatum*	Causes seed decay and damping off. The lesions that begin on the hypocotyls progress upwards and rapidly give it a grey-green wet appearance and the seedling undergoes a watery collapse.	Thiram + BHC,. Captafol, Captan thiram, duter seed treatment	207, 105, 104
5.	*Rhizoctonia solani*	Causes pre and post-emergence mortality. The reddish brown lesions are usually limited to the collar region of the hypocotyl at which point the diseased seedling topples.	Quintozene, benomyl, carbendazim, thiophanate methyl seed treatment. Varieties Ptb-1, Calicut 78, C-5.7.	207, 201, 67
6.	*Sclerotium rolfsii*	Seed rot, sprout rot where the infection is very early during germination. In germinated seedlings, the cotyledonary leaves f rot and fail to open. Soft rot symptoms are seen in the follar region which is subsequently covered by white cotton grown of the mycelium.	Resistant lines NP3, IAR 339-1 IT 82D-699, Seed treatment with thiram captafol, captan.	105, 65, 103

Table 2 : Root, collar and stem pathogens of cowpea

Sl. No.	Pathogen	Symptoms	Disease management	References
1.	*Colletotrichum lindemuthjianum*	Mainly appears on stem lesions sunken tan to brown in colour and lenticular to circular in shape. Girdling of stem, branches, peduncles and petioles is common in highly susceptible varieties. Lesions also develop on leaves and podes.	New Era, TVU 76; Benomyl or mancozeb application	108
2.	*Fusarium* oxysporum f. sp. *tracheiphilum* (Wilt)	Initially leaves appear pale green, flaccid which soon turn yellow and drop from the plant. Early infection results in premature death. Vascular discolouration is common. Reddish brown lesions are seen on the tap root.	Blackeye, Iron, TVU 990, Iron TVU 1072, Iron Tvu 1611, Tvu 1560.	69, 11
3.	*Fusarium solani* (Root rot)	Roots show dry rotting, Severely affected plant dry.		97
4.	*Macrophomina phaseolina* (Ashy stem blight)	Affected stem and branches exhibit dry, ashy blight with numerous dark pyscnidia embedded in the affected zone.	58-57, 26/4/1, V16, K 39.	20, 54
5.	*Phytophthora cactorum*	The growing tip becomes soft and light brown in colour. The infected tip portion along with the side leaves droop and finally fall down.		95, 102, 117
6.	*Phytophthora vignae* f. sp. *vignae*	The disease is characterized by lesions with reddish brown streaks. The growing tip becomes soft and light brown in colour	KV 235, Tvu 3861Caloona	102, 191

Sl. No.	Pathogen	Symptoms	Disease management	References
7.	*Pythium aphanidermatum* (Wet stem rot)	The disease is characterized by a grey-green water-soaked girdle of the steam extending from soil level upto and sometimes including the lower portion of the lower branches. The stem cortical tissue becomes packed with oospores. Infected plant quickly wilts.		73, 104, 109
8.	*Pythium butleri* (Pythium rot)	Whitish fluffy fungal growth on Petals, fruiting branches and pods. Water soaked areas develop on the leaves where infected petals fall.	Zineb or maneb sprays	66, 181
9.	*Pythium myriotylum*	Causes root rot. The pathogen severely restricts root and tap growth. Severely affected plants dry.		31
10.	*Rhizoctonia solani*	Brownish black lesions appear at soil level which later on girdle the stem. Leaves turn yellow and droop. Minute sclerotial bodies can be seen in the affected portion.	Calcut 78, S 1476, C 5-7 are resistant	139, 201
11.	*Sclerotium rolfsii*	Brown lesions near collar region. Affected portion shrinks.. White mycelial growth and brown sclerotial bodies are seen in the affected portion.	IAR 339-1 IT 82D-699, NP 3 are resistant	65, 103
12.	*Sphaceloma* sp. (Scab)	Well formed scab lesions appear on stem, leaves and pods. On leaves, white elongate 1 to 15 mm long lesions occur on veins interveinal spots are circular 1 to 5 mm diam. Flower abortion is common		38

Table 3 : Foliar fungal pathogens of cowpea

Sl. No.	Pathogen	Symptoms	Management	References
1.	*Alternaria atrans*	Irrigular spots of light grey to brown colour starting from the margins of the leaves and proceed towards the madrib and are zonate. Spots are surrounded by yellowish halo.	Mancozeb or zineb sprays	176, 200
2.	*Alternaria tenuissma*	Spots circular to irrigular, centre depressed, straw coloured with slightly raised reddish margins. In severe infection, several spots coalesce to form large spot involving the whole leaflets, spots more common on the margin of the leaf.		160
3.	*Aristostoma concentrica* (zonate leaf spot)	Folicole, causing circular spots. 2-8 mm in diameter, marked by alternating concentric zones of red and white tissue, and beneath unzoned, brownish red. Pycnidia arranged irregularly in the white zones of the spots only.		188
4.	*Ascochyata phaseolorum*	Small, circular, discoloured areas are seen. As the lesions enlarge they turn to tan in the centre and olive green in the periphery with brown margin. Spots bear concentric rings. In severe cases blightening occurs.	Vita 7 Resistant Difolatan or benomyl sprays	135, 175
5.	*Cercospora canescens*	Roughly circular, cherry red to dark red spots of up to 10 mm diameter. Severely affected leaves turn yellow and abscise. The spots on abaxial leaf surface are also coloured red.	Tvu 224, Tvu 1993, Tvu 2785, Tvx 1836-481E resistant. Mancozeb or zineb spray	30, 57, 176, 189
6.	*Chaetoseptoria vignae*	Circular, yellowish collapsed spots upto 5 mm in diameter, distinct and with raised, narrow, red-brown border above but indistinct below. Pycnidia scattered, innate.	Early black Taylor Arlington	44, 202

Sl. No.	Pathogen	Symptoms	Disease management	References
7.	*Colletotrichum truncatum*	Brown blotch symptoms common on foliage which show numerous dark spots, viz., acervuli. Die back and stem anthracnose is seen in severely affected plants under favourable conditons.	Vita 1 Resistant	106
8.	*Corynespora cassicola*	Dark reddish brown circular spots 1-2 mm in diameter, larger turn to large target spots about 15 mm in diameter. Veinal necrosis is common. Lesions are also seen on petioles and stems.		39
9.	*Curvularia lunata*	To begin with small chlorotic spots are formed on the leaves. Later on these spots coalesce forming large brown area.		195
10.	*Dactuliophoro tarri*	Leasf spots broadly zonate on upper leaf surface, usually upto 1 cm diameter, but may reach 2 cvm, rosette like with alternating whitish and tan-coloured bands. On lower surface densely covered with sclerotiophores and sclerotia. Zonation less distinct.		83, 98
11.	*Drechslera rostrata*	First appears as small, roundish pale yellow spots, 5 mm diameter, coalesce to cover large area, turns brown in colour. Conidiophores and conidia are seen in the spots.		49
12.	*Entyloma vignae* (leaf smut)	Round or irregular spots 2-12 mm in diameter, appearing as surface discolouration with diffused edges, isolated or confluent on the leaves. Brown and change to green grey with the fading of the leaf.		16
13.	*Fusarium equiseti* (top necrosis)	Necrosis of apical buds, auxillary buds proliferate, bushy appearance, petioles droop, water soaked areas on stem which later turn brown. No of formation.		122
14.	*Leptosphaerulina trifolii*	small circular spots particularly along the margins of the leaf with yellowish white centre and dark brown margins		6

Sl. No.	Pathogen	Symptoms	Disease management	References
15.	*Myrothecium roridum*	Small, circular, tan coloured spots with broad violet to brown margin, surrounded by zones of translucent area which give it the appearance of concentric rings. Fructifications of the fungus appear as dark green sporodochia surrounded by a rim of white hair like mycelia.	P911 and P944 moderately resistant. benomyl, captan or zineb sprays	56, 99, 164, 165, 169
16.	*Phakopsora pachyrhizi (Pink rust)*	Produces bright orange pustules on leaves, stem, peduncles and pods, though occasionally the pustules are confined to under surface of leaves where they are pinkish in colour.		207
17.	*phyllosticta phaseolina*	Restricted largely to mature leaves. Spots are large, brownish and cover large areas of leaf. The whole leaf gives a water soaked appearance in humid weather.		176
18.	*Pseudocercospora cruenta*	Circular to irregular, cherry-red to reddish brown spots having diffused chlorotic halos which soon become necrotic and rusty brown in colour. Lesions range from 1-16 mm and are mainly confined to leaf lamina.	Pusa Barsati, C 152, Iron gray, C 20, P426 Resistant	27, 198
19.	*Rhizoctonia solani* (web blight)	Web blight appears as small, circular, light greyish-brown spots on leaf lamina. These enlarge and become irregular in shape, surrounded by Water soaked area.	Tvu 317, 1283, 4539 are resistant	81
20.	*Sclerotium rolfsii*	Minute brown dots which expand into irregular necrotic spots 2-3 mm in size, leaf blighting, rotting. Shot hole symptoms common in severe cases.	IAR 339-1, IT 82D-699 are resistant	103, 123
21.	*Synchytrium dolichii*	Forms circular to irregular, rust coloured, crateriform galls 0.5 mm in diameter on both leaf surfaces and on the petioles.		137

Table 4 : Fungal pathogens those attack pods.

Sl. No.	Pathogen	Symptoms	References
1.	*Botryodiplodia theobromae*	The skin of the infected pod becomes soft, blackish and encrusted with pycnidia and greysh green mycelium formed as a dense weft over the skin. The infected pods on drying become ash coloured with dark black pycnidia. Seeds from infected pods are ash coloured bearing pycnidia.	86
2.	*Choanophora cucurbitarum*	Initial symptoms appear as water soaked areas on pods subsequently developing into a wet rot affecting both young and mature pods. Diseased pods bear luxuriant whitish growth of the fungus which produces black headed pin like spring structures.	210
3.	*Cladosporium vignae*	Lesions on pods conspicuous, dark purple or black, roughened or scabby spots, very irregular in shape and usually 2-6 mm in diameter.	44, 202
4.	*Colletotrichum capsici*	Maturing pods show dark pin point dots indicating the presence of acervuli. The infected pods get shrivelled prematurely and the grain size is reduced. On the grains black dots are visible which are due to black acervuli.	85, 128, 181
5.	*Fusarium equiseti*	Water soaked regions appear on pods where infection has started which is followed by the development of necrotic spots. In severe cases mycelial growth is seen around the seeds when the pod is split open.	170
6.	*Macrophomina phaseolina*	The affected portion dries up leaving around green skin which later appears blighted. Pycnidial bodies are seen in these areas.	170
7.	*Myrothecium roridum*	Small, circular, fan coloured spots with broad violet to brown margins surrounded by.	56
8.	*Phjoma vignae*	Symptoms are characterized by the production of oval to irregular, ashy brown to dark brown, slightly depressed, necrotic lesions or areas over the rind of the pods. Numerous dark gregarious pycnidia are seen in the necrotic area.	128, 129
9.	*Pythium Butleri*	Water soaked lesions on affected portion. In advanced stage rotting of pods and fruiting branches occur.	66, 181
10.	*Pythium debaryanum*	The fungus attacks pods and produces typical cottony growth. Under cool and humid environment the disease manifests as large water soaked areas on green pods, which sooner get covered with cottony growth. In some cases a dirty fluid oozes out of the infected pod.	92
11.	*Sclerotium rolfsii*	Brown lesions covered with mycellium and sclerotia are found on pods.	84

assigned it to *Uromyces phaseoli* (Pers.) Wint. var. *vignae* (Barcl.) Arth. (=*U. vignae* Barcl.). It occurs in severe form during wet periods resulting in premature defoliation causing considerable yield loss, ranging from 6 to 60% depending upon the age of the plant and severity of the decrease (26, 174).

Symptoms

The infection is first noticed on the upper surface of the leaves in the form of small raised sori. These are initially covered with epidermis which ruptures in due course releasing the powdery spore dust. Under favourable weather conditions, numerous sori appear subsequently in rings around the first sorus. Similar sori are also encountered on the lower surface of the leaf. Both urediniospores and teleutospores are observed in the sori. The symptoms are more conspicuous on the leaves, even though the disease attacks stem and fruiting branches. Serious infection results in premature defoliation and low yields (180).

The pathogen

Uromyces phaseoli var. *vignae* is a full cycled autoecious rust (41). Pycnia are epiphyllous, in small groups, globose, honey yellow becoming brownish, 100-125 μ in diameter with short ostiolar filaments. Aecia are hyphophyllous, petiolicolous, in circular groups 2-4 mm across, often with annular arrangement, low cupulate 0.2 to 0.3 mm in diameter with evanescent white peridium, margin spreading with rhomboidal peridial cells; aeciospores are ellipsoid or oblong ellipsoid, 16-20 x 20-26μ, wall colourless, 1-1.5μ, minutely and closely verrucose. Uredinia are amphigenous and petiolicolous, scattered, roundish 1 mm or less in diameter and after with ruptured epidermis; urediniospores are ellipsoid or ovate ellipsoid 18-22 x 24-30μ, thin walled, finely echinulate with two superequatorial pores. Telia are amphigenous, roundish, 0.5 mm in diameter, with ruptured epidermis; teliospores are broadly ellipsoid, 20-25 x 27-35μ, with smooth dark chestnut brown wall, moderately thick 2-2.5μ, Pedicel hyaline, fragile, rarely equalling the spore length, germinating at maturity.

Factors affecting infection and disease development

Sokhi and Sohi (180) studied the effect of rainfall, temperature and relative humidity for three seasons and found rain fall a main factor responsible for the development of rust. Similar observations were also made by Rawal *et al.* (134). Chandrashekar *et al.* (26) found high humid environment causing increased disease pressure. The germination and penetration by urediniospores occurred in the early hours when relative humidity was above 90% and temperature ranged between 12-18°C. Similar observations were also made by Wessolek and Meier (203) in field trials in Egypt. Higher disease levels were recorded in fields subjected to sprinkler irrigation than those given surface irrigation. Of the major nutrients, phosphorus and potassium had no effect on disease severity while higher dosages of nitrogen reduced the severity.

Sohi and Rawal (179) observed lower phenol contents in susceptible genotypes and significantly higher in resistant genotypes. They obtained a good correlation between phenol content and rust severity level in different genotypes. Ultrastructural comparison of susceptible and immune varieties of cowpea leaves infected with rust showed signs of detectable incompatibility in the immune variety during early stages of houstorial formation when a deposit of callose containing material was formed around the host cell wall at the point of entry of the haustorium. In absence of hypersensitive reaction, changes in the host plasmalemma, rather than sheath formation have been of primary importance in restricting further fungal growth (53).

Rarely in a resistant genotype like Queen Anne, the haustorial formation induced either rapid host cell necrosis of formation of a callose like sheath which grew up from the host cell wall to completely enclose the haustorium. In general, varietal resistance has been expressed only by more typical response of hypersensitive death of invaded cells.

Management Strategies

Chemical control and host resistance are the two major means of controlling brown rust in cowpea. In a highly susceptible variety like C 152 the chemical control generally is not economical. However, under moderate disease pressure mancozeb or mancozeb + dinocap have been found useful (180). Bayton (94) and bayleton though effective may not be economical.

The only practical means of disease control in a low value crop like cowpea is through the use of resistant varieties. Genotypes 25494-1, 488-9-1-1, 1473 (93), Cream-40, Iron (180), RC-19, Guj-2 (26) are resistant to the rust. Genotypes V-16, V-70, V-118, BG-329, COVU-85, DPC-1240, CG-218 and GC-82-7 are known to possess slow rusting resistance (26, 158).

2.2 Septoria leaf spot

Leaf spot caused by *Septoria* spp. is one of the important diseases of cowpea. Four species of *Septoria* viz., *S. kozo-polzanskii, S. vignae, S. vignicola* and *S. vigna-sinensis* are known to parasitize cowpea. *S. kozo-polzanskii* is known to occur in Russia and Zambia, *S. vignae* in Africa, Brazil and India, *S. vignicola* in Eastern Africa and India, and *S. vigna-sinensis* in Taiwan. Leaf spot caused by *S. vignicola* is quite severe during wet weather and is known to result in substantial yield loss. The loss in green pod yield due to Septoria leaf spot could be as high as 53% (173, 178).

The pathogen

Septoria vignicola Rao

Pycnidia scattered, in single, measure 63 x 113.4µ diameter. Pycnidiospores filiform, hyaline, 2-4 septate, commonly 3 septate, flexous, 21-67.2µ long, 1.26-2.1µ thick (127). The pathogen survives in the infected debris. Primary inoculum comes from this source.

Symptoms

The disease is manifested by numerous, circular to irregular, water soaked spots on the leaves. With the advance of time these spots turn brown to tan with dark tan margins. Numerous black dot like bodies can be seen in these spots which are pycnidia of the fungus. These spots measure 2 to 6 mm in diameter. Spots appear on both the surface of leaf. Severely affected leaves become yellow and with slightest touch drop down.

Factors affecting infection and disease development

The disease is severe during rainy weather and results in defoliation. It is favoured by high humidity. The cuticular thickness has a role in plant resistance. Appreciable difference in cuticular thickness of resistant and susceptible varieties has been observed. Cuticle thickness in resistant variety varied from 13.6 to 18.6µ as compared to 8.1 to 10.9µ in susceptible varieties. Thus cuticle of the resistant varieties acts as a barrier by resisting pressure exerted by the pathogen at the time of germination (131). Sohi and Rawal (177) found more of free aminoacids and total soluble proteins in resistant varieties. The pathogen is favoured by high sugar contents while phenols do not appear to have any role in resistance. The disease resistance is controlled by single dominant gene (133).

Management

Clean cultivation, removal and burning of the left out plant debris and crop rotation are useful in avoiding early disease breakout. Growing intercrop, preferably one taller than cowpea helps in preventing quick spread of the disease by avoiding rain splash dispersal of conidia. The disease can be effectively controlled by foliar application of carbendazim, difolatan, mancozeb at 15 days intervals (176).

Many varieties are known to be resistant. These include Cream 40, Iron Grey, P 426, TVu 3, EC 43819 (57, 132, 172, 176, 177). Genotypes C 152, Co-4, HG-171, RC - 48, V 16, V 276, V 385 are known to posses partial resistance to Septoria leaf spot.

2.3 Powdery mildew

Powdery mildew is known to cause significant damage in early monsoon (April-May), to late monsoon of rabi and summer crops (February-March). In addition the disease could be severe on kharif or regular monsoon crop if interspersed with long dry spells. The pathogen attacks all the above ground parts and under severe conditions is known to result in 20% loss in grain yield (193). However, it appears that the yield loss is grossly underestimated. On a variety like RC 48, which is highly susceptible to mildew, the losses could be very high. Nanjundaiah (101) recorded a yield loss of 86% under severe mildew infection in case of variety RC 48. Mildew on cowpea is prevalent in most of the cowpea growing areas of the world. In India, it is especially severe in southern parts.

The causal organism

The causal organism is *Erysiphe polygoni* DC, the conidial state of which is *Oidium*. The mycelium is superficial on plant surface and only the haustoria enter the epidermal layer. Conidiophores develop from the ectophytic mycelium and bear conidia in chains. Conidiophores septate, conidia elliptical hyaline, unicellular and measure on an average 30 x 15μ. Cleistothecia are about 90μ in diameter with myceloid appendages. Usually two to eight asci are formed which are ovate, 46-72 x 30-45μ in size. Ascospores elliptical, hyaline, unicellular and measure 19-25 x 9-14μ.

Symptoms

The pathogen attacks all the above ground parts of the plant. The first symptom is generally seen on lower leaves as small, white, circular, floury patches on upper surface. As the disease progresses, these spots enlarge and the leaves in middle and top canopy also get affected. Subsequently, whiter fluffy lighter greyish powdery growth can be seen on petioles, branches, pods and tendrils. In severe cases defoliation is common. If the infection is severe in early stage of the plant growth, cracking of stem and branches takes place. Leaves turn yellow and drop down.

Etiology

The disease, under field conditions, is generally seen on around 40 days old plants. The primary source of inoculum may come from self sown plants, early sown crop or most likely from the weeds hosts. Common weed like *Lagasca mollis* is a known as a collateral host (133).

Mildew infection results in reduction in chlorophyll content of leaves and photosynthesis. The decrease in chlorophyll content is correlated with the corresponding increase in chlorophyllase activity in the later stages of disease development (68). In subsequent studies, Satyanarayana (142) observed a significant reduction in chlorophyll a, b and total chlorophyll content in susceptible varieties even during the early stages of diseases development. The cell permeability was significantly increased in susceptible varieties due to mildew infection. There was a significant increase in the activity of catalase, peroxidase even in the early stages of disease development in susceptible varieties while such increase was noticed only in the advance stages of mildew development in resistant varieties. Susceptible healthy plants contain more reducing and non-reducing sugars which are reduced significantly due to infection. Total phenol content has been found to be significantly higher in resistant variety which remained unaltered after infection while there was a significant reduction in total phenols due to infection in susceptible varieties.

Management

Several genotypes of cowpea have been reported to be resistant (50, 57, 91, 196). These are Chinese Red, Copusa 1, Copusa 2, Copusa 3, Cream 40, G 5, HG 22, Iron Grey, Lolitha, PLS 370, TVu 310, TVu 1012-2 and 2-1.

Of the 21 varieties tested intensively over three seasons. Raju *et al.*, (120) reported V 105, V 269, V 276, V 282, and V 385 to possess slow mildewing resistance. Of the germplasm material screened under glasshouse conditions, many were found to possess good level of slow mildewing resistance (119). Patel and Kuwite (112) screened 246 lines for mildew resistance and found thirteen lines to be resistant.

Many fungicides are known to give effective control of cowpea mildew. Singh and Musyoni (163) reported triademifon to be effective. Other fungicides which are effective are cosan, benomyl, difolatan, dikar (94, 176, 180). Rajender Singh and Anilkumar (121) obtained seed treatment with carbendazim followed by foliar spray with triadimefon to be very effective.

3. VIRUS PATHOGENS

More than a score of viruses are known to attack cowpea, though many of them are not important economically. Nor all these viruses are reported to occur in any particular place together. Of these, ten viruses are known to occur on cowpea in India. The most important ones in terms of severity and losses in grain yield is cowpea banding mosaic. The other important virus diseases are, cowpea chlorotic spot, cowpea aphid borne mosaic virus, tobacco ring spot virus etc. (90). In Iran, alfalfa mosaic virus and pea leaf roll virus are important (60, 63). Cowpea severe mosaic virus is a major viral pathogen on cowpea in Venezuela (33). Cowpea mottle virus and cowpea golden mosaic virus are economically important in Nigeria (10, 136). Cowpea stunt which is due to synergistic action of blackeye cowpea mosaic virus and cucumber mosaic virus is known to cause yield reduction by over 85% in USA (115). A brief information on the viruses reported so far on cowpea is given in Table 5. In addition to those listed in the table, bean yellow mosaic viruses also occurs on cowpea (18).

3.1 Cucumo viruses

Many strains of cucumo viruses have been reported to be pathogenic to cowpea. These include cucumber mosaic virus from USA, Morocco, Spain and the Philippines (7, 36, 40, 186). Cowpea banding mosaic virus from India (143) and cowpea ring spot virus from Iran (113).

The disease could be very severe and is known to cause reduction in yield by 41.9% (145). The information on the Indian strain viz., cowpea banding mosaic virus in given below :

The Pathogen

Cowpea banding mosaic virus belongs to cucumo virus group. The virus consists of isometric particles, 35nm in diameter and contains single stranded RNA. It has 17% nucleic acid and a sedimentation coefficient of 96S. In addition to cowpea, it is known to infect frenchbean and sunhemp. The thermal death point is between 50-55°C, and dilution end point is between 10^{-3} - 10^{-4}. It remains viable up to 24 hr at room temperature. Phosphate, acetate and boric acid buffers (0.5N) maintain the maximum infectivity of the virus (143).

Symptoms

The characteristic symptoms of the disease are mosaic mottling, vein clearing and green-vein banding. The virus may also cause slight crinkling and reduction in leaf size. The diseased plants remain green for longer periods and set fewer pods under field conditions.

Table 5 : Viruses pathogenic to Cowpea

Sl. No.	Virus	Symptoms and transmission	Physical propperties	Resistant genotypes	References
1	2	*3*	*4*	*5*	*6*
1.	Alfalfa mosaic virus	Yellow mosaic of foliag, leaflet deformation and dwarfing as well as stunting of plants may be seen *Acyrthosiphon pisum* transmits in a non-persistent manner.	Particles 18 to 60 nm, isometric and bacilliform containing RNA	PI 2289209 PI 293474 tolerant to virus	61
2.	Bean common mosaic virus (Cowpea strains)	Molting, reduction in leaf size and stunting. Virus is transmitted mechanically and in non-persistent way by *Aphis craccivora, A. gossypii, Myzus persicae.* Also seed -Borne.	Flexous rods with 750-925 nm. TDP 75-80°C, DEP 10^{-3} - 10^{-4}		140
3.	Blackeye cowpea mosaic virus	Large reddish ring like lesions spreading along the veins and forming a reddish-net pattern. Where systemic infection occurs, severe mottling, distortion and reddish vein necrosis are common. Virus is transmitted through seed, mechanically and in non-persistent manner by *Macrosiphum solanifolii, A. gossypii* and *Myzus persicae.*	Flexous rods measuring upto 760 nm. TDP 57-60°C, DEP 10^{-4} - 10^{-5}	Brown gesender, Magnolia blackeye Mississippi silver, worthmore.	7, 88, 89, 115, 185
4.	Broadbean wilt virus	Systemic mosaic, brownish, necrosis, leaf malformation and distortion. Transmitted mechanically and by aphids in non-persistent manner.	Isometric, diameter 25 nm		192
5.	Cowpea aphid borne mosaic virus	Severe mosaic, show variable amount of dark green vein-banding, leaf distortion, blistering and stunting. European strain is known to cause severe distorting mosaic. African strain causes irregular angular	Flexous filamentous particles around 750 mm in length. TDP 57-60°C DEP 10^{-3} - 10^{-4}	Tvu 1948, Tvu 79, Tvu 113, Tvu 310, Tvu 1185, Tvu 1330 etc.	12, 17, 199

1	2	3	4	5	6
		broken mosaic. Seed-borne, mechanically transmitted and also by *A. craccivora, A. fabae, A. gossypii, M. Persicae* in a non-persistent manner.			
6.	Cowpea chlorotic mottle virus	Bright yellow mottle on leaves, may range from chlorotic rings, to complete and uniform yellowing. Transmitted by sap and *Diabotrica undecium punctata, Ceretoma trifurcata.*	Isometric, 25mm dia. Single strand, RNA 24%, TIP 65-70°C, DEP 10^{-4} - 10^{-5}	PI 147652, 186458 and 186465	78, 138, 184
7.	Cowpea chlorotic spot virus (Sannhemp mosaic virus-tobamo virus)	Certain isolates produce mild green mottle, certain others cause chlorotic spot, Vein cleaning, leaf size reduction, dwarfing, early defoliation are also noticed. Highly susceptible varieties show severe mosaic. Seed borne, mechanically and by aphids, beetles.	Rigid rod Shaped particles 300 x 17 nm, TIP80-85°C, DEP 10^{-5} - 10^{-7}	Early sugar crowder, sunwanne, Taylor, CM 11, V 16	21, 22, 28, 80, 143, 144
8.	Cowpea golden mosaic virus (cowpea yellow fleck)	Infected cowpea plants have b large bright yellow patches n their leaves. In severe infection the entire leaf surface turns bright yellow. Leaves are distorted and blistered. Plants stunted. Transmitted by *Bemisia tabaci* in persistent manner.	Virus particles have not been isolated	BR 10 - Piani	10, 141, 147
9.	Cowpea little leaf virus	Characterized by the presence of numerous undersized leaves, with margins rolled inwards or cupped. Bud like structures are later formed at the exils and often accompanied by profuse branching and stunting eventually causing formation of a bushy Condition. Seed borne and transmitted by *Aphis craccivora* in a non-persistent manner.	Isometric particles 25 and 30 nm diameter		186

1	2	3	4	5	6
10.	Cowpea mild mottle virus	Mild systemic mottling. Inoculated susceptible plants show necrosis. Seed-borne. Transmitted mechanically and by *Bemisia tabaci.*	Straight to flexuous, filamentous particles 650 x 13 nm TDP 65-70°C, DEP 10^{-3} x 10^{-4}		19, 100
11.	Cowpea mosaic virus (cowpea yellow mosaic)	Varies with plant variety and virus isolates. Green mottle to yellow mosaic. Infected plants get distorted and growth	Isometric particicles 24 nm diameter. TDP 55-75°C, DEP $10^{-4.7}$ - $10^{-6.7}$	Victor, K798, Vita 3, Dixielec selection	28
12.	Cowpea mottle virus	Stunting, mottling or bright yellow mosaic. Distortion and reduction in leaf size and witches broom syndrome. Seed borne, sap and by *Ootheca mutabilis*	Isometric 30 nm, single stranded, RNA, TIP 60-65°C, Dixielee Sel. DEP 10^{-6} - 10^{-7}	Bachuana, New era, Alabunch, dixielee sel.	136
13.	Cowpea severe mosaic virus	Leaves exhibit severe mosaic with blistering, distortion and necrosis. Seedling infection may result in death. Seed transmission and Chysomelidae beetles *Cerotoma ruficornia* etc.	Isometric particles, 28 nm diam., TDP 65-70°C, DEP 10^{-4} - 10^{-5}	BR 10 - Piani	2, 16, 141, 171
14.	Cowpea stunt virus	Severe stunting, leaves degenerate, pale green colour and leaves curling Transmitted by *Aphis craccivora* in a persistent manner.			55
15.	Pea leaf roll virus	Severe stunting, Transmitted by *Aphis craccivora, A. gossypi, Acyrthosiphon pisum.*	Isometric, 25 nm diameter		60, 62
16.	Peanut mottle virus	Symptomless or mild mottle, necrosis. Transmitted by seed.			34

1	*2*	*3*	*4*	*5*	*6*
17.	Southern bean mosaic virus	Vein cleaning in the primary leaves and mottling in trifoliate leaves. Seed borne sap transmitted and by *Cerotoma trifurcata Ootheca Mutabilis.*	Isometric particicles 28-30 nm. Single standed RNA. TIP 85-90°C, DEP 10^{-7} - 10^{-8}	Iron, Potmac, Brown Crowder Tardy, Georgia 21, Early Pinkeye	3, 78, 157, 167
18.	Tobacco ring spot virus	Producer concentric, chlorotic and necrotic ringspots, veinal necrosis, necrotic streaking of stem and petiole, stunting, ultimately death of the seedlings, mechanically transmitted	Isometric particles 28 nm. TIP 80-85°C in VC isolate 65-70 in Vu isolate DEP 10^{-5} - 10^{-6}	S 288, CG 5, CG 11, CG 28, Copusa 3, K 39	43, 87, 114
19.	Tomato spotted wilt virus (Bud necrosis)	Initially top necrosis is observed followed by gradual drying and death of plants. Sometimes necrotic streaks are observed on the stems of grown up plants. Leaves show veinal chlorosis followed by veinal necrosis resulting in slight downward curling.	Isometric, diameter 85 nm the particles are enveloped.	UPC 2201 and UPC 6342	32
20.	Urdbean leaf crinkle virus	Distortion and expansion of leaf lamina, crinkling of leaves. Mechanical as well as by *Henosepilachna dodecastigma.*	Isometric AV. diameter 30 nm		74, 75

Transmission

The virus is transmitted by sap, insect vector and through seed. It is transmitted in stylet borne manner by several species of *Aphis* of which *A. craccivora* is the most efficient. The acquisition feeding period though is less than one minute is best when feeding is allowed for five minutes. All the developmental forms of the aphid are equally efficient in transmitting the virus. The virus is also seed-borne and seed transmission ranges from 15 to 31% (29, 116).

Etiology

The primary source of inoculum is either seed or infected plants of cowpea, frenchbean or sunnhemp. The spread of the virus is higher in kharif than in summer (145). A significant and positive correlation is observed between the extent of seed transmission and the final incidence of the virus in both kharif and summer. There is no significant reduction in pollen viability of infected plants. Virus is known to be present in all parts of the developing seed, but maximum concentration is in plumule and radicle, and at maturity completely disappears from seed coat (47, 153). Seed lots with an initial 15% of seeds infected would result in 100% incidence in kharif and 27% in summer. The extent of seed transmission could be zero to over 30% (149), and such high level of seed transmission could be due to the localization of the virus in plumular bud (47). The extent of seed transmission varies from 15 to 31% in different varieties. Seeds from plants infected 13 days after sowing carried the virus to the extent of 32% while those from plants infected 33 and 53 days after sowing carried to the extent of 22 and 21.5%. Cowpea grown in March-May has considerably less virus infection than those sown in July. Virus spreads faster in autumn than in summer.

The rates of net and gross production as well as contents of total chlorophyll, chlorophyll a, chlorophyll b and carotenoids are reduced and the respiratory loss in higher in infected leaves than in healthy ones (190). Infected C 152 plants show a general increase in the level of phosphorus and decrease in nitrogen and potassium. Virus infection also results in a reduction in protein and free aminoacid content.

Management

Vector control is not economical, since the virus is transmitted non-persistently. As the virus is transmitted to a considerable extent through seed, use of virus free seed is very essential. The virus can be eliminated from the infected seed either by dry heat treatment (65°C for 15 minutes) or hot water (10 minutes at 50°C) treatment (146). Paraffin oil spray effectively inhibits virus transmission (151). Soaking seeds in malic hydrazide either at 40 or 100µg/ml for 90 min. eliminated seed infection. The granular insecticide such as disulfon granules at 2 kg. a.i./ha has given very effective control. Spraying plants with nucleic acid analogues 3, 10, 20, and 35 days after inoculation reduced seed transmission (116).

Crop grown in March-May generally has considerably less virus infection than those sown in July. Barrier crop of pearl millet around cowpea plants 1.5 and 2.25 M wide is known to give complete protection from virus infection. Mixed cropping with maize reduces infection. Higher sowing density and phosphorus levels reduce disease incidence (152).

Cowpea genotype V 37, K. 39, P 1475, P 304, P 109 and 1193 have been reported to posses resistance to cowpea banding mosaic virus (70, 72).

4. BACTERIAL PATHOGENS

The most widely distributed and economically important bacterial disease of cowpea is bacterial blight caused by *Xanthomonas compestris* pv. *vignicola*. Two other bacterial pathogens have been reported on cowpea. The bacterial pustule pathogen, *Xanthomonas campestris* pv. *phaseoli* which is common in Nigeria and bacterial necrosis caused by *Ceratobacterium flacumfaciens* pv. *flacumfaciens* (14, 209). These two are restricted to small geographic locations and pre relatively less important.

4.1 Bacterial blight

Bacterial blight caused by *Xanthomonas compestris* pv. *vignicola* is one of the destructive diseases of cowpea in India especially on early monsoon crop sown during April-May. Most of the cultivated varieties are susceptible and under heavy disease pressure are known to suffer over 75% loss in grain yield (107). The disease was described for the first time in 1931 from Oklahoma, USA (82) and is now known to be present in most of the cowpea growing regions of the world. It was reported from India for the first time in 1950 (111). Bacterial blight is also known to occur in Puerto Rico (197), Brazil Tanzania (37), Kenya (64), Nigeria (206) and many other African countries.

The Pathogen

The bacterium is rod shaped with average size of 1.74µ x 0.86µ, motile with single polar flagellum, gram negative and not acid fast. Capsule is distinct. The organism liquifies gelatin, produces H_2S, lacks indole formation. MR and VP tests are negative. The optimum temperature is 25-32°C and thermal death point is 50°C. The bacterium digests casein, reduces litmus milk without formation of tyrosine crystals and strict arobic (111).

Symptoms

The pathogen is both internally and externally seed-borne (48, 159). Heavily infected seeds give rise to infected seedlings where seedling mortality could be very high. The prominent symptoms are production of spots on leaves and blight on plants. The spots enlarge into brown to tan brown areas measuring 10 mm in diameter or even more. More often infection starts from the peripheral veins spreading on either side in a 'V' shape which turn brown to reddish brown in colour. In severe cases defoliation is common, wherein flower drop in high. On pods water soaked areas can be seen.

Management

The disease is favoured by warm humid weather. In studies conducted at UAS, Bangalore, blight severity was found to be very high on May sown crop and the severity gradually declined on crops sown thereafter (155). Intercropping cowpea in maize has been found to reduce the blight severity (110).

Seed treatment with Agromycin 100, streptocycline, agallol, ceresan wet significantly reduced the recovery of the pathogen from the seeds. Solar heat treatment was effective in eradicating seed borne inoculum (59). Spraying copper and copper plus mancozeb mixtures controlled the leaf spot and stem canker symptoms and significantly increased the yield (45). Chakravarti *et al.* obtained effective control by fortnightly sprays of agromycin or streptocycline (23). Three sprays of Agromycin 100 (250 ug/ml) at 10 days interval resulted in 38% increase in grain yield over unsprayed check (24).

Many workers have evaluated cowpea varieties for resistance to bacterial blight (156, 194). Allen *et al.* (5) reported Vita 4 to be resistant to bacterial blight in addition to its resistance to scab, brown blotch and *Septoria*. Abbaiah *et al.* (1) found UPC 125 to be immune and V 240 and V 701 to be highly resistant. Dhiman *et al.* (35) found Sel. 263 to possess multiple disease resistance including bacterial blight. Similarly CNC x 11-9D possesses resistance to blight, aphid borne mosaic, smut and powdery mildew (118).

Other bacterial pathogens

While *Xanthomonas campestris* pv. *vignicola* is the most widely distributed and economically important bacterial pathogen of cowpea, two other bacterial paghotens have been reported on cowpea. These are bacterial leaf spot caused by *Xanthomonas campestris* pv. *phaseoli* from Nigeria (14), and a disease from Australia caused by *Cerobacterium flaccumfaciens* pv. *flaccumfaciens* (209).

5. FUTURE LINE OF WORK

Cowpea because of its many utilities and adaptability to different farming systems can play an important role in sustainable agriculture. There are genotypes that can mature in 75-80 days and thus fit in either pre-monsoon situation as a catch crop or can be grown in rice fallows immediately after the harvest of rice, especially in coastal areas. Due to its low productivity in the above mentioned situations, plant protection measures are not economical. A useful approach to raise a healthy crop is to go for varieties with field resistance. Incidentally there is a great scope for this as genetic variability with regard to disease resistance is quite high. In most of the cowpea growing regions of the world more than one disease prevail in fairly severe form during the crop growth period and hence more and more varieties with multiple disease resistance should either be sorted out or developed. In fact a few varieties have been identified to possess good level of multiple disease resistance (3,9). In several cases their yield potentials have to increased.

Apart from the genetic improvement or selection to minimize the impact of diseases, the other way to take care of the diseases is by following useful agricultural practices. Mixed farming, especially mixing cowpea with cereals or millets helps reduce disease build-up and spread. Mixing cowpea with *Eleusine coracana* either in 1 : 2 or 1:3 proportion (rows) reduces severity of rust on cowpea significantly. Adjustment of sowing dates, wherever possible, would again help reduce diseases severity. Following crop rotation avoids fast buildup of disease. Many of the cowpea pathogens are seed-borne. Left untreated, infected seeds can result in heavy damage. Bacterial blight pathogen survives mostly in or on seed and thus is the source of primary inoculum. Simple practice of treating seeds with antibacterial substances can eliminate the pathogens. Seed treatment with fungicides should be an essential component of agronomic practices. For, an inexpensive operation like seed treatment can result in a good plant stand which is very essential for realising potential yield levels. Hence, more attention needs to be given to develop and adopt suitable agricultural practices which minimise the disease severity and increase the production.

In so far as existing scientific information on various diseases afflicting cowpea, there are many gaps. Intensive research needs to be carried out n epidemiology, physiology of parasitism, breeding for disease, particularly multiple disease resistance, and on disease management practices.

6. REFERENCES

1. Abbaiah, K., Seenaiah, P., Naidu, N. V. and Rao, M. S. 1988. Reaction of cowpea cultivars to bacterial blight. Indian J. Pulses Res., **1** : 65-66.

2. Allen, D. J. 1977. The international cowpea disease nursery programme; towards multiple virus resistance. Trop. Gr. Leg. Bull., **8** : 29-32.

3. Allen, D. J., Anno-Nyako, F. O., Ochieng, R. S. and Ratinam, M 1981a. Beetle transmission of cowpea mottle and southern bean mosaic viruses in West Africa. Trop. Agri. (Trin), **58** : 171-175.

4. Allen, D. J., Emechebe, A. M. and Ndimande, B. 1981b. Identification of resistance in cowpea to diseases of the African Savanas. Trop. Agri. (Trin.). **58** : 267-274.

5. Allen, D. J., Nebane, C.L. N. and Raji, J.A. 1981c. Screening for resistance to bacterial blight in cowpea. Trop. Pest. Manage. **27** : 218-224.

6. Anahosur, K. H., Bidari, V. B. and Hiremath, R. V. 1974. *Leptosphaerulina* leaf spot of cowpea. A new record to India. Curr. Sci., **43** : 118.

7. Anderson, C. W. 1955. Vigna and Crotalaria viruses in Florida II. Notations concerning cowpea mosaic virus (*Marmor vignae*). Pl. Dis. Reptr., **39** : 349-353.

8. Anilkumar, T. B. and Saifulla, M. 1986. Pathology Section, Annual Report for the year 1986, AICIP on Pulses, UAS, Bangalore.

9. Anilkumar, T. B. Chandrashekar, M., Saifulla, M. and Veerappa, K. B. 1989. Assessment of cowpea genotypes for multiple disease resistance. Indian Phytopath., **42** : 334 (Abstr).

10. Anno-Nyako, F. C., Vetten, H. J., Allen, D. J. and Thottapilly, G. 1983. The relation between cowpea golden mosaic and its vector, *Bemesia tabaci.* Ann. Appl. Bio.**102** : 319-323.

11. Armstrong, G. M. and Armstrong, J. K. 1980. Cowpea wilt *Fusarium oxysporum* f. sp. *tracheiphilum* race 1 from Nigeria. Pl. Dis., **64** : 954-955.

12. Atiri, G. I., Ekpo, E. J. A. and Thottappilly, G. 1984. The effect of aphid resistance in cowpea on infestation and development of *Aphis craccivora* and transmission of cowpea aphid-borne mosaic virus. Ann. appl. Biol., **104** : 331-346.

13. Atiri, G. I. and Thottappilly, G. 1984. Relative usefulness of mechanical and aphid inoculation as modes of screening cowpeas for resistance against cowpea aphid-borne mosaic virus. Trop. Agri., **61** : 289-292.

14. Bailey, A. G. 1966. A check list of plant diseases in Nigeria. Fed. Dept. agric. Res. Ibadan, Nigeria, Mem. No. 96, 37pp.

15. Barclay, A. 1891. Additional urediniae from the neighbourhood of Shimla. J. Asiatic Soc. Bengal, **60** : 211-230.

16. Batista, M. De. F., Costa, C. L., Kitajima, E. W. and Lin, M. J. 1988. Transmission of an isolate of cowpea severe mosaic virus by chrysomelid beetles. Fitopathologia Brasileira, **13** : 264-268. (Rev. Pl. Pathol., **69** : 2698).

17. Bock, K. R. 1973. East African strains of cowpea aphid-borne mosaic virus. Ann. appl. Biol., **74** : 75-83.

18. Brierly, P. and Smith, E. F. 1962. Three cowpea mosaic viruses from gladiolus. Pl. Dis. Reptr., **46** : 335-337.

19. Brunt, A. A. and Kenten, R. H. 1973. Cowpea mild mottle, a newly recognised virus infecting cowpeas (*Vigna unguiculata*) in Ghana. Ann. appl. Biol., **74** : 67-74.

20. Burke, D. W., Ditshipi, P. and DeMooy, C. J. 1986. Virus disease of cowpeas in dry land and irrigated plots in Botswana. Pl Dis. **70**: 801.

21. Capoor, S. P. and Varma, P.M. 1956. Studies on a mosaic disease of *Vigna cylindrica* Skel. Indian J. Agric. Sci., **26** : 95-103.

22. Capoor, S. P., Varma, P. M. and Uppal, B. N. 1947. A mosaic disease of *Vigna catiang* Walp. Curr. Sci., **16** : 151.

23. Chakravarti, B. P., Choudhury, S. L. and Gour, H. N. 1980. Chemical control of bacterial blight of cowpea. Indian J. Mycol. Pl. Path., **10** : 21 (Abstr.).

24. Chakravarti, B. P., Shekhawat, P. S. and Choudhury, S. L. 1976. Control of bacterial blight of cowpea. Univ. Udaipur Res. J., **14** :.89.

25. Chandrashekar, M., Anilkumar, T. B., Saifulla, M. and Salimath, P. M. 1988. Leaf rust epidemics of certain promising cowpea varieties compared by using logistic and Gompertz model. Trop. Agri., **65** : 37-40.

26. Chandrashekar, M., Anilkumar, T. B., Saifulla, M. and Yadahally, Y. H. 1989. Effect of different dates of sowing cowpea on the severity of leaf rust caused by *Uromyces phaseoli* var. *vignae*. Trop. Agi., **66**: 149-152

27. Chandrasekaran, S. and Rangaswami, G. 1960. Studies on *Cercospora cruenta* occurring in *Vigna catjang*. Indian Phytopath., **13** : 96-99.

28. Chant, S. R. 1959. Viruses of cowpea *Vigna unguiculata* (L) Walp. in Nigeria. Ann. appl. Biol., **47** : 565-572.

29. Chenulu, V. V., Sachchidananda, J. and Mehta, S. C. 1968. Studies on a mosaic disease of cowpea from India. Phytopath. Z., **63** : 381-387.

30. Chohan, J. S., Singh, R. S. and Chahal, D. S. 1971. Fungi from the Punjab state - 1. PAU J. Res., **7** : 192-196.

31. Craft, B. J. 1989. Root rot of cowpeas caused by *Pythium myriotylum* in northern Queensland. Australasian Pl. Path., **18** : 8-9.

32. Das, N. D. 1986. The occurrence of bud necrosis on fodder cowpea Indian J. Mycol. Pl. Path., **16** : 138.

33. Debrot, E. C. and de Rajas, C.E.B. 1967. El virus del mosaico del frijol *Vigna sinesis* Endl. (cowpea mosaic virus) en Venezuela. Agron. Trop, **17** : 3-15.

34. Demski, J. W., Alexander, A. T., Stefani, M. A. and Kuhn, C. W. 1983. Natural infection, disease reactions and epidemiological implications of peanut mottle virus in cowpea. Z Pl. Dis., **67** : 267-269.

35. Dhiman, J. S., Tarsem Lal, and Surjan Singh, 1989. Reaction of cowpea cultivars to bacterial blight, anthracnose and cowpea mosaic disease and their effects on economic performance, Trop. Agri., **66** : 17-20.

36. Diazo-Ruiz, J. R. 1976. Properties and electron microscopy of a Vigna virus belonging to the cucumovirus group. Microbiol. Espanola, **9** : 18.

37. Ebbels, D. L. and Allen, D.J. 1979. A supplementary and annotated list of plant diseases, pathogens and associated fungi in Tanzania. Phytopath. Paper No. 22, Commonwealth Mycological Institute, New.

38. Emechebe, A. M. 1980. Scab disease of cowpea (*Vigna unguiculata*) caused by a species of the fungus *Sphaceloma*. Ann. appl. Biol., **96**: 11-16.

39. Fazalnoor, K., Anahosur, K. H. and Narayanaswamy, B. C. 1971. *Corynespora* leaf spot of cowpea (*Vigna sinensis* (Torner) Savi) a new record to India. Curr. Sci., **40** : 116-117.

40. Fischer, H. U. and Lockhart, B. E. 1976. A strain of cowpea aphid-borne mosaic virus isolated from cowpea in Morocco, Phytopath. Z., **85** :43-48.

41. Fromme, F. D. 1924. The rust of cowpea. Phytopathology, **14** : 67-79.

42. Gaikwad, D. G. and Sokhi, S. S. 1987. Detection of seed rot and seedling infection in naturally infected cowpea seed in Senegal and their control. Pl. Dis. Res., **2** : 127-128.

43. Ganacharya, N. M. and Mali, V. R. 1981. Comparative studies on two isolates of tobacco ring spot virus from cowpea. Indian Phytopath., **34** : 112 (abstr.).

44. Gardner, M. W. 1925. Cladosporium leaf spot of cowpea. Phytopathology, **15**: 453-462.

45. Gitaitis, R. D., Bell, D. K. and Smittle, D. A. 1986. Epidemiology and control of bacterial blight and canker of cowpea Pl. Dis., **70**:187-190.

46. Gour, H.N., Gupta, U.P. and Singh, R. D. 1981. Permeability alteronations in resistant and susceptible cowpea varieties inoculated with *Xanthomonas compestris* pv. *vignicola. Proc. 3rd Intr. Syump.* Pl. Path., New Delhi, 16-18, Dec. 1981. p. 120 (Abstr.).

47. Gupta, M.D. and Sumanwar, A.S. 1980. The location of two mosaic viruses in cowpea seeds. Seed Sci. Technol., **8** : 203-206.

48. Gupta, V. P. and Chakravarti, B. P. 1980. Seed detection and transmission of bacterial blight of cowpea. Indian J. Mycol. Pl. Path., **10** : 33 (Abstr.)

49. Gupta, R.B.L., Singh, G. and Dalela, G. G. 1972. Occurrence of *Drechslera rostrataonm* cowpea in India. Indian J. Mycol. Pl. Path., **2** : 86.

50. Hanchinal, R. R., Bidar, V. B., Habib, A. V. and Goud, J. V. 1976. Varietal reaction of cowpea to powdery mildew (*Erysiphe polygoni* DC). Veg. Sci., **3** : 108-110.

51. Heath, M. C. 1971. Haustorial sheath formation on cowpea leaves immune to rust infection. Phytopathology, **61** : 383-388.

52. Heath, M. C. 1972. Ultrastructures of host and non-host reactions to cowpea rust. Phytopathology, **62** : 27-38.

53. Heath, M. C., and Heath, I. B. 1971. Ultrastructure of an immune and susceptible reaction of cowpea leaves in rust infections. Physiol. Pl. Path., **1** : 277-287.

54. Hooda, I. and Grover, R. K. 1989. Role of root exudates in altering the efficacy of fungicides used in controlling *Macrophomina phaseolina*. Crop Res. India, **2** : 78-53.

55. Iwaki, M. 1979. Virus and mycoplasma diseases of leguminous crops in Indonesia. Rev. Pl. Prot. Res., **12** : 88-97. (Rev. Pl. Path., **60** : 2857).

56. Jamaluddin, 1979. Studies on the control of *Myrothecium* rot of cowpea pods. Indian Phytopath., **32** : 313-314.

57. Jhooty, J. S., Sokhi, S. S., Sohi, H. S., Bains, S. S., Rawal, H. S. and Munsi, G. D. 1980. Evaluation of cowpea germplasm against different diseases. Crop Improv., **7** : 18-24.

58. Jindal, J. K., Patel, P. N. and Khan, A. N. 1981. Variability in Xanthomondas of grain legumes II. Pathogenic variability in *Xanthomas phaseoli* mungbean strain, *X. vignicola* and *X. phaseoli* var. *sojense.* Phytopath,. Z., **100** : 1-9.

59. Jindal, K. K., Thind, B. S. and Soni, P. S. 1989. Physical and chemical agents for the control of *Xanthomonas campestris* pv. *vignicola* from cowpea seeds. Seed Sci. Technol., **17** : 371-382.

60. Kaiser, W. J. 1972. Diseases of food legumes caused by pea leafroll virus in Iran. F.A.O. Pl. Prot. Bull., **20** : 127-132.

61. Kaiser, W.J. 1979. Natural infection of cowpea and mungbean by alfalfa mosaic virus in Iran. Pl. Dis. Reptr. **63** : 414-418.

62. Kaiser, W. J. and Danesh, D. 1974. Biology of four viruses affecting Cicer arientinum in Iran. Phytopathology **61** : 372-375.

63. Kaiser, W. J. and Mossahebi, G. H. 1975. Studies with cowpea aphid-borne mosaic virus and its effect on cowpea in Iran. F.A.O. Pl. Prot. Bull., **23** : 33-39.

64. Kaiser, W. J. and Ramos, A.H. 1979. Two bacterial diseases of cowpea in East Africa. Pl. Dis. Reptr., **63**: 304-308.

65. Karat, K.P.R., Venugopal, R. and Goud, J. V. 1985. Identification of field resistance and symptomatology for collar rot of cowpea caused by *Sclerotium rolfsii* under natural conditions. Plant Path. Newsl., **3** : 23.

66. Kataria, H. R. and Dodan, D. S. 1981. Influence of herbicides on antifungal activity of some fungicides against Pythium butleri and *Rhizoctonia solani* affecting cowpea. Pestic. Sci., **13** : 583-588.

67. Kataria, H. R. and Sunder, S. 1987. Fungicidal control of *Rhizoctonia solani* in relation to soil texture, organic matter and clay minerals. Ann. appl. Biol. **110** : 43-51.

68. Kaur, M. and Despande, K. B. 1980. Photosynthetic activities of cowpea plants infected with *Erysiphe polygoni* Indian Phytopath., **33** : 344-345.

69. Kendrick, J. B. 1931. Seed transmission of cowpea Fusarium wilt. *Phytopathology*, **21** : 979-983

70. Khalikar, P. V. 1979. Studies on seed-borne virus diseases of Pusa Vaishakhi Cowpea (*Vigna unguiculata*) cultivar. M.Sc. (Agri) Thesis, MAU, Parabhani.

71. Khan, K. R. and Sullia, S. B. 1980. *Arthrinium Phaseospermum* var. *indicum* var. nov. a new market pathogen of cowpea, gardenpea and Frenchbean. Acta Botanica Indica, **8** : 103-104.

72. Khatri, H. L. and Singh, L. 1974. Studies on a mosaic disease of cowpea. PAU J. Res., **11** : 288-294.

73. Koleosho, B., Ikotun, T. and Faboya, O. 1987. The role of oxalic acid in the pathogenicity of *Pythium aphanidermatum* on different cowpea varieties. Phytoparasitica, **15** : 317-323.

74. Kolte, S. J. and Nene, Y.L. 1972. Studies on symptoms and mode of transmission of the leaf crinkle virus of urbean (*Phaseolus mungo*). Indian Phytopath., **25** : 401-404.

75. Kolte, S. J. and Nene, Y.L. 1975. Host range and properties of urbean leaf crinkle virus. Indian Phytopath., **28** : 430-431.

76. Krishnamurthy, E. and Anilkumar, T.B. 1985. Biochemical changes in cowpea due to infection by *Xanthomonas campestris* pv. *vignicola Zbl.* Mikrobiol., **140** : 507-511.

77. Krishnamurthy, E., Joshi, S. S. and Anilkumar, T. B. 1985. Histopathological and histochemical changes in cowpea due to infection by *Xanthomonas campestris* pv. *vignicola. Zbl.* Mikrobiol., **140** : 425-430.

78. Kuhn, C.W. 1963. Field occurrence and properties of the cowpea strain of southern bean mosaic virus. Phytopathology, **53** : 732-753.

79. Kuhn, C. W. 1974. Separation of cowpea virus mixture. Phytopathology, **54** : 739-740.

80. Kulthe, K. S. and Mali, V. R. 1979. Occurrence of a tobacco virus on cowpea (*Vigna unguiculata*) in India. Trop. Gr. Zeg. Bull., **16** : 8-13.

81. Lakshmanan, P., Nair, C. and Menon, M. R. 1979. Collar rot and web blight of cowpea caused by *Rhizoctonia solani* in Kerala, India. Pl. Dis. Reptr., **63** : 410-413.

82. Larsh, H. W. 1943. Bacterial crown canker of cowpea of Oklahoma. Pl. Dis., Reptr., **27** : 555.

83. Leakey, C.L.A. 1964. Dectuliphora, a new genus of mycelia sterilia from tropical Africa. Trans. Br. mycol. Soc., **47** : 341-350.

84. Lima, E.F. and Tabosa, S.A.S. 1977. (Leaf spots and pod and branch rot of cowpea (*Vigna sinesis*) caused by Sclerotium rolfsii. *Fitopatologia Brasileira*, 2 : 187-192. Rev. Pl. Path., **57** : 4248.

85. Maholay, M.N. and Sohi, H. S. 1978a. Studies on seed-borne infection of *Colletotrichum capsici* (Byd.) B. and B on cowpea and its control. Indian J. Hort., **35** : 74-77.

86. Maholay, M. N. and Sohi, H. S. 1978b. Botrayodiplodia pod and seed rot of cowpea. Indian J. Hort., **35**: 165-167.

87. Mali, V.R. and Ganacharya, N.M. 1984. Comparative studies on two isolates of tobacco ringspot virus from cowpea. Indian Phytopath., **37** : 630-632.

88. Mali, V.R. and Kulthe, K. S. 1980a. comparative studies on three seed-borne virus isolates from cowpea. Indian Phytopath., **33** : 415-418.

89. Mali, V.R. and Kulthe, K. S. 1980b. A seed-borne potyvirus causing mosaic of cowpea in India. Pl. Dis., **64** : 925-928.

90. Mali, V.R. and Thottappilly, G. 1986. Virus disease of cowpea in Tropica. Rev. Trop. Pl. Path., **3** : 361-403.

91. Marimuthu, T. and Sivaprakasam, K. 1981. Reaction of cowpea germplasm to rust disease. Pulse Crops Newsl., **1** : 70.

92. Mathur, A. K. and Tyagi, R.N.S. 1983. A new pod rot due to Pythium debaryanum in cowpea, mung and urd. Indian J. Mycol. Pl. Path., **13** : 219-220.

93. Mathur, S. Thakore, B.B.L. and Singh, R. B. 1986. Reaction of cowpea germplasm to rust in Rajasthan. Indian J. Mycol. Pl. Path., **16** : 3418.

94. Mathur, S. Singh, R.B., Lal Thakore, B.B. and Chakravarti, B.P. 1980. Outbreaks of cowpea rust in Rajsthan and its control by fungicides. Indian J. Mycol. Pl. Path., **10** : 20 (Abstr.).

95. Milgo, J. K. 1988. Breeding for resistance/tolerance to *Phytophthora* stem rot in cowpea. TARO Newsl., **3** : 24-25.

96. Mohan, R., Alagianagalingam, M. N. and Govindaswamy, C. V. 1978. Changes in phenolics and carbohydrates in cowpea varieties infected by *Xanthomonas vignicola* Burk. Curr. Sci., **27** : 431-432.

97. Monga, D. and Grover, R.K. 1991. Chemical control of root rot of cowpea in relation to altered pathogenicity of *Fusarium solani*. Indian Phytopath., **44** : 462-469.

98. Mukibi, J. 1976. Preliminary observations on the germination of sclerotia of *Dactuliophora tarri* pathogen of cowpeas. E. Afr. Agric. For. J., **42** : 44-46.

99. Munjal, R. L. 1960. A commonly occurring leaf spot disease caused by *Myrothecium roridum* Tode ex Fr. Indian Phytopath., **13** : 150-155.

100. Muniyappa, V. and Veeresh, G. K. 1984. Plant virus diseases transmitted by whiteflies in Karnataka. Proc. Indian Acad. Sci. Animal Sci., **93** : 397-406.

101. Nanjundaiah, 1989. Development of powdery mildew (Erysiphe polygoni DC) on cowpea (Vigna unguiculata (L) Walp) M.Sc. (Agri). Thesis. University of Agricultural Sciences, Bangalore.

102. Nirwan, R. S. and Upadhayaya, J. 1972. Phytophthora blight of cowpea new to India. Indian Phytopath., **25** : 162-163.

103. Nwakpa, F.I. and Ikotunm, T. 1988. The reaction of cowpea varieties to infection by Sclerotium rolfsii. Trop. Gr. Leg . Bull., **35** : 32-34.

104. Ogundana, S. K. 1987. Control of *Pythium* wet rot of cowpea (*Vigna sinensis*) in Nigeria. Indian Phytopath., **39** : 245-248.

105. Oladiran, A.O. and Okusanya, B. O. 1980. Effects of fungicides on pathogens associated with basal stem rots of cowpea in Nigeria. Trop. Pest. Management, **26**: 403-409.

106. Oladiran, A.O. and Oso, B.A. 1983. Comparative susceptibility of some cowpea lines to brown blotch. Trop. Gr. Leg. Bull., **28** : 10-17.

107. Omotunde, C.T. 1987. A note on the effect of "*Xanthomonas* sp" on the morphology and yield of cowpea (*Vigna unguiculata* (L) Walp). East Afr. Agri. & Forest J., **52** : 298-299.

108. Onesirosan, P.T. and Barker, L.N. 1971. Stem anthracnose of cowpea in Nigeria. Pl. Dis. Reptr., **55** : 820-822.

109. Onuorah, P.E. 1973. *Pythium* seed decay and stem rot of cowpea (*Vigna sinensis* (Linn) Savi) in Southern Nigeria. Pl. Soil., **39** : 187-191.

110. Ouko, J. O. and Buruchara, R.A., 1989. Influence of different cropping systems on incidence and severity of bacterial pustule of cowpea (*Vigna unguiculata* (L) Walp) caused by *Xanthomonas compestris* pv. *vignicola* (Burkholder) Dye. Trop Pest Management, **35** : 297-300.

111. Patel, M. K. and Diwan, N. D., 1950. Bacterial blight of cowpea. Indian Phytopath., **3** : 75-80.

112. Patel, P. N. and Kuwite, C., 1983. Prevalence of cowpea aphid-borne mosaic virus and two strains of cowpea mosaic virus in Tanzania. Indian Phytopath., **35** : 467-472.

113. Pathak, H.C., 1974. Seed Borne Plant viruses : identification and diagnosis in seed health testing. Seed Sci. Technol., **2** : 3-14.

114. Patil, F.S. and Mali, V.R., 1982. Natural occurrence of tobacco ringspot virus (TRSV) on cowpea in India. Indian Phytopath., **35** : 399-403.

115. Pio-Riberio, G., Wyatt, S. D. and Kuhn, C. W., 1978. Cowpea Stunt : a disease caused by synergistic interaction of two viruses. Phytopathology, **68** : 1260-1265.

116. Prakash, J. and Joshi, R.D., 1980. Some aspects of seed transmission of cowpea banding mosaic virus in cowpea. Seed Sci. Technol., **8** : 393-399.

117. Purss, G. S., 1958. Studies on varietal resistance to stem rot (*Phytophthora vignae* Purss) Qld. J. agric. sci., **15** : 1-14.

118. Quindere, M.A.W. and Barreto, P. D., 1988. (Evaluation of cultivars and lines of cowpea, *Vigna unguiculata* (L) Walp and their reactions to diseases). *Boletimde pesqise-Empressa de Pesqisa Agropecuria do cearia do ceara* No. 11, 15 pp. Rev. Pl. Path., **69** : 5383).

119. Raju, S. G. and Anilkumar, T.B., 1990. Evaluation of cowpea genotypes for partial resistance to powdery mildew. Euphytica, **50** : 191-195.

120. Raju, S. G., Anilkumar, T. B. and Chandrashekar, M., 1991. Evaluation of certain cowpea genotypes for partial resistance to powdery mildew *(Erysiphe polygoni).* J. Phytopath., **133** : 201-208.

121. Rajender Singh, G. and Anilkumar, T. B., 1986. Efficacy of biloxazol carbendazim and triadimefon in controlling powdery mildew of cowpea. Indian. J. Mycol. Pl. Path., **16** : 30-36.

122. Ramachandran, P., Sumanwar, A. S. and Lal. S. P., 1982. Cowpea tor necrosis caused by *Fusarium equiseti* (Corda) Sacc. Curr. Sci., **51** : 475-477.

123. Ramaiah, M. and Jayarajan, R., 1976. Leaf blight of cowpea (*Vigna sinensis* Endl.) caused by *Sclerotium rolfsii* Sacc. Curr. Sci., **45** : 352.

124. Ramkishan, Srinivasan, V. R. and Ramesh Chand, 1988. Temporal and spatial spread of bacterial blight in cowpea. Indian J. Pl. Path., **6** : 31-35.

125. Rao, K.S.P. and Hiremath, R.V., 1985. Influence of host nutrition on the severity of bacterial blight of cowpea. Indian Phytopath., **38** : 81-84.

126. Rao, K.S.P. and Hiremath, R. V., 1986. Influence of growth regulators on the severity of bacterial blight of cowpea. Indian Phytopath., **39** : 512-514.

127. Rao, V. G., 1963. A New species of Septoria on an economic host. Curr. Sci., **32** : 367-368.

128. Rao, V. G., 1964. Some new records of market and storage disease of vegetables in Bombay, Maharashtra. Mycopath. et Mycol. Appl., **27** : 49-58.

129. Rao, V. G., 1966. An account of the market and storage diseases of fruits and vegetables in Bombay, Maharashtra. Mycopath et Mycol Appl., **28** : 165-176.

130. Rawal, R. D., and Sohi, H.S. 1980. Studies on inheritance of resistance to Septoria vignicola causing leaf spot disease of cowpea (*Vigna unguiculata*). Veg. Sci., **7** : 70-73.

131. Rawal, R. D., and Sohi, H.S. 1981a. Nature of disease resistance in cowpea to Septoria vignicola. Curr. Sci., **50** : 188.

132. Rawal, R. D., and Sohi, H.S. 1981b. Sources of resistance in cowpea to Septoria vignicola. Indian Phytopath., **34** : 504-505.

133. Rawal, R. D., and Sohi, H.S. and Manjunath, K. L., 1980. *Lagasca mollis* a collateral host for powdery mildew of cowpea. Indian Phytopath., **33** : 365-366.

134. Rawal, R. D., and Sohi, H.S. Sokhi, S. S., 1974. Effect of different levels of N.P. and K. on cowpea rust caused by *Uromyces phaseoli* var. *vignae*. Indian Phytopath., **27** : 405-407.

135. Rios, G.P., Zimmerman, F.J.P. and Fernandes, P.M. 1986. (Some epidemiological aspects and control of Ascochta sport (*Ascochyta phaseolorum*) on cowpea (*Vigna unguiculata*) Fitopatologia Brasileria, 11 : 911-912. Rev. Pl. Path., **67** : 3712.

136. Robertson, D. G., 1963. Further studies on the host range of cowpea yellow mosaic virus. Trop. Agri., (Trim). **40** : 319-324.

137. Robinson, R. A., 1960. Notes on Kenya agriculture VIII. Important plant diseases. E. Afr. agric. J., **25** : 131-146.

138. Rogers, K. M., Norton, J. D. and Chambliss, D. L., 1973. Inheritance of resistance to cowpea chlorotic mottle virus in southern pea *Vigna sinensis*. J. Am. Soc. Hort. Sci., **98** : 62-63.

139. Roy, A. K., 1975. Pathogenicity of Rhizoctonia solani and its control. Indian Phytopath., **28** : 184-188.

140. Sachchidananda, J., Singh, S., Prakash, N. and Verma, V.S., 1973. Bean common mosaic virus on cowpea in India. Z. pflazenkrankh. Pflanzenschutz., **80** : 88-91.

141. Santos, A. A., Friere Filho, F. R. and Cardoso, M. J., 1987. (BR 10-Piani, cowpea (*Vigna unguiculata*) cultivar with multiple resistance to viruses). *Fitopatologia Brassileira*, 12 : 400-402. Rev. Pl. Path. **67** : 5644.

142. Satyanarayana, N. 1987. Studies on physiological and biochemical changes in cowpea plants due to infection by *Erysiphe polygoni* DC. M.Sc. (Agri.) thesis, University of Agril. Sciences, Bangalore.

143. Sharma, S. R. and Varma, A., 1975a. Natural Incidence of cowpea viruses and their effect on yield of cowpea. Indian Phytopath., **28** : 330-334.

144. Sharma, S. R. and Varma, A., 1975b. Virus and Virus-line diseases in India : Proc. 62nd Indian Sc. Congr. Part III pp 74-75 (Abstr.).

145. Sharma, S. R. and Varma, A., 1975c. Three transmissible viruses from cowpea in India. Indian Phytopath., **28** : 192-198.

146. Sharma, S. R. and Varma, A., 1975d. Cure of seed transmitted cowpea banding mosaic disease. Phytopath. Z., **83** : 144 -151.

147. Sharma, S. R. and Varma, A., 1976. Cowpea yellow fleck-a whitefly transmitted disease of cowpea. Indian Phytopath., **29** : 421-423.

148. Sharma, S. R. and Varma, A., 1981a. Reaction of some cowpea cultivars and lines against three sap transmissible viruses. Prog. Hort., **13** : 127-129.

149. Sharma, S. R. and Varma, A., 1981b. Assessment of losses caused by banding mosaic (C.BMV) and cowpea chlorotic spot (CpCSV) viruses in cowpea. Proc. 3rd Intr. Symp. Pl. Path., New Delhi, p. 95 (Abstr).

150. Sharma, S. R. and Varma, A., 1982a. Control of virus diseases by oil sprays. Zb. Mikrobiol., **137** : 329-347.

151. Sharma, S. R. and Varma, A., 1982b. Effect of oils on transmission of cowpea banding mosaic virus. Phytopath. Z.,

152. Sharma, S. R. and Verma, A., 1984. Effect of cultural practices on virus infection in cowpea. Zeit. Acker und Pflazenbau, **153** : 25-31.

153. Sharma, S. R. and Varma, A., 1986. Transmission of cowpea banding mosaic and cowpea chlorotic spot viruses through the seeds of cowpea. Seed Sci. Technol., **14** : 217-226.

154. Shekawat, G. S. and Patel, P. N., 1977. Seed transmission and spread of bacterial blight of cowpea and leaf spot of greengram in summer and monsoon seasons. Pl. Dis. Reprt., **61** : 390-392.

155. Shekhariah, G., 1981. Studies on bacterial blight of cowpea. M.Sc. (Agri) thesis, University of Agril. Sciences, Bangalore.

156. Sheoraj and Patel, P. N., 1977. Studies on resistance on crops to bacterial diseases in India. IX. Sources of multiple disease resistance in cowpea. Indian Phytopath., **30** : 207-212.

157. Shepherd, R. J. and Fulton, R. W., 1962. Identity of seed-borne virus of cowpea. Phytopathology, **52** : 489-493.

158. Shoba Cherian and Anilkumar, T. B., 1992. Identification and characterization of slow rusting resistance in cowpea. Abstr, Current Trends in Diagnosis and Management of Plant Diseases, Indian Phytopath. Soc. meeting, Southern Zone, Nov. 25-26, 1992.

159. Shukla, V. D., 1978. Detecting of three Xanthomonas soo. affecting crop in Uttar Pradesh. Nat. Acad. Sci. Letters, **1** : 129-130.

160. Shukla, P., Lal., B., Singh, P. N., 1977. A new leaf spot disease of cowpea. Indian J. Mycol. Pl. Path., **7**: 159-160.

161. Singh, R. S., 1954. Wilt of lobia in Uttar Pradesh. Sci. Cult., **19** : 454-456.

162. Singh, D. and Patel, P. N., 1977. Studies on resistance in crops to bacterial diseases in India. VII. Inheritance of resistance to bacterial blight disease in cowpea. Indian Phytopath., **30** : 99-102.

163. Singh, J. P. and Musyoni, A. B. K., 1979. Control of rusts and powdery mildew by a new systemic fungicide Bayleton. Pesticides : **13** : 51-53.

164. Singh, P. N and Shukla, P., 1986a. Chemical control of Myrothecium leaf spot of cowpea (*Vigna unguiculata* (L) Walp).

165. Singh, P. N. and Shukla, P., 1986b. Effect of temperature, relative humidity and age of plant on Myrothecium leaf spot of cowpea. Farm Sci. J., 189-90.

166. Singh, P. N. and Shukla, P., 1986c. Screening of cowpea germplasm against *Myrothecium roridum* Tode ex Fr. Indian J. Mycol. Pl. Path., 10-21 (Abstr).

167. Singh, R. and Singh, R., 1974. Natural infection of sickle sena (Cassia tcra L) and cowpea (*Vegna sinensis* Savi) plants by some new strain of southern bean mosaic virus. Portugaliae Acta Biol., **13** : 87-89.

168. Singh Saroj, 1980. Field Screening of cowpea varieties to *Macrophomina phaseoli.* Indian J. Mycol. Pl. Path., **10**: 21 (Abstr.).

169. Singh Surjeet, 1980. Fungicides for the control of *Myrothecium roridum* leaf spot of russian giant cowpea. Indian Phytopath., **33** : 515-516.

170. Sinha, O.K. and Khare, M.N., 1977. Chemical control of *Macrophomina phaseolina* and Fusarium equiseti associated with cowpea seeds. Indian Phytopath., **30** : 337-340.

171. Smith, C. E., 1924. Transmission of cowpea mosaic by bean leaf beetle. Science, **60** : 268.

172. Sohi, H.S., 1975. Recent advances in the control of fungal diseases of horticultural crop plants in India. Pesticides Annu. No. pp. 101-105.

173. Sohi, H. S., 1977. Methodology of assessing crop losses due to diseases of vegetables. In : Assessment of Crop Losses due to Pests and Diseases, Eds. H.C. Govindu, G.K. Veeresh, P. T. Walker and J.F. Jenkyn, UAS Tech. series No. 33. pp. 249-258.

174. Sohi, H.S. 1987. Some important diseases of vegetables in India and losses caused by them. The Lal Baugh J., **23** : 1-11.

175. Sohi, H.S. and Nayar, S. K. 1969. New Records of fungi from Himachal Pradesh. Indian Phytopath. **22**: 462-465.

176. Sohi, H.S. and Rawal, R. D. 1974. Control of cowpea diseases. Indian Hort., **19** : 15-17.

177. Sohi, H.S. and Rawal, R. D., 1977. Role of certain physiological factors responsible for resistance to rust and *Septoria* leaf spot disease in cowpea. Symp. Physiology of Host Pathogen Interaction Proc., Madras, pp. 391-398.

178. Sohi, H.S. and Rawal, R. D., 1978. Studies on the control of leaf spot disease of cowpea caused by *Septoria vignicola* Rao with particular reference to assessment of losses. Indian J. Mycol. Pl. Path., **8** : 93-94.

179. Sohi, H.S. and Rawal, R. D., 1983. Phenols in relation to rust resistance in cowpea. Indian J. Mycol. Pl. Path., **13** : 61-62.

180. Sokhi, S. S. and Sohi, H. S., 1976. Studies on rust of cowpea caused by *Uromyces phaseoli* var. *vignae.* Indian Phytopath., **29** : 99. (Abstr.).

181. Sokhi, S. S., Munshi, G. D. and Rawal, H. S., 1984. Fungal diseases of cowpea (*Vigna unguiculata* (L) Walp) in the Punjab. Seeds and Farms, **7** : 35-37.

182. Soni, P. S. and Thind, B. S., 1987. Detection of *Xanthomonas campestris* pv *Vignicola* (Burk.). Dye from cowpea seeds and *X. campestris* pv *vignaradiatus* (Sabet *et al*) Dye from mungbean seeds by growing on tests. Phytopath. Medet., **26** : 1-6.

183 Stele, W.M. and Mehra, K.L., 1980. Structure, evaluation and adaptation to farming systems and environments in *Vigna.* In : Advances in *Legume Science* (Eds. R. J. Summerfield and A. H. Bunting), HMSO, London, pp. 393-404.

184. Sowell, G., Kuhn, C. W. and Brantley, B.B., 1965. Resistance of southern pea *Vigna sinensis,* to cowpea chlorotic mottle virus. Proc. Am. Soc. Hort. Soc., **86** : 487-490.

185. Taiwo, M. A., Gonsalves, D., Provvidenti, R. and Thurston, H. D., 1982. Partial characterization and grouping of s isolates of blackeye cowpea mosaic and cowpea aphid-borne mosaic viruses. Phytopathology, **72** : 590-596.

186. Talen, L. T., 1977. Cowpea little leaf disease in the Philippines : Possible viral etiology as detected by immunodiffusion technique. Phi. Phytopath., **13** : 43-49.

187. Talen, L. T., 1979. Cowpea viruses in the Philippines. II. Isolation and identification of a cowpea strain of cucumber mosaic virus. Phil. Phytopath., **15** : 51-57.

188. Tehon, L. R., 1933. Notes on the parasitic fungi of Illinois. Mycologia, **25** : 237-257.

189. Thirumalachar, M. J. and Chupp, C., 1948. Notes on some Cercosporae of India. Mycologia, **40** :.352-362.

190. Tripathi, O. P., Upadhayaya, P. P. and Rao, G. P., 1987. Pigments and primary productivity of cowpea leaves as influenced by cowpea banding mosaic virus infection. Nat. acad. Sci. Letters, India, **10** : 307-309.

191. Tsuchiya, S., Yanagwa, M. and Ogoshi, A., 1986. Formae specials differentiation of *Phytophthora vignae* isolates from cowpea and adzuki bean. Ann. Phytopath. Soc. Japan, **52** : 577-584.

192. Tsuchizaki, T., Yora, K. and Asuyama, H., 1970. The viruses causing mosaic of cowpeas and azuki beans and their transmissibility through seeds. Seed transmission of viruses in cowpea and azuki bean plants. I. Factors influencing seed transmission. Ann. Phytopath. Soc. Japan, **36** : 112-120.

193. Tyagi, R.N.S. and Mathur, A. K., 1978. Pathological Status of Kharif pulses in Rajasthan. Indian J. Mycol. Pl. Path., **8** : 20 (Abstr.).

194. Tyagi, R.N.S., Sankhla, H. C. and Mathur, A. K., 1979. Reaction of cowpea crosses/varieties to *Xanthomonas vignicola.* Indian Phytopath., **32** : 657.

195. Upadhyaya, P. P., 1980. Cowpea *(Vigna sinensis*)(Linn.) Endl. ex Haaska)-a new host of *Curvularia lunata* (Wak.) Boed, from India. Sci. Cult., **46** : 139.

196. Utikar, P. G., Deshmukh, R. B., More, B. B. and Shinde, P.A., 1980. Reaction of cowpea *Vigna unguiculata* (L) Walp) to powdery under field conditions. Food FMG. Agri., **11** : 218-219.

197. Vakili, N. G., Kaiser, W. J., Perez, J. A. and Cortes-Monllor, A., 1975. Bacterial blight of beans caused by two *Xamthomonas* pathogenic types from Puerto Rico. Pl. Dis. Reptr., **58** : 96.

198. Verma, P. R. and Patel, P. N., 1969. Host range, varietal resistance and epidemiological observations on *Cerospora* leaf spot disease of cowpea. Indian Phytopath. **22** : 61-66.

199. Vidano, C. and Conti, M., 1965. Transmission con afidi di un "cowpea mosaic virus: isolate da *Vignae sinensis* Endl in Italia. Alti Accad. Sci. Torino, **99**: 1041-1050.

200. Vidyasekaran, P., Balaraman, K. and Deiveekasundaram, M., 1972. First record of *Alternaria* of cowpea in India. Madras agric. J., 59-407.

201. Viswanathan, T. V. and Visweabharan, K., 1979. Screening of germplasm collection of cowpea Vigna sinensis Endl) for tolerance/resistance to collar rot caused by *Rhizoctonia solani.* Agri. Res. J. Kerala **17**: 272-274.

202. Weber, G. F., 1941. Cowpea scab. Pro. Bull. Fla. Agric. Exp. Sta. 557.

203. Wessolek, G. and Meier, U., 1984-85. (Studies on the effect of different irrigation systems on infection by *Uromyces* sp of *Vigna sinensis* subsp. *cylindricus* (cowpea) under arid conditions) *Tropen bandwirt* 85/86 : 53-60. Rev. Pl. Path., 67 : 5649.

204. Williams, R. J., 1975a. Control Control of Cowpea seedling mortality in southern Nigeria. Pl. Dis. Reptr., **59** : 245-248.

205. Williams, R. J., 1975b. The control of cowpea diseases in the IIT grain legume improvement programme. In : Tropical Diseases of Legumes, Eds. J. Bird and K. Marmorosch, Academic Press New York, pp. 139-146.

206. Williams, R. J., 1975c. Diseases of Cowpea *(Vegna unguiculata* (L) Walp) in Nigeria. *PAANS* **21**:253-269.

207. Williams, R. J., and Allen, D. J., 1976. Pathology-Grain Legume Training Course, IITA Nivgeria pp. 73.

208. Williams, R. J., and Ayanaba, A., 1975. Increased incidence of *Pythium* stem rot in cowpea treated with benomyl and related fungicides. Phytopathology, **65** : 217-218.

209. Wood, B. A. and Easdown, W.J., 1990. A new bacterial diseases of mungbean and cowpea for Australia. Aust. Pl. Path. **19** : 16-21.

210. Wilson, K. I. and Jose, P. C., 1968. Pod rot of *Vigna sinensis (L) Savi* caused by *Choanephora cucubitarum* (Berk & Rev.) Thaxter. Indian Phytopath., **21** : 130-131.

10 Wilt of Pigeonpea

Mahendra Pal

CONTENTS

Integrated Pest and Disease Management
Rajeev K. Upadhyay, K.G. Mukerji, B.P. Chamola and O.P. Dubey (eds.)
APH Publishing Corporation, New Delhi-110002, India

ABSTRACT

Wilt of pigeonpea incited by *Fusarium udum* Butler is one of the most important disease of tur crop causing 15-30% plant mortality. The disease attains economic injury level in part of M.P., Maharashtra, A.P., Tamilnadu, U.P. and Bihar, Blending of cultural, biological, chemical and host plant resistance methods of control has been found useful in avertin the menace of this disease at farmer's field.

1. INTRODUCTION

Our country has the distinction of being the largest producer of legumes with over a dozen of pulse crops grown on about 23 million hectares of land. The stagnant production of pulses over the decades against the ever increasing growth of population has led to per capita availability of less than 40 g/day as against WHO recommendation of 80 g/day for the vegetarian population. To meet the growing demand the Planning Commission has fixed the target production of pulses as 17 million tonnes with 694 kg/ha yield on an area of 24.5 million hectares in Eighth Plan against 13.5 million tonnes of the present production of pulses. Pigeonpea has to play a greater role since it is the second major pulse crop grown in India after chickpea.

India grows more than 90% of its pigeonpea crop. Outside India it is cultivated in developing countries of Asia and Africa. It is a valuable source of protein and is consumed in various forms. It is also used as a perennial forage crop for animal feed, The dried stalks are used for fuel and thatching and for making baskets for grain storage.

The crop is severely affected by a large number of diseases of which wilt, sterility mosaic and Phytophthora blight cause serious yield losses. Wilt alone causes heavy production losses and is a major hurdle in realizing full yield potentials of this crop.

2. DISTRIBUTION

The diseases is world wide in distribution wherever pigeonpea is grown. It causes production losses in Germany Italy, Uganda, Vietnam, Nepal, Kenya, Malawi, Thailand, Indonesia, Trinidad and Tanzania. In India it occurs in all the pigeonpea growing areas but is more destructive in states of Maharashtra, Andhra Pradesh, Madhya Pradesh, Uttar Pradesh and Bihar. In other states the disease is of minor importance.

3. LOSSES

The disease is responsible for 15 to 30 % plant mortality in wilt prone areas every year. Disease survey made during 1975-76 and 1976-77 by ICRISAT showed an average wilt incidence of 27.9% in Maharashtra, 5.2% in Andhra Pradesh and 1.4% in Tamil Nadu. The grain losses are almost 100% when wilt occurs at or prior to early pod formation stage and is around 30% when pods are full and the plants are close to harvest (47). Total production losses due to wilt is around 97,000 tonnes per year in India and 5 million in Malawi, Tanzania and Kenya, Annual production losses due to the combined effect of wilt and sterility mosaic has been estimated to be equivalent to US$ 113 millions (27).

4. SYMPTOMS AND SIGNS

The main symptoms of the diseases is wilting of the plants as if they are suffering from water deficiency. The pathogen is capable of causing infection to the plant at any stage of its growth but the symptoms are more pronounced when the crop is in the flowering and podding stage. The initial signs of the disease are the epinasty of the leaves followed by interveinal chlorosis, yellowing, loss of turgor and drying of the plants. The growing points droop down and the plant is killed, In some cases branches on one side of the plant dry indicating partial wilting. The disease occurs in patches in the field increasing rapidly with the dry spells that occur during the crop season.

The main root and the base of the stem tissues are blackened uniformly forming streaks. The streaks can be seen after removing the bark of the infected plant, and may be traced upto several inches high on the stem. The streaks can be traced down to the roots as well. The pathogen is specific and causes wilt disease only in pigeonpea crop. It enters through the rootless as it possesses the power of boring through the cell wall. It grows upto vascular system of the plant and plugs the vessels and become intra and intercellular. It travels from root to the stem several inches above the ground level and in some varieties the fungus can be isolated quite high in the stem becoming systemic in nature. It has been isolated from tap root, lateral roots and collar region, main stem, lateral branches and pod hels.

Phytophthora blight caused by *Phytophtora drechsleri* var. *cajani* Pal *et al.* (50), another disease of pigeonpea can be easily mistaken in the filed for Fusarium wilt, but both can be easily distinguished if the symptoms are carefully observed. There is a frequent blackening of the host stem/ tissue wholly or partially in the from of streaks in case of the wilt disease but not in case of blight. The leaves of the wilt infected plants very often turn yellow before drying but in case of Phytopthora blight the leaves of the infected plants remain light green in colour, later rolling upward and dry rapidly.

5. CAUSAL ORGANISM

Wilt of pigeonpea is caused by *Fusarium udum* Butler. The synonyms of the fungus are:

F. oxysporum Schl. emend. Snyder & Hansen f. sp. *udum* (Butler) Synder & Hansen

F. udum (Butler) var. *cajani* Padwick.

F. lateritium Nees emend, Snyder & Hansen f. sp. *cajani* (Padwixk) Gordon

F. uncinatum Wollenw.

F. lateritium Ness var. *uncinatum* (Wollenw)

The perfect stage of the fungus has been identified as *Gibberella indica* (58). The perithecia of the fungus are produced on the exposed roots, collar region and stem of the wilted plants.

Microconidia of the fungus are one celled, hyaline, elliptical, sometimes curved, 4-14 X 1-4μm. Macroconidia typically thin walled, hyaline, falcate and having distinct foot cell and apical cell with decreasing diameter towards the tip which may be curved or hooked. They are 2-3 septate (rarely 5), delicate and monthly equidistant. The apical and the basal cells may be longer than the intermediate cells. Macroconidia measure 18-64 X 2-5μm. Chlamydospores are globose intercalary or terminal, 6-14μm in diameter. Asci are mostly sub-cylindrical and ascopores are commonly tow celled, rarely 3 or 4 celled and constricted at the septa. Wollenweber (104), Wollenweber and Reinkings (105), Booth (5), Booth *et al.* (6), Subramanian (83), Gerlach and Nirenberg (19) and Upadhyay and Rai (95) described the fungus in detail.

5.1. Cultural characters

The fungus shows deep purple coloration on the agar medium. The aerial mycelium is almost absent with profuse development of pionnotes/ sporodochia. The colonies of the fungus are cottony to fluffy with the formation of concentric rings, sectors or having sticky growth (61). It has been shown that different isolates of the fungus differ in their cultural characters and production of serial mycelium, stroma, sclerotia, sporodochia as well as pionnotes (4, 12, 50, 82, 83). Fast and slow growing isolates of the fungus have also been reported.

5.2. Physiological studies

The fungus grows profusely and sporulates abundantly on potato dextrose agar and other similar media having hydrogen ion concentration of 4-9 and a temperature range of 20-30°C (42, 72, 106). Different isolates of the fungus differ in their pH and temperature requirements (31, 56, 84). The best carbon sources for better growth and sporulation of the fungus are fructose, glucose, maltose, sucrose and glycerol (44, 53, 84). Subramanian (84) demonstrated best sporulation of the fungus on lactic acid. Ammonium nitrate and asparagine are good sources of nitrogen for the luxurient growth and sporulation of the fungus, (31, 52, 53). Phosphorus (0.064%) and sulphur (0.032%) are required for its better growth. Lower concentration of these elements enhances chlamydospores production in the medium (53, 54).

Addition of pyridoxine and riboflavin in the medium stimulates growth of *F. udum* while nicotinic acid inhibits its growth. Formation of macroconidia and microconidia are depressed on addition of pyridoxene, riboflavin, nicotinic acid or calcium pantothenate in the mediu·. Their production, however, is stimulated by the addition of inositol, thiamine hydrochloride and biotin. L-ascorbic acid enhances the production of chlamydospores in the medium (55). Zinc stimulates fusaric acid accumulation resulting in the reduced production of microconidia in the medium. Each management and boron at 0.5 ppm, concentration has been found to increase growth of the fungus (66, 67).

5.3. Biochemical studies

The amount of free reducing sugars determine the level of susceptibility of pigeonpea varieties to *F. udum,* being highest in the least susceptible variety (NP-15) and the lowest in the highly susceptible variety (NP-24) (85). Chaudhary and Prasad (13) demonstrated presence of higher amount of glucose and sucrose in the highly susceptible varieties of pigeonpea, Early 269 and Early 348. Fusaric acid content has been shown to get accumulated in the roots, stems and leaves of the inoculated susceptible variety T-105. (73, 74). Amount of amino acids, glucose, fructose, maltose an raffinose decrease in pigeonpea varieties after infection by *F. udum.* (73, 74). The enzymes pectin methyl esterase, poly-galacturonase and cellulase are produced by the fungus *in-vitro* and *in-vivo* conditions (73, 74). Variation in the production of polymethyle galacturonase and cellulase enzymes in different isolates of the fungus has been demonstrated (67, 70).

5.4. Dispersal and survival of the fungus in the soil

The fungus has its highest population near the root zones of the infected stubbles after the plant dies. The population of the fungus increases saprophytically many fold in the infested plots if susceptible pigeonpea variety is grown successively in the same plot. There are conflicting reports about the period of survival of the fungus in the soil, varying from 2 to 3 years (46, 47) to many years (39). Crop sequences, use of fertilizers, micronutrients, organic amendments with different substances, types of soils and associated microflora influence the behaviour and survival of the fungus in the soil at the rate of 100 to 200 ppm (66, 67, 85). Application of zinc to the soils in which infested stubbles re burried results in the disappearance of the fungus within 5-6 weeks, (65). The competitive saprophytic ability of the fungus increases in the presence of dextrose while it decreases with the addition of urea in the soil (41). It has been shown that *F. udum* may survive in the soil as a mycoparasite on saprophytic fungi like *Aspergillus, Rhizopus, Cunninghamella, Mortierella* and *Syncephalestrum* species (96).

The fungus gets disseminated from one place to another through the infected plant debris and probably not across the soil (10). It gets dispersed through irrigation, rain water and termites which frequently food on the dead wilted plants.

Seed is another source of carrying over the pathogen from one place to other as well as from one season to the next either externally or internally (23). In the former the seed material gets infested during the harvesting and thrashing processes and is more important source for the perpetuation of the pathogen.

5.5 Disease cycle

The disease cycle of the fungus mainly occurs through its anamorphic stage though rarely the teliomorphic stage has also been reported (58). After the germination of the macroconidia/microconidia/ chlamydospores under favourable conditions of moisture, temperature and nutrients in the soil, the fungus enters the plant mostly through the roots, multiplies, travels upwards and becomes systemic in the plant. It grows inter and intracellularly in the affected pant as well as ectophytically on the roots and the collar region of the infected stems. After death of the infected pant the fungus grows saprophytically on the dead parts of the plant or other organic matter lying in the soil. The fungus is capable of remaining in the soil for several years (39). Chlamydospores are formed both in the living and dad host lying dormant in the soil.

5.6 Variability

The genus *Fusarium* has been regarded as unstable taxon by Buxton (11) due to high degree of segregation. In Uganda, while identifying *F. udum* Small (81) obtained range of variability in it in various solid and liquid media. Sarojini (67) found both cultural and pathological variability in 7 isolates of *F. udum*. Variation in different isolates of this fungus with respect to aerial mycelium, colour of the substratum on different media, texture, presence of sporodochia and sclerotia as well as conidial characters has been studied by Chattopadhayay and Sengupta (12).

It has been observe through several years that pigeonpea varieties identified resistant to wilt at one place under oneset of conditions behave quite differently at other places in the country. Variety NP (WR)-15 which is resistant to wilt at Delhi has been found susceptible at Kanpur, Hyderabad and Parbhani. The variety C-11 which is moderately resistant to wilt at Parbhani behaves as susceptible in West Bengal. This indicates the possibility of existence of physiologic races/strains of the pathogen at different places in the country.

Work conducted at IARI, New Delhi during the mid eighties showed existence of different pathological strains of *F. udum*. Out of several isolates of the fungus studies four different pathological strains were identified on the basis of morphological cultural and disease reaction on six differential varieties of pigeonpea (65). Based on pathological studies, Baldev and Amin (4) also indicated the presence of races in *F. udum*.

Working on 8 strains of *F. udum*, Sarojini (67) showed that strains I, II and III were more virulent than others. Strains V and VI caused pre-emergence damping-off of pigeonpea. Shit and Sengupta (70) reported two virulent races of the pathogen.

6. EPIDEMIOLOGY

Not much work has been done on the epidemiology of the disease. Severity and incidence of the disease in the field depends upon the soil conditions as well as saprophytic activity and survival of the fungus in the soil. The disease patches are markedly increased when susceptible variety is successively grown in the same infected soil. Wilt incidence has been observed to be more in the same infected soil. Wilt incidence has been observed to be more in the central and northern states than in the southern states of India, probably due to different physicochemical properties of the soil. Upadhyay and Rai (90) classified Indian soils into three categories; (a). Conducive soils that support growth of the pathogen and promote wilting, (b). Suppressive soils that do not favour growth of the fungus and also suppresses wilt incidence and (c). Intermediate soils that moderately suppress the pathogen growth and wilting.

A wide range of temperature, 17-20°C (42), neutral soil pH having 50% or more sand (89) and a soil moisture of 6-16% (39) have been found to favour wilt development in the field.

7. DISEASES MANAGEMENT

Increased reliance on agro-chemicals for crop protection has resulted environmental pollution, resistance in pests/pathogens and increased cost of cultivation. Therefore, the increasing public awareness for the adverse effect of indiscriminate use of pesticides has integrated pest management (IPM). To minimise adverse impact of pesticides (8, 40), IPM is adopted which is ecologically sound and environmentally safe approach of disease control. In this, several components of control methods are blended together in compatible manner so as to is maintained pathogens population below economic injury level (17, 24). Wilt of pigeonpea caused by *F. udum* is an endemic disease where pathogen remains in soil on the host plant debris for many years (39). Use of single control method may not be as effective as integrated approach comprising of two or more methods. A brief discussion in this regard is as under.

7.1 Cultural practices

7.1.1 Crop rotation and mixed cropping

Crop rotation varies with the degree of infestation of the field. Its success depends upon the sanitation which means the removal of infected stems and roots from the soil as the pathogen survives even in the deep seated roots of the host plant. The type of the crop used in crop rotation greatly influences the effectiveness of the rotation. Rotation of pigeonpea crop with tobacco has been recommended due to adverse effect of root exudates of tobacco on the pathogen (7). According to ICRISAT annual report 1986-87, one year break between pigeonpea cropping by fallowing or by growing sorghum or tobacco reduced wilt to 20, 22 and 44% respectively. A break with cotton or maize is not found effective. Upadhayay and Rai (90) found reduction in wilt by growing pigeonpea mixed with *Crotolaria medicagnae.* Intercropping of pigeonpea with sorghum reduced wilt incidence in the field significantly (16, 20, 45). Wilt incidence was only 24% when pigeonpea was sown mixed with sorghum as against 85% in a solo pigeonpea crop. Sharma and co-workers showed that intercropping of CSH-5 sorghum with H-102 pigeonpea decreased wilt incidence.

7.1.2 Soil amendment

Addition of green manure, compost of leaves of *Crotolaria medicagnea* and amendment of soil with neem, oil cakes of *Ricinus comminis,* rice husks or saw dust have been found to reduce wilt incidence in the field (25, 76, 90). Amendment of soil with roots of certain leguminous plants, molasses, sweet clover roots, groundnut and cotton cakes reduce wilt incidence through the production of more bulbiformin by *Bacillus subtilis* having antifungal activity (32, 102). Incidence of pigeonpea wilt was reduced with the addition of green manure of *Sesbania aculeata, Azadiracta indica,* mustard oil cakes as well as urea in the field (14). In the presence of nickel as soil pollutant, Khan and Salam (28) found reduction in the wilt incidence. At the same time it adversely affected the plant growth and *Rhizobium* nodulation. Cobalt was also found to suppress wilting and nodulation of pigeonpea.

7.1.3 Soil solarization

Mulching of soil with 100μm thick polythene sheeting for 6-8 weeks from April to May increases soil temperature by 6-10^{0}C in 0-20 cm soil profile and decreases *F. udum* population (14). At the same time it adversely affects *Rhizobium* population and decreases nodulation of the crop. Deep ploughing of the infected field during hot summer weather decreases pathogen population which results in the reduction of wilt incidence.

7.2 Biological control

Effect of associated microflora of *F. udum* have been demonstrated by Vasudeva and Roy (99) and Vasudeva and Govindaswamy (100). They reported lower incidence of wilt in unsterilized soil than in sterilized one and attributed it to be due to the antagonistic activity of saprophytic micro-organisms like *Aspergillus niger, Rhizopus nigricans* and *Bacillus subtilis* present in the unsterilized soil. They further concluded that *Bacillus subtilis* produced antagonistic bulbiformin which inhibited the growth of the pathogen (101). The production of antibiotic is stimulated by the external supply of magnesium, maganese and iron as well as soil amendment with certain leguminous crops, oil cakes and molasses. Upadhyay and Rai (87) described *Micromonospora globosa* as the biocontrol agent for reducing wilt incidence. They isolated a large number of saphrophytes from the soil and a few of them like *Aspergillus niger, A. flavus, A. terreus, Trichoderma viride, A. nidulans, Penicillium citrinum, P. aurtenium* and *Sacromyces gresseus* possessed the ability of biocontrol of wilt of pigeonpea (92, 93, 94). An antagonistic fungus, *Penicillium pinophilum* is found prevalent in the soil solarized by polythene mulching which reduces wilt incidence in the field (26).

Imran and Nasrath (68) reported higher population of antagonistic fungi like *Aspergillus niger* and *Trichoderma viride* in the rhizosphere of resistant pigeonpea cultivar 8858 and more *F. udum* population in the rhizosphere of susceptible cultivar 8518. Gaur and Sharma (18) isolated a large number of fungi and a few bacteria and actinomycetous from the rhizosphere of resistant NP-15 and susceptible T-21 pigeonpea varieties. They found *T. viride* to be most effective in controlling the disease followed by *Penicillium* and *Bacillus* spp. An apparent correlation existed between decrease in the wilt incidence and magnitude of population of *F. udum* per gram soil and the antagonist. Agnihothrudu (1) reported *Streptomyces* to be antagonistic to *F. udum* and was present in greater frequency in the rhizosphere of resistant than in susceptible variety. Cross protection with mild or non-pathogenic strains have also been observed by Maitra and Singh (33). They found that inoculation with *F. oxysporum* f. sp. *ciceri* or *F. oxysporum* f. sp. *vasinfectum* one week prior to inoculation by *F. udum* delayed the onset of wilt systems.

From these studies it is very much clear that if conditions suitable for the growth and multiplication of antagonistic micro-organisms are created in the rhizosphere of pigeonpea plant, wilt incidence can be kept under check.

7.3 Use of chemicals

7.3.1 In-vitro

Man has tried his efforts since centuries to control different plant diseases with the application of different fungicides. The fungicide stops the life cycle at a particular stage and thus prevents further growth and spread. Out of a large number of fungicides tested in *in-vitro* conditions, thiram, ferbam, benomyl, bavistin, captan, BAS 38601, demonsan, benlate+thiram have been found to be inhibitory to mycelial growth and spore germination of *F. udum* (65, 77, 79, 80).

7.3.2 In field

Only few workers attempted to control wilt of pigeonpea using chemicals as these do not provide an easy and economical approach to disease management, being soil borne nature of the pathogen. Sinha (80) observed reduction in the disease by bavistin applied as soil drench at 2000 ppm, 10 days before inoculation of pigeonpea with *F. udum.* Upadhyay and Rai (90) found considerable reduction in the wilt incidence with phygon, dithane Z-78 and zincop when applied as soil treatment.

Addition of superphosphate in green manured soil has been found to reduce wilt incidence by 25%. However, without green manuring superphophate has no effect on this disease (35, 36, 37, 38). Rai and Upadhayay (59) observed urea to promote activity of *F. udum* in the soil. They therefore, recommended that urea should not be used in the infested field.

Application of nematicides, aldicarb and carbofuran used for the control of *Heterodera cajani* in the soil has also been reported to reduce wilt incidence (22).

7.3.3 Seed treatment

Seed-borne infection of *F. udum* can be checked by treating the seeds with benlate+thiram at the rate of 2.5 g/kg of seed before sowing (23, 30). Since effect of fungicides applied at the time of sowing does not persist for the whole cropping season, therefore a single treatment can not provide a remedy for the disease control.

7.4 Use of resistant varieties

Use of resistant cultivars to control plant diseases in an age old phenomenon. In case of soil-borne diseases this is the most effective method of combating them and to reduce pathogen population in the soil. In pigeonpea, a large number of varieties have been screened against *F. udum* from time to time (9). Review of literature reveals that much attention has been paid on the identification of resistant sources by different workers in the country. The following are the resistant sources reported : T-80 (39); 15-3-3 (61); NP (WR)-15 (26); S-103 (26); NP-80 (16); 16-17-2; NP-41; C-38-1-2; D 419-2-4; C-11; C-28; C-36; F-18; T-17; Bori-192 (25); NP-15 (85); ST-1; ST-2; ST-3 (97); ICP-8859; ICP-8860; ICP-8862; ICP-8864; ICP-8865; ICP-8867; ICP-8868; ICP-8869; Purple-1; ICP-88663; ICP-9120; ICP-9144 (29, 47, 51, 60, 69); BDN-1; BDN-2; ICP-2376; AWR-74-15 (107, 108).

Pigeonpea variety ICP-8863 a selection from 15-3-3 from Maharashtra has been found highly resistant with highest average yield during 1985-86 trials at Gulbarga. The variety was released as 'Maruthi' in Karnataka state (51), In multilocational testing during 1978 to 1983 for broad based resistance to wilt, Nene and others (48) found ICP lines 4769, 9168, 10958, 111299, and C-11 to be resistant in all the years of testing at a large number of locations in the country. They, further found that line ICP-8863 had stable and durable resistance to wilt and had a good yield potential. Reddy and co-workers (63) identified 99 pigeonpea lines as resistant to *F. udum.* In their testing during 1980 to 1985 they found two lines ICP-9145 and ICP-10960 giving resistant reaction to wilt in Kanya and Malawi as well.

Loss of resistance in pigeonpea varieties due to seggregation has also been reported by some workers. Varieties NP-41, S-103, D-16, D-12-2, PT-12 and D-33-4-22 found resistant at one time, became susceptible to wilt after few years (43, 64, 98).

7.4.1 Phytoalexin production

Mechanism of resistance to Fusarium wilt were investigated in pigeonpea cultivars from Malawi with susceptible Malawi local and wilt resistance ICP 91-45 by Marley and Hillocks (34).

Resistant reaction appeared to depend mainly on rapid phytoalexin synthesis. Four fungitoxic isoflavonoid phytoalexins viz., hydroxygenestein, genestein, cajanin, and cajanol were isolated after 15 days of stem inoculation. They identified cajanol as the main antifungal compound being in much higher concentration (3 times) in resistant varieties than in susceptible. The cajanol content of the fungus-infected ICP-9145, 10 days after inoculation totally inhibited conidial germination of *F. udum.*

8. CONCLUSIONS AND FUTURE PROSPECTS

Wilt caused by *Fusarium udum* is the most devasting disease and is a serious concern of pigeonpea cultivators not only in India but in all other countries where pigeonpea is widely grown. It is a soil inhabitant and multiplies on the infected pigeonpea and other organic matter found in debris in the soil. The fungus causes systematic infection after entering through the root system of the plant.

Inspite of tremendous amount of studies, no suitable method for its management has been evolved. Growing resistant varieties has been considered to be the most effective method of reduce wilt incidence and also help in decreasing the inoculum potential of the pathogen population in the soil. A highly resistant pigeonpea line ICP-8863 can be utilized by the breeders for incorporating its resistance gene(s) in high yielding varieties. Number of other genotypes C-11, BDN-1, 15-3-3, AWR-74-15, ICP-9145, ICP-10960 exhibiting resistance have been identified at different research centres in the country and can also be profitably used in the resistant breeding programmes. Other possible methods of combating the disease incidence in the field are crop rotation and intercropping. Intercropping of sorghum with pigeonpea has been quite successful in reducing the incidence of wilt at farmers field.

The disease is both externally as well as internally seed-borne. The seed material collected from disease prone areas must, therefore, be treated with some suitable fungicides before sowing to minimize infection spread through seed.

Very little contribution has so far been made in the field of epidemiology which needs immediate attention of the scientists because of its soil-borne nature. Sincere efforts should, therefore, be made in this direction which can help in evolving the most effective control measures. Other areas which need further exploration are to check the menace of wilt disease by sanitation, crop rotation and intercropping. For natural suppression of wilt ecofriendly agents as neem products and microorganisms out.

Variation in virulence of different isolates of the pathogen has been of immense importance and can be worked out by using DNA finger printing with synthetic oligonucleotides. This will help in collection of a large number of isolates, preparation of homologous probes and use of restriction fragment length polymorphism (RFLP) as well as rapid amplification of polymorphic DNA (RAPD) markers to analyse the pathogen population diversity.

It is now time to focus our attention in developing and use of integrated management of plant diseases using ingredients such as host resistance, healthy seed material, ecofriendly pesticides, cultural methods, VA mycorrhizal fungi and other biological control agents, particularly against soil-borne pathogens like *F. udum.* Solarization of soil by mulching with thin plastic sheets may help in evolving a technique for the control of pigeonpea wilt but it may be detrimental as it inhibits the growth of *Rhizobium,* other beneficial microorganisms and alters the natural ecosystem of the soil which hamper the yield. If efforts to control even one devastating disease like wilt of pigeonpea succeed it is possible to reduce substantially the gap between supply and demand of pulses.

9. REFERENCES

1. Agnihothrudu, V. 1955. The density of rhizosphere microflora of pigeonpea *(Cajanus cajan* L) Millsp. in relation to the wilt caused by *Fusarium udum (F. merismoides)* Butler. Naturwissenschaften, **44** : 497.

2. Anonymous, 1954. Report of the Division of Mycology and Plant Pathology. Scientific report of the Indian Agricultural Research Institute, N. Delhi 1952-53.

3. Anonymous, 1955. Report of the Division of Mycology and Plant Pathology. Scientific report of Indian Agricultural Research Institute, N. Delhi, 1953-54.

4. Baldev, B. and Amin, K. S. 1974. Studies on existence of races in *Fusarium* udum, causing wilt of *Cajanus çajan.* SABRAO J., **6** : 201.

5. Booth, C. 1971. The Genus Fusarium. Commonwealth Mycological Institute, England, 137p.

6. Booth, C., Mordue, J. and Gibson, I. A. S. 1978. CMI description of pathogenic fungi and bacteria. Set 58 Nos. **20** : 571-580.

7. Bose, R. D. 1939. The rotation of tobacco for the prevention of wilt disease of pigeonpea. Agriculture Live-Stk. India, **6** : 653-668.

8. Burn, A. J., Coaker, T. H. and Jepson, P. C. 1987. Integrated Pest Management. Academic Press, London, 473.

9. Butler, E. J. 1908. Selection of pigeonpea for wilt disease. Agriculture India. **3** : 182.

10. Butler, E. J. 1910. The wilt disease of pigeonpea and the parasitism of *Neocosmospora vasinifecta* Smith. Mem. Dep. Agric. India (Bot. Sec), **2:** 1.

11. Buxton, E. W. 1959. Mechanism of *Fusarium* oxysporumin relation to host-parasite interaction. In Plant Pathology Problems and Progress 1908-1955, 588p.

12. Chattopadhyay, S. B. and Sengupta, P. K. 1967. Studies on wilt diseases of pulses I. Variation and taxonomy of *Fusarium* species associated with wilt diseases of pulses. Indian J. Mycol. Res., **5** : 45-53.

13. Chaudhary, S. K. and Prasad, M. 1974. Variation in sugar contents of healthy and *Fusarium oxysporum* f. *spudum* infected plants of *Cajanus cajan.* Phytopath. Z. **80** : 303-305.

14. Chauhan, Y. S., Nene, Y. L., Johansan, C., Haware, M. P., Saxena, N. P., Singh, Sardar, Sharma, S. B., Sahrawat, K. L., Burford, J. R., Pupela, O. P., Kumar, Rao, J.V.D.K., Sithawantham, S. 1988. Effect of soil solarization of pigeonpea and chickpea. Research Bulletin, ICRISAT No. **11** : 16.

15. Dasgupta, A. and Gupta, P.K.S. 1989. Effect of different soil amendments on wilt of pigeonpea *(Cajanus cajan* L.) caused by *Fusarium udum* Butler. Beitrage Zur tropischen hand wirtscha ft and veterinarmedizin, **27** : 341-345.

16. Dey, P. K. 1948. Administrative report of Agriculture Department, U. P. 1946-47 : 39-42 and 1945-46, 43-46.

17. Frisbie, R. E. and Mc Whorter, G. M. 1986. Implementing a state wise pest management program for Texas, U.S. A. In : Advisory work in Crop Pest and Disease Management. Eds. Patti and R. Ausher, Spring-Verlog, Pub., New York, pp. 234-262.

18. Gaur, V. K. and Sharma, L. C. 1991. Micro-organisms antagonistic to *Fusarium udum* Butler. Proceeding of Indian National Science Academy Part B, Biological Science, **57** : 85-88.

19. Gerlach, W. and Nirenberg, H. 1982. The Genus *Fusarium-a* Pictorial Atlas. Berlin and Hamburg : Kommissionsverlag Paul Parey, 406p.

20. Gupta, S. L. 1961. The effect of mixed cropping of arhar *(Cajanus indicus* Spreng.) with jowar *(Sorghum vulgare* Pers.) on incidence of arhar wilt. Agric Anim. Husb. U.P. **2**: 31-35.

21. Haider, M. G., Singh, R. K., Prasad, H., Nath, R. P. and Sharma, R. N. 1978. Effect of some common fungicides on the incidence of pigeonpea wilt. Indian Phytopath., **31** : 511.

22. Hasan, A. 1989. Efficacy of certain non-fumigant nematocides on the control of pigeonpea wilt involving *Heterodera cajani* and *Fusarium udum.* J. of Phytopathol., **126** : 335-342.

23. Haware, M. P. and Kannaiyan, J. 1992. Seed transmission on *Fusarium udum* in pigeonpea and its control by seed treatment fungicides. Seed Sci. and Technol. **20**:579-601.

24. Huffaker, C. B. 1980. New technology of pest control. John Wiley & Sons, 500p.

25. Indian Agricultural Research Institute (1951) Scient. Rep. of Ind. Agric. Res. Inst. for the year ended 30th Jun 1951, 120.

26. International Crop Research Institute for the Semi-Arid Tropics (1987). Annual Report-Pulse Pathology, Patancheru, A. P., India, ICRISAT, 1986-87.

27. Kannaiyan, J., Nene, Y.L., Reddy, M. V., Ryan, J. C. and Raju, T. N. 1984. Prevalence of pigeonpea diseases and associated crop losses in Asia and America. Trop. Pest. Mang. **30** : 62.3

28. Khan, M. W. and Salam, M. A. 1990. Interaction of *Meloidogyne javanica, Fusarium udum* and *Rhizobium* on pigeonpea in presence of nickel and cobalt pollutant. Annals of Biology, **116** : 549-556.

29. Konda, C. L., Parmeswarappa, R. and Rao, T. S. 1986. Pigeonpea ICP 8863-A boon to Fusarium wilt epidemic areas of Karnataka. International Pigeonpea Newsletter, **36** : 1986.

30. Kotasthane, S. R., Om Gupta and Khare, M. N. 1987. Influence of fungicidal seed treatment and soil amendment on the development of *Fusarium udum* propagules in soil and pigeonpea wilt. Indian Phytopath., **40** : 197.

31. Laha, S. K. 1978. Cultural and Pathological studies on *Fusarium udum* Butler. M. Sc. thesis, IARI, New Delhi

32. Mahmood, M. 1964. Factors governing the production of antibiotics bulbiformin and its use in the control of pigeonpea wilt. Sci. Cult., **30** : 352.

33. Maitra, A. and Sinha, A. K. 1973. Partial inhibition of Fusarium wilt symptoms in pigeonpea by non pathogenic formae of *Fusarium oxysporum.* Curr. Sci. **42** : 654.

34. Marley, P. S. and Hillocks, R. J. 1993. The role of phytoalexins in resistance to Fusarium wilt of pigeonpea *(Cajanus cajan).* Plant Pathology, **42** : 212-218.

35. McRae, W. 1924. Report of the Imperial Mycologist. Scient. Rep. Agric. Res. Inst. Pusa. 1923-24, pp, 41-51

36. McRae, W. 1926. Report of the Imperial Mycologist. Scient. Rep. Agric. Res. Inst. Pusa, 1925-26. p 54-69.

37. McRae, W. 1928. Report of the Imperial Mycologist. Scient. Rep. Agric. Res. Inst. Pusa, 1927-28. p 57-70.

38. McRae, W. 1930. Report of the Imperial Mycologist. Scient. Rep. Agric. Res. Inst. Pusa, 1928-28. pp 51-66.

39. MacRae, W. and Shaw, F.J.F. 1933. Influence of manures on the wilt disease of *Cajanus indicus* Spreng. and isolation of types resistant to the disease. II. Scient. Monogr. Coun. Agric. Res. Pusa, **7** : 37-68

40. Metcalf, R. L. 1980. Changing role of Insecticides in Crop Protection. Ann. Rev. Entomol., **25** : 219-256.

41. Mukhopadhyay, D. and Gupta, P.K.S. 1991. Factors affecting competitive saphrophytic activity of *Fusarium udum,* the causal organism of wilt of pigeonpea. Indian Phytopath., **44** : 73-79.

42. Mundkur, B. B. 1935. Influence of temperature and maturity on the incidence of sun-hemp and pigeonpea wilt at Pusa. Indian. J. Agric. Sci., **5** : 609-619.

43. Mundkur, B. B. 1946. Report of the Imperial Mycologist. Scient. Rep. Agric. Res. Inst. New Delhi, for the Triennium ended 30th June, 1944, 570.

44. Natrajan, S. 1958. Carbon nitrogen metabolism of soil fungi IV. Influence of carbon sources. Proc. Indian Acad. Sci., **47B** : 338-344.

45. Natrajan, M., Kannaiyan, J., Willey, R. W. and Nene, Y.L. 1985. Studies on cropping system on Fusarium wilt of pigeonpea. Field Crop Research, **10** : 333.

46. Nene, Y. L. and Reddy, M. V. 1981. Survival of pigeonpea wilt Fusarium in vertisols and alfisols. In Proc. Intl. Workshop i pigeonpea Vol. **2** : Patnacheru, ICRISAT (A.P.) 291.

47. Nene, Y. L., Kannaiyan, J., Haware, M. P. and Reddy, M. V. 1979. Review of the work done at ICRISAT on soil-borne diseases of pigeonpea and chickpea. In: Proceeding of the consultant groups discussion on the resistance to soil-borne diseases of legumes, ICRISAT, Patancheru (A. P.) India.

48. Nene, Y.L., Kannaiyan, J., Reddy, M. V., Zote, K. K. Mahmood, M., Hiremath, R. V., Shukla, P.P. Kotasthane, S. R., Sengupta, K., Jha, D. K., Haque, M. F., Grewal, J. S. and Mahendra Pal. 1989. Multilocational testing of pigeonpea for broad based resistance to Fusarium wilt in India. Indian Phytopath. **42** : 449-453.

49. Padwick, G. W. 1940 Report of the Imperial Mycologist. Scient. Rep. Agric. Res. Inst. Pusa, N. Delhi, 1939-40, pp. 103-115.

50. Pal, Mahendra, Grewal, J. S. and Sarbhoy, A. K. 1970. A new stem rot of arhar caused by *Phytophthora* Indian Phytopath. **23** : 583-587.

51. Parameswarappa, R., Rao, T. S., Konda, C. R., Rao, T. M., Raghumurty, M., Emminath, V. and Saguppa, K. K. 1986. Maruti-A wilt resistant red gram variety Current Research, Univ. of Agric. Sci. Bangalore, **15** : 69p.

52. Patel, M.K. and Kulkarni, Y. S. 1949. Nitrogen utilization by *Xanthomonas malvacearum* (SM) Dowson. Indian Phytopath., **2** : 62-64.

53. Prasad, M. and Chaudhary, S. K. 1965. Studies on the effect of carbon and nitrogen on macroonidial and chlamydospore production in *Fusarium udum.* Proc. Nat. Acad. Sci. India, **35B** : 171-181.

54. Prasad, M. and Chaudhary, S. K. 1967. Effect of sulphur on sporulation of *Fusarium udum* Butler. J. Indian Bot. Sci., **46** : 45.

55. Prasad, M. and Chaudhary, S. K. 1975. Influence of vitamins on growth and sporulation of *Fusarium oxysporum* f. *sp. udum*. Z. Phytopath. Z. **82** : 56-62.

56. Prasad, M. and Chaudhary, S. K. 1977. Relation of pH levels and varied nutrient media to growth and sporulation of *Fusarium oxysporum* f. sp. *udum* Butler. Res. J. Ranchi Univ., **13** : 214.

57. Prasad, M. and Chaudhary, S. K. 1977. Effect of different concentration of Dl-isoleucine, Dl-valine and Dl-alanine on growth and sporulation of *Fusarium oxysporum* f. sp. *udum.* Butler. Sny. et. Han. Zent. Bakteriol. Parasit Infect Hygiene, **132** : 735.

58. Rai, B. and Upadhayay, R. S. 1982. *Gibberella indica:* The perfect state of *Fusarium udum*. Mycologia, **74** : 343.

59. Rai, B. and Upadhayay, R. S. 1983. Competitive saprophytic colonization of pigeonpea substrate of *Fusarium udum* in relation to environmental factors, chemical treatments and microbial antagonism. Soil biol. Biochem. **15** : 187.

60. Raut, N. K. and Bhombe, B. B. 1971. A review of the work of selections of tur varieties resistant to Fusarium wilt at College of Agriculture, Parbhani (Maharashtra). Magaz. Coll. Agric. Parbhani, **12** : 37.

61. Ravi Shankar. 1936. Isolation of wilt resistant tur. Nagpur Agric. Coll. Mag., **10** : 162.

62. Reddy, P.E. and Chaudhary, K. C. B. 1985. Variation in *Fusarium udum*. Indian Phytopath., **38** : 172.

63. Reddy, M. V., Nene, Y. L., Kannaiyan, J., Raju, T. N., Saka, V. W., Davedi, A. T., Songa, W. P. and Omanga, P. 1991. Pigeonpea line resistant to wilt in Kanya and Malawi. International Pigeonpea Newsletter No., **12** : 25-26.

64. Rohewal, S. S., Joshi, B. C. and Singh, S. P. 1966. Arhar S-103, an erect type yielding high. Indian Farming., **16** : 31.

65. Sahoo, G. B. 1987. Studies on pigeonpea wilt with special reference to variability of the pathogen, *Fusarium udum.* M. Sc. thesis, I.A.R.I., New Delhi.

66. Sarojini, T. S. 1950. Soil conditions and root diseases, micronutrients elements and disease development of *Fusarium udum* on red gram *(Cajanus indicus* L.) J. Madras, Univ. Sec. B., **19** : 1.

67. Sarojini, T. S. 1951. Soil conditions and root diseases. Part II. Fusarium disease of red gram *(Cajanus indicus).* Proc. Indian Acad. Sci. Sect. BN., **33** : 49.

68. Shaik, Imran and Nasrath, M. 19887. Varietal variation in rhizosphere and rhizoplane microflora of *Cajanus cajan (L.)* Millsp., with special reference to wilt disease. Ind. J. of Botany, **10** : 126-129.

69. Sharma, H. C. 1988. Selection and development of wilt resistant pigeonpea lines for Tawa Command Area in Central India. International Pigeonpea Newsletter No. **7** : 9-10.

70. Shit, S. K. and Sengupta, P. K. 1980. Pathogenic and enzymatic variation in *Fusarium oxysporum* f. sp. *udum.* Ind. J. Microbiol. **20** : 46.

71. Singh, D. V. and Mishra, A. N. 1976. Search for wilt resistant varieties of red gram in Uttar Pradesh. Ind. J. of Mycol and Plant Pathol., **6** : 89.

72. Singh, G. P. 1968. Extracellular hydrolytic enzymes of *Fusarium lateritium* f. sp. *cajani.* Indian J. Microbiol., **8** : 95-100.

73. Singh, G. P. and Hussain, A. 1962. Production of pectic and cellulolytic enzymes by arhar wilt fungus. Curr. Sci., **31** : 110.

74. Singh, G. P. and Hussain, A. 1964. Presence of fusaric acid in wilt affected pigeonpea plants. Curr. Sci., **33** : 287.

75. Singh, G. P. and Hussain, A. 1968. Role of enzymes in pathogenesis by *Fusarium lateritium* f. sp. *cajani.* Indian Phytopath., **21** : 361.

76. Singh, N. and Singh, R. S. Effect of oil cake amendment soil atmosphere of pigeonpea wilt pathogen. Indian Phytopath., **35** : 300-305.

77. Singh, R. S. and Singh, N. 1970. Effect of cake amendment of soil on sporulation of some wilt causing species of *Fusarium.* Phytopathol. Z., **69** : 160.

78. Singh, R. S. 1975. Introduction to Principles of Plant Pathology. New Delhi. Oxford and IBH Publishing Co. 1975: 282 p.

79. Sinha, A. K. 1974. The use of benomyl fungicide against Fusarium wilt of pigeonpea. Z. Pflkrankh. Pflschutz.,**81** : 571-574.

80. Sinha, A. K. 1975. Control of Fusarium wilt of pigeonpea with bavistin, a systemic fungicide. Curr. Sci., **44** : 700.

81. Small, W. 1922. On the occurrence of a species of *Fusarium* in Uganda. Kew Bull. Miscellaneous Inform **9** : 269-291.

82. Subramanian, C. V. 1951. Is there a wilt type in genus *Fusarium*. Proc. Nat. Inst. Sci. India., **17B** : 403-411.

83. Subramanian, C. V. 1955. Studies on South Indian Fusaria IV . The wild type in *Fusarium udum* Bulter. J. Indian Bot. Soc., **34** : 29.

84. Subramanian, S. 1961. Studies on wilt of pigeonpea. Ph.D. thesis. University of Madras. Tamil Nadu, India.

85. Subramanian, S. 1963. Fusarium wilt of pigeonpea II. Changes in the host metabolism. Proc. Indian Acad. Sci. Sec B., **57** : 259.

87. Upadhyay, R. S. and Rai, B. 1978. *Micromonospora globosa* Krass. A destructive parasite of *Fusarium udum* Butler. Microbios Left., **8** : 123.

88. Upadhyay, R. S. and Ram, B. 1978. Tolerance of higher temperature by *Aspergillus nidulans* in competition with other soil fungi. Proc. Natl. Acad. Sci. India., **77.**

89. Upadhyay, R. S. and Rai, B. 1979. *Coprinus logopus* as potent and frequent saprophytic colonizer of pigeonpea in soil. Sci. Cult. **45** : 171.

90. Upadhyay, R. S. and Rai, B. 1981. Fungistatic activity of different Indian soils against *Fusarium udum* Butler. Plant soil, **63** : 407.

91. Upadhyay, R. S. and Rai, B. 1983. A new disease cycle of wilt of pigeonpea. Curr. Sci., **52:** 978.

92. Upadhyay, R. S. and Rai, B. 1983. Competitive saprophytic ability of *Fusarium udum* Butler in relation to some microfungi of root region of pigeonpea. Indian Phytopath., **36** : 539.

93. Upadhyay, R. S. and Rai, B. 1985. Antagonistic behaviour of root region microfungi of pigeonpea against *Fusarium udum.* In Ecology and Management of soil borne plant pathogens (Eds. C. A. Parker, A. D. Rovira, K. J. Moore, P.T.W. Wong and J. F. Kollmorgen). The American Phytopathological Society, St. Paul Minnesota, 131p.

94. Upadhyay, R. S. and Rai, B. 1987. Studies on antagonism between *Fusarium udum* Butler and root region microflora of pigeonpea. Plant Soil, **101** : 79.

95. Upadhayay, R. S. and Rai, B. 1989. Wilt Disease of pigeonpea and its causal organism *Fusarium udum.* In: Perspective of Phytopathology (Eds. V. P. Agnihotri, U. S. Singh, H. S. Chaube, N. Singh and T. S. Dwivedi) New Delhi : Today and Tomorrow's Printers and Publishers, 1989.

96. Upadhyay, R. S., Rai, B. and Gupta, R. C. 1984. A possible mode of survival of *Fusarium udum* as a mycoparasite. Acta Mycologia, **19** : 115-119.

97. Vaheeduddin, S. 1956. Selection of tur *(Cajanus cajan* L.) resistant varieties against wilt *(Fusarium udum* Butler). Agric. Coll. J. Osmania Univ., **3** : 12.

98. Vasudeva, R. S. 1955. The effect of associated microflora of *Fusarium udum* Butler causing wilt of pigeonpea *(Cajanus indicus)*. Proc. Sixth Intern. Cong. Microbiol., **5** : 239-242.

99. Vasudeva, R. S. and Govindaswamy, C. V. 1953. Studies on the effect of associated soil microflora on *F. udum* Butl. The fungus causing the wilt of pigeonpea *(Cajanus cajan* (L) Millsp. with special reference to its pathogenicity. Ann. Appl. Biol., **40** : 573.

100. Vasudeva, R. S. and Roy, T. C. 1950. The effect of associated microflora on *Fusarium udum* Butler the fungus causing wilt of pigeonpea *(Cajanus* cajan (L.) Millsp.) Ann. Appl. Biol. **38** : 169.

101. Vasudeva, R. S., Jain. C. and Nema, K. G. 1952. Investigation of the inhibitory action of *Bacillus subtilis* on *Fusarium udum* Butl., the fungus causing wilt of pigeonpea *(Cajanus cajan* (L.) Millsp. Ann. Appl. Biol. **39** : 229.

102. Vasudeva, R. S., Singh, P., Sengupta, P. K. and Mahmood, M. 1963. Further studies on the biological activity of bulbiformin. Ann. Appl. Biol., **51** : 415.

103. Venkateshwarulu, S., Reddy, A. R., Chauhan, V. B., Singh, O. N., Singh, R. N. and Singh, U. P. 1981. A promising resistant line of pigeonpea (Malviya arar-1) for wilt and sterility mosaic. Trop. Grain Leg. Bull., **23** : 27.

104. Wolleweber, H. W. 1938. Fusarium-produced disease of pigeonpea *(Cajanus cajan)*, Arb. Biol. Anst. Rekhsanst. Berf., **22** : 339-347.

105. Wollenweber, H. W. and Reinking, O. A. 1935. Die Fusarien, ihre Beschrejbung. Berlin Paul Parey, 335p.

106. Yogeshwari, L. 1948. The element nutrition of fungi I. The effect of boron, zinc and manganese on *Fusarium* spp. Proc. Acad. Sci. (Sect. B.), **28** : 177-201.

107. Zote, K. K., Dandanaik, B. P., and Khalikar, P.V. 1986. Sources of resistance of some pigeonpea lines to wilt diseases. Ind., J. of Mycol. and Pl. Pathol., **16** : 144-145.

108. Zote, K. K., Dandanaik, B. P. and Khalikar, P. V. 1987. Reaction of pigeonpea varieties/lines to Fusarium wilt disease. J. of Maharashtra Agri University **12** : 102-103.

11 Insect Pests of Pea and their Management

Ram Ujagir

CONTENTS

Integrated Pest and Disease Management
ajeev K. Upadhyay, K.G. Mukerji, B.P. Chamola and O.P. Dubey (eds.)
H Publishing Corporation, New Delhi-110002, India

ABSTRACT

The insect pests that attack pea, *Pisum sativum* Linn, in India and in other parts of the world are reviewed. They are grouped according to the plant growth stages at which they attack : establishment, seedling, vegetative growth, flowering and podding. The stemfly, *Ophiomyia phaseoli* Tryon, pea aphid, *Acyrthosiphon pisum* (Haris), pea leaf miner *Chromatomyia horticola* Gouveau and pod border complex are major pests from germination to maturity stages of the crop. The population ecology, quantitative sampling, assessment of losses, various control strategies, including cultural control, host plant resistance, biological control and chemical control are briefly discussed on global basis. Cost benefit, economic analysis, and IPM strategies as well as future trends have also been outlined.

1. INTRODUCTION

Pea, *Pisum sativum* Linnaeus, is widely grown throughout the world. It finds an important place in the farming systems adopted by small holding farmers in large number of developing countries. In world agriculture, pea occupies 9.267 million hectares and most of the area for various purposes in India, China, U.S.S.R., Europe and some states of USA. Grain legumes in general give low yield because of the fact that these crops have been grown under marginal conditions of both moisture and fertility and with little or no plant protection measures. Among the several factor responsible for poor yield, insect pests are the major constraints in realisation of attainable yield. Insects causes direct damage to crop by feeding on seed, seedling, foliage and pods (29, 51, 87, 110). Only few of the reported insect pests species cause sufficient loss to the crop to be regarded as major pests, however, remaining insects do not cause sufficient damage and can only be grouped as minor pests. Besides these many insects are beneficial as they feed upon the pests as parasitoids and predators. Every part and stage of plant is susceptible to pest attack. The pests acquire their food either by chewing or sucking up the plant sap. Plants that are heavily attacked at seedling stage are usually not able to compensate and even die, however, it may tolerate the pest damage in vegetative and flowering stage of the crop. Severe defoliation in vegetative stage may diminishes the photosynthesis process which may result in poor yield of the crop. Similarly, pod damage may also reduce crop yield, as plant has very little chance to compensate at this stage.

2. INSECT PEST ASSOCIATED WITH VARIOUS STAGES OF CROP GROWTH AND THEIR POPULATION ECOLOGY

About 21 insect species have been found attacking pea (Table 1), but only few species are considered to be; or will probably be of economic importance and are discussed below :

Table 1 : List of insects known to be associated with pea crop, Pisum sativum L.

Scientific name (Order : Family)	Common name	Geographical distribution	Reference
A. Root and Stem Feeders:			
1. *Agrotis (Euxoa) segetum* (Shiff) (Lep. : Noctuidae)	Cut worm	India	87
2. *Delia platura* (Meig.) (Dip. : Anthomyiidae)	Bean Seedfly	Britain	39

3.	*Ophiomyia phaseoli* Tyron (Dip. : Agromyzidae)	Pea stemfly	East and West Asia, Micronasia, N. Australia.	87
4.	*O. centrosematis* (de Meij.) (Dip. : Agromyzidae)	Stemfly	Austrilla, Tropical Asia and East Africa	80, 81, 82, 92
B. Foliage Feeders:				
1.	*Acrythosiphon pisum* (Harri.) (Hemiptera: Aphididae)	Pea aphid	South Africa, India, Europe and North America	18
2.	*Aphis craccivora* Koch (Hemiptera : Aphididae)	Black bean aphid	India	16
3.	*Bemisia tabaci* G. (Hemiptera : Aphididae)	White fly	India	73
4.	*Caliothrips indicus* Bagnall	Thrips	India	97
5.	*Chromotomyia horticola* Goureau	Pea leafminer	Europe, Asia and Africa	3
6.	*Kakothrips pisivojus* (Westw.) (Thysanoptera : Thripidae)	Thrips	Europe	50
7.	*Mythimna seprata* (Wlk.) (Lep. : Noctuidae)	Army Work	India	87
8.	*Myzus persicae* (Sulz.) (Hemiptera : Aphididae)	Green peach aphid	India	87
9.	*Seasamia inferens* (Wlk.) (Lep. : Noctuidae)	Pink borer	India	87
10.	*Spodoptera exigua* (Hb.) (Lep. Noctuidae)	Tobacco caterpillar	India	109
11.	*Tanymecus indicus* (Fst.) (Col. : Curculionidae)	Surface weevil	India	73
12.	*Thrips tabaci* Lind. (Thysanoptera : Thripidae)	Onion thrips of cotton thrips	India	97
C.	**Pod Borers:**			
1.	*Cosmolyce (Lampides) hoeticus* (L.) (Lep. : Lycaenidae)	Lycanid	India	114
2.	*Etiella zinckenella* (Treit.) (Lepi. : Phycitidae)	Pea pod borer	India	111
3.	*Euchrysops onejus* (F.) (Lep : Lycaenidae)	Blue butterfly	India	111
4.	*Heliothis armigera* (Hb.)	Green pod borer	India	111
5.	*Cydia (Laspeyresia) nigricana* Steph. (Lep. : Tortricidae)	Pea moth	Europe	28

2.1 Pests at seedling stage

The sown seeds may be damage by rats, birds like pigeon, which dig out them and eat. Cotyledons are removed and seedlings can be killed damage to the growing point or uprooting. Late sown crop in particular is not able to germinate due to attack of rats, even if the plots are treated with insecticide, specially if the field is situated near uncultivated lands and streams. The seeds not sown at desired depth are also liable to damage by these pests. Seed fly, *Delia platura* (Meig.) and millipedes may also damage pea seeds in the field. Seedling can be also damaged by rabbit and other like pests, curculionid larvae of *Sitona lineatus* (Linn.) which feed on foliage of the pea crop and are very serious in certain countries.

Cutworms *(Agrotis* spp.) seem to be endemic, particularly cause damage in low lying areas in Bihar, Orissa and Rajasthan. Termites and white grubs *(Lachnosterna* spp.) may also cause severe mortality of seedling in light soils of certain areas.

Ophiomyia phaseoli is a regular pests of early sown crop which cause mortality of the plants. It also attacks *Phaseolus* spp., *Vigna* spp., soybean and lab lab *(Dolichos lab lab)* in India. The female fly oviposits inside young leaves by making punctures on the surface of leaves. Majority of eggs are laid on the lower surface of leaves though some females do oviposit on any other tender part of the plant (17). A single female may lay between 100 to 300 eggs over two week period (64). The eggs hatch in 2 to 4 days. The larvae form short leaf mines then enter the nearest vein which is followed into the petiole and down into the stem. In young plant the main feeding take place in the lower, cortical layers of the stem but some larvae can penetrate down into the tap root. In case of heavy infestation the larvae are found more deeply inside the stem (17, 104). The larval stage is completed in about 10 days (104). In Indian condition, it passes through 3 instars. The larval period ranges from 9 to 12 days in November-December where as it is 6-7 days in June-July (2). The maggot cut an exit hole before pupation through which adult fly emerges (17). Pupation may take place in under ground stem or main stem or inside the branches. Pupal period is highly variable and it depends on climatic conditions and altitude of the place. Agrawal and Pandey (2) reported pupal period of 5 to 9 days during March-April and with an average to 7 to 19 days in November-December. The puparium remains beneath the epidermis, head upwards, normally near the base of stem (34). The lifecycle is completed in 7 to 17 days. The pest is active throughout the year and has 8-9 generations from July to April.

2.2 Pests of vegetative stage

Pen aphid, *Acyrthosiphon pisum* is widely distributed over Africa, India, North America and Southern Canada and U.S.S.R. Aphids are important pest which causes both direct and indirect damage to the crop. Nymphs and adults suck sap from the tender leaves and shoots which affect plant growth, reduces crop yield and severe attack results in stunted growth and mortality of the plants. Indirect damage are caused by acting as a vector of pea leaf-roll virus. Pea seedlings are more susceptible than older plants to pea aphid (8). Eggs are laid at the base of stem or under remain of leguminous plants. Nymph passes through five instars with 4-5 moults. Duration of nymphal period extends from 1 to 8 days with an average of 3 days (18). Even it may prolong from 20-24 days in spring at Ukraine (USSR). Pea aphid passes through varying number of generations with a range of 14 to 28 generations per year (1, 18, 21, 112).

Pea semilooper, *Plusia orichalcea* (Fabricious) is regarded as minor pest. The green larvae feed on foliage and in heavy infestaion they may defoliage the crop. Eggs are mostly deposited singly on leaves and hatch in about 3 to 4 days. Larval and pupal development may be completed in 14 and 18 days, respectively. In certain cases, the caterpillars also feed on the pod wall and grains. The humid climate is favourable for build up of population. The pea leaf miner, *C. horticola* is the most common and important pest of the pea. It is known to be common in the Mediterranean area than in northern Europe and occurs widely through Asia, including Japan. It is extremely polyphagous species whose members feed on a very large number of phylogenetically diverse plant families. Over 300 species of plants have been recorded as its host around the world (35, 104). Eggs are laid singly inside the leaf tissue in puncture made by oripositing female which may lay upto 329-358 eggs (3). The larvae feed below the epidermis, on the pallisade and spongy parenchymatous tissue. The destruction of the chlorophyll containing tissue retards the photosynthetic activity of the plant, as a result of which, growth and yield of the infested plants are adversely affected. Duration of larval stadia is highly variable which vary from 7.9 to 12.4 days from mid January to first week of April and 3 to 16 days (7). Pupariam is found at the end of third larval instar at the end of leafmine. This period is highly variable that is from 7 to 15 days and it depends on the temperature.

The pea midge, *Contarinia pisi* is a pest of pea throughout northern Europe but in Great Britain it is restricted largely to major pea growing areas like the north eastern of Cambridgeshire (119). The pest has not been reported from India.

2.3 Pests of flowers and pods

The pea thrips, *Caliothrips indicus* Bagnall has been reported to attack this crop by few workers. (97, 99, 110). Besides pea, the pest attacks several plant species like onion, groundnut, soybean, potato, linseed, brinjal, sweetpotato, barseem, coriander, khol-khol, cauliflower, chillies, cotton, cabbage and wheat (115). Singh and Mishra (91) reported that thrips can cause crumpling of leaves and retarded plant growth. Females lay eggs in the leaf tissue at the rate of 19 to 96 eggs per female. Incubation period lasts between 6-15 days during February to April. The two larval instars take about 4 to 10 days. Pupal period has two resting stages namely prepupa and pupa which is passed in the soil. Prepupal and pupal stage take 1 to 5 days and 1 to 6 days, respectively during February and April.

The other species, *Thrips tabaci* Lind., polyphagous, has also been reported to attack pea crop in European countries (89). Nymphs and adults mainly damage flowers and pods by causing either small raised bumps or scars appearing as silver mottling on the pods.

Lepidopterous species belonging to several families cause damage to peas, throughout the world. However, in most of the species are of minor importance and include mainly *Etiella zinckenella,* (Triit), *Helicoverpa armigera* (Hb.), *Lampides boeticus* Linn., *Euchrysops onejus* (Fab.) and *Cydia nigricana* (Steph). The activities of pests start when the crop enters the reproductive phase. Pea pod borer *E. zincknella* is a common pest of winter crop and causes serious damages in late sown crops. Eggs are laid singly or in groups at the junction of the calyx and pod or on pod surface. Single female may lay upto 47 to 178 eggs, which hatch in 5 to 6 days. The newly emerged larvae construct a web like structure through which they enter the developing pods and feeds on grains. The damaged pods become discoloured, watery in appearance and emit foul small and they also accumulate frass pellet in the pods (93, 107). Larval period is completed in 10 to 13 days. Pupation takes place in a cocoon in the soil or under debris. Pupal duration lasts 9 to 20 days depending upon the climate. Single generation is completed in 24 to 44 days.

Lampides boeticus is a minor pest and is active from November to March on several pulse crops. The caterpillar feeds on flowers, buds and green pods. Eggs are generally laid singly on flower buds and green pods. A single female may lay upto 30 to 120 eggs. Eggs, larval and pupal period durations last for 4 to 7, 9 to 27 and 7 to 19 days, respectively. The larvae are small, flat and oval in shape with green colour. The pest competes 2 to 3 generations in a year.

Euchrysops onejus is also widely distributed and feeds on a number of grain legumes in Kharif and Rabi seasons. The female lay 60-200 eggs singly on various parts of the plant and eggs hatch in about 2 to 4 days. The larvae feed on buds, flowers and developing pods. The entry holes made in the pods are kept closed with faecal material.Larvae undergo 4 instars within 10 to 20 days and pupate in soil or in debries at surface of ground. The pupal period lasts for 5 to 12 days.

Helicoverpa armigera becomes serious pest if podding of crop synchrosises with peak activity of the pest i.e. March-April. It is a polyphagous pest and feeds on several cultivated and wild plant species. White eggs are laid on buds, flowers, leaves and pods. The eggs hatch after 3 to 5 days and small larvae feed by scrapping on leaves and flowers. Later these bores into the developing pods and eat away grain by making large circular holes. The larvae are of different colours, with longitudinal streaks along their sides. The larva becomes full grown in about 3 weeks. It pupates in the soil and brown moth emerge from soil in about 2 weeks.

Pea moth, *Cydia nigricana* (Stephens) is the most important pest of peas in USSR, European countries and North America but absent from India and other neighbouring countries. Infestations of the pea moth are difficult to detect because the caterpillar burrow into the developing pods so their damage is unnoticed until the pea is harvested. In addition, its presence, even in small numbers, can reduce the economic value of crops, particularly those used for freezing or canning (6). Eggs are usually placed singly and hatch in about 6 to 8 days. The larvae are variable in colour from white to deep orange-yellow. Most larvae develop a yellow tinge in the third instar. There are five larval instars and larval development period is competed in 19 to 30 days. Pupation takes in a cocoon on the ground surface. Pupal period lasts from 10 to 15 days.

3. PEST LOSS ASSESSMENT

Though several species of insects feed on peas, only few cause significant yield loss. Few attempts have been made to quantify crop losses by insect pests. Stemfly has been reported to cause extensive damage of upto 70% by Agrawal and Pandey (2) in the crop season, while Bindra and Singh (17) found that stemfly could cause damage as high as 90 to 100%. Singh *et. al.*, (92) found that *O. centrosematis* may cause more than 95% damage to pea crop, if crop is sown in first week of October. Sehgal and Ram Ujagir (80) reported that more than 30% plants of susceptible cultivars yielded between 8.4 to 9.5 ton ha^{-1}, where resistant cultivars having 23 to 28% damaged plants gave yield of 1.2 to 1.8 ton ha^{-1}. Damage to pea leaves by leaf miner, *C. horticola* was reported to be as high as 90% during first week of March with upto 4 larvae and puparia per leaf (7). At Kanpur, maximum of 70% leaves damage occurred during March and 4% during December with number of larvae ranging from 169 per 200 leaves in March and 14 per 200 leaves in December (107). The number of leaf mines were found to increase gradually from 4th till 8th week after germination and there after. Higher leaf miners to 112.5 per 25 leaves and 169 leaf mines per 25 leaves were observed during 13th and 15th week, after germination respectively (41). Mavi and Singh (57) reported that more than 40% leaf infestation significantly reduced grain yield and also gave highest cost; benefit ratio of 1:3.7. Studies conducted in Yugoslavia, showed that 2 to 10 or more mines per leaf resulted in yield loss of pea (71).

Pea aphid, can cause reduction in relative growth rate and efficiency of production of new tissues (12, 37). Singh and Mishra (91) reported that *Caliothrips indicus* could cause 52% infestation to plants in early sown crops.

Saxena *et al.* (78) observed that *E. zinckenella* caused 13% and 8.7% damage to pods and seeds respectively, in unprotected crop. Singh and Dhooria (1971) found that damage caused by *E. zinckenella* was 50.9% to pod and reduction in seed weight was recorded as 24%. Delay in planting of peas led to higher damage to pods by *E. zincknella.* Percent pod damage varied from 12% to 30.8% and 3.5 to 8.5% in December and October sown crops respectively. A reduction of 1.2 ton ha^{-1} in yield was observed when crop was sown during first week of December instead of October planting. This has been found damaging in Uttar Pradesh. Varying level of pods damaged, 11 to 16% (81), 3.6 to 18.8% (95), 12.6 to 13.6% (41) during different years was observed at Pantnagar, northern India.

4. METHODS OF PEST CONTROL

4.1 Cultural control

Cultural practices against certain pests of pea may be highly effective. Different pests respond to a given cultural procedure in different ways and there may be conflicting damaged made by different pest problems. In this section, the responses of major pea pests to several categories of cultural control method are considered in order to demonstrate where integration with other components is possible.

Crop rotation

Highly mobile and polyphagous species like *H. armigera* are likely to be unaffected by crop cycling compared with endogenous pests, such as stemfly if site of sowing is not changed.

Date of sowing

Sowing date may be manipulated in order to ensure that the crop is sufficiently advanced to allow compensatory growth or to ensure that susceptible stage is not available to the pest. However, different pest react to times of sowing in different ways, and it may be difficult to design a programme that offer protection to all potential pests. Early sowing of field pea may increase the damaged caused by stemfly but at the same time reduces incidence of leaf miner and pod borer (43, 80, 92). The delayed planting may lead to higher pod damage due to *E. zinckenella* (108).

Sowing practices

A number of variables can reduce pest incidence at the time of planting, considering various practices like seed density, spacing, and depth, the use of mixture of crop, irrigation and fertilizer application. Deeply sown seed delay plant emergence reducing susceptibility to stemfly. While higher seed rate may encourage compensation for attack by stemfly. Certain weeds and cultivated hosts may act at host for several pests and these should be destroyed from field in case of leaf miner, aphid and pod borers. Systems of intercroping are unlikely to find a wide application in pea for insect pest control. Inter cropping pea with linseed and barley have reduced pest incidence (68). Gill *et al.* (31) recorded higher infestation of leaf miner when crop was irrigated after 30 and 60 mm of evaporation than when it was irrigated after each 90 or 10 mm of evaporation. Irrigation under insecticide application gave higher grain yield. Irrigation alone reduced the population of cutworm and army-worm but at the same time increased the population of other insects (70). Effects of spacing, N nutrition and weed control were studied against pea moth, *Cydia nigricana, Sitona* sp. and *Bruchus pisorum* on peas in Poland. The closest spacing reduced leaf injury by *Sitona* and pod infestation by *C. nigricana* larvae but increased aphid *(A. pisum)* number per plant. N nutrition had only minor effect on pest incidence. Weed control did not produce any significant differences (121).

4.2 Host plant resistance (HPR)

Todate very little attention has been focussed on HPR of pea and the possibility of its integration with other pest control strategies. Moderate levels of resistance have been identified for a number of pea pests. The cultivars, Asauji of mid maturity with hard and thinner stem was found resistant to stemfly attack (91). Some cultvars like KMR-83, RAU-37 and KLMR-9 were found least susceptible to stem fly damage (80). Several varieties have been identified with high and moderate level of resistance to pea leaf miner in India (4, 24, 42, 79, 82, 86, 95). The resistance in these may be mediated by antixenosis and antibiosis mechanism.

Pea cultivars like Bridger, IP3, GC 141, Meteore, Thompson, Laton, Alderman dwarf, early Badger, Bonneville, Arkel, Asauji and NLP were found relatively resistant to *Caliothrip indicus* (91).

Pea resistance to aphids attack have been reported (9, 20) and the mechanism of resistance has been reported to be a combination of preference and tolerance (84, 85). Mechanism of resistance in six resistant cultivar was attributed to antixenosis, antibiosis and tolerance (101).

Most of the pod borers are polyphagous. Therefore, it is difficult to get higher level of resistance to pest attack. However, cultivars such as JP 92A, JP 179A and P402 were found to possess resistance against both *H. armigera* and *C. horticola* and some were found to have resistance either to *Heliothis* or *C. horticola* (95). Search for pest resistance, its generally performed under extensive laboratory and field condition in germplasm, cultivars and varieties. Land races may carry greater level of resistance or tolerance to various pests. The process of screening for pest resistance is slow because it requires long term studies. Physical plant feature may be one such indicator: hard and thin stem are less susceptible to stem fly (91). The seed texture, colour and flower colour, size of leaves may play significant role in imparting resistance to leaf miner (83, 95). The small seeded varieties were relatively more resistance than larger seeded ones to *Heliothis armigera.* Chemical indicators may accelerate the process of screening but little work has been done in this direction. There is a clear inverse relationship between leaf thickness and feeding rate in case of *Sitona lineatus.* Inhibition of feeding was induced by the amino acid tyrosine. Antibiosis effect was found against aphid and leaf miner. Partial resistance with good agronomic characters may interact with biological control agents in various ways by slowing colonization, development and which expose the pest to natural enemies for longer time. Since natural enemies play an important role in the dynamics of aphid and leaf miner population on peas, it is in this group that such benefits of partial resistance are likely to be found.

4.3. Natural and biological control

The role of naturally occurring pest enemies have been studied for a number of legume pests and in number of cases they have been found effective by reducing the need for chemical control or other means of control measure. On the basis of the previous work, we know that major pests are preyed upon by several natural enemies. Many of the predators that include insects, spiders, lizard and birds feed on many pest species. Some of the pathogens i.e. bacteria, fungi, viruses are also helpful in suppressing the pest population. Several parasitoids, predators and microbial agents have been reported from India and other places. Several species of parasitoids and predators had been observed on leaf miner and stemfly on pea crop (3, 7, 32, 33, 44, 46, 47, 55, 60, 61, 83, 113). Very little information is available about parasitoids of aphid and thrips, besides the commonly aphid specific known predators like cocoinellid beetle, syrphids and apiders. Lepidopteran borers of pea are parasitized and preyed upon by several species of insects and other animals. Bhatnagar, *et. al.,* (14) had listed 27 species of parasitoids and 19 species of predators on *Heliothis.* Some natural enemies like *Trichogramma* spp. *Campolitis* spp. and *Chrysopa* spp. should be tried against *Heliothis* on pea crop. Field trials on Nuclear Polythedrosis Virus (NPV) and *Bacillus thuringiensis* under AICPIP and other parts of India, have given satisfactory control of this pest on related legumes.

Mere presence of a natural enemy in a crop system does not guarentee its potential contribution in a system where pesticide are used to get higher yield. Therefore, utmost care should be taken to preserve natural enemies. It is well accepted that injudicious use to toxicant is harmful to natural enemies to a great extent. Against pod borer entomopathogens have even greater potential with several advantages. The study of these agents constitute an entirely new area of biological control of pea pest, for which only meagre information is available. At present, the usefulness of many natural enemies in a IPM programme is limited by their unpredictable occurrence and rate of spread. It may be possible to increase their population by intercropping and to some extent by rotation with crops which carry same natural enemies. Reservoirs for aphid parasitoids may be encouraged by mixed cropping, crop diversity and early sowing etc. as has been found in case of cereal aphids (65, 66). In addition, aphid attraction can be increased by artificial food source. Some syrphid species have been found to be attracted to artificial honeydew applied in the field (36).

There is no doubt that natural enemies do play a role in preventing outbreak of peas pests and whilst it is unlikely that in near future forecasting scheme will take full consideration of these insects. Use of selective chemicals, and maintenance and extension of non crop habitat are the principal means of maximising their effectiveness at present.

4.4. Chemical control

Several studies on the evaluation of insecticides for control of pest complex on this crop have been conducted (4, 64); in India (70) and in Brazil (27). At Faizabad, several insecticides were evaluated as seed treatment, spray or basal dressing against stemfly. However, maximum grain yield was obtained with carbofuran as basal dressing (94). Among the various granular insecticides tested against stemfly, disulfoton and phorate @ 1.5 kg a.i./ha as seed treatment were found effective (118). Seed treatment with carbofuran 50 s.p. at 5% was most effective among the various insecticide tested for control of stem fly in south India (5)). In Rajasthan, carbofuran was tested as broadcast in the field before sowing. Placing of the granules in furrow, above or below the seeds any by side dressing after germination. The maximum control of stemfly was achieved when carbofuran granules at 1 to 3 kg a.i./ha, were placed below the seeds (88).

Several insecticides in various formulations had been found effective against pea aphid (21, 56, 59, 63, 76, 112). Spraying of crop with spray formulation and wettable powder of various insecticides were reported to highly effective against thrips (30, 74, 92, 99, 115). In U.K. furthiocarb and bendiocarb as polymer seed coating to pea seed gave excellent control of *Thrips angusticeps* and *Stina lineatus* (75).

Pea leaf miner, can be adequately controlled with wide range of products including phosphamidon, dimethoate, monocrotophos, methyl parathion, fenthion, chlorpyriphos, leptophos, quinalphos, endosalfan, methyl-o-demeton, dicrotophos, fenvalerate, phosalone, thiodicarb and carbaryl as spray and disulfoton, phorate, carbofuran and aldicarb as granules (7, 13, 19, 23, 77, 78, 96, 102, 105, 106, 116, 117, 118).

Pod borers can be effectively controlled with Lindane, carbaryl, dimethoate, phosphamidon, methyl parathion, monocrotophos, promecarb, bromophos, naled, triazophos, lannate, flucythrinate, primiphos-methyl deltamethrin and cypermethrin (10, 13, 15, 45, 53, 78, 107). Combined testing of mono crotophos with soil application of phorate or disulfoton gave good control (19).

Joint application of dinocarp and endosulfan, bavistin and monocrotophos and sulfex and monocrotophos provided effective control of *Erysiphe polygoni* and *E. zincknella* at Bangalore, India (90). The problem of pesticide residue on leaves, pods and grains have important consideration. Number of studies carried out in India, have examined insecticide residue to numerous products. The results have shown that by 10 to 25 days after insecticide application residue have been below tolerable limits of negligible (26, 98). Washing and boiling of pods in water can further reduce pesticide residue by 70 to 85% respectively (26).

Pest resistance to insecticide has not yet become a problem for pea growers in India and other parts of the world, and this is a major reason for the lack of interest in IPM in this crop. There is every likelyhood that pesticide resistance may develop in the near future if they are not used properly. Insecticide may have direct effect on natural enemies. The repeated application of wide spectrum pesticide will never allow population to recover. Though there are no case studies on indirect effect of pesticides on pea, there is no evidence that pesticide use in pea has led to pest outbreak. Application method, and efficient targetting of the insecticide, could play a more important role in limiting the damaging effect on polyphagous predator population. It appears that manipulation of timing of application to minimize damage to beneficial insects has received great attention.

5. MONITORING OF PEST POPULATION

Many different types of sampling techniques have been discussed and developed for estimating insect population on the host plant (38). Selection of sampling technique will basically depend upon the nature of the crop and the pest. In Germany, line sampling technique (5 plants sample/site at 5 sites at interval of 20 places along 12 lines) have been recommended to determine the level of infestation of pea aphid in large pea fields (22). Sequential plan for chemical control of *A. pisum* a field pea in Manitoba was derived from studies of aphid dispersion among plant tips in field (54). Hinz, (40) reported that the pipe methods sampling gave more number of aphids than visual counts. The economic threshold level of *A. pisum* was estimated to be 9 to 12 aphid/sweep when crop began to flower. A relationship between plant and sweep samples was equivalent to 2-3 aphid/tip of stem. For pea leaf miner, middle portion of plants (leaves) should be sampled. Fixed number of trifoliate or sweeping of adult flies are common procedure for sampling of this pest on pea. Stem flies are sampled on the basis of percent plants attacked per unit area. Traps are useful devices for mobile insects to pinpoint correct time for applying control measures. This may be important for pests such as pod borers, where treatment must be applied after eggs hatch but before the young larvae enter the pods. Against various nocturnal pests light trap had been used to monitor long term population. The principal aim is to monitor and evaluate the effects of changing agricultural practices and climatic factors on pod borer population. Chemical monitoring is largely based either on host plant or on sex attractant. Attractive plant volatile or their analogues are used to monitor various pests, notably fruit flies. In India, pheromone traps are being used for monitoring *Heliothis armigera.* In Britain, pheromone traps are used for monitoring the pea moth, *Cydia nigricana.* Two pheromone traps are placed in each pea field and are observed for 2 to 3 days. When 5 to 10 or more moths are caught per trap, then calculation based on egg development rates are used to larval hatch and hence optimum time for treatment (11, 52).

Further monitoring of migratory insects e.g. aphids, *Helicothis* can be achieved with development of remote sensing technique such as radar, genetic marking and chemoprinting (72).

6. ECONOMICS ANALYSIS OF IPM IN PEA

Integrated control programme for *C. nigricana* and pea aphid have been successful throughout Europe and America. This avoids risk of insecticide resistance to these pest. In addition, the techniques also reduces risk of phytotoxicity from pesticide applied at tender stages of the plant. This will be also cheaper in comparison to pesticide use. Where resistance in population have manifested, further development in cultural and host plant resistance technique can effect economic pest threshold and the interaction between the biological agent and its host.

7. PROGRESS TOWARDS IPM IN PEA

Exclusive reliance upon a single control measure for a pest invites vulnerability to disaster. Therefore, integration of cultural, biological and chemical control methods may be best for major insect pests of pea.

Some cultural practices measures can be used successfully against the pea stemfly; the main ones are : uniform and late planting over large area, higher plant density. This also reduces or causes to escape damage of pod borers and leaf miner attack. Early sowing reduces the attack of leaf miner and pod borers but requires use of relative insecticides in form of basal application for the control of stem fly damage. Where biological control and insect growth regulator are available, they must be used to reduce pesticide load. When crop matures, it should be immediately harvested to avoid attack of storage pest.

8. FUTURE TRENDS

The major reason behind the development of IPM has been development of resistance to pesticide so that broad spectrum pesticide strategy begin to break down. *H. armigera* in south India is an example of this. Pest will certainly continue to become resistant to some of the pesticide to which they are exposed, thus precipitating an immediate need for a different pesticide or alternative control strategy. This is most important for pea because it is consumed by both man and animals. Acceptibility of IPM depends on the availability of alternative approaches. Monitoring system and thresholds deserve more research. Interpretation of pheromone data should be refined with phenological models which take account of weather variables. Most studies on biological control of pea pests have failed to quantity the real impact of these agents with few exceptions. There is also need to emphasize that the mere presence of a natural enemy in a pea system does not guarantee its potential value as a control agent. Nevertheless, in a system in which insecticides are used to obtain high yield, care must be taken to conserve the natural enemy complex. It is generally accepted that insecticides are harmful to natural enemies of pests. In this regard, entomopathogens have even greater potential, since many are less disrupted by insecticide use. The study of these agents, constitute an entirely new area of biological control of pea pest for which only limited information is available. Microbial pesticide might be used against podborer in pea. New insecticides with an insect growth regulator effect offer the potential for some lepidopteran pests. The continuing development of newer improved components for IPM programme assume great future on pea crop.

9. REFERENCES

1. Abdulmadzhid,. A. A. 1972. Biological characteristics of pea aphid. Restritelnathaschita, **20**: 10-13.

2. Agrawal, N. S. and Pandey, N. D. 1961. Bionomics of *Melanagromyza phaseoli* Coq. (Diptera : Agronomyzidae). Indian J. Ent., **23** : 293-298.

3. Ahmad, T. and Gupta, R. L. 1941. The pea leafminer, *Phytomyza atricornis* in India. Indian J. Ent., **3** : 37-49.

4. AICPIP (All India Coordinated Pulse Improvement Project). Consolidated report on Rabi Pulses for the year 1980-1992. Entomology. Directorate of Pulses Research, Kanpur. India.

5. Ambekar, J. S., Patil, B. E., Pawar, D. B., Gandhala, D. N. and Pokharkar, R. N. 1984. A note on the control of stemfly *(Ophiomyia phaseoli* Tryon) on pea *(Pisum sativum* L.), South Indian Horticulture, **32** : 377-378.

6. Anonymous. 1985. Pea moth. MAFF Advisory Leaflet 334. HMSO, London. 6 pp.

7. Atwal, A. S., Chaudhary, J. P. and Ramzan, M. 1969. Studies on the biology and control of pea leaf miner. *Phytomyza atricornis* Mg. (Agromyzidae : Diptera). J. Res. Punjab Agric. Univ. **6** : 163-169.

8. Auclair, J. L. 1966. Dissimilarities in the biology of pea aphid. *Acyrthosiphon pisum* (Harris), (Homoptera: Aphididae) on alfalfa and pea in New Mexico. Ann. Ent. Soc. Am. **59** : 4.

9. Azarova, E. F. 1986. Nauchno, Tekhnicheskii, Byullelens Vsesoyuznogo Nanchno-Issledovalel's skogo Institute Zernobibovy khi krupyanykh Kultur No. **35** : 91-94.

10. Bailey, P. and Comery, J. 1987. Management of *Heliothis punctigera* on field peas in south-eastern Australia. Australian Journal of Experimental Agriculture, **27** : 439-443.

11. Barel, C. J. A. and Voerman, S. 1980. Vangsten in een seksval, maatstaf Voor bestrijding van de erwtepeu l boorder. Gewas bescherming, **11** : 151-154.

12. Barlow, C. A. and Messmer, I. 1982. Pea aphid (Homoptera : Aphididae) induced changes in some growth rates of plant of pea. J. Econ. Ent, **75** : 765-768.

13. Barroga, S. F. and Barrers, A. B. 1969. Evaluation of pesticides for the control of pea leaf miner, *(Phytomyza atricornis* Meign) and other insect pests attacking pea *(Pisum sativum L.)*.Phillipine J. Plant Indus., **34** : 123-130.

14. Bhatnagar, V. S., Lateef, S. S., Sithanantham, S., Pawar, C. S. and Reed, W. 1982. Research on *Heliothis* at ICRISAT. In : Proceeding of International workshop of *Heliothis* Management, 15-20 Nov., 1981 at ICRISAT, Patancheru, A.P., India.

15. Biddle, A. J. 1985. Field evaluation of timed applications of pyrethroid insecticides for the control of pea moth *(Cydia nigricana)* 371-376. Processors and Growers Research Organisation, Thornhaugh, Peterborough PE8, 6HJ-HK.

16. Bindra, O.S. 1968. Insect pests of pulse crops. Indian Fmg. **17** : 12-14.

17. Bindra, O. S. and Harcharan Singh. 1969. Pea stem borer, *Melanagromyza (Agromyza) phaseoli* Tryon (Diptera : Agromyzidae). Pesticide, **3** : 19-21.

18. Campbell, E. R. 1926. The pea aphis in California. J. Agric. Res. **32** : 861-881.

19. Chapade, H. M. 1975. Chemical control of some major pests of field pea, *Pisum sativum.* Entomologists' News Let., **5** : 17-18.

20. Chaplygina, I.N. and Amelina, K. A. 1987. Resistance of pea to diseases and pests and the search for means to increase it. Selektsiya i semenorodstva, USSR No. **4** : 27-28.

21. Cindea, E. 1982. Investigation on biology and control of some pests of edible pea crop. Problem de Protectia Plantelor, **7** : 87-96.

22. Daebeler, F. and Hinz, B. 1987. Die Besiedlung grosser Erbsenschlage durch die Erbsenblattlaus, *Acyrthosiphon pisum* (Harris). Nachrichtenblatt furden Pflanzenschutz in der DDR, **41** : 84.

23. Dash, A. N. 1990. Evaluation of some insecticide for their efficacy against the peå leaf miner, *Phytomyza atricornis* (Meign.). Indian J. Pl. Prot. **18** : 295-297.

24. Deole, J. Y., Harsana, H. S., Mishra, U. S. and Dhandere, S. Y. 1984. Varietal incidence of pea leaf miner at Gwalior. Indian J. Pl. Prot. **12** : 149-150.

25. Dikshit, A. K. 1986. Persistence pattern and waiting periods for some insecticide on pea crop. Indian J. Ent. **13** : 69-73.

26. Dikshit, A. K., Handa, S. K., Verma, S. 1988. Dissipation of carbaryel residue from pea crop. Plant Protection Bull. (Faridabad), **40** : 12-13.

27. Franca, F. H. and Giordano, L. B. 1984. Pests of peas and their control in the federal district. In XXIV congresso Brasileirode Olericultura. I, Reniao Latino-Americana de Olericultura; Jaboticabal, 16, 21 Julho. 1984, Resumose Palestras, Brazil.

28. Frohlich, G. and Sarwar, S. L. 1977. Beitrag Zur Lebensweise and Entwicklurg des olivbraumen Erbsenwicklers. Arch. Phytopathol. Pfeanzenschutz, Berlin, **13** : 117-125.

29. Fletcher, T. B. 1914. Some south Indian Insect pests. Bishen Singh Mahendra Pal singh Dehradun, 565 pp.

30. Gargar, V. P. and Vaishampayan, S. M. 1978. Relative toxicity of some chlorinated hydrocarbon and organophosphate insecticide against pea thrips *Caliothrips indicus* (Bagnall). JNKVV Res. Journal, **12** : 31-34.

31. Gill, C. K., Dhoorja, M. S., Hari Singh and Gill, S. S. 1979. Incidence of pea leaf miner, *Phytomyza atricornis* (Meign.) under different agronomic practices. J. Res. Punjab Agric. Univ. **16** ; 309-311.

32. Gokulpure, R. S. 1970. New Records of hymenopterous parasites of pea leaf miner. *Phytomyza atricornis* Meign (Diptera : Agromyzidae) J. Bombay nat. Hist. Soc. **66** : 223 - 224.

33. Gokulpure, R. S. 1972. Note on the hosts and parasites of *Phytomyza atricornis* (Meign) (Diptera : Agromyzidae). Indian J. Agric. Sci. **42** : 638-640.

34. Greathead, D. J. 1969. A Study in Africa of the bean flies (Diptera : Agromyzidae) affecting *Phaseolus vulgaris* and of their natural enemies, with description of a new species of *Melanagromyza.* Sp. Hend. Bull Ent. Res. **59** : 541-561.

35. Griffiths, G.C.D. 1967. Revision of *Phytomyza syngenesiae* Group (Diptera : Agromyzidae) including species hitherto known as "*Phytomyza atricornis* Meign" Stuttg. Beitr. Naturk. **1-77** : 28.

36. Hagen, K. S., Sawall, E. F. and Assan, R. L. 1971. The use of food sprays to increase effectiveness of entomophagous insects. Proc. Tall Tibers Conf. Ecol. Anim. Contr. Habitat Manag. **3** : 59-81.

37. Hamid, S., Shash, M. A. and Anwar, A. M. 1974. Investigations on *Acyrthosiphon pisum* (Harris) with special reference to its parasite, *Aphidius smithi* Sharma and Subha Rao. Tech. Bull. Commonwealth Inst. Biol. Cont., **17** : 69-85.

38. Heathcote, G. D. 1972. Evaluating aphid populations on plants. In: Aphid Technology (Ed. H. F. van Emden), 105-145, Academic Press, London.

39. Hill, D. S. 1973. Damage to pea seedlings and brussels sprout by larvae of bean seed fly *(Delia platura* Meig.). Plant Path. **22** : 49.

40. Hinz, B. 1989. A method for recording pea aphids in field crop. Nachrichtenblatt fur den Pflanzenschutz inder. DDR. 43-255.

41. Jamwal, S. S. 1985. Insects associated with field peas, *Pisum sativum* Linnaeus, and estimation of losses caused by them. M. Sc. Thesis, 123 pp. G. B. Pant Univ. Agri. Tech., Pantnagar.

42. Kashyap, R. K., Bharoot, J. P. and Verma, A. N. 1982. Relative susceptibility of pea cultivar against leaf miner, *Melanagromyza conavistae* Greathead (Diptera : Agromyzidae) Haryana Agric. Univ. J. Res. **12**: 128-129.

43. Kaul, C. K., Mehrotra, P. and Singh, S. D. 1988. A note on the effect of date of sowing on *Heliothis armigera* Hubner attack on pea *(Pisum sativum* L.) Indian J. Ent. **50** : 533-534.

44. Kaurava, A. S., Odak, S. C. and Dhamdhere, S. V. 1969. Preliminary studies on the biology of *Neochrysocharis* sp. (Hymenoptera : Eulophidae) A parasite of *Phytomyza atricornis* Mg. (Diptera : Agromyzidae). J. Bombay nat Hist. Soc. **66**: 396-398.

45. Khanzad, A. G. and Naqui, K. M. 1985. Preliminary studies on use of oil based ULV pyrethrold insecticide against the pod borer. Pakistan Jour. Agric. Res. **6** : 134-136.

46. Kumar, A. 1984. Incidence of pupal parasitism on *Chromatomyia horticola* (Gour) (Diptera : Agromyzidae) in some parts of Uttar Pradesh. Anu. Ent. Agra., India.

47. Kumar, A. 1985. Incidence of parasitism of *Diglyphus isaea* (Walk.) on *Chromatomyia horticola* (Gour) a pest of *Pisum sativum* in northern India. Entomon. **10** : 55-58.

48. Kumar, A. 1989. Variation in the incidence of parasitism of *Chromatomyia horticola* (Gour) (Diptera : Agromyzidae) during the different parts of the year. Entomon. **14** : 67-70.

49. Kumar, A. 1990. Records of some new hymenopterons parasites of *Chromatomyia horticola* (Gour) (Diptera : Agromyzidae) from Agra. Entomon. **15** : 135-136.

50. Lane, A. and Light, W. I. St. G. 1979. Thrips on peas. Leaflets Ministry Agric. Fisheries and Food. 170. 4pp.

51. Lefroy, H. M. 1909. Indian Insect Life. pp. 622-623. Thacker and Spink, Calcutta.

52. Lewis, T. and Stungeon, D. M. 1978. Early warning of egg hatching in pea moth *(Cydia nigracana)* Ann. App. Biol. **88** : 199-210.

53. Macaulay, E.D.M., Etheridge, P., Garthwaite, D. G., Greenway, A. R., Wall, C., and Goodchild, R. E. 1985. Prediction of optimum spraying dates against pea moth, *Cydia nigricana* (F.), using pheromone traps and temperature measurement, Crop Protection, **4** : 85-89.

54. Maiteki, G. A. and Lamb, R. J. 1987. Sequential decision plan for control of pea aphid, *Acyrthosiphon pisum* (Homoptera : Aphididae), on field peas in Manitoba. J. Econ. Ent. **80** : 605-607.

55. Mani, M. S. 1971. Some chalcidoid parasites (Hymenoptera) of leaf mining Agromyzidae (Diptera) from India. J. Bombay Nat. Hist. Soc. **5** : 591-598.

56. Mathur, A. C., Krishnaih, K. and Tandon, P.L. 1974. Comparative efficacy of some pesticide for the control of pea aphid *(Acyrthosiphon pisum* (Harris). Indian J. Ent., **31** : 386-388.

57. Mavi, G. S. and Sukhminder Singh 1988. Assessment of losses by pea leaf miner. *Chromatomyia horticola* (Goureau) on pea. J. Insect Sci. **1** : 61-64.

58. Molitas, L. P. and Gabriel, B. P. 1975. Biology of the sweet pea leaf miner, *Phytomyza horticola* Goureau (Diptera : Agromyzidae) in the Philippines. Philippine Entomologist, **3** : 89-108.

59. Mote, U. N. 1979. Efficacy of different granulated systemic insecticides against pea aphid. Indian J. Pl. Prot. **7** : 221-222.

60. Nair, M.R.G.K. 1986. Insects and mites of crops in India. Indian Council of Agricultural Research, New Delhi.

61. Narayanan, E. S., Subha Rao, B. R. and Kaur, R. B. 1956. Studies on the biology of the parasites of pea leaf miner, *Phytomyza atricornis* (Meign). Proc. Indian Acad. Sci. (B), **44** : 137-147.

62. Narayana, M. L. and Saxena, H. P. 1984. Efficacy of systemic granules alone and in combination with foliar sprays against major insect pests of pea. Indian J. Ent. **46** : 248-249.

63. Nemec, V., Havlickova, H. and Smolek, P. 1984. The effects of systematically applied dithiodiglucose on the aphid, *Achrthosiphon pisum.* Acta Entomologia Bohemoslovaca, **81** : 40-45.

64. Otanes Y. Quesales, F. 1918. The bean fly. Philipp. Agric. **1** : 2-31.

65. Powell, W. and Dewar, A. M. 1983. Alternatives to insurance spraying with aphicides on cereals. In Better British Wheat. Eds. J. Hardcastle and R. Sutherwood, 28-29. London.

66. Powell, W., Dewar, A. M., Wilding, N. and Dean G. 1983. Manipulation of cereal aphid natural enemies. 10th International congress of Plant Protection 2, 780.

67. Prasad, D., Singh, K. M., Katiyar, R. N. and Singh, R. N. 1984. Incidence of insect pests in early maturity high yielding variety of pea. *Pisum sativum* Linn. Indian J. Ent. **46** : 352-362.

68. Prasad, D., Singh, K. M., and Katiyar, R. N. 1987. Impact of intercropping on the plant growth, pest incidence and crop yield of pea, *Pisum sativum* linn. Indian J. Ent. **49** : 153-172.

69. Prasad, D., Singh, K. M., Katiyar, R. N. and Singh, R. N. 1987. Influence of irrigation on the crop growth, pest incidence and crop yield of *Pisum sativu,* Linn. Indian J. Ent. **49** : 330-334.

70. Prasad, D., Singh, K. M., Katiyar, R.N. and Singh, R. N. 1988. Interaction effects of insecticide, intercropping and irrigation on the crop growth, pest incidence and crop yield of pea, *Pisum sativum* Linn. Indian J. Ent . **50** : 116-122.

71. Pugliarini, N. and Spasic, R. 1984. Neka Zapazanja 0 podari i stetnosti muhe minera *Phytomyza horticola.* Gouraeu. Zastita Bilja, **35** : 17-22.

72. Rabb, R. L. and Kennedy, G. G. 1979. Movment of highly mobile insects, concepts and methodology in Research, North Carolina State University, Releigh, 456 pp.

73. Rehman, K. A. 1940. Some more hosts of cotton white fly. Indian J. Ent. **2** : 243-244.

74. Rawat, R. R., Saxena, D. K. and Jakhmola, S. S. 1969. Studies on chemical control of pea thrips (*Caliothrips indicus* Bagnall). Plant Prot. Bull., Indian, **21** : 1-6.

75. Salter, W. J. and Smith, J. M. 1986. Peas-control of establishment pests and diseases using metalaxyl based seed coatings. In : 1986 British Crop Protection Conference held at Brighton Metropole, England Nov. 17-20, 1986 Vol. 3. Thornton Health, U. K. British Crop Council, 1093-1100.

76. Sarup, P. Singh, D. S., Amarpuri, S. and Lal, R. 1969. Laboratory evaluation on different pesticides as contact poisons against adult of *Aphis koch* (Aphididae; Hemiptera) infesting pea crop. Indian J. Ent. **31**: 311-320.

77. Sarup, P., Sircar, P., Sharma, D. N. Singh, D. S., Dhinger, S., Dewan, R. S. and La, R. 1976. Evaluation of biological efficacy of insecticidal granular formulation against some important predators/pests of pea crop. Indian J. Ent. **38**: 153-159.

78. Saxena, H. P., Sircar, P., and Phokela, A. 1969. Studies on chemical control of pea leaf miner, *Phytomyza atricornis* Mg. and pod borer, *Etiella zinckenella* Tr. Labdev J. Sci. Tech., **7(B)**: 318-322.

79. Saxena, J. K., Srivastava, R. L., Singh, L. N. and Tripathi, R. M. 1976. Leafminer (*Phytomyza atricornis* M.) resistance in pea (*Pisum sativum* L.) Sci. Cult. **42** : 561-562.

80. Sehgal, V. K. and Ram Uhjagir, 1980. Report of entomological experiments conducted during 1980-81; AICPIP (AICAR) G.B.P.U.A.T. Pantnagar.

81. Sehgal, V. K. and Ram Uhjagir, 1982. Report of entomological experiments conducted during 1981-82; AICPIP (AICAR) G.B.P.U.A.T. Pantnagar.

82. Sehgal, V. K. and Ram Uhjagir, 1982-90. Report of entomological experiments conducted during 1982-92; AICPIP (AICAR) G.B.P.U.A.T. Pantnagar.

83. Sen, A. 1981. Field incidence, biology and host plant relationship of pea leaf miner, *Chromatomyia horticola* (Goureau), on eleven varieties of field peas, M. Sc. (Ag.) Entomology, Thesis, G. B. Pant Univ. Agri, Tech. Pantnagar, 92 pp.

84. Semenova, A. G. 1989. Evaluation of resistance of pea to *Acyrthosiphon pisum* by means of mars rearing selectsiya i semenovodstro No. **2**: 17.

85. Semenova, A. G. 1990. Resistance of pea to *Acyrthosiphon pisum* Sbornik Nauchnykh Trudov po Prikladnoi Botanike, Genetika i selekissi, **132** : 78-82.

86. Sharma, I.D. and Sharma, K. C. 1991. Screeninmg of pea germplasm for resistance to leaf miner, *Phytomyza atricornis* Meign. Indian J. Pl. Prot. **19** : 206.

87. Sharma, S. K., Batra, R. C., Mathur, Y. K. and Verma, J. P. 1970. Pests of pea in Rajasthan. Madras Agr. J. **57** : 46-48.

88. Sharma, S. K., Puri, M. K. and Shinde, V.K.R. 1981. Assessment of biological efficacy of carbofuran granules against *Melanagromyza phaseoli Cog*. in pea. Pesticides, **15** : 25-26.

89. Shelton, A. M. and North, R. C. 1987. Injury and control of onion thrips (Thysanoptera: Thripidae) on edible podded pea. J. Econ. Ent. **80** : 1325-1330.

90. Shukla, R. P. and Lal, S. S. 1988. Effect of combined application of fungicides and insecticides on the powdery mildew and pod borer of field pea. Pesticides,.**22** : 5-7.

91. Singh, Gajendra and, Mishra, P. N. 1977. Note on susceptibility of certain vegetable pea varieties to *Melanagromyza phaseoli* (Tryon) and *Thrips tabaci* Lindeman. Indian J. Agric. Sci. **47** : 11.

92. Singh, Gajendra, Mishra, P. N. and Srivastava, B. M. 1981. Effect of date of sowing of pea on the damage caused by *Opiomyia centrosematis.* Indian J. Agric. Sci. **51** : 340-343.

93. Singh, H. and Dhooria, M. S. 1971. Bionomics and pea pod borer, *Etiella zinckenella* (Treitsekhe). Indian J. Ent. **33** : 123-30.

94. Singh, H. M., Rizvi, S.M.A. and Ram Singh 1988. Insecticidal control of pea stemfly. *Ophiomyia phaseoli* Tryon. Indian J. Pl. Prot. **16** : 41-43.

95. Singh, K. V. 1984. Screening of *Pisum sativum* L. cultivars for resistance against leaf miner, *Chromatomyia horticola* (Goureau) and pod borer, *(Heliothis armigera* (Hb.) M.Sc. Thesis, 91. pp. G. B. Pant Univ. Agr. Tech., Pantnagar.

96. Singh, O. P., Dhamdhere, S. V. and Mishra, U. S. 1986. Toxicity of insecticide against pupae of pea leaf miner, *Phytomyza horticola* (Meign.) infesting vegetable pea. Indian J. Ent. **48** : 372-374.

97. Singh, S. and Sohi, G. S. 1957. A list of insect pests of economic importance in Punjab, Ludhiana. Govt. Agric. Col. Mag. **24** : 19-34.

98. Singh, M. and Yadav, P. R. 1983. Persistence of dimethoate and methyl demeton In and on pea pod shells and their movement into pea grains. Indian J. Ent.

99. Singh, Zile, and Vaishampayan, S. M. 1967. Results of trials on chemical control of *Caliothrips indicus* (Bagnall) on pea. Indian J. Ent. **29** : 21-220.

100. Singh, Zile, and Vaishampayan, S. M. 1969. Further truials on chemical control of *Caliothrips indicus* (Bagnall) on pea. JNKVV Res. J. **3** : 30-34.

101. Soroka, J. J. and Mackay, P. A. 1991. Antibiosis. and antixenosis to pea aphid (Homoptera : Aphididae) in cultivar of field peas. J. Eco. Ent. **84** : 1951-1956.

102. Sohi, G. S. and Verma, G. C. 1967. Studies on the chemical control of pea leaf miner, *Phytomyza atricornis* (Meig.) (Diptera : Agromyzidae). Indian J. Ent. **29** : 281-284.

103. Southwood, T.R.E. 1978. Ecological methods (IInd ed.) Chapman and Hall, London 524 pp.

104. Spencer, K. V. 1973. Agromyzidae (Diptera) of economic importance. Series Entomologica, 9, The Hague, W. Junk B. V. pp. 418.

105. Srinivasan, K. and Shukla, R. P. 1980. Control of pea leaf miner *Phytomyza atricornis* (Meign) and aphids *(Acyrthosiphon pisum* Harris) with certain granular insecticides. Pesticide, **14** : 19-20.

106. Srivastava, A. S. and Singh. Y. P. 1972. Bionomics and control of pea leafminer, *Phytomyza atricornis* Meig. (Agromyzidae : Diptera) Z. Angew. Ent. **70** : 437-440.

107. Srivastava, A. S. and Singh Y. P. 1974. Survey, damage, life history and control of pea pod borer, *Etiella zinckenella* (Triet.) (Pyralidae: Lepidoptera) Sci. Cult. **40** : 370-372.

108. Srivastava, A. S., Srivastava, K. M. and Singh, R. P. 1974. Effect of different dates of sowing of pwa *(Pisum sativum* L.) on the yield and incidence of *Phytomyza atricornis* M (Pea leaf miner) and *Etiella zinckenella* Tr. (Pea pod borer). Z. Angew Ent. **76** : 439-442.

109. Srivastava, B. K. 1959. Growth potential of *Spodoptera* exigua in relation to certain winter food plants. Madras Agri. J. **46** : 258-259.

110. Srivastava, B. K. 1964. Pests of pulse crops. In : Entomology in India 1938-1963. (Ed. N.C. Pant) 83-91, Indian J. Ent. New Delhi.

111. Srivastava, K. M., Singh, L. N. and Singh, R. P. 1976. Blue butterfly, *Euchrysops* sp. as pea pest in Uttar Pradesh. Entomologist Newslet. **6** : 27-28.

112. Talanov, V. P. 1967. The pea aphid in south of Ukraine Zazhchita Rastani **5** : 124-126.

113. Tandon, S. K. 1971. Notes on biology of *Phytomyza atricornis* Meign, a common leaf miner of crucifer plants (Agromyzidae : Diptera), Agra Univ. J. Res. (Sci), **20** : 47-68.

114. Thakur, J. R., Sharma, P. L. and Bhardwaj, A. K. 1972. Record of *Cosmolyce (Lampides) boeticus* Linn. (Lepidoptera : Lycaenidae) on pea pods from Simla hills (Himachal Pradesh). Indian J. Ent. **34** : 248-249.

115. Vaishampayan, S. M., Singh, Zile and Seshagiri, A. 1968. Studies on *Caliothrips* (Thrips) *indicus* Bagnall and a note on assessment of its damage to pea crop. Indian J. Agric. Sci. **38** : 295-297.

116. Verma, R. S., Prakash, N. and Nath, V. 1980. Note on chemical control of pea leaf miner, *Phytomyza atricornis* Meign and pea and pod borer, *Etiella zinckenella* Tr. Indian. J. Ent. **42** : 824-826.

117. Verma, S. and Dixit, A. K. 1982. Persistence of different insecticides against pea leaf miner, *Phytomyza atricornis* Meign. Indian J. Agric. Sci. **52** : 4-6.

118. Vyas, H. N. and Saxena, H. P. 1982. Comparative efficacy of phorate, dis Dulfoton and carboforan against leaf miner, *Phytomyza horticola (atricornis)* infesting peas *(Pisum sativum* L.). Indian J. Pl. Prot. **9** : 55-60.

119. Wall, C., Pickett, J. A., Garthwaite, D. G. and Morris, N. 1985. A female sex pheromone in the pea midge, *Contarinia pisi.* Ent. Exp. appl. **39** : 11-14.

120. Way, M. J. and Cammell, M. E. 1973. The problem of pest and disease forecasting - possibilities and limitations as exemplified by work on the bean aphid, *Aphis fabae.* Proc. 7th British Insecticide and Fungicide Conference **3** : 933-954, BCPC, Croydon.

121. Wnuk, A. and Weich, K. 1987. Effect of some cultural practices can pest occurrence and injury in green peas. Zeszyty Naukowe Akademii Rolniezej w uprawie na kollatajaw kvakowie ogrodnictwo No. **211** **(16)**: 217-229.

12 Genetic Improvement for Resistance to Insect-Pests and Diseases in Soybean

P.S. Bhatnagar, S.P. Tiwari and P.G. Karmakar

CONTENTS

Integrated Pest and Disease Management
Rajeev K. Upadhyay, K.G. Mukerji, B.P. Chamola and O.P. Dubey (eds.)
APH Publishing Corporation, New Delhi-110002, India

ABSTRACT

Losses due to biotic stresses is one of the important causes for holding the national productivity of soybean in India at low level of 1 ton/ha which is half the world average. In India major diseases of soybean are bacterial pustule, pod blight, yellow mosaic virus, rhizoctonia aerial blight, myrothecium and bud blight and major insect-pests are green semilooper, girdle beetle, stem fly and blue beetle. Disease or insect resistant varieties offer unique advantage of providing inherent control of insect-pests and diseases which involves no expense or environment pollution or destruction of useful insects and fauna. Information on sources of host plant resistance to diseases or insect-pests and their genetics are essential to select appropriate breeding methods to be effective in evolving agronomically acceptable disease/insect resistant varieties.

1. INTRODUCTION

To harness the productivity potential of a crop, it is imperative that it is insulated against the ravages of biotic aberrations. Genetic resistance against insect-pests and diseases has attained still greater importance in the context of the present ethos of eco-friendliness. Notwithstanding India is fifth and second in world and Asia for area and production of soybean, losses due to biotic stresses is one of the important causes for holding the national productivity of soybean at low level of 1 ton/ha which is about half the world average. Covering an area of over 4.5 million ha during 1993 (7, 8) India is experiencing notable growth rate. Recently enhanced invasion of the crop by insect-pests and diseases has further necessitated a renewed thrust on breeding of soybean varieties resistant to biotic stresses.

2. MAJOR BIOTIC STRESSES IN SOYBEAN

Major insect-pests and diseases of soybean are listed in Table 1 and 2. Among insect-pests, presently over 275 different species are known to feed on different plant parts throughout its growth stage (42). Moreover, many insects which seldom or did not earlier feed on soybean, have become permanent and voracious feeders. Alarming examples of such cases are white grubs, pod borer, *Cydia ptychora,* Jassids, *Apheliona maculosa;* and blue beetle, *Cneorane* sp. This list is fast increasing. The insect-pest problem in soybean in India has reached to serious proportions. This is probably due to notable area expansion as indicated in the 'species-area hypothesis' (44).

Table 1 : Major insect-pests of soybean

A. Foliage Feeders

Semiloopes/loopers

Chrysodeixis acuta (Wlk.)
Diachrysia orichalcea (Feb.)
Pseudoplusia includens (Wlk.)
Trichoplusia ni Hub.
Mocis undata Fab. *Rivula* sp.

Tobacco caterpillar
Spodoptera litura (Fab.)

Green clover worm
Plathypena scraba (Fab.)

Corn earworm
Heliothis zea (Bodd.)

Gram pod borer
Helicoverpa armigera (Hub.)

Bihar hairy caterpillar
Spilosoma obliqua (Wlk.)

Leaf folder
Hedylepta indicata (Fab.)

Leaf miner
Aproaerema modicella Deventer

Mexican bean beetle
Epilachna varivestis Muls.

Blue beetle
Cneorane sp.

B. Stem Borers

Stem flies
Melanagromyza sojae (Zehnt.)
M. obtusa Malloch
Ophiomyia phaseoli (Tryon)

Girdle beetles
Obereopsis brevis (Swed.)
Nupserha bicolor (Thompson)
ssp. *postbrunnea* Dutt
N. nitidior pic.
var. *attripennis* Brug

C. Sap Feeders

White fly
Bemisia tabaci Genn.

Jassids (Hoppers)
Empoasca terminalis Dist.
Apheliona maculosa Dist.

Thrips
Thrips tabaci Lind.
Ayyaria chaetophora Karny

Green stink bug
Nezara viridula (Linn.)

Table 2 : Major diseases of soybean

Disease	Casual organism(s)
A. Fungal Diseases	
(i) Root and lower stem diseases	
Charcoal rot	*Macrophonina phaseolina* (Tassi) Goid.
Sclerotial blight	*Sclerotium rolfsii* Sacc.
Fusarium blight or root rot or wilt	*Fusarium oxysporum* f. sp. *tracheiphilum* (E.F. Smith) Snyd. & Hans. *F. oxysporum* f. sp. *vasinfectum* (Atk.) Snyd and Hans.
Brown stem rot	*Phialophora gregata* (Allington and Chamberlain) W. Gams
Phytophthora rot	*Phytophthora megasperma* Drechs f. sp. *glycinea* Kuan and Erwin
Stem canker	*Diaporthe phaseolorum* (Cke & Ell.) Sacc. var. *caulivora* Athow and Caldwell
Sclerotinia stem rot	*Sclerotinia sclerotiorum* (Lib.) d By
(ii) Foliar and seed diseases	
Anthracnose or pod blight	*Colletotrichum truncatum* (Schw.) Andrus and W.D. Moore
Frog-eye leaf spot	*Cercospora sojina* Hara
Rust	*Phakopsora pachyrhizi* Syd. & P. Syd.
Rhizoctonia aerial blight	*Rhizoctonia solani* Kuhn
Brown spot	*Septoria glycines* Hemmi
Alternaria leafspot	*Alternaria tenuissima* (Kunze ex Pers.) Wiltshire; *A. alternata* (Fr.) Keisslor
Myrothecium leafsport	*Myrothecium raridum* Tode ex Fries
Downy mildew	*Peronospora manshurica* (Naum.) Syd. ex Gaum.
Pod and stem blight	*Diaporthe phaseolorum* (Cke. & Ell.) Sacc var. *sojae* (Lehman) Wehm., and *Phomopsis longicolla* Hobbs
Powdery mildew	*Microsphaera diffusa* Cke. & Pk.
Cercospora blight and leaf spot; purple seed stain	*Cercospora kikuchii* (T. Matsu. & Tomoyasu) Gardner
Target leaf spot	*Corynespora cassiicola* (Berk & Curt.) Wei
B. Viral Diseases	
Soybean mosaic	Soybean mosaic virus
Yellow mosaic	Mung bean yellow mosaic virus
Bud blight	Tobacco ring spot virus
Bean pod mottle	Bean pod mottle virus
Peanut mottle	Peanut mottle virus
C. Bacterial Diseases	
Bacterial pustule	*Xanthomonas campestris* pv. *glycins* (Nakano) Dye
Bacterial blight	*Pseudomonas syringae* pv. *glycinea* (Coerper) Young, Dye & Wilkie
D. Nematode Diseases	
Soybean cyst nematode	*Heterodera glycines* Ichinohe
Reniform nematode	*Rotylenchulus reniformis* Linford & Oliveira
Root-knot nematode	*Meloidogyne javanica* (Treub) Chitwood *M. incognita* (Kofoid & white) Chitwood *M. arenaria* (Neel) Chitwood *M. hapla* Chitwood

Fortunately, only few of the insect-pests visiting soybean crop are of economic importance (Table 1). Out of them, green semi-looper *(Diachrysia orichalcea)*, girdle beetle *(Obereopsis brevis)*, stemfly *(Melanagromyza sojae)* and blue beetle *(Cneorane* sp.*)* are of economic concern in the predominant soybean growing area of Central India. Some other pests like leaf miner *(Apoaerema modicella* Deventer) in Marathwada region, Bihar hairy caterpillar *(Spilosoma obliqua)* in parts of northern plains and Terai region are also becoming serious.

Over 100 plant pathogens are reported to invade soybean crop and 35 of them are economically significant (41). The important soybean diseases form the view point of occurrence and economic loss are bacterial pustule, pod blight, different mosaic diseases particularly yellow mosaic virus, rhizoctonia aerial blight, myrothecium and bud blight. Importance of different diseases however varies from region to region in India.

3. SOURCES OF HOST PLANT RESISTANCE

Genetic resistance against pests and diseases is an integral part of the Integrated Pest Management (IPM) programme. The utilization of resistant varieties in farming is the most simple, effective and the economical method in the management of biotic menaces. Besides, the resistant varieties conserve natural resources and reduce the cost, time and energy when compared with the other methods of the pathogen management.

Sources of genetic resistance to insect-pests of soybean especially for foliar feeding insects were identified in the early 1970's (12, 46). Subsequent efforts in identification and transfer of insect resistance resulted in the availability of a large number of donors. Some of these resistance sources possessed multiple insect-resistant (2, 23, 26). The progress in developing insect-resistant cultivars with acceptable agronomical characters has, however, been very slow. Two insect-resistant strains, UPIR-1 and UPIR-2, were found promising at Pantnagar in north India. At Parbhani (Marathawada region of the country), emphasis is being given to resistance breeding for leaf miner *(Bilobata subsecivella)*. Resistant donors, MACS 6613, MACS 74, are being utilised in the breeding programme. At NRCS, Indore, *G. soja* as well as germplasm line L 592 were found to have no infestation of girdle beetle. Some mutant lines were identified to be resistant to jassids. some high yielding mutant lines of Bragg and PK 472 were also identified to be tolerant to stem fly. Ram *et al.* (37) reported *G. soja* to be resistant to Bihar hairy caterpillar. The resistance appeared to be due to feeding preference rather than due to antibiosis or avoidance. The species was utilised in hybridization with the cultigen and a highly tolerant line, PK 515, was developed. During the last few years of research at AVRCD, Taiwan, three accessions, PI 171451, PI 227687 and PI 229358, were found to be resistant to various defoliators and stink bugs. All of them were not resistant to all insect species but together they covered practically all insect-pests prevalent in Taiwan. The resistance present in three multiple insect resistance donors i.e., PI 171451, PI 227687 and PI 229358 was reported to be controlled by different genes (26). Breeding advances utilizing the available sources of resistance are needed in order to develop agronomically acceptable insect resistant varieties.

Hartwig (18) and Sinclair and Backman (41) have summarised sources of disease resistance for use in breeding disease resistant soybean varieties. Resistance sources against important diseases including those prevalent in India are listed in Table 3.

Table 3 : Sources of resistance in soybean

Sl. No.	Character		Donors/Sources
1.	**Resistance to insect-pests**		
	1.1	Bihar hairy caterpillar	*Glycine soja*
	1.2	Girdle beetle	*G. soja,* L 592
	1.3	Multiple resistance to various defoliators and stink bugs	PI 171444, PI 171451, PI 227687, PI 229358
2.	**Resistacne to diseases**		
	2.1	Yellow mosaic virus	*G. soja,* UPSM 534, PK 416
	2.2	Bacterial pustules	CNS, Clark 63, Wayne, Williams, Cumberland, Lee
	2.3	Rhizoctonia aerial blight	PK 472, Jawa 16, Kalitur
	2.4	Rust	PI 200492, PI 462312, PI 230970, PI 459025
	2.5	Soybean mosaic	PI 96983, Tokyo, Ogden, Raiden
	2.6	Peanut mottle virus	CNS, Dorman, Peking
	2.7	Cowpea chlorotic mottle virus	Bragg, Hill, Lee
	2.8	Bacterial blight	Norchief, Harosoy
	2.9	Downy mildew	Kanrich
	2.10	Phytophthora rot	Mack, PI 54615-1, Kingwa, CNS, PI 171442, PI 86972-1, PI 86050, PI 91160, Altona, Mukden, FC 31745, Sanga, PI 84637.
	2.11	Brown stem rot	BSR-201, -301, -302, Chamberlain, PI 84946-2, PI 86150, PI 88820 N, PI 90138, PI 965769
	2.12	Powdery mildew	Blackhawk
	2.13	Stem canker	Kingwa, Mandarin
	2.14	Frog eye leafspot	Lincoln, Wabash, Kent
	2.15	Fusarium blight	Ripley
3.	**Resistance to nematode diseases**		
	3.1	Soybean cyst nematode	Peking
	3.2	Reniform nematode	Forrest

4. GENETICS OF RESISTANCE

Genetic behaviour of disease resistance in soybean was extensively reviewed by Palmer and Kilen (36) and Sinclair and Beckman (41).

4.1 Resistance to diseases

Soybean rust

Four independent dominant genes namely Rpp_1, Rpp_2, Rpp_3 (20) and Rpp_4 (19) confer specific resistance against soybean rust isolates.

Phytophthora rot

The resistance is controlled by six dominant Rps genes (24).

Brown stem rot

Three dominant genes independently control resistance against brown stem rot and exhibit duplicate epistasis (48).

Bacterial pustules

Resistance to bacterial pustule is controlled by a single recessive allele, 'rxp' (6).

Bacterial blight

Mukherjee *et al.* (34) reported a dominant allele (Rpg1) controlling resistance to race 1 which causes bacterial blight.

Frog-eye leaf spot

Two dominant alleles control the resistance, one of which is effective against race 1 and the other against race 2 (6).

Fusarium blight

The resistance is conditioned by a single dominant nuclear gene 'Rfs' (44).

Soybean viruses

Resistance to yellow mosaic virus is governed by a recessive gene pair, 'rym_1' and 'rym_2' (42). Resistance to soybean mosaic virus was governed by a single dominant allele 'rsv'. However resistance in variety Ogden is governed by another allele at the same locus, which is dominant to susceptibility but recessive to 'Rsv' (28). Other dominant genes controlling resistance to soybean mosaic virus have also been reported.

The inheritance of importance diseases of soybean has been reported in the Compendium of Soybean diseases (41).

Nematodes

Although seven different groups of nematodes infesting soybean have been reported in India by Bhattacharjee *et al.* (9), the parasite has not yet posed a problem for soybean cultivation in the country. 5 races of cyst nematode *(Heterodera glycines)* are known and several genes imparting resistance have been identified. Anand and Rao-Arelli (1) found that resistance to race-5 of the nematode is conditioned by two recessive genes. They further reported that at least one resistance gene was different in the three host genotypes.

4.2 Resistance to insect-pests

Van Duyn *et al.* (47) identified three soybean introductions as highly resistant to Mexican bean beetle. These were subsequently found to posses multiple resistance against several insect-pests. Kilen and Lambert (26) found that resistance in each of these three introductions, namely PI 171451, PI 227687 and PI 229358 is governed by at least one gene. Resistance to lepidopterous defoliators was found to be governed by at least three recessive genes (27). Gai *et al.* (16) studied the inheritance of resistance to the beanfly *(Melanogromyza sojae* Zehntner) in soybean. Segregation for resistance to the beanfly in crosses involving highly resistant, moderately resistant and highly susceptible varieties indicated that resistance was controlled by a single gene, 'Rms'.

Mechanisms of insect resistance in plants have been reviewed by Maxwell and Jennings (33). Painter (35) classified insect resistance in plants or : (a) Antixenosis or non-preference-where the plant is unattractive to insect for feeding, oviposition or shelter (b) Antibiosis-where the host plant has adverse effect on the growth and survival of the insect and its progeny (c) Tolerance-when the plant can withstand heavy insect attack and still produce good yield.

Pubescence in soybean appears to be one of the main resistance-imparting factors (31). Many alleles affecting the trichomes are known. Bernard and Singh (5) studied the inheritance of 5 types of pubescence and established that each of the five pubescence types differs from normal by a single gene pair and that the five loci segregate independently. Bernard (4) suggested that the two alleles (pa 1 and pa 2) are necessary for the appressed phenotype. Alleles controlling the density of pubescence are reported (3). The gentisic acid and salicylic acid produced by soybean leaves act as feeding deterent for Mexican bean beetle (15). Maturity duration appears to have considerable impact on the level of resistance in soybean (17). Generally higher level of resistance has been observed in later maturity groups (25, 29, 30, 38).

5. BREEDING METHODS

Out of several methods available, backcross method has been mostly used for resistance breeding when transfer to specific resistance gene(s) is desired in an agronomically desirable recurrent parent. Different modifications of backcross methods have been used to develop agronomically superior soybean varieties with resistance to insect-pests. Limited backcrossing has been successfully used to transfer resistance to soybean cyst nematode into populations from which desirable progeny can be selected as varieties e.g. Bedford (21). Multiple parents have been used to transfer genes for pest resistance from agronomically unacceptable parents into more desirable populations by using modified backcross method. Variety 'Kirby' resistant to soybean cyst nematode was developed by this method.

When the character to be transferred is controlled by a major gene, it can be readily monitored in backcrossing. Sufficient number of backcrossing should be effected so that the whole genetic background of the recurrent parent is transferred and the need to conduct yield trials is minimised. Several varieties have, thus, been developed especially in case of phytophthora rot resistance using 4 to 7 backcrosses (13, 47).

Induced mutagenesis has been found to be very effective in breeding soybean varieties resistant to insect-pests and diseases (7). Some mutant progeneis in soybean varieties PI 472 were identified to be tolerant to stem fly (40).

Under situations of high pressure of natural selection in the form of either inoculum of a pathogen or an insect-pest with favourable environment for infection/infestation, the bulk method of breeding could also be used for building up the frequency of resistance genes. After 4 cycles of natural selection from F2 to F5 in a bulk population the frequency of phytophthora rot susceptible genotypes in F6 generation may decrease by 19% if the susceptible genotypes yield even 75% of the resistant genotype (10). Hartwig *et al.* (22) found that the degree of field resistance to phytophthora rot, a quantitative character, influenced natural selection for individuals with specific resistance to the disease. Natural selection increased the percentage of plants with a major gene for resistance in populations derived from parents that did not have a high degree of field resistance to the disease. Natural selection was not effective when the population contained plants with field resistance that completed effectively with plants containing a major gene for resistance. Plants resistant to soybean cyst nematode had a reproductive advantage over susceptible ones in infested soil (32) and they grow in numbers over generations in segregating populations (22). However artificial selection among individuals or lines is more direct and rapid in breeding for pest resistance than relying on natural selection in a bulk population.

6. CONCLUSION

Disease or insect resistant varieties offer unique advantage of providing inherent control of insect-pests and diseases which involves no expenses or environment pollution or destruction of useful insects and fauna. In soybean, resistant varieties highly effective against twenty diseases were developed in U.S.A. (41). In India development of disease resistant varieties of soybean especially against yellow mosaic virus (Y.M.V.) has enabled successful cultivation of the crop in Y.M.V. prone areas of India. Specific resistance controlled by major genes is easy to incorporate into adapted varieties and has been used in breeding such varieties. But race-specific resistance has often been a problem in varieties used for controlling soybean cyst nematode, phytophthora root rot, rust, frog eye leafspot and downy mildew where numerous races of the causal pathogens are known. In such cases specific resistance against the predominant races can be used for breeding. Since field resistance provides protection against multiple races of pathogen, it is more desirable type or resistance but its use in breeding is difficult as it is a quantitative character. In soybean, field resistance is reported for phytophthora rot and brown stem rot diseases (11, 46). More research efforts are needed to develop varieties with sufficient level of field resistance.

Insect tolerant soybean varieties are probably the most desirable component of an ideal integrated pest management programme while for polyphagous insects resistant varieties offer a better solution. The use of an insect resistant variety over a long period of time must be integrated with other management practices. Encouraging mutation breeding results have provided bright prospects to evolve insect resistant varieties of soybean (7). However, the progress in breeding insect resistant varieties has been rather slow and the genetics of insect resistance too is less well understood. It is necessary to distinguish highly resistance sources from those with intermediate level of resistance and both field and green house screening techniques should be adapted for this purpose. Field screening should be done using appropriate plot size (14), without using any preventive sprays as generally used by plant breeders and screening observations should start as early as 3-node (V_3) stage to facilitate crossing during the same season (37).

7. REFERENCES

1. Anand, S.C. and Rao-Arelli, A.P. 1989. Genetic analyses of soybean genotypes resistant to soybean cyst nematode Race-5. Crop Sci. **29** : 1181-1184.

2. Beach, M.R. and Todd, J.W. 1988. Foliage consumption and development parameters of the soybean looper and the velvet bean caterpillar (Lepidoptera : Noctuidae) reared on susceptible and resistant soybean genotypes. J. Econ. Entomol. **81** : 310-316.

3. Bernard, R.L. 1975a. The inheritance of semi-sprarse pubscence. Soybean Genet. Newsl. **2** : 33-34.

4. Bernard, R.L. 1975b. The inheritance of appressed pubescence. Soybean Genet. Newsl. **2** : 34-36.

5. Bernard, R.L. and Singh, B.B. 1969. Inheritance of pubescence type in soybeans : glabrous, curvy, dense, sparse and puberulent. Crop Sci. **9** : 192-197.

6. Bernard, R.L. and Weiss, M.G. 1973. Qualitative genetics. In: 'Soybeans : Improvement, production and uses'. Agronomy Series, 16. Ed. B.E. Caldwell. pp. 117-154.

7. Bhatnagar, P.S. 1994. Genetic enhancement of soybean yield in tropical agriculture with particular return to India. Paper presented in World Soybean Research Conference - V, 21-27 Feb. 1994, Chiang Mai, (Thailand).

8. Bhatnagar, P.S. and Tiwari, S.P. 1993. Augmenting soybean production and soil situation in India. In: 'Research and Development of Oilseeds'. Ed. A. Saikia, NOVOD Board, Gurgaon, India. pp. 15-21 and 119.

9. Bhattacharjee, N.S., Laha, S.K. and Misra, S.D. 1979. Non-arthopod pests of soybean-incidence of nematodes. Bull. Ent., **20** : 26-33.

10. Buzell, R.I. and Haas, J.H. 1972. Natural and mass selection estimates of relative fitness for the soybean 'rps' gene. Crop Sci. **12** : 75-76.

11. Caviness, C.E. and Walters, H.J. 1966. Registration of 'Davis' soybeans. Crop Sci. **6** : 502.

12. Clark, W.J., Harris, F.A., Maxwell, F.G. and Hartwig, E.E. 1972. Resistance of certain soybean cultivars to bean leaf beetle, striped blister beetle and bollworm. J. Econ. Ent. **65** : 1669-1672.

13. Fehr, W.R., Schmitthenner, A.F., Baahrenfus, J.B., Schoener, C.S., Walker, A.K. and Tachibana, H. 1981. Registration of 'Vickery' soybean. Crop Sci. **21** : 475.

14. Fenderburk, J.E., Soffes, A.R., Burnett, R.D., Herzog, D.C. and Hinson, K. 1990. Plot size and shape in relation to soybean resistance for velvet bean caterpillar (Lepidoptera : Noctuidae). J. Econ. Entomol. **84**: 2107-2110.

15. Fischer, D.C., Kogan, M. and Paxton, J. 1990. Deterrancy of Mexican bean beetle (Coleoptera : Coccinallidea) feeding by free phenolic acid. J. Entomol. Sci. **25** : 230-238.

16. Gai, J.V., Liu, F.Y., Cui, Z.L., Xian, J.K. and Ma, Y.H., 1991. Inheritance of resistance to the beanfly (*Melanagromyza sojae* Zehntner in soybean. Soybean Genet. Newsl. **18** : 104-106.

17. Gilman, D.F., McPherson, R.M., Newson, L.D., Herzog, D.C. and Wiliams, C. 1982. Resistance in soybeans to the southern Green stink bug. Crop Sci. **22** : 573-576.

18. Hartwig, E.E. 1973. Varietal development. In 'Soybeans : Improvement, production and uses'. Agronomy Series, 16. Ed. B.E. Caldwell. pp. 187-210.

19. Hartwig, E.E. 1986. Identification of 4th major gene conferring resistance to soybean rust. Crop Sci. **26**: 1735-1736.

20. Hartwig, E.E. and Bromfield, K.R. 1983. Relationship among 3 genes conferring specific resistance to rust in soybean. Crop Sci. **23** : 237-239.

21. Hartwig, E.E. and Epps, J.M. 1978. Registration of 'Bedford' soybeans. Crop Sci. **18** : 915.

22. Hartwig, E.E., Kilen, T.C., Young, L.D. and Edwards, C.J. Jr. 1982. Effects of natural selection in segregating soybean populations exposed to phythopthora rot or soybean cyst nematodes. Crop Sci. **22**: 588-590.

23. Hatchett, J.H., Beland, G.H. and Kilen, T.C. 1979. Identification of multiple insect resistance in soybean lines. Crop Sci. **19** : 557-559.

24. Kilen, T.C. 1986. Relationships between 'Rps_2' and other genes controlling phytophthora rot in soybean. Crop Sci. **26** : 711-712.

25. Kilen, T.C. 1990. Multiple insect resistance in a soybean germplasm line. Soybean Genet. Newsl. **17** : 103-105.

26. Kilen, T.C. and Lambert, L. 1986. Evidence for different genes controlling insect resistance in 3 soybean genotypes. Crop Sci. **26** : 809-871.

27. Kilen, T.C., Hatchett, J.H. and Hartwig, E.E. 1977. Evaluation of early generation soybeans for resistance to soybean looper. Crop Sci. **17** : 397-398.

28. Kiihl, R.A.S. and Hartwig, E.E. 1979. Inheritance of reaction to soybean mosaic virus in soybeans. Crop Sci. **19** : 372-375.

29. Kraemer, M.E., Rangappa, M., Benepal, P.S. and Mebruhatu, T. 1988. Field evaluation of soybean for Mexican bean beetle resistance. I. Maturity groups VI, VII and VIII. Crop Sci., **28** : 497-499.

30. Kraemer, M.E., Rangappa, M., Mebruhatu, T. and Benepal, P.S. 1990. Field evaluation of soybean for Mexican bean beetle resistance II. Maturity groups III, IV and V. Crop Sci. **30** : 374-377.

31. Lee, Y.I. 1983. The potato leafhopper, *Empoasca fabae,* soybean pubescence and hopper-burn resistance. Ph.D. diss. Univ. of Illinois, Urbana. (Diss. Abstr. Int. **36B** : 2057).

32. Luedders, V.D. and Dulcos, L.A. 1978. Reproductive advantage associated with resistance to soybean cyst nematode. Crop Sci., **18** : 821-823.

33. Maxwell, F.G. and Jennings, P.R. 1980. Breeding plants resistant to insects. (Ed.) John Willey and Sons, New York.

34. Mukherjee, D., Lambert, J.W., Cooper, R.L. and Kennedy, B.W. 1966. Inheritance of resistance to bacterial blight (*Pseudomonas glycine* (Oerpa) in soybeans. Crop Sci., **6** : 324-326.

35. Painter, R.H. 1951. Insect resistance in crop plant. McMilan, New York. pp. 520.

36. Palmer, R.G. and Kilen, T.C. 1987. Qualitative genetics and cytogenetics. In 'Soybeans : Improvement, production and uses. Agronomy series 16 Ed. J.R. Wilcox. pp. 135-209.

37. Ram, H.H., Pushpendra, Kamendra Singh, and Ranjit, 1989. *Glycine soja* : a source of resistance for Bihar hairy caterpillar, *Spilosoma (Diacrisia) obliqua* Wallace in soybean. Soybean Genet. Newsl. **16** : 52-53.

38. Rowan, G.B., Boerma, H.R., All, J.N. and Todd, J. 1991. Soybean cultivar resistance to defoliating insects. Crop Sci., **31** : 678-682.

39. Ruffener, G.K. II, Hammod, R.B., Copper, R.L. and Martin, St. S.K. 1987. Larval antibiosis screening technique for Mexican bean beetle resistance in soybean. Crop Sci., **27** : 598-600.

40. Sharma, A.N. and Bhatnagar, P.S. 1994. Radiation induced mutants for resistance to stem fly in soybean. Mutation Breeding Newsletter (communicated).

41. Sinclair, J.B. and Backman, P.A. 1989. Compendium of soybean diseases. Third ed. APS Press. The Amer. Phytopathol. Soc. Minne sota, USA. pp. 1-106.

42. Singh, B.B. and Malik, A.S. 1978. Inheritance of resistance to yellow mosaic virus in soybean. Ind. J. Genet. **38** : 258-261.

43. Singh, O.P., Verma, S.N. and Nema, K.K. 1990. Insect pests of soybean in India. International Book Distributors, Dehradun, India 281 p.

44. Stephens, P.A., Nickell, C.D. and Kolb, F.L. 1993. Genetic analysis of resistance to *Fusarium solani* in soybean. Crop Sci., **33** : 929-930.

45. Strong, D.R. Jr. 1979. Biogeographic dynamics of insect-host communities. Annu. Rev. Entomol. **24** : 89-119.

46. Tachibana, H., Bahrenfus, J.B. and Fehr, W.R. 1983. Registration of 'BSR 201' soybean, Crop Sci., **23**: 186.

47. Van Duyn, J.W., Turnip seed, S.G. and Maxwell, J.D. 1971. Resistance in soybeans to the Mexican bean beetle-I. Source of resistance. Crop Sci. **11** : 572-573.

48. Wilcox, J.R., Bernard, R.L., Cooper, R.L., Fehr, W.R., and Neihaus, M.H. 1979. Genetic improvement of soybeans in the Midwest. Crop Sci. **19** : 803-805.

49. Willmot, D.B. and Nickell, C.D. 1989. Genetic analysis of brown stem rot resistance in soybean. Crop Sci. **29** : 672-674.

13 Integrated Management of Chickpea Diseases

Vishwa Dhar and S. N. Gurha

CONTENTS

Integrated Pest and Disease Management
Rajeev K. Upadhyay, K.G. Mukerji, B.P. Chamola and O.P. Dubey (eds.)
APH Publishing Corporation, New Delhi-110002, India

ABSTRACT

Major diseases affecting chickpea, their distribution and importance in different agroclimatic zones of India have been briefly presented. Available management practices (cultural, chemical, biological and host plant resistance) have been reviewed and their effective integration in disease management system suggested.

Significance of host-pathogen distribution and intensity maps in fixing priorities for integrated disease management (IDM) is highlighted in order to make rational use of the available infrastructural facilities and resources. Host plant resistance being the major component in IDM for this crop, needs special attention. The success achieved in developing resistant varieties against individual diseases and recent identification so source for multiple resistance encourages to intensify efforts on breeding super cultivars suited to different cropping systems and possessing resistance to multi adversities such as pathogens, insects, nematodes, weeds etc. Combining several genes in a single cultivar to achieve the target of a MAR (multi adversity resistance) genotype, is obviously not possible with conventional breeding methods. Possibilities of pyramiding resistance and developing transformed targeted genotypes by making use of recent biotechnological methods like gene cloning, protoplast fusion and recombinant DNA technology have been explored.

For Workingout more comprehensive IDM systems suiting to the requirements of different chickpea growing areas in the country, the gaps methods and use of chemicals have been indentified and approaches for future research suggested.

1. INTRODUCTION

The inevitable importance of pulses in different farming systems and in the dietary needs of the people is well realise. India has the distinction of being leading pulses producer accounting for more than 85% world's production. Chickpea is the principal pulse crop of India occupying nearly 31% of the total area under pulses and contributing over 37% to the total production. Inspite of best efforts, the crop has not gained much in terms of area and production which almost remained static with a little fluctuation in the last decade (Fig. 1). The rapid increase in population demands for nearly 24 million tonnes of pulses production by 2000 A. D., as against the present production of 14 million tonnes (1990-91). This is imperative that the production of this major pulse crop has to be doubled to meet the future demand.

Several environmental, agronomic and biotic factors constrain productivity of chickpea (9). The relative importance of these factors however, varies from region to region because of the diversity of agroecological conditions. Of the various agroecological, biological, management and socio-economic constraints, a variety of disease encountered by this crop drastically cut the yield (57). The ravages caused by disease constrain the realization of the potential yields of the new high yielding varieties. Use of chemicals and resistant varieties, though offer good protection, but these may not be a total solution to combat with the losses caused by these diseases. Only an integrated disease management system appears to be a potent and acceptable solution. Although, efforts to suggest Integrated Pest and Disease Management system have been made as early as in the begining of eighties for diseases of crops like alfalfa, corn, cotton, peanut, sorghum, wheat etc. (61), these is not much information available in this respect for chickpea diseases both within and outside the country.

This chapter attempts to outline the major disease problems of chickpea encountered in different agroclimatic regions of India and to review their management practices evolved by the National and International crop Improvement Programmes. Possible approaches for Integrated Management of these diseases based on the concepts of Integrated Pest Management have been suggested. In presenting different approaches, major consideration has been given to host plant resistance which supplemented with other management practices can offer a lasting solution to disease problems.

2. AREAS GROWING CHICKPEA AND PIGEONPEA

Chickpea (*Cicer arietinum* L.) is the world's third most important pulse crop, after dry beans and dry peas. India leads the list of chickpea producing countries with 75 and 73% as its share in area and production in the world. The current (1990-91) area and production of chickpea in India are 7.41 million ha and 5.19 million tonnes respectively (3). Of the total pulses production in the country, chickpea accounts for 37% share. Although the crop is grown almost throughout the country, the major concentration of chickpea in order of acreage is in the states of Madhya Pradesh, Maharashtra, Haryana, Karnataka, Gujarat and Bihar (Fig. 2 a, b). Both desi and kabuli type chickpea are grown the former having the maximum acreage.

3. DISEASE PROBLEMS

Chickpea is vulnerable to several diseases. It is attacked by 43 fungal, 3 bacterial and 17 viral/mycoplasmal pathogens (57). Of these, in India, chickpea is attacked by 12 fungal payhogens which are responsible for various foliage and stem infections. In addition to these, chickpea is attacked by 2 bacterial and 7 viral diseases in India, some of the important diseases and their pathogens are listed in Table 1. Among these diseases, wilt caused by *Fusarium oxysporum* f. sp. *ciceri*, Ascochyta blight (*Ascochyta rabiei*) and Gray mold (*Botrytis cinerea*) are of economic importance causing severe yield losses every year. Stunt caused by bean leaf roll virus (BLRV) is now assuming serious proportions and has become responsible for severe yield losses in chickpea growing area of Gujarat and Andhra Pradesh.

3.1. Ascochyta blight

3.1.1. Distribution

Ascochyta blight is widely distributed in the countries wherever chickpea is grown. It is reported from 32 countries (57). In India, the disease was confined to submontanceous regions of Punjab until 1981. Presently, it has assumed alarming proportions in Punjab, Haryana, Himachal Pradesh, Northern hills of U.P., Jammu & Kashmir and Western Rajasthan. Epidemics of this disease are reported in North-western parts of the country (44). The region between 28°and 36°N latitude are more vulnerable to Ascochyta blight.

3.1.2. Symptoms

The pathogen attacks all the above ground parts of the plant. The crop is more vulnerable to attack during flowering and podding stage when the weather conditions become congenial for disease development. First symptoms usually appear near the apex of young shoots and even before the necrotic lesions become consiquous, the shoots loose turgidity and droop down. Simultaneously small necrotic specks start appearing on the newly formed leaves. Under cool and wet conditions, these specks increase rapidly in size and coalesce, causing necrosis of the leaves and young buds with extensive pycnidial formation in the blighted areas. The pycnidia are generally arranged in concentric rings. On stems and petioles, the lesions are obovate or elongate and are studded with dark coloured pycnidia. The lesions often girdle the affected portions and results in killing and breaking of the branches above the giraled areas. The lesions produced on pods are usually circular and develop prominent concentric rings having dot like pycnidia. The fungus penetrates the pods and infects seeds which remain undersized and are shruken with light brown spots sometimes bearing pycnidia which are visible to naked eye.

3.1.3. Causal organism

The asexual or imperfect stage of *Ascochyta rabiei* is characterized by producing spherical or pear shaped, ostiolate pycnidia measuring upto 245 μ in diameter. The pycnidiospores are usually single celled and occasionally bicelled measuring 8. 2 to 10.0 × 4.2 to 4.5 μ. Colonies of the fungus on artificial media are flat, submerged with sparse mycelium, white initially and later turning dark. The perfect stage of the fungus *Mycosphaerella rabiei* Kovachevski was reported from Bulgaria on chickpea pods which overwintered in field (41). The perithecia were dark brown or black usually globose with a small beak and ostiole and measured 76 to 152 × 120 μ. Ascospores were produced in cylindrical clavate, pedicellate asci measuring 48 to 70x9 to 13. 7μ. Each ascus contained 8 ascospores which were mostly monostichous, ovoid, divided into two unequal cells, conspicuously constricted at the septum and measured 12.5 to 19 x 6.7 μ.

3.1.4. Pathogenic variability

Pathogenic variability in *A. rabiei* has been reported by several workers, (7, 18, 23, 62, 71, 75). Grewal (18) identified two races (1 and 2) and one biotype of race 2 from India using I-13, EC 26435 and C-235 differentials. Singh (71) studied 348 isolates of *A. rabiei* from North India states and based on morphological and cultural characters arranged them in 15 groups. Based on pathogenic reaction types of representative isolates from these groups, 12 races were identified. For identification of races, 12 chickpea cultivars and a wild species *Cicer pinnatifidum* were used as differentials. In a more recent study Singh and Pal. (75), identified only 5 races based on reaction of 11 isolates from North India on seven cultivars viz. P 1343-1, P5292-1, C235, V138, ILC 249 and I13.

Reaction of chickpea entries under multilocational screening undertaken by AICPIP also points to a great variation in the fungus. However, standardization of the differential varieties are very much needed to understand the extend of variation and for uniform results.

3.1.5. Epidemiology

The off season survival of the fungus is through infected plant debris and seed. The fungus survived in the dormant stage for over 2 years in the infected debris lying on the soil surface exposed to natural weathering conditions in north India (20). Kaiser (35), found that the fungus can survive for more than two years in naturally infected tissuses at 10 to 35°C and at 0 to 3% relative humidity at soil surface. It, however, lost viability of 10 to 40 cm. According to Bedi (6) in epiphytotic years, unharvested field may provide primary inoculum for the next crop. Infected plant bits stored with seeds in cooler areas may also help in the perpetuation of the fungus.

The seed borne nature of *A. rabiei* was first established by Luthra and Bedi (45) who found mycelium of the fungus in seed coat and cotyledons of infected seeds. Surface contamination of seed with fungal spores and their possible role in primary infection has been evidenced by Sattar. (65) Several subsiquent reports confirm the seed borne nature of the fungus. That the infected seeds serve as primary source of infection for the spread of disease in old as well as to the new races is fairly evidenced. The secondary spread of the disease in field is mainly through pycnidiospores produced in the crop debris or the infected plant from diseased seed. The disease spreads rapidly under wet and windy conditions with a temperature range of 22 to 26°C. Under north Indian plains such conditions often prevail during February and March when the crop is under flowering and active podding stages. Based on the analysis of the environmental conditions over 35 years in Pakistan, Kaushar (37) also concluded that the rainfall during chickpea season coincide with high incidence of disease. According

Table 1 : Important diseases of Chickpea in India

Disease	Pathogen	Plant Parts	Distribution	Mode of survival
Black root rot	*Fusarium solani* (Mart.) Sacc.	Root	NEPZ and NWPZ	Soil borne. Survival through chlamydospores on dead plant debris in soil.
Dry root rot	*Rhizoctonia bataticola* (Taub.) butler (*Macrophomina phaseolina* (Tassi) Goid.	Root	All zones, more frequent in CZ and SZ	Soil borne. Survives through sclerotia on dead plant debris in soil.
Wet root rot	*Rhizoctonia solani* Kuhn	Root	All zones	Soil borne. Survives through sclerotia and mycelium in colonized organic matter in soil.
Collar rot	*Sclerotium rolfsii* Sacc.	Collar region (stem base)	All zones	Soil borne. Survives through sclerotia and colonized organic matter in soil.
Foot rot	*Operculella padwickii* Kheswalla	Stem base	Punjab and Delhi in NWPZ	Soil borne. Mode of survival not known.
Wilt	*Fusarium oxysporum* Schlecht. emend Synd. & Hans. f. sp *ciceri* (padwick) Synd. & Hans.	Root	All zones	Soil brone. Survives through chlamydospores in seed and dead plant debris in soil.
Stem blight/rot	*Sclerotinia sclerotiorum* (Lin) de Bary	All aerial parts	NEPZ	Soil brone. Survival through sclerotia in soil.
Leaf blight	*Ascochyta rabiei*	All aerial parts	NWPZ	Soil and seed borne. Survival through infected debris and infected seeds.
Leaf blight	*Alternaria alternate* Keissler	All aerial parts	NEPZ	Not known
	Stemphyllium sarciniforme (Cav.) wilt	All aerial parts	EZ and NEPZ	Seed borne, also survives in infected plant debris.
Gray mold	*Botrytis conerea* pers. ex. Fr.	All aerial parts	NEPZ	Air and seed borne
Powdery mildew	*Oidiopsis taurica* (Lev.) Salmon	All aerial parts	Not specific	Air borne. Through collateral hosts.
Rust	*Uromyces ciceris arietini* (Grogu.) Jaez. & Beyer	All aerial parts	NWPZ	Possibly through collateral hosts.
Bacterial blight	*Xanthomonas cassiae* Kulkarni *et al.*	Leaf	Frequent in SZ	Seed borne
Chickpea stunt	Bean (Pea) leaf roll virus	All aerial parts	NWPZ, CZ and WZ	Through collateral hosts. Transmitted by aphids and leaf hoppers.
Mosaic	Alfalfa mosaic virus	All aerial parts	Not specific	Through collateral hosts. Transmitted by aphids.

CZ- Central Zone; WZ-West Zone; NEPZ- North-East Plains Zone; NWPZ- North West Plains Zone.

to Weltzien, and Kaack (95), rapid development of disease takes place at temperature range of 9-24°C and wetness periods of 10 hours or more. Studies at ICARDA (29) have also confirmed that minimum temperature above 5 and maximum above 15°C are essential for epiphytotics of Ascochyta blight.

3.2. Botrytis gray mold

3.2.1. Distribution

Botrytis gray mold (BGM) is an important foliar disease affecting chickpea in northern India, Nepal , Bangladesh and Pakistan. It is also reported from Argentina, Australia, Canada, Colambia, Spain, Turkey and USA. In India, epiphytoties of this disease during the last three decades have caused heavy yield losses in parts of the Indo Gangetic plains and the north west plains zone. After major outbreaks in Haryana and Punjab in the 1979-90 and 1980-81 seasons, only sporadic occurrence of the disease has been reported in these states. However, in the Tarai regions of western U. P., BGM has occurred regularly throughout the last decade.

3.2.2. Symptoms

The symptoms of BGM appear on the stem, leaves, flowers and pods as gray or dark brown lesions covered with erect hairy sporophores bearing masses of single-celled, hyaline spores. The gray fungal growth is more commonly seen on flowers and petioles if observed early in the morning. Growing tips and flowers are more susceptible to the fungus attack. In cool and high humid conditions, flower-drop and rotting of plant parts is conspicuous. On thick, hard stems, the gray mold is replaced by gray or black mass containing dark to black sporodochia.

3.2.3. Causal organism

Botrytis cinerea Pers. ex Fr. is the causal fungus of gray mold. The perfect stage is a *Sclerotinia* (*Botryotinia*) *fuckeliana* which has not been reported on chickpea (22). The mycelium of *B. cinerea* is septate, brown, the hyphae measure 8-16 μ. The conidiophores are light brown, septate with a hyline tip which are slightly enlarged bearing small, pointed sterigmata. The conidia are borne in clusters on sterigmata, hyaline, uni-celled oval of globose and measure 2-20 × 4-16 μ. The sporodochia on host are 0.5-5.0 μ is diameter. Small unicellular, round microconidia (4-8μ) develop on the sporodochia. These sporodochia soon become nonsporiferous and develop into a sclerotial mass.

3.2.4. Pathogenic variability

Pathogenic variability in *B. cinerea* is not well elucidated. Studies are needed to demonstrate the occurrence of races, if any.

3.2.5. Epidemiology

Occurrence of repeated epiphytotics of BGM in north India during the last three decades, indicate that the pathogen is able to survive in this region from season to season. The potential of the pathogen in infecting several crop species other than chickpea, contribute to its survival and multiplication. In chickpea, the pathogen is seed borne. The seed borne inoculum is influenced during storage at room conditions (43). The relative importance of seed borne inoculum and inoculum present in the environment in the epidemiology of BGM is not clearly elucidated in chickpea.

Conidia are considered as an important source of inoculum in BGM epidemics. The optimum temperature for fungal growth, sporulation and conidia germination is between 15 and 25°C. Long wet periods are prerequisite for infection. The epiphytotic of the disease usually occurs under relatively warm (Temp. range 20-28°C) and high moist conditions (Rh 95%) preceeded by intermittent rainfall. A dense foliage canopy is more conducive for infection and rapid spread of disease.

3.3. Wilt

3.3.1. Distribution

Among the major chickpea diseases, Fusarium wilt is most widely distributed in the principal growing areas of the world (57). It is a serious problem in India and some adjoining countries such as Pakistan, Myanmar and Nepal. In India, occurrence of wilt has been reported from all the chickpea growing states. The yield loss due to wilt in the country has been emperically estimated around 10% (80). At the present level of production, it amounts to approximately 520 thousand tonnes annually.

3.3.2. Symptoms

Wilt disease generally appears after 3 weeks of sowing. Affecting seedlings show drooping of the leaves and a paler colour. Adult plants show typical wilt symptoms. Leaves become chloratic which later on turn yellow then straw coloured or light brown. The roots do not show external rotting, but when split open lengthwise, the tap root display a brown discoloration of the internal tissues.

3.3.3. Causal organism

Wilt in chickpea is caused by *Fusarium oxysporum* Schlecht. emend Synd. & Hans. f. sp. *ciceri* (Padwick) Synd. & Hans. Stroma on potato dextrose-agar are pale or whitish rosy. Microconidia are borne on simple short conidiophores, oval to cylindrical in shape and measure, 2. 5-3.5 $\times$ 5-11 µ. Macroconidia are produced on branched conidiophores, fusoid with pointed ends, mostly three septate rarely four to five septate and measure 3.5-4.5 $\times$ 25-65 µ. The fungus produces chlamydospores in old cultures which are terminal and intercalary, borne singly, in pairs or in chains.

3.3.4. Pathogenic variability

The fungus exhibits physiologic specialization and occurrence of at least 4 physiologic races of *F. oxysporum cireri* in India has been reported (25). Differential reaction of chickpea genotypes at multilocation under ICAR-ICRISAT collaborative studies have indicated considerable variability across the locations. Further studies may unravel the extent of variations and help in categorisation of pathotypes.

3.3.5. Epidemiology

The fungus is mainly soil borne in nature and survives through chlamydospores produced in dead plant debris. It has ability to survive in the soil for more than 6 years (30). Survival in seed through chlamydospores has also been reported. Wilt infection has been observed to be higher in alkaline soils. Soil temperature, moisture and organic matter content are reported to influence the wilt incidence. Soil temperature around 25°C favours maximum disease development (10). Low soil moisture at sowing and flowering greatly influence wilt development (20).

4. MANAGEMENT PRACTICES AVAILABLE

Management of disease mainly aim to prevent epidemics and wide spread out breaks of disease in host plant population. This may be achieved through, reducing the initial populations of the pathogens and then decelerating the rates of increase in those reduced populations (88). The use of chemicals, resistant varieties, biological agents with modified cultural practices may help in achieving these objectives.

4.1. Cultural practices

The chickpea wilt pathogen is mainly soil borne in nature. In areas where land is not a limiting factor, the losses due to wilt may be minimised by avoiding planting chickpea in heavily infected soil. The crop rotations are not effective as the wilt fungus has the ability to survive in the soil for more than

6 years (30). Another practice of reducing the inoculum is deep ploughing during the summer months and removal of host debris from the field. At ICRISAT, it has been clearly demonstrated that solarization by covering the soil with transparent polythene sheets for 6-8 weeks during April-May effectively controls Fusarium wilt in chickpea and also improves plant growth and yield (12). Some other cultural practices suggested for reducing wilt incidence include delayed sowing (58, 85) deep sowing (15) and soil amendments with oilseed cakes (96). Roguing of wilted plants at harvest help in eliminating the number if infected seeds.

Measures to control Ascochyta blight have been sought ever since it was first described. Removal and destruction of dead plant debris, crop rotation and deep sowing to prevent infected seeds from emerging, are some methods to reduce the blight (65). In addition to sanitation, intercropping chickpea with such crops as wheat, barley and mustard reduce disease spread in the crop season (46). Burying the diseased debris 10 cm deeper and sowing infected seeds 15 cm or deeper prevent spread of the disease (28).

4.2. Chemicals

In chickpea, the wilt pathogen is also seed borne. Several reports to control the seed infection have appeared in the literature. Seed treatment with carbendazim and Benlate T (30% benomyl +30% thiram) have been reported very effective in eliminating the seed borne inoculum (26, 69, 89). For the eradication of primary seed borne infection of *Ascochyta* several fungicidal seed treatments have been recommended which include granosan, phenthiuram, thiram, benomyl, calixin -M^R (trideomorph +maneb), hexacap^R, thiobendazole, rovral^R and a combination of carbendazim + thiram (5, 36, 56, 76, 87). Carbendazim + thiram seed treatment not only eradicates the seed borne inoculum, it also improves seed germination and seedling vigour (76, 87). For the control of secondary infection, 3-5 foliar sprays of calixin-M, rovral and hexacap have been found quite effective (87).

Botrytis gray mold pathogen has been found externally as well as internally seed borne in chickpea. Infected seed may serve as the primary source of inoculum for field infections. The seed infection can be effectively controlled by treatment with fungicides such as vinclozolin (ronilam)^R carbendazim or a mixture of carbendazim + thiram (19, 48). Some other fungicides recommended for a seed treatment are, triadimefon, triadimenol, thiobendazole, iprodione and mancozeb (72). Foliar sprays with vindlozolin or carbendazim effectively control the secondary spread of disease if one prophylactic spray can be given before appearance of disease (19).

Good success has been achieved in controlling the disease in field by a combination of seed treatment with carbendazim + thiram, triademefon or mancozeb and foliar sprays of triadomefon, mancozeb or hexacap (47). A correct understanding of epidemiology of these disease is however, very much needed in order to make chemical control more remunerative and effective component of the disease management system.

4.3. Biological control

Only recently, the importance of biological control of soil borne diseases in chickpea has been realized. A limited information is available on this aspect in case of wilt and root rots where effectiveness of some antagonistic fungi such as *Trichoderma* spp. and. *Gliocladium virens* have been demonstrated (49, 50). These antagonists alongwith *Bacillus subtilis* have shown good promise against chickpea wilt under multilocational evaluation at some AICPIP centers in pot experiments (Table 2). However, before their large scale use in field is advocated, sufficient experimental data is required to be generated.

Table 2 : Effect of antagonists on chickpea wilt, rabi 1990-91 AICPIP

Treatments	Wilt incidence (%)			Disease control over check (%)		
	Dholi	Hisar	Sehore	Dholi	Hisar	Sehore
1. Sick soil with *Fusarium* (check)	95.0	80.0	64.2	-	-	-
2. 1+ *T. viride*	60.0	35.0	52.7	36.8	45.0	17.9
3. 1+ *T. harzianum*	55.0	15.0	50.0	42.1	65.0	22.1
4. 1 + *B. subtilis*	80.0	10.0	50.5	15.8	70.0	21.1

Source : Consolidated Report on Rabi pulses, Plant Pathology, (1990-91), AICPIP, p. 139.

4.4. Host plant resistance (HPR)

For developing countries like India where farmers have only limited resources, the disease management programme naturally has greater dependence on host plant resistance (HPR). Identification of HPR requires a technology to create sever disease epidemics artificially, as the natural disease incidence varies from year to year, without effective techniques to create and stimulate disease epidemics, it is difficult to screen genotypes for resistance. Pulse pathologists under National and International crop improvement programmes have made significant contributions in developing less expensive and reliable screening techniques to determine HPR in chickpea. These include use of sick plots for wilt and root rots, pot culture and field inoculation techniques for *Ascochyta* blight and *Botrytis* gray mold diseases (55). Such techniques are being used in the National and International screening programme.

Research experience have shown that resistance to *F. oxysporum* f. sp. *ciceri* is restricted mainly to the desi chickpea germplasm (24), as a result the efforts in breeding for wilt resistance concentrated mainly on the desi type. In Kabuli chickpea, the earliest example of cultivars resistant to *Fusarium* wilt are, from Mexico where resistance has been transferred from desi lines L 41 and L 1186 (79). Four kabuli cultivars, ICCVS 2, 3, 4 and 5 resistant to race 1 of *F. oxysporum* f. sp. *ciceri* have been developed at International Crops Research Institute for Semi Arid Tropics (ICRISAT), India (42). Concerted efforts made by the All India Coordinated Pulses Improvement Project (AICPIP) since 1985 through screening chickpea genotypes in sick plots at different locations against wilt have lead to identify a large number of lines with high to moderate degree of resistance against wilt (Table-3).

Table 3 : Chickpea genotypes with moderate to high level of resistance to wilt (AICPIP 1985-92)

Desi type (56 lines)	Avrodhi, BG Nos. 209, 246, 313, 323, 367, BGM Nos. 418, 419, 432, 436, 439, 443, 444, 451, GF 89-36, GL Nos. 86114, 86123, GNG Nos. 468, 469, H Nos. 82-2, 83-18, 83-23, 83-60, 83-84, 84-8, 85-65, 85-124, 86-8, 86-18, 86-20, 86-21, 86-39, 86-42, 86-72, 86-73, 86-91, 86-100, 86-142, 86-143, 86-156, 86-158, 86-170, 88-123, 89-50, JG Nos. 315, 317, 322, KPG Nos. 63, 142-1, 143-1, PDG Nos. 83-39, 84-10, 84-14, 85-7, 85-18, Phule G 87207.
Kabuli type (3 lines)	BG 287, ICCC 32, ICCC 42
ICAR-ICRISAT (82 lines)	ICC Nos. 2664, 2862, 3345, 4483, 4485, 6440, 6501, 6687, 6817, 9032, 9039, 9041, 9172, 10384, 10399, 10539, 11223, 11233, 11320, 11322, 11323, 11324, 11329, 12205, 12224, 12246, 12263, 12265, 12408, 14303, 14309, ICCL Nos. 81001, 81004, 81006, 82104,

	82108, 83142, 83145, 83225, 84225, 84303, 84311, 85102, 85107, 85108, 85111, 85116, 85117, 85221, 85225, 85310, 85311, 85412, 88001, 88002, 88004, 88104, 88106, 88108, 88109, 88112, ICCV Nos. 89219, 89224, 89234, 89306, 89344, 89402, 89445, 90041, 90201, 90301, ICCV Nos. 8, 10, 11, 17, 18, ICCC Nos. 33, 38, 48, 52, KWR 10

Source: Consolidated Reports on Rabi pulses, Plant Pathology, AICPIP. 1985-86 to 1991-92.

Ascochyta blight being a devastating disease, attracted the attention of researchers and efforts to identify genotypes resistant against the disease started as early as 1938. The pathogen has great variability, therefore, the task has been little difficult. However, since then several reports on the identification of resistance to AB have appeared. Most of these reports are from Indian Subcontinent, listing number of resistant genotypes (Table 4). Some genotypes have also been reported resistant to mixture of races of *A. rabiei* in Syria (82).

Under All India Coordinated Pulses Improvement Project, extensive screening of chickpea accessions during 1985 to 1992 at multilocation have enabled to identify some lines with moderate resistance to *Ascochyta* blight (Table 5). In addition to these, several additional resistant lines have been indentified as a result of massive screening of germ plasm and breeding material at some endemic locations like Pantnagar, Ludhiana, Palampur, Hisar and in growth chamber screening at ICRISAT, Patancheru (Table 6). Chickpea lines BG 257, BG 261, BG 300, H-75-35 (Gaurav), H83-11, H 83-16, GG 575, GG 829 and ICC 1069 have shown somewhat stable resistance to AB over a period of 2-3 years and few of these are being used as source of resistance in national breeding programmes.

Table 4 : Chickpea genotypes resistance to Ascochyta blight (1938 to 1984)

Genotype		Reference
F 8		47
C 1234		1
P 1528-1-1, I-13		21
ILC 482		27
Kabuli :	ILC 72, 196, 201, 202, 2506, 2956, 3274, 3279, 3346, 3956, and 4421 Desi : ICC 3634, 4200, 4248, 4368, 5124 and 6981	63
ILC 191, 3856		83

Table 5: Chickpea accessions moderately resistant to Ascochyta blight, identified through multilocational screening under AICPIP. India (1984-85 to 1991-92)

Year	Accessions
1984-85	BG Nos. 257, 261, 276, 300, H 75-35 (Gaurav)
1985-86	BG Nos. 261, BG 300, GG 575, H 83-11, H 83-16, ICC 41, WB 17-3
1987-88	BG 261, BG 300, GG 575, H 83-11, H 83-16
1988-89	GG 829, GL 8423, FLIP 81-26, FLIP 82-191, ICC 1069, ILC Nos. 191, 200, 3836, 3864.
1989-90	GG 829, ICC 1069
1990-91	GL 86123, H-75-35 (Gaurav), H-86-21, ICC 13816, ICCL 86447, ICCV 89445, ICCX 800677-10H, ICCX 800859-ICCX810800
1991-92	GF 89-36, GL 86152, GL 88341 GNG 600, NG 817, H 86-18, ICC 1069, ICC 13816, ICCV 90201, ICCX 800839.

Source : Consolidated Report on Rabi pulses, Plant Pathology (1984-85 to 1991-92), AICPIP.

Table : 6: Resistance to Ascochyta blight in chickpea germ plasm and breeding lines at endemic locations in India

Location	No. of lines screened	No. of lines resistant	Reference
ICRISAT	3500	18	24
Pantnagar	1000	5	68
Ludhiana	406	8	77
Palampur	125	3	38
Hisar	625	11	33
	1674	58	34
	638	27	32

Against Botrytis gray mold disease, a good level of resistance in the world germ plasm is not available. However, chickpea line Nos. GL 635, 699, 907, 926, 930, 84133, 84038, 84065, 84212, 86094, ICC 202, ICC Nos. 1069, 1903, 1905, 4000, 4018, 4065, 4950, 5033, P 1528-1-1 and C-8 have been reported as moderately resistant to gray mold at Ludhiana. Among these, ICCs 1069, 4°00, 5033, P. 1528-1-1 have been repurted as moderately resistant to both BGM and AB disease. Under multilocational evaluation at AICPIP centres in India during 1985-1992, 19 chickpea lines were identified to possess moderate level of resistance to BGM. These are BG Nos. 209, 212, 240, 244, 256, 267, 276, 297, and 298, GCP6, GNG 158, GL 769, GL 83119, HS 86-18, 86-39, 86-156, ICC 1069, PDG 83-33 and Gaurav (92). As an outcome of the multilocational resistance screening programme persued over a period of years under National and ICAR-ICRISAT collaborative projects, promising lines of chickpea have been indentified as donor parents for resistance breeding against wilt (23 lines), AB (6) and BGM (3) (Table 7). Twelve chickpea varieties possessing resistance to these major diseases have been released for cultivation in different areas of the country (Table 8).

Table 7 : Sources of resistance to major chickpea diseases identified in India rough multilocational testing under AICPIP and ICAR-ICRISAT Programme, 1985 - 1992)

1. Wilt :- ICC Nos. 2664, 3345, 4483, 6817, 6687, 9032, 9041, 10384, 11223, 11233, 11322, 11329, 12263, 12408, 81001, 85221, ICCC 32, ICCC 42, ICCV 10, H 82-2, H 86-72, H 86-156, KPG 142-1.
2. Ascochyta blight : (i) Desi type ICC 1467, ICC 2165, E 100Y (ii) Kabuli - ILC 191, ILC 202, ILC 1069.
3. Botrytis gray mold : BG 276, ILC 202, ICC 1069

Source : Consolidated Report on Rabi pulses 1991-92 - Plant Pathology, AICPIP.

Table 8 : Disease resistant varieties of chickpea released in India

Varieties	Resistant to	Zones recommended
Pusa 212	Wilt	NWPZ
JG 315	Wilt	CZ, NEPZ
H355	Wilt + root rot	CZ
ICCC 32 (Kabuli)	Wilt	NWPZ, NEPZ
BG 244	Wilt	CZ

Avrodhi	Wilt	NEPZ
Phule G 5	Wilt (Tolerant)	CZ, WZ
ICCV 10	Wilt + Root rot	CZ
Gaurav (H 75-35)	Ascochyta	NWPZ
C 235	Ascochyta	NEPZ
GNG 146	Ascochyta	NWPZ
BG 261	Ascochyta (Tolerant)	

CZ = Central Zone, WZ = West Zone, NEPZ = North East Plains Zone, NEPZ = North West Plains Zone,

Source : ICAR - DPR. 1988. Technology For Increasing Pulses Production in India, DPR, Kanpur, India.

5. APPROACHES FOR INTEGRATED DISEASE MANAGEMENT

The Integrated Pest Management (IPM) system includes management of all kinds of pests including insects, nematodes, weeds and plant pathogens encountered by crop, whereas Integrated Disease Management (IDM) is its one major component. An effective IDM system aims at minimising crop losses due to diseases through efficient management of soil, water, nutrient and plant by integrating cultural and agronomic practices, chemical application host-plant resistance, biological agents, and by the application of physical, mechanical and regulatory control methods.

The existing methods for disease control in chickpea mainly aim for the management of individual diseases through different practices. Among these, the host - plant resistance is the most worked out component with considerable success. Cultivation of disease resistant varieties is imperative for the pulse growers with limited resources. Though not impossible but combursome is the task to combine high level of resistance in a genotype with desirable attributes. Therefore, it would be worth-while to integrate available compatible control practices to develop viable and economic management strategies for diseases particularly for the genotypes which possess desirable attributes with high yields and low level of resistance to diseases. Often, the crop is damaged by more than one disease in the same field, location, region. This warrants a critical appraisal of available protection technology so as to develop integrated management practices to combat with the multiple disease encounter.

5.1 Cultural approaches

Cultural practices which may form an integrated part of the disease management systems have not gained proper attention of the pulse growers. Farmers still adopt traditional cultivation practices. The reason for this situation may be,

(1) The patterns of crop rotation and mixed cropping are highly variable in different production areas within the Indian subcontinent.

(2) The proportion of the component crops are quite variable.

(3) Use of component crops irrespective of its significance in the management of the disease prevailing in that area.

(4) Significance of crop rotation and mixed cropping in management of diseases in chickpea have not yet been demonstrated properly.

For an effective disease management system in chickpea one has to look upon these factors in relation to the disease. The reaction of host to diseases, type of crop rotation, mixed cropping, the component crop and its ratio, the time of sowing of the component crop has to be studies in detail and need to be demonstrated properly with reference to different diseases. Natrajan *et al.* (53) observed that usefulness of growing resistant varieties can be further enhanced by supplementing cultural practices including intercropping with sorghum (Table 9) known to reduce disease severity in the crop production system.

Table 9 : Effect of sorghum intercropping on wilt incidence in different pigeonpea genotypes

Genotypes reaction to wilt	Sole crop wilt (%)	Reduction of wilt (%)
Susceptible	42-73	13-24
Tolerant	20-25	6-12
Resistant	9-13	5-6

Source : Natrajan *et al.* (53)

Seed health is another important factor that needs proper care. Seeds used for commercial crop production should be free from seed borne infection of major pathogens and have high germination percentage. A sound seed production programme in disease free areas and possibly during disease free periods could make available healthy seeds to be used in disease management strategies. Since majority of diseases are influenced by agronomic practices such as irrigation, soil preparation, fertilizer and other soil amendment applications, these practices must be decided for different regions in relation to prevalent local diseases.

5.2 Use of chemicals

Although plant protection chemicals are considered as an adjuvent to modern agriculture, the marginal status of pulses has not encouraged the pulse growers for effective use of these chemicals. Out of total pesticide consumption in pulses, fungicides has less than 10% share. In chickpea, breeding for disease resistance has advanced progressively due to availability of resistant donors against wilt and root rots. However, for diseases like AB and BGM stable and high resistance sources are not yet available. While search for these needs to be vigorously persued, the available sources of resistance may be effectively utilized in breeding programmes to develop moderately resistant or less susceptible genotypes against these diseases. Such cultivars, if supplimented with judicious use of fungicides as seed dressers or foliar sprays may provide a feasible solution to combat these serious diseases. Application of seed based protection chemicals could be made more remunerative if supported with accurate disease forecasting or warning systems. Several bioclimatological, mathematical and computerised disease simulating models for predicting epidemics and disease out breaks in other crops, have been used (8, 39, 51, 52, 70, 93). It would be worth exploring to develop appropriate forecasting models for diseases like AB and BGM in chickpea. The variations in the incidence of these diseases from season to season in the same location further emphasize the urgent need of such models.

5.3 Biological approaches

There has been a considerable development in the biocontrol of plant diseases during eighties (13, 14, 59). Within the frame work of integrated disease management system, biocontrol of the diseases of pulse crops can be successfully exploited, however, till now it has not yet gained its proper place. Potential of *Trichoderma* spp. and *Gliocladium virens* as biocontrol agents has been demonstrated in chickpea against wilt and root rot diseases (49, 50). Recently, for the optimum beneficial effects from these biocontrol agents, their use supplimented with fungicides as seed dresser has been suggested (50). Thus, instead of using biocontrol agents as a single 'Shot Gun' approach, their use as one of the components of integrated disease management system may prove more effective and result oriented.

5.4 Host-plant resistance (HPR) approach

In India chickpea is cultivated in different cropping systems under a wide range of agroecological conditions. Therefore, HPR has pivotal importance in the management of diseases. It would not be worth while to consider HPR in isolation as it is influenced by several factors. For location specific as well as general disease management, understanding of host pathogen distribution, disease severity, pathogenic variability, host resistance and mechanism of resistance, genetic resistance and mode of transmission of pathogens is needed to be taken up on priority. This helps in developing varieties, with desirable traits and resistance to location specific disease problems as well as optimum exploitation of existing facilities and infrastructure in the country.

5.4.1 Host distribution map

The area of chickpea cultivation in India is known but their mapic illustration is not available. A thematic map for chickpea cultivated areas based on current statistics (1990-91) has been attempted (Fig. 3). This may form a basis to work out strategies for integrated disease management on regional basis.

5.4.2 Disease Distribution Map

Information about distribution, occurrence, frequency and severity of major diseases in different agroecological regions is of paramount importance in the management strategies, in pigeonpea in different zones of India is well realised (60). Weltzien (94) differentiated three disease intensity areas, based on geophytopathology. Using this concept of geophytopathology, and informations from AICPIP centres generated during 1985-1990 mapic illustration of present status of AB and BGM diseases has been prepared. (Figs. 4, 5). This may provide a sound base for resistance breeding strategy for the entire country based on the importance of location specific diseases (Table 10).

Table 10 : Resistance breeding strategy for chickpea

Disease	States
Wilt	Uttar Pradesh, Madhya Pradesh, Maharashtra, Andhra Pradesh, West Bengal, Karnataka.
AB	Himachal Pradesh, Jammu.
Wilt + AB	Haryana, North Rajasthan
Wilt + Stunt	Gujarat
Wilt + BGM	Bihar
Wilt + AB + BGM	Punjab, Tarai regions of West Uttar Pradesh

AB = Ascochyta blight, BGM – Botrytis gray mold.

5.4.3. Multiple disease resistance

Collections and evaluation of germ plasm and other breeding materials of chickpea under the AICPIP and ICAR-ICRISAT collaborative programme and at ICRISAT have been quite encouraging and some lines with multiple disease resistance have been identified (Table 11). Exploitation of the resistant sources should meet the location specific requirements in the country. As a out come of these efforts chickpea line ICCV 10 resistant to wilt and tolerant to root rot has been released for central zone during 1992 (4). For the development of more lines with multiple disease resistance and subsequent multiple stress tolerance (diseases, insects nematodes etc.) manifold efforts are required.

Table 11 : Examples of multiple disease resistance in chickpea

Resistance to	Cultivars/lines	Reference
Wilt +dry root rot +black root rot	ICC Nos. 12237 to 12269	54
	ICC Nos. 11322, 14303, ICCV 88101, 88106, 90039, 90041, ICCX 830240-54 830256-441	91
	H Nos. 86-20, 86-72, 86-123, 86-156, 86-158, KPG 142-1	90, 91
Wilt+Ascochyta Blight+ Botrytis Gray Mold.	ICC 1069	54
Wilt+Dry root rot+stunt	ICC10466	54
Wilt+Sclerotinia rot	ICC Nos. 858, 959, 4918, 8933, 9001.	54

AB = Ascochyta blight, BGM -Botrytis gray mold.

5.4.4. Mechanism of resistance

Proper understanding of the mechanism of genetic resistance is very important in a sound resistance breeding programme. Studies have indicated that resistance to Ascochyta blight in chickpea is controled by a single dominant gene. According to Singh *et al.* (81) a single dominant or recessive gene controls the resistance. Singh *et al.* (78) while reviewing the informations on inheritance of resistance to wilt concluded that two recessive genes and in one case, a partially dominant gene controlled late wilting. The combination of any two genes conferred complete resistance. However, there are indications that the inheritance of resistance to wilt and Ascochyta blight is not very simple as reported and needs further confirmative studies. Limited information is available on the inheriance of resistance to Botrytis gery mold. It is indicated to be governed by monogenic dominance (11, 64, 86). These reports suggest involvement of a major gene in conferring resistance to BGM in chickpea.

5.4.5. Seed transmission

The knowledge about seed transmission of the pathogens and their visible and invisible infection in the seed is essential for a successful disease management system. *Fusarium oxysporum* f. sp. *ciceri* is both internally as well as externally seed borne (26). Pigeonpea, lentil and pea are reported as its symptomless carriers. *Ascochyta rabiei* is present in the form of mycelia and pycnidia on seeds (45). Both superficial and deep infections are equally effective in disease transmission. *Botrytis cinerea* is also seed borne (43). The symptomless carriers of the pathogens and seed borne infection are the factors which have special significance in the spread of the pathogen/pathotypes to areas they do not normally occur. The possibilities of breeding for resistance to seed borne infection need to be explored. Seed treatment recommended against different diseases in chickpea is necessarily to be adopted and integrated in the disease management programme. Further, there is a need to find out still more effective seed treatment chemicals for eradication of the pathogens.

5.5 Biotechnological approaches

Disease management systems based on the integration of traditional methods of disease control such as cultural, biological, chemical and host resistance, have their own limitations and may not reduce more than 80-85% of the avoidable yield losses. Biotechnological innovations vis-a-vis genotic engineering offer an additional complement to Integrated Disease Management System to bridge this gap (16, 39, 66, 67). The major thrust areas where the genetic engineering could contribute significantly are :

(i) Developing transformed genotypes with specific resistance towards pathogens and pyramiding resistance genes.

(ii) Evolving super strains of antagonists with high survival and root colonisation potential and ability to produce wide spectrum antifungal compounds.

5.5.1 Developing transformed genotypes and pyramiding resistance genes

Recent biotechnological advances, has made it possible to transfer gene of a prokaryote or eukaryote at a desired site in the genome of an organism and produce transgenic lines with desirable attributes (17). This technology may be perfected for transfering resistance against selected diseases in chickpea without risking yield potential and other beneficial characters.

Pyramiding resistance genes is the accumulation of genes into a single line or cultivar. These genes may be major, minor, effective, non-effective, race specific and non-race specific. The ultimate aim of biological efficiency and crop management can be approached most effectively by using host resistance to botic adversities. The need is to evolve cultivars with high yield potential and possessing the gene combinations imparting resistance to multiple biotic adversities such as pathogens, nematodes, insects and weeds etc. To combine several genes in a single variety/cultivar to achieve target of a MAR (multi-adversity-resistance) genotype, is almost impossible with conventional breeding methods. Here, biotechnological methods like recombinant DNA technology, gene cloning and protoplast fusion can contribute effectively.

In chickpea, global breeding strategies fixed by the researchers (81) include combined resistance to multi adversities as indicated in table 12. Wild *Cicer* species posses a wealth of genes for resistance to various biotic stresses (84). Transfer of genes for MAR can be attempted from selected wilt *Cicer* donors to cultivated species possibly by using embryo rescue and protoplast fusion techniques.

Table 12 : Requirement of MAR (multi-adversity-resistant) genotypes for chickpea growing areas of the world

Geographic regions	Targetted MAR genotype
Indian subcontinent	Ascochyta blight + Botrytis gray mould + Fusarium Wilt and root rot + stunt virus + pod borer.
Mediterranean region	Ascochyta blight + Fusarium Wilt + other soil borne disease + leaf miner + pod borer.
The Americas	Fusarium Wilt and root rots + viruses
Australia	Botrytis gray mold + phthium root rot + stem rot + pod borer.

5.5.2 Evolving super strains of antagonists for biological control

Wild strains of some fungal antagonists such as *Trichoderma viride, T. harzianum* and *Gliocladium virens* have been successfully demonstrated as potential biocontrol agents for soil borne diseases in several crops including chickpea. However, very little attempts have been made to improve their efficiency. Future area of research in this direction could be, improving their further effectiveness by genetic manipulations brought about with the tools of genetic engineering. Development of super strains of antagonistic fungal species which are less sensitive to physical and chemical conditions of soil, have high colonisation potential in rhizosphere, and able to produce wide spectrum antifungal compounds, may be attempted.

6. CONCLUSIONS AND FUTURE OUTLOOK

Management of major diseases is a vital component of chickpea production technology. The goal of realising maximum yield depends on a series of decisions that may also determine the extent to which diseases will limit the yield. Chickpea is predominantly grown in the developing countries, and to cut down the ravages caused by diseases, the farmers with limited resources need effective and low cost protection technology which is only possible through reliable integrated disease management systems.

In the last decade, there has been considerable progress in identification of resistant sources and breeding of improved chickpea cultivars resistant to major diseases such as wilt/root rots. Chemical control of important diseases, where satisfactory level of resistance has not yet been achieved (eg. Ascochyta blight and Botrytis gray mold) have been worked out with limited success. Beneficial effect of cultural practices like oil solarization for wilt/root rots has been established. Effective fungicidal seed treatments have been worked out for better seedling emergence and control of soil borne diseases. And yet, there have been little efforts to investigate the compatibility of these practices and to develop Integrated Disease Management Systems in chickpea. While the available wealth of informations on various practices of disease management could be immediately utilized to standardise Integrated Management of individual diseases, there is an urgent need to fill the gaps in our knowledge on certain important aspects as suggested below for working out more comprehensive IDM systems suiting to the needs of different chickpea growing areas.

Often more than one disease attack the crop at the same time in some areas of the country (Ascochyta + Botrytis, wilt + root rots and wilt + root rots + stunt virus). Inspite of a limited success in developing multiple resistance against wilt + root rots we have practically little informations on the management of other diseases under such situations.

For a systematic approach to develop a long term and sustainable IDM system the foremost pre requisite will be mapping of major diseases in respect of their occurrence and severity in the different growing regions. Detailed disease distribution maps based on precise surveys need to be prepared and updated at regular intervals. This will enable to develop need based strategy for resistance breeding, exploiting the available infrastructure facilities and resources in the country.

The information gaps in the epidemiology of diseases such as Ascochyta blight and Botrytis gray mold needs to be filled in by generating needful data through well planned indepth studies. Such informations will be of great value in a successful IDM programme.

Recent informations on the identification of sources for multiple disease resistance are encouraging. The ongoing programmes should be further vigrously persued to identify stable and broad based multiple disease resistance sources and their subsiquent utilization in developing high yielding cultivars according to the requirements of the growing regions.

The potential promise of genetic engineering in understanding the host-pathogen interactions, knowledge of the structure and function of disease resistance genes and in developing of stable, non-race specific resistance against diseases and multi adversities (pathogens, insects, nematodes weeds etc.) needs to be explored and exploited at a faster speed in order to design sustainable IDM systems for chickpea. This can only be accomplished by a joint venture and team work of Breeders, Pathologists, Entomologists, Agronomists and Biotechnologists.

Active involvement of farmers at all level while evaluating the newly designed IDM system for its component is perhaps the most important factor that will ultimately stimulate the acquisition and use of technological informations.

7. REFERENCES

1. Ahmed, T., Hasanaien, S.Z., and Sattar, A. 1949. Some popular methods of plant disease control in Pakistan. Agriculture Pakistan, 1 : **18** - 22.

2. Anonymous. 1988. Technology for Increasing Pulses Production in India. Directorate of Pulses Research (ICAR), Kanpur 2080 24, India. pp. 56.

3. Anonymous. 1991. Advance estimates of area and production of total pulses, gram and tur 1990 - 91. Directorate of Economics and Statistics, Ministry of Agriculture, Govt. of India.

4. Anonymous. 1992. ICCV 10 chickpea identified for release in India. SAT News, No. 9, p. 3.

5. Bashir, M. and Ilyas, M.B. 1983. Effect of seed dressing fungicides on seed germnation, seedling vigour and seed borne fungi of chickpea seeds. Pakistan Journal of Agricultural Sciences, **20**: 65-72.

6. Bedi, K.S. 1950. Complete defence measures against gram blight. Punjab Farmer, **2** : 29-31.

7. Bedi, K.S. and Aujla, S.S. 1969. Variability in *Phyllosticta rabiei* (Pass) Trott, the incident of blight disease of gram. Journal of Research, Punjab Agricultural University, **6**: 103-106.

8. Berger, R.D. 1977. Application of epidemiological principles to achieve plant disease control. Annual Review of Phytopathology, **15** : 165-183.

9. Chandra, S. and Ali, M. 1986. Recent achievements in pulses production. Technical Bulletin 1, Directorate of Pulses Research (ICAR), Kanpur, U.P., India, pp. 44.

10. Chauhan, S.K. 1963. Influence of different soil temperatures on the incidence of Fusarium wilt of gram (*Cicer arietinum* L.). Proceedings Indian Academy of Science B, **33**: 552-554.

11. Chaturvedi, R and Singh, I.S. 1991. Genetics of resistance to Botrytis gray mold in chickpea. In: Proceedings of Golden Jubilee Symposium on grain legumes, IARI, New Delhi, India, pp. 215-217.

12. Chauhan, Y.S., Nene, Y.L., Johansen, C., Haware, M.P., Saxena, N.P., Singh, S., Sharma, S.B., Sahrawat, K.L., Burford, J.R., Rupela, O.P., Kumar Rao, J.V.D.K. and Sithanantham, S. 1988. Effects of soil solarization on pigeonpea and chickpea. Research Bulletin No. 11, ICRISAT, Patancheru, India. pp. 16.

13. Chet, I. (Ed.). 1987. Innovative Approaches to Plant Disease Control. Johan Wiley and Sons. New York, pp. 372.

14. Cook, R.J. 1985. Biological control of Plant Pathogens : Theory to Application. Phytopathology, **74**: 25-29.

15. Dahiya, S.S., Sharma, S. and Paroda, A.S. 1988. Effect of date and depth of sowing on the incidence of chickpea wilt in Haryana, India. International Chickpea Newsletter, **18**: 24-25.

16. Day, P.R. (Ed.). 1986. Biotechnology and Crop Improvement and Protection. Monograph Np. 34, Proceedings Symposium British Crop Protection Council.

17. Fincham, J.R.S. 1989. Transformation in fungi. Microbiological Review, **53**: 148-170.

18. Grewal, J.S. 1984. Evidence of physiologic races in Ascochyta rabiei of chickpea. In Ascochyta blight and winter sowing of chickpea (eds. M.C. Saxena and K.B. Singh) Martinus Nijhoff/Dr. Junk Publishers Houge, 288 pp.

19. Grewal, J.S. and Laha, S.K. 1983. Chemical control of Botrytis blight of chickpea. Indian Phytopathology, **36** : 516 - 520.

20. Grewal, J.S. and Pal, M., 1986. Fungal disease problems of chickpea. In Vistas in Plant Pathology, Eds. A. Varma and J.P. Verma, Malhotra Publishing House, New Delhi, pp. 157-170.

21. Grewal, J.S. and Vir, S. 1974. Varietal resistance of gram to Ascochyta blight. Indian Phytopathology, **27**: 643-645.

22. Groves, J.W. and Love Land, C.A. 1953. The connection between *Botryotinia fuckeliana* and *Botrytis cinerea*. Mycologia, **45**: 415-425.

23. Gowen, S.R. 1983. Pathogenic variation in *Ascochyta rabiei*. In: Proceedings, 10th International Congress of Plant Protection held at Brighton, England, November 22-25, 1993, Vol. **2** (Abst.) : 181.

24. Haware, M.P., Kumar, J. and Reddy, M.V. 1980. Disease resistance in Kabuli Desi Chickpea introgression. In: Proceedings of the International Workshop on Chickpea Improvement, ICRISAT. Feb. 28-Mar. 2, 1979. Hyderabad, India, ICRISAT, Patancheru, India. pp. 67-69.

25. Haware, M.P. and Nene, Y.L. 1982. Races of *Fusarium oxysporum* f. sp. *ciceri*. Plant Disease, **66**: 809-810.

26. Haware, M.P. Nene, Y.L. and Rajeshwari, R. 1978. Eradication of *Fusarium oxysporum* f. sp. ciceri transmitted in chickpea seed. Phytopathology, **68**: 1364-1367.

27. ICARDA. 1982. Chickpea pathology progress report 1981-92. Food legume improvement program, ICARDA, Aleppe, Syria, pp. 75.

28. ICARDA 1983. Chickpea pathology progress report 1982-83. Food legume improvement program, ICARDA, Aleppe, Syria, pp. 83.

29. ICARDA. 1984. Chickpea pathology progress report 1983 - 84. Food legume improvement program, ICARDA, Aleppe, Syria.

30. ICRISAT. 1985. Annual report 1984. ICRISAT, Patancheru, India.

31. ICRISAT. 1987. Annual Report 1986. ICRISAT, Patancheru, India. p. 177.

32. Jalali, B.L. and Harichand. 1987. Consolidated report on rabi pulses. In: Proceedings, Plant Pathology - AICPIP Workshop, Pulse Directorate, ICAR, Kanpur, p. 73.

33. Jalali, B.L. and Khirbat, S.K. 1984. Consolidated Report on rabi pulses. In: Proceedings, Plant Pathology-AICPIP Workshop, Pulse Directorate, ICAR, Kanpur, p. 73.

34. Jalali, B.L., Khirbat, S.K. and Harichand. 1986. Consolidated report on rabi pulses. In: Proceedings, Plant Pathology - AICPIP Workshop, Pulse Directorate, ICAR, Kanpur, p. 54.

35. Kaiser, W.J. 1973. Factors effecting growth, sporulation, pathogenicity and survival of *Ascochyta rabiei*. Mycologia, **65**: 444-457.

36. Kaiser, W.J. and Hannan, R.M. 1985. Chemical control of *Ascochyta rabiei* in infected chickpea seed. Phytopathology, **75**: 1308.

37. Kausar, A.G. 1965. Epiphytology of recent epiphytotics of gram blight in west Pakistan. Pakistan Journal of Agricultural Sciences, **2**: 85-195.

38. Katyar, R.P. and Sood, O.P. 1985. Screening chickpea for resistance to Ascochyta blight. International Chickpea Newsletter, **13**: 19-21.

39. Karause, R.A., Masie, L.B. and Hyre, R.A. 1975. Blitecast: A computerized forecast of potato late blight. Plant Disease Reporter, **59**: 95-98.

40. Kerr, A. 1987. The impact of molecular genetics on plant pathology. Annual Review of Phytopathology, **25**: 87 - 110.

41. Kovachevski, I.C. 1936. (The blight of chickpea, *Mycosphoerella rabiei* n.sp.) (In Russian). Issued by Ministry of Agriculture, Domains, Sofia, 80 pp.

42. Kumar, J., Haware, M.P. and Smithson, J.B. 1985. Registration of four short duration Fusarium wilt resistant kabuli (garbanzo) chickpea germ plasms. Crop Science, **25**: 576-577.

43. Laha, S.K. and Grewal, J.S. 1983. Botrytis blight of chickpea and its perpetuation through seed. Indian Phytopathology, **36**: 630-634.

44. Lal, S. and Amin, K.S. 1992. Recent Advances in Chickpea Breeding for Ascochyta blight resistance in India. In Disease Resistance Breeding in Chickpea, Eds. K.B. Singh and M.C. Saxena. ICARDA, pp 86-97.

45. Luthra, J.C. and Bedi, K.S. 1932. Some preliminary studies on gram blight with reference to its cause and mode of perennation. Indian Journal of Agricultural Science, **2**: 499-515.

46. Luthra, J.C., Sattar, A. and Bedi, K.S. 1935. Life history of gram blight (*Ascochyta rabiei* (Pass.) Lab.) on gram (*Cicer aietinum* L.) and its control. Punjab Agri. Livestock, India, **5**: 489-498.

47. Luthra, J.C., Sattar, A. and Bedi, K.S. 1938. The control of blight disease of gram by resistant types. Current Science, **7**: 45-47.

48. Mahmood, M. and Sinha, B.K. 1990. Grey mold disease of bengal gram in Bihar. Report of ICAR Scheme 1981-86, Department of Plant Pathology, Tirhut College of Agriculture, Dholi, Bihar, pp. 35.

49. Mukhopadhyay, A.N. 1988. Biocontrol efficacy of *Trichoderma* spp. in controlling damping off in tomato and egg plants and wilt complex of lentil and chickpea. Proc. 5th International Congress of Plant Pathology, Ekoto, Japan, 1988.

50. Mukhopadhyay, A.N. 1989. Microbial control of soil-borne plant pathogens by *Trichoderma* and *Gliocladium*. National seminar and 7th workshop of AICRP on biological control, Lucknow, Oct. 23-25, 1989.

51. Nagarajan, S. 1977. Meterology and forecasting of epidemics of disease. In: Symposium of Basic Sciences and Agriculture. Indian National Science Academy, New Delhi, pp. 136-143.

52. Nagarajan, S. and Josh, L.M. 1978. A linear model for seven day forecast of stem rust severity. Indian Phytopathology, **31**: 504-506.

53. Natarajan, M., Kannaiyan, J., Willey, R.W. and Nene, Y.L. 1985. Studies on the effects of cropping system on Fusarium wilt of pigeonpea. Field crops Research, **10**: 333-346.

54. Nene, Y.L. 1988. Multiple disease resistance in grain legumes. Annual Review of Phytopathology **26**: 203-217.

55. Nene, Y.L., Haware, M.P. and Reddy, M.V. 1981. Chickpea Diseases: Resistance Screening Techniques. Information Bulletin No. 10, ICRISAT, Patancheru, India, pp. 11.

56. Nene, Y.L. and Reddy, M.V. 1987. Chickpea diseases and their control. In the chickpea Eds. MC. Saxena and K.B. Singh, CAB International. Oxon, U.P. pp. 233-270.

57. Nene, Y.L., Sheila, V.K. and Sharma, S.B. 1989. A world list of chickpea (*Cicer arietinum* L.) and pigeonpea (*Cajanus cajan* L. Mill sp.) Pathogens. Legumes Pathology Progress Report-7, ICRISAT, Patancheru, A.P., India, pp. 23.

58. Padwick, C.W. 1948. Plant Protection and Food Crops of India I. Plant pests and diseases of rice, wheat, sorghum and gram. Empire journal of Experimental Agriculture, **60**: 55-64.

59. Papavizas, G.C. (Ed.) 1981. Biological control in crop production. Totowa, NJ, Allanhold, Osmun, pp. 461.

60. Pimbert, M.P. and Sehgal, V.K. 1989. The relative importance of *Helicoverma armigera* (Hubner) on chickpea in different agroecological zones of India: implications for pest management. Legumes Entomology Discussion paper No. 1, ICRISAT, Patancheru, A.P. 502 324, India.

61. Pimental, D. (Ed.) 1981. Hand book of Pest Management in Agriculture Vol. III. CRC Series in Agriculture, CRC press Inc., Boca Raton, Florida, pp 187-575.

62. Reddy, M.V. and Kabbabeh, 1984. Pathogenic variability and race establishment in *Ascochyta* rabiei in Syria and Lebanon. Plant Disease, **69**: 177.

63. Reddy, M.V. and Singh, K.B. 1984. Evaluation of a world collection of chickpea germplasm accessions for resistance to Ascochyta blight. Plant Disease, **68**: 900-901.

64. Rewal, N. and Grewal, J.S. 1989. Inheritance of resistance to *Botrytis cinerea* Pers. in *Cicer arietinum* L. Euphytica, **44** : 61-63.

65. Sattar, A. 1933. On the occurrence, perpetuation and control of gram (*Cicer arietinum* L.) blight caused by *Ascochyta rabiei* (Pass.) Lab. with special reference to Indian conditions. Annals of Applied Biology, **20**: 612-632.

66. Sen, C. 1990. Genetic engineering in plant disease management. In: Basic Research for Crop Disease Management (Ed P. Vidyasekaran), Daya Publishing House, New Delhi, pp 1-12.

67. Sen, C., Das, S. and Das Gupta, B. 1992. Potentials of Genetic Improvement in the Integrated Management of Plant Diseases caused by Fungi. In: Farming Systems and Integrated Pest Management (Eds. Jeevan P. Varma and Anupam Varma), Malhotra Publishing House, New Delhi, pp. 29-53.

68. Shukla, A., Pandya, S.P. and Rathi, Y.P.S. 1984. Screening of chickpea germ plasm against Ascochyta blight. International Chickpea Newsletter, **11**: 28-29.

69. Shukla, P., Singh, R.R. and Mishra, A.M. 1981. Search for best seed dressing fungicides to control chickpea wilt (*Fusarum oxysporum*). Pesticides, **15**: 15-16.

70. Singh, D.V., Arora, P., Srivastava, K.D., Nagarajan, S. and Aggarwal, R. 1990. Prediction of the frequency and severity of Karnal bunt occurrence in field using climatic variables. Indian Journal of Plant Protection, **18**: 225-229.

71. Singh, G. 1990. Identification and designation of physiologic races of *Ascochyta rabiei* in India. Indian Phytopathology, **43**: 48-52.

72. Singh, G. and Bhan, L.K. 1986. Chemical control of gray mold in chickpea. International chickpea Newsletter, **15**: 18-20.

73. Singh, G. and Kapoor, S. 1985. Screening for combined resistance to Botrytis gray mold and Ascochyta blight of chickpea. International Chickpea Newsletter, **12**: 21-22.

74. Singh, G. and Kaur, L. 1990. Chemical control of gray mold of chickpea. Plant Disease Research, **5**: 132-137.

75. Singh, R. and Pal, M. 1993. Pathogenic variability in *Ascochyta rabiei* (Pass.) Labr. causing chickpea blight. Indian Journal of Mycology and Plant Pathology, **23**: 53-57.

76. Singh, G. and Singh, M. 1990. Chemical control of Ascochyat blight of chickpea. Indian Phytopathology, **43**: 59-63.

77. Singh, G., Verma, M.M., Gill, A.S. and Sandhu, T.S. 1984. Screening of gram varieties against Ascochyta blight. Crop Improvement, **11**: 153-154.

78. Singh, H., Kumar, J., Smithson, J.B. and Haware, M.P. 1987. Complementation between genes for resistance to race 1 of *Fusarium*. In: Genetics and Plant Pathogenesis, Eds. P.R. Day and G.J. Gellis, Blackwell Scientific Publications, Oxford, U.K. pp. 339-342.

79. Singh, K.B. 1987. Chickpea breeding. In the chickpea, Eds. MC. Saxena and K.B. Singh, CAB International, Oxon, U.K. pp. 127-162.

80. Singh, K.B. and Dahiya, B.S. 1973. Breeding for wilt resistance in chickpea. Symposium on wilt problems and breeding for resistance in Bengal gram, Sept. 1973, IARI, New Delhi, India, pp. 13-14.

81. Singh, K.B., Kumar, J., Haware, M.P. and Lateef, S.S. 1990. Disease and Pest Resistance Breeding: Which way to go in the Nineties? In: Chickpea in the Nineties. ICRISAT, Patancheru, India and ICARDA, Aleppe, Syria, pp. 233-238.

82. Singh, K.B., Reddy, M.V. and Haware, M.P. 1989. Breeding for resistance to Ascochyta blight in chickpea. In: Disease Resistance Breeding in Chickpea, Eds. K.B. Singh and M.C. Saxena, ICARDA, Aleppo, Syria, pp. 23-54.

83. Singh, K.B., Reddy, M.V. and Nene, Y.L. 1984. International Testing of Chickpeas for Resistance to Ascochyta blight. Plant Disease, **68**: 782-784.

84. Singh, K.B., Weigand, S., Haware, M.P., Di Vito, M., Malhotra, R.S., Tahhan, O., Saxena, M.C. and Holly, L. 1989. Evaluation of wild species to biotic and abiotic stresses in chickpea. (Abst.) XII. Eucarpia Congress, 27 Feb-4 March 1989, Gottaingen, Federal Republic of Germany, Vortrage for pflanzenzuchtung 15-1.

85. Singh, U.P. and Singh, R.B. 1984. Effect of date of sowing on the incidence of Sclerotinia stem rot and wilt of gram (*Cicer arietinum* L.) Phytopathologische Zeitchrift, **109**: 254-260.

86. Tewari, S.K., Pandey, M.P., Pandya, B.P., Chaube, H.S. and Tripathi, H.S. 1985. Inheritance of resistance to Botrytis gray mold in chickpea. International chickpea Newsletter, **12**: 11-12.

87. Tripathi, H.S., Singh, R.S. and Chaube, H.S. 1986. Efficacy of some fungicides against *Ascochyta rabiei* (Pass.) Lab. International Chickpea Newsletter, **15**: 20-21.

88. Vander Plank. 1968. Disease Resistance in Plants. Academic Press, New York, pp. 206.

89. Verma, R.K. and Vyas, S.C. 1977. Effect of seed treatment with systemic fungicides in gram wilt control. Pestcides, **11**: 20-21.

90. Vishwa Dhar. 1991. Consolidated report on rabi pulses. Plant Pathology (1990-91). All India Coordinated Pulses Improvement Project, pp. 153.

91. Vishwa Dhar. 1992. Consolidated report on rabi pulses, Plant Pathology (1991-92). All India Coordinated Pulses Improvement Project, pp. 180.

92. Vishwa Dhar, Rathi, Y.P.S., Tripathi, H.S. Mahendra Pal, Singh, G., Haware, M.P. and Upadhyay, J.P. 1993. Multilocational Screening for Botrytis gray mold resistance in India. Botrytis Gray mold of chickpea-Recent Advances in Research, ICRISAT, Patancheru, India. 14-16.

93. Waggoner, P.E., Horsfall, J.G. and Lukens, R.J. 1972. EPIMAY: A simulator of southern corn leaf blight. Connecticut Agriculture Experimental Station Bulletin, 729, pp. 84.

94. Weltzien, H.C. 1989. Role of Host Plant Resistance in Integrated Disease Management. In: Disease Resistance Breeding in Chickpea, Eds K.B. Singh and M.C. Saxena, International Centre for Agricultural Research in the Dry Areas, pp. 144-154.

95. Weltzien, H.C. and Kaack, H.J. 1984. Epidemiological aspects of chickpea Ascochyta blight. In: Ascochyta blight and winter sowing of chickpea, Eds. M.C. Saxena and K.B. Singh, Martinus Nij hoff/ Dr. W. Junk Publishers, Hauge, The Netherlands.

96. Zakaria, M.A. and Lockwood, J.L. 1980. Reduction in Fusarim population in soil by oilseed meal amendments. Phytopathology, **70**: 240-243.

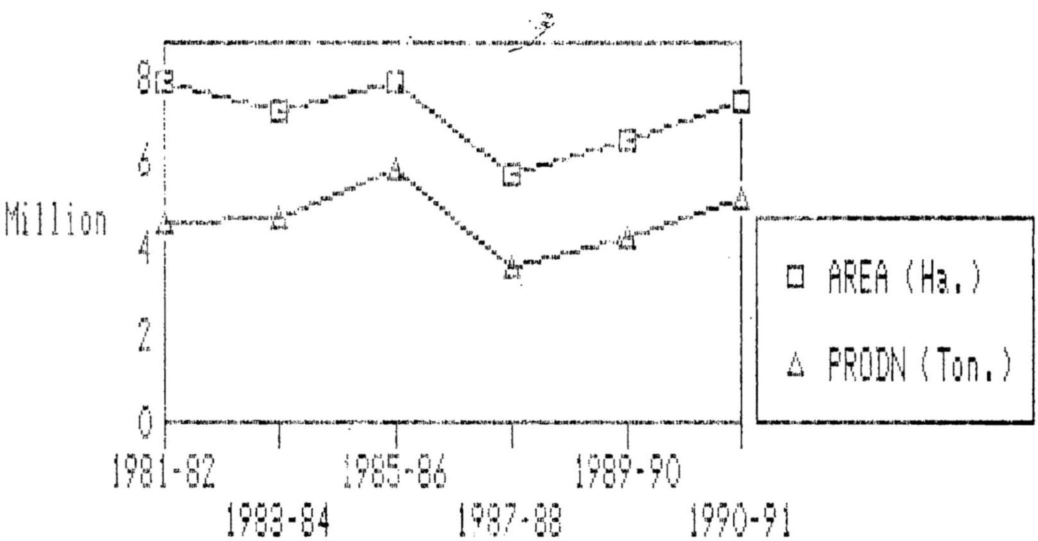

Fig. 1 : Area and Production of Chickpea in India during Eighties

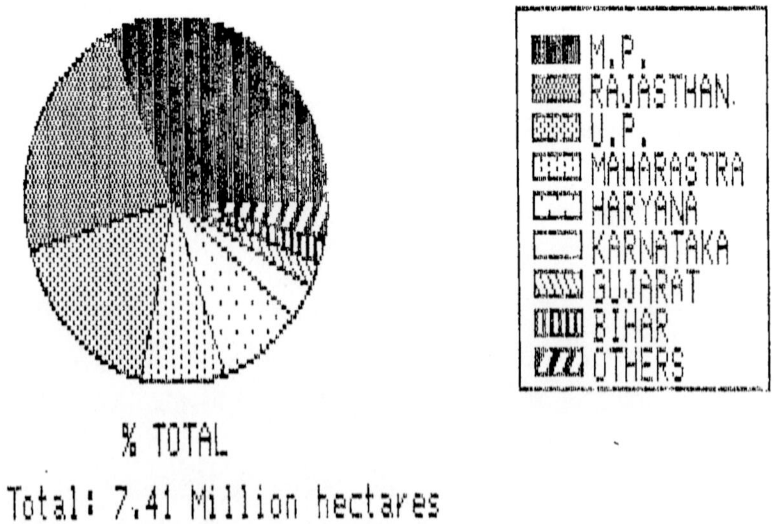

Fig. 2a : Chickpea : Principal Areas of Cultivation (1990-91)

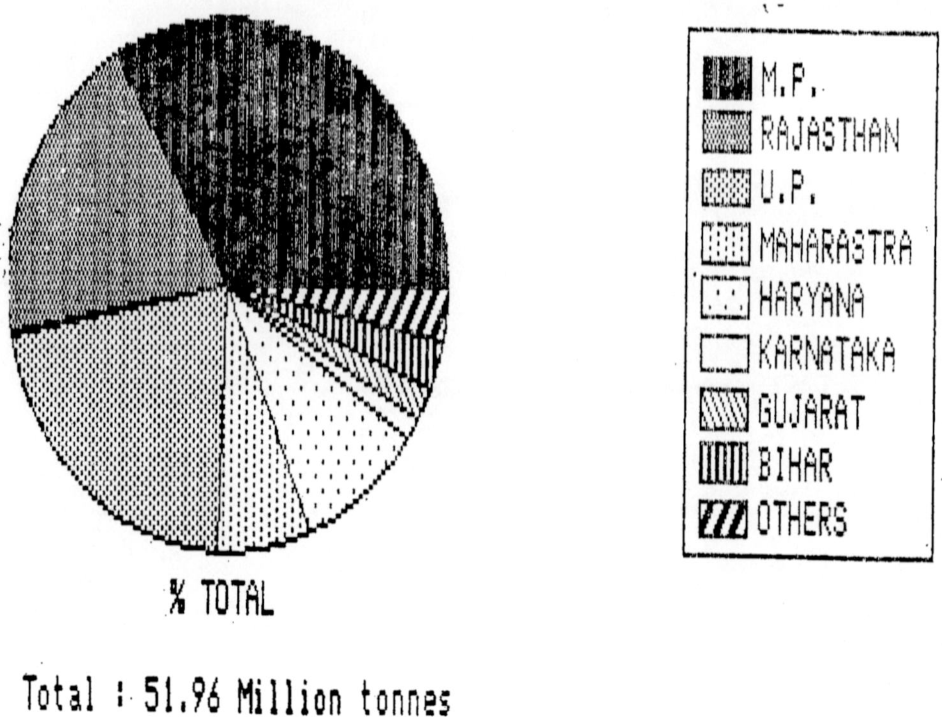

Fig. 2b : Chickpea : Production (% contribution) (1990-91)

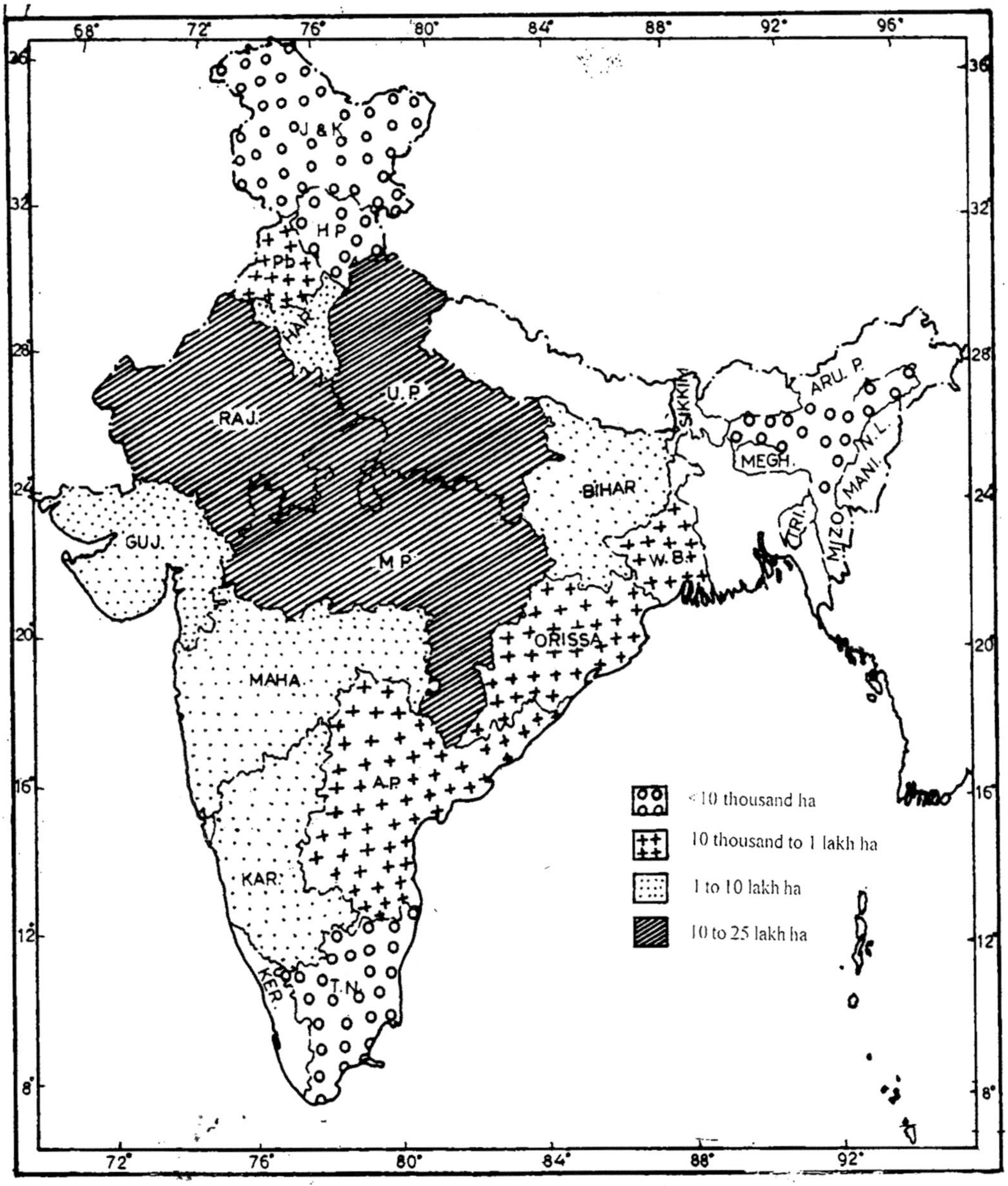

Fig. 3: Chickpea : Distribution in India (1990-91)

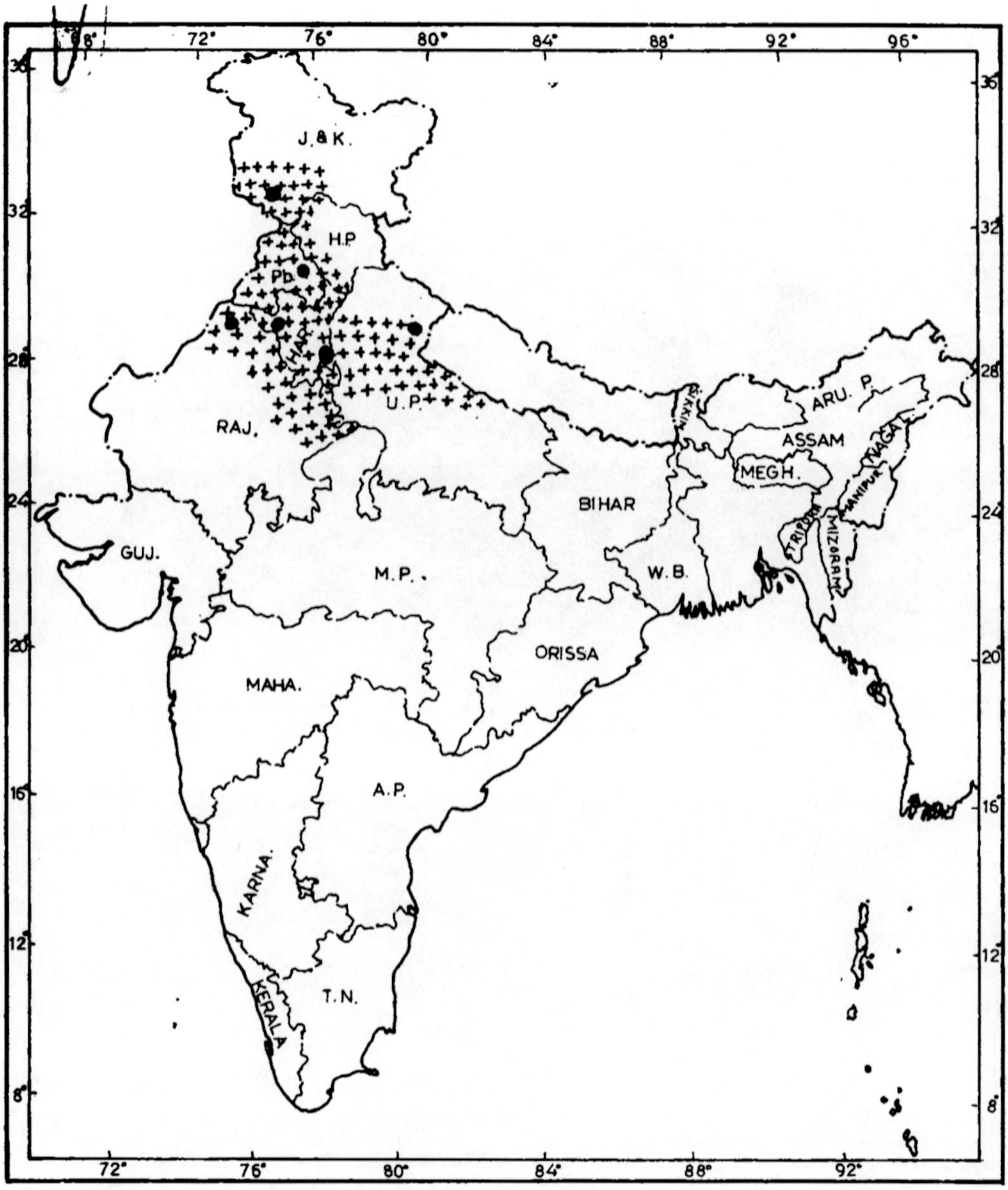

Fig. 4 : Chickpea : Prevalence of Ascochyta blight and AICPIP locations for resistance screening

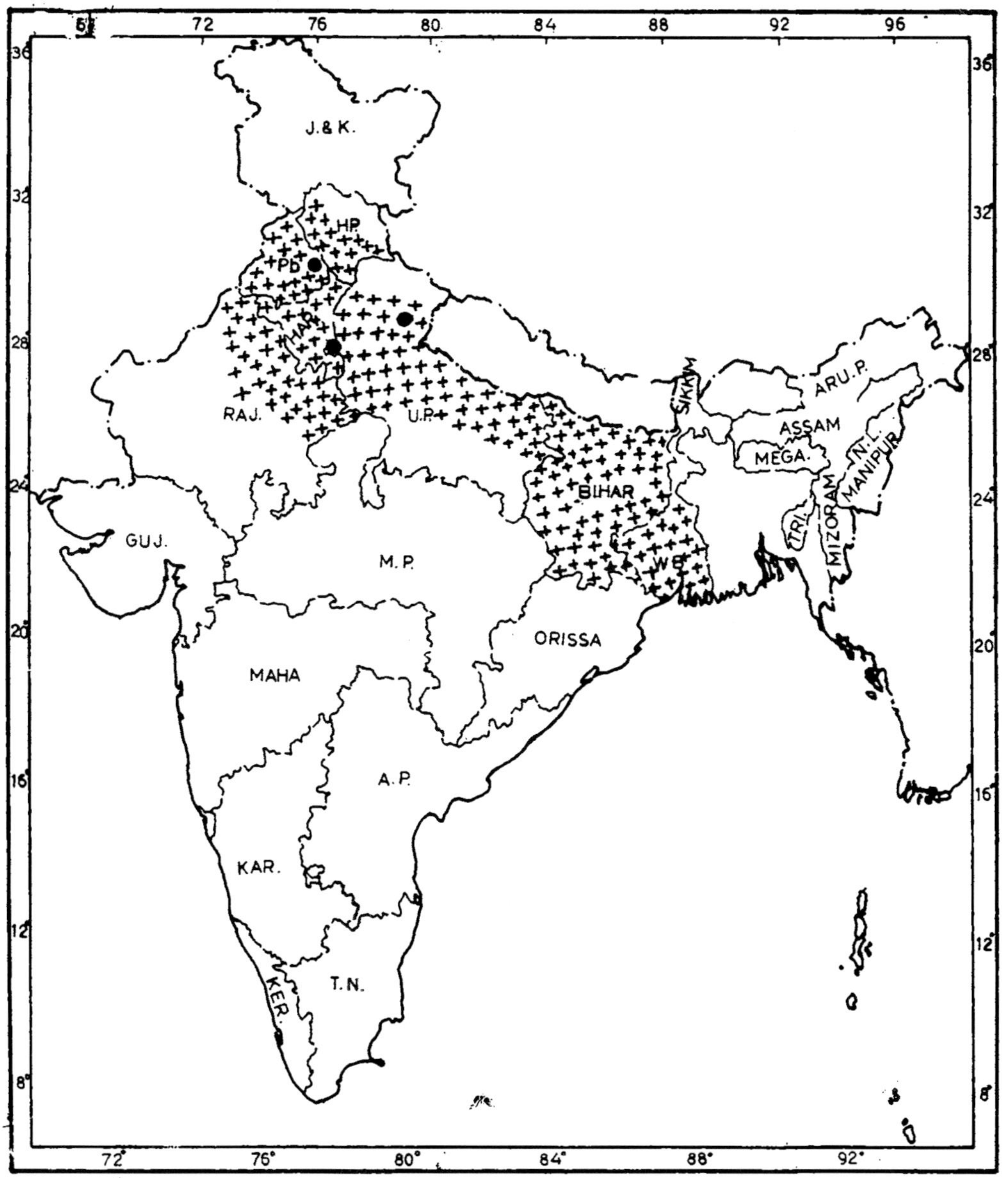

Fig. 5 : Chickpea : Prevalence of Botrytis gray mold and AICPIP locations for resistance screening

14 Integrated Pest Management in Groundnut (*Arachis hypogaea* L.)

M. P. Ghewande and V. Nandagopal

CONTENTS

Integrated Pest and Disease Management
Rajeev K. Upadhyay, K.G. Mukerji, B.P. Chamola and O.P. Dubey (eds.)
APH Publishing Corporation, New Delhi-110002, India

ABSTRACT

A large number of pests are known to be associated with groundnut crop. But all are not economically important. Different aspects like distribution, losses, symptomatology, a variety of control measures and their integration wherever possible for effective management of major pests in groundnut have been discussed. A very few disease resistant varieties, Girnar 1, ALR 1, ICG (FDRS)10, ICGV 86950 from India and Southern Runner from America have been released for general cultivation which could be used in IPM programme. A large number of advanced breeding lines possessing resistance to major diseases and insect pests are in pipe line. Biocontrol agents like NPV against *Spodoptera*, thuricides for red hairy caterpillar, *V. lecanii* and *P. islandicum* for early leaf spot, late leaf spot and rust have been successfully used. Intercrops such as pearlmillet and soybean against thrips, jassids and leaf miner, castor against jassids and *Spodoptera* and pigeonpea for early leaf spot, late leaf spot and rust are useful as traps/barriers for reduced the incidence of these pests. New approaches like pheromones for gram pod borer, tobacco caterpillar and leaf miner, and use of neem products against major insect pests and disease would go a long way in reducing the economic burden on groundnut growing farmers. If all possible pest control measures are integrated in an optimum proportion and pest surveillance activity is strengthened further, maximum returns could be realised.

1. INTRODUCTION

India is the largest producer of groundnut in the world. Out of the 9 oilseed crops grown in India, groundnut accounts for 38% of oilseeds area (8.4 m. ha) and 48% oilseeds production (9.5 m. tons). However, the average productivity is only around 900 kg/ha. Low productivity in groundnut is attributed to several production constraints. Besides abiotic stresses, biotic factors like insect pests, diseases and weeds have been recognised as the major factors limiting groundnut production. The groundnut crop suffers from a variety of insect pests pathogens and weeds, and cause heavy losses. The relative economic importance of pests (insect pests, diseases, weeds, birds and rodents etc.) in different states is variable depending upon the local cultivation practices, the environment and the cropping patterns. Some pests are widely distributed and cause economic reduction while others are restricted in distribution and are not considered to be economically important at the present time. However, there would always be shifts expected and experienced in pests situation. Therefore, timely pest surveillance is required. For this purpose qualified personel should periodically visit farmers fields, study the symptoms/injury and evaluate the damage and wherever necessary, offer advice concerning suitable control measures as well as judicious use of pesticides. This exercise has proved highly successful in case of rice and cotton in Tamil Nadu and wheat and sugarcane in North-West India. Such an effort needs to be extended to groundnut also since this activity is the key to the Integrated Pest Management strategy.

In the case of pests control the current approach of over killing through pesticides should be avoided. Although chemical pesticides have played an important role in increasing groundnut production in the country, their indiscriminate use for the control of pests has led to several environmental problems such as development of resistance in pests to pesticides, pestcides residue and the destruction of beneficial insects like parasites and predators. Therefore, the strategy should aim at economically sound Integrated Pest Management (IPM). In IPM, a series of steps are followed to keep the population level of the pests below the economic threshold level and demands a judicious use of pesticides, disease free seeds of resistant/tolerant varieties, biological suppression of the pests population usage of pheromones, timely sprays of pesticides etc. to form part of the plant production umbrella. Special attention should be given to those pests which occur consistently (Map-1) and cause heavy losses (Table 1), since growers will benefit economically from their control. With economic considerations of input versus output ratios, it would be most appropriate and feasible to develop Integrated Pest Management (IPM) programme and implement it for the effective control of pests in groundnut. In this paper the literature on major pests and their control through a variety of methods with special reference to IPM has been reviewed. Implementation and the future prospects of IPM have also been indicated.

Table 1 : Yield losses due to major insect pests and diseases of groundnut

Insect/disease	Possible yield loss (%)	References
Insect Pests :		
Leaf miner	49	5
Jassid	17	6
Thrips	17	3
White grub	20-100	153
Tobacco caterpillar	15-30	60
Hairy caterpillar	75	3
Diseases :		
Leaf spots	15-59	49
Rust	13-52	49, 53, 140
Collar rot	40-50	14, 19
Stem rot	27 & above	19
Bud necrosis disease (BND)	upto 50	34
Peanut mottle virus (PMV)	upto 30	34
Peanut clump virus (PCV)	upto 60	98

2. MAJOR PESTS OF GROUNDNUT

2.1 Insect pests

2.1.1 Floliar insects and mite pests

The foliage feeding insects which cause economic loss are belonging to Lepidoptera. These include the leaf miner *(Aproaerema modicella* Dev. Gelechidae), the red hairy caterpillar *(Amsacta albistriga* Wlk. and *A. moorei* Bult. Arctiidae), Bihar hairy caterpillar *(Spilosoma obliqua* HB; Arctiidae), tobacco caterpillar *(Spodoptera litura* (F); Notcuidae) and gram pod borer *(Helicoverpa (=heliothis) armigera* Hb; Noctuidae). The symptomatology of the damage caused by various insects are given (2,48,59).

Symptomatology

There are three or four distinct broods recorded in the leaf miner. Eggs are laid singly on the under surface of the leaflet. On the emergence of the larvae a small blister like mines in the beginning caused and later will expand in size and the major portion of the leaflet becomes brown, shrivals, some times two or more leaflets and folded and the last instar feeds from inside. In severe infestation the crop presents a burnt up appearance. In case of red hairy caterpillar on receipt of soaking showers about a month after sowing in kharif season, white moths with black marking on the hind wing emerge out from the soil during evening time. Cream coloured eggs are laid on the under surface of the leaflets. Tiny grayish caterpillars emerges and feed on leaflet leaving the upper epidermal layer. After a week the individual caterpillar with red hairs are seen feeding the leaflets voraciously leaving only the petioles and stems. They also move from one field to another in mass. The adult moth of Bihar hairy caterpillar is brown with red abdomen, wings have scattered black spots. The eggs are laid singly all over the groundnut plant or sometimes large groups on the lower surface of leaves in a single layer. On hatching, the larvae are seen with long black or black and white hairs. The young caterpillars initially feed on chlorophyll and such leaflet become papery thin and dry. Full grown larvae devour the entire foliage, flowers and growing points. The adult of the tobacco caterpillar is stout brown coloured moth with wavy white marking on the fore wing and a brown patches all along the margin of white hind wings. The eggs are laid in groups mostly on the upper surface of leaflet in mass and are covered with tuft of brown coloured hairs. The neonate larvae which are gregarious in nature feed from under surface of leaves. Heavy defoliation due to grown up larvae results in leaving the bare stem. Slender soft bodied green or brown or black stripped caterpillars can damage the crop during dusky time. While in case of gram pod borer, the adult is a stout-bodied moth, fore wings are pale brown with series of marginal markings. Eggs are laid singly and distributed efficiently over large area. On hatching, the young larvae enter the leaf bud by making a pin hole. The grown up larvae feed on foliage and flowers by preference. Red coloured spider mite *(Tetranychus telarius* Linn.) seen feeding on the leaflet. The mite infested leaves shown stippling followed by yellowing and finally almost white. In plants shown extensive webbing and tips of the plants may appear reddish from their assembly of large number of mites on them. Ultimately the whole plant wither and dries. Damage occurs always in patches. While in case of white mite *(T. utricae)* with black marking seen feeding mostly under surface of leaflets. The basel leaves get first infested and the infestation moves further upwards. Complete webbing of the plants results in withering and drying. The jassid *(Balclutha hortensis* Lindb.) belonging to Cicadellidae are light or yellowish green in case of nymphs and adults which suck and sap from under surface of the leaflets causing 'V' shaped yellowing of margin, puckering and in severe cases leading to hopper burn systems. In the case of thrips *(Caliothrips indicus* Bagnall) cream brown, yellow coloured nymphs and fringe winged black coloured adults could be seen when a wet palm is passed over the leaflets. They feed on scrapping the chlorophyll. Early stage crops suffer heavily. The aphids *(Aphis craccivora* Koch) which are dark brown adult and nymphs feed mostly on growing tips and young foliage. Under heavy infestation the plants becomes chlorotic and leaf curling expressed. They infest flower and pegs also. This is a vector for peanut mottle virus.

2.1.2 Soil insects

On the receipt of first shower during kharif season, the adults of white grubs *(Lachnosterna consanguinea* Blanch) which are dark brown in colour, emerge from soil in large numbers. During evening hours, they fly towards neem trees and by using a light source, large number of adults seen feeding on the neem leaves. The eggs are laid near the root zone of groundnut plants. The grubs feed roots and such affected plants can be easily pulled. In the case of termites *(Odontotermes obesus)* which are tiny, white with two prominant mandibles move in line. Feeds on growing roots, root hairs and tap roots and stem. The affected parts will be hollow and are filled with mud. The ear wigs *(Anisolabis annulipes)* will be brownish, blackish insect with two prominent forceps like structure at the posterior end seen moving near the pods and make holes in pods. Such pods with bore holes will be devoid of kernels.

2.1.3 Storage insect pests

These insects are of different nature in causing damage to the stored pods of groundnut. Very few are capable of damaging the entire pod thereby inviting the other secondary pests. Among the storage insects pests, the groundnut pod borer *(Carydon serratus* Olivier, Bruchidae Coleoptera) is the most devasting insect occurring through out India. The adult beetle is reddish-brown colour with dark irregular markings on the elytra. Adult females attach their eggs to the outside of the pods and kernels.

Infested pods will show a round hole, the grubs pupate inside a dull white cocoon attached to the kernel or pods. This is the largest beetle occurring in storage. Next to this pest, probably the red flour beetle (*Tribolium castaneum* Herbar, Tenebrionidae : Coleoptera) and the rice moth (*Coryra cephalonica* Station, Lepidoptera : Pyralidae) are the major pest followed by merchant grain beetle (*Orycephilus mercator* Feuvel, Silvenidae : Coleoptera), the almond moth (*Ephesita cautella* Alk, Pyralidae : Lepidoptera). The red flour beetle is the tiny brownish insect lay their eggs in cracks in the testa and make small holes in the kernels created by other adult insect. The larvae are cylindrical in shape with two prominent horns on the last abdominal segment. The pupae are formed without any cocoon. The rice moth, the adult stage is harmless and only the larval stage are capable of causing the damage to the intact kernels and feed from surface as well as within the seeds. The larvae spin a tough silken webbing. Pupation takes place inside the cocoon. Other insects such as the merchant grain beetle, almond moth and the Indian meal moth are of minor importance.

2.2 Rodents and birds

The rodents and birds are of minor importance as pests of groundnut under field and storage conditions. There are three species of rats associated with groundnut viz., *Mus booduga, Rattus meltoda* and *Tale indica.* Under field conditions the rats cause a clear runway with legs impression seen from channel to field or from field bunds to field. Holes with loose soil near the entry could be seen. Pieces of groundnut shells, cut branches and also been seen in the corners of house, storage godowns or near storage bins. The birds are generally not harmful to groundnut except crows (*Corvus splandens*), pigeons and black ibis (*Psaubis papillasa*). The former two birds generally pick up the sown seeds of groundnut leading to reduction in more than 50 percentage gaps in the sown area. The latter one is merely removes the matured pods from plant and feed the kernel. Otherwise all these birds are helpful by way of predating the insects like white grubs, grasshopper and grown up larvae.

2.3 Diseases

2.3.1 Foliar fungal diseases

Among foliar fungal diseases, only a few are economically important such as early and late leafspot commonly called as tikka disease and rust. These diseases occur wherever the groundnut crop is grown. However, the incidence and severity of each disease varies between localities and seasons, and there can be both short and long term fluctuations in their relative proportions (81). They take heavy toil on the crop in congenial environmental conditions. The magnitude of yield losses caused by these diseases is very high and ranged from 10% to 70%, all over the world, but vary considerably from place to place and between seasons (36, 41, 43, 49, 81, 139). Early leaf spot is caused by *Cercospora arachidicola* Hori (Perfect stage-*Mycosphaerella arachidis* Deighton and late leaf spot is caused by *Phaeoisariopsis personata* (Berk & Curt.) V. Arx (Perfect stage-*Mycosphaerella berkeleyii* Jenkins). Lesions caused by *C. arachidicola* are sub-circular and from 1 to over 10 mm in diameter. They are dark brown on the upper leaflet surface where most sporulation occurs and a lighter shade of brown on the lower leaflet surface. Lesions caused by *P. personata* are usually smaller, almost circular, and darker in colour than those of *C. arachidicola.* On the lower surfaces, where most sporulation occurs, the lesions are black with a slightly rough appearance. The colour of the lesion on the lower leaflet surface is light brown for early leaf spot and black for late leaf spot. The sporulation occurs on the upper leaflet surface in the cause of early leaf spot while in late leaf spot sporulation occurs on the lower leaflet surface. These are most useful characters for distinguishing between the two leaf spots in the field. These pathogens also produce lesions on petioles, stems, and pegs. On severity of infection, leaflets first become chlorotic, then necrotic, lesions often coalesce, and leaflets are shed. Temperatures in the 25°C to 30°C range and high relative humidity favour infection and disease development. The pathogens may survive from seasons to season on volunteer groundnut plants and infected crop debris. Long distance distribution on the pathogens may be by airborne canidia, by movement of the infected crop debris, or by movement of pods or seeds that are surface-contaminated with conidia. Early and late leaf spot pathogens are also soil-born (81). Rust is caused by *Puccinia arachidis* Speg. It produces orange-coloured pustules on the lower leaflet surfaces. The pustules (uredinia) rupture to expose masses of reddish brown uredinospores. The pustules are usually circular and range from 0.5 to 1.4 mm in diameter. Infected leaves becomes necrotic and dry up by end to remain attached to the plant. Groundnut rust is known to perpetuate, spread and produce severe disease out-breaks by means of its uredinospores. Groundnut crop or volunteers are available in one or the other parts of the country enabling the survival of uredinial stage round the year. Long distance dissemination for rust may occur via air-borne uredinospores, by the movement of infected crop debris, or by the movement of pods or seeds surface contaminated with uredinospores. An optimum temperature of 20°C, the presence of liquid water on the leaf surface, and high relative humidity, favour infection and subsequently diseases development. Spread of the disease within growing crops is facilitated by wind movement by rain splash, and also by insects (140).

2.3.2 Seed and soil-borne diseases

Seed and soil-bore diseases may develop from fungi already established in the seeds before sowing or may result from direct invasions of seeds or seedlings by soil fungi. Of seed and soil borne diseases, collar rot/crown rot/seedling blight *(Aspergillus niger* Van Tieghem, *A. pulverulentus* (McAlp, Thom), stem rot/Sclerotium wilt *(Sclerotium rolfsii* Sacc) dry wilt or dry root not *(Macrophomina phaseolina* (Tassi) Goid = *Rhizoctonia bataticola* (Taub., Butler) have been recognised to be the major diseases. These diseases cause severe seedling mortality resulting in patchy crop stand mostly in sandy loam soils and reduce the yields from 25-50% (19, 37). Collar rot is prevalent in almost all groundnut growing areas of the country. It is particularly serious in the sandy loam and medium black soils of Punjab, Tamil Nadu, Uttar Pradesh, Rajasthan and Haryana. Dry root rot and stem rot are sporadic in occurrence. Dry root rot is particularly serious in Rajasthan, Uttar Pradesh, Tamil Nadu, Andhra Pradesh and Maharashtra. Stem rot is becoming more prevalent in Tamil Nadu, Andhra Pradesh, Karnataka, Madhya Pradesh and Gujarat. The collaor rot disease spreads by the spores and mycelium adhering to the seed pods from season to season. In addition collar rot pathogens are saprophytic, common soil dwellers in soil humus and plant debris and capable of initial infection. The main carry over from season to season is in plant debris rotting in the soil and on the seeds harvested from a infected field. *A niger* can tolerate low soil-moisture and develops best at temperature between 30°C and 35°C. Dry root rot has wide host range. The pathogen. *M. phaseolina* is facultative saprophyte and soil dweller. Infected soil, plant debris, and pods serve as sources of inoculum. It can be obtained from kernel shell surface and shell tissue. The optimum temperature for seedling infection is 29°C and 35°C. For pods invasion the temperature range between 26°C and 32°C. The dissemination of the fungus is by sclerotia via plant debris and soil and also infected pods, shell and kernel (30). The Sclerotium wilt has also a wide host range and spreads through infected soil, rain drops, sclerotia.

2.3.3 Virus diseases

Several virus diseases of groundnut have been reported from India. However, economically important virus diseases are a few like bud necrosis diseases (BND), peanut mottle virus (PMV) and clump virus (PCV). Recently peanut stripe virus (PStV) disease has been reported on groundnut from India (106). Bud necrosis disease is widely distributed in India and has been recognised as one of the most important virus diseases of groundnut in Andhra Pradesh, Utter Pradesh and Madhya Pradesh. It has also started assuming importance in recent years in Punjab, Haryana, Rajasthan, Maharashtra, Northern Karnataka and Tamil Nadu on kharif crop (rainy season crop). It has also been reported on rabi/summer (post rainy season crop) crop of groundnut in Saurashtra region of Gujarat, Nizamabad, Nalgonda and Mehabooban nagar districts of Andhra Pradesh and Northern and Vidarbha region of Maharashtra. The incidence of the disease was observed to be as high as 92% on Spanish bunch varieties of groundnut during 1985 (summer), in Jamnagar district of Saurashtra region of Gujarat (38). Bud necrosis disease has been reported to cause yield losses upto 50% in early infected crop of groundnut in India (34). In bud necrosis disease the first sign of infection appears as chlorotic speckling or ring spots on the terminal leaf followed by necrosis of the terminal bud. Plant infected early in the season are stunted and shoots bear small leaves that are distorted and mottled. Necrosis of the bud may spread to the petiole and stem, sometimes lead to death of the plant. In late infections, pod size is reduced. Seeds are often shrivelled and have mottled and discoloured testa (115). The virus is sap transmissible. Two species of thrips, *Scirtothrips dorsalis* Hood and *Frankliniella schultzei* Trybom have been shown to transmit the disease. Both the virus and the vector have wide host ranges in crop, ornamental and weed plants. Crops like green gram, black gram, pea, soybeans, beans, groundnut, tomatoes and pea are highly susceptible to both virus and vector. Temperatures around 30°C and a wind speed of 10 km/h have been shown to favour migrations of thrips. Major spread of the disease is associated with thrips build up again during January and February and as such crops sown in December-January suffer most (2). Peanut mottle virus is reported to occur on rabi/summer groundnut mainly in the states of Andhra Pradesh, Maharashtra and Gujarat. The disease can cause upto 30% loss in yield (70). In PMV, newly formed leaves show mild mottling and vein clearing, whereas older leaves sow upward curling and intervenial depression with dark green islands. Seeds from infected pods are discoloured. PMV is reported to be seed transmitted in a range from 0.1 to 3.5% depending on the type of cultivar. PMV is transmitted in a non-persistent manner by aphids, *A. craccivora, Myzus persicae.* It occurs in nature on several important legume crops and weeds. Aphids are efficient vectors of PMV and transmit the virus rapidly to nearby plants (71). Peanut clump has been reported from Punjab, Gujarat, Andhra Pradesh and Utter Pradesh on crops grown in sandy soils. Yield losses upto 60% have been recorded in late infected plants. The virus has a wide host range and several weeds commonly occurring in groundnut fields are also infected by the virus. Indian PCV isolates are transmitted through seed (98). The virus has also been reported to be soil-borne and possibly transmitted by nematodes (34). PStV was observed in a few groundnut research centres of AICORPO located in Andhra Pradesh, Karnataka, Tamil Nadu, Maharashtra and Gujarat, during 1987 to 1992. It was also noticed in some farmers fields around Junagadh district of Saurashtra region of Gujarat during kharif, 1991 and summer, 1992 (46). The typical symptom of the PStV is the development of discontinuous dark green banding along the lateral leaflet veins. The other variations of PStV symptoms include blotches, chlorotic rings surrounding the blotching of round leaflets oak leaf pattern and vein clearing, a mosaic pattern of dark green islands and stunting of plants due to primary infection. PStV is reported to be transmitted through seed and several species of aphids. It has a wide host range. No authentic reports of yields losses due to PStV are available from India.

2.3.4 Nematode diseases

Although more than 70 species of plant-parasitic nematodes have been reported in association with groundnut diseases (126), only a few species are known to cause economically important diseases (143). Pod and root-knot diseases, caused by four species of *Meloidogyne* viz., *M. arenaria* (Neal) Chitwood, *M. hapla* Chitwood, *M. incognita* Kofoid & White and *M. javanica* (Treub) Chitwood is the most important disease of groundnut. It is almost distributed throughout the groundnut growing areas of the world. *M. arenaria* and *M. javanica* are more common in the Warmer regions, and *M. hapla* in the cooler regions, *M. incognita* occurs in the Mediterranean region (80). *M. arenaria* is the most wide spread of the groundnut root-knot nematodes. All the four species of *Meloidogyne* occurs in India. Estimates of economic losses caused by pod and root-knot disease are not available so far, however, this disease is considered to be important in India (126). Groundnut plants infected with root-knot nematode commonly develop enlarged roots and pegs which develop into galls of various sizes. *M. hapla* produces small galls but extensive root proliferation. Roots often form a dense mat on severity of infection. Pods also become infected and develop knobs, protuberances, or small warts.

Development of the root system is commonly much reduced. Plants infected with root-knot nematodes may show various degrees of stunting and chlorosis and wilting also. Pod and root lesion disease caused by *Pratylenchus brachyrus* (Godfrey) Filiipjevs Schstekh was recently reported in India from one of the major groundnut growing areas of Andhra Pradesh (148), but it is distributed in other states in India is not known. *Cricoemella ornata* (Raski) de Grisse & Loof has also been reported to be associated with groundnut in India (124). Pod and root lesion disease commonly called as Kalahasti malady caused by *Tylechorhycus brevilieatus* Williams is also important disease which damages groundnut crop in Nellore and Chittor districts of Andhra Pradesh, India. The disease caused by *P. brachyrus* is characterized by the production of small brown flecks where the nematode enters and increase in size with time. *T. brevilieatus* infection causes brownish black discolouration of the pod surface. Small brownish yellow lesions appear on pegs, and young developing pods. Affected plants are stunted and greener than foliage (113).

2.4 Weeds

Weeds are the regular menace in most part of groundnut growing area in the country during rainy seasons. They interfere with growth and development of groundnut apart from acting as alterate host for many insects and diseases. Controlling the weeds in correct time is essential to realise good yield of the groundnut. The critical periods for weeds competition in groundnut was between 3 to 6 weeks after planting. Hence the weed control practice should be taken up during this period.

2.4.1 Major weeds

The weeds occurring in groundnut ecosystem represent atleast 15 families of plant species. They can be classified broadly in to monocot and dicot weeds. Monocots : *Brachiaria eruciformis* (Graminae); *Cynotis* sp., *Commelina bengalensis* L., (Commeliaceae); *Cyodon dactylon* L. Pers, *Dactloctenium aegyptium* (L.) P. Beaue; *Diotaria sanguinalis* L. Scop.; *Echnochloa colonum* (Linn) Link; *E. cruegalli* (L.). Beanv.; *Eleusine indica* (L) Gaertn; *Eragarostis tenella* (L) Stapf, *Paspalum distichum; Setaria glauca* (Graminae). Dicots: *Abutilon indicum* (Malvaceae); *Acalypha indica* (Papavaeraceae); *Alvsicarpus* sp. (Papilionaceae); *Acalypha indica* (Euphorbiaceae); *Celosia argaentea* L. (Amaranthaceae); *Argaemone mexicana* (Tiliaceae); *Digera arvensis* Forsk (Amaranthaceae); *Portulaca oleraceae* L. (Portulaceae); *Vernonia cinerae* (Compositeae) and sedges, *Cyparus rotundus* L.; *C. defformis; C. compressus* (Cyperaceae).

3. METHODS OF PESTS CONTROL

In general, the pests, control programme involves mostly pesticides and very rarely farmers go in for other alternative methods such as cultural, physical, biolotical etc. The losses due to pests area of high magnitude all over the world. Hence, comprehensive pest management programme is most desirable. Various pest control methods adopted in India and elsewhere are given below :

3.1 Insect pests

3.1.1 Host plant resistance

Greater importance is given in India and at International level on the identification of resistant varieties to insect pest, since the resistant lines generally, support less population and lesser damage due to the feeding of insects. A groundnut line, ICGV 50 was found to be highly resistant to leaf miner and tobacco caterpillar. Who cultivars (ICGV 86031, ECGV 86699) developed by ICRISAT were found to be highly resistant to leaf miner. A variety Girnar 1 developed by the NRCG during 1988 was found to possess high level to resistance to leaf hopper *(B. hortensis)* (6). To develop resistant variety for all the insect pests occurring in a locality is rather difficult as the feeding behaviour of various insects differ considerably.

3.1.2 Cultural practices

Cultural methods involve all aspects of pest management except insecticides, natural enemies and host plant resistance. Considering the yield as well the low level of resistance of leaf miner, August Ist or 2nd week appeared to be the best season recording highly yields (1097 kg/ha) in the variety TMV 7. A significant negative correlation was observed between incidence of leaf miner and maximum-minimum temperature, while positive correlation was existed with relative humidity and maximum,-minimum temperature (78). In an another experiment, groundnut either as pure crop with 30 x 10 cm or 20 x 10 cm spacing alone with straw as mulch gave the maximum control of leaf miner as well increased the per cent larval parasitization (77). With regard to intercrop, cowpea with ground nut turn out to be better combination with least larval population. Interestingly, groundnut with pady straw mulch resulted in least weed number/unit area. Recently (86) a report indicated that low incidence (20.8%) of leaf miner, when bajra was intercropped in groundnut, apart from giving higher net income. The same authors reported 74% reduction in egg laying by leaf miner on the foliage treated with 10% concentration of bajra leaf extract in water. A highly significant positive correlation was observed between 3 generation of leaf miner and percent leaflets attacked and larvae per plant. The population of leaf miner moth catch was high in the month of October followed by August and September (77, 137) reported high leaf miner population at high plant populations levels irrespective of varieties or level of nitrogen applied. This observation was further confirmed as there was highly significant positive correlation between the plant density and leaf miner incidence (r=0.588). Yield reduction was more when groundnut was intercropped with pearlmillet (986.7 kg/ha), while in the sole crop the yield was 1300 kg/ha which was on part with yield of cowpea intercropped in ground nut (73). In case of thrips and damage by BND, the incidence was reduced when crop was sown in the onset of rainy season. Closer spacing also lessons the impact of diseases (114). Intercrop or multi crop groundnut with cereals (eg. Millet or Maize) reduce the pest incidence by 10% (152).

3.1.3 Biological control

A number of parasites, predators and pathogens are reported, however, their use under field conditions have not come into practice yet. A team of Scientists is engaged in State Agricultural University and the Tamil Nadu Agricultural University, is trying to perfect different methods on the use of Nuclear Polyherdrosis Virus of gram pod borer and tobacco caterpillar. A brief account on the natural enemies reported on the major pests of the groundnut in India is presented here under :

Aphids

Many parasites and predators are reported to cause mortality in aphids (3, 8, 145). Among the predators, such as *Menochilus sexmaculatus* Fab., *Coccinella repanda* Thumbeg. var. *transversalis* Fab., *C. maculata, C. septempunnctata; Brumus suturalis* Fab and syrphid, *Ischiodon scutellans* Fab., *I. javan* were important. Some of the bugs such as *Anthocoris* sp., chrisopidae *Chrysopa cornea* were also reported to play a dominant role in the suppression of aphids. The peak activity of Coccinellids appeared from 6th week to 10th week after sowing at Anand during 1979-1981 (145). Among Coccinellids, *M. sexmaculatus* dominated with more than 50% mean population followed by *C. septumpunctata* and *B. suturalis.*

Jassids

Not much work has been done on the use of parasites and predators on the control of jassids in groundnut. Some Lygacied bugs and spiders were observed predating on jassids, but they did not appear to cause large reduction in jassid population (3). A predatory fly *Crassopalpus* sp. belonging to Empididae, Diptera was found predating on jassids such as E. kerri Pruthi; *B. hortensis* and *Exitianus taeniaticeps* (Kirchbaum); also, this fly was reported feeding 27 jassids in 10 h when field condition was simulated (90).

Thrips

There were only two reports with regard to biocontrol of thrips. Murlidharan and Ananathakrishan (87) have reported a bug *Carayonocoris indicus* feeding on *F. schultzei*. Three species belonging to Heteroptera and four belonging to Salticidae (Spiders) were recorded predating on *C. indicus* from Junagadh (Nandagopal, unpublished).

Leaf miner

Among 28 parasites recorded from South India, Maharashtra and Gujarat (83, 154) none were put to use under field conditions. Kothai (68) reported a peak (23.5%) in the parasitism of leaf miner during October, while Khan (65) reported 38-83% parasitism during August to September from Maharashtra. Yadav *et al.* (154) reported a mean parasitism of 50% during March-May by *Gonizus* sp. and 79.2% by *Stenomesius japonicus*. Oblisami *et al.* (99) reported a fungus *Aspergillus flavus* on the larvàe of leaf miner. Occurrence of a mermithed nematode from larvae of leaf miner was also reported (68). A series of experiments at ICRISAT centre has shown that larval parasitism can reach to 90% dur to this control pressure, no increase of larval population was resulted from one generation to the next (152).

3.1.4 Chemical Control

Synthetic insecticides are inevitable in the management of devastating insect pests in agriculture. Inspite of many side effects of the synthetic insecticides, like, resurgence, outbreak of secondary pests, development of resistance in insects to insecticides apart from global concern of insecticide pollution to atmosphere, the insecticides continue to dominate as major insect control agents . Despite increased cost, handling hazards, farmers are inclined to go in for synthetic insecticides, as the quick knock-down effect of these synthetic insecticides is convincing to them. A lot of information has been generated on the use of synthetic insecticides by various workers in India and abroad, it is beyond the scope of this paper, so only which are cost efficient or proved efficient with 2 or 3 sprays (1, 3, 27, 52, 84, 88, 120, 134, 146) are presented here. Following insecticides were found to be effective against various insect pests :

Aphids

Metasystox 0.1%, trithion 0.06%, thiometon 0.08%; menazone 0.03%, BHC 0.1%; demetan - o - methyl 0.05%, 25% EC 400 ml/ha.

Jassids

Dimethoate 100 ml/ha monocrotophos 0.05%, 0.07%, 0.5 kg/ha phosalone 4%D; phenthoate 50 EC 750 ml/ha; dichlorovos 76% EC 65, 62 ml/ha; phosphamidon 375 ml/ha, demeton-o-methyl 0.3%; thiometon 0.03%.

Thrips

Carbofuron seed dress 4%; monocrotophos 0.075%, 0.5%, 0.01%; phosphamidon 0.02%; dimethoate 0.05%; demeton-o-methyl 0.03%.

Leaf miner

Dimethoate 0.03%, 0.05%, dichlorovas 0.05%, 0.1%; chloropyriphos EC 0.5%; carbofuron EC 0.025, and 0.05% WP 0.05%, Carbary 1 WP 0.2% and 10% D; BHC WP 0.5%, 0.1% monocrotophos 180 g a.i./ha, EC 1.6, 1 kg/ha, 0.04%, 0.05%, 0.07% phosalone EC phosaline 0.25 and 0.05%; quinolphos EC 0.025% and 0.05%; endosulfan 0.5 kg a.i/ha; ripcard 50 g a.i./ha.

3.1.5 New approaches for control

The use of one or two methods at a time may not help in controlling the majority of the insects occurring over time and location. Hence new approaches are being examined which can be included in the existing pest management practices. These includes moult inhibitors, pheromones, plant origin insecticides, trap cropping etc.

Moult inhibitors

The cuticular chitin in insects in degraded along with protein at the time of moulting by chitinase enzyme. If synthesis of chitin is prevented the newly formed cuticle will be devoid of chitin. The insect will be unable to withstand the muscular contraction during the moulting which may prevent the casting of excuvium and finally leading to the death of insect (9). These moult inhibitors include diflubenzuron, penfluron, triflumuron and apart from 11 coded components.

Sex pheromones

The sex pheromones in insects are the chemicals excreted outside which are species specific in attraction and in some species some components could attract subspecies of the same species(26). The sex pheromones have potential role in IPM in groundnut. The use of sex pheromones was under field use for *H. armigera* and *S. litura* (5, 103). The presence of sex pheromones in leaf miner was reported recently (92). Though these sex pheromones are under field use, no economics or its role was reported in reducing the larval population in the field. Sex phermones for *H. armigera* and *S. litura* are being explored in groundnut through the All India Coordinated Research Project on Oilseeds (5). With regard to *H. armigera* in groundnut, a mean of 3.9 males were trapped at 1 mm height of sex lure and then there was decrease with increase in height of the trap (104). In an another study when the pheromone of *S. litura* was mixed with *H. armigera*, there was 200% decrease in catch of this moth as compared to trap kept without mixing (103). Under AICORPO the adult male moths collected using sex pheromone revealed that in Ludhiana the insects continued to be present all through the year (5). Pest population based on three years datas (1986, 1987, 1988), the maximum population occurred during April, May, June, September, and October were 340, 649, 581, 246 and 308 repectively while in Jalgaon last week of July and first week of August (i.e. 125 and 108) were the peak periods (5). Pawar and Srivastava (103) reported that even though the sex pheromone of *S. litura* was mixed with pheromone of *H. armigera*, no reduction in *H. armigera* catch. The presence of sexpheromone in the virgin female moths of leaf miner was reported recently (92). The population of leaf miner at Hyderabad occurred through out the year recording a maximum of 1662, 1344 and 1100 males/week between March 3rd week to 4th week of February, 1989, respectively (Rangarao, ICRISAT, personal communication).

Use of neem products

Efficiency of neem products in insect control was reviewed recently by Saxena (121) and Schemutterer (122). According to these reviews not much work has been done on the use of neem products in the control of insect pests in groundnut. Neem oil, neam leaf extract and neem leaf decoction at 5% gave 87 to 91% cumulative mortality over 48th of exposure of the jassid *B. hortensis* (94). Under the AICORPO system, the first year trial with neem based products like Margoside CK 0.1% and aqueous leaf extract resulted significantly higher pod yield at Vridhachalam and Junagadh centres apart from reducing the leaf miner damage (5). Chandramohan and Sivapragasam (16) reported a yield of 2383 kg pods/ha in neem oil treated plot which was on par with monocrotophos. Patel (101) evaluated efficacy of neem seed extract, neem leaf extract at 10% and neem oil 2% in comparison with monocrotophos gave the highest pod yield, crude neem oil gave 1542 kg/ha pod which was significantly higher compared to control. In an another study both neem kernel suspension (1%) and neem seed extract shown growth inhibition on the *H. armigera* on *S. litura* larvae (64, 69). The neam based insecticides (Margoside OK and CK) suppressed the *S. litura* damage as there were least foliage damage in groundnut and also gave higher pod yield (5). Growth regulating hormones for *S. litura* have been recorded from 43 plants species belonging to 276 families. Among them Indian neem (*Azadirachta indica* A. Juss.), *Pongamia glabra* and *Parthenium hysterophorus* are important. Crude neem oil has been successfully utilised in the management of gram pod borer, hairy caterpillar and leaf miner in groundnut. Ovicidal property of crude neem oil was demonstrated recording 100% suppression of eggs of leaf miner (91).

3.2 Control of rodents and birds

Storage (Rodents)

Place pure popped rice in a piece of paper in different places inside the godown in the previous day. Observe the feed on the next day morning, place about 10 g of mixture given components on the same place wherever feeding was noted. Mix the following components uniformly on a plastic sheets or in a plastic container.

a. Zinc phosphide - 3 to 5 kg

b. Popped rice - 1 kg

c. Groundnut oil - 5 kg

d. Sugar or Juggary - 5 g

The bramodiolon is the single doze anti-coagulant available at 0.005% concentration. This is marketed in the name of "moosh moosh". Each cake can be divided into 4 pieces and each piece can be place inside a live borrow and the mount of the borrow should be plastered with soil.

Birds control

The current practice of birds control includes nesting the crops, birds scaring through guns, use of colour balloons and bright lighting (131).

3.3 Diseases

3.3.1 Host plant resistance

Control of groundnut diseases has been tried mostly through the use of synthetic fungicides all over the world. But repeated use of chemicals led to ecological hazards, economic problems, besides the development of resistance by some pathogens to fungicides. Hence, developing diseases resistant cultivars is one of the best means of reducing crop losses from diseases. The disease resistance breeding programme requires proven stable and easily available resistant sources for economically important diseases. Such sources have been identified both in the primary gene pool and in the wild relatives of *A. hypogaea.* Efforts are underways in many parts of the world to incorporate the genes responsible for resistance into cultivars with agronomic and quality characters suitable to different environments. Hundreds of land races, advanced breeding lines, cultivars and wild species have been screened for resistance against foliar fungal diseases by several workers, and a large number of resistant sources have been identified (15, 17, 46, 81, 82, 85, 140, 141). Sources of multiple resistance especially for late leaf spot and rust are available from ICRISAT. These are : EC 76446, (292), NCAc 17133, (RF), PI 259747, PI 350680, PI 389595 and USA 63. Recently 5 multiple disease resistant germplasm accessions namely NCAc 17149, NCAc 927, NCAc 17133 (RF), PI 393646 and PI 341879 have identified (46), which were found to be resistant to early leaf spot, late leaf spot, rust and *Alternaria* leaf spot. Genetic resistance to foliar diseases caused by *P. personata* and *C. arachidicola* has usually been associated with low yields and late maturity (105). It is not difficult to obtain lines with moderate level of resistance and produce reasonably high yields with minimal or no fungicidal applications (54). Several research organizations like groundnut research centres of the All India

Coordinated Research Project on Oilseeds (AICORPO), the National Research Centre for Groundnut (NRCG), International crop Research Institute for the Semi-Arid Tropics (ICRISAT) etc. have started disease resistance breeding programmes to evolve genetically resistant genotype of groundnut for minimizing losses caused by diseases. The programme of breeding for foliar fungal diseases resistance had received special attention during the VIIth Five Year Plan. As a result of this effort, for the first time in the country, multiple foliar disease resistant groundnut varieties viz., Girnar 1 from NRCG, ALR 1 from Tamil Nadu, ICG (FDRS) 10 and ICGV 86590 from ICRISAT were released for general cultivation in India. These varieties could be grown wherever early leaf spot, late leaf spot and rust are severe problems. In general, it is very difficult to get promising sources of resistance to many of the seed and soil-borne pathogens such as *S. rolfsii, M. phaseolina* and *A. niger* as they are not very specialized pathogens and they have a very wide spectrum of host attacking mechanism (43). However, a few genotypes have been identified as tolerant to these pathogens Chohan et al. (21) tested T 31 genotype for resistance to collar rot and found only one highly resistant genotype, EC 21115 (U4-17-7). J 11 a spanish bunch released variety and other genotypes like C No. 45-23, PI 350680, NCAc 17157, B 201 and Rau 31-3 have been reported to be resistant to collar rot (5, 14). Resistant sources like TMV 1 (123), TMV 8 (75), B 30 and B 31 (110) for root rot *(M. phaseolina)* have been reported. Genotypes, T 17, T 11-11, EC 1682, RB 4, T 25, T 9, NC 2, NC 18016, NCAc 17941 AX Elorigiant have been reported to be resistant to stem rot pathogen *(S. rolfsii)* (32,55,79). Among released varieties, Kadiri 3 (R 33-1) and ICGS 11 are tolerant to bud necrosis disease. These varieties should be used in endemic areas. At ICRISAT, ICGV 86029, ICGV 86030, ICGV 86031, ICGV 86032, ICGV 86033 and ICGV 86558 were identified to possess field tolerance to bud necrosis and also good agronomic traits (115). *Arachis chacoense, A. glabrata, A. sp.* (PI 262848), *A. pussila* were found to be immune to bud necrosis (67). while *A. chacoense* and *C. pussilla* were found to be resistant to peanut mottle virus (111). Two genotypes viz., EC 76446 (292) and NCAc 17133 RF have not shown any seed transmission of peanut mottle virus (34) and PI 261945 and IP 261946 were tolerant to PMV (72), *A.* sp. ICG 30036 was reported to be tolerant to peanut clump virus. Genotypes, PI 468141, PI 468142, PI 468144, PI 468245, PI 468174, PI 468363, PI 468366 are reported to be resistant to PStV (25). ICG Nos. 4983, 8215, 8958, 8973, 11558, 11560, 11562, and 12168 were found to be resistant to PStV. Out of these, ICG 11558 was immune (96, 107). Four genotypes JSP 1, ICGMS 2, NCAX and CGC 4 have been reported to be resistant to root knot disease caused by *M. arenaria* (56). Sakhuja and Sethi (119) identified five genotypes namely G201, Japtin 220-15, local-256, local Ambali and 4018 as resistant to *M. javanica.* PI 1290606, PI 1295233 and PI 1365553, have been found resistant to pod and root lesion disease caused by *P. brachyurus* (135, 138).

3.3.2 Cultural control

Removal of volunteer groundnut plants, removal or burial of infected crop debris, crop rotation etc. are important in reducing the leafspots and rust diseases (81, 139). Early sowing of the crop reduces incidence of rust and leaf spots (50, 66, 132, 136). Foliar disease severity could also be reduced by altering plant population (50, 66). Leafspots and rust being mainly air borne diseases, spread quickly where there is continuity and host plants over large areas. Therefore, various economically profitable crop combinations were tried to check the spread of these disease at the NRCG and elsewhere intensity of leaf spots and rust (44, 50). Muthiah *et al.,* (89) reported that the incidence of late leafspot in groundnut + cowpea was minimum followed by groundnut pearlmillet and groundnut + black gram and the rust incidence was minimum in groundnut + sorghum. Deep planting of seed should be avoided as etiolated hypocotyls are prone to infection by collar rot pathogen. Deep ploughing of fields and rotation of groundnut with gram and wheat are useful in reducing the collar rot disease incidence (120). Chohan (18) reported control of collar rot by resorting to early sowing in June. Groundnut planted with a spacing of 30 cm between the rows had lowest root rot incidence than when planted with a row spacing of 45 cm of 60 cm (125). The lower incidence of collar rot and stem rot in early sown crop and closer plant spacing have been reported (36). Cultural practices such as deep covering or burial of organic matter before planting, non-durting cultivation by avoiding movement of soil up around the base of plants, prevention of development of organic debris are useful for reducing the incidence of stem rot. Cotton is a suitable rotation crop for management of *S. rolfsii* (116). Plant density can be increased to decrease the proportion of infected plants of bud necrosis (112). Groundnut intercropped with pearl millet and early sowing showed less incidence of bud necrosis disease as compared with sole crop at ICRISAT (34). Cultivation of groundnut away from soybean, cowpea, navybeans could reduce peanut mottle virus considerably as in these crops the virus is seed transmitted and serve as foci of inoculum (28). Groundnut crop grown during rabi escapes the clump disease (111), PStV could be managed by adopting measures like use of virus free seed for sowing, sowing of suitable trap crops around groundnut fields, practice of suitable mixed cropping and intercropping to reduce secondary infection due to aphids, adopting crop rotations of groundnut with non-legumes, rouging of PStV infected plants and burning them, removal of weed hosts of PStV and avoiding of growing crops like soybean, cluster bean, fresh bean, green gram, cowpea and seasamum in the nearby areas of groundnut crop since these are hosts of PStV (46). Crops like maize, sorghum, pearlmillet and rice may be rotated with groundnut to reduce the incidence of root-knot nematode species. The crops like seasamum, castor and cotton are good crops to include in rotations for reducing the incidence of *M. arenaria* (116, 117). Nematode problems in groundnut can be effectively managed through deep ploughing during summer months, increase the use of F.Y.M., Oilcakes and other organic amendment and soil solarization (Dr. D.J. Patel, personal communication).

3.3.3 Biological control

Mycoparasites, *Dicyma pulvinata* (=*Hansfordia pulvinata*) and *Verticillum lecanii* were found to parasitize the early and late leaf spot pathogens (40,81,143). *Acrononium persicinum, A. salmonaum, Darlucu filum, Eudarluca caricis, Tuberculina costaricana, Penicillium islandicum* and *V. lecanii* were found parasitizing groundnut rust (12, 127, 128, 140, 143). *P. islandicum* and *V. lecanii* and their culture filtrates were found to be effective in controlling both late leaf spot and rust diseases under field condition (40, 42). Some soil fungi, actinomycetes and bacterial antagonistic to collar rot pathogen have been reported 919). *Trichoderma harzianum, T. auareoviride, T. viride, T. polysporum, Bacillus* spp. are reported to be antagonistic to *S. rolfsii, M. phaseolina* and *Rhizoctonia solani.* However, little efforts were made to exploit these biocontrol agents at the field level (142). For example, it is estimated that about 140 kg of *Trichoderma* granules are required to suppress *S. rolfsii* in 1 ha area (7). Both *T. viride* and *T. harzianum* were found to be capable of reducing the sclerotial population of *M. phaseolina* (125). Seed treatment with spores and mycelium of *T. polysporum* protects the seeds from invasion by *M. phaseolina* (19). Elad *et al.* (20) have proved the beneficial effects of *T. harzianum* against *S. rolfsii* and *R. solani. T. harzianum* was found to be antagonistic to *A. niger* (71). *Gliocladium virens* was found to be effective against *S. rolfsii* (100).

3.3.4 Chemical control

Several systemic and non-systemic fungicides were tried for reducing the severity of major foliar diseases viz., early leaf spot, late leaf spot and rust in India and elsewhere (12,67,81,140). However, no single fungicide was found to be very effective against all the foliar diseases. In India, leaf spots and rust were effectively managed when three sprays of carbendazim (0.05%) + mancozeb (0.2%) were given at 15 days interval commencing from the first appearance of symptoms (5,11,95). In Bangladesh, Rahman *et al.* (108) have recommended carbendazim during 10 to 70 days of plant age to manage leafspots and tridemorph (calixin) spray within 40 to 60 days of plant age against rust. Vidhyasekharan (150) reported that chlorothalonil was effective against leafspots and rust. Similarly, positive effects of chlorothalonil in controlling leafspots were reported (11,33,130). Sekhawat *et al.* (129) reported that the spray schedule consisting two sprays of mancozeb (0.2%) at 35 and 70 days after germination and one sprays of carbendazim (0.025%) at 50 days after germination was most economical. Vyas *et al.* (151) also recommended that carbendazim (0.075%) 4 mancozeb (0.15%) mixture should be sprayed around August 15 and second and third spray should be repeated depending on the intensity of both leaf spot and rust. The rust and leaf spots were substantially reduced due to the application of carbedazim (0.025%) + tridemorph (0.4%) five times at fortnightly intervals commencing 35 days after sowing during summer season (62). The application of 1.06 kg a.i./ha (118). Very recently (63) reported that five rounds of sprays, of carbenxazim (0.05%) + mancozeb (0.2%), starting from 30 days after sowing with 12 days intervals was found to have maximum control of both late leaf spot and rust disease which resulted in 57% increased in dry pod yield. Carbendazim @ 1g/kg of seed was recommended for the control of collar rot pathogen (133). Seed treatment with either thiram @ 3g/kg seed or carbaendazim @ 2g/kg seed is effective in controlling seed and soil-borne diseases (37). Seed treatments of peanut with Vitavax-captan at 3 g/kg seed and the spraying of crop with benlate at a rate of 2.5g/L caused a significant reduction in crown rot incidence (97). Good control of pre-emergence root caused by *M. phaseolina* was achieved by seed dressing with captafol (123). Soil drenching with carboxin has been reported to be effective against *S. rolfsii* (4). Flutolanil was found effective against *S. rolfsii* and *R. solani* (22). Diniconazole was effective against *R. solani* and *S. rolfsii* (23). To avoid re-infestation of thrips transmitting bud necrosis disease, an application of insecticides with strong residual effect is required (76). Amin (152) provided evidence that high and frequent doses of dimethoate have to be applied to groundnut in India before there is any reduction in the incidence of bud necrosis disease. Granular formulations of thiofanox, aldicarb acephate and disulfotan gave significant control of thrips when applied at the time of sowing. Subsequently the crop may be applied with systematic insecticides at regular intervals (115). Similarly soil application of soil biocides such as 1,2-dibromide, 3-chloropropane, carbofuran and temik reduce peanut clump disease effectively, but they are not economical (111). Aphids *(Aphis craccivora)* and other species of PStV vectors should be controlled through commonly used systemic insecticides to avoid secondary transmission of PStV (46). Application of granular nematicides like nemacur 10 g, Furadan 3g, thimet 10g, mocap 10g etc. @ 3 kg a.i. / ha is useful in reducing the incidence of nematodes in groundnut. Seed treatments with 3 and 6% aldicarb sulfone and carbofuran reduce the damage by *M. arenaria* (102).

3.3.5 New approaches for control

Use of chemicals in the control of diseases has had created a lot of environmental problems. Therefore, hunt for renewable sources of alternative chemotherapeutants is highly desirable. Green plants appeal to be the reservoirs of effective chemotherapeutants and can provide renewable sources of useful biofungicides. For examples, aqueous leaf extract (2%) of neem *(Azadirachta indica)* and mehandi *(Lawsonia inermis)* were found to be effective and economical in controlling leaf spots and rust diseases of groundnut (39). Neem oil (1%) and neerium leaf extract (10%) reduced the incidence of ringmosaic, late leaf spot and rust diseases of groundnut (31). Usman *et al.* (147) reported that foliar application of neem kernel extract (3%) reduced the late leaf spot and rust diseases of groundnut significantly and recorded the highest cost/benefit ratio. Chewande *et al.* (45) reported that the application of 3 sprays of 5% aqueous neem leaf extract reduced early leaf spot by 14% and late leaf spot by 19% with an increase in pod yield by 24%. The aqueous neem leaf extract was found to be defficaceous in controlling both late leaf spot and rust diseases and in increasing pod yield at hot spot locations (Aliyarnagar, Vridhachalam in Tamil Nadu, Dharwaad in Karnataka, Jalgoan in Maharashtra and Chiplima in Orissa) of the AICORPO. Seed treatment with neem seed extract can check the infection of *S. rolfsii* and *R. butaticola* pathogens of groundnut (149). Soil solarisation using a thin transparent plastic sheet eliminates soil-borne pathogens. This environmentally safe and efficient cultural practice obviates use of soil drenching and other granular pesticides to disinfect the fields before sowing. For example, reduction in incidence of *S. rolfsii* in peanuts, 140 days after planting was 90% and 42-64% increase in yield over control was reported in Israel (57).

3.4 Weeds

3.4.1 Mechanical/cultural control

The most prevalent methods of weed control in groundnut are hand weeding and harrowing. Experiments conducted through out the country revealed that 2-3 hand weeding or harrowing upto 45 days are optimum for groundnut. First weeding should be done at about 20-25 days after sowing and second a fortnight later.

3.4.2 Chemical weed control

A number of weedicides have been recommended to control the weeds in groundnut crop. The All India Coordinated Research Program on weed control has identified certain useful herbicides namely Nitrofen, Fluchloralin, Alachlor which were effective in controlling the weeds of importance throughout the country. Pendimithalin has also been recommended All India Coordinated Research Project on Oils seeds in groundnut. Alachlor This is applied primarily as a pre-emergence treatment at the rate of 1.5 to 2 kg a.i./ha. This is effective in controlling annual grasses and *Acalypha, Amaranthus* and *E.* species. Dephenemid : This is a water soluble wettable powder applied at the planting time @ 2.25 to 3.40 a.i./ha for the control of annual and perennial grasses and partial control of broad leaved weeds. It persists for 3 to 6 months in soil. Trifluralin : It is available at both EC and granules. It is recommended as pre-emergence application @ 0.5 kg a.i./ha in groundnut. This weedicide affected both seedling emergence of weeds and other associated phenological growth processes. Fluchloralin. This is marketed as basalin. It is selective pre-emergence weedicide controlling grasses and broad leaved weeds. Immediate irrigation is necessary after soil application @ 1.25 to 1.5 kg a.i./ha. Pendimithalin (Stomp): It is a selective , pre-emergence weedicide controlling annual grasses and certain broad leaved weeds, it is available in the form of EC. Application of the weedicide @ 1 kg a.i./ha controls most of the grassy weeds in groundnut crop Fusilade. It is a post emergence weedicide formulated as EC It is very effective against annual and perennial grasses while no effect on broad leaved weeds when applied@ 0.25 to 1 kg a.i./ha.

4. INTEGRATED PEST MANAGEMENT

In the previous pages various control measures for pests have been discussed. Majority of these methods developed by several workers have not been fully effective from economic point of view. Since most of the major pests occur throughout the world, various control measures need to be suitably integrated to cover a broad spectrum of pests. The approach, therefore, should be an integrated protection of the crop rather than integrated management of individual pests. Unfortunately, serious efforts were not made in the past to integrate the various pests control measures and as such research on integrated pests management in groundnut is still in its infancy. However, there are a few research efforts where the pests control measures have been integrated and it appears that there is a scope for an integrated pests management in groundnut (47).

Insect pests

The strategy of IPM is problem oriented and vary depending upon the pest situation and different climatic factors. So a strategy developed for a particular situation need not be necessarily relevant to other situation. Combination of different components of IPM may also vary. In groundnut, integrated management of individual pests like red hairy caterpillar *(A. albistriga)*, tobacco caterpillar *(S. litura)* (60,61) and white grub *(L. consanguinea()* has been given. A comprehensive scheme for the integrated management of pests namely leaf miner, aphid, jassid, thrips, gram pod borer and tobacco caterpillar is also proposed (84,91).

Integrated management of red hairy caterpillar

The management strategies include organising campaigns to hand pick and destroy pupaea after summer ploughing on receipt of monsoon showers, growing cowpea as an intercrop which helps to attract months to lay eggs, set up light traps or bonfires immediately after receipt of rain, collect the egg mass and destroy them, monitor the cowpea leaves for lace-like appearance due to gregarious larvae feeding, spraying of NPV at 250 larval equivalent/ha. Add wetting agents like sandovit or teepol or triton, handpick and destroy grown up caterpillar, digging a trench around the field to avoid further movements and applying any one of the insecticides namely BHC 10%, quinophox 1.5%, phosalone 4%, endosulfan 0.05%, Carbaryl 10%, 25 kg/ha each, spray of endosulfan 35 EC, quinophos 25 EC and phosalone 35 EC 750 ml/ha and chlorpyriphos at 625 ml/ha.

Integrated management of tobacco caterpillar

To manage the tobacco caterpillar, grow castor as border or intercrop to serve as trap crop, monitor the emergence of the adult moth by setting up of light or pheromone traps, mass trapping, using pheromone traps over large area, collect the egg mass on castor and groundnut and destroy the same, monitoring the larvae on leaves of castor for lace like symptoms which harbour the gregarious early instar larvae, prepare a bait with rice bran 5 kg + molasses or jaggery 1 kg+DDT 50% WP or carbaryl 50% WP 0.5% kg, to cover 0.1 ha of area. Distribute the mixture around the root zone to have effective control.

Integrated management of leaf miner, jassids, thrips and aphids

Sow groundnut early and synchronously in the rainfed and irrigated seasons, in areas where leaf miner is a major problem, sow spreading and semi-spreading varieties instead of bunch varieties which are highly susceptible, sow cowpea as an intercrop with groundnut (1 : 4) to minimise the leaf miner infestation, wherever possible, depending on the availability, mulch the soil with straw within 10 days after germination to reduce leafminer incidence, avoid water stress in irrigated crop to reduce leaf miner infestation, maintain the fields and buds free from weeds, pull out and destroy plants with bud necrosis or bud blight as thrips act as vectors of the disease. Set uplight traps upto midnight to monitor, attract and kill leaf miner moths and synchronise insecticidal application within a week after brood emergence, monitor the crop from the second week onwards and assess the leaf miner larval population. If two or more larvae are observed per plant, apply any one of the following insecticides: phosalone 4% D, endosulfan 4% D or carbaryl 10% D or fenitrothion 2% D @ 25 kg/ha or phosalone 35% EC, endosulfan 35% EC or monocrotophos 36% WSC, quinolphos 25% EC or phenthoate 50% EC at 750 ml/ha, dichlorvos 76% SC 625 ml or phosphamidom 375 ml/ha.

Integrated approach to manage the white grub

Spray of host trees of the adult beetles such as ber, neem, khejari and jular with 2% carbaryl or 0.05% monocrotophos during the evening hours just after the start of the monsoon, collect the adult beetle by jerking the tree or jerking the tree and focusing a power light down on soil surface which attract large number of beetles, soil treatment of phorate 10 G at 25 kg/ha or BHC 10% 12.5 kg/ha may effectively protect the crop, seed treatment with chlorpyriphos 20 EC may be effective, summer ploughing to expose the grubs to predatory birds or a person with a bucket behind the plough can collect and kill the grubs and flooding the field results in stagnation of water for 6 days may reduce the grub population.

Integrated management of leaf miner, gram pod borer, tobacco caterpillar, jassids, thrips and aphids

Very recently there has been a report on IPM scheme (91) which includes components such as trap crops (soybean, pearlmillet, castor), pheromones (for gram pod borer, tabacco caterpillar and leaf miner), spray application of neem oil (2%) + phosphamidon (0.02%) + endosulfan (0.04%) at every appearance of leaf miner brood. Sow 1 row of soybean after every 2 rows of groundnut, sow 3 rows of pearl millet all around the field followed by a row of castor, spray insecticide mixture at 30 to 35 days after germination followed by 2 sprays of the same of the appearance of second and third broods of leaf miner respectively. Place pheromones traps for all the 3 larval pests at the rate of 8 each/ha.

Integrated management of diseases

Collar rot and white grubs cause very high plant mortality at the seedling stage in the groundnut crop during June-July. An integration of date of sowing (10th or 20th June), seed treatment with thiram @ 5 g/ha kernel and soil application of phorate granules reduced the plant mortality and gave higher returns at Ludhiana and Samrala in Punjab (5). At Aliyarnagar and Coimbatore (Tamil Nadu) adoption of practices such as closer spacing, advancement of date of sowing by two weeks, spray of insecticides like nuvàcron and orthene, spray of antiviral principles like Jawar (JVP) and coconut (CAVP), removal of bud necrosis infected plants particularly at the early stages of crop development and removal of weeds reduced the incidence of bud necrosis disease and increased the yield from 12 to 17% (5). It is reported that when low rates of insecticides are applied for the control of an insect complex in groundnut, NC 6 a resistant variety yielded 16% more yields than a susceptible Florigiant variety of groundnut (13). Gorbet *et al.* (54) have successfully used host resistance alongwith fungicides for the control of leafspots. Bornes *et al.* (10) integrated fungicides, genotypes, cultural practices (irrigation), and environment to control *Rhizoctonia* limb rot and stem rot. They reported that application of diniconazole every 14 days at 0.28 kg a.i./ha on Florunner and New Mexico Valencia cultivars with reduced irrigations resulted in maximum reduction of the limb rot. Combining Southern Runner, a moderately resistant variety to late leaf spot, *S. rolfsii* and tomato spotted wilt virus with applications of chlorothalonil can provide late leaf spot control (24). At NRCG, various control measures were integrated for the management of early and late leaf spot during kharif 1988 to 1990. Three years data showed that most effective early and late leaf spot control was achieved where groundnut was intercropped with pigeonpea and sprayed twice (55 and 70 DAS) with fungicide mixture (Carbaendazim 0.05% + mancozeb 0.2%) or where groundnut intercropped and received one spray each of neem leaf extract (2%), fungicide mixture, and cell-free filtrate of *P. islandicum.* These treatments and the intercrops with neem leaf extract at 55 and 70 DAS and with inoculum *P. islandicum* at 55 and 70 DAS gave the highest monetary returns over three years (45). Huang Binchao (58) reported that a compound measure of integrated pest management of groundnuts including variety resistance, cultural and chemical control of groundnut diseases and pests is introduced in Gauangdong Province of South China.

Integrated approach in weed control

So far no attempt has been made to integrate all the feasible methods of weed control in groundnut. A single weed control method either chemical or cultural is not completely effective in eliminating the weed problem. For effective weed control all the feasible methods have to be used harmoniously. The known practice such as row spacing, tillage and crop rotation are worth to include in any weed control strategy. Generally, application of mixture of two or more herbicides is effective against different species of the weeds compared to application of single weedicide. For example, mixture of Nitrofen + fluchloralin @ 0.2 kg a.i. and 0.5 kg a.i./ha respectively controls various weeds in rainfed crop. The pre-emergence application of mixture of alachlor and Nitrofen is more effective in controlling weeds. Pre-sowing application of fluchloralin followed pre-emergence application of Nitrofen @ 0.7 to 1 kg a.i./ha. control weeds effectively. So the approach should be sequential application of two different herbicides which will be more effective. Pre-emergence application of fluchloralin followed by one or two hand weeding at 25 to 30 and 50 to 60 days after sowing is the most economical. Integration of herbicide and hand weeding at 30 DAS gave consistent and effective weed control and higher pod yields over two years (109).

5. THE FUTURE PROSPECTS OF IPM

It is evident from the literature that there is a great scope for integration of suitable and compatible pests control measures and thus develop cost effective and highly efficient package for the integrated pest management in groundnut. The components of IPM such as use of resistant variety possessing resistance to one and/or more pests, mass trapping of lepidopetran pests, using sex pheromones, use of biocontrol agents like NPV, thuricides, *V. lecanii, P. islandicum,* use of moulting inhibitors, use of suitable intercrops, use of plant products of neem origin, soil solarization, alellopathy for weeds and judicious use of pesticides should be integrated suitably wherever possible keeping in view of the environmental conditions, cultivation practices, cost and social factors since these components have proved to be highly successful. In USA, much emphasis is given on the interdisciplinary approach in IPM peanut programmes involving field-scouting of insect pests, plant diseases, nematodes and weeds. Such a system is also in operation on a limited scale in India which should be stregthened on a large scale for economically important pests at national level since surveillance of pests is an integral part and powerful tool of IPM.

ACKNOWLEDGEMENT

The authors are thankful to Dr. P.S. Reddy, Director, National Research Centre for Groundnut, for providing the necessary facilities and encouragement.

6. REFERENCES

1. Alexander Jesudasan, R. W., Arulsamaraj, D. and Priyakumarim Joyce. 1989. Field evaluation of new insecticides against the groundnut leaf miner, **22** : 26-27.
2. Amin, P. W. 1986. Epidemiology and control of bundnecrosis disease caused by tomato spotted wilt virus. In: Recent Adavances in Groundnut Productivity Research, Proceeding of Summer Institute, May 9-30, 1983, NRCG, (ICAR), Junagadh, India, pp. 145.
3. Amin, P. W. 1988. Insect and mite pests and their control. In: Groundnut Ed. P. S. Reddy, pp. 393-452.
4. Amma, P. A. and Shanmugam, N. 1974. Studies on the root rot of groundnut caused by *Scerotium rolfsii* Sacc. Madras Agricultural Journal, **61** : 973-978.
5. Anonymous, 1974 to 1991. Progress Report, Annual Kharif Oilseeds Work shop, All India Coordinated Research Project on Oilseeds Research (AICORPO), Directorate of Oilseeds Research, Hyderabad.
6. Anonymous, 1989. Annual Report, National Research Center for Groundnut, Junagadh, pp. 75.
7. Bakhetia, D. R.C. and Sidhu, A. S. 1976. Biology and seasonal actlvity of the groundnut aphid *Aphis craccivora* Koch. Journal of Research, Punjab Agricultural University, **14** : 299-303.
8. Balasubramanian, M. 1985. Mould inhibitors - A novel approach in pest management. In Pest and Disease Management, Ed. S. Jayaraj, pp. 224-226.
9. Barnes, J. S., Csinos, A. S. and Hook, J. E. 1990. Effects of fungicides, cultivars, irrigation and environment on *Rhizoctonia* limb rot of peanut. Plant Disease, **74**:671-675.
10. Barwal, R. N. and Gupta, D. K. 1991. Seasonal incidence and control of pod borer *(Euborellia annulipes)* and *Cercospora* leaf spots in groundnut. Indian Journal of Plant Protection, **19** : 156-160.
11. Beckman, P.A. and Rodriguez-kabana 1975. A system for the growth and delivery of biological control agents to the soil, Phytopathology, **65** : 819-821.
12. Brenneman, T. B. and Summer, D. R. 1990. Effects of tractor traffic and chlorothalonil applied via ground sprays or Center Pivot irrigation system on peanut and pod yield. Plant Disease, **74** : 277-279.

13. Campbell, W. V., Wynne, J. C. and Emery, D. A. 1977. Control of an insect complex with a resistant variety, 'NC 6'. Proceedings of APRES, **9** : 69.

14. Chahal, A. S., Dutt, S. and Chohan, J. S. 1974. Varietal resistance in groundnut *(Arachis hypogaea* L.) to collar rot *(Aspergillus niger)* in the Punjab state, Journal of Research, Punjab Agricultural University, **11**: 200-203.

15. Chohal, A. S. and Sandhu, R. S. 1972. Reaction of groundnut varieties against *Cercospora personata* and *C. arachidicola* . Plant Disease Reporter, **56** : 601-603.

16. Chandramohan, N. and Sivasubramanian, P. 1987. Evaluation of neem product against leaf miner *Aproaerema modicella* (D.) Neem Newsletter, **4** : 44.

17. Chiteka, Z. A., Gorbet, D. W., Knauft, D. A., Shokes, F. M., Kucharek, T. A. and Norden, A. J. 1988. Components of resistance to late leaf spots in peanut *(Arachis hypogaea* L.). Proceedings, American Peanut Research and Education Society, **19** : 25.

18. Chohan, J. A. 1972. Final Process Report, ICAR, Scheme for research on important diseases of groundnut in the Punjab for the period 1957-67, Department of Plant Pathology, Punjab Agricultural University, Ludhiana, India, pp. 117.

19. Chohan, J. A. 1974. Recent Advances in diseases of groundnut in India. In: Current Trends in Plant Pathology, Uttar Pradesh, India, Lucknow University Press, pp. 171-178.

20. Chohan, J. S., and Capoor, S. P. 1967. Cultural factors in relation to collar rot of groundnut. Journal of Research, Punjab Agricultural University, **4** : 536-542.

21. Chohan, J. S., Sandhu, G. S. and Dalal, J. L. 1970. Screening of world collection of groundnut *(Arachia hypogaea* L.) germplasm for resistance to collar rot disease caused by *Aspergillus niger* Van Tieghem in Punjab. Indian Journal of Agricultural Sciences, **40** : 546-551.

22. Csinos, A. S. 1987. Control of southern stem rot and *Rhizoctonia* limb rot of peanut with flutonil. Peanut Science, **14** : 55-58.

23. Csinos, A. S., Kvion, C. S. and Littrell, R. H. 1987. Activity of diniconazole on foliar and soil-borne diseases of peanut. Applied Agricultural Research, **2** : 113-116.

24. Culbreath, A. K. and Brenneman, T. B. 1992. Combining Centre Pivot irrigation applications of chlorothalonil with moderately resistant cultivar for control of late leaf spot in peanut. Plant Disease, **76**: 29-32.

25. Culver, J. N., Sherwood, J. L. and Melouk, H. A. 1987. Resistance to peanut stripe virus in *Arachis* germplasm. Plant Disease, **71** : 1080-1082.

26. David, H., Easwaramoorthy, S., Nesbitt, B. P. and Nandagopal, V. 1981. Sex attraction studies in sugarcane internode borer *Chilosacchariphagus indicus* (K.) with synthetic pheromone. Pestology, **5**: 5-11.

27. Das, B. B. 1988. Efficacy of some synthetic pyrethroids and conventional insecticides spray against groundnut pest complex. Pesticides, **22** : 10-12.

28. Demski, J. M. and Khun, C. W., 1983. Peanut mottle. In: Peanut Disease Compendium Eds. D. M. Porter, D. H. Smith and R. Rodriguez-Kabana, St. Paul, Minesota. The U.S. American Phytopathological Society, pp. 45-46

29. Elad, Y., Hadar, Y., Chat, I. and Hemis, Y. 1982. Prevention with *Trichoderma harzianum* Rifai of reinfestation by *Sclerotium rolfsii* Sacc. and *Rhizoctonia solani* Kuhn of soil fumigated with methyl bromide and improvement of disease control in tomatoes and peanuts. Crop Protection, **1** : 199-211.

30. Feakin, S. D. 1973. Pest control in groundnuts. In: Pans Manual (3rd Ed.) London U. K., Center for overseas pest control foreign and commonwealth Office, pp. 19-81.

31. Ganapathy, T. and Narayanasamy, P. 1990. Effect of plant products on the incidence of major diseases of groundnut. International Arachis Newsletter No. **7** : 20-21.

32. Garren, K. H. 1964. Inoculum potential and differences among peanut in susceptibility to *Sclerotium rolfsii.* Phytopathology, **54** : 279-281.

33. Gerlach, W.W.P., Metalanes, S. and Tuinonla, M. 1986. Evaluation of fungicides for the control of peanut leaf disease *Cercosporidium personatum* and *Puccinia arachidis* in Western Samoa, Alafua. Agriculture Bulletin, **11** : 37-39.

34. Ghanekar, A. M. 1980.Groundnut Virus Research at ICRISAT. Proceedings of the International Workshop on groundnuts Eds. R. W. Gibbons amd J. V. Martin, International Crops Research Institute for the Semi-Arid Tropics, Patancheru, India, pp. 211-216.

35. Ghanekar, A. M., Reddy. D.V.R., Iizuka N. Amin, P. W. and Gibbons, R. W. 1979. Bud necrosis of groundnut *(Arachis hypogaea* L.) in India caused by tomato spotted wilt virus. Annals of Applied Biology, **93** : 173-179.

36. Ghewande, M. P. 1983. Effect of cultural practices on the disease (bud necrosis, collar rot, stem rot) incidence and yield of groundnut. Indian Botanical Reporter, **2** : 176-177.

37. Ghewande, M. P. 1985. Seed and seedling diseases of groundnut deserve attention for better productivity. Seeds and Farms, **11** : 9-12 & 19.

38. Ghewande, M. P. 1987. Virus diseases of groundnut and their management. In: Symposium on Diseases of Oilseed and Pulse Crops and their Control., Eds. S. C. Vyas and V. N. Shroff, J. N. K. V. V. Indore, (M.P.) pp. 62-67.

39. Ghewande, M. P. 1989a. Management of foliar diseases of groundnut *(Arachis hypogaea)* using plant extracts. Indian Journal of Agricultural Sciences, **59** : 133-134.

40. Ghewande, M. P. 1989b. Biological control of late leaf spot *(Phaeoisariopsis personata)* of groundnut *(Arachis hypogaea* L.) Indian Journal Agricultural Sciences, **59** : 189-190.

41. Ghewande, M. P. 1990a. Diseases of groundnut and their management. Journal of Oilseeds Research, **7**: 78-97.

42. Ghewande, M. P. 1990b. Biological control of groundnut *(Arachis hypogaea* L.) rust *(Puccinia arachidis* Speg.) in India. Tropical Pest Management, **36** : 17-20.

43. Ghewande, M. P. and Desai, S. 1992. Integrated management of Groundnut Diseases. In: Advances in Plant Disease Management, Eds. Vyas *et al.*

44. Ghewande, M. P., Desai, S. Prem Narayan, and Ingle, A. P. 1991a. An integrated approach to the management of foliar diseases of groundnut *(Arachis hypogaea* L.). Poster Abstracts, Second International Groundnut Workshop, 25-29 November, 1991, ICRISAT, Patancheru, A. P., India.

45. Ghewande, M. P., Desai, S., Prem Narayan and Ingle, A. P. 1991b. Evaluation of fungicides and plant products to manage foliar diseases of groundnut *(Arachis hypogaea* L.). Indian Phytopathology.

46. Ghewande, M. P., Desai, S., Prem Narayan and Kamble, S. D. 1992. Some sources of resistance to early leaf spot of groundnut *(Arachis hypogaea* L.) in India. Tropical Agriculture (Trinidad), **69** : 284-286.

47. Ghewande, M. P. and Misra, D. P. 1986. Scope of IPM in increasing groundnut production. Plant Protection Bulletin, **38** : 55-56.

48. Ghewande, M. P., Nandagopal V. and P. S. Reddy. Plant Protection in groundnut (Bulletin no. 1). National Research Centre from Groundnut, Junagadh, Gujarat, pp. 33.

49. Ghewande, M. P., Shukla, A. K., Pandey, R. N. and Misra, D.P. 1983. Losses in groundnut yields due to leaf spots and rust at different intensity levels. Indian Journal of Mycology and Plant Pathology, **13** : 125-127.

50. Ghewande, M. P., Sukla, A. K. and Pandey, R. N. 1986. Management of foliar disease of groundnut-through agronomic practices. Indian Botanical Reporter, **5** : 179-181.

51. Ghewande, M. P., Singh, V. and Reddy, P. S. 1992. Occurrence of Peanut Stripe Virus (PStV) Disease in India- The possible repercussions on the groundnut production scenario. Agriculture Situation of India.

52. Ghorpade, S. A., and S. G. Thakur, 1989. Assessment of need based monocrotophos application for the control of major pest of groundnut. Pesticides, **11** : 19-20.

53. Ghuge, S. S., Mayee, C. D. and Godbole, G. M. 1981. Assessment of losses in groundnut due to rust and tikka leafspots. Indian Phytopathology, **34** : 179-182.

54. Gorbet, D. W., Knauft, D. A. and Shokes, F. M. 1990. Response of peanut genotypes with differential levels of leafspot resistance to fungicide treatment. Crop Science, **30** : 529-533.

55. Gregory, W. C. 1970. Registration of NC 2 peanuts (Reg. No. 5) Crop Science, **10** : 459-460.

56. Grewal, P. S., Chhabra, H. K. and Kaul, V. K. 1986. Screening of groundnut germplasm against root knot hematode, *Meloidogyne arcnaria.* Indian Journal of Nematology, **17** : 151-152.

57. Grinstein, A., Katan, J., Abdul - Razik, A., Zeydan, O. and Elad, Y. 1979 . Control of *Sclerotium rolfsii* and weeds in peanuts by solar heating of soil. Plant Disease Reporter, **63** : 1056-1059.

58. Haung Bingchao, 1991. Review of groundnut pests and diseases and IPM implementation in Gaungdong Province of South China. Poster Abstract : Second International groundnut work shop 25-29. November, 1991. International Research Institute for the Semi-Arid Tropics, Patnacheru, Andhra Pradesh, India.

59. Jayaraj, S., Rangarajan, A. V. and Jayarajan, R, 1986. Pest and disease surveillance in groundnut, Pest and Disease Management-Oilseeds, Pulses, Millets and Cotton, Ed. S. Jayaraj, pp. 1-13.

60. Jayaraj, S. and Rabindra, R. J. 1985. Integrated management of *Spodoptera litura* with special reference to NPV. In: Pest and Disease Management-Oilseeds, Pulses, Millets and Cotton, Ed. S. Jayaraj, pp. 30-37.

61. Jayaraj, S. and Rabindra, R. J. 1985. Management of red hairy caterpillar *Amsacta* spp. (Leipdoptera : Arctiidae) on ground nut. In Pest and Disease Mangemnt-Oilseeds, Pulse, Millets and Cotton, Ed. S. Jayaraj, pp. 38-43.

62. Jayasekar, M. and Rajasekaran, R. 1986. Appraisal of yield loss in groundnut due to rust and late leaf spot. Madras Agricultural Journal, **73** : 525-527.

63. Jayasekar, M. and Rajasekaran, R. 1991. Effect of fungicides on key diseases of groundnut in relation to crop stage. Tropical Pest Management, **37** : 270-272.

64. Joshi, B. G., Rama Paraad, G. and Satyanarayana, S. V. V. 1978. Relative hydroxide as antifeedant against tobacco caterpillar *Spodoptera litura* F. in the nursery. Indian Journal of Agricultural Sciences, **48** : 19-22.

65. Khan, M. I. and Raodeo, A. K. 1987. Importance of larval parasites in the control of *Stomopteryx subecivella* Zeller. Journal of Maharashtra Agricultural Universities, **3** : 261-263.

66. Kodmelwar, R. W. and Ingle, A. Y. 1989. Effect of sowing dates, spacing and meteriological factors on the development of tikka and rust of groundnut. Indian Phytopathology, **42** : 274.

67. Kolte, S. J. 1984. Diseases of annual edible oilseed crops. Peanut Diseases Vol. I., Florida, CRC Press Incorporated, USA, pp. 29.

68. Kothai, K. 1974. Bionomics and chemical control of the groundnut leaf miner *Stomopteryx subsecivella* (Zeller) (Lepidoptea: Gelechidae), M. Sc. (Ag.) thesis, University of Agricultural Sciences, Bangalore, India, pp. 78.

69. Kaul, D. 1985.. Azadiracht in interaction with development of *Spodoptera litura* Fab. Indian Journal of Experimental Biology, **23** : 160-163.

70. Kuhn, C. W. and Demski, J. W. 1975. The relationship of peanut mottle virus to peanut production. Georgia Agriculture Experiment Station Research Report No. 213 . pp. 19.

71. Kuhn, C. W. and Demski, J. W. 1984. Peanut mottle In: Compendium of Peanut Diseases Eds. D. M. Porter., D. H. Smith., R. Rodriguz-kabana. The American Phytopathological Society, pp. 73.

72. Kuhn, C. W., Paguio, D. R. and Adams, D. B. 1978. Tolerance in peanuts to peanut mottle virus. Plant Disease Reporter, **62**: 365-368.

73. Kulkarni, J. H. and Jojitra, V. K. 1986. Nodulation, growth and yield of groundnut under six intercropping systems. Madras Agricultural Journal, **73** : 366-369.

74. Lastin, S. M., Self El-Nasr, H. I., El-Nagar, M.A. and Nofal, M. A. 1989. Biological control of *Aspergillus niger* the causal organism of peanut crown rot by *Trichoderma harzianum*. Annals of Agriculture Science, **34** : 795-803.

75. Lewin, H. D., David, D. A., Leela, S., Natarajan, S. and Janarthanan, R. 1971. A comparative study on the pests and diseases in different strains of groundnut, Madras Agricultural Journal, **58** : 15-20.

76. Lewis, T. 1973. Thrips : Their biology, ecology and economic importance, Academic Press, London and New York. p. 349.

77. Logiswaran, G. and Mohanasundram, M. 1985. Studies on the relation between light trap catches and field incidence of groundnut leaf miner *Aproaerema modicella* (Gelechidae : Lepidoptera). Madras Agricultural Journal, **72**: 629-632.

78. Logiswaran, G., Madhava Rao, Vasudevan, G. and Kannan, V. 1982. Influence of time of sowing and weather factors on the infestation of leaf miner *Aproaerema modicella* Deventer and yield in rainfed groundnut. Madras Agricultural Journal, **69** : 359-363.

79. Mathur, R. S. and Kureel, D. C. 1965. Search of resistance in groundnut varieties to Sclerotial root rot in Uttar Pradesh. Indian Oilseeds Journal, **9** : 152-153.

80. McDonald, D. and Raheja, A. K. 1980. Pests, disease and crop protection in groundnuts. In: Advances in Legumes Science, Eds. J. summerfield. and A. H. Bunting. Royal Botanic Gardens, Kew, England.

81. McDonald, D., Subrahmanyam, P., Gibbons, R. W. and Smith, S.D. H. 1985. Early and late leaf spots of groundnut. Information Bulletin No. 21. International Crops Research Institute for the Semi-Arid Tropics: Patancheru, A. P., India

82. Melouk, H. A., Sanborn, M. R., and Banks, D. J. 1984. Sources of resistance to peanut mottle virus in *Arachis* germplasm. Plant Disease, **68** : 563-564.

83. Mahammad, A. 1981. The groundnut leaf miner *(Stomopteryx subsecivella* Zeller) (Lepidoptera : Gelechiidae). A review of world literature. Occasional paper 3, Groundnut Improvement Program,

International Crops Research Institute for Semi-Arid Tropics (ICRISAST), Patanchru, A. P., India. pp. 21.

84. Mohansundram, M. 1986. Insect pest management in groundnut. In: Pest and Disease Management-Oilseeds, Pulses, Millets and Cotton Ed. S. Jayaraj. Tamil Nadu Agril. Univ. Coimbatore.

85. Moraes, S. A. and Godoy, I. J. 1985. Evaluation of resistance to *Cercosporidium personatum* In: *Arachis hypogaea* genotypes. Summa Phytopathologica, **11** : 140-151.

86. Murali Bhaskaran, R. and Thangavelu, S. 1990. Influence of intercrop on the incidence of leaf miner *(Aproeerema modicella* Deventer) in groundnut. Journal of Oilseeds Research, **7** : 143-146.

87. Murlidharan, N. and Ananthakrishnan, T.N. 1978. Biology of four species of Anthocoridae (Hemiptera: Insecta) predacious in thrips, with key to genera of Anthocorids from India. Records Zoological Survey of India, **11** : 1-32.

88. Muthiah, C. and Hussain, H. J. 1991. Bioefficacy of insecticides against groundnut pests. Groundnut News, **3** : 6.

89. Muthiah, C., Senthivel, T., Venkatkrishnan, J. and Sivaram, M. R. 1991. Effect of intercropping on incidence of pest and disease in groundnut *(Arachis hypogaea).* Indian Journal of Agricultural Sciences, **61** : 152-153.

90. Nandagopal, V. 1988. *Crossopalpus* sp. an efficient predator of jassids of groundnut. International Arahchis News-Latter, **4** : 18-19.

91. Nandagopal, V. 1992. Studies on the Integrated Pest Management in Groundnut in Saurashtra. Ph.D. Thesis, Saurashtra University, Rajkot pp. 246.

92. Nandagopal, V. and Reddy, P. S. 1990. Report of sex pheromone in groundnut leaf miner. International Arachis Newsletter, **7** : 25-26.

93. Nandagopal, V. and Vasantha, S. 1991. Influence of plant growth regulators on the population of thrips *Caliothrips indicus* (Begnall) and damage in groundnut. Entomon, **16** : 87-89.

94. Nandagopal, V., Jaroli, R. K. Ramkumar and Reddy, P. S. 1991. Neem products : possible insecticide on the groundnut jassid *Balclutha hortensis* Lindb. International Arachis Newsletter, **8** : 22.

95. Natarajan, S. and Schitanatham, K. 1986. Combined efficacy of fungicides on the control of leaf spot and rust-disease in groundnut *(Arachis hypogaea* L.) Oleagineux, **4** : 231-233.

96. Nigam, S. N., Reddy, D. V. R., Moss, J. P., Dwivedi, S. L. and Reddy, L. J. 1991. Breeding for resistance to groundnut virus diseases at ICRISAT Centre, In: Groundnut Virus Diseases in Africa, ICRISAT, Patnacheru, A. P. India, pp. 17-18.

97. Nofal, M. A., Seif EL-Nasr, H. I., Diab, M. M., EL-Nagar, M.A.A. and El-Said, S.A.I. 1990. Effect of the systemic fungicides benlate and vitavax-captan on *Aspergillus* crown rot incidence of peanut plants. Annals of Agriculture. Sci. **35** : 407-415.

98. Nolt, B. L. and Reddy, D.V.R. 1984. Peanut Clump. In: Compendium of Peanut Diseases Eds. D. M. Porter, D. H. Smith, and R. Rodriguz-Kabana, American Phytopathological Society, pp. 73.

99. Oblisami,, G., Ramamurthy, K. and Rangaswami, G. 1969. Studies on the pathology of some crop pests of South India. Mysore Journal of Agricultural Science, **3** : 86-98.

100. Papavizas, G. C. and Collins, D. J. 1990. Influence of *Gliocladium virens* on germination and infectivity of sclerotia of *Sclerotium rolfsii,* Phytopoathology, **80** : 627-630.

101. Patel, H. K. 1968. Neem seed suspension-A sure deterrent to hairy caterpillar *Amsacta moorei* B. Madras Agricultural Journal, **55** : 509-510.

102. Patel, H. R., Vaishnav, M. U. and Dhruj, I. U. 1986. Efficacy of aldicarb sulfone and carbofuran flowable seed treatment on plant growth and against *Meloidogyne arenaria* on groundnut. Pesticides, **20** : 29-31.

103. Pawar, C. S. and Srivastava, C. P. 1988. Interaction between sex pheromones of *Spodoptera litura* Fab. and *Heliothis armigera* (Hub.) International Arachis Newsletter, **4** : 22.

104. Pawar, C. S., Sithanantham, S., Bhatnagar, V., S. Srivastava, C. P. and Reed, W. 1988. The development of sex phermone trapping of *Heliothis armigera* of ICRISAT, India, Tropical Pest Management, **34** : 39-43.

105. Pixley, K. V., Boote, K. J., Shokes, F. M. and Gorbet, D.W. 1990. Growth and partitioning characteristics of four peanut genotypes differing in resistance to late leaf spot. Crop Science, **30** : 796-804.

106. Prasada Rao, R. D., Chakrabarty, V. J., S. K., and Reddy, A. S. 1988. First report of peanut stripe virus from India. Plant Disease, **72** : 912.

107. Prasada Rao, R.D.V.J., Reddy, A. S., Chakrabarty, S. K., Sastry, K. S., Reddy. D. V.R., Ramnath Rao, and Moss, J. P. 1989. Peanut Stripe Virus research in India. In: Second Coordinators Meeting on Peanut Stipe Virus, 1-4 Aug. 1989. ICRISAT, Patnacheru, A.P., India.

108. Rahman, M.A., Ahmed, H. W. and Alam, K. B. 1986. Studies on the efficiency of fungicides and the date of commencing of spray in controlling tikka and rust of groundnut. Bangladesh Journal of Plant Pathology, **2** : 57-61.

109. Ramakrishna, A. 1991. Integrated weed management-An efficient method of groundnut weed control. Poster Abstracts, Second International Groundnut Workshop 25-29 November, 1991, ICRISAT, Patnacheru, Andhra Pradesh, India pp. 63.

110. Raj, S. A. and Prasad, N. N. 1975. Reaction of groundnut to *Rhizoctonia botaticola(M. phasolina).* Indian Phytopathology, **28** : 440-441.

111. Reddy, P.S. 1988. Genetics, breeding and varieties. In: Groundnut, Ed. P. S. Reddy, pp. 508-525.

112. Reddy. D.V.R., Amin, P.W., McDonald, D. and Ghanekar, A. M. 1983. Epidemiology and control of groundnut bud necrosis and other diseases of legume crops in India caused by tomato spotted wilt virus. In: Plant Disease Epidemiology Eds. R. T. Plumb and J. M. Tresh, J. K., Blackwell, Oxford, pp. 93-102.

113. Reddy, D.D.R., Subrahmanyam, P., Shanker Reddy, G. H., Raja Reddy, C. and Siva Rao, D. V. 1984. A nematode disease of peanut caused by *Tylenchorhynchus brevilineatus.* Plant Disease, **68** : 528-529.

114. Reddy, D. V.R. and Wihgtman, J. A. 1989. Tomato spotted wilt virus: Thrips transmission in disease vector research Vol. 5 Ed. K.F.Harris, pp. 203-219.

115. Reddy, D.V.R., Wightman, J. A., Bashear, R. J., Highland, B., Black, M., Sreenivasalu, P., Dwivedi, S. L., Demski, J. W., McDonald, D., Smith, Jr. J. W. and Smith, D. J. 1991. Bud necrosis : a disease of groundnut caused by tomato spotted wilt virus. Information Bulletin no. 31. International Crops Research Institute for the Semi-Arid Tropics, Patnachru, A. P:, India.

116. Rodriguez-kabana, R. and Morgan-Jones, G. 1987. Novel rotations and organic materials show promise for management of nematode. Alabama Agricultural Experiment Station. Highlights of Agricultural Research, **34** : 13.

117. Rodriguez-kabana, R., Roberson, D. G., Wells, L., Weaver, C.F. and King, P.S. 1991. Cotton as a rotation crop for the management of *Meloidogyne arenaria* and *Sclerotium rolfsii* in peanut. Supplement of Journal of Nematology, **23** : 652-657.

118. Salako, E. A. and Adu, J. K. 1990. Effect of foliar fungicides on foliar disease and pod, haulm and plant crude protein yield of groundnut. Tropical Pest Management, **36** : 380-383.

119. Sakhuja, P.K. and Sethi, C.L. 1985. Screening of groundnut germplasm for resistance to root-knot nematode *Meloidogyne javanica.* Indian Journal of Nematology, **15** : 129-130.

120. Sathiyanarayanamurthy, G. V., Ramasubbiah, K., Rama Subba Rao, K. and Krishnamurthy, M. M. 1988. Efficacy of certain systhetic pyrithroids and in combination against thrips and leaf miner of groundnut. Journal of Agricultural Research, Andhra Pradesh Agricultural University, **16** : 16-19.

121. Saxena, R. C. 1989. Insecticides from neem. In: Insecticides from neem Eds. J. T. Arasan., J. R. Philogene and P. Morand, ACS Symposium Series, 387, Washington, DC, pp. 110-135.

122. Schemutterer, H. 1990. Properties and potential of natural pesticides from the neem tree *Azadirachta indica.* Annual Review of Entomology, **35** : 271-297.

123. Shanmugam, N. and Govindasamy, G. V. 1973. Control of Marcrophomina root rot of Groundnut. Madras Agriculture Journal, **60** : 500-503.

124. Sharma, S. B. 1988. Survey for nematode diseases of groundnut in some coastal and Rayalaseema districts of Andhra Pradesh, India. International Arachis News Letter, **3** : 11-13.

125. Sharma, R. C. 1982. Studies on Groundnut Root rot caused by *Macrophomina phaseolina* (Tessi) Goid (=*Rhizoctoinia bataticola* (Taub) Butler. Ph. D. Thesis, IARI, New Delhi.

126. Sharma, S. B. and McDonald, D. 1990. Global status of nematode problem of groundnut. Pigeonpea, chickpea, sorghum and pearlmillet and suggestions for future work. Crop Protection, **9** : 453-458.

127. Sharma, S. K. and Agarwal, D. K. 1988. *Acremonium salmoneum* Games and Lodha-A new hyperparasite on *Puccinia arachidis* Speg. Current Science, **57** : 499.

128. Sharma, S. B., Subramanyan, P. and Starr, E. 1990. Plant parasitic nematodes associated with groundnut in niger. Tropical Pest Management, **36** : 71-72.

129. Shekhawat, P. S., Patel, V. N. and Patel, J. G. 1985. Economical fungicidal spray schedule for control of tikka and rust diseases of rainfed groundnut. Indian Journal of Mycology and Plant Pathology, **17** : 11-16.

130. Shokes, F. M. and Gorbet, D. W. 1990. Commercial fungicide use for leaf spot control on partially resistant peanut *(Arachis hypogaea* L.) cultivar. Plannary Proceedings of Soil and Crop Science Society, **50** : 26-28.

131. Shivanarayan, N. 1985. Bird pests of national importance. In: Integrated Pest and Disease Management, Ed. S. Jayaraj, pp. 258-289.

132. Siddaramaiah, A. L., Desai, S. A., Hegde, R. K. and Jayaramaiah, H. 1980. Effect of different dates of sowing of groundnut on rust development in Karnataka. Proceedings of Indian National Science Academy (B), **46** : 389-392.

133. Siddaramaiah, A. L., Prasad, K. S. K. and Hegde, R. K. 1979. Effectiveness of seed dressing chemicals against crown rot of groundnut. Pesticides, **13** : 28-29.

134. Sivasubramanian, P. and Palaniswamy, G. A. 1986. Loss in yield due to leaf hopper and thrips. Madras Agricultural Journal, **73** : 530-531.

135. Smith, O.D., Boswell, T. E. and Thomes, W. H. 1978. Lesion nematode resistance in peanuts. Crop Science, **18** : 1008-1011.

136. Smith, D. H. and Littrell, R. H. 1980. Management of peanut foliar diseases with fungicides. Plant Disease, **64** : 356-361.

137. Srinivasan, S. D. V., Siva Rao and Sundrarajan, M. 1986. Influence of crop management practices on incidence of leaf miner *(Aproaerema modicella* dev.) in groundnut. Journal of Agricultural Research APAU, **33** : 381-382.

138. Starr, J. L. 1984. Expression of resistance in peanut, *Arachis hypogaea* to *Pratylenchus brachyrus:* Impact on screening for resistance. Journal of Nematology, **16** : 404-406.

139. Subrahmanyan, P. and McDonald, D. 1983 Rust disease of groundnut. Information Bulletin No. 13. International Crops Research Institute for the Semi-Arid Tropics, Patnacheru, AP, India.

140. Subrahmanyan, P. and McDonald, D. 1987. Groundnut rust disease : Epidemiology and control. ICRISAT (International Crops Research Institute for the Semi-Arid Tropics), 1987. Groundnut rust disease proceedings of Discussion Group Meeting, 24-28, September, 1984, ICRISAT, Center, Patnacheru, AP, India.

141. Subramahmanyan, P., Rao, V. R., McDonald, D., Moss, J. P. and Gibbons, R. W. 1989. Origin of resistance to rust and late leaf spot in peanut *(Arachis hypogaea* L.). Economic Botany, **43** : 444-455.

142. Subrahmanyan, P. and Ravidranath, V. 1988. Fungal and Nematode Diseases. In: Groundnut, Ed. P.S. Reddy, pp. 453-507.

143. Subrahmanyan, P., Reddy, D.V.R., Sharma, S.B., Mehan, V. K. and McDonald, D. 1990. A world list of groundnut diseases. Legumes Pathology Progress Report 12. Patancheru, AP, India.

144. Subrahmanyan, P., Reddy, D.V.R., McDonald, D. 1990. Parasitization of rust, early and late leaf spot pathogens of peanut by *Verticillium lecanii,* Peanut Science, **17** : 1-4.

145. Upadhayay, V. R, and Vyas, H. N. 1987. Integrated pest control of major pest of groundnut in Gujarat in relation to population level and Coccinellid predators population. Proceedings of National Symposium on Integrated Pest Control-Progress and Perspectives. 15-17 October, 1987, Trivandram, Kerala.

146. Upadhayay, V. R. and Vyas, H. N. 1989. Determination of phosphamidon residues in groundnut. Pesticides, **23** : 331-332-335.

147. Usman, K. M., Durairaj, P., Karunanithi, K. and Thangavelu, S. 1991. Effect of neem products on the control of foliar disease of groundnut. Poster Abstract: Second International Groundnut Workshop, 25-29, November, 1991. ICRISAT, Patancheru, AP, India.

148. Varaprasad, K. S. and Sharma, S. B. 1990. First report of the lesion nematode, *Pratylenchus bachyurus* on groundnut in India. Indian Journal of Plant Protection , **18** : 140.

149. Venkatarao, A. and Alice, D. 1991. Neem extract checks plant diseases. Capart Press Clippigs, **7**: 23.

150. Vidhyasekaran, P. 1981. Control of rust and tikka disease of groundnut. Indian Phytopathology, **34** : 20-23.

151. Vyas, S. C., Prasad, K. V. V., Shastry, P. P. and Shukla, B. 1986. Field control of leaf spots and rust of groundnut by fungicides. Indian Phytopathology, **39** : 447-448

152. Wightman, J. A. and Amin, P.W. 1988. Groundnut pests and their control in the Semi-Arid Tropics. Tropical Pest Management, **34** : 218-226.

153. Yadav, C.P.C. 1985. White grubs and their control (Bulletin). Indian Council of Agricultural Research, Krishi Bhavan, New Delhi, pp. 15.

154. Yadav, D N., Patel, R. R. and Patel, R. C. 1987. Natural enemies of groundnuts leaf miner *Aproaerema modicella* Deveter (Lepidoptera : Glechiidae) and their impact in its infestation at Anand (Gujarat) GAU Research Journal, **13** : 13-16.

15 Early and Late Leaf Spots of Groundnut and their Management

M.P. Ghewande

CONTENTS

Integrated Pest and Disease Management
Rajeev K. Upadhyay, K.G. Mukerji, B.P. Chamola and O.P. Dubey (eds.)
APH Publishing Corporation, New Delhi-110002, India

ABSTRACT

Early leaf spot caused by *Cercopora arachidicola* and late leaf spot caused by *Phaeoisariopsis personatum* are economically important diseases all over the world. They can cause losses in pod yield of over 50%.

Symptoms and morphological characters have been described. Based on these characters, the two pathogens can be readily identified. Different aspects like, cultural characterstics, host range and physiological specialization, histopathology and pathogenesis, host parasite interaction, genetics of disease resistance, disease cycle and epidemiology and disease management strategies have been reviewed in detail and discussed. The prospects for biological control at the field level and use of plant products especially of neem origin in the control of leafspots are discussed. Research gaps and future priorities have been indicated.

1. INTRODUCTION

Groundnut is an important food legume of the world. It is grown in 20.09 million ha spread over 85 countries with the production of 22.59 million tonnes with an average productivity of 1.12 tonnes ha-^{1} (154). Though, India leads the world both in area (8.4 million ha) and production (7.5 million tonnes) of groundnut, the country ranks 8th in productivity (938 kg/ha).

Groundnut crop is prone to attack by diseases to a much larger extent than many other crops. The relative economic importance of diseases in different countries is variable depending upon the local cultivation practices, the environment and the cropping patterns. Some diseases are widely distributed and cause economic crop losses while others are restricted in distribution and are not considered to be economically important at the present time. Among the fungal foliar diseases, early and late leaf spots are the most serious diseases of groundnut on a world-wide scale. The early leaf spot is caused by *Cercospora arachidicola* Hori, while the late leaf spot is caused by *Phaeoisariopsis personata* (Berk. & Curt.) v. Arx. Worldwide, yield losses caused by these diseases range from 10% to over 50%, but vary considerably from place to place and between seasons (98).

In this review article, emphasis has been laid mainly on the distribution and economic importance, symptoms, pathogens and their characteristics, genetics of disease resistance, disease cycle and epidemiology and disease management through cultural, biological, host resistance and chemical methods.

2. DISTRIBUTION AND ECONOMIC IMPORTANCE

Both early and late leaf spots are commonly present wherever groundnut is grown (58, 98). However, the incidence and severity of each disease varies between localities and seasons and there can be both short and long-term fluctuations in their relative proportions (85, 98). The early leaf spot is predominant and wide spread in Brazil (18), Indonesia (187), Malwi (101), Mauritius (68), Rhodesia (30), Sierra leone (37), South Queenland (127), Southern U.S. (159), Taiwan (76). In India, early leaf spot is more prevalent in northern groundnut growing states (86). Recently, it started assuming serious proportion in southern and central states of India also (5, 52). The late leaf spot is more widespread in Greece (46), Israel (25), Japan (77), Philippines (138), Senegal (21). In India, late leaf spot is more severe in southern parts and central India (5, 50, 60, 168, 169). Jackson (70) reported that in Florida

(U.S.) late leaf spots were more prevalent than early leaf spots. In India also, late leaf spot is more severe than early leaf spot (5, 52).

Both early and late leaf spots cause damage to the plant by reducing the photosynthetic area, by intense lesion formation, and by stimulating leaflet abscission (45, 98). Bunting *et al.* (15) estimated that early and late leaf spot alone cause the loss of about 3 million tonnes of kernels per year. Losses in yield to the extent of 20 to 25% have been reported in Georgia alone, resulting in a monetary loss of $10 to $17 million annually (71, 103). About 60% reduction in yield of kernels due to *Cercospora* leaf spots has been reported in Nigeria (39). Whereas in other West African countries, a loss of about 15 to 20% in yield due to this disease is reported. McDonald *et al.* (98) reported that yield losses range from 10% to over 50% worldwide, but vary considerably from place to place and between seasons.

In India, losses in yield due to the leaf spots have been estimated to be in the range of 15 to 59% (16, 22, 23, 54, 130, 176, 177, 179). In semi-arid tropics, where chemical control is rare, average losses exceeding 50% are quite usual (44). The losses caused in combination with rust are generally substantial (47, 57, 65, 98, 191). Cummins and Smith (33) reported that besides the loss in yield of kernel, the value of the fodder is also adversely affected.

3. SYMPTOMS

Symptoms described by McDonald *et al.* (98) are given below:

Symptoms first appear on lower leaves as small chlorotic spots on the upper leaflet surface 10 days after infection. The spots then develop in about 5 days into mature, sporulating lesions.

Lesions caused by *C. arachidicola* are subcircular in shape and are from 1 to over 10 mm in diameter. They are dark brown in colour on the upper leaf surface where most sporulation occurs, and a lighter shade of brown colour on the lower leaflet surface.

Lesions caused by *P. personata* are usually smaller, mostly circular in shape, and darker in colour than those of *C. arachidicola.* On the lower leaflet surfaces, where most sporulation occurs, the lesions are black in colour with a lightly rough appearance.

A chlorotic halo is often present around *C. arachidicola* lesions, but its presence and prominance is altered by host genotype and environmental factors. Similar yellow halos may be found around P. *personata* lesions. Therefore, yellow halo is not a distinguishing character.

Useful characters for distinguishing between the early and late leaf spots in the field are : The colour of the lesion on the lower leaflet surface caused by *C. arachidicola* is light brown while it is black for *P. personata.* The distribution of fruiting structures is random on the upper leaflet surface in the case of *C. arachidicola* while it is in circular rings on the lower leaflet surface for *P. personata.*

In addition to causing leaf spots, the two pathogens also produce lesions on petioles, stems and pegs. These are oval to elongate and have more distinct margins than the leaflet lesions. On the severity of both the diseases, the affected leaflets first become chlorotic, then necrotic, lesions often coalesce, and leaflets are shed.

4. CAUSAL AGENT

4.1 Pathogens

Cercospora arachidicola Hori (Perfect stage : *Mycosphaerella arachidis* Deighton) and *Phaeoisariopsis personata* (Berk. & Curt.) v. Arx.

4.2 Morphological and cultural characteristics

The distinguishing morphological features of early and late leaf spots pathogens are given in table bused on the descriptions given by Kolte (85) and McDonald *et al.* (98).

Feature	*C. arachidicola (Mycosphaerella arachidis)* Deighton	*P. personata (Mycosphaerella berkeleyi* W.A. Jenkins.) (=*C. personate)*
Mycelium	Septate, initially intercellular, then becomes intracellular, no haustoria are produced.	Septate, entirely intercellular sending haustoria into the pallisade and mesophyll cells.
Stromata	Present, up to 100 μm in diameter, dark bronw in colour.	Stroma dense, globular, Pseudoparenchymatous, brown to black, upto 130 μm in diameter
Conidiospores	Mostly epiphyllous, emphigenous when old, arise from a stroma fasciculate, geniculate, 5 to many in number pale olivaceous or yellowish brown, darker at the base, continuous to several septate, unbranched, 15-45x3-6 μm in size.	Mostly hypophyllous, sometimes amphigenous, when old, arise in more or less distinctly concentric tufts from heavy stromatic bases, numerous, fasciculate, geniculate, pale to olivaceous brown with mostly hyaline tips, smooth, continuous or one to several septate, 10-100x3.0-6.5 μm in size-, conidial scars conspicuous, prominant, 2-3 μm wide.
Conidia	Subhyaline, slight olivaceous, obclavate, often curved, 3-12 septate, base rounded to truncate, tip subacute, 35-110x2.0-5.4 μm in size.	Conidia medium olivaceous, cylindrical, obclavate, usually straight or slightly curved, wall usually finely roughened, rounded at the apex, base shortly tapered with a conspicuous hilum, 1-9 septa not constricted, mostly 3-4 septate, 20-70x4-9 μm in size.
Perithecia	Scattered, mostly along lesion margins, emphigenous, partly embedded in host tissue, erumpent, ovate to nearly globoze, 47.6-84.0x44.4-74.0 μm in size, black in colour ostiole slightly papillate.	Same as described for *M. arachidis,* but the size is 84-140x70-112 μm.
Asci	Cylindrical, club-shaped, short stipitate, faciculate, aparaphysate, bitunicate, 8-spored, 27.0-37.8x7.0-8.4 μm in size.	Same as described for *M. arachidis* but the size is 30-40x4-6 μm.

Ascospores	Uniseriate to imperfectly biseriate in ascus, bicellular, the upper cell somewhat larger, slightly curved, hyaline, 7.0-15.4x3-4 μm (average 11.2x3.64 μm) in size.	Same as described for *M. arachidis* but the size is 10.9-19.6x2.9-3.8 μm (average 14.9x3.4 μm in size)

Cultural characteristics

Both pathogens grow and sporulate sparingly on culture media. However, it has been reported that the yeast extract agar (91, 125, 142), Oatmeal agar (2, 111) and glucose-asparagine-agar with thiamine (91) support good growth of these fungi. To obtain good growth, the medium should be inoculated with these fungi while it is warm. After inoculation, medium must be shaken at 30°C. Good fungal growth can also be obtained by the addition of Cu, Mn, Zn and Fe @ 1 μm/ml in the medium (125). Light is essential for sporulation of *P. personatum* but not for *C. arachidicola.* Moraes and Salgado (111) reported that *C. arachidicola* produces maximum conidia if its culture is exposed to light for eight days, followed by darkness of six days. Carrot agar medium has been reported to be the best for conidial production of *P. personatum* (125). Chohan (22) reported that both the fungi grow at an optimum temperature of 24 to 28°C and the thermal death point lies between 51 to 52°C.

4.3 Host range and physiological specialization

No authentic host species are known outside the genus *Arachis.*

There is some evidence of variation between isolates of leaf spot pathogens (34, 35, 76, 105, 178) but the races have not been clearly characterised. In India, two distinct races, (race 1 and 2) of *C. personatum* have been recognized on the basis of cultural and morphological differences and on the basis of length of incubation period and size, shape and colour of the leaf spots. Race 1 shows 15 to 16 days incubation period after inoculation to symptom development, whereas the incubation period for race 2 is reported to be only 8 to 10 days (104).

4.4 Histopathology and pathogensis

Under ideal conditions conidia germinate by producing one or more germ tubes within 3 to 8 hours. The germ tubes penetrate the leaf directly through epidermal cells or through stomata. Hydrotropic response in stomatal penetration is indicated (31). In both species the mycelium is intercellular at first, branched or botryose haustoria are formed by the mycelium of *C. personatam* and the host cells are not killed outright. *C. arachidicola* later becomes intracellular and does not form haustoria. The host cells are killed in advance as the hyphae penetrate the dead cell (197). Visutipat and Sommartya (190) reported that *P. personata* took less than 86 hr. to complete the penetration in both susceptible and resistant host foliar tissue. During the incubation period, hypha caused them to collapse while that of the resistant lines only few cells were damaged. Both direct and indirect penetrations of germ tubes into the host tissue were observed. The germ tubes pierced directly into the intercellular spaces through the cuticle layer of the epidermal tissue. Histopathological changes in groundnut leaves due to infection by *C. arachidicola* were observed to be tissue hypertrophy leading to increased leaf thickness number and length of palisade cells, and diameter of spongy cells. Inter cellular spaces were reduced in spongy tissues and the number and size of starch grains gradually decreased in palisade and spongy cells. Polysacharide grains were gradually used up by the pathogen. Proteinoplasts reduced in number and size and stained less densly infected leaves. They then became mishapen and finally disintegrated. RNA content of mesophyll cells declined (193).

4.5 Host parasite interaction

4.5.1 Physiology and biochemistry

Production of Cercosporin toxin by *C. personata* in infected groundnut leaves and its culture filtrate has been reported by Venkataraman (188). A glycopeptide nonspecific toxin has been isolated from culture filtrate of *C. personata* and from its infected host plant (129). But the exact role of these enzymes and toxins in pathogenesis is not yet known. Formation of an abscission layer resulting in defoliation due to infection is attributed to production of an ethylene-like gaseous substance (82, 180) or to some other unknown toxin. The indirect role of sugars (10), riboflavin, pyridoxine and thiamine (132) have been indicated in pathogenesis. It has been reported by Purkayastha and Mukherji (126) that 44% decrease in starch and 26% increase in total nitrogen may result from infection of *C. personata.* The severity of the disease caused by *C. personata,* at flowering stage is reported to be associated with decrease of complete disappearance of amino-acids like cystine, glutamic acid, aspartic acid, serine, glycine, threonine, alanine and histidine (115). Bourgeois and Boote (14) reported that photosynthesis was linearly reduced with increasing percentage of necrotic leaf area. A 1% increase in necrotic leaf area contributed to a reduction of 4% in the carbon exchange rate. Canopy photosynthesis was inversely proportional to total disease severity, which is an expression of both defoliation and necrotic leaf area. Ramanujan (131) reported that groundnut leaves inoculated with *C. personata* showed higher electrolyte leakage and oxygen uptake than uninoculated controls. These effects were compared with those caused by the toxins, glycopeptide and dothistromin both produced by the fungus. Glycopeptide stimulated oxygen uptake and increased electrolyte leakage from treated groundnut tissues. Dothistromin affected membrane permeability only. The effects of the toxins were less pronounced on infected leaves than on healthy tissue. Only dothistromin could be detected in infected tissues. He concluded that the glycopeptide may be involved in both the physiological responses in the early stages of the disease while dothistromin influences symptoms late in disease development only.

4.5.2 Genetic basis of resistance

It is generally accepted that race-specific (qualitative) resistance governed by major genes (monogenic) is nondurable and face non-specific (quantitative) resistance governed by minor genes (digenic/polygenic) is durable. Whether it is a case of deriving horizontal resistance or vertical resistance, knowledge of nature of pathogens is pre-requisite for rational hybridization programme.

Resistance to leaf spots in groundnut was found recessive and independently inherited (1, 145, 146, 157). Sharif *et al.* (146) reported that the resistance to leaf spots is controlled by two or more nuclear genes and quantitatively inherited. Nevill (116) in his study on resistance to foliar diseases reported that in the F_2 progenies from a cross between Robut 33-1 (Kadiri 3) and Krapovickas 16, a wide range of reactions to *C. arachidicola* and *C. personata* was observed and all components of resistance were inherited quantitatively. Later in 1982, Nevill studied the inheritance of resistance to *C. personata* and reported multiple recessive genes and non-additive gene action. Kornegay *et al.* (89) studied the inheritance of resistance to *C. arachidicola* and *C. personata* in six virginia type peanut lines (NC 3033, NC 5, NC 2, NCAc 3139, Florigiant and NC-GP 343) and reported that resistance to these pathogens is quantitatively inherited. Norden (116) has opined that resistance to leafspots may not be a dominant trait and the disease reaction may be inherited quantitatively. Company *et al.* (32) reported that resistance to early leaf spot in *A. chacoense* and *A. cardenasil* behaves like a dominant trait in interspecific hybrids. Tiwari *et al.* (185) reported that segregation for late leaf spot resistance in the F_2 populations was in the ratio of 1 resistant : 15 susceptible and resistance was found to be recessive and under bigenic control. Resistance to leafspots is influenced by cytoplasmic factors and resistance was due to additive gene effects (28, 29). Resistance to early and late leaf spots is independently inherited (6). Resistance to leafspots is due to additive gene effects (7, 62, 75, 89).

Soriano (162) reported that resistance to leafspots is recessive and controlled by 2 duplicate genes in gamma-ray induced resistant mutants. Anderson *et al.* (8) reported that the genetic nature of resistance to early and late leaf spots of groundnut is quantitative.

5. DISEASE CYCLE AND EPIDEMIOLOGY

As the common names imply, an attack by *C. arachidicola* normally precedes that of *P. personata* (130). But both diseases may appear within 3-5 weeks after sowing (98). Kucharek (90) reported that leaf spot attack is more serious where groundnut follows groundnut. Both the pathogens survive through conidia on affected plant debris in soil or through conidia being carried on the pod shell (43). The principal source of initial inoculum is probably conidia that are produced on groundnut crop residue in the soil (173). The conidia of the pathogens have sufficient longevity to carry over from one crop to another (41, 79). The pathogens may also survive from one season to another on volunteer groundnut plants and on 'ground keepers' (67, 134). It has been opined by some workers (67, 79, 134, 139, 143) that pathogens (mycelia/spores) survive in the soil and in plant debris. The production of perithecia by both the fungi on overwintering material has been studied in the U.S. (73). However, experimental data to support the possible role of either ascospores or mycelia as sources of initial inoculum are lacking (173). The perfect stages of the pathogens have not been reported so far from India. However, the role of ascospores, if formed, in initiating the outbreak can not be ruled out.

Higgins (69) reported that inoculum gets brown or splashed onto leaves giving rise to primary infection. The first lesion normally develop on the older leaves near ground level and conidia produced on them are disseminated by wind, rain splash and insects leading to secondary infections.

Studies on the aerobiology of leaf spot pathogens have shown that a diurnal periodicity exists with peak catches of conidia at dew dry-off in the morning (48, 102, 159, 164). The vertical dissemination of *C. arachidicola* to heights of 2.75m above the soil surface has been reported (159).

Lyle (95) observed the maximum number of conidia during a period of abundant rainfall and high minimal (22°C) and maximal (35°C) temperatures. Temperature in the 25 to 30°C range and high relative humidity favour infection and disease development (98). Jensen and Boyle (74) reported that severity of leaf spots can be correlated to periods of high relative humidity above 20°C. Kao and Wu (76) opined that duration of leaf wetness is another important factor in leaf spot infection. Butler and Wadia (17) reported that for late leaf spot disease, infection is greatest if leaf wetness is intermittent. After inoculation, lesions result only if the total wetness period exceeds beyond 80 hr. Maublanc (96) from Senegal and Kenknight (80) from U.S.A. attributed the rapid spread and severity of leaf spots to heavy rainfall in August-September and in spring in their respective countries. Tarr (184) observed that leaf spots were most prevalent in wet areas of Sudan with annual rainfall exceeding 500-620 mm. Davis *et al.* (36) reported that 1-day rainfall of 2.54 mm or more is an easily measured environmental variable that may be used to schedule fungicide sprays for managing peanut late leaf spot. In India, various workers (48, 177, 189, 196) have studied the leaf spot development in relation to climatic factors viz., temperature, relative humidity and rainfall.

In general, the leaf spot severity is more in kharif (rainy season) than in the rabi/summer (post-rainy season) (50, 136). The late leaf spot, in general, is more severe than the early leaf spot.

No authentic host species outside the genus *Arachis* were reported. Long distance distribution of the disease may be by movement of infected crop debris, or by movement of pods or seeds contaminated with conidia or infected crop debris. There is no evidence of disease being internally seed-borne (173). Rao *et al.* (134) reported that no collateral hosts for *C. arachidicola* and *P. personata* were found among many leguminous crop and weed plants.

6. DISEASE MANAGEMENT

6.1 Cultural

Deep burying of crop residues in the soil, destruction of crop debris by burning, removal of volunteer groundnut plants and 'ground keepers' is important in reducing the primary source of infection (38, 67, 156). One year crop rotation either with soybean or maize alternating with peanut or by not following peanut with peanut has been found to reduce the incidence of the disease early in the season by 88 to 93% (67, 81, 90, 97). The relationship between disease severity and yield increase at different spacings is reported. Chevaugeon (20) found an inverse relationship between in-row spacing of plants and intensity of late leaf spot. Ghewande *et al.* (49) reported that early planting (15 days early than normal), plant spacing of 45x10 cm and intercropping with pigeonpea and castor reduced the late leaf spot and rust by 7 to 21%. Leafspot incidence could be reduced by early planting of groundnut (136). Early planting (June, 28) and wider spacing (30-40 cm) were found to reduce tikka and rust (83). Intercropping of pearlmillet and sorghum with groundnut resulted in lowest infection of late leaf spot pathogen *(P. personata)* and *Puccinia arachidis* respectively (114). Ganpathy and Narayanasamy (42) reported that the incidence of ring mosaic, late leaf spot and rust was reduced by intercropping with sorghum and cumbu *(Pennisetum glaucum)* but not sunflower. The early leaf spot was very early and the development of the disease was rapid when groundnut followed groundnut. In groundnut followed by maize or pasture, the disease development was slower and less severe. Sowing dates also influenced the severity of early leaf spot (167).

6.2 Biological

Mycoparasites, *Dicyma pulvinata* (Berk. & Curt.) v. Arx (=*Hansfordia pulvinata* (Berk. & Curt.) Hughes (9, 106, 108, 123, 149, 151, 181, 182, 183) and *Verticillium lecanii* (Zimmerm) Viegas have been observed to parasitize leaf spot pathogens (166). These mycoparasites were found to be effective in controlling leaf spots mostly in green house studies. However no serious attempts have been made to use them at the field level (98).

Mitchel and Taber (107) reported that maximal colonization of *C. personata* by *D. pulvinata* occurred at 23-28°C under laboratory condition. The colonization of lesions of *C. personata* by *D. pulvinata* occurred within 58-65 hr. with 21-31.5 hr. leaf wetness. The mycoparasite was effective when plants were exposed to continous leaf wetness at 26°C for 5 days. In field microplot studies lesions were colonized by both mutant and wild type isolate of *D. pulvinata* within 4 days under 40 hr leaf wetness, 23-28°C temperature and 17-31 cm rainfall conditions (109). Harvey and Bailey (66) reported that *Bacillus thuringiensis* and *Pseudomonas cepacia* completely controlled early and late leaf spot pathogens in laboratory tests and provided upto 70% disease control in small scale field tests. A formulation of *B. thuringiensis* gave 74% control of leaf spots and significantly increased peanut yield.

Subrahmanyam *et al.* (174) reported that there was a significant reduction in the extent of rust and leaf spots development on peanut leaves inoculated with *V. lecani* in the glasshouse. Studies carried out at the NRCG, Junagadh showed that *D. pulvinata* and *V. lecani* and their culture filtrates inhibited the *in vitro* spore germination of *P. personata* by varying degrees (33-75%). These mycoparasites and their culture filtrates reduced *in-vitro* development of late leaf spot significantly. The spray application of culture filtrate of both the mycoparasites reduced the leaf lesions by 37.5 to 50% and lesion size by 16.8 to 37.4% of late leaf spot and gave higher pod yield of 17g/plant (50). In 2 of 3 field trials, significant reductions in the severity of early leaf spot were observed due to *Bacillus cereus* (strain 304) on chitin amended leaves (84).

6.3 Host resistance

Practical importance of host resistance is that it gives us a tool to fight diseases, or in other words, control diseases in farmers' fields. Several organizations have started breeding programmes to evolve genetically resistant genotypes of groundnut for minimising losses due to diseases. However, no considerable momentum has so far been taken in obtaining genotypes with high levels of resistance and desirable agronomic traits as genetic resistance to foliar diseases caused by *C. arachidicola* and *P. personata* has usually been associated with low yields and late maturity (120). Interestingly, there are many promising sources of resistance to early and late leaf spot diseases across the habit groups of groundnut and also in many wild *Arachis* species (3, 13, 19, 24, 27, 40, 59, 64, 88, 98, 99, 100, 110, 112, 113, 138, 141, 147, 155, 156, 163, 165, 169, 170, 171, 172). It is not difficult to obtain lines with moderate levels of resistance and produce reasonably high yields with minimal or no fungicidal applications (61). At ICRISAT, Hyderabad, a world collection of over 12,000 germplasm lines has been successfully screened in the field against rust and late leaf spot during the period 1977 to 1989 and 124 rust-resistant, 54 late leaf spot resistant and 29 rust and late leaf spot-resistant germplasm lines were identified (175). Out of 54 late leaf spot-resistant lines, 10 are highly resistant to late leaf spot disease. They are : EC 76446(292), PI 270806, PI 215696, PI 341879, PI 393516, PI 476172, PI 215696, PI 476164, PI 476195 and PI 393530. Ghewande *et al.* (53) reported 5 cultures viz., NCAc 927, NCAc 17133 (RF), NCAc 17149, PI 341879 and PI 393646 possessing resistance to early leaf spot. Genotypes resistant to early and late leaf spot available from ICRISAT are being used extensively in the hybridization programmes in the country and elsewhere. Several advanced breeding lines possessing resistance to these diseases, coupled with high yield are in pipe line. Two advanced breeding lines, viz., PPS 1 (R 33-1 x EC 76446(292) and PPS 5 (JL 24 x NCAc 17090) possessed multiple disease resistance, good quality attributes with reasonably high yields (55).

In India, 8 spanish bunch varieties viz., DH 8, TG 3, MA 16, Girnar 1, ICG(FDRS)10, ICGS 37, ICGV 86590 and GAG 24 having resistance/tolerance to late leaf spot have been released for general cultivation (137).

Resistance to leaf spot pathogens has been attributed to various anatomical and morphological characteristics of host plant (59, 67) and to different chemical constituents of the host leaves and seed.

6.4 Chemical

Several systemic and non-systemic antifungal compounds were tried for the control of early and late leaf spots in different parts of the world. Dust formulations (Copper, sulphur and copper plus sulphur) were the most commonly used fungicides upto the late 1960s although a number of spray fungicides viz., Bordeaux mixture and the dithiocarbamates (maneb and mancozeb) were also used (173). Benomyl, chlorothalonil and fentin hydroxide, the highly effective fungicides were also used in the early 1970s (160). Benomyl was very effective against leaf spots but it was not effective against rust. Since benomyl was used extensively for several years in the U.S.A., the fungicide tolerant strains of the leaf spot pathogens were developed in nature (26, 78, 94, 161). Chlorothalonil is the most widely used leaf spots fungicide in the USA. It is also very effective against rust, and some minor foliar diseases (98). However, Porter (122) found that spraying with chlorothalonil or with captafol increased levels of sclerotinia blight. Backman *et al.* (12) found an increase in levels of *Sclerotium rolfsii* Sacc. attack when Florunner groundnuts were sprayed with benomyl. Therefore, possible effects of fungicides on nontarget organisms should be considered before recommending any fungicides. In Bangladesh, Rahman *et al.* (128) have recommended carbendazim during 40 to 70 days of plant age to manage leaf spots.

In Niger at 3 locations (Sadore, Maradi and Bengon) carbendazim and chlorothalonil fungicides were found to be effective in controlling early and late leaf spots and substantially increased the pod and haulm yields during 1987 and 1988 crop season (166). Rudolph *et al.* (140) reported that febuconazole (Folicur) application at rates of only 0.187 to 0.25 kg a.i./ha as part of standard foliar spraying was extremely effective in controlling early and late leaf spot diseases for several years in the U.S.A.

Several fungicides have been evaluated for the control of leaf spot diseases in different parts of India (11, 87, 93, 133, 150, 152, 153, 191, 192, 194). Lalithakumari *et al.*, (92) reported that among the fungicides tested, baycor gave an excellent control of leaf spot. Rattan and Kang (135) reported that carbendazim and benomyl were effective in controlling leaf spots. Five sprays of carbendazim (0.05%) + mancozeb (0.2%) at 15 days interval were the best for the control of leaf spots in the late sown crops in Manipur (63).

Several fungicides and their combinations have been tested for the control of leaf spots and rust since rust and leaf spots often appear together (121). Shekhawat *et al.* (148) reported that the spray schedule consisting two sprays of mancozeb (0.2%) at 35 and 70 days after germination and one spray of carbendazim (0.025%) at 60 days after germination was most economical with a ICBR of 1:2:85 for the control of leaf spots and rust. Vyas *et al.* (195) recommended that carbendazim (0.075%) + mancozeb (0.15%) mixture should be sprayed around August 15 and second and third spray should be repeated depending on the intensity of leaf spots and rust diseases. The rust and leaf spots were substantially reduced due to the application of carbendazim (0.025%) + tridemorph (0.04%) five times at fortnightly intervals commencing 35 days after sowing during summer season (72). The leaf spots and rust are controlled effectively by the spray application of carbendazim (0.05%) + mancozeb (0.2%) at 2-3 weeks interval, 2 or 3 times starting from 4-5 weeks after planting (CBR-1:14.8 to 1:24.4) (4). Sekhar and Sekaran (142) reported that five rounds of spray, starting from 30 days after sowing, with 12 day intervals using carbensazim 0.05% plus mancozeb 0.2% was found to give maximum control of rust and late leaf spot which resulted in 57% increased dry pod yield. Foliar application of chlorothalonil (0.1%) effectively controlled late leaf spot and rust (186).

Number of plant products exhibit disease control properties. 'Neem' (*Azadirachta indica* A. Juss) its one such species which is now commanding global attention in the search for safer and more effective biopesticides. The spray of 2-5% neem leaf extract was found to be effective and economical (51, 55, 56). Foliar application of neem kernel extract (3%) reduced the incidence of late leaf spot and rust significantly and recorded the highest cost/benefit ratio (186). Greater use of these products should be encouraged becaused they are safe from the point of environmental pollution and pesticidal residues.

7. CONCLUSIONS

Both early and late leaf spots are commonly present wherever groundnut is grown. Yield losses range form 10% to over 50% all over the world, but very considerably from plae to place and between seasons. In India, losses in yield have been estimated in the range of 15 to 59%. Early and late leaf spots pathogens are both soilborne and air-borne. Temperatures in the range of 25 to 30° and high relative humidity favour infection and disease development.

The pathogens may survive from season to season on volunteer groundnut plants and infected crop debris. No authentic host species are known outside the genus *Arachis*. Long-distance distribution of the pathogens may be by airborne conidia, by movement of infected crop debris or by movement of pods or seeds that are surface contaminated with conidia or infected crop debris.

There is some evidence of variation between isolates of leaf spots pathogens but the races have not been clearly characterised. Therefore, it is necessary to make an indepth study on race flora, if any, or leaf spots pathogens since the knowledge of pathogen variability is pre-requisite for rational hybridization programme. Resistance to leaf spots is recessive and independently inherited.

A large number of germplasm lines resistant to leaf spots have been identified and are being used extensively in hybridization programme. But there is a need to look for sources of multiple disease resistance and to concentrate on "germplasm enhancement" to obtain higher level of resistance to these diseases. A very few agronomically acceptable groundnut varieties with resistance to leaf spots are available. Therefore, serious attempts have to be made to develop cultivars with multiple disease resistance. A very little information on biological control of leaf spots at the field level is available. Additional information on this aspect needs to be generated for formulating integrated disease management programme in groundnut. Direct control measures through chemicals are available. Neem products hold promise for potential biofungicides.

A disease forecasting system need to be developed to forewarn groundnut growing farmers of epidemics of leaf spot diseases. Efforts are required to be made in utilizing all available and compatible disease control measures in an integrated way wherever possible in increasing groundnut production.

8. REFERENCES

1. Abdou, Y.A.M. 1966. "The source and nature of resistance in *Arachis* species to *Mycosphaerella arachidicola* Jank. and *Mycosphaerella berkeleyi* Jenk., and factors influencing sporulation of these fungi". Ph.D. Thesis, North Carolina State University, Raleigh.

2. Abdou, Y.A.M. 1967. The source and nature of resistance in *Arachis* L. species to *Mycosphaerella arachidicola* Jenk. and factors influencing sporulation of these fungi. Diss. Abstr., **27** : (6) B, 1689.

3. Abdou, Y.A.M., Gregory, W.C. and Cooper, W.C. 1974. Source and nature of resistance to *Cercospora arachidicola* Hori and *Cercosporidium personata* (Berk & Curtis) Deighton in *Arachis* species. Peanut Science, **1** : 6-11.

4. Anon. 1981-82. Annual Progress Report, Kharif Oilseeds, All India Coordinated Research Project on Oilseeds, Directorate of Oilseeds Research, Rajendranagar, Hyderabad.

5. Anon. 1993. Annual Progress Report, Kharif Groundnut, All India Coordinated Research Project on Oilseeds, Directorate of Oilseeds Research, Rajendranagar, Hyderabad.

6. Anderson, W.F., Wynne, J.C. and Green, C.C. 1985. Potential for incorporation of early and late leaf spot resistance in peanut. Proceedings of the American Peanut Research and Education Society, **17** : 30 (Abstract).

7. Anderson, W.F., Wynne, J.C., Green, C. and Bente, M.K. 1986. Combining ability and heritability of resistance to early and late leaf spots of peanut. Peanut Science, **13** : 10-14.

8. Anderson, W.F., Holbrook, C.C., Wynne, J.C. 1991, Heritability and early generation selection for resistance to early and late leaf spot in peanut. Crop Science, **31** : 588-598.

9. Ashok Krishna and Singh, R.A. 1979. *Hansfordia pulvinata* mycoparasitic on *Cercospora* species causing "Tikka disease" of groundnut. Indian Phytopath. **32** : 318-320.

10. Aulakh, K.S. and Sandhu, R.S. 1970. Free amino acid and carbohydrate in healthy and tikka infected groundnut leaves. Indian Phytopath. **23** : 84.

11. Aulakh, K.S. and Chahal, A.S. 1973. Chemical control of leaf spot of groundnut. Indian Phytopath. **26**: 68-71.

12. Backman, P.A., Rodriguez-Kabana, R., Hammond, J.M., Clark, E.M., Lyle, J.A., Ivey, H.W. and Starling, J.G. 1977. Peanut leaf spot research in Alabama. Auburn University Bulletin No. 489.

13. Beute, M.K. and Hassan, H.N. 1977. Problems in evaluating resistance to peanut to *Cercospora arachidicola.* Proceedings of the American Peanut Research Education Association, **9** : 13 (Abstract).

14. Bourgeois, G. and Boote, K.J. 1992. Leaflet and canopy photosynthesis of peanut affected by late leaf spot. Agronomy Journal, **84** : 359-366.

15. Bunting, A.H., Gregory, W.C., Manboussin, J.C. and Rayon, J.G., 1974. A proposal for research on groundnuts *(Arachis)* by ICRISAT, ICRISAT Mimeo 1.

16. Butler, E.J. 1914. Tikka disease and the introduction of exotic groundnuts in the Bombay Presidency. Agric. J. India, **9** : 59.

17. Butler, D.R. and Wadia, K.D.R. 1992. Groundnut foliar disease, infection and leaf wetness in groundnut: a global perspective. Proceedings of an International Workshop, ICRISAT, A.P., India, 25-29 Nov. 1991 Ed. S.N. Nigam, International Crops Research Institute for the Semi-Arid Tropics, Patancheru, Andhra Pradesh, India, (Summary) pp. 475.

18. Castellani, E. 1953. Phytopathological problems in central-southern states of Brazil, Abstr. Rev. Appl. Mycol., **32** : 616.

19. Chahal, A.S. and Sandhu, R.S. 1972. Reaction of groundnut varieties against *Cercospora personata* and *C. arachidicola.* Plant Disease Reporter, **56** : 601-603.

20. Chevaugeon, J. 1952. Maladies des Plantes cultivees en Moyenne-casamance et dans le delta central Nigerien. Revue de pathologie vegetable at d'entomologie agricole de France, **31** : 3-51.

21. Chevaugeon, J. 1953. Researches on Cercosporiosis of groundnut in Middle. Casamance, Abstr., Rev. Appl. Mycol., **32** : 661.

22. Chohan, J.S. 1974. Recent advances in diseases of groundnut in India. In: current trends in Plant Pathology, pp. 171-184. Lucknow University Press, Lucknow, Uttar Pradesh.

23. Chohan, J.S. and Singh, T. 1973. Control of seed-borne pathogens in groundnut. Indian J. Mycol. Pl. Pathol., **3** : 1993-1994.

24. Chiteka, Z.A., Gorbet, D.W. and Shokes, F.M. 1988. Components of resistance to late leaf spot in peanut I. Levels and variability : implications for selection. Peanut Science, **15** : 25-30.

25. Chorin, M. and Minz, G. 1954. Leaf spots of peanuts. Hassadesh, **34** : 471.

26. Clark, E.M., Backman, P.A. and Rodriguez-Kabana, R. 1974. *Cercospora* and *Cercosporidium* tolerance to benomyl and related fungicides in Alabama peanut fields. Phytopathology, **64** : 1476-1477.

27. Coffelt, T.A., Porter, D.M., Smith, J.C. and Garren, K.H. 1977. Reaction of 45 peanut *(Arachis hypogaea* L.) lines to 5 fungal and insect pathogens. Proceedings of the American Peanut Research and Education Association, **9** : 33 (Abstract).

28. Coffelt, T.A. and Porter, D.M. 1982. Screening peanuts for resistance to Sclerotinia blight. Plant Disease Reporter, **66** : 385-387.

29. Coffelt, T.A. and Porter, D.M. 1986. Field screening of reciprocal Chico x florigiant peanut populations for resistance to leaf spot in virginia. Peanut Science, **32** : 57-60.

30. Cole, D.L. and Grimmer, C.J. 1977. The effect of foliar disease on groundnut yield, Abstr. Rhod. Agric. J., **74** : 19.

31. Cook, M. 1981. Susceptibility of peanut leaves to *Cercosporidium personatum*, Phytopathology, **71** : 787-791.

32. Company, M., Stalker, H.T. and Wynne, J.C. 1982. Cytology and leaf spot resistance in *Arachis hypogaea* x wild species hybrids. Euphytica, **31** : 885-893.

33. Cummins, D.G. and Smith, D.H. 1973. Effect of *Cercospora* leaf spot of peanuts on forage yield and quality and on seed yield. Agronomy Journal, **65** : 919.

34. Curzi, M. 1931. On *Cercospora* infections of *Arachis hypogaea.* Bolletino della stazlone di patologia vegetale, (N.S.) **11**: 84-87.

35. Das, G. 1951. 'Growth, spore germination and infection studies on a peanut leaf spot fungus. M.S. Thesis, Texas A. & M University, College Station, USA, p. 37.

36. Davis, D.P., Jacobi, J.C. and Backman, P.A. 1993. Twenty four hour rainfall, a simple environmental variable for predicting peanut leaf spot epidemic, Plant Disease, **77** : 722-725.

37. Deighton, F.C. 1953. Plant Pathology Section. Rep. Dep. Agric. S. Leone. 1951. 21.

38. De carvalho, T. and Mendis, O. 1953. A mancha castanha e a mancha preta do amendoim (*Arachis hypogaea* L.) Boletin da Sociedade Broteriana series 2, **27** : 155-162.

39. Elston, J., Harkness, C. and McDonald, D. 1976. The effects of *Cercospora* leaf disease on the growth of groundnuts (*Arachis hypogaea* L.) in Nigeria. Anon. Appl. Biol., **83** : 39.

40. Foster, D.J., Stalker, H.T., Wynne, J.C. and Beute, M.K. 1981. Resistance of *Arachis hypogaea* L. and wild relatives to *Cercospora arachidicola* Hori. Oleagineux, **36** : 139-143.

41. Frezzi, M.J. 1960. Enfermedades del mani en la provincia de cordoba (Argentina), Revista de investigaciones agricolas, **14** : 113-155.

42. Ganapathy, T. and Narayansamy, P. 1991. Effect of intercropping on the incidence of diseases of groundnut. Madras Agricultural Journal, **78** : 22-24.

43. Garren, K.H. and Wilson, C. 1951. Peanut disease. In: Peanut-The unpredictable legume, pp. 262-364. The National Fertilizer Association, Washington, D.C.

44. Garren, K.H. and Jackson, C.R. 1973. Peanut diseases. In: Peanuts-culture and uses, pp. 429-491. American Peanut Research and Education Association, Stone Printing Co., Roanoke, Virginia, USA.

45. Gerlagh, M. and Bokdam, M. 1974. Yield of groundnut in relation to the reduction of photosynthesis by *Cercospora* leaf spot or by its simulation. Abstr., Rev. Plant Pathol. **53** : 2008.

46. Georgopoulos, S.G. 1958. Some diseases of cultivated plants hitherto unreported from Greece. Ann. Inst. Phytopathol., Benaki, N.S. **1** : 334.

47. Ghewande, M.P., Shukla, A.K., Pande, R.N. and Misra, D.P. 1983. Losses in groundnut yields due to leaf spots and rust at different intensity levels. Indian J. Mycol. & Pl. Pathol., **13** : 125-127.

48. Ghewande, M.P. and Jhala, Raj. 1985. Epidemiology of foliar diseases of groundnut. Abstract Vol. Proceedings of 3rd International Conference on Aerobiology, August 6-9, 1986, Basel, Switzerland

49. Ghewande, M.P., Shukla, A.K. and Pande, R.N. 1986. Management of foliar diseases of groundnut through agronomic practices. Indian Bot. Reptr., **5** : 122.

50. Ghewande, M.P., 1989a. Biological control of late leaf spot (*Phaeoisariopsis personatum*) of groundnut (*Arachis hypogaea* L.) Indian J. of Agric. Scis., **53** : 189-190.

51. Ghewande, M.P. 1989b. Management of foliar diseases of groundnut *(Arachis hypogaea* L.) using plant extracts. Indian J. of Agric. Sci., **59** : 133-134.

52. Ghewande, M.P. 1990. Diseases of groundnut and their management. J. Oilseeds Res., **7** : 78-97.

53. Ghewande, M.P., Desai, S., Prem Narayan and Kamble, S.D. 1992. Some sources of resistance to early leaf spot of groundnut *(Arachis hypogaea* L.) in India. Trop, Agric., (Trinidad,) **69** : 284-286.

54. Ghewande, M.P., Desai S., Prem Narayan and Ingle, A.P. 1993. Evaluation of some new fungicides and plant products to manage foliar diseases of groundnut *(Arachis hypogaea* L.) Ed. Gangawane *et al.* In : Proceedings of the IVth Annual Zonal Meeting of IPS, Aurangabad, November 1-2, 1991. pp. 14-18.

55. Ghewande, M.P. 1993. Disease Resistance in Groundnut *(Arachis hypogeaea* L.). Editors, Mayee *et al.* In : Proceedings of the Vth Zonal Meeting of Western Zone, IPS, PKV, Akola, January 18, 1993, pp. 5-9.

56. Ghewande, M.P. 1994. Management of major foliar fungal diseases of groundnut *(Arachis hypogaea* L.) using aqueous leaf extract of "Neem" *(Azadirachta indica* A. Juss). In : Neem *(Azadirachta indica* A. Juss) for management of crop diseases, Ed. Mariappan.

57. Ghuge, S.S., Mayee, C. D. and Godbole, G.M. 1981. Assessment of losses in groundnut due to rust and tikka leaf spots. Indian Phytopath. **34** : 179-182.

58. Gibbons, R.W. 1966. Mycophaerella leaf spots of groundnuts. Food and Agricultural Organisation, Plant Protection Bulletin, **14** : 25-30.

59. Gibbons, R.W. and Bailey, B.E. 1967. Resistance to *Cercospora arachidicola* in some species of *Arachis*, Rhodesia, Zambia, and Malawi Journal of Agricultural Research, **5** : 57-59.

60. Gibbons, R.W. 1979. Groundnut Improvement Research Technology for the Semi Arid Tropics, presented at the Int. Symp. Development and Transfer of Technology for Rainfed Agriculture and the SAT Farmer, ICRISAT, Patancheru (India), August 1979.

61. Gorbet, D.W., Knauft D.A. and Shokes, F.M. 1990. Response of peanut genotypes with differential levels of leaf spot resistance to fungicide treatment, Crop Science, **30** : 529-533.

62. Green, C.C. 1985. Evaluation and inheritance of components of partial resistance to early leaf spot (*Cercospora arachidicola* Hori) in peanut. Ph.D. thesis North Carolina State University, Raleigh, NC, USA.

63. Gupta, D.K. 1985. Fungicidal control of tikka disease of groundnut in Manipur, Indian J. Mycol. & Pl. Pathol., **15** : 99-101.

64. Hamid, M.A., Isleib, T.G., Wynne, J.C. and Green, C.C. 1981. Combining ability analysis of *Cercospora* leaf spot resistance and agronomic traits in *Arachis hypogaea* L. Oleagineux, **36** : 605-612.

65. Harrison, A.L. 1967. Increasing peanut yields by cultural and chemical means. Plant Disease Reporter, **51** : 441-443.

66. Harvey Spurr and Bailey 1983. Biological control of peanut leaf spot. Proceedings or American Peanut Research and Education Society APRES, **15** : 93.

67. Hemingway, J.S. 1954. *Cercospora* leaf spots of groundnut in Tanganyika. East African Agricultural Journal, **19** : 263-271.

68. Hopkins, J.C.F. 1960. Plant diseases in British Colonial Dependencies. FAO Plant Prot. Bull., **8** : 109.

69. Higgins, B.B. 1956. Les maladies de l'arachide aux Etats-Unis, Oleagineux, **11** : 213-230.

70. Jackson, L.F. 1981. Distribution and severity of peanut leaf spot in Florida. Phytopathology, **71** : 324.

71. Jackson, C.R. and Bell, D.K. 1969. Diseases of peanut (groundnut) caused by fungi. University of Georgia College of Agriculture Experiment Station, Research Bulletin No. 56.

72. Jayasekhar, M. and Rajasekharan, R. 1986. Appraisal of yield loss in groundnut due to rust and late leaf spot. Madras Agric. J., **73** : 525-527.

73. Jenkins, W.A. 1938. Two fungi causing leaf spots of peanut. Journal of Agricultural Research, **56** : 317-322.

74. Jensen, R.E. and Boyle, L.W. 1965. The effect to temperature, relative humidity and precipitation on peanut leaf spot. Plant Disease Reporter, **49** : 975-978.

75. Jogloy, S., Wynne, J.C. and Beute, M.K. 1987. Inheritance of late leaf spot resistance and agronomic traits in groundnut. Peanut Science, **14** : 86-90.

76. Kao, C.Y. and Wu, L.C. 1970. The *Cercospora* leaf spots of peanut I. Early leaf spot and physiology of its causal organism *Cercospora arachidicola* Mem. Coll. Agric. Nat. Taiwan Univ., **11** : 90-100.

77. Katsuki, S. 1955. Parasitic fungus flora of Yaku island. Kyushu island. J. Jpn. Bot., **30** : 282.

78. Katan, T. 1980. Benomyl-resistant *Cercosporidium personatum* in peanut fields in the sharon region of Israel. Phytoparasitica, **8** : 213-215.

79. Karunakaran, P. and Raj, J.S. 1973. The survival of *Cercospora personata* (Berk. & Curt.) Ell. & Ev., *Cercospora hibisci* Tracy and Earle, and *Cercospora henningsii.* Allesch in the soil. Agricultural Research Journal of Kerala, **11** : 162-163.

80. Kenknight, G. 1941. Peanut diseases in certain Texas countries in 1941, with notes on occurrence of peanut rust. Plant Disease Reporter, **25** : 587-588.

81. Kenknight, G. 1942. Peanut diseases in certain South Texas countries. Plant Dis. Rep., **26** : 499.

82. Ketring, D.L. and Meloux, H.A. 1982. Ethylene production and leaflet abscission of three peanut genotypes infected with *Cercospora arachidicola* Hori. Plant Physiol., **69** : 789.

83. Kodmelwar, R.V. and Ingle, A.Y. 1989. Effect of sowing dates, spacing and meterological factors on the development of tikka and rust of groundnut. Indian Phytopath., **42** : 274.

84. Kokalis-Burelle, N., Backman, P.A., Rodriguez-Kabana, R. and Ploper, L.D. 1992. Potential for biological control of early leaf spot of peanut using *Bacillus cereus* and chitin as foliar amendments. Biological control, **2** : 321-328.

85. Kolte, S.J. 1984. Diseases of Annual Edible Oilseed Crops Vol. I Peanut Diseases, CRC Press, Inc. Boka Raton, Florida, p. 143.

86. Kolte, S.J. and Brahmachari, B.K. 1981. A comparative study of certain *Cercospora* leaf spot resistant and susceptible varieties of peanut. Abstr. 3rd Int. Symp. on Plant Pathol., New Delhi, December 14-18.

87. Kolte, S.J. and Sinha, A.P. 1976. The influence of soil and seed application of certain systemic fungicides on the severity of *Cercospora* leaf spot of groundnut. Pesticides, **10** : 31-32.

88. Kolte, S.J., Tewari, A.N. and Awasthi, R.P. 1977. Sources of resistance to *Cercospora* leaf spot of groundnut. Indian Phytopath. **30** : 421-422.

89. Kornegay, J.L., Beute, M.K. and Wynne, J.C. 1980. Inheritance of resistance to *Cercospora arachidicola* and *Cercosporidium personatum* in six virginia type peanut lines. Peanut Science, **7** : 4-9.

90. Kucharek, T.A. 1975. Reduction of *Cercospora* leaf spot of peanut with crop rotation. Plant Disease Reporter, **59** : 822-823.

91. Landers, K.E. 1964. Growth of *Cercospora arachidicola* in a glucose phosphate-aspargine-thiamine agar medium. Phytopathology, **54** : 1236.

92. Lalithakumari, D., Ganesan, T. and Nageswara Rao, M. 1984. Effect of systemic fungicides on the physiological response of groundnut plant against tikka leaf spot. Indian Phytopath., **37** : 111-114.

93. Lewin, H.D. and Natarajan, S. 1971. Fungicidal control of tikka leaf spot of groundnut in Tamil Nadu. Madras Agric. J., **58** : 480-487.

94. Littrell, R.H. 1974. Tolerance in *Cercospora arachidicola* to benomyl and related fungicides. Phytopathology, **30** : 1377-1378.

95. Lyle, J.A. 1964. Development of *Cercospora* leaf spot of peanut. Journal of the Alabama Academy of Sciences, **30** : 9.

96. Maublanc, A. 1975. The leaf spot disease of groundnut in West Africa. Agron. Colon., **13** : 126-127.

97. Mazzani, B. and Allievi, J. 1971. Effectos de la rotacion de cultivas sobre la incidencia de las manchas foliares por *Cercospora enmani*. Agronomie tropicale, Nogent-Sur-Marne, **21** : 329-332.

98. McDonald, D., Subrahmanyam, P., Gibbons, R.W. and Smith, D.H. 1985. Early and late leaf spots of groundnut, Information Bulletin No. 21. International Crops Research Institute for the Semi-Arid Tropics, Patancheru, A.P. 502324, India.

99. Mehta, P.R. and Mandal, K.K. 1978. Field screening of groundnut cultures against rust and tikka. Indian Phytopath., **31** : 259-260.

100. Melouk, H.A., Banks, D.J. and Fanous, M.A. 1984. Assessment of resistance to *Cercospora arachidicola* in peanut genotypes in field plots. Plant Disease, **68** : 395-397.

101. Mercer, P.C. 1976. Ultra-low volume spraying of fungicides for the control of *Cercospora* leaf spot of groundnuts in Malawi, PANS, **22** : 57.

102. Mercer, P.C. 1977. Pests and diseases of groundnut in Malawi I. Virus and foliar diseases. Oleagineux, **32** : 483-488.

103. Miller, L.I. 1946. Peanut leaf spot control, Va. Agric. Expect. Stn. Tech. Bull., 104.

104. Miller, L.I. 1949. Cultural and parasitic races of *Cercospora arachidicola* and *Cercospora personata.* Phytopathlogy, **39** : 17.

105. Miller, L.I. 1954. Studies on the parasitism of *Cercospora arachidicola* Hori and *Cercospora personata* (B. & C.) Eu & Ev. Virginia Journal of Science, **5** : 239.

106. Mitchell, J.K. and Taber, R.A. 1984. Comparison of mutant and wild type isolates of *Hansfordia pulvinata* for the biological control of *Cercosporidium personatum.* (Abstr.) Phytopathology, **74** : 630.

107. Mitchell, J.K. and Taber, R.A. 1986. Factors affecting the biological control of *Cercosporidium* leaf spot of peanuts by *Dicyma pulvinata.* Phytopathology, **76** : 990-994.

108. Mitchell, J.K., Jeger, M.J. and Taber, R.A. 1984. Quantitative description of colonization of *Cercosporidium* leaf spot of peanuts by the mycoparasite *Dicyma pulvinata* (Abstr.) Phytopathology, **74** : 864.

109. Mitchell, J.K., Smith, D.H. and Taber, R.A. 1987. Potential for biological control of *Cercosporidium personatum* leaf spots of peanuts by *Dicyma pulvinata.* Can. J. Bot., **65** : 2263-2269.

110. Monasterios de la Torre, T. 1980. Genetic resistance to *Cercospora* leaf spot diseases in peanut *(Arachis hypogaea* L.) Dissertation Abstracts, **41** : 1591 B.

111. Moraes, S.A. De and Salgado, C.I. 1979. Evaluation of the resistance to groundnut *(Arachis hypogaea* L.) to *Cercospora arachidicola.* Fitopatologia, **14** : 65-72.

112. Moraes, S.A. and Godoy, I.J. 1985. Differents neveis de resistencia a *Cercosporidium personatum* em genotipos de *Arachis hypogaea.* Summa Phytopathol., **11** : 74-86.

113. Moss, J.P., Rao, V.R. and Gibbons, R.W. 1989. In : The use of Plant Genetic Resources, Cambridge University Press, Cambridge, 212-234.

114. Muthiah, C., Senthivel, T., Venkatakrishnan, J. and Sivaram, M.R. 1991. Effect of intercropping on incidence of pest and disease in groundnut *(Arachis hypogaea),* Indian Journal of Agric. Sci., **61** : 152-153.

115. Narain, A. and Addy, S.K. 1970. Relationship of free amino acid and reducing sugar content of groundnut foliage in relation to tikka disease. In : Proc. Plant Disease Problems. 1st Int. Symp. on Plant Pathol. Indian Phytopathol. Soc., IARI, New Delhi, 733.

116. Nevill, D.J. 1980. Studies on resistance to foliar pathogens. In : Proc. Int. Workshop on Groundnut, ICRISAT, India, pp. 199-202.

117. Nevill, D.J. 1982. Inheritance of resistance to *Cercosporidium personatum* in groundnut, a genetic model and its implications for selection. Oleagineux, **37** : 355-362.

118. Norden, A.J. 1980. Crop improvement and genetic resources in groundnut. In : Advances in Legume Science, Eds. Summerfield, R.J. and Bunting, A.H. Royal Botanic Gardens, Kew, pp. 515-523.

119. Paningbatan, R.A. 1988. Comparative epidemiology of *Cercospora arachidicola* Hori and *Phaeoisariopsis personata* (Berk. and Curt.) von Arx on Peanut *(Arachis hypogaea* L.) Ph. D. Thesis, Phillippines University College, Laguna, Phillippines, 109 pp. (Summary, Eng.)

120. Pixley, K.V., Boote, K.J., Shokes, F.M. and Gorbet, D.W. 1990. Growth and partitioning characteristics of four peanut genotypes differing in resistance to late leaf spot. Crop Science, **30** : 796-804.

121. Ponniah, S., Sachithanantham, K. and Madhava Rao, S. 1982. Evaluation of fungicides against tikka and rust disease of groundnut. Pesticides, **16** : 27-28.

122. Porter, D.M. 1980. Increased Severity of Sclerotinia blight in peanuts treated with captafol and chlorothalonil. Plant Disease, **64** : 394-395.

123. Porter, D.M. and Taber, R.A. 1992. First report in Virginia of *Dicyma pulvinata* colonizing *Cercospordium personatum,* the causal agent of late leaf spot of peanut, Plant Disease, **76** : 1185.

124. Prasad, K.S.K., Hegde, R.K., Siddaramaiah, A.L. and Jayaramaiah, H. 1979. Screening of groundnut varieties for tikka disease resistance. Curr. Res., **8** : 104-105.

125. Purkayastha, R.P. 1968. Factors associated with conidial production and growth of *Cercospora personata* (Berk. and Curt) EU. and EV. causing tikka disease of groundnut *(Arachis hypogaea* L.) Indian J. Mycol. Research, **6** : 69.

126. Purkayastha, R.P. and Mukherji, S. 1967. Chemical changes in *Cercospora* infected leaves of groundnut *(Arachis hypogaea* L.) Sci. Cult., **33** : 341.

127. Purss, G.S. 1962. Peanut diseases in Queensland. Queensland Agriculture Journal, **88** : 540-553.

128. Rahman, M.A., Ahmed, H.U. and Alam, K.B. 1986. Studies on the efficacy of fungicides and the date of commencing of spray in controlling tikka and rust of groundnut. Bangladesh J. Pl. Pathol., **2** : 57-61.

129. Rajgopal, B.S. and Ravi, V.P. 1981. Toxic glycopeptides of *Cercospora personata.* Abstr. 3rd Int. Symp. on Plant Pathol., New Delhi, December; 14-18, p. 31.

130. Ramkrishnan, V. and Appa Rao, A. 1968. Studies on tikka disease of groundnut. Indian Phytopath., **21**: 31-36.

131. Ramanujan, M.P. 1992. Comparative physiological changes induced by infection with *Cercosporidium personatum* and toxins of the fungus on groundnut tissues. Advances in Plant Sciences, **5** : 229-236.

132. Rao, P.G. 1976. B. Bvitamin in relation to *Cercospora* infection in groundnut *(Arachis hypogaea* L.) Andhra Agric. J., (India), **21** : 158.

133. Rao, A.N.S. 1980. Control of tikka leaf spot of groundnut by carbendazim. Pesticides, **14** : 23.

134. Rao, A.S., McDonald, D. and Reddy, K.R. 1993. Perpetuation of peanut leaf spots pathogens. Oleagineux, **48** : 77-82.

135. Rattan, G.S. and Kang, M.S. 1984. Efficacy of different systemic and non-systemic fungicides against tikka disease of groundnut and their effects on host physiology. Pesticides, **18** : 30-32.

136. Ravindranath, V. and Kulkarni, L.G. 1967. Effect of different dates of sowing groundnut on the development and intensity of leaf spot disease by *Cercospora* sp. Indian J. of Agril. Scis., **37** : 362-368.

137. Reddy, P.S., Singh, V. and Singh, F. 1994. Genetic improvement for resistance to pests and diseases in groundnut. In : IPM systems in agriculture.

138. Reyes, G.M. and Romasanta, R. 1940. Varietal susceptibility of peanuts to black spot *Cercospora personata* (B. & C.) EU, & EV Phillippine Journal of Agriculture, **11** : 371-381.

139. Roldan, E.F. and Querijero, A.F. 1939. Black spot of groundnut Philipp. Agric., **27** : 669.

140. Rudolph, R.D., Noegel, K.A. and Rogers, W.D. 1989. Control of diseases of groundnut with (R) Folicur (tebuconazole) in the USA P flanzenschutz-Nachrichten Bayer, **42** : 199-214.

141. Savy Filho, A. 1980. The active bank of groundnut germplasm. (Banco ativo de germplasma de amendoim). In simposio de recursos geneticos vegetais. Sessao I. Bancos ativos de germplasma. 22-25 de maio de 1979, Brosilla, CENARGEN/EMBRAPA (1980); 121-123. Instituto Agronomico companas Brazil (fide Plant Breeding Abstracts 5120, 1982), **52** : 469.

142. Sekhar, M.J. and Sekaran, R.R. 1991. Effect of fungicides on key diseases of groundnut in relation to crop stage. Tropical Pest Management, **37** : 270-272.

143. Shanta, P. 1956. Isolation of *Cercospora personata,* its sporulation and growth in pure culture. Proc. Indian Acad. Sci., (B) **44** : 271.

144. Shanta, P. 1960. Studies on *Cercospora* leaf spot of groundnuts *(Arachis hypogaea)* Journal of Madras University, B **30** : 167-177.

145. Sharif, Y. 1972. 'The inheritance of *Cercospora* leaf spot resistance in *Arachis* species'. Ph.D. Thesis, North Carolina State University, Raleigh, USA.

146. Sharif, Y., Rawlings, J.O. and Gregory, W.C. 1978. Estimates of leaf spot resistance in three interspecific hybrids of *Arachis.* Euphytica, **27** : 741-751.

147. Sharma, L.C. and Mathur, A.K. 1979. Reaction of groundnut cultivars to *Cercospora* leaf spots. Madras Agricultural Journal, **66** : 626.

148. Shekhawat, P.S., Patel, V.N. and Patel, J.G. 1985. Economic fungicidal spray schedule for control of tikka and rust diseases of rainfed groundnut. Indian J. Mycol. & Pl. Pathol., **17** : 11-16.

149. Shokes, F.M. and Taber, R.A. 1983. Occurrence of a hyperparasite of *Cercosporidium personata* on peanut in Florida. Phytopathology, **73** : 505.

150. Siddaramaiah, A.L., Hiremath, R.V. and Krishna, K.S. 1977. Control of tikka disease of groundnut with Duter. Indian Journal of Platn Protection, **5** : 193-194.

151. Siddaramaiah, A.L., Desai, S.A. and Jayaramaiah, R. 1981. Occurrence and severity of a mycoparasite on 'tikka' leaf spots of groundnut. Current Research, **10** : 14-15.

152. Sidhu, G.S. and Chohan, J.S. 1972. Field scale testing of fungicides for the control of leaf spot (Tikka) disease of groundnut in the Punjab. Journal of Research, Punjab Agricultural University, **9** : 451-454.

153. Singh, S.D. and Naik, S.M.P. 1977. Field evaluation of some modern fungicides for the control of tikka disease of groundnut. Pesticides, **11** : 49-50.

154. Singh, F. and Oswalt, D.L. 1991. Groundnut production practices, Skill Development series No. 3 ICRISAT Centre, Patancheru, A.P., India. pp. 1-40.

155. Singh, B.P., Singh, S.P., Mohammad, A., Singh, D.K., Ahmad, Q. and Asthana, K.S. 1979. Sources of resistance in groundnut against tikka disease of groundnut. Indian Phytopath., **32** : 648-649.

156. Smart, J. 1961. The disease of groundnut in Northern Rhodesia, Emp. J. Exp. Agric., **29** : 79.

157. Smart. J. 1964. Cross compatibility relationships between the cultivated peanut *Arachis hypogaea* L. and other species of the genus *Arachis*. Ph.D. thesis, Northern Carolina State University, Raleigh, NC, USA, 83 pp.

158. Smith, D.H. 1979. Disease resistance in peanuts. In: Biology and breeding for resistance to arthropods and pathogens in agricultural plants. Ed. M.K. Harris. Proceedings of a short course entitled "International short course in Host Plant Resistance" Texas A&M University, College Station, Texas, USA, 22 July to 4 August 1979, pp. 431-447.

159. Smith, D.H. and Crosby, F.L. 1973. Aerobiology of two peanut leaf spot fungi. Phytopathology, **63** : 703-707.

160. Smith, D.H. and Littrell, R.H. 1980. Management of peanut foliar diseases with fungicides. Plant Disease, **64** : 356-361.

161. Smith, D.H., McGee, R.E. and Vesely, L.K. 1978. Isolation of benomyl-tolerant strains of *Cercospora arachidicola* and *Cercosporidium personata* at one location in Texas. Proceedings of the American Peanut Research and Education Association, **10** : 67.

162. Soriano, J.D. 1988. Resistance to leaf spot disease in peanut. Mutation Breeding News Letter, **32** : 4.

163. Sowell, G., Smith, D.H. and Hammons, R.O. 1976. Resistance of peanut plant introductions to *Cercospora arachidicola*. Plant Disease Reporter, **60** : 494-498.

164. Sreeramulu, T. 1970. Conidial dispersal in two species of *Cercospora* causing tikka leaf spot of groundnut (*Arachis hypogaea* L.). Indian Journal of Agricultural Sciences, **40** : 173-178.

165. Stalker, H.T. and Moss, J.P. 1987. Speciation, cytogenetics and utilization of *Arachis* species. Adv. Agron., **41** : 1-50.

166. Subrahmanyam, P. and Hassane, H. 1990. Response of six groundnut *(Arachis hypogaea* L.) cultivars to fungicidal control of leaf spots in Niger. Tropical Agriculture, **67** : 331-336.

167. Subrahmanyam, P. and Hildebrand, G.L. 1993. Current research on the management of early leaf spot of groundnut at SADC/ICRISAT Groundnut Project. In: Proceedings of the third ICRISAT Regional Groundnut Meeting for west Africa, Summary Ouagadougou, Burkina Faso, 14-17 Sept. 1992. pp. 49-50.

168. Subrahmanyam, P. and McDonald, D. 1986. Diseases of rabi groundnut in India. "Work-shop-cum-seminar on high yielding technology for Rabi/summer Oilseeds, Directorate of Oilseeds Research, Rajendranagar, Hyderabad-30, November 25, 1986.

169. Subrahmanyam, P., Mehan, V.K., Nevill, D.J. and McDonald, D. 1980. Research on fungal diseases of groundnut at ICRISAT. Proceedings of the International Workshop on Groundnuts held at ICRISAT Centre, Patancheru, Andhra Pradesh, October 13-17, 1980, pp. 193-198.

170. Subrahmanyam, P., McDonald, D., Gibbons, R.W., Nigam, S.N. and Nevill, D.J. 1982. Resistance to rust and late leaf spot diseases in some genotypes of *Arachis hypogaea.* Peanut Science, **9** : 6-10.

171. Subrahmanyam, P., Moss, J.P., McDonald, D., Subba Rao, P.V. and Rao, V.R. 1985. Resistance to *Cercosporidium personata* leaf spot in wild *Arachis* species. Plant Disease, **69** : 813-819.

172. Subrahmanyam, P., Ramanath Rao, V., McDonald, D., Moss, J.P. and Gibbons, R.W. 1989. Origins of resistances of rust and late leaf spot in peanut *(Arachis hypogaea,* Fabaceae), Economic Botany, **43**: 444-455.

173. Subrahmanyam, P. and Ravindranath, V. 1988. Fungal and Nematode Diseases, in Groundnut Tech. Ed. Reddy, P.S., Krishi Bhavan, Pusa, New Delhi, pp. 453-507.

174. Subrahmanyam, P., Reddy, P.M. and McDonald, D. 1990. Parasitism of rust, early and late leaf spot pathogens of peanut by *Verticillium lecanii,* Peanut Science, **17** : 1-4.

175. Subrahmanyam, P., McDonald, D., Waliyar, F., Reddy, L.J., Nigam, S.N., Gibbons, R.W., Ramnath, Rao, V., Singh, A.K., Reddy, P.M. and Subba Rao, P.V. 1991. Screening methods and sources of resistance to rust and late leaf spot of groundnut ICRISAT Information Bulletin (unpublished).

176. Sulaiman, M. 1966. Investigations on the control of tikka disease of groundnut in the Maharashtra State. Abstr. Rev. Appl. Mycol., **45** : 305.

177. Sulaiman, M. and Agashe, N.S. 1965. Influence of climate on the incidence of tikka disease of groundnut. Indian Oilseeds Journal, **9** : 176-179.

178. Sulaiman, M. and Hande, J.K. 1969. Studies on cultural, morphological, physiological and pathogenic characters of one isolate of *Cercospora arachidicola* and two of *Cercospora personata* Beitr. Trop. Sub trop. Landwirt. Tropervet. Med., **6** : 103-115.

179. Sundaram, N.V. 1965. Note on creation of tikka leaf spot of groundnut, Indian Oilseeds J., **9** : 98.

180. Swamy, R.N. 1961. Gaseous emanation from groundnut infected by *Cercospora personata* (Berk. and Curt.) Ell and Eve. Phytopath. Z., **40** : 245.

181. Taber, R.A. and Pettit, R.E. 1979. Reduction of sporulation of *Cercosporidium personata* by *Hansfordia* in Texas. Proceedings of the Am. Peanut Res. and Education Society, **11** : 45.

182. Taber, R.A., Pettit, R.E., McGee, R.E. and Smith, D.H. 1979. Reduction of sporulation of *Cercospordium personata* by *Hansfordia* in Texas (Abstr.). Am. Peanut Res. and Ed. Assoc. **11** : 45.

183. Taber, R.A., Pettit, R.E., McGee, R.E. and Smith, D.H. 1981. Potential for biological control of *Cercosporidium* leaf spot of peanuts by ***Hansfordia*** (Abstr.) Phytopathology, **71** : 260.

184. Tarr, S.A.J. 1954. Diseases of economic crops in Sudan. II. Fibers, Oilseeds, Coffee and Tobacco. FAO Plant Prot. Bull., **2** : 161-165.

185. Tiwari, S.P., Ghewande, M.P. and Misra, D.P. 1984. Inheritance of resistance to rust and late leaf spot in groundnut *(Arachis hypogaea* L.) J. Cytol. & Genet., **19** : 97-101.

186. Usman, K.M., Durairaj, P., Karunanith, K. and Thangavelu, S. 1992. Effect of neem products on the control of foliar diseases of groundnut. In: Groundnut: a global perspective Ed. S.N. Nigam. Proceedings of an International Workshop, ICRISAT Centre, Patancheru, Andhra Pradesh, India, 25-29, Nov. 1991, pp. 474.

187. Van Hoof, H.A. 1950. Disease of groundnut caused by *Cercospora personata* (B&C) E&E and *Cercospora arachidicola* Hori. Contrib. Gen. Agric. Res. Stn. Bogor, **114** : 17.

188. Venkatramani, K. 1967. Isolation of Cercosporin from *Cercospora personata,* Phytopath, Z., **58** : 379.

189. Venkataraman, S. and Kazi, S.K. 1979. A climatic disease calender for 'tikka' of groundnut. Journal of Maharashtra Agricultural University, **4** : 91-94.

190. Vetee Visutipat and Tharmmasak Sommatya, 1988. Histopathological study of *Phaeoisariopsis personata* on selected susceptible and resistant groundnut lines. In : Proceedings of the Sixth Thailand National Groundnut Meeting for 1986, Khon Kean, Thailand, 19-21 Feb. 1985. pp. 216-221.

191. Vidhyasekaran, P. 1981. Control of rust and tikka diseases of groundnut. Indian Phytopath., **34** : 20-23.

192. Vidhyasekaran, P. and Kolandaraman, P. 1968. Fungicidal control of tikka leaf spot of groundnut. Indian J. of Agric. Scis., **38** : 373-377.

193. Vijayakumar, G.C. 1990. Histopathological and histochemical changes in groundnut leaf due to infection by *Cercospora arachidicola.* Indian Phytopath., **43** : 453-455.

194. Vijayan, K.M. and Natarajan, S. 1964. A preliminary note on the fungicidal control of tikka leaf spot of groundnut. Madras Agricultural Journal, **51** : 218-220.

195. Vyas, S.C., Prasad, K.V.V., Shastry, P.P. and Shukla, B.N. 1986. Field control of leaf spots and rust of groundnut by fungicides. Indian Phytopath. **39** : 447-448.

196. Wangikar, P.D. and Shukla, V.N. 1976. Influence of prevailing temperatures and humidity on tikka disease of groundnut. Journal of Maharashtra Agricultural University, **1** : 254-264.

197. Woodroof, N.C. 1933. Two leaf spots of peanut *(Arachis hypogaea* L.) Phytopathology, **23** : 627.

16 Rust of Groundnut–An Overview

M.P. Ghewande

CONTENTS

Integrated Pest and Disease Management
Rajeev K. Upadhyay, K.G. Mukerji, B.P. Chamola and O.P. Dubey (eds.)
APH Publishing Corporation, New Delhi-110002, India

ABSTRACT

Rust caused by *Puccinia arachidis* Speg. is the most serious disease of groundnut causing upto 50% loss in suspective varieties. It perpetuate, spread and produce severe disease outbreaks by means of urediniospores. The practice of continuous groundnut cultivation without any break is an important factor in perpetuating groundnut rust in India. A variety of disease management strategies have been discussed.

1. INTRODUCTION

Groundnut is an important food legume of the world, grown in 20.09 million ha. spread over 85 countries with the production of 22.59 million tonnes with an average productivity of 1.12 t/ha^{-1} (111). India is the leading groundnut producing country since it account for 42% of world's groundnut area and 33% production (8.4 million ha. with an annual production of 7.5 million tonnes and productivity of 938 kg/ha).

An important factor contributing to low yield is disease attack. Among fungal foliar diseases, rust caused by *Puccinia arachidis* Speg. is widely distributed, economically important and can cause upto 50% losses in susceptible varieties (25, 124). Therefore, this disease has been given a high research priority all over the world. Different aspects like distribution and economic importance, symptoms, pathogen, host range and pathogenic variability, histopathology and pathogenesis, host parasite interaction, genetics of rust resistance, disease cycle and epidemiology and disease management strategies have been reviewed in this paper.

2. DISTRIBUTION AND ECONOMIC IMPORTANCE

Groundnut rust was first recorded by Spegazini in 1884 on groundnut from Caaguazu, Paraguay. Since then reports started appearing about its occurrence from many parts of the world. The disease was recorded in the USSR (43), Mauritius (115), the People's Republic of China (132), Central and South America (6, 70, 135), the Caribbean Islands (34, 120), the West Indies (11), North America (21) and the U.S.A. (34, 46, 133). In recent years, groundnut rust has spread to, and became established in, many countries in Asia, Australia, Oceania and Africa (34, 50, 97, 120, 124, 125, 127).

After the initial report of groundnut rust from Punjab in India (12) there have been records of its incidence from different parts of the country (5, 47, 68, 89, 106, 110, 136, 148). Subrahmanyam *et al.*, (120) reported that rust has been severed particularly in the Southern States of India. Surveys conducted by the National Research Centre for Groundnut (NRCG), Junagadh during kharif, 1980-81, 1981-82 and 1982-83 revealed that the rust with moderate to heavy severity was distributed in all groundnut growing districts of Saurashtra region of Gujarat (24). Rust is present in almost all rabi/ summer groundnut cultivation areas in India (128). It is now evident from the literature that the rust disease of groundnut is prevalent throughout India.

Rust caused considerable economic losses (50 and 70%) at two locations in Texas (127). However, rust as such, has not been considered a serious problem in groundnut production in the U.S.A. (36, 40). Losses as high as 40-70% were observed from rust or leaf spot alone or combined in West Africa (97). Feakin (21) reported that rust alone prevented groundnuts from becoming a commercial crop of West Indies. In the Caribbean area, rust limits or curtails production and in Central America it is as devastating as the leaf spots (34). Rust together with leaf spot has been reported to cause serious yield losses in the Caribbean Islands (14), Venezuela (66) and Central America (9). In the people's Republic of China, losses due to rust were estimated to be 49, 41, 31 and 18% at flowering, pegging, pre-pod forming, and mid-pod-forming stages respectively (151). Yield losses due to rust were less (6-31%) in the resistant (NCAc 17090, EC 76446 (292), PI 259747) than in the susceptible (40-57%) genotypes (TMV-2, Robut 33-1) (125). Significant yield loss is caused upto 100% to any unprotected crop in northern Australia in every season (67). Combined loss from rust and leaf spot was estimated in 1978 at 27-85% in Thailand (100). Estimated pod yield losses from rust alone were reported to range from 0.7 to 1.4 t/ha. in Nigeria (96). In Mauritius, 70% loss due to rust along has been reported (22). In addition to the direct yield losses rust can lower seed quality by reducing seed size and oil content (11). In India, loses in yield due to rust alone have been reported in the range of 10-52% depending upon the variety (25, 32, 45, 107, 124). The loss in oil content due to rust infection has been estimated to be about 7 to 10% (45).

Considering the magnitude of losses, and extent of prevalence, rust is now of economic importance in most groundnut growing areas of the world.

3. SYMPTOMS

Orange-red to chestnut brown coloured pustules (uredinia) appear on the abaxial (lower) surfaces of the leaves. The pustules are usually circular to elliptical, raised, and range from 0.3 to 2.00 mm in diameter. Pustules appear first on the abaxial surface. But in highly susceptible cultivars, the original pustules may be surrounded by colonies of secondary pustules. The pustules may later be formed on adaxial surface opposite to those on the abaxial surface. The pustules are formed subepidermally on compact stomata but soon burst through the epidermis and become exposed. Consequently, a reddish brown mass of urediniospores become visible on the surface of the leaves. The pustules may be formed on stipules, petioles and stem. Leaves infected with rust become necrotic and dry up but tend to remain attached to the plant (50, 128). Pods of severely affected plants are less in number and mature two or three weeks early. The seeds of such plants remain small in size (34).

4. CAUSAL ORGANISM

4.1 Pathogen

The description of *P. arachidis* L. has been consolidated as under based on the reports of Arthur (3), Jackson and Bell (41), Hennen, *et al.* (38, 39) Subrahmanyam and McDonald (124) and Kolte (50).

Spermogonia (Stage 0) and aecia (Stage I) are not known. The uredinial stage is the most predominant and commonly observed stage on groundnut plants. The uredinia are predominantly hypophyllous, subepidermal, amphigenous scattered irregularly grouped, round ellipsoid or oblong in shape, dark cinnamon brown in colour when mature. Each sorus contains numerous pedicillate urediniospores. The pedicel of the urediniospore is short, fragile any hyaline. The urediniospores measure in the range of 18.56 to 33.00 x 17.47 to 26.48 μm with an average size of 24.96 x 21.22 μm. The spore wall is brown in colour, 1.22 μ thick, finely echinulate with mostly 2, occasionally 3 or 4 germpores, nearly equetorial, in often flattened areas.

The telial stage (Stage III) is reported on groundnut from Central America, Florida and Brazil (38, 39, 114). In India, Chahal and Chohan (12) reported the occurrence of teliospores on plants growing in a green house but did not give details of spore morphology (39). Telia are chiefly hypophyllus 0.2 mm in diameter, scattered, prominent, soon naked, chestnut brown or about cinnamon brown becoming gray from germination, or almost black, raptured epidermis prominent. Teliospores are oblong, obovate, ellipsoidal, to ovate in shape, with rounded to acute and thickened apex, constricted in the middle, somewhat or gradually, attenuate at the base or more or less rounded attenuate at both ends. Teliospores are smooth walled, light yellow, golden yellow or chestnut brown, predominantly 2-celled, sometimes with 3-4 cells, 38-42 x 14-16 μ thick at sides, 2.5 -4.0 μ thick at the top, apical thickening almost hyaline, pedicel thin-walled, usually collapsing laterally, hyaline, upto 35-65 μ long but usually broken shorter or detached at spore base, germinating at maturity without dormancy. Telia are regularly produced on wild species of *Arachis* in South America (39). Metabasidia and basidiospores (Stage IV) are not known.

4.2 Host range and pathogenic variability

A large number of crop and weed species were checked for possible existence of collateral and alternative hosts of groundnut rust but none was found infected by *P. arachidis* except *Arachis* (74). Available literature does not show the existence of physiologic forms in *P. arachidis* (8, 37, 54, 126, 147). However, Cook stated that *P. arachidis* probably exists in more than one racial form. Indications about the possibility of ecotypes in the rust population are available in the literature (34, 62, 64, 71).

Based on the available literature it appears that no systematic studies on identification of races of *P. arachidis* have been done. Therefore it is necessary to make an indepth study on race flora, if any, of groundnut rust since knowledge on pathogen variability is prerequisite for rational hybridization programme.

4.3 Histopathology and pathogenesis

After deposition of the urediniospores on the surface of groundnut leaves, they germinate by giving rise to a germtube at a temperature range of 15 to 30°C, optimum range being 20 to 25°C (20, 53). For uredinospores to germinate on the leaf surface, presence of free water film is a must. Relative humidity above 80% supports the germination when spores are placed in a thin film of water (16, 57, 121). After germination, the germtube grows and forms an appressorium over stomata (16, 57). Cook (16) reported that close adhesion of the germtube on leaf surface is essential for appressorium formation. From the appressorium the infection hypha arises and penetrates the tissue through stomata. After transferring the length of the stomatal passage, the infection peg swells and forms a vesicle, and the subsequently formed mycelium become intercellular by producing knob-shaped haustoria into the cells. Intercellular hyphae are bonded to host cell walls and have pulley wheel septa. The haustoria are surrounded by double sheath (146). The stomata are not required, to open to allow successful penetration by infection hypha. The process can take place in the dark (16). Most infections take place successfully at 22 to 26°C (53). About 8 to 10 days after incubation, symptoms become visible with production of a new crop of spores, which become wind-borne and cause secondary infection (50).

4.4 Host parasite interaction

4.4.1 Physiology and biochemistry

From the changges in the physiological and biochemical processes of groundnut as a response to infection by *P. arachidis*, it was inferred that alterations in respiration, oxidative enzymes, phenol ascorbic acid, amino acid contents and nucleic acids are primarily important in resistance as they reflect the changes in metabolism that provide for a chemical and physiological environment that is either inhibitory or conducive to the growth of *P. arachidis*. The deficiency and excesses in mineral elements indirectly contribute to the altered physiology of the plant for causing the diseased condition (19, 62). Transpiration (loss of water) due to rust infection increases in infected plants compared with healthy plants (119). Changes in composition of free and protein amino acids, total sugars, and amino nitrogen have been studies (94, 118). Reddy and Khare (93) reported that inoculation with fresh urediniospores of *P. arachidis* on to the susceptible cultivary (Jyothi) and resistant line (ICG 1697) resulted in an increase in total sugars, phenols and peroxidase and phenol oxidase activity in both cultivars. Ascorbic acid increased in the inoculated susceptible cultivar and significantly decreased in the inoculated resistant genotype. Subba Rao *et al.*, (116) showed that methyl linolenate is produced in groundnut leaves infected with *P. arachidis*. Subba Rao *et al.*, isolated two new dienol phytoalexins, 9, hydroxy 10, 12 octadeca dien oic methyl ester and 13-hydroxy 9, 11 octadecdien oic methyl ester from groundnut leaves infected by *P. arachidis*. The antifungal activity of these compounds is described and their possible role in host defences against pathogenic infections is discussed. Also, alkyl bis phenyl esters were isolated as phytoalexins for the first time from groundnut leaves infected with *P. arachidis* by Subba Rao *et al.*, (117). *In vitro* antifungal activity of these compounds and their possible role in the host defence against rust infection are discussed.

An increased enzyme activity of chitinases, 1, 3-alpha and 1, 3 beta-glucanases and proteinases was reported both in resistant and susceptible cultivars of groundnut after infection by rust pathogen, *P. arachidis*. However, infected leaves of resistant cultivars showed higher activities of proteinases and glucanases than leaves of susceptible cultivars. Chitinase activity was reported to be prominent in infected leaves of susceptible cultivar, TMV-7. Possible role of these enzymes during pathogenesis is discussed (4).

4.4.2 Genetics of rust resistance

Prior to 1956 there were few investigations on genetic resistance to groundnut rust. But the rapid spread of rust disease in the late 1960s and early 1970s and increasing costs of chemicals have stimulated work in this field and a good number of sources of resistance have been identified (124, 125). High levels of resistance and immunity to rust have been found in some wild relatives of groundnut (126, 127). Bromfield and Bailey (7) reported that a natural cross between PI 298115 and an unknown pollen donor segregated for susceptibility and resistance in a ratio suggesting biogenetic control for rust resistance, with two homozygous recessive genes necessary for resistance. Cook (15) experimentally confirmed the biogenic model of inheritance postulated by Bromfield and Bailey (7). However, Nigam *et al.*, (76) reported that fourteen F3 derived rust resistant lines (FESR 1-14), from a natural cross between PI 298115 and an unknown pollen donor in the USDA rust nursery in Puerto Rico segregated for rust reaction in next generation. This indicated that resistance to rust though recessive may not be governed by duplicate loci and suggested that more than two genes were involved. Knuft and Norden (1983), studied the inheritance of rust resistance in F2 and F3 population derived from crosses between resistant lines, PI 315608, PI 314817, PI 350680 and a susceptible sister line of florunner UF-439-16-10-3. They reported the duplicate recessive gene action for inheritance of rust resistance. Tiwari *et al.*, (134) reported that the F1 populations of the crosses between resistant lines (EC 76446 (292), NCAc 17090) and susceptible cultivars (Robut 33-1, J-11, GAUG-1, TG-17) were susceptible to rust and segregation for rust resistance in the F2 populations was observed in the ratio of 1 resistant : 6 intermediate: 9 susceptible (1:6:9). Resistance to rust was thus, found to be recessive and under biogenic control. The two genes governing rust resistance acted in a polymeric fashion. Kalekar *et al.*, (44) studied the inheritance of rust resistance and found that resistance in PI 259747 was governed by a single reccessive gene in crosses with the susceptible variety, Phule Pragati (JL 24). Singh *et al.*, (109) studied the rust reaction in F1 hybrids between two rust susceptible *Arachis hypogaea* cultivars and diploid and tetraploid, *A. batizocoi* immune to rust and with aphiploids involving *A. batizocoi* and two other immune diploid wild species, *A. villosa* and *A. duranensis*. They found that all the F1 hybrids were highly resistant to rust and reported that the gene or genes controlling immunity in the wild species are partially dominant. Studies at ICRISAT on F2 plants from crosses involving three susceptible and three resistant parents suggested digenic inheritance (15 susceptible : 1 resistance) in some crosses and trigenic inheritance (63 susceptible : 1 resistant) in others (48). Zheng (150) reported that resistance to rust was found to be recessive and more than two genes were involved. Middleton and Shorter (67) from their preliminary study suggested dominant or partially dominant/additive gene action for rust resistance. Further, they concluded that genetic resistance to rust will reduce both the dependence on fugnicides and the costs of production, but will not reduce the grower's management decisions about use of fungicides. The genetics of disease resistance in groundnut has been reviewed (28). Paramasivam *et al.*, (79) studied rust resistance in the F2 generation of crosses CO2 x NCAc 17090, CO2 x PI 414331 and CO2 x PI 414332 under artificial inoculated condition. They reported that susceptibility was dominant to resistance and the F2 population segregated in the ratio of 3:1 with resistance governed by a single recessive gene. Verma *et al.*, (142) studied genetics of rust resistance involving 3 resistant and 2 susceptible genotypes. They reported that results of 6 crosses indicated that resistance to *P. arachidis* is controlled by both additive and non-additive gene action and additive gene effects were predominant. Subrahmanyam *et al.*, (131) opined that the rust resistance is of a quantitative nature and its inheritance does not seem to be simple.

The information on genetics of rust resistance in groundnut reveals that rust resistance is mostly recessive and quantitatively inherited. Although literature reveals one instance of specific resistance to rust in wild *Arachis* species, it may be wide spread.

5. DISEASE CYCLE AND EPIDEMIOLOGY

Groundnut rust is known to perpetuate, spread and produce severe disease outbreaks by means of urediniospores. The pathogen is known almost exclusively by its uredinial stage. There are no other authenticated reports of the occurrence of teliospores of groundnut rust from India. It is not known if the fungus can produce pycnia and aecia or if any alternate, alternative and collateral hosts are involved in the life cycle. It would appear that urediniospores are the main, if not the only, means of dissemination of groundnut rust disease (122, 124, 125). There is no record of the occurrence of any collateral hosts of groundnut rust outside the genus *Arachis* and in India wild *Arachis* species occur only in research centres and can hardly be involved in perpetuation of groundnut rust pathogen (125). Very rare formation of teliospores on some wild *Arachis* species and absence of an alternate host indicate over-wintering of rust through urediniospores formed on volunteer or self-sown plants or uredinospores carried on the unexposed affected crop debris (34, 77). In India, groundnut crop or volunteers are available in one or the other parts of the country enabling the survival of uredinial stage round the year. Actual surveys by the author had confirmed the survival of the uredinial stage in Saurashta region of Gujarat on Volunteers of groundnut during the off-season (24). The practice of continuous groundnut cultivation without any break appears to be an important factor in perpetuating groundnut rust in India and China. The pathogen may also survive from season to season on volunteer groundnut plants (122).

In the U.S. fungus does not overwinter but blows in from subtropical areas (138, 139). Patel and Vaishnav (81) reported that *P. arachidis* was present on monsoon, summer and premonsoon groundnut crops and on volunteer groundnut plants in wheat and mustard fields during the winter season. Thus uredinospore infection can carry the rust from one groundnut crop to another throughout the year. In Southern part of India, there is an extensive and continuous cropping of groundnut at all times of the year, and thus there is an easy availability of uredinispores as primary inoculum from one season to another (55, 89). There is no reliable evidence of groundnut rust being internally seed-borne or being spread by germplasm exchange (51, 60, 122, 143). Thus the infection appears in July-August in Souther India, in September and October in Central India and in November and December in the northeast. It appears, therefore, that the initial inoculum infecting groundnuts in northern Andhra Pradesh and Maharashtra is derived from Southern States and that in turn, Central Indian crops serve to provide the initial inoculum infecting those in West Bengal and Assam (65).

Castellani (11) reported that rust urediniospores obtained from infected leaves were viable for 3 months when stored at room temperature. Veeranjaneyulu (141) observed urediniospores viability in infected leaves for 9 weeks at room temperature; 13 weeks when burried in unsterilized soil; 15 weeks at 15°C and 33 weeks at 5°C. Mallaih and Rao (57) found that urediniospores remained viable for upto four weeks when stored at temperatures below 30°C but lost viability within two weeks when sorted at temperatures above 35°C. Mayee and Ekbote (63) reported that urediniospores remained viable for 20 days on leaf debris and at low temperature (-6°C) beyond 52 days. An optimum temperature of 20°C, availability of moisture on the leaf surface and high relative humidity favour infection (57).

Prasad *et al.*, (86) reported that intermittent rain with mean pH of 87% and average temperatures of 23-24°C for a few days favoured disease initiation. Siddharamaiah *et al.*, (108) reported that a continuous dry period characterized by high temperature (>26°C) and low relative humidity (>70%) delay rust occurrence and severity, whereas intermittent rain, high relative humidity, and 20 to 26°C temperature favour disease development. Patel and Vaishnav (82) reported that the observations during the kharif season of 1980 and 1981 indicated that severe rust infection was associated with a temperature range of 25-29°C, relative humidity of 74-89%, bright sunshine 1-4.3 hour and 10-13 mm rainfall during the preceding week. Rao *et al.*, (91) reported that after harvest of rainy and post rainy season crops, debris from groundnut plans infected with rust when kept on the field surface or burried at 5 or 10 cm depth, the fungus lost viability in 30 days, and most rapid after harvest of the post rainy season. However, when infected debris was stored indoors, the fungus survived for 30 days in a resistant genotype and for 45 days in a susceptible genotypes. Viability was also lower in the resistant genotype in the field.

Mallaiah (56) reported that during summer months (May-June) when the incidence of rust is low, the incubation period is long (18 days), while in winter when the rust is abundant, the incubation period is only 7 days. Aerial dissemination of uredinospores and diurnal periodicity with peak occurrences around noon was reported under Indian conditions (27, 58, 78). Zambettakis *et al.*, (149) reported that the most important factors were the South-West winds for contamination and relative humidity for the duration of spore generations on site in Burkina Faso. Paria and Raj (80) reported that the rust disease was favoured by warm, humid weather and spread from the lower most to the 5th leaf. Butler and Wadia (10) reported that about 70% infection occurs with a 12 hour wetness period at optimum temperatures (20-25°C). The period of wetnesses after inoculation with spore suspension must be continuous. If the continuity of wetness is broken, the amount of infection is reduced. Subrahmanyam *et al.*, (127) opined that local and long distance dissemination of the pathogen may be by airborne urediniospores, by movement of infected crop debris, or by movement of pods or seeds surface contaminated with urediniospores or infected crop debris. Plants of all ages are susceptible. The incubation period varies from 7 to 20 days depending on host genotype and environmental conditions. Spread of the disease within crops is facilitated by wind movement, by rain splash, and by insects.

6. DISEASE MANAGEMENT

Damage to the groundnut plants from rust is mainly from the loss of photosynthetic tissues to pustules and to chlorosis and necrosis that may in severed attacks cause death of the whole leaflet. Leaflets may dry up to give the plant a burned appearance. In conditions favouring rapid disease build-up, an early attack may cause wide spread death of plants and almost total crop failure (124). To minimise the losses due to rust, several methods of disease control on groundnut have been developed. However, no single method of disease control is totally effective against rust disease.

6.1 Cultural

This aspect of plant disease management was neglected and or over looked all these years. Spraying machines, chemicals and even water often become limitations to marginal and small scale farmers. In such context, adoption of cultural practices would resolve the crop losses to some extent. Cultural practices that destroy volunteer groundnut plants or crop debris are an important measure to limit primary sources of inoculum (13, 34, 124). Pods and seeds from rust-affected crops are commonly surface contaminated with urediniospores at harvest. Therefore, strict quarantine regulations should be enforced to avoid the spread of rust on olds or seeds to disease-free areas (84, 124). The early planting (15 days early than normal), plant spacing of 45 x 10 cm and inter cropping with red gram (*Cajanus cajan* L.) and castor (*Ricinus communis* L.) reduced rust intensity (26). Siddharamaiah *et al.*, (108) have recommended sowing in the first fortnight of June to avoid the incidence of groundnut rust. Kodmelwar and Ingle (52) reported that early planting (June, 28) and wider spacing (30-40 cm) were found to reduce tikka and rust. Rust progressed more slowly in treatments where 60 and 75 kg P_2O_5/ ha were applied than in those where low levels of phosphorus or no phosphorus were given (63). Inter cropping of sorghum with groundnut resulted in lowest infection of rust (72). Inter cropping of groundnut with sorghum and pearl millet could reduce incidence of *P. arachidis* (95).

6.2 Biological

Several mycoparasites, *Verticillum lecanni, Penicillium islandicum, Eudarluca caricis, Acremonium persicinum, Darluca filum, Tuberuclina costaricana* and *Hyalodendron* sp. have been observed to parasitize *P. arachidis* (5, 30, 102, 103). Mycophagus thrips, *Euphysothrips minozii* have also been reported to feed on urediniospores of *P. arachidis* (101, 140). However, no serious attempts were made to use mycoparasites in the control of rust. Attempts were made to control rust both under laboratory and field conditions using several mycoparasites and their culture filtrates. All the mycoparasites (*A. persicinum, E. caricis, P. islandicum, T. costaricana* and *V. lecanii*) and their culture filtrates inhibited urediniospores germination and reduced rust development by varying degrees. Maximum inhibition of *in vitro* development of rust was exhibited by *V. lecanni* and *P. islandicum* and their culture filtrates. The inoculation of mycoparasites and rust on the same day gave better control of rust. Sprays of culture filtrates of *V. lecanii* and *P. islandicum* significantly reduced rust disease under field conditions (30).

6.3 Host resistance

The objective of controlling the groundnut rust has two obvious facets, viz., evolving varieties with in-built resistance and direct measure of control (chemical control) till such varieties are available (24). Although rust disease can be controlled by certain chemicals (112), genetic control is decidedly the best solution. Identification of sources of genetic resistance is therefore highly important (129). Screening for resistance to rust has been intensively carried out by many workers under both field and laboratory conditions by adopting various screening techniques and a large number of sources of resistance have been reported (6, 8, 14, 34, 35, 36, 37, 66, 69, 121, 123, 124, 125, 126, 127, 129). Lin (54) screened 461 groundnut accessions, and only one genotype, PI 315608 showed resistance to *P. arachidis* under both field and rust nursery conditions. At ICRISAT, some 10, 000 groundnut germplasm lines from different parts of the world were screened in the field for resistance to rust. Out of 10, 000 germplasm accessions screened, 42 genotypes (36 *fastigiata* and 6 *hypogaea*) were resistant to rust (129). Later on, a world collection of over 12, 000 germplasm lines has been successfully screened in the field against rust and late leaf spot during the period 1977 to 1989, and 124 rust-resistant germplasm lines were identified and majority of them have originated in Peru which is believed to be one of the secondary gene centres of cultivated groundnut (130). Out of 124 rust-resistant lines, 26 genotypes had very high level of resistance to rust. They are : NCAc 17090, PI 393526, PI 393531, PI 476149, PI 476151, PI 476166, PI 476166 (B), PI 476168, (ICG 10031), PI 476177, PI 476180, PI 476182, PI 476183, PI 476186, PI 476188, (ICG 10063), PI 476189, PI 476192, PI 476197, (ICG 10073), PI 476148, PI 476168 (ICG 10935), PI 476172, PI 476173, PI 476188 (ICG 10964), PI 476195, PI 476197 (ICG 10978) and PI 476165. Most of these rust resistant sources have been used in rust-resistant breeding programme. Until recently, there were no agronomically acceptable groundnut varieties with resistance to rust. However, 5 varieties, Girnar 1 from the NRCG, ALR1 from

Aliyarnagar (Tamil Nadu), ICGV 87160, ICGV 86590 from ICRISAT and tifrust jointly developed by USDA, ARS, University of Georgia, Tifton, USA and ICRISAT has been released as cardipayne in Jumaica are available for commercial cultivation. Several wild species are also known to have immunity and high resistance to rust (8, 126, 127). Recently, Ghewande (31) reported two advanced breeding lines, PPS 1 (R 33-1) X EC 76446 (292) and PPS 5 (JL 24 x NCAc 17090) with resistance to rust and fairly good pod yield.

The rust resistance at present available in the cultivated groundnut is of the slow-rusting type, i.e. resistant genotypes have increased incubation periods, decreased infection frequency, reduced pustule size and spore production and reduced spore viability (113, 124).

6.4 Chemical

In order to ensure elimination of infection, if any, from surface-contaminated seeds, treatment of seed with thiram or captan @ 3 to 4 g/Kg seed has been recommended (13). Three to four foliar applications of mancozeb or chlorothalonil @ 0.2 to 0.3% has been found to give good control of rust disease. Chlorothalonil appears to be more effective than mancozeb (2, 13, 77, 144). Raemaekers and Preston (88) reported that three to five spray applications of chlorothalonil @ 1.5 kg/ha gave yield increase by 18 and 36% respectively as a result of control of the rust. Arneson (2) opined that the efficacy of mancozeb in controlling rust can be increased by addition of nickel sulfate in the spray suspension. Patil *et al.*, (83) reported that mancozeb was found to be very effective and economical for the control of rust giving a net profit of Rs. 543/ha followed by chlorothalonil. A systemic chemical, tridemorph, if sprayed @ 0.07% results in good control of the disease (60). Some other chemicals such as nickel chloride (99), Benodami (107) were also found effective in the control of rust disease. Natarajan *et al.*, (75) have recommended that for the control of groundnut rust, triadimefon @ 100 g/ acre can be sprayed and two sprays are necessary on 35 and 50 days after sowing (cost benefit ratio for 100 g/acre was 1:3:2). Das and Baj (18) reported that carboxin 0.1% gave the best control of groundnut rust followed by tridemorph at 0.2% over seasons.

Several fungicides and their combinations have been tested for the control of leaf spots and rust since leaf spots and rust often appear together (73, 85, 92, 105, 145). The rust and leaf spots were substantially reduced due to the application of carbendazim (0.025%) plus tridemorph (0.04%) five times at fortnightly intervals commencing 35 days after sowing during summer season (42). The rust and leaf spots are controlled effectively by the spray application of carbandazim (0.05%) plus mancozeb (0.2%) at 2-3 weeks interval, 2 o 3 times starting from 4-5 weeks after planting and gave CBR ranging from 1:14:.8 to 1:24.4 (1). Chlorothanlonil is most widely used in the U.S.A. for the control of rust and leafsports (124). A combination spray of tridemorph plus carbendazim @ 0.05% gave good control of both leaf spots and rust (59). Five rounds of spray, starting from 30 days after sowing with 12 day intervals using carbendazim 0.05% plus mancozeb 0.2% was found to give maximum control of rust and leaf spot which resulted in 57% increased dry pod yield (98). Usman *et al.*, (137) reported that foliar application of chlorothlonil (0.1%) effectively controlled late leaf spot by 37.3% and rust by 38.94%.

Although direct control measures through synthetic fungicides are available, use of botanical fungicides especially of neem origin in the control of groundnut diseases is highly desirable since they are safe from the point of environmental pollution and pesticidal residues. Ghewande (29) reported that leaf extract (2%) of neem (*Azadirachta indica* A. juss) was found useful in controlling late leaf spot and rust diseases. Foliar application of neem kernel extract (3%) reduced the incidence of rust and late leaf spot incidence significantly and recorded the highest cost benefit ratio (137). Thus neem products have good prospect in the management of groundnut fungal foliar diseases.

7. CONCLUSIONS

Groundnut rust is widely distributed and is of economic importance in most groundnut growing areas of the world. The rust of groundnut is known to occur primarily in the uredial stage. It is not known if any alternate, alternative and collateral hosts are involved in the life cycle. Therefore, critical epidemiological studies and disease surveys would enable to identify the situations of rust perpetuation. A very little information on the physiology of rust is available in the literature. It is, therefore, necessary to understand the nature and physiology of the rust as much as the host. Local and long distance dissemination of the pathogen may be by air borne urediniospores by movement of infected crop debris, or by movement of pods or seeds surface-contaminated with urediniospores or infected crop debris. There is no reliable evidence of groundnut spread by germplasm exchange. An optimum temperature in the 20-30°C range, free water film on the leaf surface and high relative humidity favour infection and subsequent disease development.

Available literature does not susbtantiate the existence of physiological races in *P. arachidis* and only indications are available. Rust resistance is mostly recessive and quantitatively inherited, although the literature reveals one instance of specific resistance to rust in wild *Arachis* species, it may be widespread. Whether it is a case of horizontal resistance or may it be vertical resistance, knowledge of nature of pathogen is pre-requisite for rational hybridizaion programme. Although a large number of sources of resistance to rust are available, durable and dependable sources of resistance are limited in number. There is a need to concentrate on "germplasm enhancement" to obtain higher levels of resistance to rust. Additional information on biological control of rust needs to be generated. Efforts are required to be made in utilizing all available and compatible disease control measures in an integrated way wherever possible in increasing groundnut production.

8. REFERENCES

1. Anon. 1981-82. Annual Progress Report, Kharif groundnut, All India Co-ordinated Research Project on Oilseeds, Directorate of Oilseeds Research, Rajendranagar, Hyderabad.

2. Arneson, P.A. 1970. Chemical control of rust and *Cercospora* leaf spot on peanuts in Honduras and Nicaragua. Phytopathology, **60**: 1539.

3. Arthur, J.C. 1934. Manual of the rusts in the United States and Canada. Lafayette, Indiana, USA. Purdue Research Foundations 438 pp.

4. Bama, M.S. and Balasubramanian, R. 1991. Chitinases and glucanases in *Arachis hypogaea* L. in detached leaves infected with *Puccinia arachidis* Speg. Zeitschrift fur Planzenkrankeeiten and Pflanzenschutz, **98**: 305 - 316.

5. Bhama, K.S. 1972. A rust on groundnut leaves near Madras, Curr. Sci., **41**: 188-189.

6. Bromfield, K.R. 1971. Peanut rust - a review of literature. Proceedings of the American Peanut Research and Education Association Inc. **3**: 111-121.

7. Bromfield, K.R. and Baily, W.K. 1972. Inheritance of resistance to *Puccinia arachidis* in Peanut. Phytopathology, **62**: 748 (Abstract).

8. Bromfield, K.R. and Cevario, S.J. 1970. Greenhouse screening of Peanut (*Arachis hypogaea*) for resistance to peanut rust (*Puccinia arachidis*). Plant Disease Reporter, **58**: 381-383.

9. Burger, O.F. 1921. Peanut rust caused by *Puccinia* (Uredo) *arachidis*. Plant Disease Bulletin, **5**:88.

10. Butler, D.R. and Wadia, K.D.R. 1992, Groundnut foliar diseases : infection and leafwetness in groundnut: a global perspective. Proceedings of an International Workshop, ICRISAT Centre, Patancheru, Andhra

Pradesh, India 25-29 Nov. 1991, Ed. S.N. Nigam, International Crops Research Institute for the Semi - Arid Tropics, Patanchru, Andhra Pradesh, India, (Summary), pp. 475.

11. Castellani, E. 1959. Rust of groundnut. Olearia, **13**:261-270.

12. Chahal, A.S. and Chohan, J.S. 1971. *Puccinia* rust on groundnut. FAO Plant Prot. Bull. **19**:90.

13. Chohan, J.S. 1974. Recent advances in diseases of groundnut in India, In : Current trends in Plant Pathology, Lucknow Univ. Press. Lucknow, India, pp. 171-184.

14. Cook, M. 1972. Screening of peanut for resistance to peanut rust in the green house and field. Plant Disease Reporter, **56**: 382-386.

15. Cook, M. 1975. Rust and leaf spot resistance in Peanut Cultivars. Ph. D. thesis, University of the West Indies, Kingston, Mamaica, West Indies, 539 pp.

16. Cook, M. 1980. Peanut leaf wettability and succeptibility to infection by *Puccinia arachidis*. Phytopathology, **70**: 826-830.

17. Cummins, G.B. 1979. Rust fungi on legumes and composites in North America, Tucson, Arizona, U.S.A. University of Arizona Press. 424. pp.

18. Das, S. and Baj, S.K. 1992. Efficacy of fungicides for the control of groundnut rust (*P. arachidis* Speg.) under field conditions in West Bengal. Plant Protection Bulletin (Faridabad), **44**: 19-20.

19. Ekbote, A.U. and Mayee, C.D. 1983. Physiological changes in rust (*Puccinia arachidis* Speg.) resistant and susceptible groundnuts after inoculation. Indian Journal of Plant Pathology, **1**: 170-174.

20. Fang, H.C. 1977. Studies on peanut rust in Taiwan. Plant Protection Bulletin, Taiwan, **19**: 218.

21. Feakin, S.D. (Ed.) 1973. Pest control in groundnuts. 3rd edn. PANS Mannual No. 2. London, U.K. : Centre for Overseas Pest Research. 197 pp.

22. Felix, S. and Ricaud, C. 1979. Occurrence and fungicidal control of groundnut rust in Mauritius. Abstr. Rev. Plant Patho. **58**: 4577.

23. Ganpathy, T. and Narayanasamy, P. 1991. Effect of intercropping on the incidence of diseases of groundnut. Madras Agricultural Jornal, **78**: 22-24.

24. Ghewande, M.P. and Misra, D.P. 1983. Grondnut Rust - A challenge to meet. Seeds & Farms, **9**: 12-15.

25. Ghewande, M.P., Shukla, A.K., Pande, R.N. and Misra, D.P. 1983. Losses in groundnut yields due to leaf spots and rust at different intensity levels. Indian J. Mycol. P. Pathol. **13**: 125-127.

26. Ghewande, M.P., Shukla, A.K. and Pande, R.N. 1986. Management of foliar diseases of groundnut through agronomic practices. Indian Bot. Reptr. **5**: 122.

27. Ghewande, M.P. and Jhala, R. 1985. Epidemiology of foliar diseases of groundnut. Abstr. Vol. Proc. of 3rd International Conference on Aerobiology, August 6-9, 1986, Basel, Switzerland.

28. Ghewande, M.P. 1989a. Genetics of disease resistance in groundnut. In : Recent advances in genetics and cytogenetics Eds. Farook and Khan, pp. 57-63.

29. Ghewande, M.P. 1989 b. Management of foliar diseases of groundnut (*Arachis hypogaea* L.) using plant extracts. Indian J. of Agri. Scis. **59**: 133-134.

30. Ghwwande, M.P. 1990. Biological control of groundnut (*Arachis hypogaea* L.) rust (*Puccinia arachidis* Speg.) in India. Tropical Pest Management, **36**: 17-20.

31. Ghewande, M.P. 1993. Management of major foliar fungal diseases of groundnut (*Arachis hypogaea* L.) using aqueous leaf extract of neem (*Azadirachta indica* A. Juss). In : Nemm (*Azadirachta indica* A. Juss) for management of crop diseases.

32. Ghuge, S.S., Mayee, C.D. and Godbole, G.M. 1981. Assessment of losses in groundnut due to rust and tikka leaf spots. Indian Phytopath., **34**: 179-182.

33. Goswami, R.N. 1974. Rust a new menace to groundnut in North East India. Indian Phytopath., **27**: 238.

34. Hammons, R.O. 1977. Groundnut rust in the United States and the Caribbean. PANS, **23**: 300-304.

35. Hammons, R.O. 1981. Breeding peanuts for disease resistance : rust and leaf spot. Proceedings of the Peanut Breeding Symposium Richmond, Virginia, North Carolian Agriculture Research Services, Research Reporter no. **80** : 52-58.

36. Hammons, R.O., Subrahmanyam, P., Rao, V.R., Nigam, S.N. and Gibbons, R.W. 1982. Registration of eight peanut germplasm lines resistant to rust (Reg. No. GP 22 to GP 29). Crop Science, **22**: 452-453.

37. Hammons, R.O. 1987. Groundnut rust research in the Americas. ICRISAT (International Crops Research Institute for the Semi - Arid Tropics) 1987. Goundnut rust disease. Proceedings of a Discussion Group Meeting, 24-28 Sep. 1984, ICRISAT Centre, Patancheru, A.P., India.

38. Hennen, J.F., Figueiredo, M.B., Ribeiro, I.J.A., and Soave, J. 1976. The occurrence of teliospores of *Puccinia arachidis* (Uredinales) on *Arachis hypogaea* in Sao Paulo State, Brazil. Summa Phytopathologica, **2**: 44-46.

39. Hennen, J.F., Subrahmanyam, P. and Figue-iredo, M.B. 1987. The toxonomy, life history, and evolution of *Puccinia arachidis* Speg. ICRISAT (International Crops Research Institute for the Semi-Aarid Tropics) 1987. Groundnut rust disease. Proceedings of a Discussion Group Meeting, 24-28 Sep. 1984, ICRISAT Centre, Patancheru, A.P., India.

40. Harrison, A.L. 1972. Observation on the development and spread of Peanut-rust in South Texas in 1971. Plant Disease Reporter, **56**: 873-874.

41. Jackson, C.R. and Bell, D.K. 1969. Diseases of peanut (Groundnut) caused by fungi. University of Georgia College of Agriculture Experiment Station Research Bulletin 56.

42. Jayasekhar, M. and Rajasekaran, R. 1986. Appraisal of Yield loss in groundnut due to rust and late leaf spot. Madras Agric. J. **73**: 525-527.

43. Jaczewski, A. 1910 Yearbook on the Diseases of Plants VI. p. 465, cited in Commonwealth Mycological Institute Distribution Map of Plant Disease No. 160, Kew Surrey, U.K.

44. Kalekar, A.R., Patil, B. and Deokar, A.B. 1984. Inheritance of resistance to rust in groundnut Madras Agri. J. **71**: 125-126.

46. Kenknight, G. 1941. Peanut diseases in certain Texas countries in 1941, with notes on occurrence of Peanut rust. Plant Disease Reporter, **25**: 587.

47. Khosla, H.K, Puranik, K.K. and Nema, K.G., 1974. Occurrence of groundnut rust (*Puccinia arachidis* Speg.) in Madhya Pradesh, JNKVV Res. Journal, **8**: 292.

48. Kishore, B. 1981. Rust inheritance studies in groundnut (*Arachis hypogaea* L.) M.Sc. (Ag) thesis, Andhra Pradesh Agricultural University, A.P., India.

49. Knauft, D.A. and Norden, A.J. 1983. Inheritance of rust resistance in peanuts. Proceedings of American Peanut Research and Education Society, **15**:76. (Abstract).

50. Kolte, S.J. 1984. Diseases of annual oilseed crops. Vol.I Peanut disease, CRC Press, Inc. USA pp. 143.

51. Kolte, S.J. and Awasthi, R.P. 1979. Some observations on occurrence of groundnut rust in the Nainital Tarai region of Uttar Pradesh. Indian Phytopath, **32**: 155-156.

52. Kodmelwar, R.V. and Ingle, A.Y. 1989. Effect of sowing dates, spacing and meterological factors on the development of tikka and rust of groundnut. Indian Phytopath. **42**: 274.

53. Kono, A. 1977. On the uredospore germination of Peanut leaf rust (*Puccinia arachidis* Speg.) on peanut leaves. Bull. Fac. Agric. Miyazaki Univ. **24**: 225-234.

54. Lin, Y.S. 1981. Effects of inoculum concentration on the resistance of peanuts to *Puccinia arachidis*. Journal of Agricultural Research of China, **30** : 324 - 331.

55. Lingaraju, S., Siddaramaiah, A.L. and Hegde, R.K. 1979. Viability and survival of uredospores of *Puccinia arachidis* Speg. in Dharwad. Curr. Res., **8**: 68-69.

56. Mallaiah, K.V. 1976. A note on the seasonal changes in the incubation time of groundnut rust Curr. Sci. **45**: 737.

57. Mallaiah, K.V. and Rao, A.S. 1979. Groundnut rust: Factors influencing disease development, sporulation and germination of uredospores. Indian Phytopath. **32**: 382-388.

58. Mallaiah, K.S. and Rao. A.S. 1982 - Aerial dissemination of groundnut rust uredinosores. Transactions of the British Mycological Society, **82** : 21-28.

59. Mathur, S. and Doshi, A. 1990. Field evaluation of some fungicides for the control of leaf spots and rust diseases of groundnut. Summa Phytopathologica, **16**: 162 - 165.

60. Mayee, C.D. 1981. Epidemiology of groundnut rust from Marathwada, Third Informational Symposium on Plant Pathology, New Delhi, December 14-18, 1981 (Abstract).

61. Mayee, C.D. 1983. Epidemiology and forecasting of groundnut rust in Marathwada region : final report India : Marathwada Agril. Univ., Parbhani, Maharashtra 76 pp.

62. Mayee, C.D. 1987. Rust disease of groundnut in Maharashtra State of India. ICRISAT (International Crops Research Institute for the Semi-Arid Tropics). 1987. Groundnut rust diseases. Proceedings of a discussion group meeting, 24-28 September, 1984, ICRISAT Centre, India. Patancheru, A.P., India.

63. Mayee, C.E. and Ekbote, A.V. 1983. Life cycle and microepidemiology of *Puccinia arachidis* Speg. rust of groundnut in Maharashtra. Indian Plant Path. **1**: 62-67.

64. Mayee, C.D., Patil, F.S. and Pant, K.G. 1979. Evaluation of groundnut cultivars for resistance to rust and possible existence of races. Indian Phytopath., **32**: 109-110.

65. Mayee, C.D., Godbole, G.M. and Patil, F.A. 1977. Appraisal of groundnut rust in India: Problems and approach. PANS, **23**: 162-165.

66. Mazzani, B. and Hinojosa, S. 1964. Varietal differences in susceptibility to groundnut rust in Venexuela. Abstr. Rev. Appl. Mycol. **43**: 314.

67. Middleton, K. and Shorter, R. 1987. Occurrence and management of groundnut rust in Australia ICRISAT (International Crops Research Institute for the Semi Arid Tropics) 1987. Groundnut rust disease. Proceedings of a discussion group meeting, 24-28 Sep. 1984, ICRISAT Centre, India, Patancheru, A.P., India.

68. Misra, A.K. and Misra, A.P. 1975. Groundnut rust in Bihar-Varietal reaction. Indian Phytopath., **28**: 557-559.

69. Mixon, A.C, Hammons, R.O. and Branch, W.D. 1983. Germplasm for use in genetic enhancement of peanut genotypes. Proc. Amer. Peanut Res. Educ. Soc. **15**: 15-38.

70. Muller, A.S. 1950. Preliminary survey of plant diseases in Guatemala. Plant Dis. Rep., **34**: 161.

71. Munde, P.N. and Mayee, C.D. 1979. Effect of temperature on incubation period of three isolates of *Puccinia arachidis*. Research Bulletin of Marathwada Agricultural University, **3**: 62.

72. Muthiah, C., Senthival, T., Venkatkrishnan, J. and Sivaram, M.R. 1991. Effect of itercropping on incidence of pest and disease in groundnut (*Arachis hyopogaea*). Indian Journal of Agricultural Sciences, **61**: 152-153.

73. Natrajan, S. 1984. Fuingicidal control of rust and tikka leaf spot of groundnut. Pesticides, **18**: 40-41.

74. Nagarajan, S. 1987. On the likelihood of pathogenic forms or virulences, in *Puccinia arachidis* Speg., that cause groundnut rust in *Arachis* species. ICRISAT (International Crops Research Institute for the Semi Arid Tropics). 1987. Groundnut rust disease. Proceedings of a discussion group meeting, 24-28 Sep. 1984, ICRISAT Centre, India. Patancheru, A.P., India.

75. Natrajan, S., Sachidanamtham, K., Rajemanickam, B. and Madhava Rao, S. 1983. Chemical control of groundnut rust with triadimefon. Pesticides, **17**: 17-18.

76. Nigam, S.N., Dwivedi, S.L., and Gibbons, R.W. 1980. Groundnut breeding at ICRISAT. pages 62-68. In: Proceedings of the International Workshop in Groundnuts, 13-17 Oct. 1980. ICRISAT Centre, India. Patancheru, A.P. India.

77. O'Brien, R.G. 1977. Observations on the development of groundnut rust in Australia. PANS, **23**: 297-299.

78. Pande, R.N. 1982. Atmospheric concentration of uredospores over groundnut field. Indian Bot. Reptr. **1**: 168.

79. Paramasivam, K., Jayasekhar, M., Rajasekharan, R. and Veerabadhiran, P. 1990. Inheritance of rust resistance in groundnut (*Arachis hypogaea* L.). Madras Agricultural Journal, **77**: 50-52.

80. Paria, T.K. and Raj, S.K. 1987. Epidemiology of groundnut rust. Puccinia arachidis Speg. Current Research, **16**: 100-102.

81. Patel, V.A. and Vaishnav, M.U. 1989 a. Perpetuation and carry-over of groundnut rust in Gujarat State, India. Journal of Oilseeds Research, **6**: 373-374.

82. Patel, V.A. and Vaishnav, M.V. 1989 b. Effect of weather parameters in relation to occurrence and development of groundnut rust. Journal of Oilseeds Research, **6**: 379-383.

83. Patil, M.B., Jadhav, B.R. and Rana, M.S. 1983. Fungicidal control of groundnut rust. Indian J. Mycol. and Pl. Pathol., **13**: 225-226.

84. Peregrine, R.G. 1971. Groundnut rust (*Puccinia arachidis*) in Brumei. PANS, **17**: 318-319.

85. Ponniah, S., Sachithanantham, K. and Madhav Rao, S. 1982. Evaluation of fungicides against tikka and rust diseases of groundnut. Pesticide, **16**: 27-28.

86. Prasad S.K.S., Siddaramaiah, A.L. and Hegde, R.K. 1979. Development of peanut (groundnut) rust disease in Karnataka State, India. Plant Disease Reporter, **63**: 692-695.

87. Puranik S.B., Bidari, V.B., Joshi M.S. and Hiremath, R.V. 1973. Rust incidence on groundnut (*Arachis hypogaea*) in Mysore State-Varietal Performance against it. Curr. Res. **2**: 81-82.

88. Raemaekers, R. and Preston, G. 1977. Groundnut rust occurrence and foliar disease control in Zambia. PANS, **26**: 166-170.

89. Ramakrishan, V. and Subbayya, J. 1973. Occurrence of groundnut rust in India. Indian Phytopath, **26**: 574-575.

90. Rao, P.V.S., Einhorn, J., Malosse, C., Geiger, J.P., Rio, B. and Ravise, A. 1990. New denial phytoalexins isolated from *Arachis hypogaea* leaves infected with *Puccinia arachidis* Speg. Oleagineux, **45**: 225-227.

91. Rao, A.S., McDonald, D. and Reddy, K.R. 1991. Perpetuation of groundnut rust disease. Oleagineux, **46**: 23-27.

92. Rawal, P.P., Jain, S.M. and Khandar, R.R. 1982. Foliar diseases of groundnut in Gujarat. Indian Farming, **32**: 17-20.

93. Reddy, N.P. and Khare, M.N. 1988. Physiology of groundnut rust disease : Changes in total sugars, phenols, ascorbic acid, peroxidase and phenol oxidase. Journal of Oilseeds Research, **5**: 102-106.

94. Reddy, M.N. and Rao, A.S. 1977. Changes in the composition of free and protein amino acids in groundnut leaves induced by infection with *Puccinia arachidis* Speg. Actaphyopathol. Acca. Sci. Hung. **11**: 167.

95. Reddy, P.M., Smith, D.H. and McDonald, D. 1991. Inhibitory effect of sorghum and pearl millet pollen on uredinospore germination and germtube growth of groundnut rust. International Arachis Newsletter, **10**: 21-22.

96. Salako, E.A. and Olorunju, P.E. 1987. Occurrence and importance of rust disease of groundnut in Nigeria. ICRISAT (Internation Crops Research Institute for the Semi-Arid Tropics). 1987. Groundnut rust disease. Proceedings of a discussion group meeting, 24-28 Sep. 1984, ICRISAT Centre, India, Patancheru, A.P., India.

97. Savary, S., Bosc, Jean - Philippe, Zadoks, J.C. and Noirot, M. 1988. Peanut rust in West Africa: A new component in a multiple pathosystem - Plant Disease, **72**: 1001 - 1009.

98. Sekhar, M.J. and Sekaran, R.R. 1991. Effect of fungicides on key diseases of groundnut in relation to crop stage. Tropical Pest Management, **37**: 270-272.

99. Seshadri, K. 1976. Effect of fungicides in the control of groundnut rust. Madras, Agri. J., **63**: 416-417.

100. Schiller, J.M. and Indhaphum, P. 1979. Economic control of *Cercospora* leaf spot and rust in rainfed peanut production. Prot. Ecol. **1**: 109.

101. Shanmugam, N., Thangavelu, P. and Vasantharaj, D. 1975. Occurrence of a myophagus thrips on the recently recorded groundnut rust. Science & Culture, **41**:80.

102. Sharma, N.D., Vyas, S.C. and Jain, A.C. 1977. *Tuberculina costanicana* syd: A new hyperparasite on groundnut rust (*Puccinia arachidis* Speg.) Curr. Sci. **46**: 311-312.

103. Sharma, N.D., Vyas, S.C. and Jain, A.C. 1980. *Euderluca caricis* (Fr.) O.Eriks, a new hyperparasite on *Puccinia arachidis* Speg., the incitant of groundnut rust. Science & Culture, **46**: 50.

104. Sharma B.D. and Mukhrji, S.K. 1972. A new record of rust on groundnut (*Arachis hypogaea* L.) in India, Curr. Sci. **41**:229.

105. Shekhawat, P.S., Patel, V.N. and Patel, J.G. 1985. Economic fungicidal spray schedule for control of tikka and rust diseases of rainfed groundnut. Indian J. Mycol. & Pl. Pathol. **17**: 11-16.

106. Shinde, P.A. and More, W.D. 1975. Out break of groundnut rust at Rahuri, The Res. Jour. M.P.K.V., **6**: 75-76.

107. Siddaramaiah, A.L., Prasad K.S.K. and Hegde, R.K. 1977. Chemical control of groundnut rust. Pesticides, **11**: 38-39.

108. Siddaramaiah, A.L., Desai, S.A., Hegde, R.K. and Jayaramaiah, H. 1980. Effect of different dates of sowing of groundnut on rust development in Karnataka. Proc. Indian Nat. Sci. Acad. (B), **46**: 389-392.

109. Singh, A.K., Subrahmanyam, P. and Moss, J.P. 1984. The dominant - nature of resistance to *Puccinia arachidis* in certain wild *Arachis* species. Oleagineux, **39**: 535-537.

110. Singh, G.P. 1978. Nematode pod rot and rust, two serious diseases of groundnut in Rachi. Indian Phytopath., **31**: 357-358.

111. Singh, F. and Oswalt, D.L. 1991. Groundnut production practices. Skill Development series No. 3 ICRISAT Centre, Patancheru, A.P., India. pp. 1-40.

112. Smith, D.H. and Littrell, R.H. 1980. Management of peanut foliar diseases with fungicides. Plant Disease, **64**: 356-361.

113. Sokhi, S.S. and Jhooty, J.S. 1982. Factors associated with resistance to *Puccinia arachidis* Peanut Sci. **9**: 96-97.

114. Spegazzini, C.L. 1884. Fungi Guaranitici. Anales de la sociedad Cientifica Argentina, **17**: 69-96, 119-134.

115. Stockdale, F.A. 1914. Annual Report of the Department of Agriculture, Mauritius (cited from Bronfield, 1971).

116. Subba Rao, P.V., Geiger, J.P., Einhorn, J., Malosse, C., Rio, B., Savery, S. and Ravise, A. 1988. Isolation of methyl linolenate, a new antifungal compound from *Arachis hypogaea* L. leaves infected with *Puccinia arachidis* Speg. Oleagineux, **43**: 173-177.

117. Subba Rao, P.V., Einhorn, J., Geiger, J.P., Malosse, C., Rio, B. and Ravise, A. 1991. Alkyl bisphenyleethers, new phytoalexins produced by *Arachis hypogaea* L. infected with *Puccinia arachidis* Speg. (Fr.). Oleagineux, **46**: 501-507.

118. Subrahmanyam, P., Gopal, G.R., Malakondaiah, N. and Reddy, M.N. 1976. Physiological changes in rust infected groundnut leaves. Phytopath. Z. **87**: 107.

119. Subrahmanyam, P., Vijayakumar, C.S.K. and Rao, A.S. 1978. Transpiration in rust infected leaves of groundnut, Curr. Sci. (India), **47**: 302.

120. Subrahmanyam, P., Reddy, D.V.R., Gibbons, R.W., Rao, V.R. and Garren, K.H. 1979. Current distribution of groundnut rust in India, PANS, **25**: 25-29.

121. Subrahmanyam, P., Gibbons, R.W., Nigam, S.N. and Rao, V.R. 1980. Screening methods and further sources of resistance to peanut rust. Peanut Science, **7**: 10-12.

122. Subrahmanyam, P. and McDonald, D. 1982. Groundnut rust its survival and carry over in India, Proc. Indian Acad. of Sci. (Plant Sci.), **91**: 93-100.

123. Subrahmanyam, P., McDonald, D., Gibbons, R.W., Nigam, S., M and Nevill, D.J. 1982. Resistance to rust and late leaf spot diseases in some genotypes of *Arachis hypogaea*. Peanut Science, **9**: 6-10.

124. Subrahmanyam, P. and McDonald, D. 1983. Rust disease of groundnut. Information Bulletin No. 13. International Crops Research Institute for the Semi-Arid Tropics, Patancheru, A.P., India.

125. Subrahmanyam, P. and McDonald, D. 1987. Groundnut rust disease : Epdemiology and control. ICRISAT (International Crops Research Institute for the Semi Arid Tropics), 1987. Groundnut rust disease. Proceeding of a Discussion Group Meeting, 24-28 Sep. 1984. ICRISAT Centre, Patancheru, A.P., India.

126. Subrahmanyam, P., Moss, J.P. and Rao, V.R. 1983. Resistance to peanut rust in wild *Arachis* species. Plant Disease, **67**: 209-212.

127. Subrahmanyam, P., Reddy, L.J., Gibbons, R.W. and McDonald, D. 1985. Peanut rust: a major threat to peanut production in the semi-arid tropics. Plant Disease, **69**: 813-819.

128. Subrahmanyam, P. and McDonald, D. 1986. Diseases of rabi groundnut in India. Workshop-cum-seminar on high yielding technology for rabi/summer oilseeds, Directorate of Oilseeds Research, Rajendranagar, Hydereabad - 30, Nov. 25, 1986.

129. Subrahmanyam, P., Ramnatha Rao, McDonald, D., Moss, J.P. and Gibbons, R.W. 1989. Origin of resistances to rust and late leaf spot in peanut (*Arachis hypogaea*, Fabaceae). Economic Botany, **43**: 444-455.

130. Subrahmanyam, P., McDonald, D., Waliyar, F., Reddy, L.J., Nigam, S.N., Gibbons, R.W., Ramnath Rao., V., Singh, A.L., Reddy, P.M., and Subba Rao, P.V. 1991. Screening methods and sources of resistance to rust and late leaf spot of groundnut. ICRISAT Information Bulletin (Unpublished).

131. Subrahmanyam, P., McDonald, D., Reddy, L.J., Nigam, S.N., and Smith, D.H. 1993. Origin and utilization of rust resistance in groundnut. In: Durability of resistance : Proceedings of a symposium International Agricultural Research Centre, Wageningen, Netherland, 24-28, Feb., 1992. pp. 147-158.

132. Tai, F.L. 1937. A list of fungi hiterto unknown from China. Sci. Rep. Tsing-Hau Univ., Sev. B., **2**: 191-639.

133. Thomson, S.S. and Smith, D.H. 1971. Distribution and significance of peanut-rust in Georgia in 1971. Plant Dis. Rep., **56**: 290.

134. Tiwari, S.P., Ghewande, M.P. and Misra, D.P. 1984. Inheritance of resistance to rust and late leaf spot in groundnut (*Arachis hypogaea* L.) J.Cytol. & Genet., **19**: 97-101.

135. Toler, R.W., Cuelluar, R. and Ferrer, J.B., 1959. Preliminary survey of plant disease in the Republic of Panama. Plant Dis. Rep., **43**: 1201.

136. Tripathi, N.N. and Kaushik, C.D. 1978. A new record of groundnut rust from Haryana, Indian Jour. Agric. Res. **12**: 273.

137. Usman, K.M., Durairaj, P., Karunanithi, K. and Thangavelu, S. 1992. Effect of neem products on the control of foliar diseases of groundnut. In Groundnut : a global perspective: Proceedings of an International Workshop, ICRISAT Centre, Patancheru, Andhra Pradesh, India, 25-29 Nov. 1991, Ed. S.N. Nigam, pp. 474.

138. Van Arsdel, E.P. 1973. Origin of South Texas peanut rust epidemics, presented at symposium: The Aerobiology of Diseases. Pests and Allergens in Western Hemisphere, Abstr., Am. Phytopathold. Soc., Mexico City, June 20 to July 4, 1973, 11.

139. Van Arsdel, E.P. and Harrison, A.L. 1972. Possible Origin of peanut rust epidemics in Texas, Abstr., Phytopathology, **62**: 794.

140. Vaishnav, M.V. and Kapadia, M.N. 1982. Occurrence of mycophagus midge, *Mycodiplosis* sp. on groundnut rust in Gujarat. Pesticides, **16**: 21.

141. Veeranjaneyulu, M., 1973. Studies on rust disease of groundnut. M.Sc. (Ag.) Thesis, A.P.A.U., Tirupati. pp. 74.

142. Verma, P.V., Ravendran, T.S. and Ganapathy, T. 1991. Genetic analysis of rust resistance in groundnut (*Arachis hypogaea* L.) Journal of Oilseeds Research, **8**: 35-39.

143. Verma, B.K. and McDonald, D. 1987. Groundnut rust disease and plant quarantine. ICRISAT (International Crops Research Institute for the Semi-Arid Tropics). 1987. Groundnut rust disease. Proceedings of a Discussion Group Meeting, 28-28, Sep. 1984, ICRISAT Centre, Patancheru, A.P., India.

144. Vidyasekaran, P. 1981. Control of rust and tikka diseases of groundnut. Indian Phytopath. **34**: 20-23.

145. Vyas, S.C., Prasad, K.V.V., Shastry, P.P. and Shukla, B.N. 1986. Field control of leaf spots and rust of groundnut by fungicides. Indian Phytopath. **39**: 447-448.

146. Wingfield, M.J. and Rijkenberg, F.H.J., 1981. A preliminary ultrastructural study of the uredial stage of *Puccinia arachidis*. Phytophylactica, **13**: 53.

147. Wongkaew, S. and Surin, P. 1984. Racevariation in groundnut rust (In Thai, Summary in Eng.) Presented at the Second Groundnut Research Workshop. 1983. Nakhon Sawan.

148. Yadav, H.L., Swarup. J. and Saksena, H.K. 1975. Occurrence of groundnut rust (*Puccinia arachidis* Speg.). A new record for Uttar Pradesh. Indian Journal of Farm Sci. **3**: 109.

149. Zambettakis, C., Picasso, C. and Bosc, J.P. 1988. Epidemiological aspects of groundnut rust in Burkina Faso. Oleagineux, **43**: 304-306.

150. Zheng Guangrou, 1987. Breeding for resistance to groundnut rust in the People's Republic of China. ICRISAT (International Crops Research Institute for the Semi Arid Tropics), 1987. Groundnut rust disease, Proceedings of a Discusion Group Meeting, 24-28 Sep. 1984, ICRISAT Centre, Patancheru, A.P., India.

151. Zhou, L.G., Huo, C., Liu, J. and Liu, Z. 1980. Studies on peanut rust in Guangdong province, People's Republic of China, (In Ch., Summary in En.). Acta Phytophylacica Sinica, **7**: 67-74.

17 Enhancing Productivity of Groundnut in India through IDM

C.D. Mayee

CONTENTS

Integrated Pest and Disease Management
Rajeev K. Upadhyay, K.G. Mukerji, B.P. Chamola and O.P. Dubey (eds.)
APH Publishing Corporation, New Delhi-110002, India

ABSTRACT

Groundnut disease are a limiting factor for successful production of the crop in India. Strategies of management based on cultural practices, chemicals and resistance are highly effective and responsive provided they take into consideration the dynamics of disease progress. The foliar diseases; rust and leaf spots, are fast multiplying pathogens and 6 to 7 infection cycles in a season lead to epidemic situation. They are governed by weather factors and forecasting would be rewarding in their management. It is estimated that nearly 25% groundnut productivity can be increased and yield of commercial cultivars can be stabilized around three metric tonnes in rainy season crops through disease management. The soil-borne diseases, though have slow multiplication rate, can be serious in patches over years. Bud necrosis virus disease caused by TSWV is yet another potential threat which can be to a large extent managed by monitoring the vector population and peaks of vector activity can be determined through weather parameters from correct prediction of the disease according to agro-ecological zones.

1. INTRODUCTION

The genus *Arachis* (Subtribe : Stylosanthineae, Tribe : Hedysareae ; Family : Papilionaceae) to which the groundnut or peanut belongs is of South American origin with centre of diversity of brazil. Nearly 22 distinct species are well described (20) which are mostly diploid. *A. hypogaea* L., the cultivated groundnut, however, is tetraploid and has two subspecies viz., *hypogaea* and *fastigiata.* In India, groundnut is one of the most important oilseed and cash crops. It occupies nearly 7 m ha area with production fluctuating from 5 to 6.4 m tonnes. Although India ranks first in production in world, the average yield is low and ranges between 700-900 kg/ha (14, 19), as against world's average yields of 1000-1100 kg/ha or on-farm yields obtained in developed countries (2500 kg/ha).

Poor realization of the potential yields has been attributed to several reasons where diseases has been considered as one of the significant factors (8, 18, 48, 66, 70). The role of groundnut diseases in reducing yields has been clearly demonstrated (18, 70). To the already established seed, soil-borne problems and leaf spots, rust, bud necrosis, and other viruses have added new dimensions in the last twenty years. Diseases render the crop uneconomical in the rainy season which is the major period of groundnut cultivation. Conservative estimates of annual loss of pod yield by foliar diseases alone is about 1 m tonne (35).

In India, fungal diseases viz., leaf spots, rust, aflatoxin problem and stem and pod rots are of great importance. Though the crop is susceptible to wide range of virus diseases, bud necrosis and stunt are considered potentially severe. Nematode diseases are restricted but cause concern under intensive cultivation. This chapter describes the major diseases of groundnut, their biology, epidemiology and integrated management practices that have direct bearing on increasing groundnut productivity.

2. BIOLOGY AND EPIDEMIOLOGY

2.1 Seed and soil borne fungal diseases

A variety of fungal plant pathogens invade ground at all stages of growth. Crown and collar rot caused by *Aspergillus niger* (55), white mold caused by *Sclerotium rolfsii* (47) and pre-emergence rot induced by *Fusarium* spp., *Rhizoctonia* spp., *Pythium* and several other fungi are important during early growth period.

Aflaroot caused by *Aspergillus flavus* is of minor importance in India (9). As the crop establishes, it is prone to root and stem rot caused by *Macrophomina phaseolina* (Tassi) Goid. (Sclarotial state-*Rhizoctronia bataticola* (Taub) Butt.), blight caused by *Sclerotium rolfsii* Sacc. (Perfect state- *Corticium rolfsii* Curzi) and to a lesser extent to *Fusarium* root-rot and wilt. The symptoms of *Fusarium* root-rot and of *M. phaseolina, Rhizoctonia* spp. and *S. rolfsii* are characterized by a progressive rotting of root system of basal portion of the stem leading to chlorosis and wilting. True wilts induced by *Fusarium* and *Verticillium* are less common than the damage induced to pegs and pods. The pathogens mostly survive in plant residue as saprophytes or in soil as sclerotia, chlamydospores, conidia or mycelium. Some of them are also carried on the seed surface or inside the seed tissue. Soil environment such as temperature, moisture and biotic factors play a great role in the disease development. Crown rot is favoured by delayed emergence and high soil and air temperatures. *Sclerotium rolfsii* is more active near the soil surface due possibly to the presence of an adequate food base and absence of competitive or antagonistic fungi. Antagonism normally increases by high soil moisture. *Macrophomina phaseolina* induced dry rot is more severe at high soil temperature and low soil moisture. Slow growing plants are quickly attacked by *Fusarium* and *Aspergillus.* In majority of the diseases, soil sickness developed over a period of time contributes largely to endemic situations. Such diseases, therefore, are often restricted to specific areas, where they occur with regular frequency.

2.2 Foliar fungal diseases

Diseases characterized by sudden and serious damage to all aerial parts of the plants include the widely distributed *Mycosphaerella* leaf spot and rust (40, 48, 66). The leaf spots caused by *Pestalotiopsis arachidis, Phyllosticta arachids hypogeae, Phomopsis* sp. and leaf blights due to *Myrothecium roridum* Tode ex. Fr. and *Alternaria alternata* (Fr.) Keissler (=*A. tenuis* Nees) are locally important. *Alternaria alternata* is fairly wide spread in coastal area where high humidity is prevalent for prolonged periods. The major foliar fungal disease problems, however, are 'tikka' leaf spots and rust.

2.2.1 Cercospora leaf spots

The *Cercospora* or 'tikka' leaf spots are generally regarded as the most important fungal diseases of groundnut occurring wherever the crop is cultivated (18, 48). Two principal species are involved : *Cercospora arachidicola*, the cause of early leaf spot and *Phaceosariopsis personatum* which causes late leaf spot (25, 26). *P. personatum* was first described in the U.S.A. in 1885 and *C. arachidicola* was reported first from Japan in 1919 (22).

The existence of two distinct species has been recognized. Their perfect stages are *Mycosphaerella arachidicola* and *M. berkeleyi* for *C. arachidicola* and *P. personatum,* respectively. Strictly these fungi should be referred by their perfect stages. However, since these diseases are commonly referred to as Cercospora leaf spots, for it is their conidial stages which are normally found in the field.

Symptoms of the two leaf spots are indistinguishable in the early stages of their development. Later, lesions caused by *C. arachidicola* tend to be lighter brown, less regular in outline and more often surrounded by a yellow halo than those caused by *P. personatum* (22, 48). *C. arachidicola* produces conidia from scattered conidiophores mainly on upper leaf surface, while *P. personatum* sporulates chiefly on lower leaf surface, conidiophores being arranged concentrically. Lesions induced by *C. arachidicola* normally first appear 3-4 weeks after sowing and those by *P. personatum* appear some 2-3 weeks later. Despite its later development, *P. personatum* is potentially more destructive because of its much higher rate of spread leading to more rapid defoliation (22, 23). Early maturing groundnut cultivars can escape the epidemic.

The main effect of both the leaf spots is premature defoliation which begins when 6% of their leaf area is diseased. Severe infections may cause complete defoliation. However, lesions are not confined to the leaves but can also occur on stems and pegs, leading to deterioration of pods. Losses to the groundnut crop may be as high as 60% under experimental conditions. The leaf spot fungi produce perithecia but their role in survival is uncertain. The fungi efficiently perennate as mycelia and conidia in plant residue. However, it seems likely that volunteer seedlings and continuous availability of groundnut in many parts of India play major hole in carry-over of the pathogens. Secondary spread of conidia takes place by rain-splash, wind and insects. Infection is favoured by temperatures between 25° to 30°C and long periods of high humidity. High humidity appears to be more critical than temperature in the germination of conidia and subsequent infection. The race situation of these pathogens is unclear although indications are available in case of *C. arachidiocola.*

2.2.2. Rust

The geographical distribution and potential economic importance of groundnut rust *(Puccinia arachidis)* have dramatically increased over the last decade, not only in India but also in many other countries of Africa, Asia, Oceania and Australia (21, 31, 33, 34, 40, 42, 70).

The first symptom of rust normally appears some 6-10 days after infection, but the incubation period varies with season (30). Whitish flecks are seen on lower surface of leaves, then yellow-green spots appear on the upper surface. Minute orange-red pustules develop within the flecks, enlarging rapidly before rupturing to expose the reddish brown urediospores. The tissue surrounding the pustules becomes necrotic and in severe infections the entire leaflet may dry up and fall prematurely. Outbreaks commonly occur in patches, often termed as 'hot-spots', but sever epidemics can be substantial, particularly where the *Cercospora* leaf spots are also involved. In India, the extent of crop loss has been estimated to be about 27% from rust infection alone, and upto 70% when leaf spot is also present (18, 66).

Groundnut rust perpetuates, spreads and produces severe outbreaks by means of urediospores. Though it infects other species of the genus *Arachis,* they can hardly be involved in perpetuation of rust in the country. Since urediospores are short lived in infected crop debris, the gap in survival is filled by the availability of host in the system of continuous cultivation of groundnut without any break in India (31-40).

Rust epidemic culminating in a speed of 0.35 to 0.40 units per day (infection rate) have often been calculated in Maharashtra during rainy season (32,33, 36, 46). The epidemic is favoured by temperatures around 22-25°C, presence of water on the leaf surface and high humidity. *C. arachidis* probably exists in racial forms but the race flora is not yet clear.

2.3 Aflatoxin problems

Species of *Aspergillus* are normal components of soil microflora. *A. niger* causes seed and seedling diseases of groundnut which may be locally important. *A flavus* occasionally causes a seedling root rot. The fungus is especially important in view of its ability to produce aflatoxin which is thought to cause liver damage in livestock and man (27). In 1960, widwapred fatalities occurred in poultry fed on groundnut meal which had been imported into the UK, the source of the toxin being traced to *A. flavus*. The consequences of consumption of *A. flavus* infected grains, are hard to appraise, but it can exert long term, particularly carcinogenic, effects many years later (5, 49, 52).

Invasion of groundnut seed by aflatoxigenic strains of *A. flavus* and subsequent aflatoxin contamination can occur both before and after harvest especially during drought stress situation. A considerable body of information is available on the conditions favouring the growth and development of *A. flavus* (49, 50, 52). Undamaged pods at maturity are commonly infected with *A. flavus* but kernels are normally free of the fungus. Damage to pods, caused either by mechanical means or by parasitic attack, tends to increase seed infection. Late season drought stress is the most important factor for *A. flavus* invasion and aflatoxin contamination. Therefore, the problem of aflatoxin is more serious in the rainfed area of India than it is thought to be. Groundnuts invaded by aflatoxigenic strains of *A. flavus* in the soil before harvest can lead to a serious contamination during drying and storage. The post harvest drying conditions tend to influence the degree of seed infection and aflatoxin contamination. Delayed harvesting also leads to increased fungal invasion.

2.4 Virus diseases

2.4.1 Bud necrosis

Amongst the virus diseases of groundnut, bud necrosis caused by tomato spotted wilt virus (TSWV), is probably the most important disease in India where crop losses upto 80% have been reported (3, 15, 29). The disease is present in every groundnut growing region of India, incidence ranging from 5-80% (16). Symptoms under field conditions develop within about six weeks of sowing. Initially chlorotic ring spots appear on leaves. Subsequently there is necrosis of petioles, stems and terminal bud. New leaves may be reduced in size and show distortion, mosaic mottling or chlorosis. Plants remain stunted. Kernels from infected plants are shrivelled and testae are discoloured. However, the virus TSWV is not seed borne in groundnut. It is transmitted by thrips *Scirtothrips dorsalis* Hood (29) and *Frankliniella schultzei* (Trybom) (16), the latter being more important vector. Primary spread is achieved by the migrant viruliferous thrips which are carried by the prevailing winds. The secondary spread is extremely limited. The nymphs are capable of acquiring TSWV and transmit the virus as adults. The disease occurs throughout the year the incidence being correlated with the vector population. It has a wide host range but the relative importance of individual crop and weed host needs to be closely assessed (4).

2.4.2 Peanut clump

The disease is caused by peanut clamp virus (PCV). Crop losses upto 60% have been reported from Punjab (15). Under natural conditions 14% seeds transmit the disease. Fungus also transmits the disease

2.4.3 Peanut mottle

Peanut mottle virus (PnMV) is probably the most widespread virus infecting groundnuts as well as various other legumes (59). In groundnut, PnMV induces a systemic mottle and necrosis. Older leaves may become distorted and show mosaic symptoms. Yield losses of 15-45% have been recorded (59). The virus is transmitted through the seed to the extent of 8.5% (2). PnMV is transmitted by aphids in the non-persistent manner.

2.5 Nematode problem

Groundnut is attracted by various type of nematodes in India of which 30% loss is due to *Tylenchorhynchus brevilineatus* and 70% losses due to *Meloidogyne arenaria* and *M. halpa* (47). The interaction of root knot with fungi, especially *Fusarium* and *Aspergillus* may lead to losses exceeding those caused by either pathogen alone. Very limited work on nematodes has been done in the country.

3. DISEASES MANAGEMENT COMPONENTS

3.1. Quarantine aspects

Quarantine plays an important role in the prevention of entry of pathogen into an area where it is not present. If precautions are not taken at right time introduction of a new disease can have serious consequences. Rust of groundnut was considered a risk through seed rust of groundnut was considered a risk through seed introduction (34). It is pertinent to know serious diseases of groundnut outside India.

Groundnut rosette is not present in India, but is a serious disease in Africa and South of Sahara (58). Similarly, witches broom is economically important in South East Asia. The bacterial wilt due to *Pseudomonas solanacearum*, *Cylindrocladium* black rot (6) and several other fungal, viral and nematode diseases have not yet been detected in the country. Therefore, a constant watch on such diseases is necessary.

3.2. Cultural and physical

Several groundnut diseases, which are less damaging to the crop, can be maintained at low levels by adopting sanitation, clean cultivation and other cultural practices. Eradication of volunteer groundnut plants and ground keepers is often advocated for reducing the primary inoculum of foliar and soil-borne pathogens.

It has been unequivocally demonstrated that rust perpetuates on groundnut itself because of continuous cultivation practices. Wherever possible, field management should include a clear break in time between successive groundnut crops. Crop rotation, destruction of volunteer groundnut plants and infected debris may limited sources of primary inoculum. It delays the onset of epidemic, but such measures alone are unlikely upto be fully satisfactory (34). Mayee (32) critically examined the effects of fertilizer application on fungal diseases of groundnut. Nitrogenous fertilizers neither reduce nor increase the disease. However, low level of nitrogen favour build up of foliar pathogens. Phosphatic fertilizers (40 to 60 kg/ha) reduce the incidence of rust and leaf spots particularly in phosphorus-deficient soils. These practices together with clean cultivation are extremely important in delaying the initial infections by foliar pathogens. Certain hyperparasitic fungi, including *Eudarluca* spp., have been found on uredia of *P. arachidis* (60) but their role in the biological control of groundnut rust is unknown.

Aflatoxin contamination can be appreciably prevented or reduced by using some cultural and storage practices. Preharvest contamination can be reduced by avoiding drought stress and insect and or fungal damage to developing pods. Harvesting the crop at proper maturity (avoiding delay), drying the produce rapidly and maintaining it in dry state, removal of damaged or moulded pods from the produce and drying the produce to a safe moisture level about 9% are the essential measures for the cultural control of the aflatoxin problem (52). Most of these management practices have been applied with considerable success by progressive farmers in the developed countries but have not been widely adopted by small growers in developing countries (51).

The most pronounced effect of cultural practices has been demonstrated on the incidence of bud necrosis disease. Early sowing of groundnut at the onset of rainy season decreases the bud necrosis incidence and losses due to the disease can be avoided. Crops sown late in the rainy season are exposed to high levels of vector influx during seedling stage and thus losses are more. It is necessary to monitor population of thrips in groundnut to determine the optimum time of sowing by avoiding the peak population influx period. Similarly, it has been repeatedly demonstrated that losses due to bud necrosis can be minimized by high plant density, closer spacing, (37.5 cm x 5.0 cm; between rows and plants) considerably reduce the disease (67-78%) and increase the yield (3). Growing of tall intercrops in groundnut, such as pearl millet, interferes in the movement of thrips and therefore, reduces disease incidence. However, it is necessary to evaluate the overall performance of groundnut in intercrops and vice versa.

The damage due to peanut mottle virus, transmitted primarily through seed can also be reduced by some cultural methods. This can be achieved through production of seed in areas free of virus in India, as has been attempted elsewhere (12).

3.3. Chemical

3.3.1. Seed and seedling diseases

Pre-emergence seed rot and post-emergence seedling mortality are of common occurrence which reduce yield. The diseases develop either from fungi already present in the seed or result from direct invasion of seeds and seedlings by soil fungi. *Aspergillus niger, A. flavus, Rhizoctonia solani, Fusarium* spp., *Sclerotium rolsfii* and *Macrophomina phaseolina* are the most predominant fungi responsible for seed and seedling diseases. It has been estimated that the fungal damage to seed is responsible for nearly 5% loss of the plant population in the best managed crop irrespective of the cultivar. The patchy stand in groundnut is partly due to such invasion. The damage to seed and seedling can be avoided by chemical seed treatment prior to sowing (13, 43). Seed dressing with thiram, mancozeb, ceptan, captafol and quintozene has been found very useful. Recently, carbendazim and carboxin have been found highly effective in protecting seeds and seedlings from fungal infections. Genotype chemical interaction exists specially in the systemic chemicals. Seed rate in groundnut varies from 80 to 120 kg/ha and as the seed coat is delicate seed treatment requires special care. The cost of seed treatment is negligible as compared to the cost of seed and the benefits are assured.

3.3.2. Foliar diseases

It is pointed out that (62) the prospects for retaining existing chemical strategies for foliar disease management in groundnut holds promise. Cercospora spots and rust (17, 35, 37, 44, 69), can be effectively controlled by fungicides. The economic benefit depends on climatic conditions the cultivar and its rate of maturity, the nature of the farming system as well as the local importance of the disease. In recent years, large number of fungi-toxicants have been tested for their effectiveness in controlling foliar diseases. Chlorothalonil is highly effective against both rust and leaf spots in USA (62). Carbendazim, tridemorph, oxathins, tin compounds, mancozeb, maneb, zeneb and inorganic sulphurs are also effective and economical (37, 41, 44). Although fungicidal control is effective yet very few farmers have adopted it. Low crop yields, fluctuating groundnut prices, non-availability of the chemical and even water for spray are some of the reasons of low adoption of chemical control. These problems are not unsurmountable. Improvement in supply of chemicals and spray machinery and information on how to use them can be efficiently generated with the help of extension agencies. In India, decision making in spraying is highly necessary because of resource constraints. Spray schedule was examined with response of tridemorph on rust development and found that the chemical applied during early disease development is highly remunerative (53). Similarly, applications of carbendazim for leaf spot control have been optimized (63).

Fungicidal control has given increased yields which commensurate with the cost of operations. However, there is a beed to integrate the management practices for all the diseases so that profitable strategy once recommended can be adopted. It is necessary to decide whether or not disease control should be undertaken. This will depend on several considerations. Once control by fungicide is planned, then decisions will have to be made on rate, timings and number of sprays. Forecasting, if available, will greatly help planning the chemical management strategy. The chemical control of foliar diseases resulted in variations in leaf area protection and in yield which were specific to each of the twenty genotypes (69). It appears that genotype chemical (and disease) interactions do occur and future strategies should take into account such possibilities.

3.3.3. Virus diseases

Among virus diseases, attempts have been made to control bud necrosis disease by controlling the vector (thrips) by insecticides. It is now well documented that insecticides are generally not effective in reducing disease incidence unless applied throughout the growing period (3). Carbofuron seed treatment is effective for a short period and the effect decreases considerably when major thrips invasion occurs, during rainy season. Dimethoate spray reduces the bud necrosis incidence in the post rainy season but it is less useful in the rainy season when thrips population is maximum. Frequent applications of insecticides, though reduce the bud necrosis incidence, are normally not economical. It is pertinent to point out that sublethal concentrations and delayed interval of application are sometimes harmful as they increase the vector activity and thereby help increase disease spread.

3.4. Host resistance

Resistance of various levels and types has been reported within cultivated groundnut to the Cercospora leaf spot, rust and *Aspergillus flavus* (1, 4, 7, 23, 26, 28, 35, 51, 53, 64-68, 71). In India, resistance in groundnut to crown rot caused by *A. niger* has been reported. Sources effective against bud necrosis have also been reported. However, despite these investigations, surprisingly few 'good sources' of resistance have been identified against the principal diseases of groundnut. Immunity against Cercospora leaf spot has not been found within *A. hypogaea*.

3.4.1. Cerocospora leaf spots

Attention has chiefly been directed at wild species of *Arachis* as potential sources of increased disease resistance. Recent evaluation of *Arachis* spp. has shown promise and despite sterility barriers, interspecific hybridization is receiving considerable attention (54, 61, 68). Nature of resistance to Cercospora leaf spots and its inheritance in groundnut is well understood. Resistance to the two species is inherited independently, and plant selections resistant to one are often very susceptible to the other. Resistance to leaf spots is associated with certain plant characters, such as, maturity plant habit and colour of foliage. Evaluation systems and screening for leaf spot resistance are fairly standardized. Longer incubation periods, reduced sporulation, less defoliation, stomatal exclusion mechanism and the absence of directed growth of germ tubes towards stomata are some of the recognized components of ' resistance to penetration' by *C. arachidicola* and *P. personatum* (56, 57). Resistance to colonization of the invaded host also occurs and is associated with the formation of a barrier around the site of infection composed of a pectic deposit on the cell wall which prevents further lesion development. Recent studies on the genetic control of leaf spot resistance suggest that several components of resistance in *A. hypogaea* are inherited in a quantitative manner. It was found that defoliation was controlled by genes with additive action (56, 57) and leaf spot resistance was determined by recessive alleles at five foci.

3.4.2. Rust

Sources of resistance to rust have also been well identified both within *A. hypogaea* and in wild *Arachis* species (54, 68). Histopathological studies revealed that *P. arachidis* is able to penetrate even immune wild species like *A. glabrata*, although development ceases after the formation of a single hypha from the sub-stomatal vesicle (56). Cultivars of groundnut differ in their susceptibility to infection, which is related to changing wettability. Germ tubes tend to pass over stomata without making successful contact (11). Close adhesion of the germ tube tip to the surface probably stimulates appressorium formation. In a resistant variety, EC - 76446, the fungus penetration, intercellular mycelial development and eruption of pustules prolonged. It is now well understood that prolonged incubation periods, low infection frequency slow rupture, less sporulation and inefficient germinability of spores produced in the uredia are major components of resistance to rust. In cultivated groundnuts, slow rusting type of resistance has been described. In all *A. hypogaea*, genes conditioning rust resistance are of recessive nature. Bigenic and trigenic inheritances occur in the interspecific crosses. However, involvement of higher order interactions due to linkages of two genes cannot be ruled out. Unlike other rust fungi, no pathogenic variants of *P. arachidis* have so far been distinguished and, therefore, the genetic stability may be on account of no known variability in the pathogen. Some measure of variation in pathogenicity occurs within certain groundnut pathogens including the Cercospora leaf spot fungi.

3.4.3. Seed and soil-borne diseases

Seed and soilborne diseases, though economically important, have been less subjected to investigations as regards plant resistance. Only a few sources of resistance have been identified against crown rot and root rot. This has restricted the progress and there is an urgent need to evolve suitable inoculation and screening procedures for identification of resistance.

3.4.4. Aflatoxin problem

Genetic resistance to seed invasion by aflatoxigenic *A. flavus* and to production of aflatoxin is possible (49-53). Resistance to penetration of pods and of the dry seed by *A. flavus* has been identified (53). The pigmentation of testa may also influence penetration by *A. flavus* (10). Resistant genotypes imbibe water at slower rates and dry out more rapidly than the susceptible ones. Genotypes, such as J-11, identified as resistant to aflatoxin, is already a recommended cultivar in India. Breeding for aflatoxin resistance, therefore, should not pose a problem. Field studies have provided evidence of resistance to insect pests and pod rots which help in reducing seed infection. There are also other possibilities of resistance to aflatoxin production in groundnut and these may or may not be correlated to seed infection and colonization by *A. flavus* . Certain pod characters are found helpful in the avoidance of pod damage by the fungus. These together with other characters of resistance will go a long way in developing aflatoxin resistant groundnut.

3.4.5. Virus diseases

Resistance to bud necrosis disease has been located (4). The lines with consistently good performance appear to be limited but have substantially less disease than the susceptible ones. The nature of resistance is not clear. However, low disease of some of the high yielding cultivars, such as, Robot 33-1, No. Ac 343 and M-13, appear to be non-preference by vector. Similarly, there is no immunity against peanut mottle virus but there are some cultivars which are tolerant (59). In some cultivars PnMV is not seed transmitted (58) and hence can be used as a source of resistance.

3.4.6. Combined resistance

Combined resistance to two or more diseases has been reported. For instance, some Peruvian accessions have resistance to both rust and Cercospora leaf spots (71). However, few commercial cultivars have adequate resistance to major diseases and there is considerable scope for incorporating available resistance to cultivated types into agronomically acceptable varieties.

Evaluation of *Arachis* germplasm including some wild species at ICRISAT (64-66) has confirmed the existence of many sources of diseases resistance. Certain wild species are resistant to rust, Cercospora leaf spot and tomato spotted wilt virus. *A. chacoense* possesses combined resistance to several diseases. Differences in ploidy level and incompatibility between certain wild species and *A. hypogaea* pose problem in employing these in resistance breeding. Other methods of gene transfer from compatible species are to make the gene pool of *Arachis* sp. available to groundnut breeders.

3.5. Biological

Recommendations are not available for control of groundnut diseases through direct biological means (40). Successful management of disease by manipulating the microbiota can be achieved by changing the soil-crop environment using diferent cultural practices. The soil-borne pathogens, such as *S.rolfsii, M. phaseolina, Rhizoctonia* spp. and *Fusarium* spp. produce dormant sclerotia and chlamydospores which are highly sensitive to soil fungistasis. These resting structures are killed by *Trichoderma* spp., *Fusarium roseum, Coniothyrum minitans* and several other fungi and bacteria. Treatments that predispose the sclerotia to microbial attack in oil include wetting and drying, flooding, sublethal fumigation, organic amendments and various forms of inorganic nitrogen. These suppress germination and growth of certain pathogens by increasing soil fungistasis.

Several hyperparasites of rust and leaf spots of groundnut have also been reported (66, 96). Although *Verticillium lecani, Penicillium islandicum, Eudarluca carisis, Darluca filum, Acremonium persicinum* and *Tuberculina costaricana* are known to parasitize pustules, serious attempts have not been made to use them in the biological control. In an air-borne disease, it is not the propagules that the killed that count but the ones, that are escaped are important in rebuilding the epidemic. In the coastal regions of Karnataka, *Hansfordia pulvinata* parasitizes tikka leaf spot pathogens in all the seasons. Further studies may reveal its potential as biological control agent.

4. EPILOGUE

Groundnut suffers heavily from several diseases in India. Losses caused by many diseases such as rust, leaf spots, bud necrosis, soil borne root and collar rots and aflatoxin can be avoided. Every effort should be made to identify the yield limiting diseases within an area/locality or agroecological zone (24). If fungicides are to be applied, they should be broad spectrum so as to cover all important pathogens. A serious endeavour is still necessary to combine resistance against major pathogens. An integrated disease management strategy that can fit comfortably with other growing practices of groundnut will go a long way in boosting production of groundnut in India.

5. SYSTEM APPROACH TO DISEASE MANAGEMENT

Management of groundnut diseases principally has to be integrated with the crop management using a system approach. Preplanting crop sanitation (eg. removal or burning of crop residues and removal of weeds through ploughing, harrowing etc.) is an important cultural measures for the reduction of seed and soil-borne diseases such as : seedling mortalities, stem rot, pod rot and dry rot caused by *Sclerotium, Rhizoctonia, Pythium, Aspergillus, Fusarium* etc. The early and late spot Pathogens are drastically reduced by such cultural practices. The success of cultural measures in the management of necrotrophic pathogens lies in depriving the causal fungus of food base and thus should be designed to burry all necrotrophic food base deep into the soil and prevent accumulation of such food base around the crown area of the plant during crop season. Rouging and rotation of groundnut with non-susceptible crops such as corn, cotton, wheat can also reduce the incidence and severity of several soil-borne problems such as ; stem and collar rots, nematode damage and also foliar diseases. Such crop rotations are extremely useful in Vertisols in reduction of stem and pod rot of groundnut. Similarly, aflatoxin preharvest field infection in normally extremely low in groundnuts grown on Vertisols than Alfisol soils and thus could be exploited for raising aflatoxin-free pods for export purposes in marked areas. Supplementary cultural practices that are designed to get better crop growth are highly essential for reduction of groundnut diseases. Soil solarization should prove useful in reducing all the soil-borne diseases including nematode damage where it involves mulching of soil with transparent polyethelene sheets in the hot season before sowing. Solar heating affects the survival ability of the pathogens to antagonists. Practices that reduce moisture retention in the canopy can discourage infection and development of not only rust and leaf spots but also of stem and pod rot.

Planting, therefore, on raised beds helps reduce moisture accumulation and assists in washings of leaf litter in the crown zone of plants. Irrigation has profound influence on the diseases. Scheduling irrigation with optimal dry periods between irrigation which are useful for crop growth are also beneficial in reduction of diseases. It is pertinent to avoid small quantities of water frequently as continuous canopy wetness will result in higher incidence of the diseases. Certain organic amendments and inorganic fertilizers can help limit the disease incidence and severity of number of groundnut diseases. High levels of calcium in plant tissue partially offset the action of oxalic acid and cell wall degrading enzymes of the pathogens. Therefore, application of gypsum is suggested as a means of better crop growth and reduced diseases. One of the classical approach to reduce bud necrosis diseases is to have higher population density of the crop. At the rate of 120 kg/ha seed rate, the groundnut can be planted at 30 cm x 10 cm rows (bunch) or 45 cm x 10 cm (spreading) for effective control of bud necrosis and obtain higher yield.

There is no substitute for selection of good seed material of a resistant or tolerant cultivar and seed dressing of the planting material prior to sowing. Cultivars with tolerance to foliar diseases like FDRS-4, FDRS-10, ICGS-76 are available similarly, J-11 has proved tolerant to aflatoxin contamination. Use of good quality, disease-free seed can help control seed rots and seedling diseases. This can be ensured by careful attention to growing, harvesting, drying and storing of sowing materials from previous season. Application of seed protectant fungicides (Thiram, Captan or Captafol) or some eradicant fungicides Bavistin (Carbendazim), Vitavax (carboxin) or mixtures of fungicides such as carbendazim and thiram, PCNB and Captan 75 WP, carboxin and thiram in suitable proportion have proved effective in preventing attack on seed and seedlings by soil fungi also protecting seedlings from seed-borne and aerial pathogens. Some chemicals are also compatible with beneficial microorganisms like, *Rhizobium* and *Trichoderma* and hence shall not pose problem while integrating biological management system with the regular practices of seed treatment.

Subsequent crop growth of groundnut is likely to be damaged by foliar diseases and diseases which attack stems, roots pegs and pod at adult stages. Diseases monitoring should be primary concern in management of these diseases. For example forecasting of rust of groundnut based on weather factors is possible. Similarly onset of late leaf spots normally follow rust in Central and South India during rainy season and post-rainy season. Also crop visits must be directed towards diagnosing early pustules or spots on the lower foliage to decide about spray schedules. Early control of the diseases in lag phases is highly remunerative and even in a most favourable diseases situation only 2 to 3 early sprays of carbendazim (0.1%) coupled with either tridemorph (0.07%) or mancozeb (0.20%) has proved very effective. Chemical control can be reduced substantially on groundnuts grown in Jan.-May, summer season as the temperatures are not suitable for disease development. However, early sprays between 30 to 60 days after planting during rainy and post-rainy seasons can ensure effective control of foliar diseases.

Aflatoxin contamination and of other pathogens can be reduced by proper harvesting the groundnuts. Harvesting the entire plants, and reversing the pods (exposing to light) in 'well' form for drying (2-3 days) before the pods are separated from plants have been shown to be extremely useful in retaining higher viability of seeds as well as low infection of kernels by pathogenic fungi. It is, therefore, very important to follow a defined system approach to disease management that can very well integrate with crop management practices. After all disease management is aimed at keeping diseases at levels that are not detrimental to yield and cultivation practices of groundnut are also aimed at achieving higher output of food and fodder.

Table 1. Important diseases of groundnut

	Diseases	Causal organism	Plant parts infected	Extent of yield loss (%)	Distribution in India	Mode of survival
	1	2	3	4	5	6
A.	**Diseases of Major Importance**					
1.	Rust	*Puccinia arachidis* pegs, stipules	leaves, petiole	50-60 India	Central and South volunteer plants	Uredinia and spores on
2.	Late leaf spot	*Phaeisariopsis personata*	leaves, petiole	upto 60	Widespread volunteers.	Spores in crop residue,
3.	Early leaf spot	*Cercospora arachidicola*	leaves, petiole	40-60 M.P., Gujarat	UP, Punjab, Orissa, M.S. in residue.	Conidi. asci mycelium
4.	Pre-emergence rots	Several fungi	seed, cotyledons, roots,	upto 40	In alll the States.	Seed and soil.
5.	Stem and pod rots	*Sclerotium rolfsii*	crown, root, stem	20-30	Endemic in some States.	Sclerotia in debris, soil soil-debris.
6.	Crown rot	*Aspergillus niger*	hypocotyl, seedling	8-10	Widespread	
7.	Bud necrosis	Bud necrosis virus (BNV)	whole Plant	upto 90	Widespread host, thrips transmitted	Collateral and alternative
B.	**Diseases of Increasing Importance**					
8.	Leaf blight	*Alternaria alternata*	leaves	2 to 10	Sporadic in coastal area	Plant residue
9.	Leaf blight	*Myrothecium roridum*	foliage	5 to 40	Orissa, Punjab, Karnataka	Residue
10.	Kalahasti malady	*Tylenchorynchus brevilineatus*	pods, pegs,	Local 50-60	Andhra Pradesh	In soil
11.	Clump	Peanut clump virus (PCV)	whole plant	More than 60	locally Punjab, A.P.	Seed, soil
12.	Stripe	Peanut stripe viruses (PStV)	leaves and aerial parts	2-10	Gujarat, Karnataka,	Seed and aphid transmitted
13.	Aflatoxin problem	*Aspergillus flavus*	pods, kernels	Serious health hazard	Widespread	Infected pods and seed
C.	**Diseases of Minor Importance**					
14.	Aflaroot	*Aspergillus flavus*	hypocotyl	2-5	Punjab, A.P.	Mycelium in seed, soil,
15.	Dry root rot	*Macrophomina phaseolina*	stem, root	20-25	Widespread	Sclerotia in soil, debris
16.	Wilt	*Fusarium oxysporium*	stem, root	1-5.	Sporadic	In soil, residue
17.	Mottle	Peanut mottle virus (PMV)	foliage	30-40	Sporadic	Seeds, aphid transmitted.
18.	Nagomeni mottle	Cowpea chlorotic mottle virus (CMMV)	whole plant	minor	Sporadic	Weeds/whitefly transmitted
19.	Bacterial wilt	*Pseudomonas solanacearum*	whole plant	minor	Sporadic	Seed, Weeds, Crop residue

Table 2 : Field diagnostic symptoms of groundnut of major importance

	Disease	Characteristic symptoms
1.	Rust	Orange coloured pustules appear on the lower surfaces of leaflets, rupture, release masses of reddish brown spores. Pustules appear subsequently on all aerial parts except flowers. Leaves dry and remain attached.
2.	Late leaf spot	Circular, dark spots occur on lower surface, slightly rough with rings of fruiting structures. Affected leaf lets become chlorotic, necrotic with coalescing lesions and premature senescence and shedding. Spots occur on stems, petioles, stipules and pegs.
3.	Early leaf spot	Sub-circular, brown lesions on appear surface of leaves, with chlorotic halo (mostly), less dark than late leaf spot and sporulation on upper surface. Spots on stems, petiole, stipules and pegs occur with distinct margins. Defoliation common.
4.	Leaf blight	*Alternaria* produces small chlorotic lesions, water soaked, distributed on leaf surfaces. Enlarge, become brown, coalesce, central portion dry and vein, veinlets, necrosed, premature senescence. *Mycothecium* lesions are round to irregular with grey-brown centres. Surrounded by chlorotic halos. Occur on both surface and rings of fungal structure develop.
5.	Seedling diseases	Infected seed and seedling reduced to dark brown spongy mass of rotted tissue covered with sporulating mycelium varied in colour as per fungal species. In crown rot, rapid desiccation of entire seedling occur and dark black mass occur at crown and extends to entire collar region. Damping off can occur due to *Rhizoctonia, Sclerotium, Rhizopus, Pythium, Fusarium* spp.
6.	Stem rot	Stem infection evident when yellowing and wilting of branches begin near the base. Sheaths of white mycelium develop near the soil line around infected area of the stem which become shreaded. Brown visible sclerotia develop. Entire plant or few branches are killed. PEG ROT, POD ROT similarly develop independently or together with stem rot. Seeds from diseased pods show plush-grey discolouration.
7.	Bud necrosis	Chlorotic spots on mottling occur on young leaf lets that develop into necrotic rings. Terminal bud necrose. Early infection results in stunting, proliferation of auxiliary shoots seeds small, shrivelled and mottled if produces.
8.	Peanut clump	Occur in patches in field. Leaves show mosaic mottling and chlorotic ring. Severe stunting of plant and dark green colour are conpicuous.

9.	Peanut mottle	Appear as irregular dark green islands. Occasionally interveinal depression, upward rolling of margins seen. Only few pods are produced.
10.	Kalahasti Malady	Appear in patches. Plants stunted with dark green foliage, small brown lesions appear on pegs and young buds. Peg length reduced. Advanced stage, the entire pod surface become black, discolouration of root.

Table 3 : Steps in the integrated management of major diseases

	The practice or step	Importance in management of
1.	Preplant sanitation (burning of crops residue, deep ploughing, harrowing)	Leaf spots, stem rot. seedling diseases.
2.	Crop rotation	Stem and pod rots, wilt, dry rot peanut clump, bacterial wilt.
3.	Organic amendments (FYM, compost, green manuring) and Gypsum addition at peg formation.	Stem, pod, dry rots and wilt, Kalahasti, leaf spots, seedling diseases.
4.	Planting on raised beds.	All soil and foliar diseases.
5.	Seed treatment	Seedling diseases.
6.	Seed treatment coupled with *Rhizobium* and *Trichoderma.*	Seedling diseases, leaf spots, stem, pod, dry rots and wilt.
7.	High density planting	Bud necrosis diseases
8.	Weeding during crop growth and prior to planting.	Virus diseases
9.	Vector control (Aphids, thrips, white flies)	Virus diseases
10.	Carbendazim+mancozeb or tridemorph sprays at 35 DAP, 50 DAP and/or 65 DAP.	Foliar diseases during rainy/post-rainy season
11.	Use of resistant or tolerant cultivars in endemic areas.	BND, Aflatoxin, foliar diseases.

* The crop should be sprayed at 0.5% Ferrous sulphate at 35 and 45 DAP to avoid iron chlorosis which is known to occur in many areas.

6. REFERENCES

1. Abdon, Y.A.M., Regory, W. C. and Copper, W.E. 1974. Sources and nature of resistance to *Cercospora arachidicola* Horiand *Cercosporidium personatum* (Berk & Curt) Deighton in *Arachis* spp.6-11.

2. Adams, D.B. and Kuhn, C.W. 1977. Seed transmission of peanut mottle virus in peanut. Phytopathology, **67** : 1126-1129.

3. Amin, P.W. 1983. Studies on arthropod vectors of groundnut viruses, their ecology and control. Groundnut Entom. Report, Gent. 3 ICRISAT, Patnacheru, India (Mimeographed).

4. Amin, P.W., 1985. Apparent resistance of Robit 33-1, groundnut to bud necrosis disease of groundnut. Plant Disease, **69**: 718-719.

5. Barnes, J.M. 1971. Aflatoxin as health hazard. J. appl. Bact., **33** : 285-298.

6. Benete, M. K. 1980. *Cylindrocladium* Black rot (CBR) disease of peanut. ICRISAT Proc. Int. Workshop. 13-17 Oct, 1980.

7. Bromfield, K. R. and Cevario, J. 1970. Greenhouse screening of peanut to rust (Puccinia arachidis) Pl. Dis. Reptr. **54** : 381-383.

8. Chohan, J. S. 1974. Recent advances in diseases of groundnut in India. In 'Current Trends in Plant Pathology' (Ed. Sp. Raychaudhury and J.P. Verma) Lucknow University, pp. 171-184.

9. Chohan, J. S. and Gupta, V.K. 1968. Aflaroot, a new disease of groundnut caused by Aspergillus flavus Link. Indian J. Agric Sci. **38** : 568-570.

10. Corker, R. D. 1980. Aflatoxin : Past, Present and future. Trop. Sci. **21** : 826-830.

11. Demski, J. W., Smith, D. H. and Kuha, C.W., 1974. Incidence and distribution of peanut mottle virus in peanut in the United States, Peanut Sci. **2**: 91-93.

12. Feakins, S.D. 1973. Pest control in groundnut. PANS Manual No. 2, Centre of Overseas Pest Control, London p. 197.

13. Fletcher, S.M., Zhang, P. and Carley, D.H. 1992. Groundnut production, utilization and trade in the 1980s. In 'Proc. IInd Int. Groundnut Workship. Nov. 25-30. ICRISAT, Patancheru, A.P. India.

15. Ghanekar, A.M. 1980. Groundnut virus research at ICRISAT. In 'Proc. Int. Workshop in Groundnuts, ICRISAT, Pattancheru, India.

16. Ghanekar, A.M., Reddy, D.V.R., Iizuka, N., Amin, P.W. and Gibbons R.W. 1979. But necrosis of Groundnut *(Arachis hypogaea)* in India caused by tomato spotted wilt virus. Ann. Appl. Biol. **93** : 17.179

17. Ghuge, S. S., Mayee, C.D. and Godbole, G.M. 1980. Development of rust and leaf spots of groundnut as influenced by foliar application of carbensdazim and tridemorph. Pesticides, **14** : 16-19.

18. Ghuge, S. S., Mayee, Godbole, G.M. 1981. Assessment of losses in peanut due to rust and tikka leaf spots. Indian Phytopath. **34** : 179-182.

19. Gibbons, R.W. 1980. The ICRISAT groundnut problem. In Proc. Int. Workshop in Groundnuts' ICRISAT, Patnancheru, India.

20. Gregory, W.C. and Gregory, M.P. 1976. Groundnut, *Arachis hypogaea* (Leguminosae-Papilionotae) In : Evolution of Crop Plants, Ed. N.W. Simmonds), Longman, London, pp 151-154.

21. Hammons, R.O. 1977. Groundnut rust in the United States and the Caribbean. PANS, **23** : 300-340.

22. Hemingsway, J.S. 1955. The prevalence of two species of *Cerocospora on groundnuts*. Trans. Br. Mycol. Soc., **38** : 243-246.

23. Hemingsway, J.S. 1957. The resistance of groundnuts to *Cercospora* leaf spot. Emp. J. exp. Agric. **25** : 60-68.

24. Ingle, R.V. and Mayee, C.D. 1986. Management of fungal diseases of groundnut. J. Oilseeds, Res., **3** : 201-204.

25. Jenkins, W.A. 1938. Two fungi causing leaf spots of peanut. J. Agric Res.., **56** : 317-332.

26. Kolte, S.J. 1985. Diseases of Annual Edible Oilseeds Crops, Vol. I. Groundnut, CRC Press Inc Boca Ration, Florida, U.S.A. p. 155.

27. Kraybil!, H.F. and Shimkin, M.B. 1964. Carcinogenesis related foods contaminated by processing and fungal metabolites. Adv. Cancer Res., **8** : 191-248.

28. Kuhn, C. W., Paguio, O.R. and Adams, D.B. 1978. Tolerance in peanuts to peanut mottle virus. Pl. Dis. Reptr. **62** : 365-368.

29. Mali, V.R. and Patil, F.S. 1979. Occurrence of tomato spotted wilt virus on groundnut in Maharashtra. Indian Phytopath. **32** : 193-197.

30. Mallaiah, K.A. 1976. A note on the seasonal changes in the incubation time of groundnut rust Curr. Sci. **45** : 26.

31. Mayee, C.D. 1982. Groundnut rust, a review. Indian Bot. Reptr. **1** : 75-83.

32. Mayee, C.D. 1983. Epidemiology and Forecasting of groundnut rust in Mahathwada Region. Final Report, ICAR, Adhoc Scheme MAU Parbhani, p. 76.

33. Mayee, C.D. 1986. epidemiology and management of groundnut rist in vistas in Plant Pathology' Ed. A Varma and J.P. Verma, MPH, New Delhi, pp. 305-309.

34. Mayee, C.D. 1987a. Rust disease of groundnut in Maharashtra. Proc. Group Discussion on Groundnut Rust, 1984. Mayee, C.D. , India, pp. 81-89.

35. Mayee, C.D. 1987b. Diseases of groundnut and their management. In 'Plant Protection in Field Crops' Ed. M.A.B. Rao and S. Sithanatham, PPAI, Hydrerabad, pp. 235-243.

36. Mayee, C.D. and Ekbote, A.V. 1983. Life cycle and microepidemiology of *Puccinia* arachidis Speg., rust of groundnut,. Indian J. Plant Pathology, **1** : 170-174.

37. Mayee, C.D. and Baheti, K. G. 1983. Influence of tridemorph on the epidemic development of rust and yield of groundnut. Trop. Plant Sci. Res. **1(4)** : 319-324.

38. Mayee, C.D. and Kokate, V. N. 1987. Environment in relation to incubation period of groundnut rust. Indian Phytopath., **40** : 67-69.

39. Mayee, C.D. and Datar, V.V. 1988. Diseases of groundnut in the tropics. Review of Tropical Plant Pathology, **5**: 169-198.

40. Mayee, C.D. 1989. Dynamics of disease progress in groundnut; an epidemiological view. In : 'Recent Res. in Ecology, Environment and Pollution' Ed. S.T. Tilak, Vol. **3** : 109-118.

41. Mayee, C.D. 1992. New approaches to disease management in Kharif oilseed crops. J. Oilseeds Res. **9**: 119-126.

42. Mayee, C.D., Godbole, G., M. and Patil, F.S. 1977. Appraisal of groundnut rust in India. Problems and Approach. PANS, **23(3)** : 162-165.

43. Mayee, C.D., Mali, V. R. and Pawar, K. R. 1985. Stabilizing productivity of rainfed groundnut through disease management. In 'Oilseed Production" Constraints and Opportunities. Ed. Srivastava *et al.* HLRF, OF & IBH Pub, New Delhi pp. 447-459.

44. Mayee, C.D. Sondge, V. D. and Rodge, R.P. 1985. Spray schedule decision model for rust of groundnut. J. MAU, **10(2)** : 150-161.

45. Mayee, C.D. Sandhikar, R. N. and Bulbule, S. V. 1990. Comparison of Gompertz and logistic equations to describe groundnut rust disease progress, Indian Phytopath, **43** : 531-534.

46. Mayee, C.D. Asghari, M. R. 1991. Comparative efficiency of management practices on stem and pod rots of groundnut. Indian Phytopath. **44** : 329-332.

47. Mc. Donald, D. and Rahji, A. K. 1980. Pests, diseases, resistance and crop protection in groundnut. In 'Adv. in legum. Sci. Ed. R. J. Summer Field and A.H. Bunting. HSMO Londoon, pp. 501-514.

48. Mc Donald, D., Subrahmanyam, P., Gibbons, R.W. and Smith, D. H. 1985. Leaf spots of groundnut. Information Bull. No. 21, Patnacheru, A.P. ICRISAT.

49. Mehan, V.K. and Mc Donald, D. 1984. Research on aflatoxin problem in groundnut at ICRISAT, Plant and soil, **79.** : 255-260,

50. Mehan, V. K. Mc Donald, D., Nigam, S. N. and Lalitha, B. 1981. Groundnut cultivars with seed resistance to invasion by *Aspergillus flavus* Oleagineux, **36** : 501-507.

51. Mehan, V. K., Mc Donald, D. and Rajagopalan, K. 1987. Resistance of peanut genotypes of seed infection by *Aspergillus flavys* in field trials in India. Peanut Sci. **14** : 217-221.

52. Mehan, V. K., Mayee, C. D., Jayanthi, S. and Mc Donald, D. 1991. Preharvest seed infection by *Aspergillus* group of fungi and subsequent aflatoxin contamination in groundnuts in relation to soil types. Plant and Soil, **136** : 239-248.

53. Mixon, A. C. and Rogers, K. M. 1973. Peanut accessions resistant to seed infection of *Aspergillus flavus* Agron. J., **65** : 560-562.

54. Moss, J.P. 1980. Wild species in improvement of groundnuts. In : 'Advances in Legume Science. Ed. R. J. Summerfield and A.H. Bunting, HmSO, London, pp. 525-533.

55. Nema, K.G., Jain, A. C. and Asthana, R.P. 1955. Further studies on Aspergillus blight of groundnut seedlings : its occurrence and control. Indian Phytopath., **8**: 13-21.

56. Nevill, D. J. 1980. Studies on resistance to foliar pathogens. In : "Proc. Int. Workshop of Groundnut. ICRISAT, Hyderabad, pp 199-202.

57. Nevill, D. J. 1981. Components of resistance to *Cercospora arachidicola* and *Cercosporidium personatum* in groundnuts. Ann. appl. Biol. **99** : 77-86.

58. Reddy, D.V.R. 1980. International aspects of groundnut virus research. In 'Proc. Int. Workshop in Groundnut. ICRISAT, Patancheru, India.

59. Reddy, D.V.R., Iizuka, N., Ghanekar, A. M., Murthy, V.K. Kuhn, C. W., Gibbons, R.W. and Chohan, J. S. 1978. The occurrence of peanut mottle virus in India. Pl. Dis. Reptr. **62** : 978-982.

60. Sidaramaiah, A. K., Desai, S. A. and Jayaramaiah, R. 1981. Occurrence and severity of a mycoparasite on tikka leaf spots of groundnut. Curr. Res., **40** : 14-15.

61. Singh, A. K., Shastri, D. C. and Moss. J.P. 1980. Utilization of Wild *Arachis* species at ICRISAT. In : 'Proc. Int. Workshop in Groundnut'. ICRISAT, Patancheru, A.P India, pp. 82-90.

62. Smith, D. H. and Littrell, R. 1980. Management of peanut foliar diseases with fungicides. Plant Disease, **64** : 356-361.

63. Sondge, V. D., Mayee, C.D. and Rodge, R.P. 1985. Spray schedule decision model for management of leaf spots of groundnut. J. Oilseeds Res. **2** : 72-77.

64. Subramanyam, P., Gibbons; R. W., Nigam, S. N. and Rao, V. R. 1980. Screening methods and further sources of resistance to peanut rust. Peanut Sci., **7** : 10-12.

65. Subrahmanyam, P., Mammons, R.O., Nigam, S. N., McDonald, D. and Gibbons, R, W. 1983. International cooperative screening for resistance of peanut to rust and late leaf spot. Pl. Disease, **67** : 1108-1111.

66. Subramamyam, P. and Mc Donald, D. 1983. Rust disease of groundnut. Inf. Bull 13 , ICRISAT, Patancheru, A.P. India.

67. Subrahmanyam, P., Gibbons, R.W., Nigam, S.N. Nevill, D.J. 1982. Resistance to rust and late leaf spot disease in genotyps of *Arachis hypogaea.* Peanut Sci., **9** : 6-10.

68. Subramahmanyam, P., Moss, J.P., Mc Donald, D., Subba Rao, P. V. and Rao, V. R. 1985. Resistance to *Cerocosporidium personatum* leaf spot in wild *Arachis* species. Plant Dis. **69** : 951-954.

69. Subrahmanyam, P., Williams, J. H., Mc Donald, D. and Gibbons, R. W. 1984. The influence of foliar diseases and their control by selective fungicides on a range of groundnut genotypes. Ann. Appl. Biol. **104**: 467-476.

70. Subrahmanyuam, P., Reddy, L. J., Gibbons, R. W. and Mc Donald, D. 1985. Peanut rust : a major threat to peanut production in the semiarid tropics. Plant Dis., **69** : 813-819.

71. Subrahmanyam, P.; Ramanatha Rao, V. Mc Donald, D., Moss, J. P. and Gibbons, R. W. 1989. Origins of resistance to rust and late leaf spot in peanut. Economic Botany, **43** : 444-455.

18 Biological Control of Sugarcane Pests : Present Status and Future Thrusts

S. Easwaramoorthy

CONTENTS

Integrated Pest and Disease Management
Rajeev K. Upadhyay, K.G. Mukerji, B.P. Chamola and O.P. Dubey (eds.)
APH Publishing Corporation, New Delhi-110002, India

ABSTRACT

In sugarcane pest management, biological control holds a pivo position, because of the minimum use of pesticides, availability of all stages of the crop at a given time in a location, succession of different host insect species throughout the growth period of the crop, relatively high economic threshold values for the pest owing to the compensating ability of the plant, nature of canopy and favourable microclimatic conditions. The present status of biological control of sugarcane pests is dealt with under three headings: conservation of natural enemies, importation and colonisation, and augmentation of native natural enemies. A large array of natural enemies is active in sugarcane ecosystem and their conservation and enhancement should form the primary component in sugarcane pest management. Though several attempts have been made to import and colonise exotic natural enemies for the control of sugarcane pests in India, no successful colonisation has resulted. Augmentation of certain native natural enemies like *Trichogramma* spp., *Isotima javensis* Rohw., *Sturmiopsis inferens* Tns., *Epiricania melanoleuca* F., predatory coccinellids and granulosis virus of shoot borer *Chilo infuscatellus* snell has resulted in appreciable pest suppression. The importance of other promising natural enemies like parasites of scale insect, pyrilla and whitefly, bacterial and fungal pathogens are also discussed. Future thrust areas in conservation, colonisation and augmentation are highlighted.

1. INTRODUCTION

Sugarcane, which forms the basic raw material for the second largest agro-based industry in India, is cultivated under diverse agro-climatic conditions ranging from tropical peninsular India with a moderate climate to the sub-tropics characterised by extremes of weather conditions. The cane and sugar yields are affected by various factors like variety, environmental stresses, poor crop husbandry and occurrence of pests and diseases. Pests inflict considerable losses, which are estimated to be around 20% in cane yield and 15% in sugar recovery (7). Various pest management tactics are being employed to mitigate the losses due to pests and among them biological control holds a pivotal position, for the following reasons. The cryptic habits of the injurious stages of borers and certain sucking and subterranean pests make them less amendable for chemical control. The scope of evolving varieties resistant to key pests combined with desirable agronomic traits and disease resistance appears to be only a remote possibility in sugarcane. Several cultural and mechanical control methods are useful in reducing the pest density but difficult to be implemented over large areas.

2. SCOPE OF BIOLOGICAL CONTROL

Sugarcane is a long duration crop and it remains in the field for more than one or two years as plant and ratoon. To meet the demands of sugar industry, the crop is planted in a staggered manner. This makes available crops of different ages in contiguous areas, which support several generations of the pests. Moreover except a few pests, all others can infest the successive stages of the sugarcane crop.

Sugarcane being a C4 plant, is able to compensate for the damage inflicted by several pests, to a certain extent over a period of time. For example in case of shoot borer attack there is extra tillering, the order of effective compensation varying from 1.84 to 116.61% (50). The economic threshold levels for several pests remain at a higher level due to compensating ability of the plant and market value of the pest affected plants.

Sugarcane being the tallest among the field crops with dense canopy, it is difficult to apply insecticides once the canopy becomes dense. As a result, the insecticide pressure is less in sugarcane compared to other cash crops like cotton. In fact, only 2 to 3% of the insecticides used in India has been utilised in sugarcane pest management (8).

The microclimatic conditions are favourable almost throughout the year for the activity of entomophagous and entomopathogens in tropical parts of India, except probably during summer. In sub tropical parts of India also favourable microclimatic conditions are available during most part of the year.

The above factors weigh in favour of biological control of sugarcane pests. In addition, the sugar factories have an organisational set up which helps in cane development activities. The cane departments in the factories actively involve in the production/procurement and distribution of biocontrol agents to farmers in the factory zone. All these factors make biological control practicable in sugarcane insect pest management.

3. PRESENT STATUS

3.1 Conservation of natural enemies

Although more than 200 species of insect pests are reported on sugarcane in India (17), only about a dozen of them are economically important, whereas other species are kept well under check by natural control. For various reasons mentioned in the earlier section, large scale use of insecticides and other disturbing cultural practices is minimum in sugarcane. As a result large number of parasites, predators and pathogenic micro-organisims (Table 1) are active in the sugarcane ecosystem. Some of these entomophages and entomopathogens occur throughout the cropping season, while others occur in certain seasons (13).

Table 1 : Number of species of natural enemies reported on important pests of sugarcane

Pest	Number of species		
	Parasites	Predators	Pathogens
Chilo infuscatellus Snellen	36	5	10
Chilo auricilius Ddgn	31	*	2
Chilo sacchariphagus indicus (Kapur)	17	15	4
Sciroophaga excerptalis Walker	61	*	5
Pyrilla perpusilla Walker	13	12	6
Melanaspis glomerata (Green)	17	6	1
Aleurolobus barodensis Mask	8	11	2
White grubs	2	2**	4

* Not recorded

** Does not include avian predators.

Life table studies of shoot borer (2) internode borer (33), top borer (65) and pyrilla (11) and the earlier studies on the role played by natural enemies as reviewed by David and Easwaramoorthy (13) indicate the importance of natural enemies in the suppression of key pests of sugarcane (Table 2).

Table 2 Extent of mortality caused by natural enemies

Pest	Natural enemy	Extent of kill (%)
C. infuscatellus	*Trichogramma* spp.	4 - 5
	Telenomus spp.	15-22
	Cotesia flavipes Cameron	4-14
	Sturmiopsis inferens Tns.	6.5 - 22.2
	Stenobracon decsae Cameron	5-7
	Egg predator	13.4
	Granulosis virus	2.6 - 14.6
C. auricilius	*C. flavipes*	10-14
	Campyloneurus mutator F.	60-78
	S. inferens	20-50
C. sacchariphagus indicus	*Trichogramma* spp.	2-78.5
	Telenomus spp.	15-90
	C. flavipes	4-14
	Rhaconotus spp.	25
	Egg predators	9.4-22.4
	Granulosis virus	3.8-31.5
S. excerptalis	*Trichogramma* spp.	4-5
	Telenomus spp.	5-85.5
	C. flavines	4-14
	Elasmus zehntneri Ferr.	5-15
	Rhaconotus sciroophagae Wlkn.	7-12.2
	Isotima javensis Rohw.	5-40
P. perousilla	Egg parasites	2.7-97.6
	Epiricania melanoleuca F.	15-100

In pest management programmes that emphasize biological control, conservation and enhancement of native natural enemies should form the primary component. If these are properly exploited in sugarcane cropping system, the need for other management practices could be greatly reduced. Disruption of naturally occurring biological control in sugarcane ecosystem by poor choice and or application technique of insecticides and/or employing harmful cultural practices is well documented. For example, the scale insect *Melanaspis glomerata* (Green), though reported to occur in Tamil Nadu since 1903, suddenly flared up at Pugalur during 1956, when large scale spraying of endrin was done to suppress top borer *Sciroophaga excerptalis* Walker. The outbreak of scale insect was mainly due to the heavy mortality of the parasites and predators which are normally abundant in Tamil Nadu and maintain this pest at low densities (14). The comparatively low incidence of the leafhopper, *Pyrilla perousilla* Walker in eastern Uttar Pradesh was attributed to the reduced usage of pesticides as compared to Western Uttar Pradesh (9), where, during the sixties, large scale aerial spraying was resorted to whenever the pest incidence was noticed to increase. The tactics of reduced usage of pesticides and redistribution of the parasite, *Epiricania melanoleuca* Fletcher have reduced the epidemics of pyrilla in the north Indian sugarcane belt (60). Insecticides like endosulfan and phosalone do not seem to reduce the natural enemy population. Malathion has also been reported to be less harmful to *E. melanoleuca* when ULV spraying from ground is made at a dose of 0.567 to 1.132 kg/ha (10). However, observations on the aerial spraying showed that the parasite larvae are adversely affected. The larvae are dislodged from the host body and fall to the ground and could not form viable cocoons (80).

The impact of inundative releases of *T. chilonis* used for the suppression of internode borer has been found to vary with the insecticides sprayed prior to parasite release (98). Sugarcane foliage treated with 0.1% endosulfan, endrin, endrin + thiometon, leptophos, phosalone and chlorpyriphos showed residual toxicity to the adults of *Trichogramma chilonis* Vigg. for 1, 7, 6, 7, 7 and 9 days respectively (95). Toxicity depended upon the developmental stage of the parasite at the time of insecticide application (96).

Under field conditions, soil application of granules of lindane, chlorpyriphos and sevidol and foliar spraying of endosulfan does not affect the activity of *Sturmiopsis inferens* Tns. In pot culture experiments, soil application of carbofuran, chlorpyriphos and sevidol granules and whorl application of lindane, chlorpyriphos and sevidol granules does not have adverse effect on this parasite, while soil application of lindane EC is highly toxic (38). Spraying of decamethrin 0.005% or cypermethrin 0.02% also does not seem to have any adverse effect on *S. inferens* (82).

Studies on the effect of insecticides on mortality of scale insect and activity of its principal parasite, *Adelencyrtus mayurai* (Subba Rao) under field conditions showed that Ripcord, Malathion and Marshall do not adversely affect the parasite, while Mitac, Volatan, Ekalux and Metacid cause adverse effect (66).

Trash burning has deleterious effect on the natural enemy complex of *M. glomerata* (93). Similarly, detrashing also has an adverse effect on the population of *Adelencyrtus mayurai* (4).

3.2 Importation and colonization

Several attempts have been made to import and colonise exotic natural enemies for the control of sugarcane pests in India. A total of 40 species of parasites and predators were imported (25, 87) and colonised but none of them established permanently and/or controlled the target pests successfully. This is primarily due to the fact that most of the sugarcane pests are native to the Indian subcontinent and it is difficult to find suitable exotic species of natural enemies which can colonise in a new agro-ecosystem and compete with the existing local natural enemies. Another important factor which contributed to the failure of colonization of exotic natural enemies is that in most of the cases, the natural enemies introduced were those which were readily available with little or no cost or effort. In these introductions the basic attributes of the natural enemies like ecological compatibility, temporal compatability etc. were not generally taken into account. Furthermore, the introduced natural enemies were not given a fair trial and the number of individuals released and frequency of releases were generally low and the testing was limited to one or two locations in the country (88).

3.3 Augmentation of native natural enemies

In earlier years, it was generally felt that use of native natural enemies may not provide successful results in biocontrol. But this view has gradually undergone changes due to rarity of success in classical biological control programmes especially against native insect pests (86). Mass multiplication and release of native natural enemies for the control of sugarcane pests has received considerable attention in India, since 1930s, and some of the native natural enemies have provided appreciable pest suppression when their populations were augmented. Some notable successes are discussed below.

3.3.1 Parasites

Trichogramma spp.

Trichogramma spp. are widely used for the control of moth borers of sugarcane in several countries. Of the seven species of *Trichogramma* and one species of *Trichogramma toidea* reported in India on sugarcane pests (68, 69), *Trichogramma chilonis* Ishii is most widely used for control of shoot borer, *Chilo infuscatellus* Snellen, internode borer, *C. sacchariphagus indicus* (Kapur) and stalk borer *C. auricilius* Ddgn., while *T. japonicum* Ashmead is generally recommended against top borer, *Scirpophaga excarotalis* Walker. Recently *T. chilotraea* has been tried against stalk borer.

Field testing of *Trichogramma* against sugarcane borers was initiated during 1930 in Karnataka (52) and later continued in different states (70, 81, 105).

Encouraging results were reported by various workers (91, 100, 101, 102, 104, 105). However, the results of early experiments on inundative release of *Trichogramma* for the suppression of sugarcane borers were conflicting and inconsistent. Narayanan and Mookherji (70) after reviewing the work done between 1930 and 1950 recommended release of this parasite for the control of sugarcane stem borers and/or root borer in Tamil Nadu, Maharashtra, Orissa and Bihar. However, Gupta (40) on the basis of five years' experiments in Uttar Pradesh and review of work done in other states, reported that the parasite releases were ineffective except in Bihar. So the claims of success and failure of this method remained a matter of debate for several years mainly because of unsatisfactory experimentation and presentation of data (88). But studies carried out during the 1970s and later indicated that *Trichogramma* may be useful in the suppression of various sugarcane pests (Table 3).

Isotima javensis

T. javensis (Ichneumonidae) is native to north India, and was introduced into Pugalur in Tamil Nadu from Muzaffarnagar in Uttar Pradesh (41, 42, 77, 78, 79) for the control of *S. excerptalis.* During March and April 1958, a total of 500 wasps were released in an area of 1.2 ha. in a special season (August-September) planted crop. The parasite got immediately established and by May 1958 itself it accounted for 3% parasitism. Parasitism by *Isotima* increased to 22% in four months and at harvest it increased to 67.1% (Table 4). Consequent to the parasite release there was only 3% increase in the borer incidence. The parasite also dispersed over a distance of 1600 m from the release site and in the neighbouring fields 28-81% parasitism was observed in the next season crop (77).

During 1959 to 1964 a total of 1, 22, 128 parasites were released (78). The performance of the parasite in these years is presented in Table 5. Due to the parasite activity, top borer incidence had declined from 37.4 and 33.0% in main and special season during 1958 to 10.8% in 1960. Since then, the top borer has become a minor pest in this area. In the surveys conducted during 1985 and 1986, only 6.3 and 10.3% incidence of top borer was noticed in this sugar factory area (92) indicating the permanency of suppression offered by this parasite.

Similar successful results were obtained from Vadapathimangalam in Tamil Nadu state (47) and Mandya in Karnataka state (74).

Unlike in sub-tropical India where the pest and parasite undergo hibernation, the continuous availablity of the host, and short life cycle and high searching ability of the parasite made the introduction of the parasite an outstanding success in tropical India (25).

Table 3 : Impact of inundative release of Trichogramma chilonis. T. japonicum1 and T. Ehilotraeae2 in the suppression of sugarcane pests

Pest	Dosage of parasite per ha.	Details of release	Pest incidence (%)		Intensity of infestation (%)		Egg parasitism (%)		Reference
			Treated	Control	T	C	T	C	
1	2	3	4	5	6	7	8	9	10
Chilo infuscatellus	1,25,000	2 to 4 times	7.1	16.7	-	-	-	-	48
	50,000 to 1,50,000	Single	4.0 to 11.4	21.0 to 31.4	-	-	28.9 to 52.9	1.5 to 3.1	103
Shoot borer complex	50,000	Weekly	0.4	10.2	-	-	-	-	62, 63
	50,000	Weekly	0.5	8.3	-	-	-	-	61
Chilo sacchariphagus indicus	50,000	Weekly (4th to 11th month)	69.9	75.6	8.3	12.2	76.2	64.1	97
	1,50,000	per season	39.0 to 61.1	64.4 to 88.0	4.6 to 8.0	14.0 to 17.0	-	-	99
	37,500	28 fort nightly releases at Sirugamani Pugalur	49.1 65.6	71.3 73.3					106
	12,500	12 releases (5th to 11th month) Kancheepuram Chengalpattu	40.1 39.9	78.0 70.1					
C.sacchari phagus indicus &	1,25,000	2 to 4 releases	27.6	38.6	6.6	6.9	-	-	48
C.infusca tallust	50,000	6 releases 10 days interval	15.8	62.3	-	-	86.8	5.2	5
C. auricilius	50,000	Weekly	8.8	14.2	8.9	19.6	-	-	62
	50,000	3 releases October at 10 days interval	6 to 12	34 to 50	-	-	-	-	5
	50,000[2]	7 releases at 10 days interval	7.9	22.2	-	-	-	-	5
Acigona steniellus	1,25,000	4 times coinciding with brood emergence	1.9	9.3	-	-	-	-	110
	50,000 to 1,60,000	Single release	6.1 to 9.1	9.9 to 21.7	-	-	25.4 to 71.0	1.6 to 3.1	103
Scirpophaga excerptalis	50,000[1]	Weekly	4.4	32.8	-	-	-	-	62
	50,000[1]	Weekly	0.9	6.5	-	-	-	-	61

1. *T. japonicum* against *S. excerptalis*
2. *T. chilotraeae* against *C. auricilius*

@ includes *C. infuscatellus, C. auricilius, S. inferens. Rhapimetopus ablutella*

Table 4 : Progressive increase in parasitism by I. javensis in the first year of release (after Raja Rao, 1961)

Month (1958)	Top borer incidence (%)	Parasitism by Isotima (%)
March	30.0	-
April	33.5	-
May	32.07	3.0
June	31.8	3.0
July	Not recorded	6.0
August	Not recorded	22.0
October	33.3	67.1

Table 5 : Reduction in top borer incidence due to parasitism by Isotima javensis (after Raja Rao, 1964a)

Main Season			Special Season		
Year	Top borer incidence (%)	Parasitism (%)	Year	Top borer incidence (%)	Parasitism (%)
1957-58	37.4	0.0			
1958-59	20.8	78.6	1958	33.0	67.0
1959-60	12.5	70.9	1959	12.7	44.7
1960-61	10.8	70.4	1960	10.8	46.7

Sturmiopsis inferens

The indigenous tachinid parasite *S. inferens* parasitises *C. infuscatellus, C. auricilius, Acigona steniellus* Hampson and *Sesamia inferens* Wlk. It has high reproductive potential, good dispersing capacity and searching ability which offer much promise in biological suppression of target pests. In addition, its larvipositional behaviour is a distinct advantage over the hymenopterous parasites which face considerable difficulty in ovipositing through the thick stemmed sugarcane (67). It distributes its off spring efficiently (Table 6) in or near borer tunnels (20). In addition, the parasite tolerates fairly high range of temperatures (19). Field releases of 7922 gravid females in Chengalpattu district and 87 mated females and 2000 *S. inferens* parasitized shoot borer larvae in Thanjavur district of Tamil Nadu showed successful establishment of this parasite (16). Recent studies on large scale liberation of *S. inferens* in Simbhaloi Sugar Mills in Uttar Pradesh have given encouraging results (Grewal. Pers. Commu). However, in Haryana, release of 1370 mated females in 0.4 ha field did not enhance the level of parasitism in *C. auricilius* (12).

Epiricania melanoleuca F.

Redistribution of *E. melanoleuca*, the nymphal and adult parasite of *Pyrilla perpusilla* has resulted in the successful suppression of the leafhopper population in several states. During 1975 to 1977 the parasite population was distributed in Haryana, Uttar Pradesh, Rajasthan, Delhi, Karnataka and West Bengal by the Central Biological Control stations (56). Subsequently, the parasite was established in Ahdhra Pradesh, Gujarat, Madhya Pradesh and Maharashtra (43, 57, 58, 59, 60, 73). Release of several thousands of the parasite cocoons and egg masses in different sugar mill zones of Haryana by the Haryana Agricultural University has resulted in the successful suppression of the pest population (54).

Table 6 : Distribution of maggots by S. inferens at shoot borer bore hole

Number of maggots/fly	Number of shoots alighted	Maggots laid	
		at bore hole	near bore hole
340	314	170	170
312	303	254	58
306	303	252	54
521	417	419	102
338	242	203	135
191	147	141	50
123	38	71	52
286	229	189	97
126	59	87	39
307	304	283	24
2850	2356	2069	781

Generally, release of 4000 to 5000 cocoons and 0.4 to 0.5 million eggs per ha. provides effective suppression of the pest (72). Better establishment of the parasite can be obtained by stapling 203 egg clusters and 5-7 cocoons per leaf in close proximity at each release point (60). Effective suppression of the pest occurred when the host. parasite ratio reached to a level of 1: 3.5 (94). The ratio advocated for field release is 1:4 when the nymphal population is more and 1:5 when adult population is predominant (53). According to this estimate, 2333 cocoons are required when the host density is 1.25 individuals per leaf whereas 7000 cocoons are required when the host density is 7.5 per leaf.

High fecundity, short developmental period, adaptability to wide range of agro-climatic conditions, susceptibility of all stages of nymphs and adults for parasitization, good searching ability of the parasitic larvae and fairly high rate of spread (6-8 km per year) (60), are the man reasons for the success of this parasite. In addition, parasitization by *Epiricania* reduces the fecundity and hatchability of eggs of pyrilla (53). The effect of parasitization on fecundity is more pronounced when both the sexes are parasitized.

Other parasites

Laboratory multiplication techniques have been developed for the scale insect parasites, *Adelencyrtus mayurai* (Subba Rao) and *Botryoideclava bharathiva* Subba Rao (36). *A. mayurai* is being multiplied and field released on a limited scale in Tanjore district of Tamil Nadu by the sugar factories. Techniques have also been developed for the multiplication of three species of egg parasites of pyrilla - *Ooencyrtus papilionis* Ashmead, *Cheiloneurus pyrillae* Mani and *Tetrastichus pyrillae* Crawford (113). The effectiveness of the parasite in relation to host population density and temperature has also been studied (114). However, evaluation of the efficacy of these parasites under field conditions is yet to be taken up. Redistribution of the whitefly parasite, *Amitus minervae* Silvestry showed establishment in new areas. The level of parasitization of whiteflies is also slowly increasing in the released areas (3).

3.3.2 Predators

A large number of oligophagous and polyphagous predators occur in the sugarcane ecosystem (39). They play an important role in the suppression of pest populations.

Coccinellids

Among indigenous predatory coccinellids, *Chilocorus nigritus* (F.) and *Pharo scymnus horni* Weise are found to feed on the scale insect, *Melanaspis glomerata* (Green). The grub of *C. nigritus* on an average consumes 46 scale insects per day whereas the adult feeds on 125 per day (90.75). The larval period varies from 9 to 14 days and the adults live for 45-60 days. The grub of *P. horni* consumes 24.2 scale insects on an average and adult 43.2 per day. One individual is capable of preying 1890 scale insects during its life span (76). Field releases of these predators were made during 1976 to 1983 in the coastal districts of Andhra Pradesh. The total number of *C. nigritus* beetles released was 2, 12, 400 while that of *P. horni* was 2, 03, 450. The predators have got established in Chagallu and Tanuku areas (64). However, the adults migrate from one field to another and prefer to settle down only there a heavy scale insect population exists (90) . A minimum release of 1500 beetles per hectare, every season, at the first appearance of the pest is suggested for suppression of scale insect (55).

3.3.3 Entomopathogens

A number of entomopathogens have been reported to infect sugarcane pests (22). However detailed studies have been carried out on only a few of them.

Virus

Granulosis virus

The granulosis virus infecting shoot borer (24) is found widely distributed in Tamil Nadu and Pondicherry (26). The virus is active throughout the year in nature (21) and is highly virulent in laboratory studies. In field evaluation, four sprays of the virus at 10^6—10^8 inclusion bodies per m! applied at fortnightly intervals starting from the 30th day of crop age have given promising results (21, 35). In recent studies, it is observed that two rounds of virus application is as effective as insecticides and virus + insecticide combinations (71), in the suppression of shoot borer population. Large scale trials in various sugar factory areas have shown successful suppression of the pest. The virus is more effective when applied as high volume spray compared to low volume spray (32).

The GV is host specific (27) harmless to parasites and predators of sugarcane pests (28), silkworms (30), honey bee (29) and higher animals. Another advantage is that the virus gets vertically transmitted to offspring through adults (31). The virus infection was also found to alter the sex ratio in pupae in favour of males which will have significant influence on the population build up of the pest. The persistence of the virus on sugarcane foliage is inversely related to the exposure time and the 1/2 is found to be 6.8 days (21).

Bacteria

Bacillus thuringiensis

Application of 0.4% 'Thuricide' a commercial preparation of the bacterium, at weekly intervals during July-October reduced the infestation of Gurdaspur borer, *Acigona steniellus* Hmpsn (6). The bacterium has been reported to be promising in killing the grubs of *Holotrichis* spp. in pot tests when applied as dust at 10 g containing 3 x 109 viable spores (89).

Bacillus papillae

The milky disease bacterium has given promising results in the control of white grubs in laboratory studies (15, 18, 111). Field application of spore powder in Kapadvaj area in Gujarat, an endemic area for *H. consanguinee* resulted in 35-60.7% larval infection in 1976 and 20-75% in 1977 (112).

Fungi

Metarhizium anisopliae

The green muscardine fungus infects pyrilla and whitegrubs and causes epizootics in the former under favourable weather conditions (51). In a field experiment conducted at Annamalainagar, Tamil Nadu, the fungus caused 100% mortality of nymphs and adults (84). In Uttar Pradesh, overwintering population of pyrilla could be readily infected by the fungus application and the infected individuals were capable of spreading the infection and inducing an epizootic (109). The fungus is safe to *E. melanoleuca* (107) the principal parasite of pyrilla.

Beauveria brongniartii

B. brongniartii reported from white grubs (83) is found widely distributed in Karnataka (44). It is highly pathogenic to all the stages (45). Under field conditions, broadcasting of the fungal spores mixed with pig manure at 10^9 spores/g of soil provides good control of white grubs (44).

B. bassiane

It is reported from shoot borer (34) stalk borer (108) and white grubs (46, 85). The fungus is found to have potential in the suppression of overwintering population of stalk borer and black bug, *Cavelerius sweeti* Slater and Mugomoto. The inoculum level and efficacy of the fungus can be increased by releasing laboratory infected adults of black bug (107).

Nematodes

The indigenous species *Heterorhabditis indicus* at 13 x 10^5 infective juveniles/pot (40 x 42 cm) caused 62.5 and 100.0% mortality in third instar grubs of *H. serrata* and *Leucopholis lenidephora* Brum respectively (49).

4. FUTURE THRUST

4.1 Conservation

The significant role played by natural enemies in pest suppression in an ecosystem like sugarcane, where there is minimal disruption by insecticidal applications has to be understood well in the context of integrated pest management. Species complexity, seasonal abundance, influence of abiotic factors and natural enemy - host relationships have to be worked out for different agroclimatic zones to know the impact of naturally occurring biological control. A knowledge of the disruptive effects of insecticides and harmful cultural practices is also essential so that the management practices can be manipulated in such a way that they can be beneficial to natural enemies but harmful to the pests. Role of intercropping, influence of weeds and interaction of other fauna in maximising natural control also merit detailed investigation.

4.2 Colonisation

Planned introductions of selected natural enemies, detailed laboratory evaluation and repeated releases in large numbers in different agroclimatic zones may enhance the possibilities of establishment of exotic natural enemies.

4.3 Augmentation

In India, more than 270 species of indigenous natural enemies have been reported on sugarcane pests. But, at present, only a few of them are being utilised in applied biological control. So the first step involved in augmentation is the identification of other candidate natural enemies that can be utilised in sugarcane pest suppression. Preliminary studies in this direction revealed the importance of *Telenomous beneficiens* Zehnt on internode borer (37) and *Telenomus* sp. on top borer (65) as important mortality factors in the egg stage of these pests. Similarly *Rhaconotus* spp. appears to be the predominant larval parasite of top borer both in tropical (92) and subtropical (65) India. *Amitus minervae* has been recognised as the predominant parasite of white flies (3). The role played by entomopathogens is being documented in recent years. Another important aspect which needs consideration is the existence of races/strains with differential ecological adaptability as reported in *Trichogramma chilonis* (1). There was evidence of heterogenecity in the granulosis viruses of *C. infuscatellus* and *C. sacchariphagus indicus* (23) probably due to strain variation. A knowledge of these aspects will be useful in utilization of suitable species/strains of natural enemies.

Development of laboratory culturing and mass production techniques forms the next important step in the augmentation programmes, various aspects like designing of laboratory facilities, development of artificial diets for host insects like shoot borer and top borer, identification of alternative laboratory hosts (for the multiplication of *Isotima, Epiricania*, GV etc.) which can be cultured with least difficulty and at low cost, development of artificial media for production of parasites such as *Trichogramma, Sturmiopsis*, Coccinellid predators, maintenance of genetic purity and virulence need perfection.

Evaluation of the biocontrol agents under diverse agroclimatic situations to determine the conditions under which they can work, determination of dosage in relation to weather conditions and pest densities and frequency of releases and prediction of effectiveness of biocontrol agents are necessary to make proper pest management decisions.

Studies on chemical ecology to know the role played by semio-chemicals, as evident in the preliminary studies with shoot borer and its natural enemies, are essential to use these chemicals in improving the effectiveness of selected biocontrol agents.

Use of genetic engineering techniques to obtain ecologically better adaptive strains of entomophages and a planned genetic reassortment and gene site specific modification in entomopathogens like GVs of shoot and internode borers to increase the host range and virulence are essential to improve the effectiveness of biological control.

Establishment of biocontrol agent production centres to mass produce and make readily available the required quantity of quality biocontrol agents is very essential for the successful adoption of this method. The production centres, instead of concentrating on the multiplication of a single biocontrol agent, should have diversification depending upon the requirements. For example during summer months, *S. inferens* and granulosis virus of shoot borer may be produced and during cooler months *T. chilonis* may be mass cultured by the centres in Tamil Nadu to make them economically viable. There should be close coordination between research and production centres.

A net work of marketing system should be developed to distribute the biocontrol agents to every needy farmer as is done with chemical pesticides. In addition, advisory facilities should be made available to advise the farmers about the pest problems, kind of biocontrol agents needed, dosage and frequency of release.

5. CONCLUSION

Biological control is an important component in sugarcane pest management and it receives much more attention in recent years than in the past.

The sugarcane ecosystem is comparatively less disturbed than other agroecosystems and consequently diversity and abundance of natural enemies is greater. Avoidance of disruptive practices and suitable manipulation in the ecosystem will enhance the scope of naturally occurring biological control.

Earlier attempts on colonisation of exotic natural enemies have failed for various reasons. Planned long term efforts are necessary to obtain success in colonisation of exotic natural enemies.

Augmentation of native natural enemies like *Trichogramma* spp. *E. melanoleuca, I. javensis* and granulosis virus have given promising results. Identification of other potential natural enemies and their use in augmentation will increase the scope of biological control of major pests of sugarcane. There is a greater scope for the exploitation of entomopathogens in the management of selected pest species. Modern tools like biotechnology and genetic engineering can be utilised to obtain better adaptable natural enemies.

For adopting biological control methods, ready and timely availability of quality biological agents is an essential prerequisite. At present the biocontrol laboratories in our country are at the most equipped for mass production of *Trichogramma* spp. and redistribution of *E. melanoleuca.* Unless biocontrol agents are made as readily available as chemical pesticides, this method will not become practicable on farmer's holdings. This emphasizes the urgent need to create additional facilities and to upgrade the existing facilities to make biological control as an important component of sugarcane pest management.

ACKNOWLEDGEMENT

The author is thankful to Dr. K. Mohan Naidu, Director, Surgarcane Breeding Institute for the facilities provided.

6. REFERENCES

1. Abraham, C.C. 1970. Exploration of the feasibility of developing races of *Trichogramma australicum* Girault (Trichogrammatidae: Hymenoptera.) suitable for different environments. Ph.D. Thesis, Indian Agric. Res. Inst., New Delhi 126 pp.

2. Ananthanarayana, K. and David, H. 1982. Construction of life table for shoot borer. Ann. Rep. Sug. Cane Breed. Inst. Coimbatore. pp. 163-164.

3. Ananthanarayana, K. and Salin, K.P. 1992. Parasite complex of sugarcane whiteflies. Ann. Rep. (1991-92) Sug. Cane Breed. Inst., Coimbatore. p. 62.

4. Anonymous. 1984. Ann. Rep. All India Coordinated Res. Proj. Biol. Control Crop. Pests Weeds. Bangalore pp. 34-42.

5. Anonymous, 1992. Ann. Rep. All India Coordinated Res. Proj. Biol. Control. Crop Pests Weeds. Bangalore pp. 77-80.

6. Atwal, A.S. and Sohi, B.S. 1969. Effectiveness of *Bacillus thuringiensis* Berliner against *Bissetia steinellus* Hampson (Lep: Pyralidae). J. Res. Punjab agric. Univ., **6**: 174-176.

7. Avasthy, P.N. 1983. Insect pest management for sugarcane in India. In: Sugarcane Pest Management in India, Eds. M. Balasubramanian and A.R. Solayappan. Tamil Nadu Co-op. Sugar Federation, Madras pp. 71-77.

8. Balasubramanian, M. 1988. Strategy for insecticidal efficiency in sugarcane pest management. In Biocontrol Technology for sugarcane pest management Eds. H. David and S. Easwaramoorthy. Sug. Cane Breed. Inst., Coimbatore. pp. 328-344.

9. Banerjee, S.N. 1974. Epidemic of sugarcane pyrilla in India. Pl. Prot. Gleanings, **3**: 1-3

10. Bindra, O.S. Harcharan Singh, and Chaudhary, J.P. 1970. **Ground application of ultra low volume sprays** for the control of sugarcane leaf hopper, *Pyrilla perousilla.* **Indian J.** Ent., **32** : 4-6.

11. Brar, R.S. and Bains, S.S. 1979. Population dynamics of *Pyrilla perpusilla* Walker - mortality factors. Indian J. Ecol. **6**: 110-121.

12. Chaudhary, J.P., Yadav, S.R. and Bhardwaj, S.C. 1981. Biological control of sugarcane stalk borer, *Chilo suricillius* Ddgn. Indian J. Sug. Technol.

13. David, H. and Easwarmoorthy, S. 1986a. Biological control. In : "Sugarcane Entomology in India" Eds. David, H., S. Easwaramoorthy and R. Jayanthi, Sug. Cane Breed. Inst. Coimbatore p. 383-422.

14. David, H. and Easwaramoorthy, S. 1986b. Resugarcane of sucking pests in sugarcane. In : "Resugarcane of sucking pests - Proc. Natl. Sym. Eds. S. Jayaraj. Tamil Nadu Agric. Univ. Coimbatore. p. 1-11.

15. David, H. and Easwaramoorthy, S. 1988. An evaluation of some microbial agents for the control of white grubs infesting sugarcane. In : "Advances in Management and conservation of soil Fauna" Eds. G.K., Veeresh, Rajagopal, D. and C.A. Viraktamath, Oxford IBH Publ. Co. New Delhi p. 223-229.

16. David, H. and Kurup, N.K. 1991. Tachinid parasites for management of sugarcane borers. In : "Biocontrol Technology for sugarcane pest management" Eds. H. David, and S. Easwaramoorthy, Sug. Cane Breed. Inst. Coimbatore p. 51-64.

17. David, H. and Nandagopal, V. 1986. Pests of sugarcane - Distribution, symptomatology of attack and identification. In "*Sugarcane Entomology in India*" Eds. David, H., S.Easwaramoorthy and R. Jayanthi Sug. Cane Breed. Inst. Coimbatore p. 1-30.

18. David, H., Alexander, K.C. and Ananthanarayana, K. 1973. Milky disease bacterium of the sugarcane whitegrub. Curr. Sci., **42**: 695-696.

19. David, H., Easwaramoorthy, S., Nandagopal, V., Kurup, N.K., Shanmugasundaram, M. and Santhalakshmi, G. 1981. Influence of different temperatures on the tachinid parasite, *Sturmiopsis inferens* (Dip). Entomophaga **26** : 333-338.

20. David, H.,.Easwaramoorthy, S., Nandagopal, V., Kurup, N.K., Sharmugasundaram, M. andSanthalakshmi, G. 1989. Larvipositional behaviour of *Sturmiopsis inferens* Tns. (Tachinidae : Diptera) on sugarcane shoot borer in India. *Trop.* Pest Mgmt. **34** : 267-270.

21. Easwaramoorthy, S. 1984. Studies on granulosis viruses of sugarcane shoot borer, *Chilo infuscatellus* Snellen and internode borer, *C. sacchariphagus indicus* (Kapur). Ph.D. thesis, Tamil Nadu Agric. Univ. Coimbatore. 262 pp.

22. Easwaramoorthy, S. 1986. New approaches in sugarcane pest control In : "Sugarcane Entomology in India", Eds. David, H., S.Easwaramoorthy and R. Jayanthi. Sug. Cane Breed. Inst., Coimbatore p. 437-458.

23. Easwaramoorthy, S. and Cory, J.S. 1990. Characterization of the DNA of granulosis viruses isolated from two closely related moths, *Chilo infuscatellus* and *C. sacchariphagus indicus*. Arch. Virol. **110** : 113-119.

24. Easwaramoorthy, S. and David, H., 1979. A granulosis virus of sugarcane ashoot borer, *Chilo infuscatellus* Snell., (Lepidoptera : Crambidae). Curr. Sci., **48** : 685-686.

25. Easwaramoorthy, S. and David, H., 1992. Augmentation of native natural enemies for the control of sugarcane pests - Prospects and problems. In : "Emerging trends in Biological control of Phytophagous Insects" Ed. T.N. Ananthakrishnan, Oxford & IBH Publ. Co. New Delhi. p. 199-212.

26. Easwaramoorthy, S. and Jayaraj, S. 1987a. Survey for the distribution of granulosis virus infection in sugarcane borers, *Chilo infuscatellus* Snellen and *C. sacchariphagus indicus* (Kapur). Trop. Pest Mamt. **33** : 200-201.

27. Easwaramoorthy, S. and Jyaraj, S. 1987b. Cross infectivity of granulosis virus infecting *Chilo infuscatellus* Snell. and *C. sacchariphagus indicus* (Kapur). J. Ent. Res. **11** : 170-174.

28. Easwaramoorthy, S. and Jyaraj, S. 1987c. Safety of two granulosis viruses infecting sugarcane borers to certain parasites and predators. J. Biol. Control, **1** : 106-110.

29. Easwaramoorthy, S. and Jyaraj, S. 1987d. Lack of susceptibility of Indian honey bee, *Apis cerana indica* Fabr. to two granulosis viruses infecting sugarcane borers. Indian Bee J., **49** : 1-2.

30. Easwaramoorthy, S. andJyaraj, S. 1988. Effect of two granulosis viruses on the silkworms, *Bombay mori meridionalis* P. and *Philosoma ricini B.* Entomon. **13** : 57-59.

31. Easwaramoorthy, S. and Jyaraj, S. 1989. Vertical transmission of granulosis virus of sugarcane shoot borer, *Chilo infuscatellus.* Trop Pest Mgmt., **35** : 352-353.

32. Easwaramoorthy, S. and Jyaraj, S. 1991. Influence of spray fluid volume and purity on the effectiveness of a granulosis virus infecting sugarcane shoot borer, *Chilo infuscatellus* Snellen. Trop. Pest Mgmt., **37**: 134-137.

33. Easwaramoorthy, S. and Nandagopal, V. 1986. Life tables of internode borer, *Chilo sacchariphagus indicus* (K) on resistant and susceptible varieties of sugarcane. Trop. Pest Mgmt. **32** : 221-228.

34. Easwaramoorthy, S. and Santhalakshmi, G. 1987. Occurrence of a fungal disease on sugarcane shoot borer, *Chilo infuscatellus* Snell. Entomon., **12** : 394-395.

35. Easwaramoorthy, S. and Santhalakshmi, G. 1988. Efficacy of granulosis virus in the control of shoot borer, *Chilo infuscatellus* Snellen. J. Biol. Control, **2**: 26-28.

36. Easwaramoorthy, S. and M. Sharmugasundaram, 1988. Techniques for culturing hymenopterous parasites. In : "Biocontrol Technology for Sugarcane Pest Management" Eds. H. David and S. Easwaramoorthy, Sug. Cane Breed, Inst. Coimbatore p. 170-179.

37. Easwaramoorthy, S., Nandagopal, V. and David, H. 1983. Seasonal importance of parasites and predators of sugarcane internode borer, *Chilo sacchariphagus indicus* (K). Proc. Symp. Ecol. Resource Mgmt., Muzaffarnagar p. 55-61.

38. Easwaramoorthy, S., David, H., Santhalakshmi, G., Sharmugasundarm, M., Nandagopal, V. and Kurup, N.K. 1990. Toxicity of certain insecticides to *Sturmiopsis inferens* a larval parasite of sugarcane moth borers. Entomophaga, **35** : 130-136.

39. Easwaramoorthy, S., Kurup, N.K. and Santhalakshmi, G., 1991. Arthropod predators in sugarcane pest management. In :"Biocontrol Technology for sugarcane pest management Eds., H. David and S.Easwaramoorthy . Sug. Cane Breed. Inst. Coimbatore p. 201-248.

40. Gupta, B.D. 1951. A resume of the trials conducted during 1941-44 with *Trichogramma evanescens minutum* Riley against the stem borer, *Argyria sticticraspis* Hmpsn. and the root borer, *Emmalocera depresella* Swinh. in Uttar Pradesh. Proc. Bienn. Conf. Sug. Cane Res. Dev. Wkrs. India, **1** : 15-23.

41. Gupta, B.D. 1958. Some friends of the sugarcane farmer. Part I - Parasites of the top borer, *Sciroophaga bivella* F. Indian Sug. **8** : 439-444.

42. Gupta, B.D. 1960. An indigenous parasite succeeds in controlling the sugarcane top borer. Proc. A. Conv. Sug. Technol. Assoc. India, **28** : 24.

43. Jagdish Prasad, Sharma, R.S., Ravindra Singh and Pawar, A.D. 1981. Introduction of an ectoparasite, *Epipyrpps melenoleuca* Fletcher (Epipyropidae : Lepidoptera) in Chittauragarh district (Rajasthan) for the bio-control of sugarcane Pyrilla, *Pyrilla perpusilla* Walker Co-op. Sugar. **12** : 459-460.

44. Jayaramaiah, M. 1981. Studies on the entomogenous fungus, *Beauveria brongniartii* (Sacc.) Petch in relation to white grub, *Holotrichia serrata* Fab. and silkworm, *Bombyx mori* Linn. and possibility of its use in the management of white grubs. Ph.D. thesis, Univ, Agric. Sci. Bangalore.

45. Jayaramaiah, M. and Veerash, G.K. 1983a. Studies on the symptoms caused by the silkworm white muscardine fungus, *Beauveria brongniartii* (Sace.) Petch to different stages of the whitegrub, *Holotrichia serrata* Fab. (Coleoptera : Scarabaeidae). J. Soil Biol. Ecol. **3** : 7-12.

46. Jayaramaiah, M. and Veerash, G.K 1983b. Fungal pathogens of white grubs in Karnataka. J. Soil. Biol. Ecol., **3** : 83-87.

47. Kalra, A.N. and David, H. 1964. Introduction of *Isotima javensis* Roh., a parasite of top borer of sugarcane in Tanjore area of Madras State. Proc. All. India Conf. Sug. Cane. Res. Dev. Wors., **5** : 602-605.

48. Kamalakara Rao, C. 1980. Biological control of sugarcane shoot borer and internode borer in the factory area of the Jeypore Sugar Company Ltd., Chagallu, W.G. District, Andhra Pradesh. In :"Biological control of sugarcane pests in India", Eds. S. Sithanantham and A.R. Solayappan, Tamil Nadu Co-op. Sugar Federation, Madras - p. 37-38.

49. Karunakar, G. 1990. Studies on entomophilic nematodes for the control of two species of sugarcane white grubs. M.Sc. (Ag) thesis, Tamil Nadu Agricultural Univ. Coimbatore.

50. Khanna, K.L. 1956. On tiller mortality and compensation due to borers in sugarcane. Proc. Inst. Soc. Sug. Cane Technol., **9** : 936-941.

51. Kulshreshtha, R.S. and Gursashani, K.A. 1961. An observation on entomogenous fungus, *Metarhizium anisopliae* (Metsch) Scr. Indian J. Sug. Cane Res. Dev., **5** : 163-164.

52. Kunhikanna, K. 1931. The mass rearing of the egg parasites of sugarcane moth borer in Mysore (Preliminary experiments). J. Mysore Agric. Exp. Un., **12** : 57-61.

53. Madan, Y.P. 1985. Bionomics, effectiveness and mass multiplication of *Epipyrops melanoleuca* Fletcher, an ectoparasitoid of *Pyrilla perousilla* Walker, Ph.D. thesis, Haryana Agric. Univ. Hisar p. 138 pp.

54. Madan, Y.P. and Singh, R.A. 1979. Augmentation of *Epipyrops melanoleuca* Fletcher cocoons in the sugar mills zone of Haryana. Ann. Rep. Sug. Cane. H.A.U. Hisar (1978-79).

55. Manjunath, J.M. 1984. Role of commercial insectaries in promoting biological control. Proc. First Regional Symp. Biol. Control. Univ. Pertanian Malaysia, Selengor, Malaysia p. 287-294.

56. Misra, M.P. and Pawar, A.D. 1980. *Epipyrops melanoleuca* Fletcher (Lepidoptera : Epipyropidae) an effective parasite of sugarcane pyrilla, *Pyrilla perpusilla* Walker (Hemiptera : Fulgoridae) in northern India. Entomophaga **32** : 125-130.

57. Misra, M.P. and Pawar, A.D. 1981. Release and establishment of *Epipyrops melanoleuca* Fletcher (Lepidoptera : Epipyropidae) on sugarcane pyrilla in West Bengal. J. Pl. Prot. **9**: 128-131.

58. Misra, M.P. and Pawar, A.D. 1982a. Colonisation and recovery of *Epipyropos melanoleuca* Fletcher (Lepidopera, Epipyropidae) against sugarcane pyrilla, *Pyrilla perpusilla* (Walker) (Hemiptera : Fulgoridae: Lophopidae) at Navsari, Valsad, Gujarat State, India. Indian Sug. **31** : 673-674.

59. Misra, M.P. and Pawar, A.D. 1982b. Colonisation and recovery of *Epipyrops melanoleuca* Fletcher (Lepidoptera : Epipyropidae) against *Pyrilla perpusilla* (Walker) (Hemiptera : Fulgoridae) from Raolgaon, Maharashtra State. Indian Sug. **32** : 593.

60. Misra, M.P. and Pawar, A.D. 1984. Use of *Epipyrops melanolueca* Fletcher (Lepidoptera : Epipyropidae) for the biocontrol of sugarcane pyrilla, *Pyrilla perousilla* (Walker) (Hemiptera : Fulgoridae). Indian J. agric. Sci., **54** : 742-750.

61. Misra, M.P. and Pawar, A.D. 1987. Biological control of early both borer of sugarcane by *Trichogramma* in North Bihar. J. Biol. Control. **1** : 70-71.

62. Misra, M.P., Pawar, A.D. and Srivastava, U.L. 1986. Biocontrol of sugarcane moth borers by releasing Trichogramma parasites at Harinagar, West Champaran, Bihar. Indian J. Pl. Prot., **14** : 89-91.

63. Mokha, K.S. and Singh, S. 1988. Biological control of sugarcane moth borers in Harinagar area of north Bihar. Barathiya Sug. **13** : 53-57.

64. Mukthilingam, B., Gangaram, M.N. and Appa Rao, M. 1983. Mass multiplication of predators of sugarcane scale insect (*Melanaspis glomerata* Green) in Andhra Pradesh. Fourth All India Workshiop on Biocontrol., Univ. Agric. Sci., Bangalore, Feb. 22-24 pp. 6

65. Mukunthan, N. 1989. Life tables of sugarcane top borer, *Scirpophaga excerotalis* Wlk. Insect Sci. Applic. **10** : 269-276.

66. Muralikrishna, Y. 1984. Studies on the insecticidal control of sugarcane scale insect *Melanaspis glomerata* (Green) M.Sc. (Ag.) Thesis, Tamil Nadu Agric. Univ., Coimbatore.

67. Myers, J.G. 1935. Biological control of sugarcane pests in the British West Indies and British Guiana. Proc. Int. Soc. Sug. Cane Technol. **5** : 381-385.

68. Nagarkatti, S. 1980. Trichogrammatid parasites of sugarcane borers in India - Distribution and identification. In :"Biological control of sugarcane pests in India", Eds. S. Sithananthan, and A.R. Solayappan. Tamil Nadu Co-op. Sugar Federation, Madras p. 13-19.

69. Nagarkatti, S. and S. Nagaraja. 1977. Biosystematics of *Trichogramma* and *Trichogrammatcidea* species. A. Rev. Ent., **22** : 157-176.

70. Narayanan. E.S. and Mookherjee, P.B. 1953. Two decades of field trials in India with *Trichogramma* against the sugarcane borer, *Argyria* sp., and the root borer, *Emmalocera depressella*. Proc. Indian. Sol. Congr. **40** : 157-176.

71. Parameswaran, S., Sundara Babu, P.C., Suresh, P.J. and Jayaraj, S. 1992. Field efficacy of granulosis virus and insecticides in the control of sugarcane shoot borer, *Chilo infuscatellus* Snellen. J. Biol. Control, **6**: 46-47.

72. Pawar, A.D. 1983. National efforts in biocontrol of sugarcane pests. In : "Sugarcane Pest Management in India". Eds. M. Balasubramanian, and A.R. Solayappan. Tamil Nadu Co-op. Sugar Federation, Madras p. 34-45.

73. Pawar, A.D., Jagdish Prasad, Ravindra Singh, Yadav, K.P. and Ram Asre. 1980. Introduction and establishment of an ectoparasite, *Epipyrops melanoleuca* Fletcher (Epipyropidae: Lepidoptera) in Bundi (Rajasthan) for the biocontrol of sugarcane Pyrilla, *Pyrilla perousilla* Walker. Indian Sug., **30** : 745-749.

74. Puttarudriah, M. and Usman, S. 1961. Possibilities of successful colonisation of *Melcha ornatipennis* Cameron for the biological control of the sugarcane top borer, *Sciroophaga nivella* Fabricius in V.C. tract, Mandya district, Mysore State. Mysore Agric. J., **35** : 220-222.

75. Raghunath, T.A.V.S. and Rao, B.H.K. 1980. Studies on the feeding potential of *Chilocorus nigritus* Fabr. (Coccinellidae) on seale insect, *Melanaspis glomerata* Green of sugarcane. Co-op. Sug., **12** : 191-192.

76. Raghunath, T.A.V.S. and Rao, B.H.K. 1982. Studies on the biology and feeding potential of *Pharoscymnus horni* (Weise) (Coccinellidae: Coleoptera), a predator on the scale insect of sugarcane, *Melanaspis glomerata*. Maharashtra Sug. **7** : 41-45.

77. Raja Rao, S.A. 1961. Introduction of indigenous parasite for the control of top borer in sugarcane at Pugalur (Madras state). Proc. All India Conf. Sug. Cane Res. Dev. Wkrs., **4** : 504-512.

78. Raja Rao, S.A. 1964a. Mass production of *Isotima* sp. (javensis) an Ichneumonid parasite of top borer of sugarcane, *Scirpophaga nivella* F., at Pugalur (Madras State). Proc. All India Conf. Sug. Cane Rest. Dev. Wkrs. **5** : 597-602.

79. Raja Rao, S.A. 1964b. Biological control of top borer, *Scirpophaga nivella* F., of sugarcane at Pugalur (Madras State). Proc. All India Conf. Sug. Cane Res. Dev. Wkrs., **5** : 588-592.

80. Rajk, R.L., Pawar, A.D., Misra, M.P., Prasad, J., Varma, A. and Singh, G.P. 1987. Sugarcane pyrilla epidemic a case study. Pl. Prot. Bull. **39 (3)** : 1-9.

81. Ramachandra Iyer, S., Krishnamurthy, C. and Basheer. 1951. Biological control of the early shoot borer, *Argyria sticticeraspis* Hampson by *Trichogramma minutum* Riley in the Madras State. Proc. Bienn. Conf. Sug. Cane Res. Dev. Wkrs. India, **1** : 31-39.

82. Ramachandran, S. 1986. Chemical control of shoot borer *Chilo infuscatallus* Snellen for sugarcane production in Cauvery deltaic area in Thanjavur district. M.Sc. (Ag.) thesis, Tamil Nadu Agric. Univ. Coimbatore.

83. Ranganathiah, K.G., Veeresh, G.K. and Govindu, H.C. 1973. A new entomophagas fungus of the root grub, *Holotrichia serrata* F. from Mysore. Curr. Sci., **42** : 4323-433

84. Rangaswami, G., Ramamoorthi, K. and Oblisami G. 1968. Studies on microbiology and pathology of insect pests of crop plants. Final Rep. P.L. 480 Res. Proj. Univ. Agric. Sci., Bangalore.

85. Rao, G.N. and Vijayalakshmi, U. 1959. A note on the occurrence of certain parasitic fungi on insect pests of sugarcane. Curr. Sci., **28** : 297.

86. Bidgway, R.L. and Vinson, S.B. 1976. Biological control by Augmentation of Natural enemies. Plenum press, New York. pp. 480.

87. Sankaran, T. 1974. Natural enemies introduced in recent years for biological control of agricultural pests in India. Indian J. Agric. Sci. **44** : 425-433.

88. Sankaran, T. and Manjunath, T.M. 1988. Biological control : Progress and problems. In :"Fourty years of Plant Protection in India 1946-1986". Ed. B. Reddy. Plant Prot. Assoc. India, Hyderabad, p. 75-83.

89. Sekhar, P.S. and Venkataramiah, G.H. 1964. Susceptibility of the Cockchafer, *Holotrichia* to *Bacillus thuringiensis* Berliner. Curr. Sci., **23** : 405-406.

90. Seshagiri Rao, C. 1980. Biological control of sugarcane scale insect, *Melanaspis glomerata* (Green) in West Godavari district (Andhra Pradesh). In : "Biological control of sugarcane pests in India" Eds. S. Sithanantham, and A.R. Solayappan. Tamil Nadu Co.op. Sugar Federation, Madras p. 67-70.

91. Seshagiri Rao, D., Puttarudriah, M. and Sastry, K.S.S. 1956. Biological control of sugarcane stem borer in Mysore - A brief review of work done in regard to the production and liberation of *Trichogramma evanescens minutum* Riley in Mysore. Proc. Int. Soc. Sug. Cane. Technol., **9** : 810-814.

92. Shanmugasundaram, M. 1989. Studies on natural enemies of larval and prepupal stages of sugarcane top borer, *Scirpophaga excerptalis* Wlk. in Tamil Nadu. M.Sc. thesis, Univ. Bombay, Bombay pp. 156.

93. Shukla, G.S. and Tripathi, N. 1981. Effect of trash burning on the management of sugarcane scale insect, *Melanaspis glomerata* G. Indian J. Ent., **43** : 437-438.

94. Singh, O.P. and Dayal, B. 1975. Some observations on *Epipyrops melanoleuca* Flet. an ectoparasite of Pyrilla. Cane Grow. Bull. **2** : 1-2.

95. Sithanantham, S. and Navarajan Paul, A.V. 1977. Toxicity of some insecticidal sprays to the parasitoid *Trichogramma australicum* Gir. (Hymenoptera : Trichogrammatidae) Proc. Second Orient. Ent. Symp. Madras (Abstract).

96. Sithanantham, S. and Navarajan Paul, A.V. 1980. Toxicity of some insecticidal sprays to the parasitoid *Trichogramma australicum* Gir. Pestol.; **4** : 21-24.

97. Sithanantham, S., Muthusamy, S. and Durai, R. 1973. Experiments on the inundative release of *Trichogramma australicum* Gir. in the biological control of sugarcane stem borer, *Chilo indicus* (Kapur). Madras agric. J., **60** : 547-561.

98. Sithanantham, S., Varadarajan, G. and Rajan, S.D. 1974. Experiments on releasing parasites in combination with insecticidal application for the control of sugarcane stem borer, *Chilo indicus* (Kupur). Madras agric. J., **60** : 547-561.

99. Solayappan, A.R. and Krishanmoorthy, V. 1973. Role of biological control of sugarcane borers in improving sugar recovery in factory areas in Tamil Nadu State. Proc. Ann. Conf. Prod. Council. Trichirappalli, **4** : (Abstract).

100. Subramaniam, T.V. 1937. Preliminary experiments on mass production of *Trichogramma* parasite for control against sugarcane borers in Mysore : Indian J. agric Sci., **7** : 149-155.

101. Subramaniam, T.V. 1941. Control of the sugarcane borers by the egg parasite *Trichogramma minutum* in Mysore. J. Mysore Agric. Exp. Un., **19** : 183 - 187.

102. Subramaniam, T.V. and Seshagiri Rao, D. 1940. Biological control of sugarcane borers in Mysore. Techniques for the mass breeding of the flour moth (*Corcyra cephalonica* St.) as host material for the bulk production of the borer egg parasite, *Trichogramma minutum.* J. Dept. Agric. Mysore (Ent. Ser), **12** : 1-12.

103. Tuhan, N.C. and Pawar, A.D. 1983. Life history, host suitability and effectiveness of *Trichogramma chilonis* (Ishii) for controlling sugarcane borers in the Punjab. J. Adv. Zool. **4** : 71-76.

104. Usman, S., Shivashankara Sastry, K.S. and Puttarudriah, M. 1957. Report of work done on the control of the sugarcane borers in the Visvesvaraya Canal tract. Dept. Agric. Mysore State. 69 pp.

105. Usman, S., Puttarudriah, M. Appanna, M. and Shivasankara Sastry, K.S. 1964. Report of work done during the period from 10th September. 1948 to 31st March, 1965. Dept. Agric., Mysore State. pp. 102.

106. Varadharajan, G. 1976. Studies on the utilization of *Trichogramma australicum* Gir., for the control of sugarcane internode borer. Madras agric. J., **63** : 569-570.

107. Varma, A. 1988. Effects of entomogenous fungi on some pests of sugarcane and other possible value in controlling the pests. Ph.D. thesis, Lucknow Univ. Lucknow.

108. Varma, A. and Mitra, Krishna, 1981. Recent researches leading to evolve a suitable control programme of stalk borer, *Chilo auricilius* Dudg. Proc. natn. Symp. Stalk borer. Karnal p. 63-65.

109. Varma, A. and Singh, K. 1987. *Metarhizium anisopliae* (Metschaikoff) Sorokin in the management of *Pyrilla perpusilla* Walker, the sugarcane leaf hopper. IISR Ent. Newslet., **18** : 12.

110. Verma, G.C., Bindra, O.S. Sukhdev Singh and Baljinder Singh, 1979. Comparative efficacy of biological, mechanical and integrated methods for controlling sugarcane borers in Punjab. J. Res. Punjab Agric. Univ., **16** : 45-48.

111. Veenakumari, K. and Veeresh, G.K. 1982. Evaluation of different methods of inoculation for the production of the milky white disease caused by the bacterium, *Bacillus papillae* Dutky on the white grub, *Holotrichia serrata* Fabricius. J. Soil. Biol. Ecol., **2** : 1-7.

112. Vora, V.J. and Ramakrishanan, N. 1978. Studies on the milky disease of whitegrub, *Holotrichia consanguinea* Blanchard (Coleoptera : Scarabaeidae). J. Ento. Res., **2** : 136-141.

113. Yadav, R.P. 1983. Bionomics, effectiveness and basic requirements for mass multiplication of common egg parasitoids of *Pyrilla perousilla* Walker. Ph.D. thesis, Haryana Agric. Univ. Hisar.

114. Yadav, R.P. and Chaudhary, J.P. 1987. Relative effectiveness of three important egg parasitoids of the sugarcane leafhopper (*Pyrilla perousilla* Walker) under controlled temperature conditions. Co-op. Sug., **18** : 309-314.

19 Sugarcane Diseases, Host-Parasite Relationships and Integrated Management System

U.S. Shukla and O.K. Sinha

CONTENTS

Integrated Pest and Disease Management
Rajeev K. Upadhyay, K.G. Mukerji, B.P. Chamola and O.P. Dubey (eds.)
APH Publishing Corporation, New Delhi-110002, India

ABSTRACT

Sugarcane is one of the important cash crops and is attacked by many important diseases like red rot, smut, wilt, leaf scald, ratoon stunting, grassy shoot diseases and mosaic viruses. These diseases have been described. Information partaining to Integrated Management of these diseases has also been discussed.

1. INTRODUCTION

Sugarcane (*Saccharum* interspecific hybrid) is cultivated in both tropical and sub-tropical regions of India. It is a principal source of sugar, gur and khandsari. In our country, during 1990-91 crop season nearly 240 million tonnes of sugaracane was produced from 3.7 million hectares of land and sugar production was recorded 12 million tonnes. As per projections of National Commission of Agriculture by 2001 AD sugar production target has been kept at 18 million tonnes and sugarcane production 300 million tonnes.

In meeting out this projected target, beside a-biotic factors disease alone constitute a serious drain on sugarcane production. According to an estimate of UN Food and Agriculture Organisation, the world sugarcane production in 1982 was only 811 million tonnes as against 1802 million tonnes of potential production. Thus, 991 million tonnes (55%) of sugarcane was lost chiefly due to diseases, insects and weeds. From diseases alone, the loss was estimated to be 19.2%, whereas insects and weeds incurred 20.1 and 15.7% loss, respectively (1). In India crores of rupees are lost annually on account of diseases of sugarcane.

By the turn of last century, a few diseases of sugarcane were recognised in the country, of these the most dreaded fungal diseases-red rot was identified in 1901 followed by smut in 1906 and wilt in 1913. Subsequently, two important bacterial diseases, ratoon stunting disease (RSD) and leaf scald were recognised in 1956 and 1961, respectively. The mycoplasmal disorder, i.e., grassy shoot disease (GSD) was first observed by Dr. C.A. Barber in 1919. In 1921, mosaic disease was detected in a number of local and exotic varieties. These diseases have been found to cause substantial economical all losses by scaling down the cane yield and recoverable sugar.

The serious outbreaks of red rot epidemics in the early part of this century had threatened the sustenance of several sugar mills in subtropical India. In the process, a large number of promising varieties had to be discontinued due to their susceptibility to red rot only. The most dominant variety of whole of subtropical India after sixties i.e., Co 1148 and later CoJ 64, the wonder variety of Punjab have been withdrawn due to red rot disease alone. In view of absence of any chemical control measures, the situation prompted sugarcane scientists to breed sugarcane varieties resistant to red rot, thus saving the sugar industry from near collapse. Massive sugarcane varietal development programme started by the two ICAR Institutes, viz., Indian Institute of Sugarcane Research, Lucknow and Sugarcane Breeding Institute, Coimbatore, various State Agricultural Universities and State Sugarcane Research Stations, has resulted in the development and release of several promising varieties in tropics as well as subtropics (Table 1.).

Table 1 : Some promising varieties of sugarcane released for various states (118)

Sl. No.	State	Early Ripening	Mid season & late tipening
1.	**Uttar Pradesh**		
	Normal conditions	CoS 687, CoS 8436 CoPant 84211	CoS 767, CoS 802, BO 91, UP 12, CoLK 8001, CoS 85272, CoS 8118, CoS 8119, UP-1, CoLK 8102
	Water logging	-	BO 91, CoS 767, CoS 8118, CoS 8119
2.	**Bihar**		
	Normal conditions	BO 102, BO 120	BO 104, CoS 767, BO 91
	Water logging	BO 90	BO 109, BO 91, BO 110
3.	**West Bengal**	BO 90	Co 1148, Co 7224
4.	**Orissa**	Co 7704, CO, 8401, Co 62175	Co 7706, Co, 8402, Co 62175
	Water logging	-	Co 62175, Co 6304
5.	**Assam**	Co 1008	Co 8314, Co Jor 1 Co Jor 2
6.	**Punjab**	CoJ 89	CoJ 67, CoJ 81
7.	**Haryana**	CoJ 58, Co7717	CoS 767
8.	**Madhya Pradesh**	Co 7314, CoC 671	Co 1307, Co 62175, CoLK 8001, Co 8304, CoS 767
9.	**Gujarat**	CoC 671	Co 7219, CoM 7125, Co 7527, CoC 671
10.	**Maharashtra**	Co 7219, CoC 671	Co 7219, CoM 7125, Co 7527, CoC 671
11.	**Karnataka**	IC 225, KHS 3296	Co 62175, Co 8021, KHS 2045
12.	**Andhra Pradesh**	CoA 7701, CoA 7601, CoA 77-1, CoT 8201, CoA 8401	Co 62175, CoA 7602, CoA 8402, Co 7219, Co SO14, CoR 8001
13.	**Tamil Nadu**	CoC 671, CoC 771 CoC 8001, CoC 850061	Co 6304, Co 62198, CoSi 776, CoSi 86071

Red rot, smut, wilt, RSD, leaf scald, GSD and mosaic diseases are transmitted through setts or sets (cane cuttings used for gegetative propagation of sugarcane). Although infection of most of these diseases can be eliminated to a great extent by heat therapy or chemicals the secondary spread of these diseases under favourable environment culminates into reinfection of large plant population which poses a serious constraints in profitable cane cultivation. Under such a situation, disease resistant varieties provide the alternative and effective method as no single treatment is effective in controlling the disease. Some of the practices like rouging, seed selection, intercropping, crop rotation and sett treatment with chemicals have been effective in minimizing the diseases. It is, therefore, obvious that effective disease control can be accomplished by integration of available physical, chemical and biological methods against one or more diseases of sugarcane with minimum use of chemicals so that there is least disruption of the sugarcane ecosystem.

This chapter provides a current review on sugarcane diseases *vis-a-vis* host-parasite relationships, various methods of disease control and a proposed integrated disease management system for managing the diseases. For greater details the readers may consult earlier reviews (3, 52, 64, 92, 93).

2. MAJOR DISEASES OF SUGARCANE

Out of a large number of sugarcane diseases only economically important ones are briefly described below :

2.1 Red rot (Causal fungus : *Glomerella tucumanensis* (Speg.) Arx & Muller)

This is the most dreaded disease of sugarcane in India. Outbreaks of red rot in Pusa (Bihar) and Intiathok (U.P.) in 1932 and 1936, respectively created a havoc and forced the growers to abandon cane cultivation. The epidemics of this disease that occurred during 1930s to 1960s wiped out several excellent varieties like Co 210, Co 213, Co 312, Co 453, Co 997, BO 3, BO 10, BO 11, BO 17, BO 54 and CoS 510. In tropical India, Co 449 and Co 658 were amongst most popular varieties, but could not survive due to this disease.

Symptoms : Both leaves and stalk display characteristic symptoms of the disease. Leaves show red elongated lesions on the midrib and reddish patches on leaf sheath. A number of acervuli are produced in the central region of the lesion. Lamina may also exhibit reddish brown specks.

The affected canes reveal purplish discolouration on rind. In the nodal region, root primordia produce black acervuli. On splitting the cane longitudinally, the pith shows dull red colour interrupted with white patches. (Fig. 1). The affected tissue emits odour of vinegar. In severe cases the fungus traverses whole length of the cane and may extend to spindle which may also show rotting symptoms. The affected stool dries up in the later part i.e., September-October.

2.2 Smut (Casual fungus : *Ustilago scitaminea* Syd.)

Culmicolous smut is the next important disease of sugarcane. It is more serious in ratoon crop in which the incidence increases manyfolds. In 1942-43, the epidemic of smut in Bihar affected nearly 66% of canes. Co 1158, a superior variety in subtropical India and Co 740 in tropical India (especially in Maharashtra) were worst affected by this disease and have gone out of cultivation. Varieties like Co 419, Co 975, CoS 510, CoLK 7701 and CoLK 7901 are highly susceptible to the disease.

Symptoms : The disease can be readily identified by the black whip-like structure coming out of the top of the cane. The size of whip may range from a few centimetres to more than a metre. Whip is covered by a white membrane of the host origin. Millions of teliospores are produced in numerous sori that are formed on it. In addition to this conspicuous structure, the top leaves may become stiff narrow and pointed.

2.3 Wilt : (Causal fungi - *Acremonium furcatum, A. implicatum, Fusarium maniliforme*)

In recent years, wilt has become a serious problem in certain pockets of the country, particularly Bihar, U.P., Punjab, Haryana and Gujarat. This disease is responsible for the withdrawal of excellent varieties, viz. Co 527, Co 951, Co 975, Co 1007 and Co 1223. Wilt often occurs in association with red rot, pineapple disease, scale insects, root borer or plant parasitic nematodes. Such an association enhances magnitude of losses.

Symptoms : Wilt occurs during and after monsoon. The affected plants show chlorotic leaves which finally dry from base upwards. In extreme cases, the crown dries. In split open canes the pith display dull red colour tissue. A cavity is formed in the internode because of desiccation of the tissue. A foul odour emits from the affected tissue.

2.4 Ratoon stunting disease (Causal bacterium - *Clavibacter xyli* subsp. *xyli*)

This disease is most insidious in nature and has potential to cause significant loss of cane yield in a ratoon crop. In India, it has been recorded in varieties like CoS 510, Co 421, Co 738, Co 1148, Co 1158, Co 1336, BO 17 and BO 70.

Symptoms : RSD is not easily recognized in the field as it causes inconspicuous symptoms like stunting of plants with reduced vigour both in plant as well as ratoon crops. In addition, the tillering is reduced. In ratoon crop, the diseases plants grow slowly and the highly susceptible ones may die. Affected field may show infestation of weeds. Since stunting of plants is not uniform, the crop shows 'up and down' appearance.

In young plants with immature cane, pinkish discolouration appears in the pith tissue below the apical meristem. In mature diseased cane, on slicing through basal nodal area, orange-red dots, commas and short lines are observed. These symptoms have been considered unreliable because all the diseased canes do not show such discolouration and, therefore, are of little diagnostic value.

RSD can be diagnosed by light microscopy. Crude juice, taken out by applying negative or positive pressure on basal portion of diseased stalk, contains large number of bacteria. Under phase-contrast or dark-field microscopy, RSD bacterium can be observed. Electron microscopy and serodiagnosis are also helpful.

2.5 Leaf scald (Causal bacterium - *Xanthomonas albilineans* Ashby Dowson)

This disease is assuming important in the country. Till date, several commercial varieties like CoS 767, CoJ 64, CoLK 8001, Co 1148, Co 1158 and BO 70 have been found susceptible to the disease.

Symptoms : Of two phases of the disease, viz., chronic and acute, the later one is more injurious as it causes wilting, particularly during dry weather. In India, acute phase of the disease is uncommon, whereas the chronic phase is common occurrence. The leaves display white pencil-line streak along the entire length of the lamina. Leaves start drying from margins. The buds on the standing stalk exhibit premature sprouting in acropetal fashion. The side shoots (lalas) show white stripes on all the leaves. Occurrence of leaf scald during grand growth phase results in shortening of internodes. On cutting the internode, transversely, red vascular strands appear as red dots. On germination, root system becomes poorly developed.

2.6 Grassy shoot disease (Causal organism - Mycoplasma)

Grassy shoot disease has been reported earlier under different names such as 'chlorotic disease', 'yellowing', 'albino' and 'bunchy diseases' or leaf tuft due to variable symptoms at different stages. GSD inflicts heavy yield losses because negligible millable canes are found. The disease assumes serious proportions in ratoon crop if the preceding crop had been infected in the early stages of crop growth. In India, it occurs in almost all the sugarcane growing states, but in Maharashtra it is serious in old varieties like Co 419, Co 740 and Co 658, although new releases also show some infection.

Symptoms : The disease is characterized by premature and excessive tillering, with narrow leaves having varying degree of chlorosis. Loss of chlorophyll may be negligible, partial or complete. In a clump, only one or two lean and thin canes are produced and these do not bear inflorescence. The buds become papery and elongated which often sprout *in situ*. Occasionally, aerial roots develop from the lower nodes. Root system is poorly developed. The symptoms of GSD resemble to those of white leaf disease which is common in Taiwan.

2.7 Mosaic (Casual agent- Sugarcane mosaic potyvirus)

This disease is wide spread in India. Its incidence is more or less similar in tropical as well as sub-tropical region. Mosaic does note cause economical loss in cane yield in the country. But is has potential to cause significant losses when it occurs in combination with other diseases. At one time in Louisiana (USA), this diseases had threatened the survival of sugar industry. The disease occurs in a severe form in variety Co 740 where not a single cane free from mosaic is available.

Symptoms : The symptoms can best be seen on the basal portion of young leaves of the crown against the light source. The chlorotic areas alternate with the green ones and this imparts the mosaic pattern hence known as mosaic. The chlorotic areas may be accompanied by varying degrees of reddening or necrosis. Occasionally, elongated yellow stripes are formed on leaves. Leaf sheath may also display chlorotic areas. Symptoms are not produced on stalks, but necrotic areas may be sometimes present inside the stalk tissue.

3. HOST-PARASITE RELATIONSHIPS

For understanding of disease development, it is imperative to know the source of primary inoculum. It is ths small inoculum which builds up in disease development.

3.1 Sources of inoculum

3.1.1. Seed cane or sett

Sugarcane is a vegetatively propagated crop. The seed cane itself harbours the inoculum and becomes the source of primary infection. The sett consists of at least one node and an internode. The node comprises of a leaf scar, a bud which is located in root band region and a growth ring. In root band region, numerous root primordia are present.

The diseased sett carries incipient infection of *C. falcatum* in the nodal region. Mycelium of smut fungus lies dormant in the meristematic tissue of the bud, whereas wilt fungi are present in vascular bundles. The bacterial pathogens of leaf scald and RSD as well as sugarcane mosaic virus are found in the xylem bundles. The GSD pathogen is confined in the phloem tissue.

Soon after germination and elongation of shoot, these pathogens pass on to the mother shoot and consequently to the tillers which are formed after 2 to 3 months of germination.

3.1.2. Stubble

In commercial cultivation of sugarcane, ratooning (or stubbling) is invariably practised by the growers. Usually one or two and occasionally three ratoons are taken after the harvest of plant crop.

The infected stubble readily serves as the source of primary inoculum for the successive ratoon crop. It is, therefore, not uncommon to observe enhanced disease incidence in the ratoon crop in comparison to the preceding crop.

3.1.3. Soil

Following harvest of the crop, the incoculum survives in soil and crop debris for varied periods of time depending upon the type of resting structures, soil moisture and temperature as well as physical and biotic factors of the soil.

Colletotrichum falcatum can survive for 6 months in a fallow land (25). Singh *et al.*, (89) reported its survival up to 63 and 34 days during winter and autumn season, respectively. The fungus may survive in the form of chlamydospores, setae, thick-walled appressoria and hyphae. Teliospores of *Ustilago scitaminea* remain viable for 16-64 days in dry soil and up to 8 days in wet soil (41). Wilt fungus can survive in soil for 27 to 31 months (31). Possibility of leaf scald bacterium to survive in soil is remote as it remains viable for two days only (61). So far, no study has been made to estimate the survival period of RSD bacterium in soil.

Some soilborne pathogens like species of *Pythium, Rhizoctonia, Sclerotium* and *Fusarium* when get opportunity to infect the setts, cause rotting.

3.1.4. Insects and insect-vectors

Although red rot fungus has been reported to enter through borer holes of *Diatraea saccharalis* in USA (27), this mode of entry appears to be uncommon in India. Teliospores of *U. scitaminea* frequently adhere on the body of mycophagous insects viz., *Phalacrus immaginatus* Champ., *Diphyllus* sp. nr. *egens* Grouvella, *Trocetes bostrychophillus* Badonnel and *Blattiscius keegani* Fox and are disseminated to healthy plants (7, 30).

Earlier reports claimed that GSD pathogen can be transmitted through aphids (19), and also by the hopper *Proutista moesta* Westwood (29). However, recent reports (66) do not corroborate transmission of GSD pathogen either through aphids or hopper. Sugarcane mosaic potyvirus has been reported to be transmitted through a number of aphis and bugs (1, 9, 13, 16, 22, 36, 37, 43, 65, 119). So far, leafscald and RSD pathogens have not been demonstrated to be insect-borne.

3.1.5. Collateral hosts

Sugarcane belongs to family Gramineae. Therefore, most of the pathogens of sugarcane can infect related grasses which become constant source of inoculum. Collateral hosts of important pathogens are listed in table 2.

Table 2. Collateral hosts of sugarcane pathogens (3, 64)

	Fungus/Disease	Collateral hosts
1.	*Colletotrichum falcatum* (Red rot)	*Leptocholoafiliformis, Miscanthus* sp., *Saccharum spontaneum, Sorghum halepense, S. vulgare*
2.	*Ustilago scitaminea* (Smut)	*Cyperus dilatatus, Erianthus saccharoides, Imperata arundinacea, Narenga* sp., *Saccharum barberi, S. munja, S. robustům, S. sinense, S. spontaneum, Sclerostachya fusca, Sorghum vulgare.*
3.	*Clavibacter xyli* sub sp. *xyli* (Ratoon stunting disease)	*Brachiaria milliformis, B. mutica, Chloris gayana, Cynodon dactylon, Echinochloa colonum, Imperata cylindrica, Panicum maimum, Pennisetum purpureum, Rhynchelytrum repens, Sorghum verticilliflorum, S. helepense, Sporobolous capensis, Zea mays.*
4.	*Xanthomans albilineans* (Leaf scald)	*Bambusa vulgaris, Brachiaria piligera, Coix lacryma jobi, Cymbopogan citratus, Imperata cylindrica* var. *major, Panicum maximum, Paspalum paniculatum, P. dilatatum,, P. commerconii, P. conjugatum, Pannisetum, purpureum, Sorghum verticilliflorum, Thysanolaena maximum, Zea mays, Z. mays* var. *saccharata.*

5. Sugarcane mosaic potyvirus (Mosaic)	*Arundinaria gigantea, **Brachiaria platyphylla**, Dactyloctenium **aegyptiacum**, Digitaria abscendens, D. didactyla, **Echinochloa** sp., E. crusgalli, Eleucine indica, **Eragrostis ferrudinea**, Gynerium sagitattum, Imperata cylindrica* var. *koenigii, Miscanthus sinensis, Panicum dichotomiflorum, Paspalidium flaridum, Paspalum boscianum, Pennisetum glaucum, P. purpureum, P. typhoides, Setaria lutescens, S. magna, S. sulcata, S. aurea, Sorghum vulgare, Stenotaphrum secundatum, Syntherisma, sanguinalis, Urochloa mosambicensis.*

3.1.6. Contaminated agricultural implements

During harvest and seed cutting operations, implements often get contaminated with the inoculum. This situation is commonly observed in case of diseases like leaf scald and RSD. This mode of transmission, is, however, uncertain in case of GSD pathogen.

3.2 Mode of penetration

Under a set of environmental conditions within favourable range, the pathogen when comes in contact with the host surface, penetrates and establishes infection. Some of the modes of penetration of important sugarcane pathogens are described below :

3.2.1. *Direct penetration*

This type of penetration occurs in red rot and smut fungi. Conidia of *C. falcatum* germinate to produce appressoria from which infection hyphae emerge out and penetrate the cell wall through pits. The infection thread penetrates through the tissues of bud scales, leaves, leaf sheath, growth ring or root primordia. Besides, wounds caused by mechanical injury, insect damage, growth cracks and leaf scar are the vulnerable sites for penetration.

Telioospores of *U. scitaminea* present in soil or on the bud, germinate to produce sporidia or dikaryotic infection hyphae which enter through the basal region of the innermost scale-leaf and penetrate the meristematic tissue of the buds and lie dormant (15).

3.2.2. By cultural operations

Conventionally, cane is cut into three bud setts for planting purpose. The cut-ends of the sett provides direct avenue for the entry of soilborne or setborne (RSD and leaf scald bacteria) pathogens. Similarly, the expowred cut-ends of stubble facilitate the entry of pathogens. The wilt fungus, *Acremonium* sp., enters through the exposed vascular strands at the cut ends (10, 107).

3.2.3. Through injuries caused by nematodes infestation

The plant parasitic nematodes mechanically puncture the root cell by stylet while feeding (20, 90). Ectoparasites (e. g., *Xiphinema, Trichodorus* and *Paratylenchus)* insert the stylet into the host cell while their body remains outside. They feed on epidermal cells in root elongation zone, root hair or root cap cells. Semi-endoparasites (e.g., *Radopholus* and *Meloidogyne)* fully enter the host tissue. A number of nematodes, *(Meloidogyne, Hoplolaimus* and *Longidorus* etc.) provide courts for entry of other pathogens, like *Acremonium* spp., that incite wilt disease (54, 124), The entry points of these nematodes provide easy passage to the infection hyphae of the pathogen.

3.2.4. By insect-vector feeding

Sugarcane mosaic potyvirus (SCMV) has been demonstrated to be carried to the host through the insect-vectors that feed on sugarcane leaves. In India, *Melanaphis sacchari,M. indosacchari, Lipaphis pseudobrassicae, Rhopalosiphum rufiabdominalis* and *Assamia moesta* (13, 22) have been reported to transmit SCMV. The stylet of the aphid feeding on a diseased leaf gets contaminated with virus particles. When such aphids happen to feed on the other plants, the virus particles are infected in the healthy plant.

3.3 Alterations in the host

Consequent upon successful entry and establishment of infection, the pathogen brings about cellular and biochemical changes in the host. These are summarized below :

3.3.1 Red Rot

Structural changes: The invading mycelium of *Colletotrichum falcatum* advances intracellularly in the parenchymatous tissue. The damage to cell depends on the nature of pathotype and susceptibility of the host. In moderately resistant genotype, the colour of the protoplasmic contents of the cell changes before hypha reaches there. A gummy dark red substance is secreted by the host and deposited in the intercellular space. The tracheary vessels become occluded with gum deposition and cell wall turning red. Gum like formation is considered to be a resistant reaction of the host and restricts the advancement of the fungus which may later lyse (106). In highly susceptible genotypes there is no gum formation. The central portion of the lesion displays off-white patches or spots which harbour the mycelium of *C. falcatum* (108). The rotten tissue which turns red emits the odour of vinegar and is formed by fermentation process.. In resistant genotypes, the response of host cells is rapid , whereas lesion development is slow and restricted.

Depending upon the degree of host resistance, the infection may remain confined to single internode or may extend to whole of the cane. In severe cases the fungus may reach the apical portion and cause spindle rotting also. In the pith region, where the tissue is shrivilled sometimes, fungus sporulates to produce abundant mycelium with conidia acervuli, appressoria, chlamydospores and thick-walled hyphae. On rind of the internode portion and also on nodal region, the fungus produces abundant acervuli in which conidia are embedded in visible pinkish mass. Rarely, the rind exhibits long 'cirrhus' consisting of millions of conidia embedded in gelatinous matrix.

The fruting bodies or conidia are washed away by rain or irrigation water and thus the inoculum spreads and infects fresh canes nearby healthy plants. For this reason, red rot disease build-up takes place rapidly during and after monsoon rains and also under inundated conditions of the field. When the weather is dry, detachable flakes rind containing acervuli and conidia may become air-borne and infect healthy plants (99).

Biochemical alterations : *C. falcatum* produces cellulolytic and pectolytic enzymes for tissue disintegration (84, 85). Invertases, located at the outer layer of plasmalemma of fungal cell, are produced by the fungus the break and sucrose molecule into glucose and fructose. This inversion of sucrose results in enormous loss of recoverable sucrose.

The affected tissue manifests reduction in carhobydrates, chlorophyll, carotenoids, total organic nitrogen and activity of nitrate reductase. RNA content and total protein, however, increase with no appreciable change in DNA (6, 123). Sinha *et al.* (102) reported marked increase of free proline in and around affected tissue. Such an increase is perhaps necessary to meet the high energy requirement during the pathogensis.

The extracted juice from diseased cane contains high quantum of total salts, gum and reducing sugars as well as titrable acidity (96).

3.3.2. Smut

Structural changes : Under high humidity and temperature around 28°C, the teliospores of *Ustilago scitaminea* germinate and produce infection hyphae. It takes 20-25 days to successfully colonize the meristematic tissue of the nodal bud (100) and the hyphae can be detected by a staining technique developed by Sinha *et al.* (103). As the nodal bud germinates and shoot elongates, the mycelium keeps pace with the meristem and remnants of hyphae are left behind in each bud primordium. The hyphae present in the young tissue of internodes, gradually lyse, but remain dormant and alive in the meristem of the nodal bud and apical shoot. Till fungus remains dormant to no visible structural changes are observed. Under moderately high temperatures (28-30°C) that favour vigorous growth of sugarcane, the smut fungus usually keep pace with the growing point and manifests no external symptoms. At higher temperatures ranging from 34-40°C or more than mycelium present in the apical meristem becomes sporogenous and under the influence of growth regulators the apical shoot elongates rapidly to form a whip-like structure. A whip consists of numerous sori where millions of teliospores are produced. Several morphological modifications are observed in the host plant (74), e.g., premature flowering, tissue proliferation at the base of whip multiple bud formation, elongation and sprouting of nodal buds, whip production from the lateral buds and the bud groove becoming prominent.

Teliospores from the whip are liberated and air current helps their dissemination causing secondary infection. Spores that fall on ground are carried through rain or irrigation water.

Biochemical changes : In smut affected canes, there is sharp decline in sucrose continent (67, 117). In susceptible genotypes, total phenols, reducing sugars and free amino acids are increased, while o-dihydroxy phenols and total sugars decreased (56). The fungus synthesizes erythritol and mannitol with smaller quantities of arabitol, inositol and trehalose under *in vitro* conditions. It appears that the fungus probably utilizes suncrose of the host system and produces these carbohydrates. Sucrose is rapidly inverted (60).

In leaf and apical meristem of infected cane, invertase activity increases two-folds (57). Sankpal and Nimbalkar (68, 69) reported reduction in non-reducing sugars, polyphenols, potassium, phosphorus, iron, manganese and chlorophyll content. They noted enhancement in reducing sugars, sodium, calcium and magnesium. Sulphur uptake was shown to increase in smut affected plant and more sulphate was partitioned to leaf laminea (75).

3.3.3. Wilt

Structural changes : *Fusarium moniliforme* and *Acremonium* spp., develop inside the xylem vessels. *F. moniliforme* can also infect parenchymatous tissue which turns dark muddy-red near the node. The walls of xylem bundles become dark. Concomitantly, gum is produced in the vessels, intercellular spaces of parenchymatous cells and xylem parenchyma. It appears that presence of gum and hyphae obstructs the transport of water an causes wilting. Production of vivotoxin has also been suspected in inducing wilt fungus proliferates and sproulates profusely (32).

Biochemical changes : The analysis of foliar tissue reveals decrease in protein and total chlorophyll content. Nitrate reductase activity gets retarded (91). Such changes become prominent after three months of infection and lead to low recoverable sugar and higher reducing sugars, gums, titrable acidity, soluble salts and flavonoids. Reduction in sugar recovery has been estimated to be 3-29% (35).

3.3.4. Ratoon stunting disease

Structural changes : The bacterium is non-aggressive and after establishment of infection, it colonizes the xylem. It multiplies and moves to leaf sheath, leaf laminae and root and shoot of underground portion. The bacteria do not cause any recognisable structural changes except reddening of small areas at nodes.

Biochemical changes : In case of RSD, biochemical changes have been studied in RSD affected canes and these have been compared with infected but moist hot air treated canes. Moist hot air treatment (54°C for 7 hr) kills the RSD bacterium to a greater extent (81). Shukla *et al.* (79) and Madan *et al.*, (48) have recorded very low acid invertase activity in RSD affected buds, whereas it was quite high in heat treated ones. Neutral invertase was absent in the treated and untreated buds. The inhibition of acid invertase activity by RSD bacterium was attributed to some thermolabile factor and the effect of which was reversible.

Sulphur nutrition studies have shown that RSD bacterium disrupts metabolic functions in roots and cause increase in accumulation of non-protein nitrogen (48). It was also observed that ammonium sulphate application masked the effect of delayed tillering caused by RSD infection. Similarly, stunting effect of the disease was reduced by sulphur aplication (71).

3.3.5 Leaf scald

Structural changes : The bacterium is restricted to the xylem elements of the vascular bundles and caused chlorosis of adjacent parenchymatous cells. In chlorotic tissue, bacterium is not found. The chlorosis is caused by a toxic metabolite of pathogen origin. It acts on plastics before chlorophyll synthesis. The white pencil-line stripes appears by the destruction of plastids and chlorophyll (63). Chlorosis including isolates of *X. albilineans* produce a number of antibacterial compounds, the major one was identified as albicidin (14). There was positive correlation between albicidin production and ability to cause chlorosis. Albicidin affects both bacterial and chloroplasts.

The bacterium has been detected in fluff (true seed) of sugarcane (26) but the nature of damage is seed tissue is yet to be ascertained.

Biochemical changes : Pronounced increase in the activity of acid invertase has been noted in leaf scald affected buds (47). This is evident by the fact that the affected cane displays early sprouting of buds. Amylase increases two-folds in affected buds. Agnihortri and Madan (4, 5) found increase in quantity of reducing sugars in sprouted buds. There appears to be breakdown of protein into amino acids which show increasing trend in the infected buds.

In leaf scald affected leaves, there was reduction in nitrate reductase activity, protein and chlorophyll, whereas protein content increased (91). Chlorophyll-a was affected more than chlorophyll-b.

Juice quality gets impaired due to leaf scald infection. It has been observed that protein and sucrose contents decrease, while quantum of reducing sugars and polysaccharide increases (2).

3.3.6. Grassy shoot disease

Structural changes : In deceased plant, shape, size, structure and orientation of stem cells are altered (70). Electron microscopy of GSD affected leaf reveals the presence of mycoplasma-like organisms (MLOs) in the sieve tubes of the phloem cells. The size of MLOs ranges from 50 nm to 700 nm. The following structural and cellular changes were observed (80).

(a) Chloroplast : Chloroplast in diseased leaf tissue reduces in number and size. The lamellae of the stroma are poorly developed which may be due to the action of enzyme chlorophyllase. The stroma and grana become almost indistiguishable.

(b) Mitochondria and nuclei: Number of mitochondria per cell increases several folds (77). The nucleus undergoes internal changes in the affected leaf. The chromatin material forms dark aggregates towards the peripheral side of the nucleus which contains lesser number of osmophillic granules.

Biochemical changes : As consequence of GSD infection the plant undergoes following biochemical changes. Quantity of free amino acids and amides enhances in the leaves with concomitants reduction in proteins (39, 82). This shows that protein synthesis is altered in diseased leaves. Organic acids accumulate in the diseased leaves with increase in citric acid and malic acid ratio.

Carbohydrate metabolism gets disturbed in the diseased plant. Dhumal and Nimbalkar (23) recorded increase in reducing sugars and reduction in non-reducing sugars, total sugars and starch. Activity of sucrose synthetase and sucrose phosphate synthetase is adversely affected, while invertase activity gets stimulated. Shukla *et al.* (78) reported higher quantum of water soluble non-structural carbohydrates in GSD affected leaves.

The respiration rate of diseased plant increases 1.5 times. The enzymatic activity gets altered in the leaves. Activity of peroxidase, polyphenol oxidase and ascorbic acid oxidase enhances considerably after GSD infection (24, 78). Increase in peroxidase activity converts phenols into quinones.

Analysis of mineral composition of diseased leaves showed higher amount of K,P,Na, Fe, Zn and Cu and lower quantity of N, Mg, Mn and Si and an inverse relationship between B and Ca was noted (76). It was also observed that there was increase in the rate of sulphur uptake and decrease in partitioning of S to leaf lamina.

3.3.7. Mosaic

Structural changes : Following entry and successful infection, the SCMV particles move both upwards and downwards to almost all parts of the plant. The virus being systemic, all the leaves in a diseased plant, carry virus particles, but symptoms are conspicuous on young leaves only.

In mosaic affected leaves Lin (44) observed intra-cytoplasmic crystalline inclusions. These were present near the nuclei and contained structures like mitochondria, proplastids and crystalline inclusions. Virus particles were randomly distributed in the cytoplasm. In diseased leaf, there is redution in the growth of cells in chlorotic region. Such cells show little or no differentiation. The chloroplasts become smaller in size and reduce in number. The nuclei of cells are more prominent and the nucleoli may be more in number.

Biochemical changes : Biochemical alterations in mosaic affected plant have not been studied in much detail. In mosaic affected leaf, photosynthetic rate on a unit chlorophyll basis gets increased (38). Ghorpade and Joshi (33, 34) observed reduction in chlorophyll content, carbon assimilation and activity of photosynthetic enzymes. There was high activity of acid invertase, while neutral invertase was low. High activity of sucrose synthetase and low of sucrose phosphate synthetase was recorded and appears to be due to limited availability of uridine diphosphate glucose. Madan *et al.* (46) reported absence of acid invertase activity in the buds, whereas protein and total carbohydrate content was lower in diseased plant. Quantity of non-reducing sugars was decreased. Srinivasulu *et al.* (105) found that activity of starch synthetase was higher and resulted into more synthesis of starch in mosaic infected plant. Amylase activity was lower. The activity of starch synthetase and amylase, and starch content enhanced with the maturity of internode. In mature stalk, quantum of non-reducing sugars was higher.

3.4 Host resistance

The mechanism of resistance of disease in sugarcane has been well studied in case of red rot and smut mainly because these are most destructive diseases.

3.4.1. Resistance to red rot

Of five species of *Saccharum*, most of the clones of *S. sinense* and *S. robustum* are susceptible to red rot, whereas those of *S. spontaneum* and *S. barberi* are resistant ones. In breeding programme for development of red rot resistant varieties. *S. spontaneum* and *S. barberi* are the main source of resistance. Sugarcane offers two types of resistance to red rot disease, viz., mechanical and biochemical.

Mechanical : This type of resistance is also referred to as morphological or static resistance. The entry of *C. falcatum* is resisted by a number of morphological structures like greater thickness of epidermis, presence of thick cuticle, hard rind, presence of thick bud scales, relative abundance of vascular bundles in the stalk just below the rind and presence of thick bundle sheath. In rind portion, the lignified sclerenchyma restricts entry of the pathogen. Atkinson (11) reported that in some resistant varieties, comparatively a few vessels extend continuously from one internode to another and, therefore, offer resistance to movement of spores in the xylem vessels. The presence of cross walls or septa in the xylem vessels also impedes the movement of spores.

Thus, the varieties in which xylem vessels and board, continuous with a few cross walls, favour fast spread of the fungus and such a variety is rated susceptible.

It has been experienced that varieties which are susceptible to red rot by plug method (18) (a method of inoculation in which rind of cane is broken and inoculum is placed in contact with the pith) and possess morphological barriers to fungus invasion, show field resistance. For example, Co 312 survived in field for several years mainly because of morphological resistance and was otherwise susceptible to red rot by plug method.

Biochemical : It is also referred to as physiological, protoplasmic or dynamic resistance. This kind of resistance is due to either preformed chemicals that are present in the host before infection or post-infectionally formed chemicals.

In resistant varieties, gum formation checks the spread of the pathogen. Copious gum formation may lead to hypersensitive reaction. Enzymes like peroxidase, polyphenol oxidase, phenylammonia lyase (PAL) and tyrosine ammonia lyase (TAL) have been implicated in red rot resistance, but their precise role is yet to be established. Peroxidase activity which is linked with lignin synthesis has been found higher in moderately resistant mutants of a susceptible donor variety CoJ 64 (110). Polyphenol oxidase (PPO) is responsible for oxidation of phenols into quinones which are known to impart disease resistance. PPO activity has been reported to be higher in moderately resistant varieties and low in susceptible ones (98, 110). Some workers consider that the amount of phenols present in a variety is linked with the resistance to red rot (62, 73, 122, 127). But Singh *et al.* (91) did not find any correlation of phenols with red rot resistance.

PAL and TAL enzymes are known to synthesize phenolics, phytoalexins and lignins. Madan *et al.* (49) recorded higher activity of these enzymes in resistant and moderately resistant varieties (CoS 767 and Co 1158) as compared to susceptible ones (Co 312 and Co 1148).

3.4.2. Resistance to smut

A number of clones of *S. spontaneum* are resistant to this disease, but some clones have been found susceptible. Similarly, many clones of *S. sinense* and *S. barberi* are resistant to smut. The variety Co 6806 is highly resistant to smut. These resistant clones have been used as sources of resistance in the breeding programme.

In this case too, morphological and physiological resistance has been identified in sugarcane against smut.

Morphological : Tight bud scales offer resistance to invading hyphae of *U. scitaminae.* Removal of bud scales in a resistant variety may turn it susceptible to smut (88, 126). Muthusamy (53) reported that bud shape and size are controlled by environmental factos and, therefore, the latter may play important role in contributing resistance to a variety. James, (40) however, did not obtain correlation between the resistance and degree of tightness of bud scales.

Biochemical : Failure of certain varieties like Co 997 (88) to show resistance to smut even after removal of bud scales, is indicative of biochemical resistance that operates against smut. It was further substantiated by the fact that extract from fresh, homogenized bud scales of several varieties contained a glycosidic substance in concentrations highly correlated with resistance to *U. scitaminea* (45). Sinha (101) recorded higher germination of teliospores in callus suspension of susceptible varieties like Co 1158 and CoLK 7901. In case of resistant varieties, teliospore germination was inhibited and not a single teliospore germinated in callus suspension of Co 6806 which is highly resistant to smut. Bud scale diffusates of Co 6806 and Co 449 (resistant clones) have been reported to inhibit teliospore germination, while those of Co 1287 and Co 740 (susceptible clones) did not cause adverse effect (55). Probably, the resistance factor (a glycosidic substance) is less active in immature buds and subterranean buds as compared to mature ones because the former kind of buds are more susceptible than the latter ones (120).

4. INTEGRATED PEST MANAGEMENT

4.1 The concept

During 1950s, organic pesticides, like DDT, proved extremely successful in controlling arthropods of pest status. Soon these pesticides became widely popular and formed integral part of public health and agricultural pest control programmes. Stern *et al.* (116) observed the following problems that arose from indiscriminate use of organic pesticides :

(a) Arthropod resistance to insecticides

(b) Outbreak of non-target arthropods

(c) Resurgence of target species necessitating repetitive insecticide application

(d) Toxic residues on food, feed and forage crops.

(e) Hazards to insecticide handlers

(f) Legal complications from suits.

In view of these problems, the authors suggested integrated approach utilizing both biological and chemical measures and, for the first time, furnished the concept of integrated control for insect-pests. In such an approach, the chemical applied for control must be least disruptive to biological control which forms the part of natural control that governs the population density of pest species. Chemical control offers only immediate and temporary decimation of localized pest population and does not contribute to permanent density regulation.

Definition : Stern *et al.* (116) defined integrated control as applied pest control which combines and integrates biological and chemical control.

Later in 1960s, the integrated control became popularly known as Integrated Pest Management (IPM). In 1987, Food and Agricultural Organization (FAO) defined IPM as a pest management system that, in the context of the associated environment and population dynamics of the pest species, utilizes all suitable techniques and methods in as compatible a manner as possible and maintains the pest populations at levels below those causing economic injury (121).

It is obvious that IPM makes use of alternative methods of pest management and involves the process of selection, integration and implementation of pest or disease control based on predicted economic, ecological and sociological consequences.

IPM strategy against plant diseases has received lesser attention probably because of the fact that chemicals (funguicides and antibiotics) are relatively less toxic than most insecticides. Of late, increasing use of chemicals for disease control has been observed. The adverse effects of such chemicals have also been well recognized. Oganomercurial fungicides have been banned and their indiscriminate usage has been restricted. Such a situation has forced the plant pathologists to adopt integrated disease management system involving alternative methods of controlling diseases so that sugarcane ecosystem is not disturbed.

4.2 Components of Integrated Disease Management system

The main components in context of controlingsugurcane diseases are discussed below:

4.2.1. Regulatory

Since prehistoric times sugarcane seed material is being exchanged within the country and between the countries. This is true because from its place of origin in New Guinea, sugarcane is now grown world over. Development of superior hybrid varieties has been the major source of attraction for seed material and thus resulted in its migration from one place to another. In this process, many seed transmitted diseases like red rot, smut, wilt, ratoon stunting, leaf scald, grassy shoot and mosaic have been introduced in areas where the diseases were absent. For example, subtropical India was almost free from red rot until hybrid canes were introduced from tropical India. Similarly, when smut was first recorded in Hawaii (USA) in 1971, the mainland USA was free from smut. In 1978, smut appeared for the first time in Florida and spread to other parts of the mainland (21). Ratoon stunting bacterium is considered to have spread to new areas through exchange of germplasm. In the same manner, other diseases have been inadvertently introduced in disease-free areas.

It is, therefore, important that the quarantine procedures are strictly followed so that seed material containing pathogens is checked and destroyed before it goes to the yield. It is necessary to make available precise and rapid techniques of detecting pathogens in the seed material. Serology, Elisa and DNA recombinant technology would be of immense use at quarantine stations.

4.2.2. Physical

For the control of sugarcane diseases, various types of thermotherapy have been attempted, viz., hot water treatment (HWT), Hot air treatment (HAT), moist hot air treatment (MHAT) and aerated steam therapy (AST). The primary inoculum which is present in the seed material gets killed and, therefore, disease build-up is reduced in the field.

Each of these methods is briefly described in relation to disease control.

Hot water treatment : For this purpose, different designs of hot water plant are available. The seed material is kept in baskets that can be lowered or lifted from the tank. In India, Joshi (42) obtained good control of red rot and smut by HWT at 52°C for 18 min. Chona (17) claimed complete cure of smut when seed pieces were treated at 55 to 66°C for 10 min. HWT has also been found effective against wilt (109), RSD (114), and leaf scald (115). SCMV was eliminated from Co 975 by repeated HWT of bud chips for 4 days at 55, 56.5 and 57.5°C for 7 and 10 min. with 24 hr interval between each treatment. Benda *et al.* (12) obtained good control of mosaic by serial HWT of infected shoots at 51.5, 59, 61, 63°C for 0-20 min. at 20 to 24 hr interval.

In hot water tank, fungicides have been incorporated by many workers for deeper penetration of the chemical. This has increased the efficiency of HWT.

Hot air treatment : Dry hot air is passed into the chamber which contains seed material stacked on trays. In U.S.A., HAT (54°C/8 hr) is used on a commercial scale for controlling RSD (113). In India, it was tried against red rot (86) and other disease, but had to be abandoned as it drastically reduces but germinatoin.

Moist hot air treatment : MHAT plant was designed and developed at the Indian Institute of Sugarcane Research, Lucknow (111). It is an improvement over HAT as the dry hot air was replaced by moist hot air protecting bud desiccation. At 54°C for 2-2.5 hr, most of the pathogens of red rot, smut, leaf scald, RSD and GSD present in seed material are killed. The time/temperature combination may vary with the titre of inoculum and kind of genotype.

Aerated steam therapy : AST was designed and developed at the T.N. Agricultural University, coimbatore (29). This was first used against GSD. At sugarcane Breeding Institute, Coimbatore, this therapy is being, used to treat seed material for raising foundation seed. In USA, AST is used against RSD. AST has been found effective against smut (54), but not against red rot (58).

4.2.3. Cultural

Various cultural methods have been recommended for keeping the diseases at low level. These are as follows :

Trash burning : Field sanitation is important to reduce the initial inoculum. After harvest of sugarcane, crop residue like dry foliage, dead and dry canes and stubble are left in the field. Inoculum of a number of diseases survives in the crop debris and, therefore, immediately after harvest the crop debris should be burnt. This practice is helpful in reducing the insect-pest population also.

Use of healthy seed : Sugarcane being vegetatively propagated crop, many important pathogens are transmitted through seed-pieces. Incipient infections of these pathogens are sometimes not visible. Therefore, seed material should be obtained from apparently disease-free crop, preferably the plant crop. For seed purpose, short seed crop nurseries should be raised from heat treated seed material. The seed crop should be free from diseases, insect-pests, drought or waterlogging and should be harvested after 8-12 months age.

At the Indian Institute of Sugarcane Research, Lucknow Singh (87) proposed a "three-tier" seed programme for the production of healthy seed cane.

There are three important components of this programme and is completed in three stages.

Foundation seed : Seed is derived from apparently healthy and genetically pure plant crop free from abiotic stress. For ensuring the purity of seed, it should be obtained from research stations. The seed cane is given heat treatment (moist hot air treatment at 54°C for 2 hrs or aerated stem at 50-52°C for hr. After the treatment the seed should be treated with fungicides (e.g., Dithane M-45 @ 0.25%).

The crop thus raised should be inspected at monthly intervals and diseased clumps, if any, should be uprooted and removed as soon as they appear. The seed cane thus obtained from this crop is called 'Foundation seed'.

Certified seed : Foundation seed cane is multiplied in the second year. All cultural operations are repeated. At planting time, foundation seed is not heat treated. The crop is inspected after 4 and 7 months of planting. The last inspection should be 15 days before harvesting. The seed obtained from this crop is called 'Certified seed'.

Three-tier seed programme is being carried out at many sugar factory farms for the last two decades. It ensures healthy and continuous source of seed cane to farmers.

Drainage in fields : Sugarcane fields should have good drainage system as waterlogging during monsoon favours rapid secondary infection and disease development of red rot and smut. In areas which are prone to these diseases, upland fields should be selected.

Roguing : Removal of infected clumps, as soon as symptoms appear, has been found very effective in checking the build-up of diseases like red rot, smut and mosaic. In case of smut, rouging should be strictly followed because millions of spores are liberated from a single mature whip. In case of mosaic, rouging is effective till disease incidence is 2%. Removal of collateral host manually or by herbicide treatment is necessary for checking spread of mosaic.

Discontinuation of ratooning : After the harvest of plant crop, the inoculum of many diseases survives in the underground portion of the stubble. Since in the ratoon crop, tillers are formed from the sprouting of subterranean buds, the inoculum in the bud or onto the bud passes into the tiller. In case of smut, the dormant mycelium present in the buds of stubble keep pace with the growing point of tillers and, therefore, a number of whips appear in a clump. In case of red rot affected stubble, the inoculum spreads to other healthy clumps through irrigation or rain water. Thus, infected stubble serve infection foci in a ratoon crop and resultantly, disease incidence becomes higher than the preceding crop. It is, therefore, advisable that ratooning of infested crop be discouraged.

Crop rotation : Sugarcane is a 10-12 months crop in subtropical India and 12-18 months in tropical India. Monoculture of sugarcane, i.e., planting sugarcane after sugarcane has allowed many diseases to build-up in the field. Under such a situation, it is important that crop rotation should be followed so that the pathogens are killed in the absence of host crop. In a sugarcane based cropping system, several crop rotations have been recommended (125). For example, in sub-tropical India the following rotations are practised.

1. Maize-wheat-sugarcane-ratoon-wheat
2. Rice-potato-sugarcane-ratoon-wheat
3. Rice-wheat/mustard-sugarcane-ratoon-wheat
4. Green manuring-lahi-sugarcane-ratoon-wheat
5. Green manuring-potato-sugarcane-ratoon
6. Maize (early) - potato-sugarcane-ratoon
7. Rice (early) - sugarcane (autumn)-ratoon-moong
8. Maize+arhar-green manuring-sugarcane-ratoon

In tropical India, sugarcane is grown generally in rotation with rice. Other crops like groundnut or green manure crops can also be taken in rotations. In Maharashtra, cotton-sugarcane-lahi-jowar sequence is followed (112).

4.2.4. Intercropping (companion cropping)

In case of smut and wilt diseases, beneficial effect of intercropping has been observed. In autumn planted sugarcane, secondary spread of teliospores of *U. scitaminea* was interrupted by the intercrops like wheat, Indian mustard and coriander, whereas in spring planted sugarcane by maize, sorghum and pigeon pea (51). Wilt incidence is reduced when sugarcane is intercropped with coriander or mustard (95).

4.2.5. Biological

Use of a biological agency in disease control is the safest means of control provide the biological agency is established in the sugarcane ecosystem. Use of disease resistant varieties in sugarcane cultivation has been found most effective and cheapest mean of controlling diseases like red rot and smut (Table 3). Availability of large number of red rot resistant varieties in various parts of the country in sixties onward has checked recurrent red rot epidemics.

Table 3. Disease resistant varieties of sugarcane

Disease	Resistant and moderately resistant varieties
Red rot	BO 32, BO 47, BO 70, BO 90, BO 91, BO 99, BO 109
	Co 449, Co 527, Co 975, Co 1007, Co 1157, Co 1158, Co 1261, Co 1287, Co 1336, Co 6808, Co 7314, Co 7704, Co 7708, Co 7801, Co 7810, Co 62101, Co 62399
	CoA 7601, CoA 7602, CoA 7701,
	CoC 671, CoC 9161
	CoL 9
	CoLK 7710, CoLK 7901, CoLK 8001, CoLK 8002, CoLK 8005, CoLK 8102
	CoPant 84211
	CoS 510, CoS 654, CoS 658, CoS 687, CoS 776, CoS 770, CoS 771, CoS 802, CoS 7918, CoS 8315, CoS 84321, CoS 8436
	UP 25
Smut	BO 11, BO 22, BO 24, Co 449, Co 527, Co 658, Co 974, Co 1148, Co 6806, Co 7108, Co 62174, CoLK 8102
Ratoon stunting disease	CP 29-116, Q 50
Leaf scald	Q 42, Q 50, Q 98, Q 813, PoJ 36, PoJ 2725, CP 807, CP 29/116, Co 290, Co 301, Co 331, Co 421, B 4908

Scientists have produced a synthetic gene for resistance to sugarcane mosaic virus. This gene produces coded viral protein when induced into sugarcane plant cells by electroportion (104).

4.2.6. Forecasting

In advanced countries, prediction models have been developed for the outbreaks of certain diseases like potato blight, apple scab, barely leaf blotch, brown rust of wheat, etc. On regional or national basis, computer services are available for knowing the disease and pest situation.

In India no work has been carried out on this aspect in relation to sugarcane diseases. In view of rapid technological development, forecasting models can be developed to make the disease control programme more effective.

4.2.7. Chemical

Currently, emphasis is being laid upon minimum use of chemical so that agroecosystem is not disturbed. In sugarcane crop, vast number of insect-pests attack at various phases of growth and a number of insecticides are recommended for their control. In such a condition switching over to biocontrol is a must. In contrast to insect-pests, chemicals are rarely used against sugarcane diseases as they are least effective.

In view of some success that has been achieved with chemicals to control primary inoculum of sugarcane disease, chemicals can be applied as sett treatment or may be incorporated in hot water tank during hot water treatment of seed cane as very little quantity of chemical is required in comparison to its field application.

Some of the chemicals that have been shown effective against sugarcane diseases are listed in Table 4.

Table 4. Chemicals used to control sugarcane diseases (3)

Red rot	Sett treatment	1.	Benomyl (100 or 200 ppm) under air pressure of 10 lb psi for 15 min.
		2.	Benomyl (0.25%) for 1 hr.
		3.	Vitavax (0.05%) for 1 hr.
		4.	Agallol (0.5%) for 15 min.
		5.	Bavistin (0.5%) for 1 hr.
	Spray	1.	Blitox-50 (0.25%)
		2.	Dithane Z-78 (0.25%)
		3.	Perenox (0.25%)
Smut	Sett treatment	1.	Aretan (0.1%)
		2.	Aretan (0.5%) for 15 min.
		3.	Dithane Z-78
		4.	Agrosan (1%) for 7 min.
			Bayletan (0.2%)
			Bayletan (0.1%) for 5 min.
			Bauletan EC (0.05% a. i.) for 2 hrs.
	Soil application	1.	BAS 3192 (5% a.i.)
		2.	BAS 327 (5% a.i.)
Wilt	Sett treatment	1.	Agallol (0.1%)
	Soil treatment	1.	Boron or manganese (40 ppm)
		2.	Boric acid (15 kg/ha) to soil + Aretan (0.1%) as sett treatment + Bavistin (0.2%) drenching in soil.

Leaf scald	Sett treatment	Streptomycin (100 pp.) for steeping one-bud setts for one day. The setts are desiccated for 20 hr before treatment.
Grassy shoot	Sett treatment	1. Tetracycline (250 ppm) applied under negative pressure. 2. Achromycin (250 ppm) 3. Terramycin (250 ppm) 4. Tylan (250 ppm) 5. Erythromycin (250 ppm) 6. Ledermycin (500 or 1000 ppm)
Mosaic	Spray (for checking insect-vectors)	1. Insecticides available for killing aphids & bugs.

4.3 Integrated management system for sugarcane diseases

For successful and economical control of plant diseases, information pertaining to the crop, its pathogens, the environmental conditions prevailing in the area, locality, availability of materials and costs, is taken into account in developing the control programme (8). Although control programme is usually aimed against all the diseases affecting a crop, occasionally it may be aimed against an important disease prevalent in the region.

The following integrated management schedule is suggested for sugarcane diseases.

4.3.1. Pre-planting measures

Selection of seed material : In crop like sugarcane in which important pathogens are transmitted through seed cuttings, the health of seed is of paramount importance and, therefore, the control programme must start with planting of healthy seed. Three-tier programme is the primary source of healthy seed. In many sugar factory farms, seed crop is grown for availability of seed material to growers.

In red rot prone areas, seed of resistant or moderately resistant variety should be used. Similarly, if any other disease is important, disease resistant variety should be selected. Before, planting, setts must be treated with fungicides like Dithane M-45 or Dithane Z-78 for protection against soilborne fungi. Bayletan (0.1%) should be used for sett treatment in those areas where smut is a serious problem (see table 4)

Selection of land : Field selection is important if the area is prone to diseases like red rot and wilt. Waterlogging favours rapid red rot development, while moisture stress may promote wilt incidence. Upland fields are suitable where inundation is . recurrent problem.

Field sanitation : It is of utmost importance as the soil and crop debris harbous the inoculum of many diseases. It is recommended that the trash must be burnt immediately after the harvest of sugarcane crop.

Crop sequence : In many parts of the country, monoculture of sugarcane is practised. If the area is prone to diseases, monoculture should be discontinued and recommended crop rotation and intercropping be taken. In case the crop suffers from disease, the ratooning should not be practiced as the diseases like smut and ratoon stunting assume severe form.

4.3.2. Post-planting measure

Roguing: When the infected sett gets access into the field or soil inoculum infects the sett, the symptoms of smut. GSD and mosaic are observed at the post-emergence stage. Under high temperature conditions, smut whips emerge out. GSD symptoms are readily observed after two months of planting. Mosaic symptoms appear on young foliage. As soon as disease symptoms appear, the plants should be uprooted and burnt to check the secondary spread of the disease. At the onset of monsoon when grand growth phase of sugarcane crop starts, crop canopy become dense and survey for diseased clumps becomes difficult. At the same time mortality of canes occurs due to red rot. In a red rot prone area, surgery must be continued to locate and destroy the affected clumps.

Adoption of these measures during cane cultivation will prove effective to keep the diseases away or at a low-key. In case of seed transmitted diseases, reduction of primary inoculum is most important to keep the rate of disease progress slow. With this in view, the integrated disease management schedule has been formulated.

5 EPILOGUE

Sugarcane diseases, viz., red rot, smut, wilt, ratoon stunting, leaf scald, grassy shoot and mosaic, are destructive as these inflict losses of millable cane and recoverable sugar. Both structural and biochemical changes in the host have shown that loss in the quantum of sugar occurs due to lesser photosynthetic rate in the diseased plant and conversion of sucrose into monosaccharide mediated by the invading pathogen. The reduction in number and weight of millable cases occurs due to adverse influence of pathogens on the plant metabolism and disruption of water supply in the xylem vessels.

The diseased plant or clump in the field serves as infection focus for the secondary spread of the disease. After harvesting of the crop, the infected dry leaves and cane pieces make available inoculum for the successive crop. Since the important diseases of sugarcane are seed transmitted, for effective disease control, use of healthy seed and field sanitation are critical factors. Reduction of initial inoculum permits slow rate of disease progress. In view of hazards of using agrochemical in crop cultivation, in integrated pest management emphasis has been laid on the use of all those methods that are least disruptive to the agroecosystem with least use of chemicals.

The areas which need more research and attention include development of disease forecasting, models, identification of antagonists of plant pathogens, parasites and predators of insect-vectors of SCMV and plant parasitic nematodes, implementation of regulatory measures for checking inter-regional movement of seed material, inspection of healthy seed or nursery crop by trained personnel, development of chemicals to eradicate sett-borne pathogens, biotechnological methods to introduce resistance in superior clones and cross protection against sugarcane mosaic potyvirus.

6. REFERENCES

1. Abott, E.V. and Charpentier, L.J. 1963. Additional insect-vectors of sugarcane mosaic. Proc. Intern. Soc. Sugarcane Technol. 11th Congr., pp. 735-760.

2. Agnihotri, V. P. 1984. Changes in juice quality in leaf scald affected sugarcane varieties. Indian Sugar Crops J. **12** : 21-23.

3. Agnihotri, V. P. 1990. Diseases of Sugarcane Oxford & IBH Pub., New Delhi, p. 483.

4. Agnihotri, V. P. and Madan, V. K. 1987. Changes in carbohydrates and nucleic acids in leaf scale affected sugarcane genotypes. Indian. J. Plant Pathol. **5** : 155-157.

5. Agnihotri, V. P. and Madan, V. K. 1989. Changes in carbohydrates and nucleic acids in leaf scale affected sugarcane varieties infected with both leaf scald and red rot. Sugarcane (Spring Suppl.) : 11-12.

6. Agnihotri, V.P., Madan, V.K. and Lal. Ramji, 1989. Changes in carbohydrates and nucleic acids in sugarcane genotypes affected by *Colletotrichum falcatum:* Intern. Sugar J. **4** : 7-8.

7. Agrawal, R.A., 1956. Some mycophagous insects and their parasites associated with sugarcane. Curr. Sci. **25** : 332.

8. Agrios, G. N. 1988. Plant Pathology (3rd Ed.). Academic Press Inc., (London) Ltd., pp. 803.

9. Anzaone , L. Jr. and Pirone, T.P. 1964. Transmission of sugarcane mosaic virus by *Myzus persicae.* Plant Dis. Reptr. **48** : 984-985.

10. Arber, C.A. 1928. The Havana 1927 conference and cane diseases. Intern. Sugar J. **30** : 575-582.

11. Atkinson, R.E. 1939. On the nature of resistance of sugarcane to red rot. Proc. 6th Congr. Intern. Soc. Sugarcane Technol., 1938, pp. 684-692.

12. Benda, G.T.A.M., Mock, R. C., Gillaspie, A.G. Jr. 1989. Control of sugarcane mosaic by serial heat treatment. II. The pattern of cure at high temperatures. Sugarcane No. **2** : 6-10, 16, 19.

13. Bhargava, K.S., Joshi, R.D. and Rizci, S.M.A. 1971. Some observations on the insect transmision of mosaic virus in India. Sugarcane Pathol. Newsl. **6** : 20-21.

14. Birch, R. G. and Patil, S.S. 1985. Preliminary characterization of an antibiotic produced by *Xanthomonas albilineans* which inhibits DNA synthesis in *Eschericia coli.* J. Gen. Microbiol. **131** : 1069-1075.

15. Bock, K. R. 1964. Studies on sugarcane smut *(Ustilago scitaminea)* in Kenya. Trans. Brit. Mycol. Soc. **47** : 403-417.

16. Brandes, E.W. 1920. Artificial and insect transmission of sugarcane mosaic. J. Agric. Res. **19** : 131-138.

17. Chona, B.L. 1943. Sugarcane smut and its control. Indian Fmg. **4** : 401-404.

18. Chona, B. L. 1954. Relative resistance of sugarcane varieties to ret rot. Indian J. Aggric. Sci. **24** : 301-315.

19. Chona, B.L., Capoor, S. P. Varma, P.M. and Seth, M.L. 1960. Grassy shoot disease of sugarcane. Indian Phytopath. **13** : 37-47.

20. Chona, B.L. and Swarup, G. 1960. Nematode root injury of sugarcane. Indian Phytopath. **13** : 173-175.

21. Comstock, J.C., Ferreira, S.A. and Tww, T.L. 1983. Hawaii's approach to control sugarcane smut. Plant Dis. **67** : 452-457.

22. David, H., Alexander, K. C. and Ananthanarayana, K. 1972. Insect transmission of mosaic virus in sugarcane. Sugarcane Pathol. Newsl. **9** : 15-16.

23. Dhumal, K. N. and Nimbalkar, J. D. 1982. Physiological studies on grassy shoot disease infected sugarcane cultivars. Indian Phytopath. **35** : 345-348.

24. Dhumal, K. N. and Nimbalkar, J.D. 1983. Oxidative enzymes and some organic constituents in grassy shoot disease affected sugarcane leaves of clone Co 419 and Co 740. 33rd Sugarcane Technol. Assoc. India Conv.

25. Dastur, J. F. 1946. Report of the Imperial Mycologist. Sci. Rep. Agri. Res. Instt. New Delhi, 1944-45, pp. 66.

26. Duttamajumder, S.K. 1991. Fluff transmission of *Xanthomonas albineans,* the incitant of leaf scald disease of sugarcane. Curr. Sci. **59** : 744.

27. Edgerton, C.W. 1911. The red rot of sugarcane. La. Agri. Expt. Sta. Bull., p. 133.

28. Edison, S. and Ramakrishnan, K. 1972. Aerated steam therapy for the control of grassy short disease (GDS) of sugarcane. Mysore J. Agric. Sci. **6** : 491-494.

29. Edison, S. 1973. Investigation on the grassy shoot disease of sugarcane. Ph.D. Thesis, T.Nadu. Agric. Univ., Coimbatore.

30. Francis, C. B. 1968. Sugarcane smut. Maras Agirc. J. **26** : 468-474.

31. Ganguly, A. and Chand, J. N. 1963. Longevity of *Cephalosporium sacchari* Butler causing wilt disease of sugarcane. Sci. & Cult. **29** : 247-348.

32. Ganguly, A. 1964. Wilt. In : Sugarcane Diseases of the World Vol. 2. Eds., C.G. Hughes *et al.* Elsevier Amsterdam.

33. Ghorpade, L.N. and Joshi, G.V. 1980. Studies on photosynthesis in sugarcane var. Co. 740 infected with sugarcane mosaic. Indian Phytopath. **33** : 33-37.

34. Ghorpade, L.N. and Joshi, G.V. 1982. Effect of sugarcane mosaic virus on photosynthetic enzymes in sugarcane Co. 740. Sugarcane Pathol. Newsl. **28** : 42-46.

35. Gupta, M.R. and Gupta, S.G. 1976. Assessment of losses caused by wilt disease of sugarcane in Western U.P. Indian Sugar. **26** : 151-152.

36. Ingram, J.W. and Summers, E.M. 1936. Transmission of sugarcane mosaic by the rusty plum aphid *Hysteroneura setraiae* (Thomas) J. Agric. Res. **52** : 879-887.

37. Ingram, J. W. and Summers, E.M. 1938. Transmission of sugarcane mosaic by the green bug *(Toxoptera graminum* Rond) J. Agric. Res. **56** : 537-540.

38. Irvine, J. E. 1972. Photosynthetic rate in sugarcane. Effect of virus diseases, genetic disorders, and freezing. Proc. Intern. Soc. Sugarcane Technol. **14** : 1131-1138.

39. Jaiswal, S. P. and Bhatia, S. 1971. The metabolic changes in sugarcane associated with grassy shoot disease. Sugar Y. Azucar. **66**: 15-16.

40. James, G.L. 1969 . Current programme in sugarcane-Disease Research Centres (5) Rhodesia 'Smut'. Sugarcane Pathol. Newsl. **4** : 48.

41. James, G. L. 1969. Viability of sugarcane smut spores in the soil. Sugarcane Pathol. Newsl. **3** : 10-12.

42. Joshi, N.C. 1954. Effect of hot water treatment of setts for the control of red rot and smut disease of sugarcane. Indian Sugar N.L. **4** : 1-2.

43. Lawas, O.M. and Fernaxez, W.L. 1949. A study of transmission of the corn mosaic and some of the physical properties of its virus. Philip Agric. **32** : 231-238.

44. Lin, S.C. 1969. Cytoplasmic inclusions in plants affected with sugarcane mosaic virus. Sugarcane Pathol. Newsl. **3** : 26-27.

45. Lyoyd, H.L. and Naidoo, G. 1983. Chemical assay potentially suitable for determination of smut resistance of sugarcane cultivars. Plant Disease, **67** : 1103-1105.

46. Madan, V. K. Agnihotri, V.P., Singh, K., Pande, H.P. and Saxena, Y.R. 1981. Biochemical studies on sugarcane diseases. Sugar J. **44** : 19-20.

47. Madan, V.K., Agnihotri, V.P., Singh, K., Pande, H.P. and Saxena Y.R. 1982. Biochemical studies on diseases of sugarcane III. *Xanthomonans albilineans*. Sugarcane Pathol. Newsl. **29** : 36-40.

48. Madan, V.K., Singh, K., Shukla, U.S. and Saxena, Y.R. 1986. Biochemical studies on Sugarcane *(Saccharum officinarum L.)* affected with ratoon stunting disease. Indian Phytopath. **39** : 506-568.

49. Madan, V.K., Soni, N., Nigam, M., Solomon, S. and Agnihotri, V.P. 1991. Enzyme activity and cane genotype resistance to red rot. Sugarcane No. **2** : 6-8.

50. Mirza, M.S., Ahmad, M. and Anwar, M.S. 1986. Elimination of sugarcane mosaic virus from diseased sugarcane buds with thermotherapy J. Agric. Res. (alore). **24** : 207-210.

51. Misra, S.R., Ram, R.S. and Singh, K. 1989. Effect of intercrops on dispersal of smut *(U. scitaminea)* spores in sugarcane fields. Indian J. Agric. Sci. **59** : 114-117

52. Misra, S. R. and Ram, R. S. 1993. Major Sugarcane diseases and their management. In : Sugarcane Research and Development in Subtropical India, Eds. G. B. Singh and O.K. Sinha, pp. 191-221, Indian Institute of Sugarcane Research, Lucknow.

53. Muthusamy, S. 1974. Varietal susceptibility to smut *(Ustilago scitaminea* Sydow) in relation to bud characters. Proc. Intern. Soc. Sugarcane Technol. **15** : 289-291.

54. Natarajan, S. and Muthusamy, S. 1981. Effect of aerated steam, Sugar. J. **83** : 39-40.

55. Padmanabhan, P., Alexander, K.C., Shanmugan, N. 1988. Mechanism of Smut resistance in sugarcane. Sugarcane No. **6** : 14-16.

56. Padmanabhan, P., Alexander, K.C. and Shanmugum, N. 1988. Some metabolic changes induced in sugarcane by *Ustilago scitaminea.* Indian Phytopath. **41** : 229-232.

57. Padmanabhan, P., Mohanraj, D. and Aledander, K.C. 1988. Tissue invertase as a possible index of smut infection. Curr. Sci. **57** : 902-903.

58. Padmanabhan, D., Natrajan, S., Kalaimani, T. and Rajasekaran, S. 1990. Disease build up in sugarcane setts treated with aerated steam,. SISSTA Sugar J. **16** : 31-34.

59. Patel, D.J., Patel, H.R. and Patel, C.C. 1988. Epidemic of sugarcane wilt associated with plant parasitic nematodes in South Gujarat. Curr. Sci. **57** : 858-859.

60. Peros, J.P., Piombo, G. and Dumar, J.C. 1986. Carbohydrate metabolism of the sugarcane smut fungus, *Ustilago scitaminea.* Curr. Microbiol. **13** : 303-306.

61. Persley, G. J. 1971. A progress report on leaf scald investigations. Proc. Qld. Soc. Sugarcane Technol; 38th Conf., pp. 163-168.

62. Rao, K.C., Krishnamurthy, T. N., Lalitha, E. and Rajalakshmi, V.K. 1968. Phenols in relation to resistance of sugarcane varieties to red rot. Curr. Sci. **37** : 532.

63. Ricaud, C. and Ryan, C.C. 1989. Leaf Scald. In : Diseases of Sugarcane : Major diseases, Eds. C. Ricaud, B.T. Egan, A.G. Gillaspie and C.G. Hughes, Elsevier : Amsterdam.

64. Rigoud, C., Egan, B.T., Gillaspie, A. G. Jr. and Hughes, C. G. 1989. Diseases of Sugarcane. Elsevier, Amsterdam, p. 399.

65. Rizvi, S.M.A. and Bhargava, K. S. 1973. Prevalence of vectors of sugarcane mosaic. Sugarcane Pathol. Newsl. **10** : 40-41.

66. Samaddar, K., Singh, K. and Varma, A. 1977. Studies on the vectors of grassy shoot disease of sugarcane. Annual Rept. Indian Institute of Sugarcane Research, Lucknow, pp. 108.

67. Sandhu, S.A., Bhatti, D.S. and Ratan, B.K. 1969. Extent of losses caused by red rot *(Physalospora tucuanensis* Speg.) and smut *(Ustilago scitaminea* Syed.) J. Res. (PAU) **6** : 341-344.

68. Sankpal, S.D.and Nimbalkar, J.D. 1979. Effect of smut infection on sucrose metabolism in sugarcane. Indian Phytopath. **32** : 306-307.

69. Sankpal, S.D. and Nimbalkar, J.D. 1980. Physiological studies on smut infected sugarcane variety Co 740. Indian J. Expt. Biol. **18** : 95-96.

70. Saosh Misra, S.R., Singh, K. 1986. Anatomical studies on sugarcane affected with grassy shoot disease. Indian Phytopath. **39** : 499-502.

71. Saxena, Y.R., Shukla, U.S., Pande, H.P. and Madan, V.K. 1981. Biochemical studies on sugarcane I. Effect of sulphur application on the roots of sugarcane affected with RSD. Sugarcane Pathol., Newsl. **26**: 3.

72. Sharma, D.D. and Sinha, S.K. 1980. A survey of antatonistic antinomycetes from the soils of Agra region. Indian Phytopath. **33** : 581-585.

73. Sharma, K.P., Bhatia, S.K. and Kanwar, R.S. 1982. Phenolic contents of red rot resistant and susceptible varieties of sugarcane. J. Res. (PAU) 19-21.

74. Sharma, S.L. 1956. Morphological modifications in sugarcane plant by systemic infection of smut *(Ustilago scitaminea* Syed.) Proc. Intern. Soc. Sugarcane Technol. 9th Congr., pp. 1134.

75. Shrivastava, A.K., Singh, K., Ghosh, A.K. and Saxena, Y.R. 1987. Effect of smut and grassy shoot disease on sulphur uptake and partitioning in sugarcane. Sugarcane (Sprint suppl.), pp. 20-22.

76. Shrivastava, A. K., Singh, K., Srivastava, S. N., Ghosh, A. K. and Saxena, Y. R. 1987. Ion uptake and its partitioning in sugarcane settlings raised from grassy shoot disease and red rot infected planting material. J. Nuc. Agric. & Biol. **16** : 197-200.

77. Shukla, U.S. and R.R. Bhansali. 1985. Increase in mitochondria, respiration and oxidative enzymes in grassy shoot diseased sugarcane plants. Indian Phytopath. **38** : 740-742.

78. Shukla, U.S., Singh, V. and Mohan, P. 1988. Histopathological and biochemical changes in grassy shoot diseased of sugarcane plant. Indian J. Plant Pathol. **6** : 164-170.

79. Shukla, U.S., Singh, K., Madan, V.K. and Saxena, Y.R. 1979. Effect of moist hot air treatment on germination and invertase activity in RSD affected sugarcane buds. Sugarcane Pathol. Newsl. **23** : 21-22.

80. Shukla, U.S., Singh, K., Varma, A. and Prakash, N. 1984. Changes in chloroplast ultrastructure and chlorophyll in relation to mycoplasm infected sugarcane leaves. *Indian* J. Plant Pathol. **2** : 114-119.

81. Shukla, U. S., Ram, R. S., and Tripathi, R.C. 1974. Effect of moist hot air treatment on the control of GSD and RSD. Annual Rept., Indian Institute of Sugarcane Research, Lucknow, pp. 72-74.

82. Singh, A. and Singh, O. S. 1966. Studies on the metabolism of albine and healthy leaves of sugarcane. Proc. All India Conf. Sugarcane Res. & Dev. Workers, Coimbatore, **5** : 337-339.

83. Singh, G.R. 1967. Hot water treatment for the control of grassy shoot disease. Sugar J. **29** : 14-16.

84. Singh, G. P. and Husain, A. 1963. Role of enzymes in red rot disease of sugarcane. Curr. Sci. **32** : 81-83.

85. Singh, G.P. and Hussain, A. 1964. Relation of hydrolytic enzyme activity with the virulence of strains of *Colletotrichum falcatum.* Phytopathology **54** : 1100-1101.

86. Singh, K. 1973. Hot air therapy against red rot of sugarcane. Plant Dis. Reptr. **57** : 220-222.

87. Singh, K. 1977. Sugarcane diseases and three tier seed programme. Sugar News. **9** : 81-89.

88. Singh, K. and Budhiraja, R. R. 1964. The role of bud scales as barriers against smut infection. Proc. Sugarcane Res. & Dev. Workers **5** : 687-691.

89. Singh, K., Budhraja, T.R. and Agnihotri, V.P. 1977. Survival of *Colletorichum falcatum* in soil, its portals of entry and role of inoculum density in causing infection. Interim. Sugar, J. **79** : 43-44.

90. Singh, K. and Misra, S.R. 1976. Pathogenicity and histopathology of *Hoplolaimus indicus* on sugarcane. Nematologia, *22* : 433-436.

91. Singh, O., Waraitch, K. S. and Kanwar, R. S. 1981. Leaf scald disease stress mediated losses and alteration in some metabolities of sugarcane. Sugarcane Pathol. Newsl. **26** : 7-11.

92. Singh, K., Sinha, O.K., Srivastava, S. N. and Srivastava, P. 1990. Sugarcane disease epidemiology and management with emphasis on quarantine. In : Recent Advances in Sugarcane Ed. P.N. Rao, pp. 382-415, The KCP Ltd., Vuyyuru, India.

93. Singh, K., Srivastava, S.N., Sinha, O.I. and Srivastava, P. 1992. Red rot of sugarcane. In : Plant Disease of International Importance, Diseases of Sugar, Forest and Plantation Crops, Vol. IV. Eds. A. N. Mukhopadhayay, J. Kumar, H. S. Chaube and U.S. Singh, pp. 1-15 Prentice Hall, New Jersey.

94. Singh, N. 1993. Potential biocontrol fungi in the management of red rot disease of sugarcane. Biocontrol.

95. Singh, N., Singh, K., Misra, S. C. and Singh R. 1983. Natural incidence of sugarcane wilt in companion cropping. Indian Sugar Crops. J. : 16-17.

96. Singh, O.N. and Waraitch, K.S. 1977. Metabolic changes induced by *Colletotrichum falcatum* Went in sugarcane. Sugarcane Pathol. Newsl. **19** : 7-9.

97. Singh, K., Singh, R.P. and Agnihotri, V.P. 1976. Phenolics in relation to sugarcane resistance against red rot disease. Sugarcane Pathol. Newsl. **15/16** : 37-41.

98. Singh, R.P. and Lal, S. 1984. Comparative changes in host following red rot infection. Annual Rept. Indian Institute of Sugarcane Research, Lucknow.

99. Singh, R. P., Lal, S. and Singh, K. 1983. Development of dormant nodal infections of *Colletotrichum falcatum* Went in sugarcane. Proc. Annu. convention, Sugarcane Technol. Assoc. India. **47** : 79.

100. Sinha, O.K. 1982. Studies on epidemiology and control of sugarcane smut. Annual Rept., Indian Institute of Sugarcane Research.

101. Sinha, O.K. 1991. Studies on resistance expression in tissue cultures against red rot and smut diseases of sugarcane. Annual Reprt. 1990-91, Indian Institute of Sugarcane Research, Lucknow, pp. 53, 55.

102. Sinha, O.K., Bhansali, R.R. and Singh, K. 1984. Free proline accumulation in response to infection by *Colletotrichum falcatum* Want in sugarcane. Curr. Sci. **53** : 793-494.

103. Sinha, O.K., Singh, K. and Misra, S.R. 1982. Stain technique for the detection of smut hyphae in nodal buds of sugarcane. Plant Dis. **66** : 932-933.

104. Smith, G. 1991. Transgenic sugarcane-application of biotechnology to sugar industry needs. BSES Bull. No. 36 (October), pp. 6-7.

105. Sreenivasulu, K.V., Raju, B.C. and Nyudu, M.V. 1985. Carbohydrate metabolism in mosaic virus infected sugarcane (*Saccharum officinarum* L.). Indian Bot. Rept. **4** : 129-133.

106. Srinivasan, K. V. 1958. Report of the mycologist. Annual Rept. Sugarcane Breeding Institute, Coimbatore, 1957-58, pp. 69.

107. Srinivasan, K. V. 1964. Some observations on sugarcane wilt J. Indian. Not. Soc. **43** : 397-408.

108. Srinivasan, K.V. and Bhat, N.R. 1961. Red rot of sugarcane-Criteria for grading resistance. J. Indian Bot. Soc. **40** : 566.

109. Srinivasan, K.V. and Rao, J.T. 1968. Heat treatment of sugarcane seet material. Indian Sugar, **18** : 683-690.

110. Srivastava, B. L. and Solomon, S., 1990. Biochemical analysis of induced mutant of sugarcane (*Saccharum* sp.). International Symposium Genetic Engineering and Plant Tissue Culture in relation to disease and pest resistance of crop plants, TNAU, Coimbatore.

111. Srivastava, N.S.L., Singh, K.and Sharma, M.P. 1977. The I.I.S.R. moist hot air treatment plant for seed cane. Sugar News, **9** : 96-102.

112. Srivastava, S.C. and Seth, S.R. 1991. Sugarcane Agronomy and Cropping. In : Recent Advances in Sugarcane Eds. P.N. Rao, pp. 207-247, The KCP Ltd. Vuyyuru, India.

113. Steib, R.J. and Forbes I.L., 1958. Hot air for the control of the ratoon stunting disease of sugarcane in Louisina. Phytopathology, **48** : 3298.

114. Steindl, D.R.L. 1950. Ratoon stinging disease. Proc. Intern. Soc. Sugarcane Technol. 7th Congr., pp. 457-465.

115. Steindl. D.R.L. 1972. The elimination of leaf scald from infected material. Proc. Intern. Soc. Sugar Cane Technol. 14th Congr. pp. 925-929.

116. Stern, V.N., Smith, R.F., Van den Bosch, Robert and Hagen, K.S. 1959. The integrated control concept. Hilgardia, **29** : 81-101.

117. Subha Raja, K.T., Natrajan, S., Ahmed, J. and Padmanabhan, D. 1972. Effect of smut incidence on plant. Sugarcane Pathol. Newsl. **9** :9.

118. Sugarcane Production Technology of Sugarcane, All India Coordinated Research Project of Sugarcane, Indian Institute of Sugarcane Research, Lucknow, p. 57.

119. Tate, H.D. and Vandenburg, S.R. 1939. Transmission of sugarcane mosaic by Aphids, J. Agric. Res. **59**: 73-79.

120. Toffano, W.B. 1977. Study of the interrelation between resistance to *Ustilago scitaminea* Syd. and state of maturation of sugarcane buds. Arquivos do Inst. Biol. **44** : 55-58.

121. U.N.F.A.O. 1984. Analysis and Design of Integrated Crop Management Programmes, Working Paper No. AGP PEST/84/WP/314, FAO Committee of Exports on Pest Control, 3rd Session, Rome, Oct. 24-26.

122. Verma, A. K., Jaiswal, S. P., Bajaj, K. L. and Bhatia, I. S. 1971. A Study of polyphenols in sugarcane in relation to red rot disease present in the stem of sugarcane varieties. Sugar Y. Azucar, **66** : 11-13.

123. Verma, R. K., Rao, G.P., Varma, K.P. and Singh, H.N. 1991. Some biochemical changes in red rot infected sugarcane leaves. Indian Sugar. **40** : 817-820.

124. Vijaylakahmi, U. and Rao, J. T. 1961. Screening sugarcane varieties to root-knot nematode, *Meloidogyne javanica.* Curr. Sci. **30** : 349-350.

125. Yadav, R.L. 1993. Cultivation practices of sugarcane in sub-tropical India. In : Sugarcane Research & Development in Subtropical Indian, Eds. G.B. Singh & O.K. Sinha, pp. 77-112, Indian Institute of Sugarcane Research, Lucknow.

126. Waller, J.M. 1970. Sugarcane smut *(Ustilago scitaminea)* in Kenya II. Infection and resistance. Trans. Brit. Mycol. Soc. **54** : 405-414.

127. Wilson, K. I. and Srivastava, D.N. 1969. Chromatographic detection of chlorogenic acid and a histochemical test for phenols in nodal tissues of sugarcane. Indian Phytopath. *22* : 306-308.

Fig. 1 : Red rot disease of sugurecane

20 Genetic Improvement for Resistance to Insect Pests and Diseases in Cotton

A.K. Basu

CONTENTS

Integrated Pest and Disease Management
Rajeev K. Upadhyay, K.G. Mukerji, B.P. Chamola and O.P. Dubey (eds.)
APH Publishing Corporation, New Delhi-110002, India

ABRSTRACT

Cotton in India is damaged by a number of insect pests and diseases causing sizable losses in yield. Pesticides are principal method of management of the diseases and insect pests, so much so that cotton occupying 5% of the area under cultivation consumes 53% of the pesticides used annually in India. Almost exclusive reliance on pesticides has increased the cost of cultivation, polluted environment, caused upsurge of minor pests and made some insects resistant to insecticides. As such, here is an urgent need to adopt integrated pest management wherein resistant/tolerant cultivars are an important component.

Work has been done to develop genetic resistance to key pests and diseases of cotton. Development of reliable and efficient screening techniques, identification of lines with genetic resistance and utilisation often of backcross method for incorporation of genetic resistance in adapted varieties have contributed to successful resistance breeding for some diseases and insect pests. Resistance breeding for some pests such as bollworms is difficult because of the absence of reliable source of resistance to the pest. Genetic engineering with incorporation of gene from bacterium *Bacillus thuringenesis* into cotton may impart resistance to bollworms.

1. INTRODUCTION

Cotton is an important commercial crop of India providing almost 80% of the raw material of the domestic textile industry of more than 1100 textile mills. It is grown in 7.5 million hectares in northern, central and southern regions of India, producing annually 12-13 million bales of cotton. This needs to be augmented to meet the increased needs of the future. Damage by pests and diseases is an important constraint for improvement of cotton production. We have relied heavily on pesticides for the management of pests and diseases so far. 53% of pesticides used annually in Indian agriculture costing about 300 crores of rupees is consumed to protect cotton crop occupying only 5% of the total area from vagaries of pests and diseases. Excessive use of pesticides has increased the cost of cultivation, made the insects resistant to pesticides, caused upsurge of minor pests apart from environmental pollution in cotton growing niches. It is therefore necessary to adopt Integrated Pest Management Technology for the control of pests and diseases of cotton. Resistant/tolerant genotypes (varieties/hybrids) is an important component of IPM, requiring least amount of pesticides. This has become crucial particularly farmers are switching over from cotton to other crops.

2. DISEASES OF COTTON

Cotton crop is affected by a number of diseases. More than 25 diseases caused by fungi, bacteria, viruses or nematodes have been reported in cotton in India (21), of which bacterial blight (*Xanthomonas campestris* pv. *malvacearum*), grey mildew (*Ramularia areola*), leaf spots (*Alternaria* sp. and *Myrothecium roridum*), root rots (*Macrophomina phaseolina* and *Rhizoctonia solani*), fusarium wilt (*Fusarium oxysporum*), Verticillum wilt (*Verticillium dahliae*) are economically important in different cotton growing regions of the country.

Increased area under tetraploid varieties and hybrids (Table 1), and irrigations, indiscriminate use of pesticides, and possible change in virulence of pathogens have aggravated disease problems including species composition of the pathogen in cotton. While the diploid cottons (*Gossypium arboreum* and *Gossypium herbaceum*) are affected by Fusarium wilt, root rot and grey mildew the bacterial blight and leaf spotting diseases like *Alternaria* and *Myrothecium* leaf spots take a heavy toll of the tetraploid cottons (*Gossypium hirsutum, G. barbadense)* under ideal situations of pathogenicity. Grey mildew which used to be a disease of diploid cotton only now affects the tetraploid cottons as well. The bacterial blight has now become an important disease of *G. hirsutum* in Rajasthan (55) contrary to its earlier presence in traces only.

Table 1 : Species composition over different periods

Species	Pre-War	Area in 100,000 hectares		
		1947-48	1955-56	1993-94
G. hirsutum	1.5	1.4	16.3	31.4
	(2)	(3)	(20)	(42)
G. barbadense	Nil	Nil	Nil	0.1
				(0.1)
G.arboreum	52.1	27.9	44.0	13.2
	(64)	(65)	(54)	(18)
G.herbaceum	27.1	13.9	20.5	9.5
	(34)	(32)	(25)	(12)
Hybrid	Nil	Nil	Nil	21.3
				(28)
Total	80.7	43.2	80.8	75.5

Figures in parentheses are percent of total area.

Losses caused by diseases of cotton vary depending on the cultivars used, extent of application of water and fertilisers, climate and disease management. Loss in yield of cotton due to various diseases is estimated to be around 10-15%.

2.1 Breeding for resistance to major diseases

Host plants are designated as disease resistant or susceptible in accordance with their reaction to the pathogen. Plants/varieties inherently less damaged by a pathogen than other plants/varieties under comparable environmental conditions are called as resistant (8). The host reactions in most cases show graded variation where they are classified in a scale (score) starting from 0 (with no disease) to 4 (most susceptible) (53). Host may be immune where it does not show the symptoms of the disease or tolerant where it is affected by the disease like a susceptible line but withstands the attack and shows no loss in yield. Breeding objectives in cotton include development of varieties resistant/tolerant to major diseases so that they can be cultivated with no or minimal usage of fungicides/antibiotics.

2.2 Sources of disease resistance

Germplasm consisting of cultivated species or their wild relatives constitutes an important source of resistance for genetic improvement of cotton for resistance to diseases. (Table 2).

Table 2 : Lines Resistant/Tolerant to Diseases

Disease	Resistant/Tolerant Lines	Reference
Bacterial blight	BJA 592, Reba B50, 101-102B, XAB 5 x Tanguis 126-5-6 AR4, B56-181, B61-2037, EL 156E, K4005, PK1572, Wilds Early, 1301 DD,	8, 9, 20, 21, 41, 42, 66

	AC 84, AR15, B56-247, B56-286 CT, B57-819, B57-960, CB2493-5, Coker 413, CP 15/2, EL 389, EL 592, EWLS, Tidewater, GRS 60-5, GRS 60-6, ISC 84, Lankburn, Mysore 14-66, Paymaster, PK1592, Reba TK, Saenz Pena Toba, Stoneville 7A-12, T85-2, UK53, UPA (57)-17, UPA (62)-31, VLI, 21-1-4-3, 134 xCO2, 134 CO2, M21, 150-3-1-1, 561, IC 1430, Project 38, Tamcot SP 21, Tamcot SP23, Tamcot SP37, Tamcot-SP 215, Tamcot CAMD-E, TxBonham, TxMaroon 2-78, Coker 304, Reba PVT 9, Demeter III (I), Demeter III (2), Super Okra leaf, 9-1487, CB 2628, Coker 413-68, NHH 300, NA1266, DS 28, 39F Coker 41768 JK 44-42, KW 61-240, 127F, TxORS 80, ORS 75-75, CNH 142, LP 371, GS 27, HC7069, BJR 97-16-4, P 95-2, race punctatum, CRH 71, RKR 4145, 1412, JR 80, Stoneville 20, B 31. (*G. anomalum*)	
Grey mildew	G135-49, 30805, 30814, 30826, 30838, 30856, AC24, AC631, AC 655, AC727 MH, AC727 DH, Desi 1, G 112, Gao CB3-NLL, H574, K53-519, AKA 54, AKA 81, AKA 102, AKA 137, AKA 148, AKA 843, AKA 8419, Malvi 20, 30802, 30815, 30821, 30845 (*G.arboreum*) ERB 4492, SB289 E (*G.barbadense*)	43, 53
Myrothecium leaf spot	Akansas green, GP 197 Kirghis K2, Line F(Hyb 330), M DH66, Mex 685-3, Pee Dee 113, Pee Dee 259, Pee Dee 875, SR1 F4, AC 241, AKH 8246, B163, B1007, FJA, GA231, G. Cot14, GP187, IMH4041, JK97, JK97 FBRN, JK97 FORN, Kirghis K 3, M-DH84, NHS 1412, NIS4, Surat Dwarf, WH25, 79-4303 P1 79-4385-P2, 1083, 433, 0133, 8216 (*G.hirsutum*) Early Pima restorer (*G. barbadense*), AK8410, Bangladesh - 1, NA 80, 231R, M60, PA 65, PA59 (*G. arboreum*), GH 458, GH 529, SM 61 (*G.herbaceum*)	53, 71
Alternaria leaf spot	ABT4, ABT 6, ABT 8, ABT 10 ABT 11, ABT 19, AHO 66-113, CPD 8-1-22, CRH 71, DS 28, HC 122/60, 5153-C, AC 79, Hansi, G-2-1-17-904, 49/60/231, Big boll, Triumph, PH 144-34-7, J-34, 418/49-4-6, RRD 275-2, RRD 276, RRD 280-1, RRD 381-2, RRD 392, RRD 418-5, RRD 419, DP 354, DP 452, DP 954, NA 585, NA 920, JK 97-CAC-77-1-1, JK 260, KG 1128, MESR 27, RAMPBS 22, RAMPBS 43, RAMPBS 108, RAMPBS 127, RAMPBS 159, RAMPBS 220, RAMPBS 223, RAPBS 261, RAMPBS 293, RAMPBS 296, TCH 665, TKHH 2A, 1-10 B, 101-102 B (*G. hirsutum*)	47, 53

Verticillium wilt	SeaIsland and other *G. barbadense* culture; 108F, 137F, 138F, Tashkent-1, Tashkent-4, Hopicala, Varmexicanum, f. nervosum (*G.hirsutum*), Paymaster 266-999 (*G.hirsutum*)	54
Fusarium wilt	2-2-4-2-2, H407, H487, AKH496 AKA 91, AKA 154, AKA 156, AKA 8 416, 8183-2, NA 40/80, PA 78/82, NA 305, 306, 259, 141, 85/86, 85/139, PBN 564, 244, 3286, 814, 834, 844, RDS 164, SM6, SM 150B, JLA 69, JLA 118, JLA 122, JLA 128, JLA 137, JLA 163, 7181-1-1, MDS 56, SM 102, HD 85, LD 312, LD 113 etc. (*G. arboreum* & *G. barbadense*)	53
Multiple Disease Resistance		
Bacterial blight Alternaria leaf and Grey mildew	IRH 1-10, NA 1369, CNH 148, VHB 150, RAMPBS 261	53
Bacterial blight, and Alternaria leaf spot	ABT 4, ABT6, ABT8, ABT10, ABT11, ABT19, AHO 66-113, KG 11-28, RAMPBS 22, RAMPBS 43, RAMPBS 159, RAMPBS 223, RAMPBS 261, RAMPBS 293, TKHH 2A, 1-10 B, 101-102 B	
Bacterial blight and Grey mildew	DHB 220, DHB 270, S1-44/27 RAMPBS 128	

2.3 Genetics of disease resistance

Resistance to balckarm disease in *G. hirsutum* cotton is due to the presence of two dominant cumulative factors B1 and B2 together with minor genes constituting the modifier complex (35). Additional genes for blackarm resistance in *G. hirsutum* identified are B2, B3, B4 and B7 which do not have selective value in the absence of the disease as they have no effect on the plant apart from giving resistance to blackarm (32, 34, 36). Studies conducted in India on inheritance of resistance to bacterial blight indicate dominance for resistance (53). Singh *et. al* (59) from their studies suggested on the other hand additive and additive x additive type gene effects for resistance.

Resistance to verticillium wilt is partially dominant except in one environment where overdominance is observed (54). Single dominant gene with modifiers govern the resistance of *G. barbadense* and mexicanum strain of *G. hirsutum* (65). Studies on genetics of Fusarium wilt in diploid cotton show resistance to be governed by dominant genes (31, 53). Temburnikar (73) reports both dominance and recessive inheritance depending on genotypes involved. Quantitative inheritance of resistance has also been suggested (54).

2.4 Disease resistant varieties

Varieties and hybrids have been evolved with resistance to Verticillium wilt in tetraploid cotton and Fusarial wilt in diploid cottons. Varieties and hybrids tolerant to bacterial blight have also been developed.

In view of the high susceptibility of the existing *G. hirsutum* varieties to Verticillium wilt, a high degree of resistance is difficult to achieve and 'tolerance' has been accepted as a more practical goal (65). Individual selection in a highly infested background in MCU 5 variety led to the selection of MCU 5 VT with a 10% tolerance limit (65). The selection was entered in All India Coordinated Cotton Improvement Project and on the basis of its performance in yield and quality vis-a-vis MCU 5 was released for commercial cultivation in Tamil Nadu in 1982 as MCU 5 VT (10). The variety is currently grown in large areas in Tamil Nadu and Andhra Pradesh.

Fusarium wilt which was a serious disease of both *G. herbaceum* and *G. arboreum* has now been overcome with the development and introduction of commercial cultivation of resistant varieties in both the species. Laboratory and field screening of diploid cottons is a prerequisite for the release of diploid cottons for commercial cultivation. Breeding for wilt resistance produced varieties like Jarila, H420, Suyog, Vijalpa, Jayawant, Gaorani 12 in earlier times. Prominent diploid varieties and hybrids with resistance to Fusarial wilt are listed in Table 3.

Table 3 : Fusarium wilt resistant varieties/hybrids

Sr. No.	Name of species	Variety	State	References
1.	*G. arboreum*	Jarila	Maharashtra	(5, 53)
		Jyoti	Maharashtra	
		Virnar	Maharashtra	
		PA 32 (Eknath)	Maharashtra	
		G Cot 15	Gujarat	
		AKA 8401	Vidarbha	
2.	*G. herbaceum*	Jayadhar	Karnataka	
		Sujay	Gujarat	
		G. Cot 11	Gujarat	
		Vijalapa	Maharashtra	
		DB 3-12	Karnataka	
		Dharwar-2	Karnataka	
		G cot 13	Gujarat	
3.	Interspecific Desi Hybrid	G Cot DHy7	Gujarat	
		G Cot DHy9	Gujarat	
		DDH 2	Karnataka, Tamilnadu & Andhra Pradesh	

Inspite of the availability of sources with immune reaction to all the races of *Xanthomonas malvacearum* no variety except MCU 10 and Sujata has been developed in tetraploid cotton which are completely free from the disease.

Culture CRH 71, is a gamma ray induced mutant developed by Tamil Nadu Agriculture University and released as MCU 10 *(G. hirsutum)* and Sujata (*G. barbadense*) is released from Central Institute for Cotton Research Regional Station (30). MCU 10, however was completely replaced by a more higher yielding variety LRA 5166 and Sujata by Suvin. However, there are varieties which are tolerant to bacterial blight disease namely, Bikaneri Nerma, Khandwa 2, SRT 1, H 777, F 414 etc. Work is in progress to breed *G. hirsutum* varieties with resistance to bacterial blight by incorporating genes from *G.anomalum* and immune cultures of *G. hirsutum, G. arboreum* and *G.herbaceum* are resistant to bacterial blight.

North eastern parts of India which are favourable for the disease and where *arboreum* cotton is under cultivation for more than thousand of years is likely to have been the host species on which bacterial blight first appeared (64). It is suggested that *G.arboreum* race cernnum and race burnanicum grown in the area evolved a high degree of resistance through natural selection over the centuries. It is suggested by Srinivasan (64) that races indicum and bengalase of *G.arboreum* and races grown elsewhere in India have acquired resistance through crossing with races grown in north east India.

Suvin, Sujata (*G.barbadense*), Laxmi, Sargam (*G.hirsutum*) and Varalaxmi (interspecific hybrid) are reported to be resistant to grey mildew (54).

2.5 Biochemical aspects of resistance

On analysis of the three pectinolytic (PMG, PTE and PME) and the cellulose enzyme activities of the races of *Xanthomonas malvacearum* it was seen that virulence of the races could be correlated to their PMG and PTE activities. Races 10 and 18 which are considered to be most virulent had highest PMG and PTE activities. Race 13 which is least virulent had the lowest enzyme activity. This has a bearing on their ability to overcome bacterial blight (16).

Analysis of the three amino acids in three varieties (Arkansas green, Kirghis K2, ED 875) resistant to *Myrothecium roridum* and healthy as well as diseased sample of three susceptible varieties (H777, L147, SRT 1) revealed that two amino acids (Leucine and Tryptophan) were invariably present in all the three resistant varieties but not in susceptible ones. This indicates that these two amino acids are responsible for imparting resistance to this pathogen (70).

3. INSECT PESTS OF COTTON

Insect pests constitute one of the major limiting factors in cotton production. Among 162 species of insects that are associated with this crop in India about 12 species are of major importance accounting for sizable loss in production. They are sucking pests, viz., jassids, (*Amrasca biguttula* Ishida), aphids (*Aphis gossipy* Glover), whitefly (*Bemisia tabaci* Gennadius) and bollworms spotted bollworms, (*Earias vitella* Fabricius and *E. insulana* Boisd); pink bollworm, (*Pectinophora gossypiella* Saunders) and American bollworm, (*Heliothis armigera* Hubner). Besides these, leaf-eating caterpillars, (*Spodoptera litura* Fabricius) and stem weevil, (*Pempherulus affinis* Fab.) also damage the crop in some areas in south. The early vegetative phase of the crop from the date of emergence of seedlings upto 50 days is associated with the sucking pests such as jassids, aphids and thrips. These insects, by sucking the sap of the growing plants, interfere with photosynthesis and reduce the vigour and further growth of the crop. The spotted bollworms appear in the crop from 45 days and bores into the stem when no fruiting parts are available. The early stage attack, if not checked, serves as a source for its further spread which damages squares, flowers and bolls throughout the cropping season. In southern parts of the country where cotton is sown in September, the pink bollworm though appears in the middle of November, shows peak activity after January. It occurs in norther zone from September onwards. The leaf feeding insect such as the leafworm occurs during the months of October and November. *Spodoptera litura* sometime bores into the reproductive parts of the plant. American bollworm has become a serious pest of cotton in all the three cotton growing regions of the country and appears in August - October (north), September - November (central) and October - January (south). Hybrids and extra long staple cotton of longer duration are more vulnerable to pests as is evident from the amount of insecticides used vis-a-vis varieties (Table 4). Indiscriminate use of pesticides particularly synthetic pyrethroids in recent times has caused upsurge in minor pests like whitefly and aphids. Genotypes with hairly leaves developed for resistance to jassids harbour whitefly and bollworms.

Table 4 : Requirement of insecticides in varieties and hybrid of cotton (67)

	Cultivars				
	MCU 5	MCU 9	Varalaxmi (Hybrid)	Suvin (long duration variety)	CD (P=0.01)
Insecticides ai/kg/ha	2.335	2.715	5.639	5.695	1.019
Natural enemies Mean No./sample	33.94	99.27	13.92	9.74	27.64
Seed cotton Q/ha	18.10	23.28	21.17	17.32	4.52
Stationed cotton Q/ha	2.15	2.29	1.75	1.64	0.86

4 BREEDING FOR RESISTANCE TO MAJOR INSECT PESTS

Evolution of genotypes with tolerance/resistance to sucking pests and bollworms forms an important component of Integrated Pest Management. Plant resistance is defined, as the collective heritable characteristics by which a plant species, race clone or an individual may reduce the probability of successful utilisation of that plant as a host by an insect species, race, biotype or individual (8). Breeding for resistance requires dependable and efficient screening technique, identification of lines with genetic resistance and often backcrossing for transfer of resistance to adapted lines. Resistance to insects may be due to (i) non preference (plants not attractive for colonization or oviposition (ii) antibiosis where feeding adversely affects development and or reproduction of the insects (iii) tolerance, i.e. ability of the variety to produce larger yield than a susceptible variety at the same level of infestation and (iv) avoidance i.e. escape from infestation (62). Resistance is associated with certain morphological, physiological and biochemical factors.

Considerable work has been done in breeding for resistance to sucking pests (jassid, aphid, thrips and whitefly) and bollworms (spotted, American and pink bollworms) in India.

4.1 Sucking Pests

4.1.1.. Jassids

Sources of resistance

Germplasm collections of both cultivated species, and its wild relatives forms an important source for breeding for resistance to jassids and other sucking pests (Table 5). Jassid is a pest of major importance to *Gossypium hirsutum* and its hybrids which occupy 70% of cotton area in India and therefore receives considerable attention. The diploid cotton (*G. arboreum* and *G. herbaceum*) are generally resistant to jassids and other sucking pests. The tetraploid wild species *G. tomentosum* is resistant to attack of jassids. Several lines of *G. hirsutum* have also been reported to be resistant/tolerant to jassids (27).

Table 5 : Lines resistant/tolerant to sucking insect pests

Insect Pest	Resistant/Tolerant Lines	References
Jassid	AS 122, A74, J 132, SH467, SH 169, PL 405, Aboharia, EC 601, BC 68-24, Surat, U-4, 193, BP 52, HC 122/60, H 297, EL 123E, 4716 SR, CCH 3-5195, 3628 Pakistan, ACH 101, CT 1421, KH 33/1146, AC 738, DHY 287,	5, 22, 27, 46

	H667, H854, CA 8745, CNH 108, F671, RS 454, SH 381, LH 1078, B57-619, 53-3-1, NH124, KW 74-3168, 69KH 62-1470, JLH 2, KH 95/2018, NH 262, JLH 323, IC 439, IC 601, ACH9, ACH 99, J 260, AV 2804-111/2 (*G.hirsutum*)
	G. tomentosum
	G. herbaceum
	G. arboreum
	G.anomalum
	G. armourianum
Whitefly	H 814, H 922, RS 510, RS 513, SH281, KOP498, RFS 8402, D53, NHY 1 (*G. hirsutum*)
Aphid	HCS 321729, ACP 71-12-3,
Thrips	AHO 61-38-2, NCBr 2-3/4, LH 299, C50, DS 35 (*G.hirsutum*).

Breeding and mass rearing of insect pests helps in screening lines for resistance to pests under field and artificial zoophytic conditions. Growing a susceptible crop like Lady's finger in alternate rows with experimental material is also used for screening for jassid resistance.

Resistance to jassid has been found to be highly correlated with hairiness of leaves. Degree of resistance depends on the degree of hairiness, length and density of hairs per unit area of leaf (45), thickness of leaf lamina and toughness of the vein (2, 3, 45, 57). Breeding hairy cottons that are resistant to jassid attack has, however, brought other problems hairy plants are preferred host for egg laying of bollworms' moths (76). This brings in the need for study of biochemical aspects of resistance to jassids. Results of study of chemical composition of leaves of *G. hirsutum* cottons vis-a-vis jassid resistance reveal that tannin, phenol and silica contents of the leaf play an important role in defending the plant against jassid attack (17, 72).

Genetics of jassid resistance

Genetics of jassid resistance has been studies by several workers in India and abroad (1, 33, 37, 57). Some reported monogenic inheritance (1), others found jassid resistance due to two factor pairs (57). Resistance in two perennial *barbadense* types has been reported to be controlled by a partially dominant hairiness gene known as H1 accompanied by a number of minor genes (33, 37). Another hairiness gene H2 is reported to be responsible for dense tomentosums of wild tetraploid *G. tomentosum* (51).

Jassid resistant varieties

Breeding for jassid resistance utilising germplasm resources for resistance have resulted in the development of resistant varieties of *G. hirsutum* (Table 6). The CTI genetic material developed at the Institute of Plant Industry, Indore from three way cross involving *G. tomentosum* with H_2 gene, CO_2 and Indore 2 has shown useful segregants with resistance to jassids, eg. Badnawar-1 (1961), Khandwa-2 (1971) and Vikram (1981) in M.P. SRT-1 (1974) in Gujarat and Maharashtra and B 1007 (1967) and DHY 286 (1973) in Maharashtra. The varieties released for commercial cultivation in the norther zone states of Punjab, Haryana, Rajasthan and Uttar Pradesh in seventies and eighties have tolerance to jassids, the exception being Ganganagar Ageti in Rajasthan. Several intra-hirsutum popular hybrids such as AHH468, NHH-44, JKHY-1, G. Cot. Hyb. 8 etc. are resistant/tolerant to jassids because one or both parents of the hybrids have jassid resistance.

Table 6 : Varieties and hybrids resistant/tolerant to jassids

G.hirsutum	Varieties/Hybrids	State grown
Jassids	BN, H777, F414, F505, Pramukh,	Punjab, Haryana, Rajasthan, UP
	B 1007, DHY 286, L147, AK 32, SRT 1, AHH468, NHH 44	Maharashtra
	SRT1, G.Cot 12	Gujarat
	Badnawar 1, Khandwa 2	Madhya Pradesh
	Krishna, NHH 44	Andhra Pradesh

4.1.2. Whitefly

Whitefly which was hitherto a minor pest has become a serious pest in several cotton growing niches in India in recent years. There had been severe crop losses in A.P. and Gujarat due to whitefly attack in mid and late eighties. Management of whitefly has therefore received much attention of researchers. Breeding for resistance has been considered as an important aspect of whitefly management. Morphological basis of resistance to whitefly has been studied by many workers (7, 15, 23). One of the reasons for whitefly flare up is the cultivation of pubescent genotypes such as MCU5, MCU 9, LRA 5166, SRT-1 etc. which are conducive for higher rate of oviposition and population build up. Hair density and leaf thickness are positively correlated with the population of *Bemisia tabaci.* The thinner and glaborous leaved genotypes confer resistance to whitefly (15). Varieties with light green narrow lobed (okra) thin glabrous leaves as well as early maturity withstand better the attack of whitefly (46). Based on the above observations genotypes including Kanchana, 83 KTT 217-218, L57 X 1128-43, AET-5, Fregobract 407-14-44, ICMF 4, NA1280, LK 861, Supriya, and USA 22 (5, 15) have been isolated. Kanchana (LPS 141) and LK 861 were released for commercial cultivation in A.P. to counter the menace of whitefly in mid eighties. They served the farmers well in the days of crisis. However, whitefly subsided but jassids reappeared due to glabrous leaves causing the two cultivars unpopular with the farmers. Therefore the need is to work for biochemical basis of resistance to whitefly to evolve varieties tolerant to both whitefly and jassids. Tannin and gossypol may play a role in resistance to both the sucking pests simultaneously.

4.2. Bollworms

While spotted (*Earias* spp.) and pink bollworms (*Pectinophora gossypiella*) are the principal cotton bollworms in the northern cotton zone, spotted bollworms dominate in the central and southern cotton zones. American bollworm (*Heliothis armigera*) has assumed status of a serious pest in all the three zones in recent times probably due to changes in the cropping system and species composition. Moreover, *Heliothis* spp. have also developed resistance to most of the chemicals in certain areas.

Studies on population dynamics reveal that spotted bollworms appear during August-October (North), September-February (Central) and October-November (South), pink bollworm appears in September-December (North) and October-onwards (Sough) and *Heliothis* spp. appears in September-November (Central) and September-December (South).

4.2.1. Breeding for resistance to bollworms

Development of inbuilt tolerance/resistance to the bollworms as a component of integrated management is a goal for which work is being carried out at different Indian cotton research stations.

4.2.2. Resistance factors

Several factors have been reported to be contributing to resistance to bollworms in cotton. They are crop maturity, morphological characters and biochemical factors.

Preferred sites of oviposition in the plant for spotted bollworm (*Earias vittella*), American bollworm (*Heliothis armigera*) and pink bollworm (*Pectinophora gossypiella*) as revealed from studies are shoot terminal and upper surface of leaves of upper half of the plant and base of bracts respectively (8). Knowledge of the preferred sites of oviposition may help in breeding for modification of the sites which will be deterrent for oviposition. Thus glabrous leaved genotypes and frego bract genotypes may detract *Heliothis* and *Pectinophora* respectively since the most preferred sites of these bollworms are leaves and base of bracteoles respectively. However, oviposition on less preferred sites of the plant in that event cannot be ruled out.

Crop maturity

Early crop maturity has been seen to have a definite advantage in reducing the incidence of bollworms. It escapes the attack of bollworms particularly pink bollworm which appears late in the season. Development and introduction of early maturing varieties in the northern cotton zone namely, F414, F286, LH900, H777, SH 131, Ganganagar Ageti in place of the older long duration varieties, 320F, LSS and Pramukh have reduced the pink bollworm incidence considerably. Similarly, the early maturing hybrids namely, H6, AHH 468 and GHH 334 in place of Hybrid -4 reduced the incidence of bollworms in the central cotton zone. The early maturing variety, AKH 081 released in Vidarbha recently also attract less incidence of bollworms. Table 7 shows bollworm incidence of early maturing varieties vis-a-vis their late maturing predecessors.

Table 7 : Mean bollworm infestation (per cent) in the early varieties vis-a-vis their longer(8) duration predecessors (checks)

Variety/Hybrid	Percent bollworm infestation Locule basis
F414	32.7
J 205 (Check)	37.6
LH 900	40.8
F 286 (Check)	52.2
SH 131	16.3
Pramukh (Check)	23.6
Vikram	29.8
Khandwa-2 (Check)	42.8
NH 239	32.5
SRT 1 (Check)	40.2
NHH 44	16.0
H 4 (Check)	20.1

Morphological characters

Bhat and Basu (13) studied the role of morphological traits in bollworms incidence in cotton in F2 population of cross between SH 131, a variety of UP and an exotic strain red stem with contrasting characters including red stem, frego bract and nectariless leaves to impart tolerance to bollworms in cotton. Red pigmented plants having okra lead character, and frego bract have been found to be tolerant to bollworms (4, 12, 61, 74, 80). Thickness of boll rind also contributes to resistance to bollworms. The lower incidence of bollworms in frego bract is attributed to increased insecticidal efficacy as a result of better penetration of insecticides and also increased activity of larval predators. Less hairy leaf genotype like FBRN is less preferred for egg laying by *E. vittella* (Table 8) (68). Frego, nectariless and red stem (FBRN) are also less preferred for oviposition (Table 9). Glabrous leaves have been found to be less susceptible to bollworms (*Earias* and *Heliothis*) since they do not prefer it for oviposition (29).

Table 8 : Number of eggs as influenced by hair density in two genotypes of cotton (68)

Lines	Hair density in leaf *		No. of eggs
	Lower surface	Upper surface	deposited
FBRN 2-6	4.6	0.4	2.7
L 92	34.8	19.7	51.7

* Hair density in microscopic field in 2nd leaf from top.

Table 9 : Ovipositional preference of spotted bollworm (E. vittella) (68)

Lines	Mean No. of eggs laid/plant	
	Actual value	+ 0.5 values
FBRN 2-6	1.00	1.18
IC 563	4.00	1.98
Frego bract I	2.75	1.76
HG IPS 625	4.50	2.22
DF 64-6	2.00	1.54
MCU 5 (Check)	6.50	2.64
CD at 5% level		0.74

Studies on the role of individual characters on bollworms incidence on isogenic lines of JK 97 background indicate that it is the combination of characters rather than one character which is useful in imparting tolerance to bollworms (28). JK cultures viz., JK 276-4, JK 260, JK 344, JK 345 having green less hairy leaves, less hair on stem tips, small bract, thick boll rind with early maturity have proved to be less susceptible to bollworms and show higher yield compared to controls (Table 10). Wilson (78) in his studies in USA showed that combination of morphological traits in a resistance genetic background (AET 5) would increase the level of resistance to pink bollworm (Table 11). Singh (58) reported better tolerance to bollworms with combination of characters in cultures derived from backcross involving ROIL 3 as recurrent parent and Acala glandless, nectariless and frego bract as donor parents (Table 12). In a study conducted at CICR Regional Station, Coimbatore DF 64-6, a frego bract culture recorded less bollworm damage and higher yield compared to MCU 5 and LRA 5166 under schedule protection and need based protection (69).

Genetics of many of these morphological characters have been studied. While frego bract and okra leaf are controlled by single gene (monogenic recessive), nectariless is controlled by duplicate recessive genes (6, 11, 12, 50). Red plant body has been found to be monogenic dominant (25, 38, 49, 77). Basu and Bhat (12), however recorded interplay of two dominant factors giving a ratio of 9:6:1: for red plant body. Investigation on joint segregation revealed no association but independent assortment of most of the qualitative characters (12). Genetics of morphological traits has helped in incorporation of these traits through breeding programme.

Table 10 : Shed squares and bolls (percent) due to bollworm and seed cotton yield (Mean of 3 years) (29)

Strains	Shedding due to bollworms (percent)	Bollworm attacked bolls/plant	Unattacked good open bolls/plant	Seed cotton yield (Q/ha.)
JK 276-4	53.1	2.75	4.91	12.1
JK 260	52.2	4.09	5.36	12.5
JK 344	64.5	3.91	3.78	10.7
JK 345	66.1	3.53	3.48	11.3
Sharada (Check)	75.5	4.34	2.43	6.9
Laxmi (Check)	78.4	2.66	1.76	5.4

Table 11 : Seed damage in AET-5 cotton; in its nectariless (N), smooth leaf (S) and okra leaf (L) isolines and in nectaried (DPL 61) and nectariless (DPL-N) semismooth leaf check cultivars (78)

Lines	Resistance Traits combined	Actual	Seed damage AET 5 observed	Percent of DPL 61 observed	Expected
AET-5	1	32.9b*	100	79	-
AET-5N	2	25.6c	78	62	62
AET-5S	2	34.3b	104	83	82
AET-5L	2	24.2c	74	58	58
AET-5NS	3	28.4bc	86	69	64
AET-5NL	3	22.9c	70	55	46
AET-5SL	3	28.6bc	87	69	61
AET-5NSL	4	24.5c	74	59	47
DPL-61	0	41.4a	126	100	-
DPL-N	1	33.4b	102	81	-

* Means with letter(s) in common are not significantly different at the 0.05 level of probability, according to the FLSD test (2).

Table 12 : Influence of combination of morphological characters on bollworm incidence (56)

Lines	Morphological Character(s)	Bolls infested (percent)	Locules infested (percent)
1035	Okra leaf, frego bract	52.2	30.2
957	Normal leaf, frego bract	65.1	38.2
ROIL 3	Okra leaf	80.4	58.2

1014	Okra leaf, nectariless	58.6	36.3
1001	Normal leaf nectariless	69.6	44.4
Check	Normal leaf normal bract	73.4	50.8
CD at 5% level		8.5	7.8

Biochemical factors

Phenolic compounds namely tannin, gossypol, heliocide etc. present in the cotton plant detract bollworms (14, 19, 39, 52, 60, 79). Chakravarty *et al.* (19) in their studies on *G. hirsutum* and *G. arboreum* varieties showed that varieties with higher concentration of tannin in seed coat, bud or other fruiting parts and lesser quantity of protein in anthers/bud alone or along with seed were more tolerant to pink bollworms. G 27 (*G. arboreum*) with high tannin and low protein was found to be least affected by bollworms (Table 10). Singh *et al.* (61) also reported lower incidence of bollworms (pink and spotted) in G 27 as compared to Virnar (28% as against 72%) and this was attributed to higher glandular area in the former besides containing more phenolic compounds.

Kadapa *et al.* (29) reported higher tannin content (50-100% more) in boll rinds of bollworm tolerance *G. hirsutum* cultures compared to susceptible cultures. Volume of glandular area (containing phenolic compound) in *G. arboreum* cottons was highly and significantly correlated with the incidence of bollworm i.e. higher the gandular area, lower is the bollworm incidence. They indicated scope for selection of genotypes for resistance/tolerance to bollworms by selecting types with higher glandular area. Heliocide 1 and Heliocide 2 are the other two phenolic compounds present in pigment glands which also impart resistance to bollworms. *Heliothis* tolerant/resistant genotypes have been found to have higher concentration of heliocide in both flower buds and leaves (63). Sharma *et al.* (52) studied the variability of gossypol and tannin in different species of cotton and found sanguineum (parent of G 27) to have the maximum percentage of gossypol. Relationship between phenolic compounds and glandular area in the plant was found to be useful for screening genotypes for bollworm tolerance (60).

Singh *et al.* (61) studied the genetics of bollworm resistance using a susceptible and resistant genotype of *arboreum* cotton and found the gene action to be mainly additive, indicating ease with which resistant character can be fixed.

4.2.3. Host plant resistance

Host plant resistance can be evaluated under natural field conditions with optimum insect population. Such screening of germplasm, breeding material, entries in the various coordinated trials conducted at multi locations have shown several cultivars to be tolerant to bollworms. For field evaluation the following scale for resistance/tolerance/susceptibility to bollworms is adopted in All India Coordinated Cotton Improvement Project.

Percent bollworm infestation	Rating
0 to 30	Resistant/tolerant
31 to 40	Moderately susceptible
41 to 50	Susceptible
Above 50	Highly susceptible

Tolerant lines identified by screening done during three seasons are listed in Table 13.

Table 13 : Lines tolerant to bollworms (8, 27)

North Zone
F 572, LH 900, LH 947, HS 2, LH 892, B.N. PUSA 31, CA 10-2, H938, LH 1004, LH 1078, CH 1006, F 907, LH 1141, HS 131, H 83-1, RS 650, HS 84-1, KOP 498, HS 139, LH 972, LH 892, RST 9, F 806, LH 922, H 999, 081, F 670, SH 281, SH 583, H 83-3, H 83-5 (*G. hirsutum*).
SD-6-9, SD 9-1, HD 106, LD 262, H084, LD 272, LD 286, LD 293, HD 86, LD 306, HD 37, HD 40, HD 88, NA 114, RG 10, DS 5, Pusa Desi, LD 26, HD 86 (*G. arboreum*)
Central Zone
KOP 495, KOP 498, EC 142764, EC 137592, AKH 081, LRA 5166, SRT 1, Purnima, PH 36, G 2637, G 4222, NH 347, NH 358, PH 23, PH 54, NH 290 (*G. hirsutum*)
JLA 2, Y1, Eknath, NA 80, NA 139, NA 189, NA 39, NA 89, NA 128, NA 130, NA 202, NA 212, NA 216, NA 114, NA 106, NA 40, NA 48, JLA 97, G 27, Eknath, Rohini, NA 116, NA 269 (*G. arboreum*)
Southern Zone
JK 276-4, JK 260, JK 344, JK 305, RAS 301, MESR-17, FBRN 2-6, ASD 7-70, L 21, RAMPHS 22, PKH 496, JK 119-25-24, L 1245, JK 78, B.N, DF 64-6, KG 5-29, KG 11-26, JK 78, MESR-6 (*G.hirsutum*)
BCS 10-5, BCS 9-67, BCS 10-1 (*G. barbadense*)

There is need to confirm the results obtained at each centre by testing them under multilocations in bigger plots under favourable conditions for optimum insect load. For example crop such as Ladys finger can be grown as trap crop to maximize infestation by *E. vittella*. Further confirmation of tolerance/resistance to bollworms needs to be made in the laboratory (Net house/cage studies) involving studies on non preference for ovipositing, non preference for infestation of squares, flowers and bolls and feeding. Morphological, phenological and biochemical factors responsible for resistance are to be identified. Resistance for individual bollworm rather than the bollworm complex is our aim. Breeding programme to transfer the resistant gene(s) to high yielding genotypes follows.

(iv) Source of resistance from wild relatives of cotton.

The role of wild species of cotton in imparting resistance to bollworms has been studied by various workers (40, 45). Some amount of resistance to bollworm was reported in cross involving *G. thurberi* (24). They reported retarded growth of 1st instar larva of *E. vittella* in a sterile cross in *G. arboreum, G. barbadense* and *G. armouranium* thus indicating presence of a resistance factor. *G. armouranium, G. anomalum, G. raimondii, G. somalense* and primitive race Rozi of *G. arboreum* are considered to have genes for resistance to bollworms (44). However, reports of their specific and planned evaluation for bollworm resistance are rare. Planned studies as to their utility as donors for resistance to bollworms is therefore needed.

Diploid cottons are more tolerant to bollworms and therefore the areas of their cultivation are to be preserved. Their quality can be improved to compete with the tetraploids.

5. ROLE OF GENETIC ENGINEERING IN DISEASE AND INSECT PEST RESISTANCE

There has been increased awareness on the role of molecular genetics in understanding the molecular mechanisms of pathogenicity. Proper understanding of the molecular basis of virulence of different races of *Xanthomonas malvacearum* can pave the way for the control of bacterial blight. Distinct reproducible and conserved nature of plasmids of the bacteria are indicative of their epidemiological importance (18). The plasmids can possibly be used as a diagnostic tool in the identification of this pathogen and its differentiation from other pathogens of *X. campestris*. Transfer of genes governing antagonism in the phylloplane/phillosphere microflora may help in controlling foliar diseases of cotton (56). Somaclonal variation through tissue culture can be exploited for disease resistance particularly seedling diseases.

Genetic engineering offers an unique opportunity for resistance to insect pests. This hopefully will be an important tool for breeding for resistance to lepidopteran pests (bollworms) for which there are no reliable sources of resistance in the cultivated species of cotton or its wild relatives. Genes from *Bacillus thuringenesis* (BT) which produces crystal containing one or more insecticidal proteins have been incorporated into cotton through genetic engineering. The crystal proteins when ingested by susceptible insects interact and lyse their midgut epithetial cells eventually causing death (75). Integration and expression of alien (BT) genes in cotton is carried out by TI plasmid of *Agrobacterium tumefaciens*. Such transgenic cottons have shown high levels of field resistance to tobacco budworm, cotton bollworm, pink bollworm and leaf perforator (26). In India several Institutions including Central Institute for Cotton Research, Nagpur have embarked upon programme of developing transgenic cottons with resistance to bollworms.

6. CONCLUSION

The future of cotton in India and the world is bleak unless we manage the diseases and pests of the crop with due consideration of economics and safety of environment. Integrated Pest Management system which is an economically sound and environmentally safe system and of which we are talking about for many years now should reach the farmers. Host plant resistance is one of the most important components of IPM. Genotypes which mature early escape pests and are therefore of immense importance in host plant resistance. Morphological characters such as pubescent and glabrous leaves have helped in resistance to jassids and whitefly respectively. Other morphological characters frego bract, nectariless, red plant body, thicker rind of bolls singly or collectively have been reported to confer some degree of resistance to bollworms. Biochemical factors such as tannin, gossypol heliocide etc. have also role in resistance to bollworms. Breeding approach to combine earliness as well as morphological and biochemical parameters for pest resistance will be useful Genetic engineering as a novel tool in resistance breeding appears to have great potential and should be exploited. For cotton to stay as a crop in competition with food and other cash crops, importance of IPM with resistant/tolerant genotypes can hardly be overemphasized.

7. REFERENCES

1. Afzal, M. 1941. Present position as regards breeding for jassid resistance in cotton. 2nd Conf. Cott. Gr. Prob. ICCC Bombay, pp. 54-58.

2. Afzal, M and Ghani, M.A. 1953. Cotton jassid in the Punjab. Sci. Monogr, No.2. Pakistan Asscn. Ad. Sci. Lahore.

3. Ahmad, N., Afjal M. and Ghani, M.A. 1950. Studies on cotton jassid *Empoasca devastans* Dist. in the Punjab. XII Apparatus for measuring toughness of leaf veins of cotton. Pakistan J. Sci. **2** : 117-120.

4. Anonymous, 1989a. Annual Report Central Institute for Cotton Research, Nagpur, p. 52.

5. Anonymous, 1989b. A catalogue of cotton genetic resources in India. Central Institute for Cotton Research, Nagpur, India.

6. Balasubramanyam, R. 1951. A mutant in Asiatic cotton. Curr. Sci. **20** : 73.

7. Basu, A.K. 1986. Resurgence of whitefly in cotton and strategies for its management. In: Proc. Natl. Symp. 'Resurgence of Sucking Pests', Tamil Nadu Agri. Univ., Coimbatore, pp. 129-133.

8. Basu, A.K. 1988a. Prospects of breeding for resistance to bollworms in cotton. In : Proc. Group Discussion on bollworm in Cotton, Central Instt. for Cott. Res., Nagpur 8-9 March, 1988, pp. 208-231.

9. Basu, A.K. 1988b. Glimpses of history of varietal improvement of cotton in India. J. Indian Soc. Cott. Imp. **13** : 79-92.

10. Basu, A.K. 1989. Technology for increasing cotton production in India. All India Coordinated Cotton Improvement Project, Central Instt. for Cott. Res. Reg. stn., Coimbatore, pp. 9-63.

11. Basu, A.K., Duhoon, S.S. and Singh. 1974. Gamma ray induced okra leaf mutant in cotton (*Gossypium hirsutum* L). In : Proc. Symp. Use of radiations and radioisotopes in studies of plant productivity, G.B. Pant Univ. Agri. and Tech., Pantnagar, pp. 129-133.

12. Basu, A.K. and Bhat, M.G. 1984. Inheritance of certain qualitative characters in cotton. (*Gossypium hirsutum* L.). Indian J. Genet. **44** : 153-158.

13. Bhat, M.G. and Basu, A.K. 1984. Effects of certain morphological characters on bollworm incidence in cotton. J. Indian. Soc. Cotton. Imp. **9** : 64-67.

14. Bottger, G.T., Sheehan, E.T. and Lukefhr, M.J. 1964. Relation to gossypol content of cotton plant to insect resistance. J. Econ. Ent. **57** : 283-285.

15. Butter, N.S. and Vir, B.K. 1989. Morphological basis of resistance in cotton to the whitefly *Bemisia tabaci*. Phytoparasitica, **17** : 251-261.

16. Chakravarty, P.K. 1989. Annual Report. Central Instt. for Cott. Res. Nagpur 1988-89, pp. 35-36

17. Chakravarty, S.C. and Sahni, V.M. 1972. Biochemical basis of resistance to jassid (*Empoasca* spp.) in *G. hirsutum* cotton. Indian Agri. **16** : 45-48.

18. Chakravarty, P.K. and Sheoraj 1994. Plasmids of *Xanthomonas campestris* pv. *malvacearum* and their role in most virulent race 18. In : Abst. of papers 1st World Cott. Res. Conf. Brisbane, Australia 13-17 Feb. 1994, p. 96.

19. Chakravarty, S.C., Basu, A.K. and Sahni, V.M. 1981. Variability in tannin and protein contents in fruiting parts in relation to the incidence of pink bollworm in cotton. Indian J. Agri. Sci. **52** : 767-773.

20. Chalam, T.V., Sudarshan, M.R. and Ranga Reddy, M. 1991. A note on bacterial blight resistance in cotton. J. Indian. Soc. Cott. Imp. **16** : 65.

21. Chopra B.L., Randhawa, L.S. and Singh, T.H. 1994. Evaluation of *Gossypium hirisutum* strains for developing bacterial blight resistant hybrids in cotton. In: Abst. of paper presented at Natl. Seminar on Cot. Prodn. Challenges of 21st century, Haryana Agri. Univ. Hissar, April 18020, 1994, p. 90.

22. Das, B.B., Bhutai, D.K. and Basu, A.K. 1975. Screening of cotton varieties for resistance to jassids. Sci. & Cult. **41** : 130-133.

23. David, B.A., Jesudasan, R.W.A. and Winstone, A.A. 1986. On the outbreak of *Bemisia tabaci* (Gennadius) on cotton and Brinjal in South India. In : Proc. Natl. Symp. on 'Resurgence of sucking pests', Tamilnadu Agri. Univ. Coimbatore, pp. 116-124.

24. Ganesan, D. 1947. Interspecific hybridisation in cotton. 3rd Conf. Cott. Gr. Prob. India, ICCC Bombay.

25. Harland, S.C. 1939. The genetics of cotton. Jonathan Cape, London, 193 p.

26. Jenkins, J.N. 1994. Host plant resistance in cotton to insects. Abst. of papers 1st World Cot. Res. Conf. Brisbane, Australia, 13-17 Feb. 94, p. 11.

27. Jayaswal, A.P. and Sundaramurthy, V.T. 1992. Achievements in insect management in cotton. In : Achievements, AICCIP Silver Jubilee Souvenir, Central Instt. for Cott. Res. Nagpur, pp. 117-151.

28. Kadapa, S.N. 1980. JK 97- forms a multiple genetic method in *G. hirsutum* cotton. Curr. Res. **10** : 16-17.

29. Kadapa, S.N., Thimmaiah, G. and Khadi, B.M. 1987. Breeding for bollworm tolerance in *G. hirsutum* cotton IV. Experimental findings, plant characters and yield. Paper presented at Biennial Workshop of AICCIP, Andhrapradesh Agri. Univ., Hyderabad, Jan 21-24, pp. 1-10.

30. Kandaswamy, T.K. 1980. Diseases of cotton and their control. In : Monograph on cotton, Tamilnadu Agri. Univ. Coimbatore, pp. 103-122.

31. Kelkar, S.G., Chowdhuri, S.P. and Hiremath, N.B. 1947. Inheritance of *Fusarium* resistance in Indian cottons. 3rd Conf. Cott. Gr. Prob. India, ICCC, Bombay, pp. 125-162.

32. Knight, R.L. 1944. The genetics of blackarm resistance IV. *Gossypium punctatum* cross. J. Genet. **46** : 1-27.

33. Knight, R.L. 1952. The genetics of jassid resistance in cotton. I. The genes H1 and H2. J. Genet. **51**: 47-66.

34. Knight, R.L. 1953. The genetics of blackarm resistance IX. The gene B6m from *Gossypium arboreum*. J. Genet. **38** : 139-159.

36. Knight, R.L. and Hutchinson, J.B. 1950. The evolution of blackarm resistance in cotton. J. Genet. **50**: 36-58.

37. Knight, R.L. and Sadd, J. 1954. The genetics of jassid resistance in cotton III. The kapas purao, kawanda and Turgus on group. J. Genet. **52**: 186-198.

38. Kokuev, V.I. 1935. Inheritance of certain agronomic and morphological characters in cotton. Pl. Breed. Abst. **6**: 176-177.

39. Lukefahr, M.J., Owman, C.B., Pfrimmer, T.R. and Noble, L.W. 1966. Resistance of experimental cotton strain '1514' to the bollworm and leaf hopper. G. Econ. Ent. **59**: 395.

40. Mehta, N.P., Patel, C.T., Chaudhuri, P.P., Vashi. D.R., Patel, B.K. and Maisuria, I.P. 1988. Role of introgressive hybridisation in bollworm resistance breeding. Proc. Gr. Discussion on bollworm resistance in cotton. Central Instt. for Cott. Res. Nagpur, March 8-9, pp. 169-196.

41. Meshram, M.K. and Sheoraj. 1986. Reaction of the germplasm of *Gossypium hirsutum* Linn to bacterial blight. Indian. J. Agri. Sci. **56** : 236-237.

42. Meshram, M.K. and Sheoraj. 1991. Resistance of upland cotton (*Gossypium hirsutum*) lines to bacterial blight caused by *Xanthomonas camplestris* pv. *malvacearum*. Indian. J. Agri. Sci. **61** : 954-955.

43. Mukewar, P.M., Sheoraj and Singh, V.V. 1993. Screening for resistance against grey mildew in *Gossypium arboreum*. Proc. Indian. Phytopath. Soc. W. zone, Punjabrao Krishi Vidyapeeth, Akola.

44. Narayanan, S.S., Singh, and Verma. P.K. 1984. Introgressive gene transfer in *Gossypium* goals, problems, strategies and achievements. Cott. Fibre. Trop. **39** : 123-125.

45. Narayanan, S.S., Singh. J and Verma. P.K. 1985. A brief review of the genetic welath in *Gossypium* with special reference to wild species. J. Indian. Soc. Cott. Imp. **10**: 51-64.

46. Patel, C.T. 1987. Some considerations for breeding varieties tolerant to whitefly. Cott. Dev. **17** : 53-54.

47. Padagmur, G.M. and Basavaraj, M.K. 1984. Reaction of cotton cultures to *Alternaria* blight caused by *Alternaria macrospora* zimm. J. Indian. Soc. Cot. Imp. **9** : 36-37.

48. Parnett, F.R., Macdonald, D., Rustom, D.F. and King, H.E. 1945. Breeding and genetics, Leaft hairiness and jassid resistance. Emo. Cott. Gr. Rev. Corp. Rep. Exptl. Stn. (1943-44), 23-24 & 26-30.

49. Ramchandran, C.K., Nagarajan, C. and Peter, S.D. 1961. Inheritance of a marker gene in Cambodia Cotton. Indian. Cott. Gr. Rev. **15** : 180-181.

50. Shoemaker, D.N. 1908. Rep. An. Breed. Assn. **5**: 116-119.

51. Sethi, B.L., Sikka, S.M., Dastur, R.H., Gadkarni, P.D.. Balasubramaniam, R., Maheshwari, P., Rangaswamy, N.S. and Joshi, A.B. 1960. Cotton in India. A monograph Vol. I. Indian Cent. Cott. Comm., Bombay.

52. Sharma, H.C., Agarwal, R.A. and Singh, M. 1982. Effect of some antibiotic compounds in *Gossypium* on post embryonic development of spotted bollworm (*Earias vitella*). Proc. Indian. Acad. Sci. (Animal science) **91** : 67-77.

53. Sheoraj. 1992. Achievements in Plant pathology. In : Achievements of AICCIP Silver Jubilee 1967-92. Eds. A.K. Basu and S.S. Narayanna, Central Instt. for Cott. Res. Nagpur, pp. 152-176.

54. Sheoraj and Verma, J.P. 1988. Diseases of cotton in India and their management. Rev. Trop. Pl. Path. **5**: 207-254.

55. Sheoraj, Meshram, M.K., Ajmra, B.D., Narula, A.M. and Chauhan, M.S. 1991. Outbreaks of bacterial blight of cotton in Rajasthan and Haryana causes and future strategies for its management. J. Cott. Res & Dev. **5**: 158-162.

56. Sheoraj, Chakravarty, P.K., and Monga, D., 1994. Future options for the management of cotton diseases. 1st World Cott. Res. Conf. Brisbane, Australia, 13-17 Fbe, 1994, p. 100.

57. Sikka, S.M. and Singh, A. 1953. Inheritance of jassid resistance in some upland American cotton. ICGR. **7** : 113-117.

58. Singh, J. 1982. Ann. Report. Central Instt. for Cott. Res. Nagpur, pp. 20-22.

59. Singh, I.P., Mor, B.R., Nagwekar, S.N. and Paroda, R.S. 1987. Genetics of resistance to bacterial blight in upland cotton. J. Indian. Soc. Cot. Imp. **12** : 112-116.

60. Singh, M. 1987. Bollworm resistance in *G. arboreum* cotton. Paper presented at 11th Biennial workshop AICCIP. Andhrapradesh Agri. Univ. Hyderabad, Jan. 21-24.

61. Singh, M., Joshi, A.B. and Agrawal, R.A. 1976. Genetics of cotton bollworm (*Pectinophora gossypiella* Saunder) Resistance in intervarietal crosses of *G. arboreum* cotton. Et. Fibres Tropicals **31** : 369-372.

62. Singh, B.D. 1983. Plant Breeding, Kalyani publishers, Ludhiana. pp. 397-413.

63. Stipanovic, R.D., Bell, A.A. and Lukefahr, M.J. of Singh. M. 1987.

64. Srinivasan, K.V. 1991. The geolocation of the origin of cotton bacterial blight and development of resistance in Indian Cotton. J. Indian. Soc. Cott. Imp. **16** : 73-80.

65. Srinivasan, K.V., Krishnamurthy, R. and Kannan, A. 1976. Progress report of *Verticilium* wilt control. J. Ind. Soc. Cott. Imp. **1** : 77-83.

66. Sivasubramaniam, V., Srinivasan, S., Ramanatha Pillai, O., Nagarajan, C. and Panneerselvam, S. 1977. Studies on the varietal resistance of rainfed *G. hirsutum* L. to blackarm ((*Xanthomonas malvacearum*). Cott. Dev. **7** : 30-31.

67. Sundaramurthy, V.T. and Basu, A.K. 1985. Management of cotton insect pests in polycrop systems in India. Outlook on Agri. **14** : 79-82.

68. Surulivelu, T. and Sundaramurthy, V.T. 1987. Natl. Symposium, Insect Pest Control. Trivandrum Oct. 15-17, 1987 (Of. Basu, 1988).

69. Surulivelu, T., Sundaramurthy, V.T. and Basu, A.K. 1987. Natl. Symp. Int. Pest Control. Trivandrum Oct. 15-17, 1987.

70. Taneja, N.K. 1989. Studies on myrothecium leaf spot disease of cotton. Ph.D. thesis Y.S. Parmar Univ. Hort. & Forestry, Solan (HP) pp. 100.

71. Taneja, N.K., Sheoraj and Seth, P.K. 1991. Reaction of the germplasm of *Gossypium* species to *Myrothecium roridum*. Indian. Phytopath. **44** : 404.

72. Taneja, A.D., Sharma, J.C., Singh, D.P., Sharma, A.P. and Kairon, M.S. 1988. Biochemical changes in *Gossypium hirsutum* cotton in relation to attack by jassid (*Emposca* spp). J. Indian. Soc. Cott. Imp. **13** : 33-36.

73. Temburnikar, S.T. 1994. Inheritance of resistance to Fusarium wilt in diploid cotton. Abst. of papers presented at Natl. Seminar on Cotton Prodn. challenges in 21st century, Haryana Agri. Univ. Hissar, April 18-20, 1994, pp. 16-17.

74. Thombre, M.V. 1980. Association of some morphological characters to confer resistance to bollworms in *G. hirsutum* cotton. Andhra Agri. J., **97** : 19-20.

75. Tripathi, B.K. and Basu, A.K. 1992. Genetically engineered insect resistant cotton. Paper presented at Silver Jubilee Seminar of AICCIP, Tamilnadu Agri. Univ. Coimbatore.

76. Vedamoorthy, G. and Reed, W. 1976. Integrated Control of Cotton Pests in India. J. Indian. Soc. Cott. Imp. **1** : 25-30.

77. Ware, J.O. 1929. The inheritance of red plant colour in cotton. Bull Ark. Agri. Expt. Stn. 220, p. 80.

78. Wison, F.D. 1987. Pink bollworm resistance, lint yield and earliness of cotton in a resistant genetic background. Crop Sci. **27** : 957 - 960.

79. Wilson, F.D. and Lee, J.A. 1971. Genetic relation between tobacco budworm feeding response and gland number in cotton seelings. Crop. Sci. **11** : 419-420.

80. Wilson, R.L. and George, B.W. 1982. Effects of okra leaf, frego bract and smooth leaf mutant on pink bollworm damage and agronomic properties of cotton. Crop. Sci. **22** : 778-801.

21 Integrated Management of Cotton Diseases and the Approach for Tomorrow

Sheo Raj, N.K. Taneja, M.K. Meshram and O.M. Bambawale

CONTENTS

Integrated Pest and Disease Management
Rajeev K. Upadhyay, K.G. Mukerji, B.P. Chamola and O.P. Dubey (eds.)
APH Publishing Corporation, New Delhi-110002, India

ABSTRACT

In India, cotton is grown under three diverse agroecological zones, namely, northern, central and southern. Besides ecological and climatological diversities, many other aspects like genetic variability, cultural practices, use of pesticides etc. make the cotton cultivation a unique object of management.

A number of disease such as, bacterial blight, *Alternaria* and *Myrothecium* leaf spots, grey mildew, anthracnose, boll rots (on aerial plant parts), root rot, *Fusarium* and *Verticillium* wilts (soil borne) affect cotton crop at various stages of growth with varying severity in different zones over the seasons. It is, therefore, difficult to estimate the true magnitude of losses caused due to diseases. However, a sizeable loss is caused every year to warrant management as a regular feature.

As on today, we do not have perfect package of recommendations for the management of diseases, however, a good data base has been generated with regard to epidemiology, host resistance and chemical control measures for important diseases. The breakdown of host resistance due to evolution of new races of the pathogen and the development of chemical mutants are important risks in the present management system. Now, there is a need to take advantage of this knowledge for integrated management and also to think about the frontline of researches currently available, such as computer simulation and forecasting, multiadversity resistance, biological control and biotechnological approaches. Finally, a strong network of extension activities needs to be built up to make the integrated disease management a reality.

A complete account on different aspects of cotton diseases is given in this article alongwith the likely approaches to be followed in developing the IDM methodology.

1. INTRODUCTION

Cotton has an important place in Indian agriculture, textile and handloom industry because it sustains a large number of farmers, provides vegetable oil and generates vast employment opportunities both on farm and in industry. For all these reasons and also because of the unique properties of the fibre it has withstood stiff competition from the synthetic fibres.

India ranks top in the world in terms of area under cotton (24%) but only fourth (11.42%) in production (4). The chief reason for this poor performance is obviously that 70% of the total of 7.5 million ha under cotton is rainfed and therefore, depends on monsoon. Other reason is that about 65% of the cotton cultivators belong to small and marginal group, holding less than 1 ha. For these resource-poor farmers, the high cost modern cotton technologies are difficult to be followed. Out of the total cotton area only 16% is sown with quality seed and only 10% of the farmers make appropriate use of pesticides (201). Irrigation is thus the first priority with the farmers followed by fertilizer use and plant protection comes only next. Because of the resource crunch, most of the farmers have no choice but to rely on pesticide dealers for their credit requirements and have to accept the dealers' advice in the selection of a particular pesticide.

There is much diversity in conditions of cotton cultivation in India which has been divided into three main ecological zones : North, Central and South (Fig. 1). Within each zone there are many differences in the sub-zones with regard to climatic conditions, soils, varieties/hybrids grown, cultural practices, cropping systems, etc (Table 1). Because of these, during the last 45 years, a lot of changes have taken places in cotton disease scenario (Table 2) in India (187).

Table 1 : Diversity of cotton cultivation in India

Aspects	North zone	Central zone	South zone
States	Punjab, Haryana, Rajasthrn and Western U.P.	Maharashtra, Madhya Pradesh and Gujarat	Andhra Pradesh, Karnataka and Tamil Nadu
Soil types	Alluvial	Black	Black
	Sierozem	Alluvial	Red
	Sandy	Red	Alluvial
Area (lack ha)	12.3	46.7	16.1
% area irrigated	98	12	26
Production* (lakh bales)	26	56	25
Productivity* (kg lint/ha)	359	204	264
Predominant vars. *G. arboreum*	DS 5, G 27, LD 327, RG 1, RG 8	AKH 4, AKH 8401, Jyoti, Virnar, AKA 5,	Jaydhar
G. herbaceum	-	V 797, Y 1	
G. hirsutum	B.N., F 414, F 505, LH 900, LH 1134, RST 9, G. Ageti, Zurret	DHY 286, Khandwa 2, LRA 5166, SRT 1, AKH 081, CNH 34	Abadhita, LRA 5166, MCU 5 VT, Savitha
G. barbadanse	-		Suvin
Hybrids (H x H)	-	H 6, H 8, JKHy 1, Kriti, MECH 4, NHH, 44, AHH 468, NHH 302	JKHy 1, H 6
(H x B)	-	NHB 12	DCH 32, HB 224, TCHB 213, Varalaxmi
Sowing time	May	June-July	August-Sept

* Based on 1989 yield data

Under the current system of cotton production, fungicides/antibiotics are sparingly used. The general lack of any direct control measures represents a major void in our system of managing cotton diseases. Several factors are responsible for this limited use of direct control measures. For example, farmers generally fail to reckon the true nature of disease development which is an interplay between the host, pathogen, environment and time. Absence of the disease or its appearance in low levels in one year does not indicate that it is harmless. The disease may cause devastation under favourable conditions in another year even before the farmer becomes aware of the dangers and contemplates taking control measure against it. It is, therefore, essential to gauge the real importance of various diseases attacking cotton by the level of losses caused by them, identify key variables affecting their development to epidemic proportions and then develop strategies based on this knowledge to contain them.

Several reviews have been written on cotton diseases and their management (190, 191, 270, 282). The present article focuses on integrating the knowledge available on managing important diseases of the three agro-climatic cotton zones in India and also endeavours to find out the gaps in our knowledge of the diseases which can be streered for improving the existing management practices. Various cotton diseases here reported from India alongwith their etiology (Table 3).

Table 2 : Changing Disease sitution in relation to *Gossypium* species composition

	Till 1947	Diseases prevalent After 1947	1980
Gossypium sp.	Mainly diploids	Diploids to tetraploids	Mainly tetraploids and hybrids
Bacterial blight	Minor	Got established and major in 1960s	Major
Anthracnose	Major	Minor	Minor
Myrothecium leaf spot	Unknown	Reported in 1960 and remained sporadic	Endemic in Haryana and sporadic throughout
Grey mildew	Major	Major in diploids	Sporadic throughout and endemic in Karnataka, Tamil Nadu, and Maharashtra
Alternaria leaf spot	Minor	Minor	Sporadic throughout and endemic in Karnataka
Root rot	Major	Major	Major
Fusarium wilt	Major	Major on diploids	Major on diploids in north India. A potential danger
Verticillium wilt	Unknown	Reported in 1967 and major in 1970s	Important in Tamil Nadu
New Wilt	Unknown	Reported in 1978	Major in hybrids

Table 3 : Major diseases of cotton, their causal agents, species attacked and distribution in India

Disease	Causal agent	*Gossypium* species attacked	Distribution
Bacterial blight	*Xanthomonas compestris* pv. *malvacearum*	All the four cultivated species	Throughout India
Alternaria leaf spot	*Alternaria macrospora*	-do-	-do-
Myrothecium leaf spot	*Myrothecium roridum*	-do-	-do-
Cercospora leaf spot	*Cercospora gossypina*	-do-	Sporadic in south zone
Helminthosporium leaf spot	*Helminthosporium gossypii, H. spicifer*	-do-	Sporadic in north zone
Anthracnose	*Colletotrichum indicum*	-do-	Sporadic in central zone
Grey mildew	*Ramularia areola*	-do-	Mainly in central and south zones
Rust	*Phakopsora desmium*	-do-	Sporadic in south zone

Root rot	*Rhizoctonia solani* *R. bataticola* (*Macrophomina phaseolina*)	-do-	Major in Rajasthan Haryana, Gujarat, M.P. and Punjab
Fusarium wilt	*Fusarium oxysporum* f. sp. *vasinfectum*	*G. arboreum* *G. herbaceum*	Rajasthan, Haryana, Punjab, M.S.
Verticillium wilt	*Verticillium dahliae*	All the four cultivated spp.	Major in south zone
New wilt	Unknown	do	Major in central and south zone

2. DISEASES OF COTTON

Most of the diseases occur in all the three zones with varying intensities. Bacterial blight, Myrothecium leaf spot, boll rot, Fusarium wilt and root rot are moderate to severe in the north zone. Bacterial blight, Myrothecium and Alternaria leaf spots, grey mildew, boll rot, Fusarium wilt and New wilt are moderate to severe in the central zone. Bacterial blight, Alternaria blight, grey mildew, rust, Verticilium wilt and new wilt appear in moderate to severe form in the south zone. Other diseases are of minor importance or appear sporadically.

2.1 Bacterial Blight

Losses

Bacterial blight is the most important disease of cotton and responsible for reduction in stand, defoliation, stem blighting and gridling, stem breaking during strong wind/storm, boll shedding, boll rotting, and reduction in grade due to lint staining. In United States, an average annual loss of 1.03% was recorded (56). Critical studies of actual crop losses have been made in several countries, both in irrigated and rain grown cotton, and losses in crop yield upto 50% have frequently been recorded (93).

In Sudan yield losses of 20% were common in irrigated varieties of *G. barbadense,* and even it led to 77% reduction in one trial of partially resistant variety (53).

The losses of 10 to 30% are of common occurrence (93, 202) and complete crop failure is also known (125). In India, the experimental data showed that under natural infection, bacterial blight could cause a loss of 6.0% in hybrid Varalaxmi and 14.3% in JKHy-1 (156). A loss of 1 to 27% was reported (144) depending upon variety, stage of the crop when infected and prevailing weather during diseases development. A linear relationship and significant positive correlation was observed between bacterial blight intensity and losses in seed cotton yield (150). While adopting the control measures, the diseases in susceptible cultivars should be brought down to the level of moderately resistant grade for avoiding the economic losses (146). The potential loss in susceptible cultivars varied from 9.10 to 32.29% depending upon the nature of genotypes, severity of the disease and weather conditions (148, 195).

Epidemiology

Xanthomonas campestris pv. *malvacearum* (Smith) Dye (Xcm) is carried by seed externally and internally (31, 250). Infected seed acts as a primary sources of inoculum but the bacterium remains alive in dried leaves for 17 years. It could survive for 8 days in water suspension, both in moist and air dried soil at 21-33°C and thus could be disseminated by dust. Covering the trash or chopped stalks with most soil makes the pathogen non-infective. The cultures on yeast extract-glucose-calcium carbonate-agar under mineral oil at 5-10°C remains viable and pathogenic, at least, for 20 months (273).

Climatic conditions strongly influence the infection and the disease development. Primary infection requires a lower temperature 30°C but secondary infection is favoured by higher temperature 35°C. Infection occurs most readily when air temperature is high ranging from 30-35°C coupled with high atmospheric humidity (above 75%) are prevalent whereas dry and hot weather retards the development of the disease (145). In the field, rain splash or sprinkler irrigation is the main agency by which the disease is spread and regular heavy wind driven rainstorms followed by a period of sunshine favour the build up of the disease into epiphytotic form. There is an evidence that two cotton pests (*Earias* spp. and *Dysdercus koeningii*) transmit the disease at least mechanically (277). A large number of collateral hosts are known, however, their role in the disease cycle is not very definite.

Races/strains

The pathogen *Xcm* is variable in pathogenicity, became increasingly evident when the resistance of cotton strains and varieties broke down. Probably biological specialization occurs in this species (115). In India, the presence of definite biological races was described which could break the resistance of BAR type with B2 and B3 genic make-up (11). Later, six races A,B,C,D (168) and E,F (275) identified.

A set of 8 upland cotton lines having different bacterial blight resistant genes was developed (82) and 15 races identified (30). Two new races were described, from India as 16 and 17 (274). Race-18 from Pakistan (84) and race-19 from Brazil (209) were reported subsequently.

New races designated as 20 (which has broken down the resistance of 101-102 B with combined genes B2+B3 in Acala background and other combined genes for resistance from Western and Central Africa (62) and Post-Barcket from Sudan (233) have also been reported.

In India, as many as 17 races, namely, 1-16 and 18 have been described so far (197, 273). Race 18 was most prevalent followed by race 10 which possess the virulence factors for five (B7, B4, B2, BIn and BN) and four (B7, B2, BIn and BN) resistance genes respectively.

Host resistance

Several resistant sources to bacterial blight of cotton are found in India e.g. 101-102 B, BJA 592, P-14-T-128, Reba B-50, HG-9, Tamcot CAMD-E, TxBonham, BJR etc. (143, 147, 149, 273). Many genotypes identified as immune or highly resistant against the most virulent race 18 of Xcm are being utilized in breeding programme for developing durable bacterial blight resistant cultivars. Nineteen *G. hirsutum* lines with high degree of resistance have been developed at Central Institute for Cotton Research (CICR), Nagpur (231). A variety BJR serves as a combined source of resistance to bacterial blight and jassid (230). In Pakistan, 38 of the 54 bacterial blight resistant lines were homozygous immune without visible symptoms to the most virulent and common race 18 which include 101-102 B, Im 216 and Reba B-50 (32). In Sudan, it was shown that the key to resistance in the important Albar stocks from East Africa was the major gene B2, fortified by genes of small effect (90). Im 216, a blight immune line developed by Brinkerhoff at Oklahoma was selected as an immune plant from a segregating population of Bird's B2, B3 Empire, one of the parental populations of 101-102 B (34).

The first rainfed upland cotton of Sudan resistant to bacterial blight was Bar SP 84, a selection from Uganda SP 84 R that was homozygous for B2 (116). The first commercial variety of long staple barbadense cotton was Bar XLI (a Lambert type homozygous for B2 and also resistant to leaf-curl disease) which was most popular in irrigated areas (79). By 1970s the variety Barkat (a Lambert type homozygous for B2, B3) had eventually replaced Bar XLI (224). Four Reba varieties had the maximum resistance to *Xcm* and of these Reba-W-296 with genes B2, B3 exhibited uniform resistance (35).

The flexibility in the approach for breeding a resistant variety by backcrossing (88) to transferring the known gene combination B2 B6, of susceptible variety Acala 4-42 in A (57) 12, which was extremely variable in its reaction to blight was successfully practised (91, 113). The B2, B6 Acala 4-42 variety Barac (67)B is widely grown in the Sudan Gezira and in several rainfed areas such as Gedaref.

Among the varieties from the IRCT Breeding work that are grown on a large scale in Central Africa and other cotton growing countries are Reba B-50 (from Allex x Stoneville) and BJA 592 (208). The variety Reba P 279, derived from the cross (Reba B 50 x Deltapine Smoothleaf) is grown in Paraguay and Argentina. Other IRCT bacterial blight resistant varieties 1 1422, BTK 12, SRI F4-71, IRCO 5028 and 3372, Y 1422 and BKA 592 are widely grown in different countries. /Tamcot varieties developed by Bird (22, 23, 24) and B2, B3 and B7 genes and are highly resistant to the 18 established races of the bacterium. During 1967 and 1968, two MAR varieties Tamcot SP 21 and Tamcot SP 37 were released in areas of Texax (21). Three MAR varieties, namely, Tamcot SP 215, Tamcot SP37H, and Tamcot CAMD-E were released in 1975 in USA (22, 23, 24).

Heritable resistance to bacterial blight was first demonstrated in Sudan (119). Ten major genes were identified to which the symbols B1-B10 were ascribed, of which 8 were dominant or partially so (117, 118). The symbols B9 and B10 for bacterial blight resistance genes were found in the upland variety Allen (122). These genes are not homologous with Knight's B9 and B10 and to avoid the confusion all four were redesignated B9L, B10L and B9K, B10K in that order (89). In USA, three major genes in upland varieties were identified and given the symbols BIn, BN and BS (69). Polygenic complexes for resistance in the upland varieties Stoneville 2B and Deltapine have been given symbols B SM and B DM, respectively (27). A dominant gene for resistance in a mutant plant from irradiated susceptible upland variety Westburn 70 was identified (33).

The importance of minor and modifier genes and the interaction of B genes with each other in enhancing the resistance conferred by major genes has been established by a number of workers. It was evidently proved (7, 92, 94) to indicate that high level of resistance could be achieved by the accumulation of polygenes. The *G. harknessii* cytoplasm slightly enhanced the expression of blight resistance in F1 hybrid (133). Cytoplasm from *G. herbaceum* also enhances the resistance to bacterial blight.

A major part of the total expression of resistance was attributable either to additive gene effects or to an interaction of additive of dominance effects (54). In Africa, the resistance of 101-102 B, showing outstanding resistance to bacterial blight is controlled by a complex of polygenes and two major resistance genes, B2 and B3 (94). The resistance was observed to be incompletely dominant and additional genes were responsible for imparting greater resistance (175).

Immunity in Im-216 was inherited as a completely dominant triat of two dominant independently inherited genes. The F2 data also fit a ratio for two dominant and one recessive independently inherited genes (33). Individual blight resistant genes have not yet been identified in lines derived from crosses of immune and susceptible parents.

Resistance was correlated with a high glucose level in the leaves (80). A substantial level of free threonine but no detectable free lysine in cotton cultivars susceptible to *Xcm* was reported (279). Serine and alanine were not detected in cotton plant infected by less virulent isolate of the pathogen but in case of infection with highly virulent isolate, these were present proline was absent and there was a three to four fold increase in glumtamic acid (169). Serine was most inhibitive and probably competitive inhibition was operative (167, 170, 276). Phenolics and Leucoanthocynins probably cause resistance in bacterial blight (95).

Resistance is associated with low content of citric acid and high contents of sugars and amino acid like serine, which inhibits the growth of the pathogen and interfere with its utilization of glutamic acid and alanine (276). The amino acids extracted from cotyledonary leaves of Im 216 were more inhibitory to *Xcm* than those from Acala 44 (289). The oxidation of catechin, catalysed by peroxidase, was suggested as the mechanism of resistance (271).

A direct relationship was observed (17) between the speed of induced terpenoid synthesis and host resistance. However, during the first 48 hours the susceptible cultivars showed rapid synthesis similar to resistant, but beyond 48 hours the additionally induced terpenoid synthesis was almost totally suppressed suggesting that the pathogen actively interfered with induced terpenoid synthesis. Thus, the induced but not the constitutive terpenoids were essentially associated with resistance. The concentrations of fluorescent phytoalexins in leaves adjacent to colonies of *Xcm* were studied and it was observed that the bacterium multiplied to reach much lower populations in the immune cultivar than the susceptible one (182), almost totally suppressed suggesting that the pathogen actively interfered with induced terpenoid synthesis. The rate of darkening of internal pigment glands was negatively correlated to the degree of compatibility (172) and the size of the lesion could be a reflection of the rapidity of leakage from the pigment glands.

Management

The main source of primary infection is seed which may carry the pathogen both externally as well as internally (250). The pathogen is also present in debris. Hence, use of delinted seed and removal of infested plant debris should invariably be practised. Where the disease is very severe, the seeds should be treated with a solution of 0.1% streptomycin before sowing. The experiments conducted throughout India showed that the disease was effectively controlled by spraying the mixture of streptomycin+copper oxychloride preparations (188).

Since drug resistant mutants of *Xcm* have been observed (165), it is better not to use streptomycin frequently. A protective effect of preinoculation of avirulent strains, heat-killed races of *Xcm* and phylloplane bacteria was demonstrated when the challenge dose of virulent strain of *Xcm* was given at 8-24 hours (278). The phylloplane bacteria possessing this preinoculative protective effect belonged to the genera *Flavobacterium, Pseudomonas* and *Aeromonas.* Field trials have already indicated that a spray of these bacteria could effectively reduce the severity of bacterial blight and could be used as an effective input in the integrated management of the disease (232).

2.2. Alternaria leaf spot

Losses

Losses caused by *Alternaria* leaf spot have not been systematically estimated. However, under severe infection it affects leaves, stem, petiole and even bolls and damages the crop heavily.

Epidemiology

Alternaria macrospora Zimm., the causal agent of the disease, was isolated from the seeds when the opened bolls were exposed to the atmosphere where the fungus was present but it could not penetrate the boll wall and reach the seed (12). The pathogen is internally seed borne in *G. herbaceum* (diploid) and the infected seeds act as primary source of inoculum (176). It survived on commercial cotton seeds and on post season plants left growing at the field edges. Survival in plant debris under field conditions was minimal and was considered to be less important.

The disease development was 5.5-8.5 times higher in cotyledons than in leaves (14). For infection, the minimum, optimum, and maximum temperatures were less than 10, 20-25 and 35°C, respectively. For cotyledons infection, the required wetting period was lower than 4 h. and temperatures of 20°C and 25°C whereas for leaves infection the requirement of wetting period was more.

Host resistance

Eleven lines viz., RRD 275-2, RRD 276, RRD 280-1, RRD 381-2, RRD 392, RRD 418-5, RRD 419, DP 354, DP-954, NA 585 and JK-97-CAC-77-1-1 were observed as resistant for *Alternaria* leaf spot (177). Three lines namely 260-2, PA 122-18 and IC-1230 were reported to be field immune (166). Several resistant lines have been identified under field test at Dharwad in Karnataka where the disease appears in severe form (188).

The infection of the leaves by *A. macrospora* reduced the chlorophyll (both a and b) content. In healthy cotton seedlings phenol content was higher in leaves than stem and in resistant (Acala) than susceptible (Pima) plants (13). Phenols were oxidized by polyphenoloxidase (PPO) rather than peroxidase and catalase during diseases development; PPO activity and phenol accumulation being higher in leaves of resistant than susceptible plants. Out of the seven different PPO isoenzymes detected in diseased plants, two were found in healthy plants also, 2 were induced in plants during disease development and 3 were of fungal origin. This confirms that the phenol oxidative activity in diseased plants is caused by the plant, the pathogen and their interaction.

Management

The most effective treatment for the control of this disease was found to be 1-2 hour soaking of seed either in ceresan (phenyl mercury acetate) or duter (fentin hydroxide) (178). RH 2161 (fenapanil) and Panoctine (guazatine) were most effective followed by Cuman L (Ziram), Dithane M-45 (mancozeb), Blitox (copper oxychloride) and Duter under *in vitro* evaluation (48). Dithane M 45 was also found to be effective (20). In field test, the best control was achieved with Blitox and Dithane Z-78 (zineb) which reduced disease incidence and increased yield (228). Cuman and Blitox gave the best control in upland cotton cultivar MCU 5 (178).

2.3. Myrothecium leaf spot

Losses

A loss upto 15% due to Myrothecium leaf spot with disease intensity of 37% was recorded in the field experiment (253).

Epidemiology

The pathogen (*Myrothecium roridum* Tode ex Fr. was isolated from the seed of H4 from Gujarat (239). Contaminated seeds with *Myrothecium roridum* (239) reduced germination and delayed emergence (45) and soaking in suspension of the pathogen resulted in reduction in germination (252). High relative humidity and fairly low temperature favour the spread of the disease (242). The toxin of the fungus at high concentration inhibited the germination, retarded the growth of young seedlings and caused wilting when introduced in the young plant through roots and cut ends of the stem (39).

Races/strains

Strain variation in *M. roridum* attacking different host species was reported earlier (8). However, pathogenic variation in the isolates of cotton and *Vigna* spp. was observed and grouped into 5 pathotypes, namely, MRI, MRII, MR III, MR IV and MR V (256).

Host resistance

Two varieties of diploid cotton *(G. arboreum)* viz., 231 R and M 60 A-2 (40) and ten of G. *hirsutum* namely, Arkansas green, GP 197, Kirghis K2, Line F (Hyb 330), M-DH-66, Mex 685-3, Pee Dee 113, Pee Dee 259, Pee Dee 875 and SRI F 4 were observed as resistant to Myrothecium leaf spot (257).

Two amino acids leucine and tryptophan were present only in resistant varieties and were ascribed to be responsible for resistance (255).

Management

The disease can be managed either by cultural or chemical means. Early sowing resulted in less incidence of disease under north Indian conditions (38). Captafol, maneb, ziben and thiram reduced the disease incidence considerably under field conditions (236). Bavistin, Blitox 50 and Delsan were found to be the best in controlling the disease and increasing the yield (254).

2.4. Grey mildew

Losses

In the United States and most other countries, grey mildew causes little economic loss. In fact, defoliation caused by the disease in late fall may favour opening of bolls and access of air to the plant. Thus reducing boll rots. The early outbreak of the disease in a few countries, such as Madagascar, Uganda and India (Maharashtra state), makes it very severe. In such case, it causes heavy defoliation which is responsible for economic losses (223).

Epidemiology

In temperate climate, the pathogen *Ramularia areola* Atk. over winters in dead infected leaves while in tropical and sub-tropical climate the pathogen survives on perennial plants or on the off season plants and serve as primary inoculum (187). Wind, irrigation water, workers and machinaries passing through the fields help in secondary infection. Host penetration is only through stomata under cycles of night wetting and day drying. For maximum infection four cycles are required (225).

Races/strains

Different isolates of *R. areola* vary in growth rates depending upon their virulence the fastest growing is most virulent. The pathogen is restricted to *Gossypium* species only and all the four cultivated species are afflicted. The isolates from China and India are believed to infect diploid cottons but the Indian isolates now infect *G. hirsutum* also (76). This demonstrates the possibility of races in *R. areola.*

Host resistance

G. barbadense varieties, namely, ERB 4492, Suvin, Sujatha, SB 289 E, *G. hirsutum* cultivars Laxmi and Sangam and interspecific tetraploid hybrid Varalaxmi are resistant to grey mildew (46). Some *G. arboreum* lines also have been identified as resistant (42, 77, 124, 161).

Management

The disease can be managed by dusting sulphur or spraying wettable sulphur (225, 270). The best control of this disease was achieved by sprays of tridemorph or carbendazim (2).

2.5. Boll rots

Losses

Boll rot is responsible for losses in all cotton growing regions of the world but extent of loss varies widely with climate. Damage is generally much greater in areas of high rainfall. In the United States, the average loss of 2.36% is caused by boll rot but in the Louisiana-Mississippi delta region the losses of 50% regularly occur due to high rainfall and fertile soil which result in luxuriant growth. In the Oklahoma Texas, plant growth is less and the air in the canopy is not excessively humid and annual losses sustained in this region have usually been under 2%. In India, the losses have been reported from 10 to 30% (120).

Epidemiology

Cotton boll rot is predominant in the humid areas and depends upon the moisture, plant population/density and plant type. Bollworm damage is reported to aggravate the disease as it paves way for the entry of secondary invaders. Changes in microclimate within the canopy indicated that reduced light intensity and longer period of 95% or more relative humidity and free moisture favoured boll rot development (206).

Management

The beneficial effect of sulphuric acid treatment on fungi and bacteria present on cotton seed from diseased bolls is well known. It was suggested to spray a mixture of fungicide and insecticide, which gave better control than fungicide alone (41). Best results were achieved by combining insecticide with bactericide and fungicide (188, 229, 251).

2.6. Anthracnose

In India, it was first reported from Pusa (Bihar) and since then has been recorded from Madras, Madhya Pradesh, Bombay and Bengal (270).

Losses

In Madhya Pradesh, the disease though of minor importance may at times become quite destructive, particularly in wet years. The disease, however, is of minor importance in other parts of the country and occurs there only occasionally (270).

Epidemiology

The mycelium of the causal organism *(Colletotrichum capsici / C. gossypii* Southw. survives in the seed whereas spores on the seed surface. Old rotten bolls and other plant parts in the fields are responsible for perpetuation of the fungus. The two common weeds *Aristolochia bacteata* and *Hibicus diversifolius* are also affected by the pathogen (270).

Humid weather is much favourable for the disease. Rains after seedling emergence cause severe seedling blight (266). Boll rot also increases if it rains during boll development.

Management

Use of disease free seed from healthy crop is good method of control. Ploughing the plant debris just after harvesting eliminates the primary source of infection. One year rotation also helps in minimising the inoculum. Defoliation of the plants by dusting calcium cyanamide was recommended (234)) to reduce humidity and thereby bollrot phase.

Acid delinting of seed helps in removing the surface borne spores of the fungus. Seed treatment with ceresan, ceresan M, zinc trichlorophenate and sulphur was found to be effective (270). Spraying the seedlings with Bordeaux mixture has also been recommended.

2.7. Helminthosporium leaf blight

In India, it occurs sporadically in Karnataka and Punjab. The disease caused by *Helminthosporium speciferum* (Bain) Nicot is of minor importance.

Host resistance

Twelve lines of *G. hirsutum,* namely, AC 738, Acala SJ-1, Deltapine 14, JR 8, JR 78, K 4004, LH 751, PRS 72, Ps 10, SRT 1, 9/11, Nigeria and 9/17 Nigeria are resistant to the disease (16).

Management

Karathane was found to be most effective fungicide in reducing conidial germination and germtube growth of *H. gossypii* (164), while the growth and sporulation of *H. spiciferum* were best reduced by vitavax (43).

2.8. Root rot

Losses

A loss of about 3% in general and upto 90% in certain cases due to root rot has been reported (270).

Epidemiology

The thermal death points of *Rhizoctonia bataticola* (Taub.) Butler and *R. solani* Kuhn, the causal organisms of root rot, are 68°C and 60°C, respectively, therefore these are able to survive during hot summer. The soil moisture of 15 and 20% and temperatures of 39°C and 35°C respectively are most suitable for infection (270). Growing the susceptible cultivar Digvijay under the regimes of 25 and 50% available soil moisture reduced the incidence of *M. phaseolina* without significantly affecting the seed cotton yield but the yield was reduced at 75% soil moisture (261). The root rot of indigenous cotton caused by *M. phaseolina* in Maharashtra and Gujarat differed from the root rot of cotton in Punjab and maximum development of the disease was observed at 28°C and 20% soil moisture (160).

Races/strains

Two races have been identified on the basis of cultural characters and pathogenicity of 19 isolates if *M. phaseolina* on three host variety (246). Race-1 formed olive green mycelium in culture, remained mainly vegetative and caused 100% mortality of host plants while races-2 had dark grey mycelium and many sclerotia causing 59.4% mortality.

Host resistance

Three lines viz., KH-33-146, 15-KW-2 (MB) and 9-KW-2 (MB) were observed as resistant to both *R. solani* and *R.bataticola* (264).

The higher contents of gossypol and phenol and lower sugar content in older seedlings as compared to younger ones were attributed for resistance to *R. bataticola* (204). Toxin produced by *R. bataticola* in culture reduced seed germination, retarded root extension and induced wilting in cuttings of cotton plants (49) and loss in carbohydrates and electrolytes from toxin treated seeds and leaves was also observed.

The concentrations of terpenoids desoxyghemigossypol, desoxy-6-methoxyhemigossypol, hemigossypol, 6-methoxyhemigossypol, 6, 6-dimethoxyhemigossypol and catechin are considered to be involved in age related resistance to *R. solani* in seedlings of cotton (81, 83). The lowest *Rhizoctonia* infection on cotton seedlings was observed when complete nutrient without magnesium was applied, whereas lack of potassium, sulphur and phosphorus appeared to increase disease intensity (203).

Management

Marked reduction in the disease incidence was obtained in the sowings of first week of April to the end of June as compared with May sown crop (270). Intercropping with mothbean (*Vigna aconitifolia*) was also effective in reducing the root rot incidence (270).

Soil treatment with Brassicol was found to be effective in bringing down the incidence. Seed as well as soil treatment with Brassicol was recommended (238) for its control. Soil treatment with vitavax, Brassicol+captan and Demosan reduced the disease incidence (222). Soil drenching with benlate, bavistin or TBZ has also been found useful (205). Very good control was obtained with soil application of brassicol+intercropping with mothbean under field conditions (198). Seed treatment with carbendazim followed by quintozene provided maximum disease control (110).

A strain of *Pseudomonas fluorescens* antagonistic to *R. solani* was isolated from the rhizosphere of cotton seedlings (78). *Penicillium* sp., *P. petuletum, Aspergillus* sp., *A. terreus, A. fumigatus* and *Chaetomium* sp. are antagonistic to *R. solani* (2119). Seed coating with *Trichoderma* spp. reduced disease incidence upto 83% in greenhouse (52). A strain of *Baccilus megaterium* from the rhizosphere of cotton seedlings was also found effective against *R. solani* (269). Seed pelleting with isolates of *T. viride* and *T. harzianum* increased the germination rate and reduced post emergence mortality in cotton variety LRA 5166 (3, 159).

2.9. Fusarium wilt

Losses

Prior to introduction of tetraploid *G. hirsutum* and *G. barbadense*, the losses in diploid cultivars of *G. arboreum* and *G. herbaceum* due to *Fusarium* wilt was 40 to 60% in India (121). Due to cultivation of resistant diploid and immune tetraploid cultivars to Indian race of the pathogen, the disease incidence has been reduced to significant level.

Epidemiology

The pathogen *Fusarium oxysporum* f. sp. *vasinfectum* (Atk.) Snyder and Hansen, the incitant of *Fusarium* wilt survives for long periods in soil and on cotton debris or other organic materials in the soil. The fungus grows at a wide range of temperature from 11 to 35°C. The thermal death point is about 50°C whereas pH range for optimum growth is between 5.2 to 7.2 (157). Aeration is helpful for mycelial growth but not for sporulation of the fungus (214). Low germination of conidia was observed in the soil amended with microelements as compared to untreated soil (247). In unsterilized soil it remains viable for more than 24 months (245). Under controlled conditions the disease manifests at soil temperature between 22-30°C, however, the most optimum temperature is 24-28°C, decreases at 32°C and is completely inhibited at 35°C (266). Hot and dry periods of long duration followed by rains favour the maximum disease development (234).

Although *Fusarium* wilt occurs in the absence of nematodes their presence in cotton rhizosphere aggravates the wilt incidence. Besides, *Meloidogyne incognita,* other nematodes eg. *Belonolaimus gracilis, B. longicaudatus, Pratylenchus brachyurus, P. pratensis, Rotylenchulus reniformis* and *Tylenchorhynchus* spp. have also been reported to be associated with *Fusarium* wilt of cotton (139, 151).

When both the organisms were inoculated simultaneously, wilt was more in wilt resistant variety as compared to susceptible variety (211). The same plant stand was observed in *Fusarium* susceptible but *Meloidogyne* resistant (N-6072) and *Fusarium* tolerant but *Meloidogyne* susceptible (Auburn-56) varieties (85).

Races/strains

Tetraploid cottons are immune to Indian strain of *F.oxysporum* f. sp. *vasinfectum* (162). American strain of the fungus did not infect American cottons on the Indian soils which were highly susceptible to that strain on the American soils.

Four races are known on the basis of differential host reactions (5). Race-1 is widely distributed. Races 1 and 2 are present in USA. Race 2 has got very limited distribution i.e. two sites in South Carolina. Race 3 is pathogenic to *G. barbadense* variety Sakel and *G. arboreum* variety Rozi in Egypt. This race is also reported from USSR (142) on *G. barbadense.* Three races viz., 1, 2 and 3 virulent on fine staple cotton varieties were identified by inoculating 9 cotton cultivars (226). Race 4 is known to infect only diploid cottons in India. Race 5 infects *G. barbadense* variety Ashmouni in Sudan (86) and race 6 was isolated from cotton plants in Paraguay (6). Besides these distinct 6 races, variations are known in the isolates of the fungus in pathogenicity on different cultivars (59, 123, 163, 227, 262) in isozyme bands of enzymes (132), , serological relationship (37, 100) and in morphological and physiological characteristics (141).

Host resistance

Tetraploid cotton varieties are immune to Indian race of *F. o.* f. sp. *vasinfectum.* High degree of resistance was obtained from *G. herbaceum* var. *frutescens, G. arboreum* var. *neglectum* f. *bengalensis* and *G. arboreum* var. *neglectum* f. *indicum* (267). There was comparatively less resistant material in *G. arboreum* than in *G. herbaceum* indicating higher potential source of resistance in the latter (74, 114). Ten strains viz. C-C-1-1-35, JLA-101, AKH-590, NA-3(77) *(G. arboreum)* and 321, 6455, SM-133, SM-143, SM-3-3-131, EZ-13-02 *(G. herbaceum)* have been developed as highly resistant/immune to *Fusarium* wilt (73).

Several varieties resistant to *Fusarium* wilt in *G. hirsutum* and *G. barbadense* have been identified (105, 174). Varieties, namely, Rivers, Dillon, Dixie, Lighting Express, Super Seven and Coker 100 with resistance have been developed (235, 281). Auburn-56, Auburn-M, Dixie King-II, Coker-310 and Stoneville-603 possess resistance to both Fusarium wilt and root knot nematode (103). Cultivars resistant to wilt-nematode complex were more resistant to wilt and Auburn 634 was highly resistant to nematode (243). Two interspecific diploid hybrids G. CotDH-7 and G. cot DH-9 released for commercial cultivation are tolerant to *Fusarium* wilt in India (188). In China, variety 86-1 was evolved as resistant to various races of this pathogen. A large number of resistant varieties have been developed in USSR also (15, 18, 19).

Resistance to *Fusarium* wilt is likely to be polygenic in *G. hirsutum* while in *G. barbadense* it may be controlled by relatively few genes (51). In Egyptian immune varieties of *G. barbadense* a cumulative gene effect was demonstrated (60). A single dominant factor was responsible for the inheritance of resistance (158). In Sea Island cotton *(G. barbadense*), two major genes inherited the resistance (235). Quantitative inheritance of resistance has also been suggested in certain lines (99). Inheritance of resistance was also suggested as predominantly additive (104). Studies on resistance in

G. arboreum and *G. herbaceum* revealed the presence of two dominant complementary genes but the third dominant gene when present completely inhibited the action of other two genes (111). Susceptible cotton cultivars had more cross reactive antigenes than resistant ones (1).

Management

The association of nematodes aggravates the wilt, thereby complicating the disease situation. Plants deficient in K suffered more nematodes damage than the plants high in K content, although it enhanced nematode reproduction. Potash reduced the wilt incidence while nitrogen and phosphorus increased the disease to certain extent. Soil amendment with zinc also reduced the wilt but farm yard manure had adverse effect (248).

Hot water treatment at 58°C can kill even the internally seed borne infection but it causes injury to the embryo (270). Seed treatment with systemic fungicides like Daconil, Tecto and Vitavax provided resistance to infection by inducing increased phenolic enzyme activities (61).

The systemic fungicide Algin and Arilon have given very good control of wilt (210). Good results were achieved by seed treatment with Bavistin followed by Agallol (221). Best control, however lies in the cultivation of resistant varieties. In India, several resistant varieties have been evolved which are being grown in various tracts of diploid cottons (187).

2.10. Verticillium wilt

Losses

Average yearly loss due to *Verticillium* wilt in the cotton growing states of USA during 1952-81 has been estimated at 2.37% with a range of 1.01-4.4% (72). In India, it affects mostly hirutum cottons and at times even upto 90% crop is damaged.

Epidemiology

Some models have been developed for *Verticillium* wilt (70) and *Phymatorichum* root rot (96) which can forecast and project, if and when the disease will be severe. The information on epidemiological studies will be useful to growers in management of diseases. The best strategy to manage diseases is to restrict the pathogen at the initial stage of its primary source. Next comes the secondary spread of the disease where epidemiological studies are helpful in selecting suitable management strategy.

Races/strains

Six races of *V. dahliae* have been identified in different countries. Out of five races found in USA, two races SS-4 and T-1 are most common (218). In USSR, Races 1 and 2 (183) and race 0 (258) were identified. Race 2 is represented by two biotypes namely 2 V and 2 A, the former is aggressive and does not cause severe leaf fall (286). Later on 6 races were described (68). Out of the six races, three races have been identified in India (102, 241).

Biotypes have also been reported (108, 140). Morphological mutants (88 in number) of races 1 and 2 were categorised as highly virulent, moderately or mildly virulent and avirulent (184). Somatic recombination was reported as one of the causes of new races (107).

Host resistance

American upland cottons are highly susceptible to *V. dahliae* (44) but many Sea Island resistant sources like Sea Brook were useful in breeding programme (283). Two lines viz., Hopicala (a derivative of a cross between the semiwild Hopi and the American cultivated Acala) and Leningrad strain of the var. *mexicanum* f. *nervosum*, very few accessions of *G. herbaceum* and *G. arboreum* possessed acceptable degree of resistance, whereas out of 101 *G. barbadense* lines, 52 proved to be highly resistant (237, 240).

Twelve derivatives of Tashkent 4 were more resistant to wilt than 108-F and Tashkent 1 (154). Wild *mexicanum* form as a good donor for most resistant types. Five accessions of *G. hirsutum* var. *mariegalante* were identified as highly resistant (71). Several *G. arboreum* and *G. herbaceum* lines were reported resistant (10, 101, 106).

Many varieties and lines from USA and Peru tested in USSR exhibited resistance (199, 215, 259, 260). Sixty eight accessions belonging to *G. hirsutum* ssp. *latifolium, punctatum, morilli* and *palmeri* originating in Mexico and Guatemala were identified as resistant to wilt (97). Lines with greater resistance have been developed through mutation breeding (50, 65, 179).

Some Paymaster series and the open bolled varieties (Acala B-3080 and New Mexico-B-8040) had low percentage of wilt whereas Paymaster -266-999 was the most resistant variety (152). Varieties 108-F, 137-F, 138-F, 147-F, 149-F, 152-F, 153-F, 159-F, 162-F, 163-F, 164-F, 167-F, 172-F and 137-F have been bred from *G. hirsutum* spp. *mexicanum* var. *nervosum* (265). In India, a variety MCU-5 VT has been developed by selection from MCU-5 (241).

The resistance to Verticillum wilt was partially dominant except in one environment where overdominance was observed (272) while tolerance was reported as recessive character (207). Monogenic dominance with modifiers was observed to govern the resistance in *G. barbadense* and *mexicanum* strain of *G. hirsutum* (153, 241). The existence of polygenic resistance was also reported in some resistant lines (134). Three dominant genes located in genomes A-1 and D-1 of autotetraploid cotton was observed (288).

Dominant genes for resistance in *G. hirsutum* sub sp. *paniculatum* var. *contextum* and *punctatum* Hopi A-2-A and the Indian line K-4640 have been reported (200). Nine varieties with susceptible to resistant reactions showed that 2 pairs of genes gave a ratio of 13 susceptible : 3 resistant (67). Crosses between susceptible and tolerant showed quantitative control dependent on polygenes with additive effects outweighing those of dominance. Completely dominant and intermediate resistance in serial lines have been observed (134).

Management

Denser sowings (9, 216), sub-surface irrigation (171), flooding the field (185) or covering the field with thick polythene sheet (186) helped in reducing the wilt incidence and increasing seed cotton yield.

Application of ammonium sulphate and calcium cyanamide (155), Nitrogen and phosphorus in a ratio of 1:0.7 or 1:1 at 150 or 250 kg N/ha (287), 250 kg K_2O/ha in addition to 100 kg N+100 kg P_2O_5 (173) and the fertilizers containing Cu, Mo and carboammophos (249) decreased the wilt incidence and enhanced cotton yields. Incidence and development of wilt were reduced by urea also (135, 219). Rotation with puddle rice reduced the incidence to 3.3-15.6% and two years rotation involving lucerne or chrysanthumum reduced incidence by 54 to 76% (241). Similarly ten years of rotation in USSR is reported to reduce the wilt incidence (189).

Benzimidazol fungicides reduces infection from 82 to 6-7% in field trial depending upon the amount of chemical present in the plant tissue (217). Incorporation of paraformaldehyde in the soil twenty days before sowing cotton decreased incidence from 90 to 30% (280). When ammophos with 0.25% copper was introduced with the soil, disease incidence was reduced by 55% (268). Spraying cotton with 5% uzgen together with butiphos as defoliant considerably reduced wilt (258).

Addition of lignin of gluacontile to the soil alongwith *Trichoderma* before cotton sowing increased resistance to wilt (112). Various trichodermins prepared from *T. viride* when introduced in the soil were highly antagonistic to *V. dahliae* (263). Besides *T. viride,* several fungi, namely, *T. harzianum, Penicillium petulum, Aspergillus sulphureus, A. ochraceus, A. flavus, Gliocladium virens, Myrothecium verrucaria* and *M. roridum* have been reported as antagonistic to *V. dahliae* (219).

2.11. New wilt

Losses

A disease of unknown etiology was observed on all the four *Gossypium* spp., however, *G. hirsutum* varieties were more prone but the economic damage is caused in certain intra-hirsutum and inter-specific (hirsutum x barabadense) hybrids. The loss in affected plants was observed to be 14.76 to 94.5% depending upon the genotype and the time of manifestation of wilting (192). Fibre quality from the wilted plants is also affected (137, 194).

Epidemiology

The prolonged drought conditions followed by rain is considered to be an important factor for the onset and development of new wilt (193). Similarly, the soil environment and the parameters affected by soil type, has profound influence on disease development.

Host resistance

Although all the four *Gossypium* spp. are affected, many varieties/lines are not susceptible. Bacterial blight resistant lines are more prone to new wilt (196) however, resistant line T x ORSC-78 was found free of new wilt incidence. The inheritance of new wilt was studied in hybrid JKHy 1 (Khandwa 2 x Reba B 50) and a duplicate factor model was suggested (136).

3. INTEGRATED MANAGEMENT OF DISEASES IN DIFFERENT ZONES

3.1 North zone

The crop debris must be destroyed, either by burning or burying in a compost pit. Acid delinted seeds should be treated with captan/thiram or dipped in 0.01% streptocycline solution before sowing. The crop must be sprayed with streptocycline (0.1%) + copper oxychloride (0.2%), 2-3 times at 10-12 days intervals. These measures restricts bacterial blight incidence and leaf spots. In case of boll rots, carbendazim along with recommended insecticide is to be sprayed.

For root rot management, the seeds are to be treated with carbendazim (2 g /kg seed) or captan (3 g/kg seed). FYM should be added @25 tonnes/ha in the affected patch. Cotton may be intercropped with *Vigna aconitifolia.*

3.2 Central zone

The same measures as recommended for north zone are to be adopted with regard to bacterial blight, leaf spots and boll rot diseases.

For grey mildew, the crop must be sprayed with wettable sulphur (0.2%) or carbendazim (0.1%). Sulphur should not be sprayed on deshi cotton as they are sensitive. For Fusarium wilt, more than a dozen varieties have been released and recommended for cultivation in arboreum / herbaceum tract of cotton growing areas.

3.3 South zone

This zone is specific for Verticillium wilt appearance. MCU-5 VT, a resistant variety, must be cultivated in wilt prone areas. Two compost @ 25 tonnes/ha may be applied in the affected patch. The fields should be well drained to avoid water logging. The cotton crop may be rotated with lucerne or paddled paddy in wilt affected areas.

For foliar diseases, the recommendations for other zones will be applicable. *Alternaria* blight, a serious problem in Karnataka, can be managed with the application of Dithiocarbamate and copper fungicide.

4. RELATIVE EFFICACIES, BENEFITS AND RISKS OF PRESENT MANAGEMENT TACTICS

The current approach to managing cotton diseases can best be described as effective, economical and having minimal impact on the environment. In no way, however, can it be considered to be final or complete. Since drug resistant mutants of *X. c.* pv. *malvacearum* have been observed (165) and copper resistance in *X. c.* pv. *vesicatoria* is also reported (138), the use of copper and streptomycin in cotton needs thorough investigation. With cotton, as with many other crops, we lack sufficient knowledge to apply established principles of IDM.

The present day plant disease control is based mainly on the use of chemicals, selected cultural practices and host resistance and a number of disease management methods. The complexity of crop protection tactics makes it difficult for the farmer to integrate piece meal recommendations into a workable management plan and too often the farmer must integrate crop protection and production technology on his own farm. We should pursue research and advisory activities on a coordinated basis in developing management regimes for the North-zone, Central-zone and the South-zone cotton agroecosystems.

5. APPROACH FOR TOMORROW

Some of the possible refinements and immediate requirements in different aspects for improving existing methods of cotton disease managemet in India are discussed here.

5.1. Understanding the cotton agroecosystem

The relationship of a disease flare up to the agroecosystem is shown in Fig. 1. The concept of IPM has thus evolved into a holistic, multi-disciplinary system that integrates control methods on the basis of ecological and economic principles, for pests of all classes (pathogens, insects and weeds) that co-exist in an agroecysystem (66).

Better understanding of the population dynamics of the pathogens and of the natural and man-imposed factors that regulate disease intensity of economic damage would improve the management systems. When practical and economical, we should maximise use of biological control which at present is of academic interest only.

The utilisation of management principles leads to the development of a framework for the study of mechanisms involved in biological interactions in an agroecosystem. Such studies would further refine the existing management strategies as also would lead to novel ideas in management.

It is the degree of refinement, based on accumulated information about host, pathogen and environment, that would be utilised to make the disease management of the future a new approach and the success of a DM programme will totally depend upon the completeness and accuracy of the knowledge on which it is based.

5.2. Identification of problem

Identifying the problem is the first step in developing disease management strategies. We know fairly well about the relatively important diseases of different zones. Within each zone, however, we need to define a field, a farm or a tract at which our strategy must be aimed.

5.3. Loss appraisal

The magnitude of yield losses due to cotton diseases in India is such as to warrant utilisation of management tactics. What we lack is systematic information on the extent of total losses caused by various diseases. Consequently, it is difficult to appreciate the magnitude of losses and take appropriate management decisions. It is worth emulating the Cotton Disease Council organised in 1936 in USA which has done a good service by providing loss assessment for the entire cotton belt in USA from 1952-1984 (56). Though the extent of average loss by individual diseases in USA seldom went beyond 2.84%, the overall loss was estimated at 12.73% which is alarming and needs control strategies. In India, we need urgently to develop such a system.

5.4. Monitoring

A basic premise of the disease management is that it will be utilised only when actual or predicted diseases loss reaches the economic threshold level (except resistant varieties or cultural practices where no additional cost is involved). Our lack of crop loss and diseases severity data, for most of the cotton diseases in India has resulted in the sparse application of the economic thresholds in managing cotton diseases. Once an economic threshold has been established, reliable disease and pathogen monitoring techniques must be developed.

Monitoring data would allow proper timing of control treatments and would be a basis for determining if, when and where treatments will be required. It may eventually be possible to use remote sensing technology to monitor disease progress and estimate crop losses with satellite imagery. This would help us in the development of economic threshold values for the principal cotton diseases.

5.5. Predictive, conceptual and integrated crop management models

The ultimate objective of disease management is the development of predictive models for agroecosystems through which management decisions can be optimized resulting in maximum benefits to the producer. This process entails the collection and integration of several sets of complex data, with each set relating to a dynamic biological, physical, meterological or socio-economical sub-system. This process can be facilitated by system analysis and mathematical modelling. Forecasting models mostly do not involve pathogen monitoring - they assume that inoculum is always present. Better methods to measure inoculum potential must be developed and should be coupled with environmental data.

A conceptual model (Fig-2) developed by El-Zik (55) for integrated management of *Verticillium* wilt integrates and links the available knowledge of system components to plant health and quality. The system adjusts biological activities to environment and management tactics. *Verticillium* wilt of cotton exemplifies the Integrated Crop Management System (ICMS). An ICMS must consider the biology and phenology of the crop, phenologies and dynamics of pests, rhizoplane and phylloplane microorganisms, soil characteristics, cultural and management practices and environment (Fig - 3). The system, as with all production systems, is dynamic in that it can vary according to the production region and environmental conditions. This ICMS can be used as a production model and is a checklist of key disease control tactics and options that should be considered throughout the season.

The field of conceptual and simulations modelling can also provide analytical tools in cotton breeding. Models have a role in integrating and qualifying the inter and multi-disciplinary knowledge, concepts and approaches needed in developing new cotton cultivars. The use of modelling in cotton breeding will enhance efforts in identifying useful traits and provide additional decision aids to genetic improvement programmes.

5.6. Disease resistance

Disease resistant varieties offer control with no input from the farmer, so incorporating resistance seems the best way to help small holders control disease problems, Future strategies must incorporate components that complement and add to host resistance in a manner to achieve realistic results.

The choice of cultivar probably is the single most important management decision the grower makes in an ICMS. The cultivar sets the framework for the level of susceptibility to pests, the tactics applied to management of the crop and the production cost (58). Resistant cultivars provide the cornerstone for successful IPM System (57).

Although a considerable amount of information on resistance to a number of cotton diseases such as Fusarium wilt, Verticilium wilt, bacterial blight, Alternaria leaf spot, and Myrothecium leaf spot etc., has been generated in India, only a few cultivars have been specifically released as having resistance to a particular disease. The disease resistance in the current cultivars is not sufficient to obviate the requirement for alternative chemical and cultural disease management practices.

The development of cultivars with multiple resistance is a high priority research need in India. Elsewhere, multiple disease resistant cultivars (213) and cultivars resistant to pathogens, insects and environmental stresses (25, 26, 28, 29, 58) have been developed.

The multi-adversity resistance (MAR) is a system of direct and indirect selection for genes and traits that provide broad resistance to adversities in cotton production. It is a monumental task to select for resistance to every adversity affecting cotton. A system of selecting directly for four or five key traits, which indirectly improves levels of the remaining traits, would simplify genetic improvement. The MAR concepts and selection procedures for genetic improvement of cotton were developed by Bird (25) beginning in 1963.

Direct selection for improvement traits indirectly provides genetic gains for resistance to other major pathogens, insects and stresses alongwith higher yield potential and earliness (Fig. 4). Application of the system has been very successful in developing new superior cottons (58).

MAR concept could be profitably used in India in the context of large rainfed cotton belt exposed to various biotic and abiotic stresses. Resistant cultivars, however, are not a panacea for all the pest problems. The use of resistant cultivars alone should not be expected to control pests under all conditions or in all locations and, therefore, these should be used in concert in IPM programme.

5.7. Biocontrol

To date the search for disease control organisms applicable to cotton has been limited to those associated with soil and the rhizosphere. Control of leaf spots such as bacterial blight and others continue to rely on chemicals or genetic resistance.

5.7.1 Novel biocontrol approaches

There are considerable microflora in soil with biocontrol potential apart from the common soil saprophytes such as *Trichoderma, Gliocladium, Penicillium* and *Aspergillus*. These microflora are comparatively difficult to isolate as they are fastidious and/or they may be difficult to culture on a large scale. Isolates of *Dendrastibella* and *Cladorrhinum* not easily isolable by conventional methods as also *Lactisaria arvalis*, a basidiomycete were shown recently to reduce cotton damping off caused by *Rhizoctonia solani* (129).

Biocontrol agents (BCAs) that do not show antagonism in culture might act by inducing host resistance. Non-pathogenic isolates of *R. solani* suppressed damping off in cotton caused by virulent isolates (87).

A wide variety of unique resident microflora exists in soil which may be exploited by generating additional information on their occurrence, ecology and applicability.

5.7.2 Commercial uses

Two commercial preparations of bacteria as cotton seed treatments for control of seedling diseases are currently available. *Bacillus subtilis* A-13, is manufactured by Abbott Laboratories and sold as "Quantum-4000" for use against *Rhizoctonia.* A new strain of *B. subtilis* (GB-03) has been isolated that colonizes cotton roots much more efficiently than the original strain. In 1988 Ecogen marketed a second seed treatment for cotton seedling diseases, called "Dagger-G" prepared by firm Ecogen from a strain of *Pseudomonas fluorescens* has been used.

Only about ten BCAs are currently registered for use against soil borne pathogens and many of these are still not widely used. This situation needs to be addressed urgently to make biocontrol more feasible option.

5.7.3 BCAs in IPM

There has been considerable interest to use BCAs in combination with reduced amounts of pesticides and an impetus has been added from recent developments of strains of biocontrol fungi resistant to certain pesticides (180). A good example of this is the successful integrated control of egg plant wilt caused by *V. dahliae*. This system involved the introduction of *Trichoderma viride* in fields treated with either dichloropropene plus methyl-isothiocynate (fumigant) or thiabendazole.

An innovative approach in the formulation of BCAs, involves immobilization or entrapment of biomass within cross-linked matrix of organic polymers such as alginate, polyacrylamide or carrageenan (284). The method consists of mixing biomass and a carrier (bulking agent) with a sodium alginate solution. The mixture is then dripped into a calcium salt gellant solution to form insoluble gelatinous beads which dry into stable, hand pellets of uniform size. The preparations are easily handled, can be applied to soil with conventional fertilizer equipment, store well at cool temperatures and are generally dust free. Alginate pellets containing biomass of *Gliocladium virens* prevented damping off of cotton (128). Three patents have been awarded to USDA during 1987-1989. The future development of this technology and its commercial adaptation by industry seem certain.

The recent developments as experimental formulation of biocontrol fungi, such as the alginate pellets (126) and the inclusion of small amounts of pesticides and nutrients in such delivery systems will provide great stimulus to use biocontrol fungi together with pesticides or nutrients for integrated control. Technology for encapsulation has been developed (63, 127, 128) and licenced to industry.

There has been limited interest in combining biocontrol fungi with cultural practices for integrated control (181).

5.7.4 Constraints and restrictions on use of BCAs

Many serious problems and constraints exist on the road to a successful establishment of a biocontrol programme.

Many biocontrol systems with introduced fungi are successful in the laboratory and green house, but fail when tested in production system. This is one of the greatest constraints to biocontrol on a large scale (130).

The complex microbial community of soil imposes a buffering effect on population increase by individual species. It is because of fungistasis and competition, and prevents production of antibiotics and enzymes or inactivates them once formed. It is necessary, therefore, that the environment to which a prospective biocontrol agent is added should be studied carefully and modified accordingly by cultural practices such as addition of reduced amounts of pesticides, solarization, nutrients and organic amendments, to assist in the establishment and proliferation of BCAs.

Specificity is another limitation in the implementation of biocontrol because of limited market potential. For example, an isolate of *Trichoderma*, effective in the prevention of damping-off of cotton caused by *R. solani*, did not effectively prevent damping-off of sugarbeet caused by the same pathogen isolate (131).

Liquid and semi-solid formulation technologies are available. Whereas the liquid formulation procedure is expensive and accurate, the semi-solid formulation does not require elaborate facilities and is especially appropriate for small scale, labour intensive units in developing countries. Such small scale units need to be promoted in India.

Some other problems are listed below :

1. Lack of large scale arsenal of highly effective, stable, long living BCAs and questionable adaptability of such agents of industrial processing.
2. Low efficiency of preparations compared with cost.
3. Unknown health and environmental risks and consequent difficulties in registration process.
4. Apprehension by the farming community about efficacy.

At this point it is critical that successful cases of biocontrol be demonstrated and published to encourage efforts to register and commercialize the products. Future research will undoubtedly show biocontrol to be safe, effective and economically feasible.

5.7.5 Botanical pesticides

The present day pesticide approach in agriculture had its beginning in the use of botanical materials. Plants provide richest and inexhaustible source of organic chemicals. Tropical plants provide a rich and intriguing source of natural products with potent biological activity.

Recent reports on higher plants have indicated their potentiality in providing non-toxic and biodegradable pesticides, with several plant species exhibiting antimicrobial activity. The extract of *Rheum officinale* was effective against *X. c.* pv. *malvacearum.*

Currently, work is in progress at CICR, Nagpur on evaluating the medicinal properties on indigenous plant species against cotton pathogens. So far, 45 plant species have been evaluated. Extracts from *Datura inermis, D. metel, Lawsonia inermis, Sesamum indicum* (wild), *Lagerstomia parviflora, Leucas aspera* and *Punica granata* were efective against two or more of the cotton pathogens viz., *Alternaria macrospora, Myrothecium roridum* and *X. c.* pv. *malvacearum.*

Although many plants are known to have pesticidal properties, few have gained importance as pesticides, perhaps due more to a lack of scientific attention than to a lack of pest control properties. In future, medicinal plants are likely to command more attention as potential candidates in management of cotton diseases.

5.8. Biotechnology in control of cotton diseases

5.8.1. Biocontrol

With the exception of mutagenesis and protoplast fusion, no biotechnology involving gene splicing and cloning have been done with biocontrol fungi. Even conventional genetics has only been used occasionally to improve biocontrol fungi (109).

A strain of *T. harzianum* (KRL-AG-2) has been developed at Cornell University as a bioprotectant against seedling diseases (98). It was selected from protoplast fusions of two parental strains. The KRL-AG-2 grew faster at 11° and 25°C than the parental strains, and it was a competitive rhizosphere colonizer. The fungal strain seems to be particularly effective in providing protection against *Pythium*. In field trails, cotton yielded more when the seeds were treated with the bioprotectant than when treated with effective combination of fungicides (75).

Pseudomonas fluorescens strain HV 37A has been used as a seed treatment for control of *Pythium, Rhizoctonia* and *Thielaviopsis*. Through use of mutants deficient in antibiotic production, they found that 60-70% of disease suppression was due to the antibiotic produced by the organism. With this knowledge, and by genetically engineering a stronger promoter into a gene critical for antibiotic synthesis, they were able to increase antibiotic production by over 50%. This resulted in additional 10-20% suppression of diseases.

Fungi are thought to be susceptible to chitinase because of the presence of polymer, chitin, in the hyphal wall. Promising results have been obtained with chitinase in the control of *R. solani* on cotton (220). A chitinase gene was obtained from *Serratia marcescens* and cloned into *E. coli* with a strong promoter to obtain high expression of the gene. A crude preparation of the enzyme reduced disease on cotton by 62% under greenhouse conditions. Similar control was obtained where viable cells of genetically engineered *E. coli* were applied. These results indicate that the enzyme chitinase may have an important future role in the biocontrol of pathogenic fungi. As yet, its influence on beneficial fungi has not been assessed. Because of this lack of discrimination or specificity, general use of chitinase as a biopesticide may be limited. This enzyme, however, may have a more suitable role in engineered host plant resistance.

The mechanisms of action of biocontrol agents at the biochemical, molecular and micro-environmental level must be understood. This would lead us to :

(i) identify metabolites that can be quickly monitored to evaluate potential biocontrol strains,

(ii) determine the gene codes for a genetic improvement programme of such agents, and

(iii) meet the requirements of knowledge of mechanisms for registration.

5.8.2. Plant resistance to pathogens

The field of biotechnology offers opportunities to enhance our efforts in developing new cotton cultivars. The new techniques include cell and tissue culture, recombinant DNA and gene cloning, protoplast fusion, and plasmid insertion. New molecular genetic technology provides the prospect for conferring resistance to insects and plant pathogens and tolerance to broad-spectrum herbicides, through the introduction of genetic material from sources whose resistance cannot be incorporated through sexual hybridization.

The amount of current effort directed towards genetic engineering of resistance to pathogens is much less than directed toward resistance to herbicides and insects. This is primarily related to the low-value added potential for pathogen resistant seed since getting resistance through conventional breeding is available.

Two genes that are receiving attention are those that code for chitinase and glucase. The anticipation is that these will attack cell walls of invading fungi and render them more susceptible to natural plant defences.

Considerable biochemical and molecular work is directed toward bacterial blight resistance in cotton. A series of cotton isolines with single resistant genes have been developed (244). The lines are being utilized, at the biochemical level, to know, how they regulate the expression of the phytoalexin which imparts resistance.

A set of cloned avirulence genes have been developed (47, 64) from *Xcm* that have gene-for-gene recognition by specific resistant genes in the cotton isolines. In addition, some avirulence genes do have strict gene-for-gene interaction rather some resistant genes recognize more than one avirulence gene. This type of interaction is indicative of broad resistance. In India also, the work on characterization of *X. c.* pv. *malvacearum* races on the basis of plasmid profile has been initiated (36).

The cotton model system, to which molecular techniques are being applied, is providing basic answers to the nature of pathogenicity and resistance in plant-microbe interactions.

6. CONCLUSIONS

The development of agriculture over last 10,000 years or more has had great impact on mankind and his environment. During the evaluation of early agriculture man developed stable agroecosystems. The stability on the other hand has taken us to unnatural farming thereby creating new challenges. Cotton is being cultivated since early human civilization and has faced all facets of changes in agroecysystems. During the process of agricultural development the present cottons have got selected against the vagaries of weather, insect pests, disease and so on. This has also resulted in the emergence of various new problems. These, individually may not be responsible for enormous losses in the crop but when combine together make us realize about great reduction in lint quantity and quality.

The management of cotton diseases with diverse nature like seed, soil, debris and air borne affecting root systems, seedlings and all aerial plant parts needs skill. For making the management system successful, the conceptual models, on the basis of studies on factors affecting disease development, have to be developed. In India, such models have not been developed as in USA. Cultural practices, the important component of integrated disease management system, have not been adopted in India due to diverse agroecosystems and socio-economic conditions of the farmers. Although sincere efforts have not been made to breed varieties resistant to foliar diseases, success has been achieved in case of soil borne diseases like *Fusarium* and *Verticillium* wilts. Root rot, a great menace throughout India and more particularly in north India has remained untackled. Seed treatment and intercrop have little impact on disease amelioration. Although very little chances of getting good sources of resistance to both the pathogens - *R. solani* and *R. bataticola* causing root rot of cotton, because of their being highly polyphagous in nature, in cultivated species of *Gossypium.*, lot is needed to be done in this direction including searching the genes for resistance in wild *Gossypium* species and transferring the same through conventional breeding procedure if possible or through biotechnological options. The multi-adversity resistance (MAR) system has much scope in the integrated disease management system (25).

Biocontrol of diseases has been studied in detail in USA and other countries. The commercial products have also been prepared but the use of the technology is yet limited. In India, the studies have just been initiated and are being tried under field conditions also. The bioagents when applied to a phylloplane rhizosphere and rhizoplane are challenged by the environment, co-habitants and several unknown factors. The success of biocontrol systems lies in overcoming these problems. Transferring the genes of bioagents to the cultivars is currently gaining momentum. In certain cases, such genes have been cloned. If this technology becomes successful, as has been achieved in case of insect-pests, it will be a new dimension in disease management. Significant co-operative efforts are urgently needed among the scientists in various disciplines and the Government agencies to achieve success in IDM.

7. REFERENCES

1. Abd El-Rahim, M.A., Abou-Taleb, E.M. and Tohamy, A. 1988. Common antigen(s) to cotton in *Fusarium oxysporum* f. sp. *vasinfectum*. J. Phytopath. **121** : 217-223.

2. Alagarsamy, G. and Jeyarajan, R. 1986. Efficacy of fungicides against grey mildew disease of cotton. Madras Agri, J. **73** : 651-652.

3. Alagarsamy, G., Mohan, S. and Jeyarajan, R. 1987. Effect of seed pelleting with antagonists in the management of seedling disease of cotton. J. Bio. Control, **1** : 66-67.

4. Anonymous 1991. Cotton International, pp. 250.

5. Armstrong, G.M. and Armstrong, J.K. 1960. American, Egyptian and Indian Cotton-wilt Fusaria, their pathogenicity and relationship to other wilt Fusaria. USDA Tech. Bull., : 1219.

6. Armstrong, G.M. and Armstrong, J.K. 1980. Race 6 of the cotton wilt Fusarium from Paraguay. Plant Dis. **64** : 596.

7. Arnold, M.H. 1963. The control of bacterial blight in rain-grown cotton. I. Breeding for resistance in African Upland varieties. Journal of Agricultural Sciences, **60** : 415-427.

8. Arya, H.C. 1959. On the occurrence of physiological strains in *Myrothecium roridum* Indian Phytopath. **12** : 164-167.

9. Avrarov, A., Gubanov, G. and Faizillafev, M. 1976. Wilt and dense sowing. Kholopkovodstvio No. **10**: 29-30.

10. Babamuratov, Kh., Gesos, K. and Arutyunova, L. 1977. Methods of improving wilt resistance and fibre quality. Khlopkovodstvo No. **12** : 35-36.

11. Balasubrahmanyan, R. and Raghvan, A. 1950. Bacterial Blight on cotton in Madras. Indian Cotton Growing Review, **4** : 118-123.

12. Bashan, Y. 1984. Transmission of *Alternaria macrospora* Zimm. in cotton. Phytopathogische Zeitschrift, **110**: 110-118.

13. Bashan, Y. 1986. Phenols in cotton seedlings resistant and susceptible to Alternaria macrospora. Phytopathology, **116** : 1-10.

14. Bashi, E., Roten, J., Pinnschmidt, H. and Kranz, J. 1983. Influence of controlled environment and age on development of *Alternaria macrospora* and on shedding of leaves in cotton. Phytopathology, **73** : 1145-1147.

15. Bazhanova, A.P. 1979. Ash-167 B, a new fine fibre variety of cotton, obtained through distant hybridization. Biologicheskikh Nauk No. 241-244.

16. Bedi, P.S. and Chopra, B.L. 1984. Source of resistance against Helminthosporium blight of cotton in Punjab. Indian Phytopathology, **37** : 173-174.

17. Bell, A.A. and Stipanovic, R.D. 1978. Biochemistry of disease and pest resistance in cotton. Mycopathologia, **65** : 91.

18. Berdymuradov, R. 1979. Interspecific distant hybridizaation in the breeding of fine-staple cotton. Biologicheskikh Nauk No. **2** : 27-33.

19. Berdymuradov, R. and Akmuradov, Sh. 1981. Genetic characteristics of low-growing varieties on fine-staple cotton. Biologicheskikh Nauk No. **5** : 42-45.

20. Bhaskaran, R. and Shanmugam, N. 1973. Laboratory evaluation of some fungicides against *Alternaria macrospora* zimm. [on control]. Madras Agr. J. **60** : 646-647.

21. Bird, L.S. 1976. Registration of Tamcot SP 21, Tamcot SP 23, and Tamcot SP 37 cottons (Reg. No. 61, 63 and 64). Crop Science, **16** : 884.

22. Bird, L.S. 1979a. Registration of Tamcot SP 21S cotton (Reg. No. 73) Crop Science, **19**: 410-411.

23. Bird, L.S. 1979b. Registration of Tamcot CAMD-E cotton (Reg. No. 74). Crop Science, **19** : 411-412.

24. Bird, L.S. 1979c. Registration of Tamcot SP 37H cotton (Reg. No. 75). Crop Science, **19** : 412.

25. Bird, L.S. 1982a. The MAR (multi-adversity-resistance) system for genetic improvement of cotton. Plant Dis. **66** : 172.

26. Bird, L.S. 1982b. Multi-adversity (diseases, insects and stresses) resistance (MAR) in cotton. Plant Dis. **66** : 173-176.

27. Bird, L.S. and Hadley, H.H. 1958. A statistical study of the inheritance of Stoneville 20 resistance to the bacterial blight disease of cotton in the presence of *Xanthomonas malvacearum* races 1 and 2. Genetics, **43** : 750-767.

28. Bird, L.S., El-Zik, K.M., Percy, R.G., Thaxton, P., Benedict, J.H., Reyes, L., Creelman, R.A., Clark, L.E., Heald, C.M. and Bird, L.S., El-Zik, K.M. and Thaxton, P.M. 1986. Registration of ITAMCOT CAB-CS) Upland cotton. Crop Sci. **26** : 384-385.

29. Bird, L.S., El-Zik, K.M. and Thaxton, P.M. 1987. TAMCOT CD3H a Multi-adversity Resistant Cotton Variety, Tex, Agric. Exp. Stn. L-2240, 13 pp.

30. Brinkerhoff, L.A. 1970. Variation in *Xanthomonas malvacearum* and its relation to control. Annual Review of Phytopathology, **8** : 85-110.

31. Brinkerhoff, L.A. and Hunter, R.E. 1963. Internally infected seeds as a source of inoculum for the primary cycle of bacterial blight of cotton. Phytopathology, **53** : 1397-1401.

32. Brinkerhoff, L.A. and Hussain, T. 1978. Evaluation of American and African bacterial blight resistant upland cotton lines in Pakistan in 1977. Plant Disease Reporter, **62** : 1082-1085.

33. Brinkerhoff, L.A. and Verhalen, L.M. 1976. Inheritance of immunity to bacterial blight in an upland cotton cross. Proc. Beltwide Cotton Prod. Res. Conf., Natl. Cotton Council, Memphis, Tenn. 31 .

34. Brinkerhoff, L.A. Verhalen, L.M., Mamaghani, R. and Johnson, W.M. 1978. Inheritance of an induced mutation for bacterial blight resistance in cotton. Crop Science, **18** : 901-903.

35. Carvalho, P.P. 1970. Susceptibility of cotton cultivars to bacteriosis. Methos of investigating the disease. Agronomia Mocambicana, **3** : 1969.

36. Chakrabarty, P.K., Mahadevan, A., Raj, S. and Meshram, M.K. 1992. Plasmid profiles and curing of plasminds of *Xanthomonas compestris* pv. *malvacearum* stains isolated from cotton. *In Symposium* on Frontiers in Plant Biotechnology, Biotechnology Centre, IARI, New Delhi, November 25-27, 1992. Abstr. 19.

37. Charudattan, R. 1970. Studies on strains of *Fusarium vasinfectum* Atk. II. *In vitro* production of toxin and enzymes and immunoserology, Phytopath Z. **67** : 129-143.

38. Chauhan, M.S. and Suryanarayana, D. 1970. Effect of sowing cotton on different dates on the incidence of the Myrothecium leaf spot disease of cotton caused by *Myrothecium roridum* (Tode ex Fr.) in Haryana state. H.A.U. Journal Res. **7** : 580-584.

39. Chauhan, M.S., Surayanarayana, D. and Jalali, B.L. 1970. Toxin production by the fungus *Myrothecium roridum* attacking cotton in Haryana, India, Cotton Gr. Rev. **47** : 292-297.

40. Chauhan, M.S., Suryanarayana, D. and Singh, V.P. 1972. Myrothecium leaf spot disease of cotton in Haryana State, India. PANS, **18** : 329.

41. Chin-chyu, Tu and Yeng-Hsiung C. 1970. Experiments for the control of cotton boll black rot. Plant Dis. Rep. **54** : 29.

42. Chopra, B.L., Sharma, J.R. and Singh, T.H. 1980. Studies on Ramularia leaf spot of cotton in the Punjab. Indian J. Mycol. & Pl. Pathol. **10** : 96.

43. Chopra, B.L., Bedi, P.S. and Dutt S. 1985. Studies on Helminthosporium leaf blight of cotton in Punjab. Indian J. Mycol. & Pl. Pathol. **15** : 155-158.

44. Cotton, J. R. 1965. Breeding cotton for tolerance to *Verticillium* wilt. USDA Agric. Res. Bull. pp. 34-80.

45. Dake, G.N. 1980. Effect of *Myrothecium roridum* on the germination of cotton seeds. Indian Phytopath. **33** : 591-593.

46. Dake, G. N. and Kannan, A. 1981. Relative resistance and susceptibility of cotton cultivars to *Xanthomonas campestris* pv. *malvacearum* (E. F. Smith) Dye and *Ramularia areola* Atk. *In* Third Intern. Symp. on Plant Pathology, New Delhi, December, 14-19, 1981. 260-261 (Abstr).

47. De Feyter, R. and Gabriel, D. W. 1991. At least six avirulence genes are clustered on a 90-kilobase plasmid in *Xanthomonas campestris pv.* malvacearum. Molecular Plant Microbe Interactions **4** : 423-432.

48. Desai, S. A. and Patil, S. A. 1982. *In vitro* evaluation of different fungicides against *Alternaria macrospora* Zimm; incitant of leaf spot disease of cotton. Current Res. **11 (7/8)** : 72-73.

49. Deshpande, K. S. and Wadje, S. S. 1978. Toxic metabolites of *Rhizoctonia bataticola*. Indian J. Exptl. Biol. **16** : 137-138.

50. Dzhabarov, Kh. 1978. Wilt resistant mutants of cotton developed under the action of chemical mutagenes. *In* Genit Izmenchivost Vozbuditelya Vilatai puri povysh. Vitoustoichivosti Khlopchatnika, Dushambe, USSR, 164-168.

51. Ebbels, D. L. 1975. Fusarium wilt of cotton : /A review, with special reference to Tanzania. Cott. Grow, Rev. **52** : 295-339.

52. Elad, Y., Kalfon, A. and Chet, I. 1982. Control of *Rhizoctonia solaniin cotton by seed coating with* Trichoderma spp. spores. Plant and Soil, **66** : 279-281.

53. El Nur, E. 1970. Bacterial blight of cotton. *In.* Cotton Growth in the Gezira Environment. (eds. M. A. Sidding and L. C. Hughes) pp. 179-188., Sudan Agricultural Research Corporation.

54. El-Zik, K.M. 1968. Diallel analysis of the inheritance of resistance in cotton. *Gossypium hirsutum* L., to races of *Xanthomonas malvacearum* (E. F. Smith) Dowson. Dissertation Abstracts, **8** : 3959 B.

55. El - Zik, K. M. 1985. Integrated control of Verticillium wilt of cotton. Plant Dis. **69** : 1025-1032.

56. El - Zik, K. M. 1986. Half a century dynamics and control of Cotton disease : Dynamics of Cotton disease and their control. The 1986 proceedings of the Beltwide Cotton production Research Conferences. 2 pp.

57. El - Zik, K. M. and Frisbie, R. E. 1985. Integrated crop management systems for pest control and plant protection. *In* CRC Handbook of Natural Pesticides : Methods. Vol. I. Thjeory, Practice and Detection. Ed. N. B. Mandava. CRC Press, Inc., Boca Raton, FL. pp. 21-122.

58. El - Zik, K. M. Thaxton, P. M., Wallace, T. P. and Cook, C. G. 188. Improvement in fiber quality, yield, potential and resistance to pests of MAR cottons. Proc. Beltwide Cotton Prod. Res. Conf. National Cotton Council of America, Memphis, TN. pp. 556-557.

59. Fahim, M. M., Youssef, B. A. and Sharkawy, T. A. 1973. Host range of *Fusarium oxysporum* f. sp. *vasinfectum* (Atk.). Egyptian J. Phytopathol. **5** : 37-42.

60. Fahmy, T. C. 1934. The selection of wilt immune strains of long staple cotton. Min. Agric. Egypt. Tech. & Sci. Service, Bull. 130.

61. Fayed, M. A. M. 1981. Thirty-third International Symposium on Phytopharmacy and phytiatry. Rijksuniversiteit Gent, **46** : 923-932.

62. Follin, J. C. 1983. Races of *Xanthomonas campestris* pv. mavacearum (Smith) Dye in Western and Central Africa. Cotton et Fibres Tro. **38** : 274-279.

63. Fravel, D. R., Marois, J. J., Lumsden, R. D. and Connick, W. J., Jr. 1982. Encapsulation of potential biocontrol agents in an alginate-clay matrix, Phytopathology, **75** : 774-777.

64. Gabriel, D. W., Burges, A. and Lazo, G. R. 1986. Gene-for-gene recognition of give cloned avirulence genes from *Xanthomonas anmpestris* pv. *malvacearum* by specific resistance genes in cotton. Proceedings of the National Academy of Sciences (USA), **83** : 6415-6419.

65. Gaibullaev, I. Kh., Avazkhodzhav, M. Kh. and Egamberdiev, A. G. 1975. Reaction of chemical mutants of cotton to infection with *Verticillium dahliae.* Genetika USSR, **11** : 47-50.

66. Glass, E. H. 1975. Integrated pest management : Rationale, potential, needs and Implementation. Spec. Publ. 75-2, Entomol. Soc. Am.

67. Godoy Avila, S. and Mungoz Orozco, A. 1979. Cotton resistance to *Verticillium dahliae* K. I. The inheritance of resistance. Agroceincia, **37** : 171-183.

68. Gorkovtswa, E. A., Kasyanenko, A. G. and Ryabova, I. M. 1978. The problem of cotton wilt Note 3. Specialization of *Verticillium dahliae,* The pathogen of cotton wilt. Izv. AN. Tadzh. SSR., Biol. N. No. **3** : 3-10.

69. Green, J. M. and Brinkerhoff, L. A. 1956. Inheritance of three genes for bacterial blight resistance in upland cotton. Agronomy Journal, **48** : 481-485.

70. Gutierrez, A.P., Devay, J. E., Pullman, G. S. and Friberstshauser, G. E. (1983.) A model of Verticillium wilt in relation to cotton growth and development. Phytopathology, **73** : 89-95.

71. Haley, A. B. and Wilhelm, S. 1975. *Gossypium hirsutum* race *marie-galante,* a new source of *Verticillium* wilt in resistance in cotton. Proc. Am. Phytopathol. Soc. **2** : 114.

72. Halloin, J. M. 1983. Thirty year summary of cotton disease loss estimates : crop year 1952-1981. pp. 3-4. *In* Proc. Beltwide Cotton Prod. Res. Conf., National Cotton Council, Memphis, T. N.

73. Hande, Y. K. and Rane, M. S. 1985. Development of deshi cotton lines immune to Fusarium wilt. Paper presented in National Seminar on Rainfed Cotton Production and Technology, M.A.U., Parbhani on 24 May, 1985 pp. 146-149.

74. Hande, Y. K., Padule, D. N. and Lokhande, S. B. 1983. Screening of some diploid strains against Fusarium wilt. J. Maha. Agric. Univ. **8** : 191-192.

75. Harman, G. E., Taylor, A. G. and Stazz, T. E. 189. Combining effective strains of Trichoderma harzianum and solid matrix priming to provide improved biolotical seed treatment systems. Plant Dis. **73** : 631-637.

76. Holey, N. R. and Moghe, P. G. 1977. Incidence of grey mildew or dahiya of H4 cotton in Vidarbha. J. Maharashtra Agric. Univ. **2** : 68-69.

77. Holey, N. R., Satpute, G. N., Ghoderao, B. N. and Peshney, N. L. 1992. Evaluation of diploid cottons (*Gossypium* species) for resistance to grey mildew disease caused by *Ramularia gossypii.* Indian J. agric. Sci. **62** : 293-294.

78. Howell, C. R. and Stipanovic, R. D. 1979. Control of *Rhizoctonia solani* in cotton seedlings with *Pseudomonas fluorescens* and with an antibiotic produced by the bacteriu. Phytopathology, **69** : 480-482.

79. Hughes, L. C. 1970. Varieties in the Gezira environment. I. Long staple varieties. *In* Cotton growth in the Gezira Environment (eds. M. A. Siddig and L. C. Hughes), pp. 1654-172. Sudan Agricultural Research Corporation.

80. Hughes, L. C. and Fowler, H. D. 1953. Resistance to a plant disease associated with high glucose content of leaf. Nature, London, **172** : 316.

81. Hunter, R. E. 1978. Effects of catechin in culture and in cotton seedlings on the growth and polygalactoronase activity in *Rhizoctonia* solani Phytopathology, **68** : 1032-1036.

82. Hunter, R. E., Brinkerhoff, L. A. and Bird, L. S. 1968. The development of a set of cotton Lines for differentiating races of *Xanthomonas malvacearum.* Phytopathology, **58** : 830-832.

83. Hunter, R. E., Halloin, J. M., Veech, A. J. and Canter, W. W. 1978.. Terpenoid accumulation in hypocotyls of cotton seedlings during ageing after infection by *Rhizoctonia solani.* Phytopathology **68** : 347-380.

84. Hussain, T. and Brinkerhoff, L. A. 1978. Race 18 of the cotton bacterial blight pathogen, *Xanthomonas malvacearum,* identified in Pakistan in 1977. Plant Disease Reporter, **62**: 1085-1087.

85. Hyer, A. H., Jorganson, E. C., Garber, R. H. and Smith, S. 1979. Resistance to root-knot nematodes in control of root-knot nematode Fusarium wilt disease complex in cotton. Crop. Sci. **9** : 989-901.

86. Ibrahim, F. M. 1966. A new race of cotton wilt *Fusarium* in the Sudan Gezira, Emp. Cotton Grow. Rev. **43** : 296-299.

87. Ichielvich-Auster, M., Sneh, B., Koltin, Y. and Barash, I. 1985. Suppression of damping-off caused by *Rhizoctonia* species by a nonpathogenic isolate of *R. solani.* Phytopathology, **75** : 1080-1082.

88. Innes, N. L. 1964. Bacterial blight resistance breeding. Progress Report from Experiment Stations. Republic of the Sudan, 1963-64. Empire Cotton Ciorporation, London, pp. 21-27.

89. Innes, N. L. 1965a. Resistance to bacterial blight of cotton : The genes B8 and B10. Experimental Agriculture I. 189-191.

90. Innes, N. L. 1965b. Inheritance of resistance to bacterial blight of cotton : I. Allen *(Gossypium hirsutum)* derivatives. Journal of Agricultural Sciences, **64** : 257-271.

91. Innes, N. L. 1966. Promising selections from Albar A (57) 12 in the Sudan. Empire Cotton Growing Review, **43** : 383-272.

92. Innes, N. L. 1969. Inheritance of resistance to bacterial blight of cotton. 4 Tanzania selections. Journal of Agricultural Science, Cambride, **72** : 41-57.

93. Innes, N. L. 1970. Breeding for bacterial blight resistance. *In* Cotton Growth in the Gezira Environment (ed. M. A. Sidding and L. C. Hughes), pp. 144-152. Sudan Agricultural Research Corporation.

94. Innes, N. L., Brown, S. J. and Walker, J. T. 1974. Genetical and Environmental variation for resistance to bacterial blight of upland cotton. Heredity, **32** : 53-72.

95. Jalali, B. L., Singh, G. and Grover, R. K. 1976. Role of phenolics in bacterial blight resistance in Cotton. Acta phytopathological Academiae Scientarium Hungaricae, **11** : 81-83.

96. Jeger, M. I. and Lyda, S. D. 1986. Epidemics of phymatotrichum root rot *(Phymatotrichum omnivorum)* in cotton : environmental correlates of final incidence and forecasting criteria. Annals of Applied Biology, **109** : 523-534.

97. Jenkin, J. N. and Parrott, W. L. 1978. Field evaluation of primitive races of *Gossypium hirsutum* L. for resistance to cercospora leaf spot and verticillium wilt. Mississipi Agric. expt. Sta. Tech. Bull. **92** : 6 pp.

98. Jin, X., Hayes, C. K. and Harman, G. E. 1990. Principles in the development of biological control systems employing *Trichoderma* species against soil borne plant pathogenic fungi. In *Trichoderma* species against soil borne plant pathogenic fungi. *In.* Symposium on Industrial Mycology. Ed. G. F. Latham, Mycological Society of Americ. Brock-Springer, New York.

99. Jones, J. E. 1961. Inheritance of resistance to fusarium wilt in upland cotton. Ph. D. thesis, Lousiana State Univ. Diss. **22** : 34-35 (Abstr).

100. Kalyanasundaram, R. and Charudattan, R. 1969. Serological studies in the genus *Fusarium,* a comparison of strains. Phytopath. Z. **64** : 28-31.

101. Kandov, A. I. and Bakhramov, K. B. 1976. Earliness, wilt resistance and yield in new cotton varieties and lines. Uzbekistan Biologija Zbernail No. **4** : 67-69.

102. Kannan, A. and Srinivasan, K. V. 1984. A new race of *Verticillium dahliae* affecting cotton. Curr. Sci. **53** : 104.

103. Kappelman, A. J. 1971a. Fusarium wilt resistance in commercial cotton varieties. Plant Disease Reporter, **55** : 896-897.

104. Kappelman, A. J. 1971b. Inheritance of resistance to Fusarium wilt in cotton. Crop Sci. **11** : 672-674.

105. Kappelman, A. J. 1982. Resistance to Fusarium wilt pathogen in currently used cotton cultivars. Plant Disease, **66** : 837-839.

106. Kasyanenko, A. G. and Alyanov, A. 1976. Donors of genes of wilt resistance. Kholpkovodstvo, **26 (12)** : 30.

107. Kasyanenko, A. G. and Portenko, L. G. 1978. Somatic recombination as source of new races in *Verticillium dahliae* on cotto. Viltoustoichivosti Khlopchatnika, Dushanbe Tajik SSR, 107-109.

108. Kasyanenko, A. G., Gorkovtswa, E. A. and Ryabova, I. M. 1978. Physiologic races of the pathogen of Verticillium wilt of cotton and their biotypes. *In.* Genet. izmenchivost vozbuditelya vilta i puti pouysh viltoustoichivosti khlopchatnika, Dushanbe, USSR, 32-46.

109. Katan, T., Dunn, M. T. and Papavizas, G. C. 1984. Genetics of fungicide resistance in *Talaromyces flavus*. Can. J. Microbiol. **30** : 1079-1087.

110. Kathpal, V., Chauhan, M. S. and Gangopadhayay, S. 1987. Chemical Control of fungal root rot complex of cotton caused by mixed inocula of *Rhizoctonia solani* an *Macrophomina phaseolina*. J. Cotton Res. & Dev. **1** : 47-60.

111. Kelkar, S./ G., Chowdhuri, S. P. and Hiremath, N. B. 1947. Inheritance of Fusarium resistance in Indian cottons. Proc. 3rd Conf. Cott. Grow. Probl. India, ICCC, 125-157.

112. Khakimov, A., Saryamaskova, R., Yuldashev, A. and Yumusov, M. 1977. Effectiveness of *Trichoderma* with lignin in controlling wilt. Khlopkovodstvo No. **7** : 42-43.

113. Kheirala, A. I. 1970. Varieties in the Gezira environment. 2. Medium staple varieties. In : Cotton Growth in the Gezira Environment (eds. M. A. Sidding and L. C. Hughes), pp. 173-178.

114. Khetmalas, M. B., Kadam, V. C., Hande, Y. K. and More, W. D. 1989. Reaction of certain deshi cotton cultivars to Fusarium wilt. J. Maha. Agric. Univ. **14** : 369-371.

115. Knight, R. L. 1946. Breeding cotton resistant to blackarm disease *(Bact. malvacearum)*. Emp. J. Exptl. Agriculture, **145** : 153-174.

116. Knight, R. L. 1954. Cotton breeding in the Sudan. Part I. Egyptian cotton. Empire Journal of Experimental Agriculture, **22** : 68-80.

117. Knight, R. L. 1957. Blackarm disease of cotton and its control. N Plant Protection Conference, 1956, p. 53, Butterworth, London.

118. Knight, R. L. 1963. The genetics of blackarm resistance. 12. Transference of resistance from *Gossypium herbaceum* to *Gossypium barbadense*. Journal of Genetics, **38** : 328-346.

119. Knight, R. L. and Clouston, T. W. 1939. The genetics of blackarm resistance. I. Factors B1 and B2. Journal of Genetics, **38** : 133-159.

120. Krishnamurthy, V. and Verma, J. P. 1974. Preliminary studies on boll rot of cotton in India. Cotton Grow. Rev. **51** : 226-229.

121. Kulkarni, G. S. and Mundkur, B. B. 1928. Mem. Dept. Agric. India, Bot. Ser. **17** : 7-27.

122. Lagiare, R. 1960. La bacteriose due contonnier *(Xanthomonas malvacearum)* (E. F. Smith) Dowson dans le monde et en Re'publique centrafricaine (Oubangui : Chari), Paris, I.C.R.T., p. 252.

123. Lakshminarasimhan, C. and Kalyanasundaram, R. 1979. Strains of *Fusarium vasinfectum* VI. Pathogencity and *in vitro* production of pathogen factor. Proc. Indian Acad Sci. B, **88** : 1-11.

124. Lanjewar, R. D., Sapka, K. N. and Shukla, V. N. 1975. Reaction of some cotton varieties against grey mildew disease. P. K. V. Res. Jnl. **3** : 93-95.

125. Last, F. T. 1960. Effect of *Xamthomonas malvacearum* (E. F. Sm) Dowson on cotton yields. Emp. Cotton Grow. Rev. **37** : 315-.

126. Lewis, J. A. 1990. Formulation and delivery systems of biocontrol agents with emphasis on fungi. *In* Rhizosphere and Plant Growth (eds. D. L. Keister and P. B. Cregan)). Dordrecht. The Netherlands Kluwer Academic Publ..

127. Lewis, J. A. and Papavizas, G. C. 1985. Characteristics of alginate pellets formulated with *Trichoderma* and *Gliocladium* and their effect on the proliferation of the fungi in soil. Plant Patho. **34** : 571-577.

128. Lewis, J. A. and Papavizas, G. C. 1987. Application of *Trichoderma* and *Gliocladium* in alginate pellets for control of *Rhizoctonia* damping off. Plant Pathol. **36** : 438-446.

129. Lewis, J. A. and Papavizas, G. C. 1988. Biocontrol of *Rhizoctonia solani* (Rs) by some novel soil fungi. Phytopathology, **78** : 862.

130. Lewis, J. A., Barksdale, T. H. and Papavizas, G. C,. 1990. Greenhouse field studies on the biological control of tomato rot caused by *Rhizoctonia solani.* Crop Prot. **9** : 8-14.

131. Lewis, J. A., Papvizas, G. C. and Lumsden, R. D. 1989. A new biocontrol formulation for *Trichoderma* and *Gliocladium.* Phytopathology, **79** : 1160.

132. Lu, C., Nium, S. and Kan, L. 1982. Determination of esterase iso-enzymes of cotton wilt pathogen *Fusarium oxysporum* f. sp. *vasinfectum* (Atk) Synder & Hansen in China. Acta Phytopathologica Sinica, **12 (2)** : 41-45.

133. Mahill, J. F. and Davis, D. D. 1978. Influence of male sterile and normal cytoplasm on the expression of bacterial blight in cotton hybrids. Crop Science, **18** : 440-443.

134. Makhbubov, M. 1977. Genetic of resistance to physiological race 1 of *Verticillium dahliae* in Gossypium hirsutum. Ed. N. I. Vavilova, Leningrad, USSR, 331.

135. Makhmudov, A. M., Sidorova, S. F. and Gubaidullina, M. G. 1978. Effect of nitrogen fertilization of cotton wilt infection. Khimiya V Selskom Khozyaistve, **16(1)** : 24-27.

136. Mandloi, K. C. and Krishna, A. 1988. Duplicate factor model for New Wilt of cotton. Paper presented in All India Cotton Production and Utilization Conference. Bombay, April 23-24., 1988. p. 35 (Abstr.)

137. Mandloi, K. C., Nagwekar, S. N., Pai, S. D., Oka, P. G., Krishna, A. and Kotwani, S. G. 1987. The new wilt and its effect on fibre quality in cotton. J. Indian Soc. Cotton Improv. **12** : 44-46.

138. Marco, G. M. and Stall, R. E. 1983. Control of bacterial spot of pepper initiated by strains of *Xanthomonas campestris* pv. *vesicatoria* that differ in sensitivity to copper. Plant Dis. **67** : 779-781.

139. Martin, W. J., Newton, L. D. and Jones, J. E. 1955. Relationship of nematodes of the general *Meloidogyne, Tylenchorhynchus, Helicotylenchus* and *Trichodorus* to the development of Fusarium wilt in cotton. Phytopathology, **45** : 349.

140. Matveev, G.G. and Sidorova, S. F. 1978. Heterogenecity of Physiological race 1 of *Verticillium dahliae* in cotton. Mikologia i Fitopatologiya, **12** : 37-39.

141. Menlikiev, M. Ya. 1962. Fusarium wilt of fine staple cotton and a study of *Fusarium oxysporum* f. sp. *vasinfectum* strs. as the causal agent of the disease in conditions of the Vaksh Valley. Izv. Akad. Nauk. Tadzh. SSR. Set. Biol. Sci. **4** : 48-59.

142. Menlikiev, M. Ya., 1977. The composition of the population of *Fusarium oxysporum* f. *vasinfectum.* Izv. AN Tadzh. SSR. Biol. N. No. **1** : 34-37.

143. Meshram, M. K. and Raj, S. 1986a. Reaction of the germplasm of *Gossypium hirsutum* Linn. to bacterial blight. Indian J. agric. Sci. **56** : 236-237.

144. Meshram, M. K. and Raj S. 1986b. Assessing yield losses due to bacterial blight in cotton. *In* Advances in Research on Plant Pathogenic Bacteria (Eds. Gnanamanickam and A. Mahadevan). Today & Tomorrow's Printers and Publishers, New Delhi.

145. Meshram, M. K. and Raj S. 1987. Development of bacterial blight on cotton in relation to genotypes, their age and weather under rainfed conditions. Indian J. Plant Prot. **15** : 19-26.

146. Meshram, M. K. and Raj S. 1988. Seed cotton yield and fibre quality as influenced by different grades of bacterial blight under rainfex conditions. Indian J. Plant Protection, **16** : 257-260.

147. Meshram, M. K. and Raj S. 1989. New Sources of resistance to bacterial blight in xotoc lines of upland cotton *Gossypium hirsutum*. Indian Journal of Agricultural Sciences, **59** : 169-170.

148. Meshram, M. K. and Raj S. 1991a. Influence of bacterial blight infection on seed cotton yield of upland cotton. J. Cotton Res. & Dev. **5** : 158-162.

149. Meshram, M. K. and Raj S. 1991b. Resistance of upland cotton *(Gossypium hirsutum)* lines to bacterial blight caused by *Xanthomonas compestris* pv. *malvacearum*. Indian Journal of Agricultural Sciences, **61**: 954-955.

150. Meshram, M. K. and Raj S. 1992. Effect of bacterial blight infection at different stages of crop growth on intensity and seed cotton yield under rainfed conditions. Indian J. Plant Protection, **20** : 54-57.

151. Michell, R. E. and Powell, W. M. 1972. Influence of *Pratylenchus brachyuras* on the incidence of Fusarum wilt in cotton. Phytopathology, **62** : 336-338.

152. Minton, E. B., Ray, L. L. and Fest, G. A. 1975. Verticillium wilt of cotton cultivars. Proc. Am. Phytopathol. Soc. **2** : 59.

153. Mirakhmedov, S. M. 1977. Results and prospects of breeding cotton for resistance to *Verticillium dahliae* Sel'skokh ozyaistvennaya Biologiya, **12** : 685-693.

154. Mirakhmedov, S. M., Senoidov, V. P., Akhmedzhanov, A. A., Khalkhodzhaev, Khidyaev, K. H., akhimova, M. and Gafurava, Ra 1975. Interspecific remote hybridization of cotton G. hirsutum L. for breeding wilt resistant early maturing varieties with high quality fibre. Uzbekskii Biologicheskii Zhurnal, No. **5** : 28 - 35.

155. Mirpulatova, N. S., Akbarov, S. and Shadmanova, N. A. 1978. The development of *Verticillium dahliae* Kleb. In the soil and infection of cotton by wilt with different sources of nitrogen nutrition. Mikologiya i Fitopathologiya, **12** : 341-343.

156. Mishra, S. P., Krishna, A. and Mandloi, K. C. 1984. Assessment of losses due to *Xanthomonas campestris* pv. *malvacearum* in cotton. Indian Phytopath. **37** : 415.

157. Mitra, M. and Kheswalla, K. F. 1935. The effect of temperature on the growth of *Fusarium vasinfectum Atk.* Proc. Indian Acad. Sci. **2** : 495-499.

158. Mohmed, H. A. 1963. Inheritance of resistance to Fusarium wilt in some Egyptian cottons. Emp. Cott. Grow. Rev. **40** : 292-295.

159. Monga, D. and Raj, S. 1992. Biological control of root rot of cotton. AICCIP Silver Jubilee Souvenir, September 17-19, 1992, abstr. 142.

160. Moniz, L. and Bhide, V. P. 1963. Root rot of cotton in Maharashtra and Gujarat states caused by Macrophomina phaseoli (Maubl) Asbhy. var *indica* N. Indfian Cotton Grow, Rev. **12** : 292-302.

161. Mukewar, P. M., Raj, S. and Singh, V. V. 1993. Screening for resistance against grey mildew in *Gossypium arboreum*. Paper presented during zonal chapter meeting of the Indian Phytopathological Society, P.K.V. Akola, January 18, 1993.

162. Mundkur, B. B. 1936. Resistance of American Cotton to Fusarium wilt in India. Proc. Indian Acad. Sci. **3** : 498-501.

163. Muratnmukhamedov, Kh. 1975. Effect of the type of soils and of the resistance of cotton on the variability of *Fusarium oxysporum* f. sp. *vasinfectum* In.: Izuch. rastitelnosti Turkmenii, Ashkhabad, USSR, Ylym, 181-186.

164. Murthy, B. R., Rao, P.S.S. and Malliah, K. V. 1985. Effect of fungicides on conidial germination and germ tube growth of *Helminthosporium gossypii* Tucker. Indian Bot. Reptr. **4 (1)** : 79-80.

165. Nafade, S. D. and Verma, J. P. 1985. Drug resistant mutants of *Xanthomonas compestris* pv. *malvacearum*. Indian Phytopath. **38** : 74-80.

166. Narula, A. M., Raj, S. and Meshram, M. K. 1989. Reaction of *Gossypium hirsutum* lines to leaf spot caused by *Alternaria macrospora* Zimm. and *Myrothecium roridum* Tode ex Fr. Cotton Development, **18 (3 & 4)** : 59-60.

167. Nayudu, M. V. 1963. *In vitro* nutrition of *Xanthomonas malvacearum*. Phytopathology, **53** : 300.

168. Nayudu, M. V. 1964. Variation in *Xanthomonas malvacearum*. Indian Cotton Drow. Rev. **18** : 350-355.

169. Nayudu, M. V. 1969. Stress physiology in plant bacterial diseases, Proc. Indian Acad. Sci. Ser. B. **69** : 146.

170. Nayudu, M. V. 1984. Physiology of infection and disease development to plant bacterial diseases, Proc. Symp. Host-Parasite Relations in Plant Pathology, Budapest, 1964, 113.

171. Nerpin, S. V., Stipanog, V. N. and Khudocrov, T. Kh. 1979. Doklady Vsezsoyumai Ordena Lenina Akademii Sel Skokhozyaistvennykh Nauk Imeni v. I. Lenina No. **11** : 13-14.

172. Novocky, A. 1972. Influence of *Xanthomonas malvacearum* infection on cotton pigment glands. Plant Dis. Reptr. **56** : 765.

173. Novruzov, M. 1976. *In* Rezultaty issledovanii malodykh uchenkh Aznikhi po Khlopkovodstvu, Kirovabad, Zzerbaijan, SSR : 121-126.

174. Orton, W. A. 1900. Wilt diseases of cotton and its control. Bull. U. S. Div. Veg. Physiol. Path. 27 pp.

175. Owen, D. F. 1967. A genetic study of bacterial blight resistance in five lines in American Upland cotton. Dissertation Abstracts, **28** : 24-B.

176. Padaganur, G. M. 1979. The seed borne nature of Alternaria macrospora Zimm. in cotton. Madras Agric. J. **66** : 325-326.

177. Padaganur, G. M. and Basavaraj, M. K. 1984. Reaction of cotton cultivars to blight caused by *Alternaria macrospora* Zimm. J. Indian Soc. Cotton Improv. **9(1)** : 36-37.

178. Padaganur, G. M. and Basavaraj, M. K. 1987. Fungicidal control of Alternaria blight of cotton in transition belt of Karnataka. Indian J. Agric. Sci. **57** : 445-447.

179. Paiziev, P. 1977. Induction of mutants in a wild cotton species by means of ethyleneimine. *In* 3 ii s" czd Vses ob-va genetikov i Selektwnerov in N. I., Vavilova, Liningrad, USSR, 389-390.

180. Papavizas, G. C. 1987. Genetic manipulation to improve the effectiveness of biocontrol fungi for plant disease control. In : Innovative Approaches to Plant Disease C ntrol. ed. I. Chet. N. Y. John Wiley & Sons, New York. pp. 193-212.

181. Papavizas G. C. and Lewis, J. A. 1988. The use of fungi in integrated control of plant diseases. *In* Fungi in Biological Control Systems, Ed. M. N. Burge. Manchester University Press, Manchester, U. K. pp. 235-253.

182. Pierce, M., cover, E. C. and Essenberg, M. 1983. Camparison of phytoaloxin concentrations near sites of bacterial blight infection in leaves of resistant andimmune cotton. Proc. Int. Congr. Plant Pathol., Abstr. **4** : 757.

183. Popov, V.I., Vunusov, M. and Tribunski, A. N. 1977. Problems of wilt control. Zashchita Rastenil No. **8** : 28-29.

184. Portenko, L. G.and Kasyanenko, A. G. 1978. Virulence and pathogenicity in induced mutants of *Verticillium dahliae.* viltoustoichivosti khlopchatnika, Dushanbe tajik, SSR,. 74-83.

185. Pullman, G. S. and De Vay J. E. 1981. Effect of soil flooding and paddy rice culture on the survival of *V. dahliae* and incidence of verticillium wilt in cotton. Phytopathology, **71** : 1285-1289.

186. Pullman, G. S., De Vay, J. E., Garber, R. H. and Weinhold, A. R. 1981. Soil solarization : effects on verticillium wilt of cotton and soil borne populations of *V. dahliae. Puthium* spp. *R. solani* and *Thielaviopsis basicola.* Phytopathology, **71** : 954-959.

187. Raj, S. 1992a. Changing cotton disease scenario in India. Proc. Natl. Seminar on Changing Pest Scenario. Jan. 31-Feb 1, 1992, Hyderabad.

188. Raj, S. 1992b. Achievements in Plant Pathology. In : Achievements of AICCIP (1967-1992). Eds. A. K. Basu and S. S. Narayanan, pp. 152-176.

189. Raj, S. and Butter, N. S. 1987. Integrated insect pest and disease management in U. S. S. R. Report of a study tour under Indo-USSR Protocol, July 4-24, 1987..

190. Raj, S. and Taneja, N. K. 1987. Diseases of cotton and their management. *In* Plant Protection in Field Crops (Eds. M. Veerbhadra Rao and N. Sitanathan) Plant Protection Association of India, 301-317.

191. Raj, S. and Verma, J. P. 1988. Diseases of cotton in India and their management, Rev. Trop. Pl. Path. **5**: 207-254.

192. Raj, S., Chakrabarty, P. K. and Meshram, M. K. 1987. The New Wilt of Cotton-remains a challenge. Paper presented in Beinnial Workshop, AICCIP, Hyderabad.

193. Raj, S., Chakrabarty, P. K. and Meshram, M. K. 1990. Epidemiology of and losses due to new wilt of cotton Indian J. Plant Pathol. **8** : 93-77.

194. Raj, S. Chakrabarty, P. K., Meshram, M. K. and Bambawale, O. M. 1991a. Investigations on new wilt of cotton. Cot. et. Fib. Trop. **46** : 207-215.

195. Raj, S., Meshram, M. K., Ajmera, B. D., Narula, A. M. and Chauhan, M. S. 1991b. Outbreak of bacterial blight of cotton in Rajasthan and Haryana-causes and future strategies for its management. J. Cotton Res. & Dev. **5** : 35-42.

196. Raj, S., Meshram M. K. and Mukewar, P. M. 1984. Some observations on new wilt of cotton. Proc. Group Discussion on New wilt in cotton pp. 14-19.

197. Raj, S., Meshram M. K. , Narula, A. M. and Taneja, N. K. 1988. Present status of races of *Xanthomonas compestris* pv. *malvacearum* in India and their utilization for durable resistance in Gossypium hirsutum. Paper presented in All India Cotton Production and Utilisation conference, Bombay, April 23-24, 1988. pp. 35 (Abstract).

198. Rajpurohit, T. S. 1983. Control of root rot of cotton caused by *Rhizoctonia bataticola* (Taub) Butler, Madras Agric. J. **70** : 751-752.

199. Rakhimzhanov, A. G. 1977. Resistance of varietal population of cotton to *Verticillium dahliae.* Mirov, rastitelth resursy V. Srendnei Azii No. **3** : 128-131.

200. Rammoorthy, K. 1990. Economics of cotton production in India. *In* Cotton Scenario in India. A souvenir, International Hybrid Cotton Consolation, CICR, Nagpur pp. 103-108.

202. Ramapandu, S. Sitaramaraju, K., Subbarao, K. and Prasadrao, M. P. 1979. Screening of cotton germplasm against bacterial blight caused by *Xanthomonas malvacearum.* Indian Phytopathol. **32** : 486.

203. Ramasami, R. and Shanmugam, N. 1976. Effect of nutrients on the incidence of Rhizoctonia seedling disease of cotton. Indian Phytopath. **29** : 465-466.

204. Ramasami, R. and Shanmugam, N. 1977. Possible roe of sugars, phenols and gossypol in cotton seedling and disease resistance. Indian J. Mycol. & Plant Path. **7**: 52-54.

205. Ramasami, R., Shanmugham, N. and Kandaswamy, T. K. 1976. Studies on the control of seedling disease of cotton caused by *Rhizoctonia solani* (Taub). Butl. Madras Agric. J. **63** : 389-392.

206. Ranney, C. D., Hursh, J. S. and Newton, O. H. 1971. Effects of bottom defoliation on microclimate and the reduction of boll rot of cotton. Agran. J. **63** : 259-263.

207. Roberts, G. L. and Staten, G. 1972. Heritability of Verticillium wilt tolerance in crosses of American upland cotton. Crop Sci. **12** : 63-64.

208. Roux, J. B. 1977. Selection de varieties resistance a certains maladies du cottonier. Seminaire Cotonnier & Akp. (October, 1977). I.R.C.T., Paris.

209. Ruano, O. and Mohan, S. K. 1982. A new race of *Xanthomonas compestris* pv. *malvacearum* (Smith) Dye in Parana State. Fitopatologia Brasileria, **7** : 439-442.

210. Sadykov, A. S. 1980. More attention to the control of wilt. Zaschita Restenii No. **6** : 24-26.

211. Salem, A. A. M. 1980. Effect of *Meloidogyne incornita* and *Fusarium oxysporum* f. sp. *vasinfectum* on cotton. Egypt J. Phyutopath. **12** : 27-30.

212. Santhanam, V. 1976. Breeding Procedures for cotton. Technical Bulletin (Agriculture) Series 10 : 10-12, Indian Council of Agricultural Research, New Delhi.

213. Sappenfied, W. P., Baldwin, C.H., Wrather, J. A. and Bugbee, W M. 1980. Breeding multiple disease resistant cotton for the North Delta. Proc. Beltwide Cotton Prod. Res. Conf. National Cotton Council of America, Memphis, TN pp. 280-283.

214. Sarojini, T. S. and Yogeshwari, L. 1947. Aeration affecting growth and sporulation of some soil fusaria in liquid cultures. Proc. Acad. Sci. Sect. B, **26** : 69-76.

215. Sattarov, B.K. 1981. *Gossypium barbadense* L. Specimens resistant to *Verticillium dahliae* Kleb. Byulleten Vsesoyuznogo Nauchno-Isledovatelskogo Instituta Rastenievodstvaimeni N. I., Vavilova No. **115** : 35-37.

216. Savov, S. G. 1978. Effect of some agrotechnical measures on verticillium wilt of cotton. Rasteniev dni Nauki, **15(4)** : 122-127.

217. Sboeva, Zh. N., Nematov, G. N. and Yumusov, M. R. 1977. Antiwilt effectiveness of benzimidazole derivatives and their behaviour in plants and soil, Byulleten Vsesoyuznogo Nauchno Isledovatelskogo Institute Zashchity Rastenii No. **40** : 23-25.

218. Schanthorst, W. C. 1981. Life cycle and epidemiology of *Verticillium*. *In* Fungal Wilt Diseases of Plants. (Eds. M. E. Mace, A. A. Bell and C. H. Beckman). Academic Press, New York, 81-111.

219. Sezigin, E., Karcilioglu, A. and Ycmiscioglu, U. 1982. Investigations on the effects of some cultural applications and antagonistic fungi on *R. solani* and *V. dahliae* in the Aegean region. J. Turkish Phytopathology, **11(3)** : 79-91.

220. Shapira, R. Ordentlich, A., Chet, I. and Oppenheim, A. B. 1989. Control of plant diseases by chitinase expressed from cloned DNA in *Escherichia coli.* Phytopathology, **79** : 1246-1249.

221. Sharma, J. R. and Bedi, P.S. 1986. Seed treatment to control Fusarium wilt of cotton in Punjab. Pesticides, **20(2)** : 19-20.

222. Sharma, J. R. and Bedi Sharma, K. B. and Sharma, N. D. 1976. Chemical control of root rot cotton. Indian J. Mycol. & Pl. Pathol. **6** : 190-191.

223. Shivankar, S. K. and Wangikar, P. D. 1992. Estimation of crop losses due to grey mildew disease of cotton caused by *Ramularia areola.* Indian Phytopath. **45** : 74-76.

224. Sidding, M. A. 1973. Barakat, a new long staple cotton variety in the Sudan. Cotton Growing Review, **50**: 307-315.

225. Siddiqui, M. R. and Sahni, V. M. 1962. Problem of Grey Mildew (Dahiya) in Vidarbha. Indian Cotton Grow. Rev. **16** : 343-345.

226. Sidorova, S. F. and Akmuradov, B. 1980. Intraspecific differentiation of the pathogen of Fusarium wilt of cotton in Turknenia. Miklogia i Fitopatologiya, **14** : 511-521.

227. Sidorova, S. F. and Akmuradov, B. 1981. Virulence of Turmkenian strains of the pathogen of Fusarium wilt of cotton to foreign differential varieties. Mikologia i Fitopatologia, **15** : 314-317.

228. Singh, I. M. and Srivasta, M. P. 1980. The chemical control of Alternaria leaf blight of cotton in Maharashtra State. J. Maha. Agric. Univ. **5** : 87-88.

229. Singh, I. M. and Srivastava, M. P. 1982. Chemical control of boll rot complex of cotton. Indian Phytopath. **35** : 138-140.

230. Singh, Munshi, Verma, J. P,., Agarwal, R. A., Singh, V. P., Singh, R. P. and Katyar, K. N. 1973. SABRAO General Congr. New Delhi, **2** : 190.

231. Singh, V. V., Meshram, M. K., Raj, S. and Narayanan, S. S. 1988. Evolution of upland cotton *(Gossypium hirsutum)* lines with multiracial resistance to bacterial blight caused by *Xanthomonas compestris* pv. *malacearum.* Indian J. Agric. Sci. **58** : 255-258.

232. Sinha, P. P. and Verma, J. P. 1983. Role of Phylloplane bacteria in bacterial blight disease of cotton. International J. Tropical Pl. Dis. **1 (2)** : 125-133.

233. Sippell, D. W., Khalifa, H., El Hilo Omer, M. and Bindra, O. S. 1986. A method of accumulate horizontal resistance in *Gossypium hirsutum* to *Xanthomonas malvacearum.* Int. Congr. Plant Pathol., Abstr. **4** : 830.

234. Smith, A. L. and Dick, J.B. 1953. Anthracnose and some blights. Plant Diseases USDA, Yearbook of Agriculture, 303-311.

235. Smith, A. L. and Dick,, J. B. 1960. Inheritance of resistance to Fusarium wilt in Upland and Sea Island cottons as complicated by nematodes under field conditions. Phytopathology, **50** : 44-48.

236. Sohi, H. S. and Kalra, J. S. 1983. *In vitro* evaluation of different fungicides against *Myrothecium roridum* Pesticides, **16** : 20-21.

237. Srinivasan, K. V. 1971. Verticillium wilt of cotton-some observations on resistance, Proc. Intern. Symp. on pathological wilting in plants, Madras, 1971.

238. Srinivasan, K. V. 1975. Indian Farming January.

239. Srinivasan, K. V. and Kannan, A. 1985. Aspects of the seed pathology of cotton. ZJ. Indian Soc. Cotton Improv. **10** : 1-7.

240. Srinivasan, K. V. and Santhanam, V. 1973. Breeding for resistance to verticillium wilt. Proc. SABRAO General Congr. New Delhi, Indian J. Gen. Pl. Breed. **34** : 876-882.

241. Srinivasan, K. V., Krishnamurthy, R. and Kannan, A. 1976. Progress Report of Verticillium wilt control. J. Indian Soc. Cotton Improv. **1** : 73-83.

242. Srivastava, M. P. and Singh, A. 1973. Myrothecium disease of cotton. HAU J. Res. **3** : 221-223.

243. Starr, J. L. and Martyn, R. D. 1991. Reaction of cotton cultivars to Fusarium wilt and root knot nematodes. Nematropica, **21** : 51-58.

244. Stewart, J. Mc D. 1992. Biotechnology of cotton achievements and perspectives. ICAC Review articles on Cotton Production Research No. 3.

245. Subramanian, C. V. 1950. Soil conditions and wilt diseases in plants with special reference to *Fusarium vasinfectum* on cotton. Proc. Indian Acad. Sci. Sect. B, **31** : 67-102.

246. Sulaiman, M. and Patil, B. C. 1966. Existence of physiological races of *Macrophomina phaseoli* causing root rot of cotton. Beitr. trop. Subtrop. Landwirtsch. Tropenveterinarmedizin, **4** : 291-298.

247. Sulochna, C. B. 1952 a. Soil conditions and root disease VI. Germination of conidia of *Fusarium vasinfectum* in micro-elements amended soils. Proc. Indian Acad. Sci. Sect. B, **26** : 229-233.

248. Sulochna, C. B. 1952 b. Soil conditions and root diseases, Proc. Indian Acad. Sect. B, **36** : 234-242.

249. Sultanov, B., Urunov, I S. and Bicherov, E. V. 1977. Complex fertilizers with microelements and cotton wilt. Tr. Srednaez N II Zaschity Rast. No. **11** : 74-76.

250. Taneja, N. K. and Raj, S. 1988. Seed borne infection of bacterial blight *(Xanthomonas compestris* pv. *malvacearum)* in cotton *(Gossypium* species) Indian J. Agric. Sci. **58** : 323-324.

251. Taneja, N. K. and Raj, S. 1989. Studies on boll rot complex in cotton. Proc. Natl. Seminar on Futuristic Approaches in Cotton Improvement, HAU, Hisar, Feb. 24-25, 1989.

252. Taneja, N. K. and Raj, S. 1990. Role of contaminated seed and infected plant debris on the development of Myrothecium leaf spot in cotton. J. cotton Res. & Dev. **4** : 70-72.

253. Taneja, N. K. and Raj, S. and Seth. P. K. 1989 a. Estimation of losses due to Myrothecium leaf spot of cotton. J. cotton Res. & Devt. **3** : 46-48.

254. Taneja, N. K. and Raj, S. and Seth. P. K. 1989b. Chemical control of Myrothecium leaf spot disease of cotton. J. Cotton Res. & Dev. **3** : 169-173.

255. Taneja, N. K. and Raj, S. and Seth. P. K. 1990a. Estimation of amino acids in resistant and susceptible varieties of cotton to Myrothecium leaf spot. Indian Phytopathology, **43** : 101-102.

256. Taneja, N. K. and Raj, S. and Seth. P. K. 1990b. Existence of pathogypes in *Myrothecium roridum.* Indian Phytopath. **43** : 464-466.

257. Taneja, N. K. and Raj, S. and Seth. P. K.1991. Reaction of the germplasm of *Gossypium* species to *Myrothecium roridum.* Indian Phytopath. **44** : 404.

258. Tellyaev, R. and Ligai, A. 1987. Use of anti-wilt preparations jointly with defoliation. Khlopkovodstvo No. **8** : 31.

259. Ter-Avanesyan, D. V. 1977. Breeding wilt resistant cotton varieties. Biologicheskaya No. **5** : 776-780.

260. Ter-Avanesayn, D. V., Babaev, F A and Aldinov, F. A. 1977. Wilt resistance of cotton from the USA under Azerbeijan conditions. Vestnik Selsko-Khozaistvennoi Nauk No. **2** : 39-41.

261. Thakar, N. A. 1984. Influence of moisture on root rot of cotton. Madras Agric. J. **71** : 529-630.

262. Tian, F. 1981. Studies on specialized biotypes of cotton Fusarium wilt fungus in Xijiang, Acta Phytopathologica Sinica, **11** : 27-30.

263. Tillaev Kh. T. and Velitskaya, I. S. 1981. Methods of producing different forms of the biopreparation trichodermin and their effectiveness against cotton wilt. Mikologiya i Fitopatologiya, **15 (4)** : 317-321.

264. Tiwari, A. and Shroff, V. N. 1982. Differential reaction of cotton genotypes to root rot. Indian Phytopath. **35** : 514-515.

265. Tribunski, A. N., Kim, A. V. and Butkova, V. Ya. 1976.. Brief results of research into the genetics and breeding of wilt-resistant cotton *In.* Sb. Nakch. tr. Andizhan. fil. VN II Khlopkovodstva, Tashkent, Uzebek SSR, 3-21.

266. Upal, B. N. 1948. Diseases of Cotton in India. I.C.C.C., Bombay.

267. Uppal, U. N., Kulkarni, Y. S. and Ranadive, J. D. 141. Further Studies in breeding for wilt resistance in cotton. I. Isolation of wilt-resistant types. II. A preliminary note on the genetics of wilt resistance in Indian cottons. 2nd Conf. Sci. Res. Workers Cott, India.

268. Urunov, I., and Sultanov, B. 1978. Copper containing ammophos reduces wilt infection of cotton. Khlopkovodstvo No. **4** : 30.

269. Vargas, R. and Ramirez, C. 1983. Biological control of damping off caused by *Rhizoctonia solani* on cotton. Agromia costarricense, **7(1/2)** : 73-75.

270. Vasudeva, R. S. 1960. Diseases. In : Cotton in India - a monograph Vol. II. ICCC, 164-216.

271. Venere, R. J. 1980. Role of peroxidase in cotton resistant to bacterial blight. Plant Science Letters, **20** : 47-56.

272. Verhalen, L. M., Brinkerhoff, L. A., Fun, K. C. and Morison, W. C. 1971. 1971. A quantitative genetic study of *Verticillium* wilt tolerance in crosses of American Upland Cotton. Crop. Sci. **11** : 407-412.

273. Verma, J. P. 1986. Bacterial blight of cotton. CRC Press. Florida, USA, 278 pp.

274. Verma, J. P. and Singh, R. P. 1970. Two new races of *Xanthomonas malvacearum,* the cause of blackarm of cotton. Cotton Growing Review, **47** : 203-205.

275. Verma, J. P. and Singh, R. P. 1974a. Recent Studies on the bacterial disease of fibre and oilseed crops in India, 134-145, *In* Current Trends in Plant Pathology. Eds. Raychaudhury, S. P. and Verma, J. P., Univ. of Lucknow, India, 342.

276. Verma, J. P. and Singh, R. P. 1974b. Studies on the nutrition of *Xanthomonas malvacearum* and its relationship to the free amino acids, organic acids and sugars of resistant and susceptible cotton cultivars. Acta Phytopathological Academiae Scientiarum Hungaricae, **9** : 55-63.

277. Verma, J. P. Borkar, S. G. and Singh, R. P. 1981. Transmission of *Xanthomonas compestris* pv. *malvacearum* (Smith) Dye through red cotton but. An. Agric. Res. **2** : 57.

278. Verma. J. P., Chowdhury, H. D. and Singh, R. P. 1978. Interaction between phylloplane bacteria and *Xanthomonas malvacearum.* Proc Int. Conf. Plant Pathogenic bacteria, **4** : 795.

279. Vora, S. and Chand, J. N. 1971. Relation of free amino acids to the development of angular leaf spot (black arm) of cotton. Phytopathol. Z. **70** : 177.

280. Vunusov, S. 1978. *In.* Sintz i primenenic novykh khinicheskikh preparatov protiv vilta khlopchatnika, Tashkent, Uzbek, SSR 10-12.

281. Wate, J. O. and Young, V. H. 1934. Control of cotton wilt and "rust". Ark. Agr. Crop. Sta. Bull. No. 308.

282. Watkins, G. M. (Ed.) 1981. Compendium of Cotton Diseases. Pub. Am. Phytopathol. Soc. 87 pp.

283. Wilhelm, S., Sagen, J. E. and Tritz, Helga 1969. Dominance of Sea Island resistance to Verticillium wilt in F1 Progenies of upland X Sea Island Crosses, Proc. Beltwide Cot. Prodn. Res. Conf. 31 p. (Abstr.).

284. Woodward, J. 1985. Immobilized Cells and Enzymes, IRL Press, Washington D. C. 178 pp.

285. Yakutkin, V. I. 1978 a. Physiological race 0 of *Verticillium dahliae* on cotton. Mikologiya i Fitopatologiya **12** : 48-50.

286. Yakutkin, V. I. 1978 b. Biotypes as structural units of physiological races of *Verticillium dahliae* Kleb. Mikologiya i Fitopatologiya, **12** : 145-147.

287. Yerovenko, A., Kadyrkhozhaev, P. 1975. *In* Sinetz i Primenenic novykh khimcheskikh Preparatov protiv vilta khlopchatnika Tashkent, Uzbek, SSR : 81-82.

288. Yuldashev, A., Gafiullina, G., Sultanov, K. and Malanina, E. 1976. Aspects of the genetics and breeding of wilt resistant cotton. In : Vopr. genet. Selektsii. i. semenovodsta khlopchatnika i. lyutserny, 13, Tashkent Uzbek, SSR 31-41.

289. Ziegler, M. G. and Essenberg, M. 1978. Evaluationof the role of free amino acid imbalance in the inhibition of *Xanthomonas malvacearum* in immune cotton leaves. *In* 3rd International Congress of Plant Pathology, Munchen, 16-23 August, 1978.

Fig. 1 : Cotton growing zones and important diseases in India

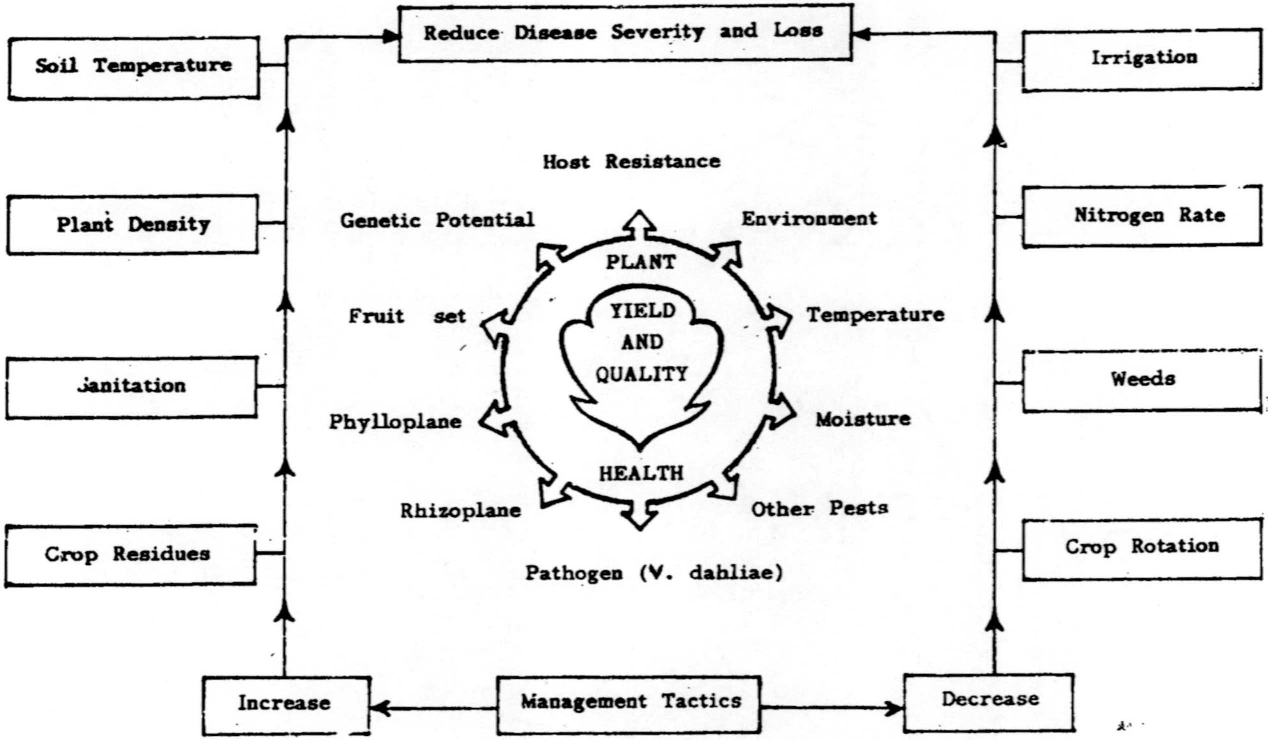

Fig. 2 : Conceptual model of an integrated management system for controlling *Verticillium* wilt of cotton. After El-Zik (55)

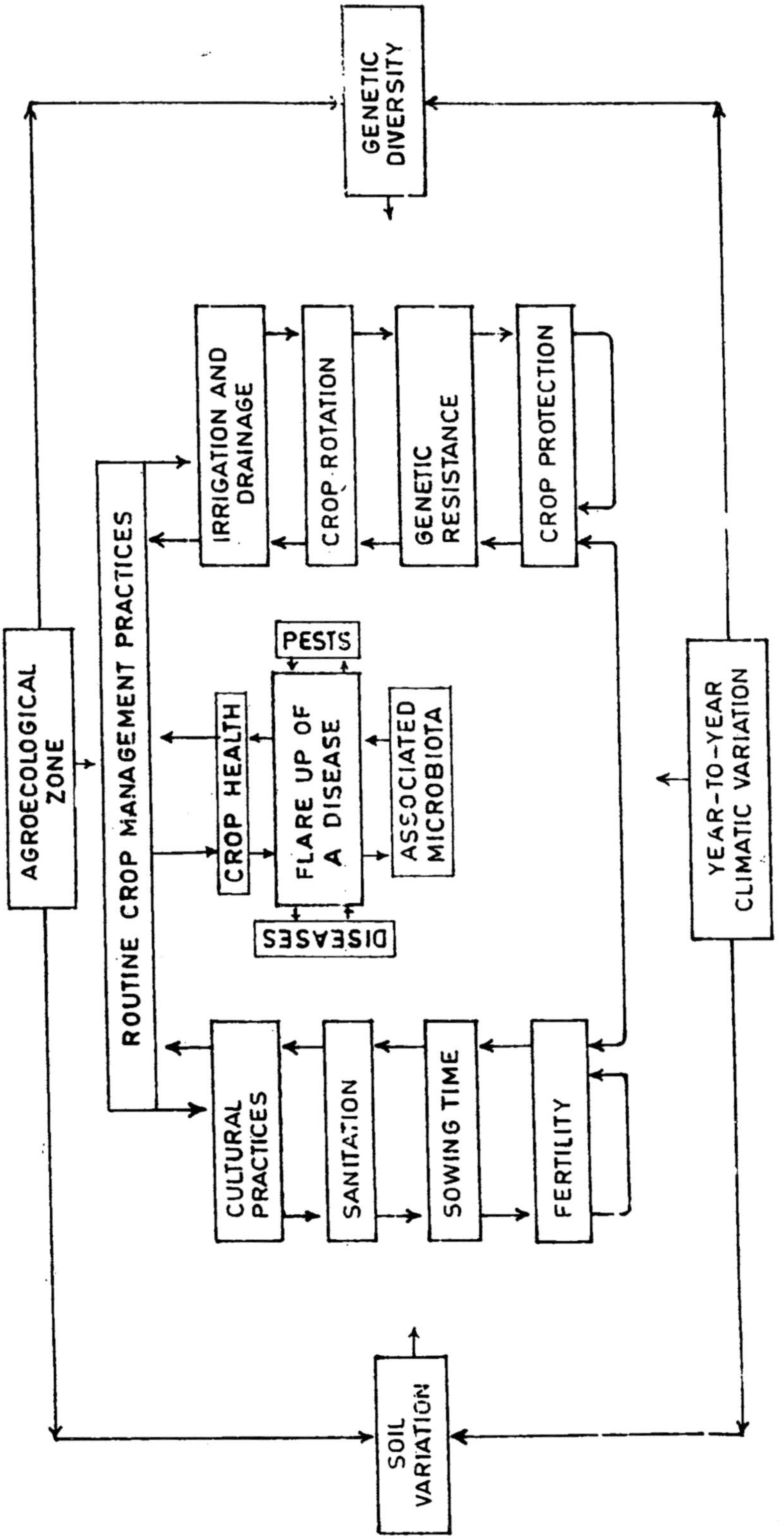

Fig. 3 : A conceptual model of holistic influence of factors beyond immediate control (outer ring) and within control (inner ring : crop management practices) on flare up of a disease

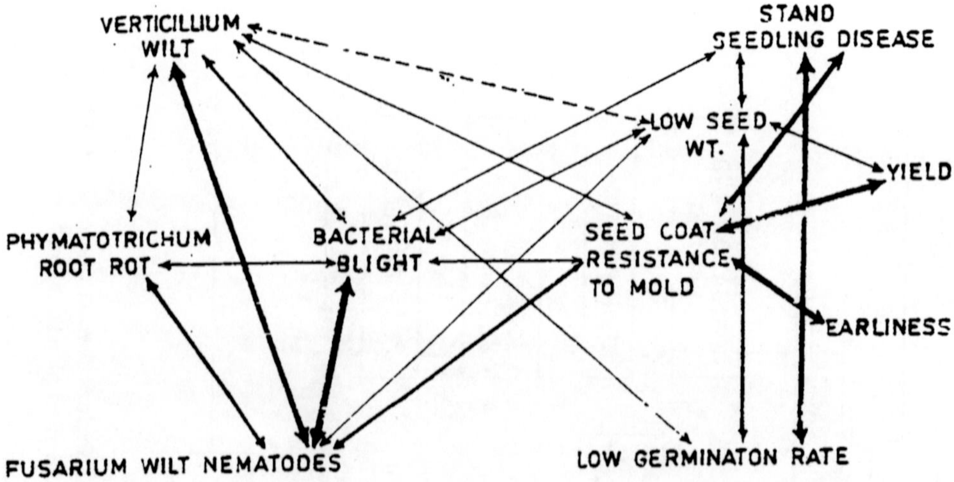

Fig. 4 : Genetic interrelationship pathways among genes for disease resistance and performance traits in the MAR system. The width of the pathway line indicates the relative strength of the relationship. Favourable gains for all traits in the system are aided by selecting for seed coat resistance to mold, low germination rate and bacterial blight resistance in the laboratory and greenhouse. After Bird (25)

22 Apple Scab and Powdery Mildew of Apple and their Management

G.K. Gupta

CONTENTS

Integrated Pest and Disease Management
Rajeev K. Upadhyay, K.G. Mukerji, B.P. Chamola and O.P. Dubey (eds.)
APH Publishing Corporation, New Delhi-110002, India

ABSTRACT

Apple scab and powdery mildew are the major limiting factors in harvesting the full genetic potential of the appole crop. These two diseases not only reduce the yield but also affect quality of the fruit. Disease management utilizing forecasting system developed for apple scab disease and cultural operations to reduce inoculum load by spraying urea etc. on leaf litter has been suggested. Besides, safer and more effective chemical molecules recently registered to combat the disease has been recommended as a last resort of control operation.

1. APPLE SCAB *(VENTURIA INAEQUALIS)*

1.1 History and distribution in India

In India, apple cultivation had been started by European settlers and missionaries as early as 1870 (52). However, rapid expansion of this industry into commercial organisation came only during the last three to four decades. Apple scab was first noticed in the Kashmir valley in 1930 on an indigenous cultivar "Ambri" (47). For nearly next four decades, no notice was taken of this disease as its spread was sporadic and of not much consequence. After 1960, Red Delicious was gaining importance and a large area had come under this cultivar. In 1973 scab affected this cultivar badly. No doubt other cultivars were also affected. This was an odd experience for the growers as scab had hit hard at a commodity which was Khashmir's pride and had enjoyed a high place in its economy and trade. The unholesome thing was that the prices of apple afflicted by scab slumped heavily. This problem promptly engaged the attention of the Indian Council of Agricultural Research, and the Directorate of Plant Protection, Quarantine and Storage, Government of India. As a result of organisation of various seminars and group discussions held at several places (New Delhi, March, 1974; Srinagar, October, 1974; New Delhi, April, 1975 etc.), apple scab was considered a disease of national importance and hence on a priority list. These media took paid to pool down all the relevant literature on various aspects of the disease, mainly from foreign countries, and suggested steps to counter the menace through the application of long term and short term measures, including spray schedules and their cost economics. Subsequent investigations on this disease at the Indian Agriculture Research Institute from 1974 to 1976 (28, 29, 30) provided clues to the distribution of apple scab in the valley, the response of various cultivars to the disease, the pathogen and its variability, role of certain environmental factors in disease development, sources of inoculum, *in-vitro* efficacy of fungicides against scab fungus, and exploration of methods for inhibiting pseudothecial formation and spore release so as to break the life cycle of scab fungus.

During 1977, scab was also detected in Himachal Pradesh in an orchard in Kotkhai tehsil of Shimla district and in a few localized pockets of Kullu, Mandi and Chamba districts (13). In 1983, there was an epidemic of the disease whereby almost entire apple growing area had been affected by the disease (36). In subsequent years, apple scab has been recorded in other apple growing areas of the country which include UP hills, Sikkim, Arunachal Pradesh and Nilgiri hills of Tamil Nadu (24). It seems that the scab fungus has got introduced into these areas through the apple-planting material.

1.2 Economic importance and losses accured

Apple scab is a disease which devalues the fruit quality, weakens the tree by mid-season defoliation and failure of fruit bud formation in the tree, and involves heavy expenses to combat it.

In favourable wet season, early scab infection of fruit pedicels leads to no fruit setting. Leaves with infection fall prematurely and weaken the tree considerable as there may be complete crop failure and no fruit bud formation for the coming season. Year to year, premature leaf fall in young, non-bearing trees leads to stunted and reduced growth of the tree.

Reduced quality of fruit due to scab attack provides very poor price or such fruits are not marketed. Another burden scab creates is the application of fungicides repeatedly at short intervals which escalates the cost of production.

First epidemic of apple scab had occurred in Kashmir valley in 1973 which had resulted in a loss of Rs. 54 lakhs to the apple crop in a single season. In 1982 and 1983, this disease had hit hard at the apple crop in Himachal Pradesh, and all the potential apple growing areas (40,000 out of 47,354 ha) had been engulfed by the disease. Out of 10 to 50% scabbed fruit, nearly 10% of the apples (30,000 of 300,000 metric tons of production) were unfit for market and were destroyed on the spot. The state suffered a loss of Rs. 1.50 crores (17).

1.3 Symptoms

Scab symptoms commonly appear on leaves and fruits, and rarely on twigs, petioles, pedicels and sepals of the flower also exhibit symptoms.

New and tender leaves in spring and summer are more susceptible to infection. As the fruit buds open, the lower surface of the tender leaves may develop scab lesions but at later stages, both sides of the leafy may be infected, light brown to olive green lesions turning mousy black with age develop on the leaf lamina. They may be feathery and indefinite later on developing distinct limits. Lesions are dark due to sporulation but in resistant cultivars such as Versified (**Maharaji**) and Cox's Orange Pippin (**Kesri**) these are light in colour due to scanty or no sporulation and are chlorotic or reddish-brown (28). As the leaf becomes old, the neighbourhood of the scab lesion thicken and result in an upward bulging of the affected area with a corresponding cup like depression on the lower surface. Severe leaf infection leads to curling, dwarfing and distortion of the leaf, and premature yellowing and defoliation. In very humid weather, the scab fungus covers the large portion of leaf lamina with black powder of growth known as sheet scab. Infection of petioles and pedicels provides similar scab lesions and leads to premature defoliation and fall of fruits, respectively.

On young fruits, scab lesions develop around the calyx end but in little developed fruits, these may appear any where on the fruit epicarp. Initially scab lesions are small, circular and olive coloured but with age, become dark, brown and corky. Young lesions have grey appearance as these remain covered partially by the fruit cuticle. In older lesions, rupturing of cuticle exposes black, powdery mass in the centre. Appearance of fruit is distorted by large, dark and wasty lesions, severe early infection may result in misshapen, knotty fruits. As the fruit enlarges deep cracks develop in the corky scabbed areas and serve as avenues for other fruit rotting organisms to cause fruit rot, under moist conditions in the orchard, pink fungus *Trichothecium roseum* covers the scab lesions and may result in fruit rot.

Infection of twig is restricted to the Kashmir valley (27) Scab lesions develop in the form of small, raised, reddish - brown blisters on terminal growth as well as on 1 to 3 years old twigs. In spring, the blisters rupture the upper surface, with silvery-grey peeling and an olive green to black powdery mass underneath. The blisters on coalescion give a scurfy appearance to the bark and wood.

Long wetting periods occurring in late summer months near to fruit harvest may result in fruit infection commonly referred to as pin point scab. Small, rough, black and circular lesions may be apparent on the fruits at harvest or in the storage after some time. Apparently healthy fruits picked from infected trees exhibit scab lesions even in cold storage (1°C) and may not last long. The lesions on the fruit grow larger, and the areas surrounding these lesions show depression, and lead to fruit shrinkage (34), the affected fruits develop rot and become source of infection to the healthy fruits in storage.

1.4 Causal organism

Apple scab is caused by the fungus *Venturia inaequalis* (Cke) Winter with the imperfect stage as *Spilocaea pomi* Fr. The fungus produces pseudothecia (ascocarps) in a stroma in overwintered leaves on the orchard floor. The pseudothecia are negatively geotropic, separate, dark brown to black, and spherical. They have a short beak with a distinct ostiole and single-celled bristles at the apex. The centrum is pseudoparenchymatous with pseudoparaphyses.

Asci about 50-150 per pseudothecium are fasciculate, cylindrical, short-stipitate and eight spored. The ascus wall is thin and bitunicate. Ascospores are yellowish green to tan and unequally two celled, with the upper cell shorter and wider than the lower cell.

Conidiophores arise from the ribbon-like strands or from the more compact stroma as simple, cylindrical structures, pale to mild olivaccous brown and sometimes swoollen at the base. Conidia are produced sequentially and leave abscission ridges on the conidiophore. They are obypyriform to obelavate, pale to mild olivaceous brown, smooth, with a truncate base and somewhat pointed apex, Conidia are produced on the surface of leaf and fruit lesions throughout the growing season. Conidial pustules on the fruit show the dark olive conidia surrounded by a fringe of silvery white, torn cuticle.

The fungus is bipolar heterothallic. Fungal strains of opposite mating types produce ascogonia and antheridia, which following plasmogamy, give rise to ascogenous hyphae. Ascus formation occurs following the formation of croziers in the usual manner. The stroma and the pseudothecia develop concurrently with the ascogenous hyphae.

Among five physiologic races identified so far, the Indian isolates of the scab fungus seem to belong to Race 1, which is commonly encountered in the United States of America and various other countries (23). Besides this, the fungus consists of at least two strains, the virulent one infecting Red Delicious and the less virulent one infecting Ambri cultivar of apple in the Kashmir valley. This was based on morphological and biological characteristics (29). Another hot-specific strain on *Contoneaster affinis* var. *bacillaris* (Wallich ex Lindl.) Schreid (14) and on *Cotoneaster aitchinsonii* Schneid (50), is known. The fungus in its *Spilocaea* state has also been recorded on a crab apple, *Docynia indica* (6), which seems to be another host-specific strain.

1.5 Epidemiology

Perennation of apple scab fungus-The scab fungus perennates in the perfect state, as pseudothecia, in fallen leaves. The scab fungus enters the saprophytic phase soon after leaf fall in autumn. Most of the pseudothecium development takes place within 30 days of leaf fall mainly in leaves which fall in September, October and November as the leaves of the earlier months do not survive and get decomposed quickly (31). Cool and moist climatic conditions either due to rain or plenty of snow as are available at higher elevations and in shady portions of the orchard are congenial for the development of pseudothecia in the overwintering leaves. When moisture is not a limiting factor, temperature has a major influence on the development of pseudothecia, Fungus starts producing initials of pseudothecia at low temperature of 4-8°C but maturity is attained at optimum temperature of 15°C. Prolonged dry weather conditions in the winter and spring do not allow normal development of pseudothecia and even the ascospore maturity gets delayed by a month or so.

Pseudothecia start maturing at silver tip stage of apple tree in spring and continue to mature till early summer (15). Each pseudothecium contains 100 to 150 asci, each with 8 ascospores, thereby yielding 800-1200 ascospores (51). On a single leaf there may be a production of nearly 2000 pseudothecia with a total load of 2 million ascospores. The ascospores on maturity get discharged from the pseudothecia upon wetting of over-wintering leaves on the orchard floor and are then wind dispersed within the orchard, maturation and discharge of ascospores usually last 4-9 weeks (12, 18). In addition to ascosporic infection in spring, conidia produced in scabbed lesions on 1-3 years old twigs may cause primary infection.

Infection of the host plant-The spore (ascospore or conidia) on germination produces the germ tube with disc like appressorium at the tip, and peg like structure emerging out of appressorium punctures the cuticle of plant tissue to start primary infection depending on the availability of Mills infection periods (17). After infection has occurred, the fungus grows under the cuticle and after 9-17 days comes out and produces olive-brown lesions. Each lesion produces 50,000-100,000 conidia, and these being washed by rain or sometime even detached by wind (40) serve to spread the disease to uninfected tissues throughout the summer and early autumn. Humidity and temperature conditions favouring conidia production in scab lesions are 60-100% R.H. and 4-28°C temperature with optima being 90% and 16-20°C respectively (40).

Spore production on scab lesions stop generally within 30-36 days, and lesions become necrotic or bronzed and reddish brown. In summer months - particularly when weather is congenial for scab development, application of 1% urea increases the susceptibility of plant tissues to scab pathogen, and needs to be avoided (41). In monsoon months, secondary infection to fruits can occur but symptoms may not appear till stored for some time (34). In valley areas, dew is of sufficient length and in absence of rain, may result in initiation and build up of new infection (36). Such late infection on leaves may result in production of large number of pseudothecia to create trouble in the coming season.

1.6 Disease prediction

Three main criteria, viz, quantity and relative maturation of pseudothecia as source of primary inoculum, tree phenology, and occurrence of infection periods form the basis of disease prediction in Himachal Pradesh (23). Such prediction provides timely warning to the growers for initiation of spray and adoption of nee-based spray materials.

Appraisal of primary inoculum is made by studying either the (i) ascospore discharge (productivity) based on the number of ascospores/cm^2 over-wintered leaf area, and/or (ii) ascospore dose as the number of spores/volume air (19). In studying ascospore discharge, examination of crushed pseudothecia picked up from overwintered apple leaves is commonly used to categorize ascospores/asci maturity level. Usually examination of asci of 25 pseudothecia randomly picked up from the overwintered leaves and carefully crushed so as to spread the asci in a thin layer, and record the total number of asci per pseudothecium, Rotored sampler is a convenient trap for measuring ascospore dose in the air. The 'I' rods of this sampler are easy to remove and require less time to examine than the melinex tape of Burkard trap (33, 35).

Ample evidence has been accumulated about the maturity of ascospores coinciding with the break of dormancy of apple trees (17, 22). Thus it is clear that temperature conditions which accelerate apple host development also accelerate ascospore maturity. Much reliance is given to beginning the spray programme at silver tip to 1/4" green tip stage of apple tree and continuing fungicidal spray at short interval until the primary scab season is over (17). Empirical model for predicting quantum of inoculum, in spring is lacking. Among three biological components (leaf lesion, pseudothecial and leaf litter density), leaf litter density has the major role to influence the build up of the scab fungus (42). As such, adoption of 5% urea spray at leaf fall helps in reduction of disease status for the coming season, thereby requiring few fungicidal application in the season and keeping orchard floor free from such leaves.

Suitable temperature and moisture conditions for germination of spore, penetration of the cuticle and bringing about infection are termed as infection periods. These have been defined by Mills (45). Weather monitoring devices such as dewit leaf wetness meter, Belfort hygrothermograph or apple scab predictor (Reutor stokes or Metos) do help in identifying infection period and carrying on necessary spray. How far Mills infection periods are correct to different regions, are still required to be worked out. At Bajaura (Kullu valley), it has been observed that minimum 9 hours wetting does not provide light infection and one more hour wetting would be necessary for the purpose. As such, validity of minimum hours of wetting for light, moderate and heavy infection is still to be established for different apple growing areas.

Scab warning service is vogue in Himachal Pradesh since 1981 (19). Initially such prediction has started at Bajaura (Kullu valley) and Mashobra (Shimla hills), but during the last 5 years, wider areas of apple growing belts have been covered by utilizing all the three parameters of disease prediction and issuing scab warnings from the field laboratories at Thanedhar (Kotgarh apple belt), Kotkhai and Sarahan of Shimla district. Apple growers are kept abreast by siren signalling, phone and personal contacts, and also informed by special warnings carried on 2-3 times in a day through All India Radio, Shimla. Recording of leaf wetness in an individual orchard either by making direct observation of leaf wetness in the shady portions of the orchard or by the use of humidity data following rain can bring about precision in this system. Such process will be still easy if leaf wetness recorder or apple scab predictor is placed in each locality.

1.7 Disease management

Chemical control - Apple scab is best managed by chemical (fungicide) sprays. Fungicide spray schedules have either of, these three basic strategies: protective, after-infection (curative and pre-symptom), and irradiative (post-symptom) (17). The strategy differs from the point of fungicidal action and its timing of application.

Protective sprays - Protective sprays start at bud break stage, and are regularly carried out a 12-14 day interval. At times, sprays are carried on in routine as the weather is not favourable for disease development. This way, unnecessary wastage of fungicide, manpower and money is involved.

After-infection (curative and pre-symptom) sprays - To make protective sprays economical and need based, best course is to adopt after-infection spray strategy. For adoption of such a spray programme, basic need is the availability of predictive devices (weather monitoring and spore trap) (19). Most of the conventional fungicides (like mancozeb, captan and carbendazim) have limited after-infection activity and are to be applied within 18-24 h from the start of infection period (26, 58). S.I. fungicides have a longer curative activity upto 96 h while dodine can be utilized in between 36-48 h depending on temperature (26, 57). Once an infection period occurs, a fungicide with longer curative activity can be applied.

In case, it is not possible to carry on spray even after 3-4 days due to continuous rainy weather, certain fungicides can be applied late in the incubation period to mask the developing lesions but not beyond 2-3 days prior to symptom expression as ascertained from Mills table (17, 23). S.I. fungicides, dodine, carbendazim/thiophanate methyl have excellent pre-symptom activity, and do not allow appearance of typical scab lesions. Infact, the lesions are either chlorotic or reddish borwn and have no or limited sporulation (26, 55). Among S.I. fungicides, there is difference in activity as bitertanol (Baycor) inhibits sporulating lesions temporarily and need is to spray it again at 14 day interval. Thus it acts as fungistatic. On the other hand, mycobutanil (Systhane) inhibits lesions permanently and is fungitoxic (59).

The curative spray strategy as worked out and tried in the orchard with bitertanol and myclobutanil does provide reduction of 3 sprays in comparison to protective spray programme (56). Such strategy is of practical value to the growers falling within 3-5 km. range area of apple scab prediction centre.

Besides chemical spray programme as stated above, the other methods of managing/reducing scab severity are adoption of pre-leaf fall spray, orchard sanitation, biological control strategy and resistant apple cultivars.

Pre-leaf fall spray - Single spray of 5% urea just before general leaf fall provides efficient control of the overwintering stage of the fungus and helps in proper management of disease in the following season (14,18,20). This could be a part of biological control, as urea enhances the bacterial population and helps in quick decomposition of leaves thereby depriving the production of overwintering stage of the fungus (9).

Orchard sanitation - Reduction of overwintering leaves (leaf litter) from the orchard floor by way of urea spray or sweeping does bring down disease level in the coming growing season (42). Moreover keeping trees away from each other to accelerate proper ventilation, and not allowing bushy formation of the tree do help in quick drying of trees and inhibiting new scab infection.

1.8 Biological control

To reduce dependence on chemicals the need is to explore possibility of non-pathogenic microbes for effective elimination of the overwintering stage of scab pathogen. Two fungal organisms viz. *Athelia bombacina* Pers and *Chaetomium globosum* Kunze ex Fries have proved antagonist to scab fungus, alongwith 5% urea, *A. bombacina* is being utilized for enhanced and quick decomposition of leaves, and inhibiting the saprophytic phase of the scab fungus in Wisconsin, USA (38). Similar efforts are required to be made in India.

Resistant cultivars - The most used apple species and cultivars bearing genes of resistance to apple scab are the following (48):-

Some resistant species/cultivars	**Resistant gene**
Malus floribunda clone 821	Vf
Malus micromalus	Vm
Malus atrosanguinea	-
Malus pumila R 127-40-7A	Vr
Malus baccata	Vf
Malus baccata jackii	vbj
Antonovka	va poly.

Vf earlier thought to be a single dominant gene represents a genetic system with a group of closely linked genes.

This most used scab resistant "gene" vf has proved to be very strong, because after more than 40 years. It is still effective against all known races of apple scab.

The genes from *Malus micromalus* (Vm) and *Malus atrosanguinea* 804 have been overcome by race 5 of *Venturia inaequalis*. The new Canadian cultivars, Nova Easygro (*M. pumila*, Vr), Murray (Vm) and Ronville (Vm) are therefore not completely resistant to apple scab, and stand discarded from breeding programmes.

Among 48 scab resistant cultivars evolved so far in Europe, USA and Canada (8), 11 have been under investigation for their performance at Bajaura and Mashobra (17, 53), two cultivars namely, Emra (local name given to co-op 12) and Red Free (co-op 13) have been found promising in low and mid-altitude hills of Himachal Pradesh. These are early ripening apples with scarlet-red colour and stripes, sweet in taste and having good keeping quality (23). Priscilla, though still in trial has been found to be a mid-season promising cultivar (22). There is much demand for the planting material of Emra and Red free in newly establishing orchards of Kullu valley. Last year another promising scab resistant European cultivar, Florina has been introduced at Mashobra for observing its performance.

Locally bred hybrids (Ambstarking, Ambroyal, Ambrich and Ambred) (4, 5) though were bred for improvings storage quality of Delicious apples, have also been observed to be field resistant to apple scab (20). However these hybrids still lack preference by growers as these are late maturing and best suited for the upper altitudes (above 2400 meter mean sea level).

Efforts are on at Mashobra to breed a cultivar similar to Delicious but being early maturing, sweet in taste, red with good keeping quality and resistant to scab. Once such Delicious type cultivar comes out, there won't be any need of chemical spray. From integrated pest management point of view, best approach is to go for curative sprays and adopt single pre-leaf fall spray of urea besides proper orchard management.

2. POWDERY MILDEW *(Podosphaera leucotricha* (Ell. and Everh) Salm

2.1 Distribution and economic importance

Apple powdery mildew is widely distributed in all the apple growing areas of the country. It is primarily damaging to apple seedling stock and nursery plants while in established plantations. It is severe on susceptible cultivars, which include Cox's Orange Pippin, Versifield or White Dotted Red (**Maharaji**), Benoni (**Hazratwali**), Jonathan, Ambri, Granny Smith, Buckingham, Winter Delicious and Golden Delicious (10,21). Cultivars of Delicious group which comprise of about 60-80% of the plantings in major apple growing areas of India are moderately resistant as infection in them remains restricted to current season's growth only.

In nurseries and young plantings, mildew stunts tree growth and causes poorly formed, misshapen plants. In hearing orchards in susceptible cultivars, powdery mildew causes reductions in tree vigour and blossom bud production, aborted blossoms, and fruit russeting (at it is common in Jonathan). According to Vender and Scheer (60), and Yoder and Hickey (64), infection can reduce trunk growth, fruit size, crop weight and value. Severe infection may reduce the amount of bloom and almost eliminate the crop the following season (65). Tree vigour and bloosom production are mainly affected due to reduced photosynthesis, transpiration and carbohydrate content of the host (11).

2.2 Symptoms

Powdery mildew attacks foliage, growing twigs, blossoms and fruits. The disease appears as small, greyish or white, felt like patches on the underside of the leaves. The infected leaves become crinkled and curled. Whitish or greyish felt like patches often cover the entire leaf. Severely infected leaves become brittle and sometimes fall from the plants in mid-season. The pathogen spreads from leaves to current seasons's growth. Infected shoots are stunted, have shortened internodes, and are covered with greyish patches of mycelium, the colour of which may persist until the following season.

Petals of infected flowers are pale yellow or green and covered with mycelium. The flowers are shrivelled, may fail to set fruit, and are more susceptible to spring frosts. Severe blossom infection also involves the flower cluster leaves.

Severely infected blossoms emerging from infected buds give rise to small, severely russetted fruit, less severe blossom infection occurring later results in milder, net like russetting. The growing fruits often show cracks.

2.3 Causal organism

Apple powdery mildew is caused by an ascomycetous fungus, *Podosphaera leucotricha* (Ell. and Ev.) Salm. (anamorph *Oidium* cooke). Fungal conidia are produced in long chains and contain distinct fibrosin bodies, cleistothecia bear both apically and basally inserted appendages. The apical appendages are three to seven times as long as the diameter of the ascocarp and are usually straight and undivided. The basal appendages are rudimentary, rarely well developed, short, and more or less tortuous. The cleistothecium contains single ascus which has eight ovate to elliptic ascospores. The fungus is heterothallic (7).

Pear is another host for *P. leucotricha*. Other powdery mildew reported on apple is *Erysiphe herachei* (DC) St. Am. from Kashmir (49).

2.4 Epidemiology

Spore production and dispersal Fruit buds emerge first and provide the source of inoculum. Terminal shoot buds constitute on important inoculum source because of their larger infected area and the continued growth, and spore production on new growth well into the season (1). Infected buds of all types usually appear slightly later than their healthy counterparts. Spore dispersal usually takes place at pink bud and during bloom. A diurnal periodicity of spore release is observed in this fungus as peak of spore collection takes place at mid-day or early afternoon. In Himachal Pradesh, certain mildewicides which are also anti-scab in action do form part in 2nd spray at pink bud. These are wettable sulphur, Rubigan and Baycor.

2.5 Spore germination

Optimum temperature for germination of conidia is 20-22°C with minimum and maximum being 4 and 30°C. Germination of conidia does not take place in water but high relative humidity (90%) favours it. However germination of conidia can occur at as low as 70% R H on apple leaves, apple fruits and rose petals (46).

2.6 Infection process

The conidium on germination produces a germ tube which penetrates the article. The hypha upon penetration of the article becomes thin and peg like, penetrates the epidermal cell and forms a haustorium. When the leaf surface is covered with a water film, no injection takes place.

2.7 Time of bud infection

Most fruit ends remain briefly susceptible to primary and secondary infection (1). Lateral and terminal buds remain susceptible to infection for about one month after initiation. Terminal but susceptibility may stop during rainy weather but dry period in early summer may restrict terminal shoot growth. Young shoots having checked growth due to insect attack may also have midew susceptibility (62).

2.8 Role of cleistothecium

Evidence provides clue about ascospores being non-significant in spread of the disease as no germination takes place.

2.9 Mathematical models

For mathematical models are available on various aspects of disease epidemiology. These include incidence and severity relationships of secondary infection (52), relationship between disease progress and cumulative numbers of trapped conidia effects of leaf age and fungicidal control (44), and the functional relationship between primary inoculum and control by the fungicide bitertanol (44).

2.10 Disease management

General management strategy of powdery mildew is based on reduction of primary inoculum and protection from secondary inoculum (63).

Cultural practices and chemical control - In young orchards, primary inoculum can be reduced by removing infected terminal shoots during the winter pruning operation (32) but in larger plantations, the most practical approach is by using a surfactant based fungicide for eradication of powdery mildew mycelium in dormant season. Results obtained at the Long Ashton Research Station, Bristol (England) reveal the excellent control of powdery mildew by a single dormant season spray of 'Dormakil' (1.75% alkylphenolethoxylate PP 222) plus bupirimate (60g/100L). Such an spray reduces the number of routine mildew sprays required in the growing season. This treatment is now of practical utility in England (personal communication from Everard O' Donnen of Tiverton). Similar approach is required to be tried in India.

In the growing season, protection from secondary inoculum can be had by fungicide sprays applied throughout the period from tight cluster stage until the end of extension growth. Thorough spray coverage must be achieved on all parts of the trees especially the growing tips, and the fungicide application should be repeated at shorter spray internal of 5 to 7 days than 10 to 14 or 15 days intervals (3, 61, 63). Starting from bud break stage till the late summer or setting in of monsoon rains, there may be eight or more protective sprays for nursery plants and for trees of highly susceptible vigorous growing cultivars. Under severe disease prevalence such an approach is effective but cost prohibitive.

Fungicides and their rate of application efficaceous in growing season spray programmes include 0.05% benomyl, 0.05% carbendazim 50 WP, 0.05 to 0.10% dinocap, 0.05% thiophanate - methyl 70 WP, 0.30 to 0.50% wettable sulphur (Thiovit, Wetsulf, Sulfex etc.), 0.05%, Oxythinonox, 0.10% binapacryl, 0.05% triadimefon and 0.04% fenarimol. Among these fungicides, binapacryl (Morocide) and oxythioquinox (Morestan) stand withdrawn from Indian market, Triadimefon is excellent in its activity against powdery mildew (61) and can be employed in 15 days interval spray programme. In less susceptible cultivars, even bitertanol (0.05 - 0.10% Baycor) provides excellent control of powdery mildew (25) besides being effective against apple scab (16). Bitertanol (Baycor) and fenarimol (Rubigan) do form place at pink bud stage in apple scab spray programme adopted for Himachal Pradesh during 1993 to knock down the active powdery mildew fungus on emerging tissues besides taking care of apple scab.

Integrated management - In order to make powdery mildew control economical, best approach is to adopt an orchard- scouting method for supervised control decisions based on summer mildew ratings (2). The method involves assessment of infected leaves in summer months. Four shoots are rated on each sampled tree. Only the five leaves below a reference leaf near the shoot tip are assessed at each sampling. The proposed action thresholds and suggested actions based on this sampling technique are as follows : light disease rating (less than 8% mildewed leaves) - extend spray intervals particularly when weather is cool and wet or there is slow growth of shoot after petal fall; moderate rating (8-20% incidence) - plenty of short interval sprays are required; severe infection (more than 20% incidence) - adopt short interval sprays with 2 or 3 different fungicides to achieve effective control of powdery mildew. Such research needs to be undertaken in India.

Disease resistance - Like delicious cultivars of apple, Tydeman's Early Worcester is moderately resistant to powdery mildew and should have preference over Golden Delicious and Jonathan as pollinizer in Indian orchards. In the United States of America, some cultivars bred for scab resistance also have resistance to powdery mildew. Common ones are Prima, Sir Prize, Liberty, Emra and Red Free (23). Since last two cultivars have much more preference in lower - and mid - zones of Himachal Pradesh, adoption of these would be a practical approach to combat powdery mildew along with scab without the need of chemical spray.

3. REFERENCES

1. Burchill, R.T. 1960. The role of secondary infections in the spread of apple mildew (*Podosphaera leucotricha* (Ell. & Ev.) Salm.). J. Hortic. Sci. **35** : 66-72.

2. Butt, D.J. and Barlow, G.P. 1979. The management of apple powdery mildew : a disease assessment method for growers, Proc. 1979 Br. Crop. Prot. Conf. Pests and Diseases, 77-86.

3. Byrde, R.J.W., Hucheon, J.A., Coyla, J. and Holgate,M.E. 1984. Diseases of fruit crops. Apple mildew (caused by *Podosphaera leucotricha*). Long Ashton Res. St. Rep. 1984 : 42-43.

4. Chadha, T.R. and Sharma, Y.D. 1978. Breeding of apple varieties in Himachal Pradesh. Ind. J. Hort. **35**: 178-183.

5. Chadha, T.R. and Sharma, Y.D. 1980. New Cultivars of apple. In: Some Research Contributions on Apple. Eds. H.S. Verma, R.L. Sharma and G.K. Gupta, Farmers communication No. 101, pp. 6-7, Reg. Fruit Res. Stn. Mashobra, Simla.

6. Chohan J.S., and Kaul, S. 1976. Occurrence of Indian crab-apple scab in the plains of the Punjab. Indian Phytopath. **29** : 440-441.

7. Coyier, D.L. 1973. Heterothallism in the apple powdery mildew fungus, *Podosphaera leucotricha.* Phytopathology, **64** : 246 - 250.

8. Crosby, T.A., Janick, J., Pecknold, P.C., Korban, S.S., O' Cannor, P.A., Ries, S.M., Goffreda, J. and Voordeckers, A. 1992. Breeding apples for scab resistance. Fruit varieties J. **46** **(3)** : 145-166.

9. Crosse, J.E., Garrett, C.M.E., and Burchill, R.T. 1968. Changes in the microbial population of apple leaves associated with the inhibition of the perfect stage of *Venturia inaequalis* after urea treatment. Ann. Appl. Biol. **61** : 203 - 216.

10. Dar, G.N. and Kaul, R.N. 1981. Powdery mildew disease of apple and its control in Kashmir. Pestology, **5 (9)** : 9-10.

11. Ellis, M.A., Ferree, D.C. and Spring, D.E. 1981. Photosynthesis, transpiration, and carbohydrate content of apple leaves infected by *Podospharra leucotricha,* Phytopathology, **71** : 392-395.

12. Ghulam - Ud-Din, S., Ambradar, V.K., Najar, A.G. and Kaul, R.N. 1987. Maturing and discharge of ascospores of *Venturia inaequalis* in Kashmir. Indian J. Mycol. Pl. Path., **17** : 271-275.

13. Gupta, G.K. 1978. Present status of apple scab (*Venturia inaequalis*) in Himachal Pradesh and strategy for its control. Pesticides, **12 (5)** : 13-14.

14. Gupta, G.K. 1979. Role of on-season, post harvest and pre-leaf fall sprays in the control of apple scab (*Venturia inaequalis*). Indian J. Mycol. Pl. Path. **9**: 141-149.

15. Gupta G.K. 1981. Maturation and discharge of ascospores of *Venturia inaequalis.* Indian Phytopath. **34**: 502-504.

16. Gupta, G.K. 1983. Behaviour of fungicides and various spray schedules in the control of apple scab (*Venturia inaequalis*). Int. J. Tropical Pl. Dis. **1**:181-185.

17. Gupta, G.K. 1985. Apple Scab (*Venturia inaequalis*) E. Merck (India Ltd., Agrochemical Division, Bombay, 36 pp.

18. Gupta, G.K. 1987. Investigations on the effect of urea and fungicides in suppressing the ascigerous stage of apple scab pathogen. Int. J. Tropical Pl. Dis. **5** : 93-97.

19. Gupta, G.K. 1988. Effects of weather on the perennation and prediction of apple scab (*Venturia inaequalis* (Cke) Wint. In: Perspectives in Mycology and Plant Pathology, Eds. V.P. Agnithotri, A.K. Sarbhoy and D. Kumar, Malhotra Publishing House, New Delhi , 353-369.

20. Gupta, G.K. 1989. Diseases of pome-fruit orchards in India and research objectives. In: Integrated control of Pome Fruit Diseases. Vol. II, Eds. C. Gessler, D.J. Butt and B. Koller, WPRS Bulletin 1989/XII/6, Institute fur Pflanzenweissenchaften/Gruppe Pathologie, ETH, Zurich, 272-285.

21. Gupta, G.K. 1990. Major temperate fruit diseases in India and their management strategies. Pl. Dis. Res. **5** (special) 136 - 142.

22. Gupta G.K. 1991. Research findings on apple scab in Himachal Pradesh during 1990-91. 2nd Workshop ICAR's AICP on Apple Scab, pp. 1-8. YSPUHF, Nauni, Solan, Oct. 28, 1991.

23. Gupta, G.K. 1992. Apple Scab (*Venturia inaequalis* (Cke.) Wint.). In: Plant Diseases of International importance. Diseases of Fruit Crops. Vol. III, Eds. J. Kumar, H.S. Chaube, U.S. Singh and A.N. Mukhopadhyay, Prentice Hall, New Jersey (USA), 1031.

24. Gupta, G.K. 1993. Important temperature fruit diseases in India and current trends in their management. In: Important Diseases of Orchard and Field Crops Eds. K.S. Bhargava, V.P. Agnihotri, Om Prakash, Ram Kishun, A.K. Misra).

25. Gupta G.K. and Gupta, S.K. 1987. Post symptom antisporulant activity of chemicals against apple powdery mildew. Pl. Dis. Res. **2** : 131-132.

26. Gupta, G.K. and Kumar, J. 1985. Studies on the curative (after infection) and eradicant (post symptom) activity of fungicides against apple leaf infection by *Venturia inaequalis*. Pesticides, **19 (5)** : 18-20 & 25.

27. Gupta, G.K. and Lele, V.C. 1976. The scab fungus (*Venturia inaequalis* (Cke) Wint.) on apple twigs in Kashmir. Curr. Sci. **45** : 565.

28. Gupta, G.K. and Lele, V.C. 1980. Prevalence, distribution and intensity of apple scab in Kashmir valley. Indian J. agric. Sci. **50** : 45-50.

29. Gupta, G.K. and Lele, V.C. 1980 a. Morphology, physiology and epidemiology of the apple scab fungus, *Venturia inaequalis* (Cke) Wint. in Kashmir valley. Indian J. agri Sci. **50** : 51-60.

30. Gupta, G.K. and Lele, V.C. 1980 b. Role of urea in suppression of ascigerous stage and comparative *in vitro* efficacy of fungicides against apple scab. Indian J. Agric. Sci., **50** : 167-173.

31. Gupta, G.K. and Raj, M. 1988, Studies on the factors affecting development of pseudothecia of *Venturia inaequalis*, the apple scab pathogen. Pl. Dis. Res. **3** : 22-31.

32. Gupta, G.K. and Verma, K.D. 1980, Influence of cultural and chemical practices on the incidence of powdery mildew of apple. Indian. J. Mycol. Pl. Path. **10** : LXI.

33. Gupta, G.K. and Verma, K.D. 1984. Studies on apple scab. Measurement of ascospore dose in the orchard air. A. Rep. Reg. Fruit Res. Stn., Mashobra, Shimla, 1983-84 : 37-39.

34. Gupta, G.K. and Verma, K.D. 1986. Studies on the keeping quality of apple fruits infection with scab (*Venturia inaequalis*) in late summer season. Advances in Temperate Fruits, YSPUHF, Solan, 307 - 309.

35. Gupta, G.K. and Verma, K.D. 1989. Comparative performance of two different spore traps in measuring the ascospores of *Venturia inaequalis* in orchard air, Natl. Symp. Epidemiology and Forecasting of Pl. Diseases. IARI, New Delhi, Feb. 28 - March 2, 1989, C-33,

36. Gupta, G.K., Verma, K.D. and Pal, J. 1984. Some observations on the prevalence and severity of apple scab in Himachal Pradesh during the year 1978 to 1983. Indian J. Mycol. Pl. Path. **14** : XIII-XIV.

37. Gupta, S.K. and Gupta G.K. 1991. Note on post-symptom antisporulant influence of fungicides against powdery mildew. Scientific Hort. **1** : 57-60

38. Heye, C.C. and Andrews, J.H. 1983. Antagonism of *Athelia bombacina* and *Chaetomium globosum* to the apple scab fungus, *Venturia inaequalis*. Phytopathology, **73**: 650-654.

39. Jeger, M.J. 1984. Relating disease progress to cummulative numbers of trapped spores : Apple powdery mildew and scab epidemics in sprayed and unsprayed orchard plots. Plant Pathology, **33** : 517.

40. Kumar, J. and Gupta, G.K. 1986. Influence of host response and climatic factors on the development of conidial stage of apple scab fungus (*Venturia inaequalis* (Cke) Wint.). Indian J. Mycol. Pl. Path. **16** : 123-135.

41. Kumar, J. and Gupta, G.K. 1986 a. Influence of NPK fertilizers on the susceptibility of apple plants to scab pathogen (*Venturia inaequalis*), Indian Phytopath. **39** : 480-481.

42. Kumar, J. and Gupta, G.K. 1990. Effect of leaf litter density on potential ascospore dose of *Venturia inaequalis* the incitant of apple scab. Pl. Dis. Res. **5** : 7-11.

43. Lalacette, N. Jr. and Hickey, K.D. 1984. Apple powdery mildew disease progress on sections of shoot growth: an analysis of leaf maturation and fungicide effects. Phytopathology, **75** : 130-134.

44. Lalacette, N., and Hickey, K.D. 1986. An apple powdery mildew model based on plant growth, primary inoculum, and fungicide concentration. Phytopathology, **76** : 1176-1182.

45. Mills, W.D. 1944. Efficient use of sulfur dusts and sprays during rain to control apple scab. Bull. N.Y. Agric. Exp. Stn. 630.

46. Molnar, J. 1971.Cleistothecia of the fungus *Podosphaera leucotricha* (Ell. ec. Ev.) Salm. under the conditions of (Zechoslovkia. Ceska Mykologia. **25** : 211-218.

47. Nath P. 1935. Studies on the disease of apples in Northern India. II. A short note on apple scab due to *Fusicladim dendriticum* Fcl. J. Indian Bot. Soc. **14** : 121-124.

48. Overhofer, H. 1993. New developments in breeding. Disease resistant apple cultivars. Indo-Italian temperate Fruit Development Project, Shimla, H.P., India, April, 1993 : 1-15

49. Puttoo, B.L. 1972. Two new fungi on apple (*Malus sylvestris*) in Kashmir Indian J. Mycol. Pl. Path. **2**: 193.

50. Qasba, G.N., Dar, G.N. and Shah, A.M. 1982, Scab on *Contoneaster aitchinsonii* in India caused by *Venturia inaequalis (Spilocaea pomi)*. Indian Phytopath. **33** : 698-699.

51. Raj, M. 1985. Studies on the factors affecting the development of pseudothecia of *Venturia inaequalis* the apple scab pathogen. M.Sc. thesis, 80 pp, HPKVV, Solan.

52. Seem, R.C. and Gilpatrick, J.D. 1980. Incidence and severity relationships of secondary infections of powdery mildew on apple. Phytopathology, **70** : 851-854

53. Sharma, R.L., Kumar, J. and Ram, V. 1988. Performance of some scab resistant cultivars of apple in Kullu valley of Himachal Pradesh. J. Tree Sci. **7** : 45-49.

54. Singh, R. 1969. Fruits. National Book Trust, New Delhi.

55. Thakur, V. and Gupta, G.K. 1990. Evaluation of pre-symptom activity of fungicides on symptom expression, conidia production and viability of *Venturia inaequalis*. Indian Phytopath. **43** : 520 - 526.

56. Thakur, V. and Gupta, G.K. 1990 a. Note on development of spray schedule for controlling apple scab in Himachal Pradesh. Ind. J. Hort. **47** : 60-63.

57. Thakur, V.S. and Gupta, G.K. 1991. Studies on the effect of sterol inhibitors on apple scab *Venturia inaequalis* (Cke). Winter pathogenesis. Indian J. Plant Prot. **19** : 185 - 190.

58. Thakur, V.S. and Gupta, G.K. 1992. Evaluation of preventive and curative activity of organic fungicides against apple scab (*Venturia inaequalis* (Cke) Wint.). Pestology, **16 (7)** : 38-44.

59. Thakur, V.S., Gupta, G.K. and Garg, I.D. 1992. Effect of post-infection application of fungicides on the apple scab pathogen, *Venturia inaequalis* (Cke) Wint. J. Phytopathology, **135** : 160-166.

60. Vander, H.A., and Scheer, T.R. 1980. Threshold of economic injury for apple powdery mildew and scab. In: Integrated Control of Insect Pests in the Netherlands, 49.

61. Verma, K.D. and Gupta, G.K. 1986. Field evaluation of fungicides for control of powdery mildew (*Podosphaera leucotricha* (Ell. and Ev.) Salm.) of apple. Advances in Research in Temperate Fruits, YSPUHF, Solan : 407-409.

62. Woodward, R.C. 1927. Studies on *Podosphaera leucotricha* (Ell. and Ev.) Salm. Trans. Br. Mycol. Soc. **12** : 173-204.

63. Yoder, K.S. 1992. Powdery mildew of apple. In: Plant Diseases of International Importance. Diseases of Fruit Crops. Vol. III, Eds. J. Kumar, H.S. Chaube, U.S. Singh and A.N. Mukhopadhyay, Prentice Hall, New Jersey (USA), 66-89.

64. Yoder, K.S., and Hickey, K.D. 1983. Control of apple powdery mildew in the Mid-Atlantic Region. Plant Disease, **67** : 245-248.

65. Yoder, K.S., Byers, R.E., Cochran, A.E. II and Warren, J.R. 1985. Comparison of morning and evening concentrate applications of sterol-inhibiting fungicides on Jonathan apple, 1983-84. Fungicide and Nematicide Tests 40, Am. Phytopathol. Soc. 30-31.

23 Integrated Pest Management in Fodder Crops

C. R. Hazra, K. C. Pandey, R. B. Bhaskar and N. C. Sinha

CONTENTS

Integrated Pest and Disease Management
Rajeev K. Upadhyay, K.G. Mukerji, B.P. Chamola and O.P. Dubey (eds.)
APH Publishing Corporation, New Delhi-110002, India

ABSTRACT

This review article deals important insect pests and diseases of economic significance of major fodder crops including cowpea, guar, lucerne, sorghum, mustard and grasses. Insect pests particularly transmitting viral diseases and diseases incited by bacteria, fungi and viruses have been described in brief. Control options based on economic injury levels specifically non-chemical control methods have been described as a system approach.

1. INTRODUCTION

Indian livestock population numbering about 435 million in 1990, representing 15% of the total livestock population of the world thriving in merely 2% geographical area of the world. This itself is an indicative of the poor forage availability to the livestock, which is an important component of the economy of Indian villagers. The livestock population is estimated to be 494 million by the turn of this century. The requirement of green fodder and dry forage is estimated to be 837 and 529 million tonnes respectively against the estimated availability of 40% only. About 4% of cultivated area is under cultivated forages in India and about 24% of the total area comprising of natural grasslands, grazing lands, culturable wastes and degraded forests. The opportunities for increasing the cultivated area for cultivated forages are remote due to human food pressures. However, possibility exists for lessening the gap of 60% forage deficit through appropriate management practices. Pests (including plant pathogens, insect, nematodes and weeds) incur yield loss upto 25-30% in cultivated forages. For increasing the effectiveness and stability of forage production, pest management can help to a great deal.

A number of workers have outlined the ingredients and steps necessary to conduct IPM (47, 72, 108). Botterll (15) noted that IPM is the selection, integration, and implementation of pest control based on predicted economic, ecological and sociological consequences. The approach is ecologically oriented that incorporates the use of control techniques based on an understanding of the biology and life history of the pest (26).

The information available on various aspects of pests in forage crops is not very large. Although efforts have been made at this place; and few other places in the country regarding identification of important pests, their biology and management for some of the major crops like lucerne, cowpea berseem, oats, sorghum, Brassicas and some of the grasses.

In India though some attempts have been made regarding partial IPM, on insects have been made by Shri Ram *et al.*(89) in lucerne, cowpea, surghum and mustard. In this chapter an efforts has been made to present available information on various aspects of IPM and possibilities have been explored to integrate them.

2. PESTS OF FORAGE CROPS

The forages consist of an array of plants and almost four dozen of plant species are grown for forages (Table 1) and so is the diversified nature of insect-pests problem. In general, the problem of pests (including pathogens, insects and weeds) is less conspicuous on range grasses and legumes and also on trees and shrubs. These pests are of vital importance on cultivated fodders both leguminous and cereals. The fodder crops provide a better environment for development of insect and pathogens

Table 1 : Important forage crops and their botanical name

Sl. No.	Common name	English name	Botanical name
1.	Berseem	Egyptian clover	*Trifolium alexandrinum* L.
2.	Lucerne	Alfalfa	*Medicago sativa* L.
3.	Senji	Sweet clover	*Melilotus indica* L.
4.	Shaftal	Persian clover	*Trifolium resupinatum* L.
5.	Methi	Fenugreek	*Trigonella foenumgraceum* L.
6.	Lobia	Cowpea	*Vigna unguiculata* (L.) Walp.
7.	Guar	Cluster bean	*Cyamposis tetragonoloba* (L.) Toub.
8.	Rice bean	Rice bean	*Phaseolus calcaratus*
9.	Jaie	Oats	*Avena sativa* (L).
10.	Jau	Barley	*Hordeum vulgare* (L).
11.	Chari (Jowar)	Sorghum	*Sorghum bicolor* (L).
12.	Bajra	Pearl millet	*Pennisatum americanum* (L).
13.	Makka	Maize	*Zea mays* L.
14.	Makchari	Teosinte	*Euchlaena mexicana* Schrad.
15.	Chara sarson	Chinese cabbage	*Brassica pekinensis* (Lour.) Rupr.
16.	Napier grass	Elephant grass	*Pennisetum purpureum* Schum.
17.	Guinea grass	Guinea grass Green panic	*Panicum maximum* Jacq.
18.	Nandi grass	Setaria	*Setaria anceps*
19.	Deenanath grass	Pennisetum	*Pennisetum pedicellatum*
20.	Para grass	Bracharia	*Bracharia mutica*
21.	Ruzi grass	Ruzi	*B. ruziziensis*
22.	Signal grass	Signal grass	*B. decumbens*
23.	Dharaf grass	Chrysopogon	*Chrysopogon fulvus* (Spring) Chiov.
24.	Anjan grass	Buffel	*Cenchrus* ciliaris (L)
25.	Yellow anjan grass	Birdwood	*Cenchrus setigerus* Vahl.
26.	Rhodes grass	Rhodes grass	*Chloris gayana*
27.	Marvel grass	Marvel grass	*Dicanthium annulatum* Stapf.
28.	Sabi grass	Urochloa	*Urochloa mosambicensis*
29.	Sudan grass	Sudan grass	*Sorghum sudanense* Stapf.
30.	Shem	Dolichos bean	*Dolichos lablab*
31.	Kulthi	Horse gram	*Dolichos biflorus*
32.	Moth	Moth bean	*Phaseolus aconitifolius* Jacq.
33.	Subabool	Leucaena	*Leucaena leucocephala* (Lamk) de Wit.
34.	Shevri (Jayanti)	Sesbania	*Sesbania sesban* (L.) Merr.
35.	Agati	Agati sesbania	*Sesbania grandiflora* (L.) Poir.
36.	Dasrath grass	Hedge lucerne	*Desmenthas virgatus*
37.	Aparjit	Clitoria	*Clitoria ternatea* (L.)
38.	Siratro	Siratro	*Macroptylium atropurpureum*
39.	Stylos	Stylos	*Stylosanthes hamata, S. guyanensis S. humilis, S. scabra*
40.	Axillaris	Axillaris	*Macrotyloma axillare*
41.	Desmodium	Desmodium	*Desmodium intortum*
42.	Centro	Centrosema	*Centrosema pubescens*
43.	Glycine	Glycine	*Glycine wightii*

Table 2 : Pests of major fodder crops

Crop	Pests
Lucerne (*Medicago sativa* L.)	**Diseases** - Downy mildew (*Peronospora trifolii*), Rust (*Uromyces striatus*), Common Leaf Spot (*Pseudopeziza medicaginis*), Anthracnose (*Colletotrichum trifolii*), Crown wart (*Physoderma alfalfa*), Leaf spot (*Cercospora mechicaginis*), *Stemphylium botryosum*, *Alternaria medicaginis*), Powdery mildew (*Erysiphe polygoni*), Mosaic (*Alfalfa* mosaic virus), Wilt (*Fusarium* sp.), Root rot (*Rhizoctonia* and *Fusarium* sp.).
	Insects - Leaf hoppers (*Exitianus indicus* Dist., *Empoascanara* sp., *Austroagalia* sp., *Macrosteles* sp., and *Empoasca* sp., *Pentatomid* bugs, *Andrallus spinidens* Fabr., *Antestiopsis cruciata* Fabr., *Bagrada cruciferarum* Kirk, *Dolycoris indicus* Stal., *Nezara viridula* Linn., *Lygacid* bugs, *Nysius* sp., *Graptostethus* sp., *Mirid* bugs, *Creontiades pallidifer* Walk., *Campylomma* sp., Aphids, *Aphid craccivora* Koch., *Therioaphis maculata* Buck., Cow bugs, *Tricentrus* sp., *Leptocentrus* sp., Thrips, *Taeniothrips distalis* Karney, *Caliothrips indicus* Bagnall., Pumpkin beetle, *Aulacophora foveicollis* Lucas, *Galerucid* beetle, *Monolepta signata* Mots., *Alfalfa* weevil, *Hypera postica* Gyll., Grey weevil, *Myllocerus mosolosus*, Hairy caterpillars, *Euproctis lunata* Walk., *E. subtonata* Walk., *E. virguncula* Walk., *Psalis pennatula* Fabr., Alfalfa caterpillar, *Spodoptera exigua* Hub., Tobacco caterpillar, *S. litura* Fabr., Lucerne butterfly, *Terias hecabe* Linn., Semilooper, *Plusia nigrisigna* Walk., *P. orichalcea* Fabr., Grasshoppers, *Hieroglyphas banian* Fabr., *Oxya obneri* Will., *Chrotogonus trachypterus* Blanch, *Attractomorpha crenulata* Fabr., *Phanerophtera gracilis* Burm., Cut worms, *Agrotis spinifera* Hub., *Heliothis armigera* Hub., Termites, *Microtermes anandi* Holmgren.
	Weeds - *Anagelis arvensis, Spergula arvensis, Medicago denticulata, Meliolotus alba.*
Berseem (*Trifolium alexandrinum)*	**Disease** - Root rot complex (*Rhizoctonia solani, Fusarium semitactum, Tylenchorrhynchus vulgaris*), Stem rot (*Sclerotinia trifoliorum*), Leaf spots (*Alternaria, Myrothecium, Epicoccum* sp.)
	Insects - *Heliothis armigera,* hairy caterpillars(*Euproctis virguncula* and *E. lunata*), Semilooper (*Plusia nigrisigna, P. orichalcea)*, Pod borer (*Heliothis armigera*), Red pumpkin beetle (*Raphidopalpa favcicollis*), Red cotton bug (*Dysdercus koenigii*), Aphid (*Aphis craccivora*).
	Weeds - Chicory (*Chiclorium intivus*).
Oats (*Avena sativa* L.)	**Diseases** - Leaf blight (*Helminthosporium avenae*), Crown rust (*Puccinia coronata avenae*), Stem rust (*P. graminis avenae*), Leaf blotch (*Pyrenophora avenae).*
	Insects - Aphid
	Weeds - *Cyperus rotandus, Chenopodium album, Anagallis arvensis, Spergula arvensis.*
Mustard (*Brassica* sp.)	**Disease** - Leaf spot (*Alternaria brassicae, A. brassicola*) White rust (*Albugo candida*).

Insects - Sawfly (*Athalia legens*), Aphid (*Lipaphis erysimi*), Painted bug (*Bagrada hidaris*).

Weeds - *Chenopodium album, Medicago denticulata, Melilotus alba, Spergula arvensis.*

Cowpea *(Vigna unguiculata* (L.) Walp)

Diseases - Leaf spots (*Cercospora canescens, C. cruenta*), Anthracnose (*Colletotrichum lindemuthianum*), Bacterial blight (*Xanthomonas compestris* pv. *vignicola*), Root rot (*Mactophomina phaseolina*), Seedling rot (*Rhizoctonia, Pythium, Sclerotium phytoptithora* sp.,), Top necrosis (*Fusarium equiseti*), Wilt (*Fisarium* spp.), Mosaic (Cowpea mosaic virus), Powdery mildew (*Erysiphe polygoni*).

Insects - *Leafhoppers (Empoasca kerri)*, Flea beetle (*Pagria signata, Modurasia* spp.), Tobacco caterpillar (*Spodoptera litura*), Semilooper (*Plusia nigrisigna*), Aphid (*Aphis craccivora*), Grasshoppers (*Colemania sphenarioides* ; *Chrotogonus trachypterus* and *Attractomorpha crenulata*), Blister beetle (*Mylabris sp.*).

Weeds - *Celosia argentea, Cyperus rotandus, Solanum rurottense, Dactyloctenium aegypticum.*

Sorghum, Maize, Pearl millet

Diseases - Anthracnose (*Colletotrichum graminicola*), Sooty strips (*Ramulispora sorghi*), Zonate leaf spot (*Gloeocercospora sorghi*), Grey leaf spot (*Cercospora sorghi*), Rough leaf spot (*Ascochyta sorghi*), Leaf blight (*Helminthosporium fareicum*), Downy mildew (*Pseudopernospora sorghi, Sclerospora graminicola*), Charcoal rot (*Macrophomina phaseolina*), Rust (*Puccinia penniseti*), Ergot (*Claviceps fusiforms*), Leaf spot (*Ramulispora* sp., *Pyricularia* sp.).

Insects - Shoot fly (*Atherigona varia soccata*), Cut worm (*Agrostis* sp.), Army worms (*Mythimna separata, Spodoptera exigua*), Aphids (*Rhophalosiphum* sp.), Leaf roller (*Marasmia trapezalis*), Stem borer (*Chilo partellus, Sesamia inferens*), Grasshopers (*Hieroglyphus nigrorepletus, Colemania sphenarioides, Chrotogonus trachypterus*), Blister beetles (*Zonobris pustulata, Lytta tenuicolis*), Ear head bug (*Calocoris angustatus*), Ear head caterpillar (*Heliothis armigera*).

Weeds - *Striga, Celosia argentea, Cyperus rotandus, Solanum surattense, Dactyloctenium aegypticum, Echinochloa colonum, Digitaria arvensis, D. sanguinensis, Commelina benghalensis, Trianthema porchulacastrum.*

Guar (*Cyamopsis tetragonoloba* (L.) Taub.)

Diseases - Leaf spot (*Alternaria cyamopsidis*), Blight *(Xanthomonas campestris* pv. *cyamopsidis*), Powdery mildew (*Erysiphe polygoni*), Root rot (*Rhizoctonia solani*), Anthracnose (*C. lindemuthianum*) viruses.

Insects - Flea beetle (*Pagria signata*), Leaf roller (*Anarsia epipias*), Leafhoppers (*Empoasca* sp. and *Exitianus indicus*), Yellow mite (*Polyphagotarsonemus latur*).

Weeds - *Cyperus* group, *Echinochloa colonum, Corchorus capsularis, Cynodon dactylon, Digitaria sanguinalis, D. arvensis, Tridex procumbene, Amaranthus virdis.*

Grasses

Diseases - Rust (*Puccinia* and *Uromyces* spp.), Leaf spot (*Phyllachora, Curvularia, Pyricularia, Sorosporium, Colletotrichum* spp.).

	Insects - Yellow hairy caterpillars (*Psalis pennatula* Fabr.), Hairy caterpillar (*Euproctis* sp., *Creatonotus gangis* Linn.), Red hairy caterpillar (*Amsacta* sp.), Army worm (*Cirphis unipuncta* Haw., *C. albistigma, Spodoptera mauritia* Boised, *Euxoa spinifera*), Grass hoppers (*Hieroglyphus nigrorepletus* Bolivar, *Catantops pinguis* Walk., *Oedaleus abruptus* Thun., *Chrotogonus trychypterus* Blanch., *Aelopus tamulus, Colemania* sp., *Oxya* sp., *Locusta migratoria, Melanoplus spretus, Atractomorpha* sp.), Termites (*Microtermes* sp., *Odontotermes* sp.), Aphid (*Rhapalosiphum maides* Fitch.), Leafhoppers (*Nephotattix* sp., *Empoasca* sp., *Empoascanara* sp., *Exitianus indicus* Dist.).
Stylo	**Diseases** - Anthracnose (*Colletotrichum gloeosporioides*), Root rot (*Sclerotium rolfsii*).

as they are grown densely and by nature they are listed in Table 2. A description of major disease and insects is given below:

2. 1. Cowpea

2. 1. 1. Root rot

This disease incited by *Macrophomina phaseolina* produces primary symptoms of water soaked lesions on the rootlets, into decaying of entire root system and drying up of the plant. The disease is prevalent in all cowpea growing areas and its intensity is influenced by drought spells occurring in rainy season. The pathogen survives with seed and in soil with plant debris, which later serve as the source of primary infection.

2. 1. 2. Mosaic

A number of viruses infect this crop, however, cowpea mosaic virus is the real problem. The virus is seed borne in nature and primary infection occurs through seeds. The symptoms appear in the form of chloratic patches on the cotyledons, which later advance with the growth of the plant, covering entire foliage with mosaic symptoms. The secondary spread of the disease occurs by insect vectors particularly aphids, *Aphis craccivora, A. gossypi* and *Myzus persicae*. The host range of this virus is very wide and it occurs on a large number of leguminous and other crops.

2. 1. 3. Leafhopper (*Empoasca kerri*)

The leafhopper female lays 80-142 eggs mainly on the leaf veins, which hatch in about 9 days. Five nymphal instars are completed in about 10 days and adults survives for 34-52 days. The symptoms of damage are characteristic yellow discoloration of leaf edges followed by cupping, curling or crumpling of leaves. The pest appears at an early stage of crop growth and attains the peak during 3rd week of August (89).

2. 1. 4. Flea beetle (*Pagria signata*)

The adult flea beetle feeds on leaf epidermis. The leaf resembles tissue paper after severe infestation, which starts from germination and continues till harvest. The beetle is a small, metallic brown, with whitish spots on its elytra.

2. 1. 5. Semilooper (*Plusia nigrisigna*)

The female moth lays 76-183 creamy white eggs singly on the leaves. These hatch in 4-5 days. Young larvae are white, but soon becomes light green after feeding on foliage subsequently, discoloured white stripes develop laterally throughout the body. There are 5 larval instars completed in 11-28 days. The pupae develop within a white cocoon. The pupal period is 4-11 days during April-October and 14-23 days during January-March. The moths are brownish grey with forewings yellowish brown with characteristic silvery spots. The hing wings are straw or buff coloured. The incidence starts in the advance stage of crop growth i. e. in the month of August and reaches its peak in late September or October with almost all the top leaves of crop devoured.

2. 2. Guar

2. 2. 1. Bacterial blight

This disease is incited by *Xanthomonas campestris* pv. *cyamopsidis*. The symptoms appear in form of small water soaked lesions which later becomes necrotic covering large area of the leaves, these appear on flowers and pods also. The pathogens are seed borne (101). The primary infection occurs through infected seeds. In field the disease is favoured by rains.

2. 2. 2. Stylo anthracnose *(Colletotrichum gloeosporioides)*

The disease is prevalent all over stylo growing countries of the world. The symptom appears in the form of water soaked lesions of pin head size which later enlarge and form necrotic spots with characteristic anthracnose symptoms i.e. light brown centre and dark margins. In later stages the infection reaches to the stem, flower and pods. In the centre of the spots enormous black bodies are seen, which are the acervuli of the pathogen containing spores. The disease is internally seed borne in nature and also survives with plant debries buried in the soil. The intensity of the disease is greatly influences by the environmental factors like temperature and humidity. The optimum temperature and RH for the development of the disease is 25-30°C and 85-100% respectively. The fungus also infects other pasture species like *Aschynomine, Bauhinia, Indigofera* etc.

2. 3. Lucerne

2. 3. 1. Downy mildew (*Peronospora trifolii*)

The primary symptoms of the disease appears in the from of chlorotic patches on the leaflets. The reverse side of the patch is full of light brown mycelium spores of the pathogen. At the later stages of the disease the leaves of plants are dried and shed-off leaving bare stems and branches. The disease is greatly influenced by environmental factors like temperature and humidity. The optimum temperature for the development of the disease is 20-30°C with 85% humidity. The pathogen survive with the fallen leaves buried in the soil which serves as the source of primary infection. The sporulation of the fungus occurs during dark and the spores shed off by the time sun rises.

2. 3. 2. Rust (*Uromyces striatus*)

The disease is well distributed in the country. It is generally high in parts of Maharshtra and Gujarat. In Uttar Pradesh and Rajasthan severity of disease occurs occasionally. The fungus produces characteristic rust pustules on the leaf blade, though often they may be found on the petiole and stems. The reddish brown uredia and telia develop on the stem late in the season. In severe infection the leaf may wither resulting into considerable damage to the crops. The fungus is an obligate parasitic and heteroecious in nature. The uredial stage repeats several times, which further spread the disease through wind. The disease is favoured by humid climate. The survival of pathogen is through perennial lucerne plants.

2. 3. 3. Lucerne weevil (*Hypera postica*)

The adults are active during early morning and late evening, rest of the day they remain hidden under the leaves or in leaf buds. The female weevil lays eggs inside the stem singly or in clusters, numbering 600-1300 eggs. The eggs are oval in shape and bright yellow in colour, which hatch in 6-

14 days. There are four larval instars completed in 10-19 days. When the larvae is about to pupate, it stops feeding and spins whitish silken threads between two leaves or two folds of a leaf or in soil. This pre-pupal stage lasts for 2-3 days. Pupa is light brown in colour about 5 mm in length, with a darker strip extending down their backs more than half of the length of body. During April to October the adult weevil aestivates in various trees, plant stubbles or in soil. The eucalyptus tree scales are good medium for its survival (27).

Incidences of lucerne weevil larvae are maximum during middle of February to first week of March. The weevil has only one generation in a year. Temperature and relative humidity plays a major role in regulating weevil population. The weevil development is rapid at a temperature range of 11.5-31.6°C and 63=76% relative humidity (68).

2. 3. 4. Spotted alfalfa aphid (*Therioaphis maculata*)

The aphid is pale yellow with six or more rows of black spots alongwith the back. Nymphs and adults suck juice from leaves and stem and are usually most numerous in the lower parts of the plant. The feeding causes leaves to turn yellow, die and drop off. The aphid is more severe on seedlings causing death and older plants remain stunted. Maximum population is observed during third week of January. The most suitable temperature and humidity for development of the aphid are 20-25°C maximum, 9-10°C minimum and R.H. 80-85% (30).

2.4 Sorghum

2.4.1 Anthracnose (*Colletotrichum graminicola*)

It has a world wide distribution and prevalent wherever sorghum is grown. In India leaf spot phase of the disease is common. The symptoms of the disease appears as small coloured spot on the leaves of lower part of the plant and spread upward. The colour depends on the variety affected being red, purple or brown. The centre of the spot is light brown and dark margins. In older spots acervuli appear as black dots in the centre. Infection of the disease is seed, soil and air borne. The pathogen has a wide host range. The disease is favoured by low or unbalanced soil fertility.

2.4.2 Sooty stripe (*Ramulispora sorghi*)

The disease is prevalent in Haryana, Delhi, Uttar Pradesh, Andhra Pradesh and Tamil Nadu. Elongate elliptic, oblong or irregular lesions are formed on the lamina. At first the spores are grey to tan in the centre with a distinct red margin. Later, the surface becomes powdery and sooty coloured as the sclerotic lesions are formed. The lesions are small in beginning but enlarge rapidly, especially during rainy season. Several of them coalesce and form bigger patches. The leaves dry up prematurely. In humid weather the disease spreads rapidly. The pathogen is seed and soil borne. The optimum temperature for growth is 28°C. The pathogen survives in the soil with fallen leaves season to season.

2.4.3 Zonate leaf spot (*Gloeocercospora sorghi*)

The disease symptoms appears as semicircular zonate spots with alternating dark and light boundaries on the lamina. The lesions are red or brown in colour. The lesions enlarge becoming irregular or semicircular and extending across the lamina with a distinct red margin. In larger lesions, zonation becomes conspicuous. The disease is favoured by rains as the severity of the disease occurs during rainy season.

2.4.4 Downy mildew (*Sclerospora sorghi*)

This disease is prevalent in many countries in Asia and Africa. In India it is common in Tamil Nadu, Karnataka, Andhra Pradesh, Maharashtra, Madhya Pradesh and to some extent in Uttar Pradesh. The incidence of the disease is higher in fodder sorghum than grain crops. Different types of symptoms have been noticed depending on the stage of growth and the nature of infection. Infected seedlings have pale yellow narrow leaves, covered on both surfaces by soft, white, downy growth of the conidial stage of the fungus. The infected plants remain stunted with withered leaves which dry up prematurely. The pathogen survive in soil or on seeds during off seasons.

2.4.5 Shootfly (*Atherigona variasoccata*)

Grey coloured shootfly is about 6 mm in length with 3 pairs of black spots on dorsum of its pale pink abdomen. The first to seventy leaf stage of plants is preferred by fly to lay eggs on the underside. The maggot hatch within 48 hours and tunnels towards the base of the plant behind or through the leaf sheath, ultimately cutting the growing point, causing dead heart. The shootfly is prevalent throughout the sorghum growing tract of the country, it is most severe on late planted kharif and early planted rabi crops.

2.5 Mustard

This crop is seriously attacked by two major insect pests, the mustard saw fly (*Athalia legens* Klug) at seedling stage and mustard aphid (*Lipaphis erysimi* Kalt.) which appears right in the month of December and remains till the harvest of the crop infesting almost all parts of the plant. Heavy infestation results in cessation of growth and depletion of stand. The aphids also secrete honey dew which invites black sooty mould fungus (*Polydesums exituosus*). This renders the crop unpalatable for animals.

2.6 Grasses

Almost all the important cultivated and range grasses namely *Dicanthium annulatum, Cenchrus ciliaris, Chloris gayana, Pennisetum pedicellatum, Brachiaria* sp., *Pennisetum* sp., *Panicum maximum, Chrysopogon fulvus, Heteropogon contortus, Iseilema laxum, Sehima nervosum* etc. are attacked by a number of grasshopper species. *Hieroglyphus nigrorepletus* appears in the second fortnight of June and remains active till the first half of November, attaining highest population in second fortnight of August. *Oedaleus abruptus* appear in the first half of August reaching maximum in the first half of September remaining till October end. *Catantops pinguis* appears in the second half of August and remains in the field till the first half of November with maximum population in first fortnight of October.

Table 3 : Quantitative losses in forages due to pests

Crop	Pest	% loss	Reference
Lucerne	Insect	37.70	89
	Diseases	72.40	3
Cowpea	Insects	30.20	85
	Disease	74.00	19
	Weeds	62.80	28
Sorghum	Insects	75.20	84
	Diseases	50.00	2, 105
	Weeds	54.00	28
Maize	Insects	28.00	88
Bajra	Diseases	30.00	2, 105
Mustard	Insects	36.80	89
Stylo	Diseases	35.90	21
Cluster bean	Diseases	75.00	19
	Weeds	44.00	93
Oats	Diseases	55.00	105
Grasses (*C. ciliaris*)	Insects	19.75	80

3. ASSESSMENT OF LOSS

Reichelderfer *et al.* (78) considered information on pest induced losses as necessary, but often neglected component in the design of pest management programme. Pest in general reduce the quality and quantity of forages and may contribute to the rate of stand decline in case of perennial crops. The extent of losses depends on many factors *viz.*, the incident and degree of infestation/infection, variety of crop and yield potential of crop due to agronomic and other related reasons. Assessment of crop losses can be determined from statement by experts, questionnaire data, pesticide sales, surveys, field experiments and simulation models. Out of these, survey, field experiments and simulation models only can provide consistent and repeatable approaches for assessment of losses. Information available on quantitative losses in various forage crops is presented in table 3. Most of the studies relating to pest losses reported here have been made on the basis of survey and field experiments. Simulation studies is reported by Fick (34).

4. ECONOMIC INJURY LEVELS

The pests of forage should be controlled when their collective injuriousness to the cop causes economic loss equal to or exceeding the cost of control. This is expressed as Economic Injury Level (EIL). Four basic variables for calculating EIL as suggested by Pedigo *et al.* (71) are control costs, market value, injury per unit pest and damage per unit of injury. Out of these control costs, injury per unit pest and damage per unit of injury can be calculated as per procedures for other agricultural crops, but for market value of forages, frequently clearly defined market does not exist. Craven and Hasbargen (1979) suggested market value of forages calculated on the basis of replacement costs with a substitute feed.

As has been discussed earlier a number of pests cause economic damage in forage crops. Pedigo *et al.* (22) pointed out limitations of EIL, out of those two are discussed here. He pointed out that use of EILs or ETs cannot describe the quantitative relationship to yield of a number of pathogens as sampling and quantification of these pathogens is frequently impractical and control of many pathogens are preventive and not curative. He further points out that use of EILs or ETs is relatively suitable for multiple best situations. Use of an EIL or ET for management decisions involving many pests is considerably removed from their original use for which they are developed. He adds, however, the capacity for making appropriate management decisions with many pests or a pest complex is important goal of Integrated Pest Management (IPM).

Most of the information on EILs or ETs are based on the single species or a groups of insects doing similar type of injury viz., for lucerne, in California (USA), the EIL has been indicated by 25 larval per sweep for *H. postica*, 40 aphids in spring and 20 aphids in summer per stem for *T. maculata* and 15 larvae (longer than 1.27 mm) per sweep for *S. exigua* ; in USSR 3-6 beetles/m^2 for Hypera, 5-8 beetles/m^2 for Sitona weevil, 10 beetles/5 sweep for Apion (29); in India for lucerne 12 grubs/0.5 m^2 for *H. postica*, for cowpea defoliators damage is 9.89% foliar damage, for mustard aphid 50 aphid/ 4 cm twig (89), for sorghum shoot fly 3.8-15% dead hearts in different varieties (76).

For developing a sound IPM programme in forages, there is a need to develop comprehensive threshold. Pedigo *et al.* (22) considered best method for approaching comprehensive thresholds through the EIL is by examining the host response to pest injury. Practically an improved understanding of physiological injury may be incorporated in EILs by developing standard equivalents for guilds of species with similar injuries. Thus a single EIL in standard equivalent could serve for a complex of pests having the same type of injury.

The physiological response of pest damage has been studied in case of insect-pests of lucerne. Hutchins *et al.* (50) reported that in lucerne both leaf-mass consuming insects and assimilate removing insects causing yield and quality losses, consequently reflects a delay in maturity of the crop. Buntin and Pedigo (16) and Hutchins and Pedigo (49) utilized days of regrowth delay as the estimate of injury per insect when calculating EILs for leaf-mass consuming or assimilate removing insects.

5. COMPONENTS OF IPM

5.1 Cultural control

Cultural control is a prophylectic method of pest control, which is achieved by manipulation of the environment in such a way as to render it unfavourable for the pests. Coaker (20) considered it as the first-ditch defence around which to build other control options. These methods are unlikely to reduce pest infestations to desirable levels, but within the framework of IPM system these are used to reduce overall levels of infection/infestation to below economic thresholds.

Tillage practice

Soil tillage is one of the important and widely used method of pest suppression. Though pest control may not be the primary goal of cultivation, but it markedly influence the soil environment, either by creating conditions not suitable for pest survival, by exposing pests to their natural enemies, by inflicting direct physical injury, by exposing to sun and other abiotic factors.

Cultivation of soil markedly affect the number of eggs and nymphs of grasshoppers, *Kraussaria angulifera* and *Oedaleus senegalensis* by exposing egg pods to desiccation, reducing the level of food, shelter and vegetation and by making the soil rough and unsuitable for egg laying (5). The population of Sweetclover weevil (*Sitona cylindeicollis*) can be reduced upto 99% by ploughing before the plants of sweet clover (*Melilotus officinalis*) are in full bloom (60). Weed population in berseem can be effectively controlled by pre-sowing irrigation to reduce germination and destruction of weeds through preparatory tillage operations (41).

Planting date

Varying the planting time of crops works as a means of pest control by creating asynchrony between crop phenology and the pest phenology which can retard the rate of colonization (32). Cowpea crop sown in third week of June and last week of July had least incidences of leafhoppers (90). Summer cowpea sown in March to May suffers less from viral diseases (82). In lucerne, early sowing in third week of October escapes the incidences of leafhoppers and lucerne weevil, their population increased as sowing is delayed (90). In sorghum insect-pests are more severe on late sown Kharif (rainy season) and early Rabi (winter season) crops. The kharif sorghum crop should be sown immediately after the onset of monsoon.

Intercropping

In modern agriculture, tendency towards the use of monocultures have led to proliferation of pests in crops. The poly cultures like mixed intercroppings, row intercroppings, strip intercropping, relay intercropping and alley intercropping can help in reducing pests. Intercropping of cowpea and sorghum reduces leafhoppers and defoliators damage in cowpea crop (90). Cowpea intercropped with pearl millet or maize have less viral infection (82). Introduction of pea (*Pisum sativum*) @ 30 kg/ha in lucerne crop and cowpea in maize and sorghum crops showed smothering effect and thus significantly lowered the weed infestation (1, 41, 107).

Crop fertilization

This is a well known fact that any factor that modifies that development, in turn, alter the response of pests to the plants. Properly selected mixes of fertilizers must have an adverse affect on pests and disease pathogens and increase the tolerance of plants to them. In lucerne, phosphorus and potassium deficiency causes heavy attack of spotted alfalfa aphid (62). Shri Ram *et al.* (89, 91) reported combination of 50 kg K_2O and 100 kg P_2O_5 per hectare in cowpea and lucerne makes the crop less attractive for insect pests. Fertilizer combination of 75 kg N, 75 kg P_2O_5 and 45 kg K_2O per hectare enhanced the fodder yield with less insect attack in cowpea and lucerne. Nitrogen concentration in mustard plants shows a direct correlation with the abundance of mustard sawfly and aphids, whereas, potassium and phosphorus are inversely correlated (89). Similarly the bacterial blight (*Xanthomonas campestris* pv. *cyamopsidis*) increased with increase in nitrogen level from 20 to 40 kg per hectare in guar (35).

Irrigation

Presowing irrigation is advantageous for controlling weeds in maize, soybean and sorghum because of their good growth and canopy profile built up inducing smothering effect on weeds (64). Frequency of irrigation plays a vital role on the aphid incidence in mustard. Moderately irrigated crop (4 irrigations) harbours reasonably lower aphid population, higher frequency of irrigation (six) increases succulancy of the plants, which attract higher aphid population (89).

Crop rotation

From an agronomist point of view, a good crop rotation is one that ensures optimal production of the fodder per unit time per unit area. Nutrients in the soil fully, alongwith the water and applied fertilizers, but from plant protection point of view it must serve the purpose of pest reduction i.e. reduce the nutritions of pests in every subsequent year or totally deprive them of food, reduce the initial infection by diseases and ensures unfavourably stratum for the development of pests. Suitable crop

Table 4 : Crop rotations in various locations/regions

Region/Crop rotations
Northern region
Palampur (H.P.) (Sub-temperate, humid, red soils)
1. White clover + rye grass
2. Napier-bajra hybrid + velvet bean - berseem + sarson
Pantnagar (U.P.) (Tarai region, red & yellow soils)
1. Dinanath grass - berseem - maize + cowpea
2. Napier-bajra hybrid +Subabool
Hissar (Haryana) (Semi-arid, sandy soils)
1. Napier-bajra hybrid _ berseem
2. Napier-bajra hybrid + lucerne
Cenral and Western region
Jhansi (U.P.) (Semi-aird, mixed & black soils)
1. Napier-bajra hybrid + cowpea - cowpea - berseem
2. Cowpea - sorghum + cowpea - berseem
Jabalpur (M.P.) (Sub-humid, black soils)
1. Napier-bajra hybrid + cowpea = cowpea = berseem
2. Sorghum + cowpea - berseem + sarson - Jowar + cowpea.

Anand (Gujarat) (Semi-arid, balck soils)

1. Napier-bajra + guar - lucerne
2. Maize + cowpea - maize + cowpea - oats - maize + cowpea

Eastern Region

Kanke (Bihar) (Sub-humid, red acid soils)

1. Bajra + cowpea - maize + cowpea - oats
2. Maize + cowpea - jowar + cowpea - berseem + sarson

Kalyani (West Bengal) (Sub-humid, alluvial soils)

1. Maize + cowpea - Dinanath grass - oats
2. Maize + rice bean - berseem + sarson

Jorhat (Assam) (Humid, acid alluvial soils)

1. Napier-bajra hybrid alone
2. Maize + cowpea - maize + cowpea - maize + cowpea - oats.

Southern region

Coimbatore (Tamil Nadu) (Semi-arid, balck soils)

1. Mapier-bajra hybrid + hedge lucerne
2. Sorghum + cowpea - maize + cowpea - maize + cowpea

Vellayani (Kerala) (Humid, red soils)

1. Guinea grass
2. Congosignal under coconut garden

rotations of forages recommended for various regions of the country are presented in table 4 as has been worked out through the aegis of All India Coordinated Research Project on Forage Crops (AICRPFC).

5.2 Host plant resistance

Host plant resistance (HPR) as defined by Beck (11) that the collective heritable characteristics by which a plant species, race, clone or an individual may reduce the possibility of successful utilization of that plant as a host by an insect species, race, biotype or individual. Accordingly, Kumar (58) reiterated that HPR as the inherent ability of crop plants to restrict, retard and overcome pest infestation. In forage crops, particularly the HPR has a special significance because of the relatively high cost of chemical application, comparatively low acreage value and the possibility of harmful residual effects. The importance of pest resistant varieties in IPM is very high as this provides a solid base on which the complementary forms of control are based.

Mechanism of resistance

The mechanism of resistance to insects has been divided into three categories, i.e. (i) non-preference, (ii) antibiosis and (iii) tolerance. However, Kogan and Ortman (57) preferred the term antitoxins in place of non-preference to explain the means of resistance mechanism employed by the plants to deter colonization of insect. Antitoxins resistance in terms of 'choice' and 'no choice' situations has been worked out in lucerne and other *Medicago* species (7, 66). Another criteria for assessing antitoxins is in terms of insects leaving the plants. Dent (25) observed that alate forms of grass aphid, *Metopolophiumfestucae* did not colonize in the grasses. *Festuca rubra* and *F. arundinacea*, 94% and 84% flew from them, respectively. On the other hand this pest aggressively colonized in *Lolium multiflerum*, *L. perenne*, oats and wheat where only 36-64% of the alates left. Antitoxins is evaluated by insect behaviour in terms of latency, duration, frequency and intensity. Latency is the time from some specified event to the onset of the first occurrence. The aphid, *Aphis craccivora* when transferred to cowpea plant, it took more time to start feeding in resistant lines (37). Duration is the length of time for which a single occurrence of the behaviour pattern last and in their studies it was observed that the total probing time was longer on resistant lines. The frequency is the number of occurrence of the behaviour per unit time and the criterion used in case of aphids in number of probes made per unit time. Intensity is a subjective assessment like the extent of restlessness.

Antibiosis is the measure of development, reproduction and survival of host on the host plant. In case of different *Medicago* species/cultivars least larval weight gain was observed in species, *M. rugosa, M. scutellata* (66). Spotted alfalfa aphid could not reproduce in *M. scutellata* (66). This indicates resistance potential of *M. scutellata* to insect-pests. Host plant resistance to insects is controlled by morphological as well as biochemical mechanisms. The morphological factors include plant texture, colour, anatomical features, surface and sub-surface factors. Glandular hairs in *Medicago* species are not conducive to egg laying by lucerne weevil (54). This weevil egg laying increases in lucerne with increase in stem diameter beyond 1.00-1.33 mm and crown width 4.67-6.00 cm (69). Insect feeding on lucerne leaves is inhibited on leaves coated with wax easters and on this basis resistant plants have been identified (12).

The biochemical basis of resistance can be divided into two broad categories, those influencing, behavioural responses and those influencing the physiological responses of insects. Insect behaviour influencing chemicals has been further classified as attractants, arrestants, stimulants, repellents and deterrents. Adult lucerne weevils, showed a strong olfactory attractancy to fresh alfalfa plant material (17). Nitrate is a deterrent to feeding by the sweet clover weevil (*Sitona cylindricollis*) in sweet clover. Plant chemicals affecting the physiological processes may be classified as nutrients, physiological inhibitors and toxicants. High saponin concentration in lucerne retarded growth of *Sclerotium rolfsii*, pea aphid (*Acyrthosiphon pisum/Acyrthosiphum pisum*) and potato leafhopper (*Empoasca fabae*) (46).

Genetics of resistance

Inheritance of resistance studies are presented in table 5.

Table 5 : Inheritance of resistance to pests in forage crops

Crop	Pest	Inheritance	Reference
Lucerne and clovers	Sweet clover aphid (*Therioaphis riehmi*)	Single dominant gene An additional complementary gene in some resistant clovers	61
	Pea aphid (*Acyrthosiphum pisum*)	One dominant gene	39
	Spotted alfalfa aphid (*Therioaphis maculata*)	1. Quantitative 2. Resistance could be both vertical and/or horizontal	38 63
	Wilt (*Fusarium oxysporum*)	Monogenic, oligogenic	48

	Root rot (*Phytophthora megasperma*)	Oligogenic	51
Cowpea	Pulse beetle (*Collosobruchus maculatus*)	Additive dominance and maternal components	31
	Bean Yellow Mosaic	Monogenic	77
	Cercospora leaf spot (*Cercospora cruenta*)	Monogenic	33
	Bacterial blight (*Xanthomonas vignicola*)	Monogenic	92
Dolichos	Bacterial leaf spot (*X. campestris* pv. *phaseoli*)	Oligogenic	104
	Shootfly (*Antherigona soccata*)	1. Quantitative 2. Heritability 15-25% under high to moderate levels of infection	102 44
	Stem borer (*Chilo zonellus*)	Polygenic	74
	Green bug (*S. graminum*)	Single incompletely dominant gene	112
	Rust (*Puccinia purpurea*)	Oligogenic	75
Pearl millet	Downy mildew (*Sclerospora graminicola*)	Polygenic	9
Oat	Crown rust (*Puccinia coronata*)	Monogenic, Oilgogenic	48

Breeding for resistance

Pedigree breeding has been used by Johnson and Teetes (54) for breeding resistance to sorghum midge and shootfly in sorghum. Line selection has been used in breeding pest resistance in lucerne cultivars but as a method it is less preferred to individual plant selection because there is less control of pollen sources. The individual plant selection involve an initial population screening of 1000 to 10, 000 plants, usually from a cross variety or plant introduction. Selection cycles are made, where resistant material is selected. At each cycle the number of plants is greatly reduced until a desired number of superior plants remain and then progeny tests be used. For recurrent selection the number of selected plants or lines in each cycle should be at least 25-100 to ensure that the frequency of detrimental characters is maintained at a level, not higher than in the original population. These selected plants/lines from one cycle are interpollinated and seeds obtained undergo another cycle of selection for the same characters. This method of recurrent phenotypic selection is well suited to cross pollinated forage crop like lucerne, for concentrating highly heritable genes by use of large population. This has added advantage that plant selection and backcross methods can also be incorporated into it. The phenotypic recurrent selection has been used to breed lucerne cultivars with multiple pest resistance to rust, common leaf spot, bacterial wilt, anthracnose, spotted alfalfa aphid and potato leafhopper (45). A number of pest resistant lucerne cultivars like Williams, Cherokee, Team, Arc, Hagmark, WARP have been bred by this method.

For development of superior forage sorghum varieties with resistance to pests through hybridization have been attempted by Solomon and Premkishore (99). Promising hybrids PCH-101, PCH-102, PCH-103, PCH-105, and PCH-106 with insect pest resistance have been developed by crossing PC-112 and PC-115 with male sterile lines separately.

Source of resistance

A number of varieties resistant to scores of pests have been identified through the aegis of All India Coordinated Research Project on Forage Crops (Table 6).

Table 6 : Disease/pest resistance status of some released varieties of important forage crops.

Crop	Variety	Resistance against	Resistance status
Cowpea	HFC-42-1, EC-4216, GFC-1, GFC-2, GFC-3, GFC-4 and Russian giant	Leaf hopper, flea beatle & semi looper	Reasonably tolerant
	UPC-287	Yellow mosaic virus, Anthracnose wilt, stem and root rot	Resistant
	UPC-5286 and UPC-5287	Yellow mosaic virus complex, Anthracnose, hairy catter pillar, pod and seed borer	Resistant
Lucerne	CO-1	Anthracnose	Fairly resistant
	LLC-3 and LLC-5	Downy mildew	Fairly resistant
	S-244, Sirsa-8	Downy mildew, Aphids & weevil	Fairly resistant
	Sirsa-9/T-9	Weevil	Fairly resistant
Berseem	BL-1, BL-10, BL-22, JB-1, Mescavi and Pusa giant	Root and stem rot	Moderately resistant
	S-991-1, BL-2, JB-3 and UPB-110	Bacterial wilt	Tolerant

Crop	Variety	Resistance against	Resistance status
Field bean	JLP-4	Yellow mosaic virus	Moderately resistant
Rice bean	K-1	Yellow mosaic virus	Moderately resistant
Cluster bean	HFG-119 and HFG 156	Leaf blight	Tolerant
	Guara-80	Leaf blight and bacterial blight	Moderately tolerant
	IGFRI-212-1	Leaf blight	Moderately resistant
	FS-277	Alternaria blight	Tolerant
Sweet clover	YSL-106	Downy mildew, aphids, lucerne weevil	Moderately resistant
Persian clover	Ch.-20 and Ch.-48	Fusarium wilt	Resistant
Methi (Fenugreek)	T-8	Leaf blight	Moderately resistant
Brassica	Chinese cabbage	Downy mildew	Moderately resistant
Teosinte	Improved	Leaf blight	Moderately resistant
Sorghum	PC-1	Red leaf spot and stem borer	Moderately resistant
	SL-44	Red leaf spot	Moderately resistant
	HC-171	Foliar diseases, stem borer, shoot fly and midge	Highly resistant
	HC-260	Stem borer	Moderately tolerant
	PC-6	Foliar diseases	Susceptible
	PC-9 & PC-23	Major diseases and pests	Reasonably tolerant
	Raj. Chari-1 and Raj. Chari-2	Stem borer	Resistant
	U.P. chari-2	Zonate leaf spot, Athracnose, shoot fly and stem borer	Moderately resistant
Maize	African Tall	Leaf blight, rust, downy mildew, stalk rot and stem borer	Resistant

Crop	Variety	Resistance against	Resistance status
	Vijay comp.	Leaf blight rust, downy mildew, stalk rot and stem borer	Moderately resistant
	Moti comp.	Downy mildew	Moderately resistant
Pearl millet	Giant bajra, L-72, L-74, TNSC-1, UUJ-IV-M-1	Downy mildew	Resistant
Naper-bajra hybrid	NB-21, Co-1, IGFRI-3, IGFRI-6, IGFRI-7, IGFRI-10	Major pests and diseases	Tolerant
Oats	Kent, OS-6, OS-7, JHO-822 and UPO-212	Major pests and diseases	Resistant
	UPO-94	Stem rust, crown rust, leaf blight and smut	Resistant
	JHO-851	Leaf blight, leaf blotch and crown rust	Resistant

Pesticides

Pesticides have been the backbone of pest control from almost last 50 years. Subsequently the widespread use of these pesticides resulted to several problems. Now the use of these chemicals have been questioned with the advent of the ecologically sound IPM approach to pest control. Callender based approach of the pesticide use have to be changed with the judicious use of these in appropriate situations.

Fungicides

Lucerne downy mildew can be effectively controlled by three sprays of rixomil (0.1%) or propamocarl (0.15%) (56). Four sprays of mancozeb (0.2%) at 10-12 days interval starting from forty days after crop emergence increased grain and fodder yield of sorghum by controlling foliar disease incited by *Colletotrichum graminicola, Puccinia purpures* and *Gloeocercospora sorghi* (6). Use of banomyl or bavistin spray in cowpea effectively controlled anthracnose (98). The cowpea root rot caused by mixed inoculum of *Rhizoctonia solani, R. bataticola* and *Fusarium solani* can be controlled by seed treatment and soil drenching of carbendazim and thiophonate methyl (36). *Macrophomina phaseolina* root rot of cowpea incidences are reduced by application of 0.1% carbandazim or quintozene or 2% thiram (73). The incidences of root rot of berseem can be minimized by seed treatment with bavistin (13).

Insecticides

In sorghum, maize and pearl millet soil treatment with carbofuran 3 G or phorate 10 G @ 1.5 kg a.i./ha can control shootfly effectively, however, to control shoot borer also, this treatment should be followed by two sprays of endosulfan 35 EC (0.1%) at 40 and 60 days after sowing. Application of aldicarb 10 G 1 kg/ha at sowing of cowpea followed by sprays of endosulfan (0.7%) or monocrotophos (0.04%) or malathion (0.05%) or metasystox (0.03%) or dimecron (0.04%) or lindane (0.01%) or dimethoate (0.03%) at 40 and 60 days after crop emergence reduced insect-pest infestation and enhanced forage and seed yield. Some dust formulations like malathion (5%) or carbaryl (5%) @ 20-25 kg/ha applied twice at 30 and 60 days crop reduced insect pests in cowpea. In lucerne, soil treatment with granular insecticides like aldicarb, endosulfan and phenthoate @ 750-1000 g a.i./ha with spray of endosulfan (0.08%) gives good control of leaf hoppers and lucerne weevil. Spray application 20 days after each cut of lucerne of monocrotophos (0.04%) or dimecron (0.04%) or fenitrothion (0.05%) or dimethoate (0.1%) or quinolphos (0.05%) or methyl demeton (0.1%) or malathion (0.04%) or dichlorvos (0.04%) or carbaryl (0.1%) (83). In grasses for control of grasshoppers neem kernal extract spray (2%) is useful.

Weedicides

Pre-emergence application of trifluratin (2 kg a.i./ha) or linuron (0.5 kg a.i./ha) controlled the emerging weeds in oats (41, 70, 97). Of the different herbicides, use of some chemicals i.e. pre-emergence application of atrazine (1.0 kg), simazine (0.75 kg), atrazine (0.5%) and alchlor (3.0 kg) or linaron (2.0 kg) were found selective for weed control in M.P. chari, maize, bajra and sunflower, respectively. Weedicide trifluralin (0.5 kg a.i./ha) was found selective for weed control in cowpea (52, 95). In mixed cropping system, pre-emergence application of alachlor (2 l/ha) was found suitable in maize + cowpea, maize + guar and maize + soybean (4, 94). Pre-planting application of fluchloralin (0.6 and 1.0 kg/ha) significantly reduced the weeds in berseem and black gram (96, 106). In lucerne annual and broad leaved weeds were controlled by 0.6 kg HDB and 0.50 kg paraquat (114); dandelian perennial sowthistle and dodder with 2.2. kg dichrobenil/ha (23, 110).

5.3 Biological control

Although there is a renewed interest in biological control of soil borne plant pathogens through nematodes (8), some fungi like *Trichoderma viride, T. herzianum, Gliocladium virens* and bacterium *Bacillus subtactum* (59, 109), but this approach has rarely been implemented. Biological control of weeds in forages has not been seriously pursued. A number of entomophagous insects and pathogenic diseases have been reported on insect-pests of forages. A list of parasitic insects in major forage crop insect-pests have been given in table 7. Here we will limit our discussion on insect-pests of lucerne and clovers (including genus *Medicago, Trifolium* and *Melilotus*). Main insect-pests of lucerne and clovers are aphids and weevils. Four aphid species, namely, pea aphid, spotted alfalfa aphid, sweet clover aphid and cowpea aphid attains severe outbreaks if their natural enemies are in reduced populations, may be due to biotic, abiotic or human interferences. In general all over the world predators like coccinellids, syrphid flies are primarily responsible for much of the biological control of these aphids, but parasites have their own role to play.

Hagen *et al.* (43) briefly reviewed classified biological control of lucerne pests. An aphidid, *Asphidius smithi* an introduction from India to California in 1957, later on to Poland in 1960 and Czechoslovakia in 1967 is a parasite of pea aphid. Further this species has spread to central Mexico and Hawai (10). Another species of *Asphidius, A. ervi* has been most effective under mild, wet conditions, whereas *A. smithi* is most effective in hot and dry areas (18).

For spotted alfalfa aphid most important predator is *Hippodamia convergens.* Three exotic parasites, *Aphelinus asychis, Praon exsoletum* and *Trioxy complantue* were introduced from old world to California in 1955-1956. These species established by 1959 and had given significant control. These three parasites also control *Therioaphis riehmi*, an aphid pest of *Melilotus* species. In cowpea aphid, *Aphis craccivora*, aphidid parasite *Lysiplebus testacipes* appears to be most common.

Table 7 : Parasites of forage insect-pests

Systamic position	Major species	Common host	Host stage
Fam :. Braconidae Order : Hymenoptera	*Apanteles* spp.	*Chilo partellus, Mythimna separata, Amsacta* sp.	Larval
	Bracon sp.	*Chilo partellus, Spodoptera exiqua, Agrotis* spp.	Larval
	Chelonus spp.	*Chilo partellus, Spodoptera litura, Heliothis armigera, Agrotis* spp.	Larval
	Rogus sp.	*Mythimna separata*	Larval
	Disophruys sp.	*Spodoptera litura*	Larval
	Microctonus aethiops	*Hypera postica*	Larval
	Praen sp.	*Therioaphis maculata*	Nymphal
Fam : Chalcidae Order : Hymenoptera	*Invreia* sp.	*Chilo zonellus*	Larval
Fam : Chamaemyiidae Order : Diptera	*Leucopis* sp.	*Aphis craccivora*	Nymphal
Fam : Eulophidae Order : Hymenoptera	*Crataepiella* sp.	*Atheriogona soccata*	Pupal
	Necremnus leucarthros	*Hypera postica*	Larval
	Tetrastichus ayyari	*Chilo zonellus*	Pupal
	T. ingertus	*Hypera postica*	Larval
Fam : Ichneumonidae Order : Hymenoptera	*Enicospilus* sp.	*Chilo zonellus*	Larval
	Methopius sp.	*Spodoptera exigua*	Larval
	Xanthopimla stemmator	*Chilo partellus*	Larval
Fam : Tachinidae	*Bathyplectes curculionis*	*Hypera postica*	Larval
	Carcelis sp.	*Chilo partellus*	Larval
	Chelonus sp.	*Spodoptera exigua*	Larval
Fam : Trichogrammatidae	*Trichogramma* sp.	*Spodoptera litura, Plusia* spp.	Larval
Order : Hymenoptera	T. australicum	*Atherigona soccata*	Egg
	Trichogrammatoidea spp.	*A. soccata*	Egg

A number of species have been released for biological control of lucerne weevil in USA. The success of biological control of this weevil has been attributed to *Bathyplectes anurus* and *Microctonus aethiopoides*, out of several species which have been released (24). Lucerne weevil is a widely distributed pest of lucerne, occurring throughout the world, wherever lucerne is grown. In USA the experience of successful biological control of this pest reveals that not only release of a complex of suitable parasite against a widespread introduced pest, but also release of a complex of biotypes of a single species from appropriate climatic zone and of biotypes physiologically attuned with the pest population at source of origin helps in achieving successful biocontrol.

Other weevil species like Egyptian alfalfa weevil, *H. bruneipennis*, clover leaf weevil, *H. punctata* and sweet clover weevil, *Sitona cylindricollis* are not important in India.

Integration of IPM components

Several projects on IPM has been taken up in USA on insect pest management in lucerne and the total IPM programme have been described by Good (40). In the erstwhile USSR system of Plant Protection, IPM has been worked out and are under actual application for clovers, lucerne, cereal grasses, corn and forage root crops evolving agronomical, biological, and chemical methods (55). Use of combination of insecticide and fungicide benomyl and permethrin gave effective and economic rate of application, benomyl @ 1.0 kg a.i./ha and permethrin 0.08 kg a.i./ha reduced brown blotch incidence and pod borer infestation to less than 10% in cowpea (65).

In India integration of various components of IPM for insect pest management have been worked out at Indian Grassland and Fodder Research Institute, Jhansi for cowpea (86) by growing least susceptible variety IGFRI-450, sowing of the crop in first week of July and fertilizer doses of 30 kg N, 100 P_2O_5, 80 kg K_2O per hectare; in lucerne (87) by growing least susceptible variety, IGFRI-244, sown in last week of October with fertilizer dose of 30 kg N, 80 kg P_2O_5 and 50 kg K_2O per hectare; in mustard (87) using Chinese cabbage sown in first week of November with fertilizer dose of 50 kg N, 50 kg P_2O_5 and 30 kg K_2O per hectare.

However, studies on complete on complete IPM on forage have not been attempted in India. Attempts have been made through All India Coordinated Research Project on Forage Crops, to collect information on various pest factors like pest status, damage thresholds in various agroclimatic regions of the country, which may throw light on the problem, incidence and required thrust to control them effectively through integrated pest control measures, an effective step towards IPM on forages.

6. REFERENCES

1. Abraham, C.T. and Singh, S.P., 1984. J. Agric. Sci. **103 (1)** : 103-115.
2. Ahmad, S.T., 1969. Ph.D. Thesis, Aligarh Muslim Univ.
3. Ahmad, S.T., Sreenath, P.R. and Gupta, M.P. 1977. Indian Phytopath. **30** : 466-468.
4. Akobundu, I.O., 1982. Weed Sci. **30 (3)** : 331-334.
5. Amatobi, C.I., Apeji, S.A. and Oyidi, O., 1988. Tropical Pest Management, **34** : 173-179.
6. Anahosur, K.H., 1986. Indian Phytopath., **39 (4)** : 526-528.
7. Barnes, D.K. and Ratcliffe, R.H., 1967. Crop Sci. **9** : 640-642.
8. Barnes, G.L., Russell, C.C., Foster, W.D. and McNew, R.W., 1981. Plant Dis. **65** : 423-424.

9. Basavrajan, R., Saffee Ullah, K.M. and Murty, B.R. 1981. Ind. J. Genet. & Pl. Breed, **41** : 144-249.

10. Beardsly, J.W. 1961. Proc. Hawai Entomol. Soc. **17** : 333-336.

11. Beck, S.D. 1965. Ann. Rev. Entomol. **10** : 207-232.

12. Bergman, D.K., Dillwith, J.W., Zarrabi, A.A., Caddel, J.L. and Berberet, R.C., 1991. Environ. Ento. **20** **(3)** : 781-785.

13. Bhaskar, R.B. and Ahmad, S.T. 1990. Indian Phytopath. **43 (4)** : 589-590.

14. Borikar, S.T. and Chopde, P.R. 1980. Maydica, **XXV** :175-183.

15. Bottrell, D.R. 1979. Conc. Environ. Qual. Washington DC, GPO, 120 pp.

16. Buntin, G.D. and Pedigo, L.P. 1986. J. Econ. Entomol. **79** : 1507-1512.

17. Byrne, H.C. and Steinhaur, A.L. 1966. Ann. Entomol. Soc., America, **59 (2)** : 303-309.

18. Campbell, A. and Mackauor, M. 1973. Z. Angew. Entomol. **74** : 47-55.

19. Chester, K.S. 1950. Pl. Dis. Rep. Suppl. **193** : 190-362.

20. Coaker, T.H. 1987. Integrated Pest Management, Eds. A.J.M. Burn, T.H. Coaker and P.C. Jepson, Academic Press, London, pp. 69-88.

21. Cordoso, C.C.I., Gutirrez, P.J.A., Lenne, J.M. 1983. Acta Agron. **33 (2)** : 49-55.

22. Craven, R.H. and Hasbarge, P.R. 1970. Univ. Minnesota Agric. Ext. Serv. Fact. Sheet, 560.

23. Dawson 1989. Weed Res. **28(5)** : 607-611.

24. Day, J.H. 1981. In : Biological control in crop production, Ed. G.C. Papavizas, Allanheld Osmum, Montclaire, N.J., pp. 361-374.

25. Dent, D.R. 1986. Ann. of Applied Biol. **188** : 577-583.

26. Dent, D.C. 1991. Insect Pest Management, CAB International UK, 604 pp.

27. Dhaliwal, J.S. 1988. Ind. J. Ecol. **15(1)** : 87-91.

28. Dutta, T.R., Panwar, O.P.S., and Vinod Shankar 1972. Indian Agriculturist, **10** : 33-39.

29. Fadeev, I.U.N. and Novozhilov, K.V. (Eds.) 1987. Integrated Plant Protection, Oxonian Press Pvt. Ltd., New Delhi.

30. Faruqui. S.A., Pandey, K.C. and Patil, B.D. 1986. Indian J. Ecol. **13(1)** : 120-122.

31. Fatunla, T. and Badaru, K. 1983. Agric. Sci., **101** : 423-426.

32. Ferrow, D.N. 1987. In : Insect Outbreaks, Eds. P. Barbosa and J.C. Shultz, Academic Press, San Diego and London, 195-205.

33. Fery, R.L., Dukes, P.D., Cuthbert, F.P. Jr. 1976. J. Am. Soc. Hort. Sci., **101** : 148-149.

34. Fick, G.W. 1982. Agron. J., **74** : 835-840.

35. Gandhi, S.K., Chand, J.N. 1985. Ind. J. Agric. Sci., **55(5)** : 326-327.

36. Gangopadhyay, S. and Grover, R.K. 1985. Ind. J. Mycol. and Pl. Patho. **14(1)** : 57-68.

37. Givovich, A., Weiball, J. and Petersson, J. 1988. Entomologia Expt. et. Appl., **49** : 259-264.

38. Glover, C. and Melton, B. 1966. N.M. Agric. Exp. Stn. Res. Rep., **127** : 4.

39. Glover, D.V. and Stanford, E.H. 1966. Crop Sci. **6** : 161-165.

40. Good. 1973. Paper presented at US-USSR Pest Manag. Conf. 10-18 Sept., 1973 Kiev.

41. Gupta, S.D. 1992. Subject Matter Workshop cum Seminar on Forage Development and its role in livestock, IGFRI, Jhansi (Jan 2-16, pp. 43-47).

42. Gupta, M.P. and Shri Ram 1989. Trop. Pest Management, **35 (4)** : 399-402.

43. Hagen, K.S., Viktorov, G.A., Yasumatsu, K. and Schuster, M.F. 1976. Biological control of pests of range, forage and grain crops (pp. 397-442). In : Theory and practice of biological control. Eds. C.B. Huffakev and P.S. MessengerAcademic Press, New York.

44. Halalli, M.S., Gowda, B.D.S., Kulkarni, K.A. and Goud, J.V. 1982. SABRAO J. **14** : 165-170.

45. Hanson, C.H., Busbice, J.H., Hill, R.R. Jr., Hunt, O.J. and Oakes, A.J. 1972. J. Environ. Qual. **1** : 106-111.

46. Hanson, C.H., Pederson, M.W., Berrang, B., Wall, M.E. and Davis Jr., K.H., 1973. In : Anti-quality components of forages. Crop Sci. Soc. of America, Madison, Wisconsin.

47. Haynes, D.L., Bradenburg, RK. and Fisher, P.D. 1973. Environ. Entomol. **2** : 889-899.

48. Hijano, E.H., Barnes, D.K. and Frosheiser, F.I. 1983. Crop Sci. **23** : 31-34.

49. Hutching, S.H. and Pedigo, L.P. 1990. J.Econ. Entomol. **83** : 1587-1594.

50. Hutchins, S.H., Buntin, G.D. and Pedigo, L.P. 1990. Agron. J. **82** : 1035-1044.

51. Irwin, J.A.G., Maxiwell, D.P. and Bingham, E.T. 1981. Crop Sci. **21** : 271-276.

52. Johnson, W.A. and Amling, H.J. 1964. Proc. South Weed Sci. Soc. **17** : 221-227.

53. Johnson, K.G.R., Sorensen, E.L. and Horber, E.K. 1980. Environ. Entomol. **9(1)** : 133-136.

54. Johnson, J.W. and Teetes, G.L. 1980. Proc. of an International Short Course in Host Plant Resistance. Texas A & N University, 22 July - 4 Aug., 168-180.

55. Karavianskii, N.S. 1987. In: Integrated Plant Protection, Eds. I.V.N. Fadeev and R.V. Noruzhilow, pp. 250-268.

56. Karwasra, S.S., Chand, J.N. and Lodhi, G.P. 1982. Forage Research, **8(2)** : 155-157.

57. Kogan, M. and Ortman, E.F. 1978. Bull. of Entomol. Soc. Ame. **24** : 175-176.

58. Kumar, R. 1984. Insect pest control with special reference to African Agriculture, Edward Arnold, London.

59. Maiti, D. and Sen, C. 1985. Ind. J. Agric. Sci. **55(7)** : 464-468.

60. Manglitz, G.R. and Calkins, C.O. 1963. J. Econ. Entomol. **6(5)** : 716-717.

61. Manglitz, G.R. and Gorz, H.J. 1968. J. Econ. Entomol. **61** : 90-93.

62. McMurty, M. 1962. Hilgardia, **32** : 501-539.

63. Nelson, M.W. and Kuchl, R.O. 1982. Environ. Entomol. **11** : 989-996.

64. Okoli, P.S.O., Dralsom, P.N. and Schall, J.M. 1984. Agron. J. **76** : 363-366.

65. Oladiran, A.O. 1990. Trop. Pest Mgmt. **36(4)** : 397-402.

66. Pandey, K.C. Singh, A. and Faruqui, S.A. 1984. Ind. J. Genet. & Pl. Breed. **44(1)** : 253-258.

67. Pandey, K.C., Singh, A. and Faruqui, S.A. 1984. Ind. J. Genet. & Pl. Breed. **44 (1)** : 1-6.

68. Pandey, K.C. and Faruqui, S.A. 1988. Ind. J. Ecol. **15(2)** : 199-201.

69. Pandey, K.C., Singh, A. and Faruqui, S.A. 1990. Ind. J. Genet. & Pl. Breed., **50(1)** : 77-80.

70. Pandey, R.K. and Gupta, S.D. 1979. Curr. Agric. **3(3-4)** : 147-150.

71. Pedigo, L.P., Hutchins, S.H. and Higley, L.G. 1986. Ann. Rev..Entomol. **31** : 341-368.

72. Rabb, R.L. and Guthrie, F.E. (Eds.) 1970. Concept of Pest Management, Cong. Proc. N.C. State Univ., Raleigh.

73. Ram Dass, S. and Sivaprakasam, K. 1988. Madras Agric. J. **75(3-4)** : 133-138.

74. Rana, B.S. and Murty, B.R. 1971. Indian J. Genet. & Pl. Breed. **31** : 521-529.

75. Rana, B.S., Tripathi, D.P. and Rao, N.G.P. 1976. Indian J. Genet. & Pl. Breed., **36** : 244-249.

76. Rai, S., Jotwani, M.G. and Jha, D. 1978. Indian J. Ent., **40(3)** : 254-260.

77. Reeder, B.D., Norton, J.D. and Chambliss, O.L. 1972. S.J. Am. Soc. Hortic. Sci. **97** : 235-237.

78. Reichelderfer, K.H., Carlson, G.A. and Norton, G.A. 1984. FAO Plant Protection and Protection paper 58; FAO Rome.

79. Ridgway, R.L., Silverstein, R.M. and Inscoe, M.N. (Eds.) 1990. Behaviour modifying chemical for Insect Management, Application of Pheromones and other attractants. Marcel Dekker Inc. New York and Basel.

80. Shah, N.K., Pandey, K.C. and Faruqui, S.A. 1991. Range Mgmt. and Agroforestry, **12(1)** : 67-68.

81. Shah, N.K., Pandey, K.C. and Faruqui, S.A. 1991. Range Mgmt. and Agroforestry, **12(2)** : 159-163.

82. Sharma, S.R. and Verma, A. 1984. Zeitschrift fur Acker and Pflanzenbau, **153(1)** : 23-31.

83. Shivanker, V.J., Tomer, A.S. and Tomer, P.S. 1989. Agric. Rev. **10(4)** : 183-194.

84. Shri Ram and Gupta, M.P. 1971. Sorghum Newsletter, **14** : 71-72.

85. Shri Ram and Gupta, M.P. 1977. Ind. J. Ento. **39(1)** : 95-99.

86. Shri Ram and Gupta, M.P. 1989. Trop. Pest Mgmt. **34(4)** : 348-351.

87. Shri Ram and Gupta, M.P. 1990. Trop. Pest Mgmt. **36(3)** : 258-262.

88. Shri Ram, Patil, B.D. and Shivanker, V.J. 1983. All India workshop on crop losses due to insect pests held at A.P.A.U., Hyderabad (7-9 Jan), pp. XIV. 1

89. Shri Ram, Gupta, M.P. and Shivanker, V.J. 1987. Integrated pest management of some fodder crops. IGFRI-ICAR, New Delhi.

90. Shri Ram, Gupta, M.P. and Patil, B.D. 1989. Trop. Pest Mgmt., **35(4)** : 345-347.

91. Shri Ram, Gupta, M.P. and Purohit, M.L. 1990. Indian J. Ent. **52 (4)** : 627-636.

92. Singh, D. and Patel, P.N. 1977. Ind. Phytopath. **30** : 99-102.

93. Singh, F. and Sinha, N.C. 1981. Ind. Soc. Agron. Nat. Symp. Hissar, pp. 290-293.

94. Singh, F., Sinha, N.C. and Patil, B.D. 1984. Ann. Conf. of Indian Soc. Weed Sci. (Feb. 27-28), B.H.U., Varanasi.

95. Sinha, N.C. 1980. Ann. Report Weed Eco. & Control, IGFRI, Jhansi.

96. Sinha, N.C. and Singh, R.P. 1985. Ind. J. Weed Sco. **17(3)** : 61-64.

97. Sinha, N.C. and Singh, R.P. 1987. Indian J. Plant Physol. **30(3)** : 266-271.

98. Sohi, H.S. and Rawal, R.D. 1984. Pesticides, **18(5)** : 33-34.

99. Solomon, S. and Prem Kishore 1989. J. Ent. Res. **13(1)** : 21-25.

100. Springer, T.L., Kindler, S.D. and Sorenson, E.L. 1990. Environ. Ento. **19(5)** : 1614-1617.

101. Srivastava, D.N. and Rao, Y.P. 1963. Indian Phytopath. **16** : 390-391.

102. Starks, K.J., Eberhart, S.A. and Doggett, H. 1970. Crop Sci. **10** : 519 - 522.

103. Starks, K.J., Muniappan, R. and Eikenbary, R.D. 1972. Ann. Entomol. Soc. Am. **65** : 650-655.

104. Sulladamath, V.V., Shivshankar, G., Anil Kumar, J.B. and Veerappa, K.B. 1977. Ind. J. Genet. and Pl. Breed. **37**: 101-102.

105. Sundaram, N.V. 1970. 1st Workshop, Forage Production and Utilization, ICAR, Hissar, India.

106. Tiwana, M.S., Tarlok Singh and Brar, L.S.1985. Proc. XI Int. Grassland Cong. (Aug. 24-31, 1985), Kyoto, Japan.

107. Tiwari, N.P., Raghuwansi, N.K. and Tiwari, S.P. 1985. Ann. Conf. Ind. Soc. Weed Sci. 77-78.

108. Tummala, L.T. and Haynes, D.L. 1977. Environ. Entomol. **6**: 339-349.

109. Upadhyay, J.P. and Mukhopadhyay, A.N. 1986. Trop. Pest Mgmt. **32** (3) : 215-220.

110. Waddington, J. 1980. Weed Sci. **28** : 164-167.

111. Walker, P.T. 1983. Agric. Ecosystems and Environ. **9** : 119-158.

112. Weible, D.E., Starks, K.J., Wood, E.A. and Morrison, R.D. 1972. Crop Sci. **12** : 334-336.

113. Wong, L.S.L., Mckenzie, R.D.H., Harder, D.E. and Martens, J.W. 1983. Can. J. Genet. Cytol. **25** : 329-335.

114. Wilson, R.G. 1986. Weed Sci. **34(3)** : 423-426.

24 Mycoherbicides : An Overview

Rajni Gupta, K.G. Mukerji and Rajeev K. Upadhyay

CONTENTS

Integrated Pest and Disease Management
Rajeev K. Upadhyay, K.G. Mukerji, B.P. Chamola and O.P. Dubey (eds.)
APH Publishing Corporation, New Delhi-110002, India

ABSTRACT

Weeds are wild plants, grow at places where they are unwanted. These plants compete with the crop plants and inflict considerable losses. So far, chemical and mechanical methods of weed control have been adopted, though both are quite expensive. Besides, chemicals have got residual effects in soil, water and food stuff, and ill effects on non-target organisms and environment, and induce herbicide resistance in crop plants. As a result, perception of the farmers, scientists and administrators has changed particularly with reference to environmental safety. Therefore, biological control has received new momentum and is being used successfully throughout the world as an economical, effective and environmentally safe method for controlling weeds.

1. INTRODUCTION

The harmful effects of weeds in and near cultivated fields are many (24). The most widespread and serious of these effects is yield loss by weed competition which is still not well understood because of the many factors involved. A quantitative approach to weed competition can be helpful to integrated weed control practices. The yield of crops (Y) infested with a specific weed can be expressed as $Y=1/(a+bx)$, where x is the population density of weeds. The values of parameters a and b, however, vary with changes in factors influencing competition, so the variation in Y has to be elucidated under a wide range of field conditions. Through the analysis of weed competition, valuable information which contributes to an integrated approach to weed control is obtained i.e. information about target weeds to be controlled, desirable time to weed control, choice of weed control measures and balance sheets of gain and cost of weed control.

Biota in natural environment are essential for the maintenance of environmental quality and for the survival of life on the earth. Over and misuse of agrochemicals including herbicides can put many species at risk and so jeopardize the preservation of functional ecosystems, and also allow some herbicide resistant weeds to proliferate in croplands (48, 81). However, rapid and extensive use of herbicides, 29% of the total world pesticide production, has brought about a drastic reduction in the use of labour and the cost of weed control in many countries including Japan (18, 44). It may therefore be of merit to select the most appropriate chemical, cultural, mechanical or biocontrol measures and combine them according to the crop situation involved.

1.1 Phases of weed damage

Weeds affect many phases of crop culture, cause serious losses in yield and quality of crops, and increase the cost of crop production (24). Direct damage due to the presence of weeds in and near cultivated fields can be summarized as follows :-

i) Weeds in cultivated fields reduce crop yield and quality by competing for growth requirements, such as nutrients, water and light.

ii) Weeds intensify the problems of diseases, insects and other pests by serving as hosts (41). For instance, cut grass *(Leersia oryzoides)* is the well known host of bacterial leaf blight of rice (39). It has also been reported that grass weed infestation causes increase in the population of rice stink bug (65).

iii) Weeds reduce the efficiency of harvesting or other machines because trailing weeds get entwined in machines and the rank growing weeds often cause crop lodging.

iv) Poisonous weeds in pasture or forage crops cause injuries to live stock. Many kinds of perennial broad leaf weeds in grazing lands are known to be poisonous (41). Bracken *(Pteridium aquilinum)* is one of the most extensively known weeds to be poisonous to live stock in Japan (62).

v) Acquatic weeds reduce efficiency in irrigation systems (78). The most troublesome and widely distributed is water hyacinth *(Eichhornia crassipes)* Water is wasted through evapotranspiration as well as reduced flow of water. Water loss through evapotranspiration by water hyacinth was reported to be about 7.8 times more than that of open water (37).

Of these different types of damages due to weeds, yield loss by weed competition in cultivated fields is the most wide spread and serious.

1.2 Factors determining the degree of competition

Weed competition is complicated because various factors affect the extent to which it occurs. The magnitude of the contribution of any one factor can be expected to vary greatly because of its dependence on other factors. For example the density of weeds can be expected to have little effect on a crop if the weeds are only short lived.

1.3 Approaches to the better method of weed control

Though current knowledge is rather limited, much useful information for the integration of weed controls can be expected to be obtained through the type of analysis described. The information would assist in :

1) improving cultural methods of weeds control such as planting time, spacing and fertilizer management,
2) confirming which target weeds are to be controlled,
3) determining the optimum time of weed control,
4) selecting desirable weed control measures according to the level and consistency of their effect, and
5) forecasting yield loss due to weed competition

1.4 Target weeds to be controlled

A large, tall weed such as barnyard grass may reduce the rice yield even at a density of one plant per ha, though the yield reduction may be negligible. On the other hand, it is possible that a small short weed, such as *Monocharia, Rotala indica* or slender spike rush *(Eleocharis acicularis)*, grows without any damage to rice when its infestation is very slight. Less than 20 gm/m^2 of such seeds at the maturing stage could be empirically expected to have "no effect" on rice and so need not be controlled. Weeds which are introduced into fields after rice plants are well established need not be controlled because their growth is strongly suppressed by rice. Target weeds are those which emerge early during the season, roughly within 3 or 4 weeks after transplanting of rice, though this period will vary depending on weed species, weed population, cultural methods of rice and locality. Weed control, accordingly, should aim to prevent these target weeds from emerging or to kill them before they affect the panicle number of rice.

1.5 Forecasting yield loss due to weed competition

Forecasting yield loss due to weed competition is important for determining whether the practice of weed control is necessary or not, how much money can be profitably spent on control and the extent to which weed control is necessary. When the relation between weed population and yield loss is known, an estimate for yield loss could be easily obtained in advance. For instance, the formula Y=1/(a+b x) can be used for competition between the crop and weed. The values of parameters vary with the changes in factors influencing competition. Accordingly, the average yield and its standard deviation for each density of weeds were calculated using the results obtained from several years experiments which covered a number of the major factors influencing competition.

2. SUITABILITY OF A WEED FOR BIOLOGICAL CONTROL

There may be several reasons, why a weed may be selected for biological control; it may be difficult to control by conventional methods, may have developed resistance or tolerance to the chemical herbicides, or the use of herbicides may not be allowed or be undesirable or uneconomic. However, it is advisable to carefully analyse the weed problem to determine whether it is suitable for biological control (27, 60).

It is important to correctly identify the weed species, and to know its biology, mode of dispersal and mode of reproduction. In the case of skeleton weed *(Chondrilla juncea)*, an apomict, three clones were found to occur in Australia. Searches were therefore made in Europe for strains of the rust *Puccinia chondrillina* for each individual clone (30,31,33). Whether a weed is a single plant species or several species together is important. European weedy black berrier in Australia are composed of several species of *Rubus,* also known as taxa of *Rubus fruticosus* aggregate (1). It was therefore necessary to collect several strains of the rust *Phragmidium violaceum* with different degrees of pathogenicity on individual taxa and make a pool of spores from these strain, before, testing against other members of Rosaceae (8).

Information should be gathered on the origin of the weed-whether it is introduced or endemic. In the case of the former the country of origin and other areas where it commonly occurs should be located. An important point in selecting a weed for biological control is to evaluate not only the economic losses it causes, but also any beneficial aspects it has. If the weed presents some beneficial properties such as flowers being useful for the honey industry, it should not be processed for biological control. This will obviously be affected by the extent of the weed's distribution and the economic importance of the industry it affect. In cases of conflict of interest the cost and feasibility of other means of control is relevant. However, in some situations biological control may be the only solution to a weed problem.

3. STRATEGIES IN BIOLOGICAL CONTROL OF WEEDS WITH PLANT PATHOGENS

The management of natural enemies (mainly insects and fungi) is the basis of biological control of weeds. It is based on the ecological observation that natural enemies can be of the prime and that some of these natural enemies have restricted host range (14, 16, 28, 34, 60, 61, 87).

Traditionally the "classical strategy" of biological control of weeds has been practised by introducing exotic arthropods from the original habitat of the weed or else where to control perennial naturalised weeds commonly occupying wider areas of undisturbed land (38, 82). In principle an introduction on one occasion is adequate to allow the build up of a sufficiently large population to have a long term impact on the weed density. This strategy has more recently been applied also to annual weeds (2) and extended mostly by the use of plant pathogens (32, 33). Thus plant pathogens have been searched for in the original habitat of the weed and introduced into the new environment thereafter relying on their ability to self perpetuate and reduce the host population below the economic there hold. Despite the traditional use of the classical method against weeds in large stable situations of low economic return and with single dominating weed species, the successful biological control of *Chondrilla juncea* with *Puccinia chondrillina* in wheat crops in South Eastern Australia (20, 35) has shown that the method is also effective in annual cropping systems.

Recently a new strategy of biological control of weeds has been developed using virulent strains of pathogens already occurring on the weeds and enhancing their destructed action. Pathogens are mass produced and used against weeds in a manner similar to conventional herbicides. In this "bioherbicide or inundative strategies" both endemic and exotic pathogens can be used as "mycoherbicides" though to date mostly endemic fungi have been so employed (32). According to Templeton (75) the reasons for the lack of effect of such endemic pathogens on indigenous weeds under normal conditions are multiple, including cultivation and the use of selective pesticides or management practices that disrupt or eliminate stages of pathogens life cycle. In order to overcome these constraints the pathogens are collected from undisturbed areas and multiplied in large fermentation tanks. However, some of the basic considerations in the selection of pathogens are outlined here.

3.1 Selection of pathogens

All kinds of plant pathogens including viruses, bacteria, mycoplasmas, fungi and nematodes that can cause reduction in plant growth can be considered as biocontrol agents for weeds. So far fungi have received most attention probably because they are numerically more important, have a well defined taxonomy, in many cases their host specialisation is well known, and at the same time more is known of the mechanism and stability of that specialisation. Moreover most viruses require insects for the transmission, many bacteria are not specific and relatively few nematodes with potential are known. Additionally, of course, in classical programmes only those pathogens not already present in the target area or only represented by ineffective host specialised races would be candidates for selection.

3.2 Survey of weed diseases

A literature survey should enable one together all possible information on the occurrence of the weed and its diseases and their geographical distribution. It is necessary if spores and mycelia are applied so as to inundate the host plant with inoculum at the most propitious time for infection to eliminate the weed in its early stages of growth. This technique is mostly applied in the annual cropping system but it has also been proved efficient in orchards (55) and water ways (19).

Among other strategies can be cited the integration of biological control into pest management systems, which will be referred to here as "integrated weed control". It is a relatively new field compared with integrated control of arthropods. Attempts are being made to combine the destructive effect of pathogens with compatible herbicides and mechanical or cultural practices (53).

4. WEED BIOLOGICAL CONTROL WITH EXOTIC PLANT PATHOGENS

4.1 The classical strategy

As the success of control depends on the ability of the pathogen to self perpetuate, disperse and reduce the host population in the infested region, special care is taken to select agents which are strongly pathogenic to the plant and are eco-climatically suitable for the target area. Also pathogens should be specific to the weed or at least to a small group of related plants as long as these include to determine whether certain organisms are already present and the extent of their damage to the target weed in the area where control is desired. This information helps to select areas to be surveyed. During subsequent field surveys, collections are made of all pathogens causing damage to the weed not only those already known, and if this can be done at different times of the season it will enable organisms attacking different plant stages to be collected. Priority should be given to pathogens attacking all stages of development, from very young seedlings to the mature plants. Those pathogens with high virulence and wide spread occurrence are normally considered as suitable for further studies for their biological control effectiveness.

Plant pathogens can only be used in biological control of weeds if there is sufficient evidence that these will not be harmful to useful plants. Most of this evidence can be obtained by glasshouse tests for host specificity, but field observations can also be supporting evidence. Further indications regarding host specificity of a pathogen can also be had from its taxonomic position. A literature survey will indicate if it is recorded only on the weed or on other plants, and also the level of specificity characteristic of its close relatives. Plants are selected on the basis of their phylogenetic relationship to the target weed. This method is most commonly used in the current testing programmes and includes, in order of relationship (i) other forms of the target species, (ii) other species of the same genus, (iii) other members of the tribe, (iv) other members of the sub family, (v) other members of the family, (vi) other members of the order. In some cases where a pathogen has never been recorded on plants outside a family, the list of test plants may include only members of that family, is not being necessary to include members of the order. Thus *Phragmidium violaceum* from blackberry was tested only on the plants belonging to the Rosaceae, as the genus *Phragmidium* has been recorded only on plants of this family (8). When determining the host specificity of a fungal pathogen, a visual examination is made to assess the external symptoms produced by the pathogen. Detailed microscopic observations must also be made to determine if the plant is immune or resistant to the pathogen. In the latter case, there is some development of a mycellium with in or outside the host tissue and this can vary according to the environment or physiological conditions of the plant.

4.2 Inundative method

The inundative method consists of applications of massive doses of inoculum to the weed population to create a fast and high level of epidemic when conditions conducive for disease development and weed control are present (74). An inundative agent is manufactured, formulated, standardised, packaged, and registered as a herbicide. It is applied to the weed by conventional chemical application methods and tools, and the application schedules (67, 73), are usually integrated with other pesticides schedules. Because an innudative agent can produce rapid and complete weed control similar to the fast acting chemical herbicide, it is the best option for agricultural and intensively managed agroecosystems (12). Agricultural soils hold considerable seed reserves to initiate new weed infestations each growing season, making it necessary to reapply the inundative agent, because of these similarities with chemical herbicides, the inundative tactic has been called the "Mycoherbicide" or the "microbial herbicide strategy" (74). Although any pathogen can be applied in a massive inoculum dose, the term 'inundative agent' and inundative control should be reserved only for a pathogen that can be mass produced *in vitro* and applied as herbicide preparations. According to the above definition, obligate parasites such as rusts, smuts and viruses cannot be classified as 'herbicidal' agents since they can not be cultured in a practical sense and therefore it would not be possible to treat large weed infestations over hundreds or even tens of hectares. Nor can they be used to produce a rapid and high level of weed control with in a short time.

4.3 Augmentative method

In this method, weeds are inoculated with obligate parasites using inoculum gathered from naturally infected plants. Phatak *et al.* (53) have demonstrated the feasibility of controlling Canada thistle with *Puccinia obtegens*, yellow nut sedge with *P. canaliculata* and Johnson grass with *Sphacelotheca holci*. Scheepens & Van Zon (59) have cited similar use of *Puccinia punctiformis* to control Canada thistle, *Cirsium arvense*, and of cynophages to control blue green algae. Unlike the obligate parasites used as classical agents, these pathogens are native or naturalised in their respective regions and occur on native or naturalised weeds, generally causing endemic diseases. But, similar to the inoculative agents, they are capable of self dissemination and of causing epidemic build up after an application of inoculum. In terms of the quantity of inoculum used, the types of inoculum deployment, and the management actions necessary to ensure the occurrence of the epidemic, the augmentative strategy can be viewed as being the middle of a continuum between the inoculative and the inundative approach; they are generally large in the case of the inundative method.

5. DEVELOPMENT OF BIOHERBICIDES FOR WEED CONTROL

Even though bioherbicides are usually pathogens already present in the infested region, inundation of the environment with massive dose of inoculum still requires them to be host specific, and to be effective they should obviously still be sufficiently virulent. The general procedure for the development of bioherbicides is therefore broadly similar to the classical strategy.

5.1 Scope of fungi in biocontrol of weeds

As for classical biological control any plant pathogen can be considered as a bioherbicides. However, Deuteromycetes or fungi imperfecti have received the major attention, as fungi belonging to this group are mostly facultative and can be easily grown on artificial media. This facilitates the mass production of the infective stage of the fungi, either in the form of reproductive spores or as vegetative mycelium.

Pathogens selected to serve as inundative control agents must posses certain essential features. They must be culturable *in vitro* and readily produce spores or other infective propagules that can be used as standarised inoculum. The inoculum must maintain viability and be readily capable of infecting and killing the host in environments with reasonably wide latitude (23). It should be possible to grow the agents in submerged liquid cultures because the current industrial production facilities can meet only this type of fermentation need. These features are shared by facultative parasites, which can be mass produced *in vitro*. There are general pathogenic genera among the facultatives that will satisfy both the efficacy and host specificity requirements. However, the fungi have received maximum attention (75) as they offer a wide choice of being biocontrol agents. Most importantly, unlike bacteria and viral pathogens which need insect vectors, natural openings, or wounds for entry into hosts, fungi are capable of active penetration, which makes them the desired candidates for spray applications. Although both indigenous and exotic pathogens may be equally effective as inundative control agents the choice is clearly with native pathogens. Wilson (90) listed three advantages that the use of plant pathogens would have over herbicides.

1) they can be specific to the weeds,
2) residue and toxicity problems would be reduced or eliminated altogether,
3) there would be no accumulation of the herbicides in the soil or ground water

The degree of success of an exotic fungus imported into a new region depends on its ability to adapt the climate of the new region, which may influence factors affecting disease cycle.

5.2 Control of specific weeds

Wapshere (86) outlined a strategy for demonstrating the safety of new biological control agents for weeds. The new agent should be tested against a group of Phylogenetically related plants plus those socially important plants which may be considered at risk for one of several reasons including the fact that (1) the plants are related to the weed, (2) the mycology related to the new agent and the 'at risk' plant is inadequate, (3) the plant is attacked by a close relative of the new agent, (4) the plant has never been ecologically exposed to the new agent.

This method of testing for safety is replacing the older method of exposing a broad spectrum of crop plants to the new agent. The new method considers the safety of not only crop plants but also other desirable plants whether they are introduced or native species. Caution is well warranted. For example there are two rusts of Knapweed, *Centauria diffusa*, which would be highly desirable as biological agents to help control this rapidly spreading weed in North America. In California there is a race of this rust which appears to be host specific to the slender flower thistle (88). A biological control for *Cirsium arvense* has been sought for many years. Rust diseases of this aggressive weed have been proposed for its control since 1894 (26). *Puccinia obtegens* for example has been observed to kill individual plants or small stands of this weed (80).

Other economically important weeds which are subjects of biological control using fungi include water hyacinth, yellow nutsedge and giant ragweed. Spread of *Eichhornia crassipes* has been controlled mostly using mycoherbicides. Rust fungus *Uredo eichhorniac* which has been found in Argentina, once was considered for release in Florida, USA, as a classical biocontrol agent (15). Control of giant ragweed, *Ambrosia trifida* using *Puccinia xanthii.* Halst has been proposed using isolates from North America (4). The control of yellow nutsedge, *Cyperus exulentis* and *C. rotundus* has also been proposed (11, 82). The first international use of a plant pathogenic fungus to control a weed was probably attempted in Australia when Hasan (29) suggested the possibility of introducing *Puccinia chondrillina* into the country for control of rust skeleton weed. After much research and deliberation, the rust was finally released through quarantine in the summer of 19871-72 at Wagga (21).

The rust appears to exist in the form of several races which are very specific to different types of weeds which are dissimilar either phenotypically or in terms of resistance or a combination of both. *Chondrilla juncea* is an apomictic plant and populations of the weed in a region are in reality clones of the original introduction.

Another weed which has been successfully controlled by an introduced fungal pathogen is the black berry. The rust *Phragmidium violaceum* was released in Chile in 1973 on two species of black berry, *Rubus constrictus* (49). The rust disseminated rapidly and reduced weed populations. Another successful example of a rust is the use of *Uromyces galega* saccardo, which is reported to be established and spreading on goatsrue. Another rust *P. carduorum* Jacky, attacks the musk thistle (54). The foremost commercial production of a fungal pathogen as a mycoherbicide is the use of endemic anthracnose disease *(Colletotrichum gloeosporioides* f. sp.*aeschynomene* for the biocontrol of northern jointvetch, *Aeschynomene virginica*, a leguminous weed in rice fields, and to some extent in soybean in Arkansas (68). The weed is responsible for reduction in rice yield and in quality. *Colletotrichum gloeosporioides* f. sp *aeschynomene* occurs naturally on its host, but the infection is not very severe and the disease rarely kills the plant.

Phytophthora rot of milk weed vine-A strain of the stem and root rot fungus initially identified as *Phytophthora citrophthora* was isolated from dying milk weed vine, *Morrenia odorata*, a serious weed in the citrus groves of Florida (9, 10). The pathogen was evaluated as a mycoherbicide. (9, 10). The pathogen was found to be stable even after manipulation for several years and proved to be host specific when tested against 58 members of 12 families. A few plants other than milk weed were found to be infected, but only with a very high dose of the fungus (55). Morin *et. al.*, (47) reported the presence of pathogenic fungi on the *Noogoora burr* complex in Australia. Shivas (63) studied infection caused by *Phomopsis emicis* on the *Emex australis.*

6. GENETIC MANIPULATION OF HOST PATHOGEN SYSTEMS

It is possible to harvest protoplast from the cells of a number of fungal genera (51). The procedure involves enzymatic digestion of cell wall in the presence of an osmotic stabilizer. Conditions are optimized to provide the greatest yield, health and uniformity of protoplasts. Somatic hybridization resulting from protoplast fusion can be manipulated both within and between species. In cases where natural events of hyphal fusions or cells fusions of compatible mating types are difficult to achieve, protoplasts provide a means of producing diploids or multinucleate heterokaryons (52). Protoplast fusion crosses between closely related species of *Aspergillus* or closely related species of *Penicillium* resulting in hybrid heterokaryotic and diploid colonies upon regeneration. Upon haploidization a range of nonparental segregant are recovered (52). Leslie (42) induced mutation by UV irradiation in *Gibberella zeae* and identified morphological, auxtrophic, and temperature sensitive variants protoplast fusion was performed with four of the auxotrophic mutants, resulting in prototrophy by heterokaryosis as long as selective pressure was maintained. It is clear that protoplast fusions could facilitate the combining of desirable traits of potential mycoherbicide fungi. Enhanced virulence, toxin production, sporulation, tolerance to environmental stresses and shifts in specificity could be the benefits of these techniques. The production and regeneration of protoplasts from mycelium of *Alternaria crassa* a potential biocontrol agent of Jimson-weed, has been demonstrated (7).

Charudattan (12) identified the areas of mycoherbicide development that could benefit from genetic alteration of the pathogen. Charudattan (12) believes that immediate need for improving fungal strains for weed control can be met by random mutation and selection. Pesticide resistance is one example. The efficacy of Collego northern jointvetch is reduced if benomyl is applied within 3 weeks after Collego treatment for control of rice blast caused by *Pyricularia oryzae* or for control of soybean anthracnose diseases (68). Te Beest (72) induced mutation in spores of the Collego fungus for benomyl resistance by treatment with ethylmethane sulphonate. The mutants may eventually be used in a new and improved Collego with pesticide resistance.

Mutation and selection might be adequate for recovering isolates with preferred capabilities in particular environments. For example an ability to germinate and penetrate the host more rapidly when moisture is available would alleviate much concern over the duration of dew periods following inoculation. Midseason control of prickly sida by *Colletotrichum malvarum* could be achieved if mutants could be selected that express high virulence at high temperatures.

Virulence of pathogens might be increased by mutation and selection. Candidates for improve virulence include the Collego fungus and Indian Jointvetch and *Colletotrichum cocciodes* on velvet leaf. The current problem is that too little is known about the virulence factors of these fungi and consequently it is not known whether genetic material exists in the nuclei or cytoplasm that could contribute to dramatic changes in virulence. Recombination DNA technology provides a more elegant means of alterning the DNA of a pathogen. Transformation of filamentous fungi requires the use of protoplasts (79). Several vectors for gene transfer are available and have been transformed, including several plant pathogens (57). Although incorporation of genetic material by these techniques is still somewhat random, greater control over the site and extent of genetic changes can be exercised compared to random mutation. It is possible to select isolates with desirable characteristics and evaluate the DNA chemically and physically. The accurate and subtle manipulations favour the prospects of taking virulence genes from one fungus and inserting it into another while still retaining specificity of the second. However, as far as we know, no genes for virulence have yet been cloned in plant pathogenic fungi.

Turgeon & Yoder (79) postulated that mycoherbicides may be improved using genetic engineering techniques to increase virulence by manipulation of genes controlling production of host non specific toxins, along with genes for resistance to the toxin to protect recipient cells from self destruction, could result in enhanced virulence in a pathogen that is specific for a target weed. So far, no genes controlling toxin production have been cloned successfully in plant pathogenic fungi. These techniques may also be useful for increasing sporulation capacity and directing adaptation to adverse environment. Pesticide resistance has already been stably incorporated into pathogens by transformation techniques that did not alter pathogenicity. The β-tubulin genes cloned from *Colletotrichum graminicola* and *Neurospora crassa* benomyl resistant mutants were used by Panaccione *et al.* (50) to transform

C. graminicola, a pathogen of corn and sorghum to benomyl resistance. Extensive changes in an organism may not be worth while, since research and development costs of DNA cloning can be expensive. However, it should be more practical to insert genes from one organism into another organism that already produces the product, by passing a rate limiting step or improving growth characteristics of the producing organism (22).

Loss of virulence in culture is a pervasive problem in plant pathology. Undoubtedly, some reports of loss of virulence actually involve loss of viability or improper handling of pathogens and plant material.

Some aspects may be controllable by manipulating physiological influences such as media, temperature, and illumination under which cultures are maintained. Reversible loss of virulence has been attributed to changes in characters determined by cytoplasmic inheritance. Irreversible changes in isolates may be due to chromosomal mutations. The degree to which mycoviruses or plasmids are involved in loss of virulence is unknown. The problem of loss of pathogenicity or virulence in culture can be solved with certainty if the followings are known.

a. particular characters that contribute to pathogenicity of virulence,

b. which of those characters have been lost,

c. how those characters are regulated,

d. what genes are directly or indirectly involved in altering those characters,

e. how the genes have been changed or regulated.

The paradox of mycoherbicides is that the specificity sought is often too narrow for the market aspirations. Luboia, Collego, Biomal and *Colletotrichum orbiculare* would all be more valuable if they controlled additional weeds in the same genus or family as the target weed. Transformation technology may allow scientists to broaden selectively the number of hosts a pathogen can control or at least enhance disease of hosts it naturally infects. Once again, success depends on an understanding of the mechanisms involved in resistance and susceptibility of the particular weed-pathogen system. A pathogenicity factor could be the absence of a molecule or process that might otherwise trigger active defence responses (36).

Another problem with potential mycoherbicides is a host range that is too broad. For example *Pyricularia grisea* on Crabgrass may affect rice, and *Gloeocercospora sorghi,* a pathogen of shatter cane *(Sorghum bicolor)* and Johnson grass *(S. halepense)* can cause disease in corn and sorghum (45). Species such a *Rhizoctonia, Sclerotium* and *Pythium* have little host specificity, killing weed and crop plants without discrimination. However, in the cases of *Rhizoctonia, Pythium* and *Sclerotium*, it does require a commitment of the land to the planning of crops that have incorporated resistance and a genetically manipulated inability of the fungi to overwinter. Durable resistance might be achieved by introducing several new defence mechanisms into the plant that are unmatched by fungal pathogenicity factors or by introducing one or two defence mechanisms that the fungus is incapable or evolving a means to over come (36).

7. COMMERCIAL APPROACHES TO THE USE OF BIOLOGICAL CONTROL AGENTS

The success of affluent countries in meeting the increasing food demands of their populace has largely been achieved by increased agricultural production and the control of losses during storage and distribution. New chemicals and new technologies have played an indispensable role in achieving this agronomic 'miracle' (85).

Currently, the only reliable and relatively cheap solution of problems of pests and diseases in large scale agriculture is chemical treatment. Environmental and organic food groups have lobbied for change but there is little to offer as an alternative to chemical treatment if large scale agriculture is to continue to provide stable food supplies in the future. Agriculture, therefore, has a problem. High yields depends upon effective use of agrochemicals but the costs of registration of new products are becoming increasingly expensive.

Biological control agents offer a potential solution to the dilemma from both the environmental impact of BCAS assumed to be less than that of agrochemicals. BCA can also be used in integrated that of agrochemicals. BCA can also be used in integrated pest management (IPM) schemes, wherein combination with good crop husbandry and hygiene, a high quality product can be produced with low chemical input. On the industrial side, attraction includes cheaper discovery and registration of BCAs relative to chemical pesticides, with a consequently shorter lead time. At present, the total global market for BCA's and other microbial based products amounts to about $ 48 million per annum, of which about 50% is accounted for by *Bacillus thuringiensis* and 18% by *Rhizobium* (43). The bulk of the remaining section of the market is accounted for by *Agrobacterium radiobacter* and silage additives. Even with the relatively reduced costs of developing BCA permitting the exploitation of smaller markets there is still a lower limit for commercial investment. For example the production of Upjoin and the USDA of Collego *(Colletotrichum gloeosporioides* f. sp. *aeschnomene)* to control northern jointvetch, will never recover development costs of approximately $ 2 million (76) as the market for collego is estimated at $ 132000 per annum. Requirements for the successful discovery of BCA which include the standard types of methodology for isolation, screening and mode of activity studies are well documented (64). To date, only 5% of deliberate releases of BCAs have actually achieved their aim (40). Bacteria, viruses and fungi have all been used as microbial insecticides and the greatest commercial impacts of BCA have been made in the insecticide market.

The early evaluations of myco insecticides considered them to be impracticable because of the adverse effects of the environment on the activity of the fungus (5, 69, 70). At presents there are few plant pathogens that are being considered as microbial herbicides, and as yet there are no commercially available viral or bacterial preparations. All interest is currently on the use of fungi in this respect. This reflects the fundamental differences in the physiologies of the other groups, in that viruses tend not to be specific and they require a vector, and that plant pathogenic bacteria normally require moulds to enter their hosts. In contrast, fungi can actively penetrate a healthy plant. An extensive list of projects has been compiled by Scheepens & Van Zon (58) but to date there are only three commercial mycoherbicides preparations available; Collego *(Colletotrichum gloeosporioides* f. sp. *aeschynomene)* against *Aeschynomene virginica* in rice or soyabean, *Cercospora rodmanii* against water hyacinth and Devine *(Phytophthora palmivora)* against milk weed vine in Citrus groves.

The mycoherbicide "Collego" has been used commercially since 1982 in Arkansas to control Northern jointvetch *(Aeschynomene virginica)* a leguminous weed in rice *(Oryza sativa)* fields (77). It is marketed as a dry formulation consisting of 15% viable, dry conidia of *Colletotrichum gloeosporioides* f. sp. *aeschynomene* and 85% inert ingredients. Each formation batch is assayed or packaged to contain 7.54 x 10^4 viable conidia per bag, the amount required for treatment of 4.05 hectares at the rate of 93.5 ha^{-1} (6). This amounts to 1.8 x 10^7 viable spores m^{-2}. The pathogen infects but does not adequately control, a closely related species, Indian jointvetch *(Aeschynomene indica)* a serious localized weed in the gulf coast rice production area of Louisiana and Texas.

Mycoherbicide Devine has been used commercially since 1981 in Florida citrus (Rutaceae) groves to control milkweed vine, *Morrenia odorata* (55). It is marketed as a wet formulation of chlamydospores of *Phytophthora palmivora* with a shelf life of six weeks in cold storage. It is applied at the rate of 8 x 10^4 chlamydospores m^{-2} to the surface of moist soil under citrus tree. A safety zone of 1.6 km must be maintained between site of application and planting of several susceptible vegetable crops and ornamental plants. It is feasible to distribute and market this relatively liable product because of the limited marketing area. No damage to non target crops or citrus have been noted in the seven years since first commercial use of this mycoherbicide despite mild infection of citrus in controlled environments with high inoculum rates (200 x field rates).

The persimmon wilt fungus, *Acremonium diospyri,* is not commercially available but is routinely used as a mycoherbicide to control Persimmon trees (*Diospyrous virginiane*) in rangeland in South Central Oklahoma. It has been used since 1960 to control trees up to 10 cm in diameter (89). All trees in a grove may be killed with in three years following hand inoculation of 80% or more of the trees with a conidial suspension of the fungus (25).

Velgo is a potential mycoherbicide for velvet leaf *(Abutilion theophrasti)* control in corn *(Zea mays)* and soybeans *(Glycine max)* in the U.S. corn belt and southern Ontario, Canada. It is a strain of *Colletotrichum coccoides* (91). Commercialization is anticipated in 3 to 5 years as a combination treatment with reduced rates of several chemical herbicides that will not control velvet leaf alone. The fungus alone at the rate of 1 x 10^9 spores m^{-2} kills 46% of the plants inoculated at the two or three leaf stage. Lack of complete kill is due to premature shedding of infected leaves and regrowth of new leaves. The host range of this isolate of *C. coccoides* is restricted compared to other isolates, and although it infects several other plants, it does not cause significant damage to them.

Luboa 2 is a selected strain of *Colletotrichum gloeosporioides* f. sp.*cuscutae* that is used as a mycoherbicide in The People's Republic of China to control dodder *(Cuscuta chinensis* and *Cuscuta australis)* parasitic on broadcast-planted soyabeans. It has been used since its discovery in 1963 for practical control of this parasitic weed. Spore concentrations of 2 x 10^7 spores ml^{-1} are applied with a hand sprayer until run off. Best results are obtained when spraying is done at 10.00 to 17.00 h on days when humidity is high, usually in late July to early August. Luboa 2 is a improved strain necessitated by loss of virulence in Luboa 1. Virulence is positively correlated with spore size greater use is expected when new formulations with longer shelf life are available.

Biomal is a potential mycoherbicide for control of round leaved mallow *(Malva pusilla)* in wheat *(Triticum aestivum)* and lentils *(Lens culinaris)* in the provinces of Manitoba and Saskatchewan, Canada and the northern tier of wheat producing states in the United States. It is a selected strain of *Colletotrichum gloeosporioides* f. sp. *malvae.* It is applied in spore suspension containing 2 x 10^9 spores at the rates of 5 x 10^2 1 ha^{-2} (46). This amounts to 6 x 10^7 viable spores m^{-2}. The pathogen infects, but doesn't adequately control, velvet, leaf, a serious malvaceous weed widely distributed in corn and soybeans.

Casst is a potential mycoherbicide for control of sickel pot *(Cassia obtusifolia)* and Coffee senna *(Cassia occidentalis)* in soybeans and peanuts *(Arachis hypogaea)* in the Southern States of the United States (83). It is a strain of *Alternaria crassiae* and is applied at the rate of 1.1 kg ha^{-1} in 76.7 1 of water with an oil-based adjuvant. The spore concentration in the spray tank is 7.5 x 10^4 spores ml^{-1}, amounting to application of 1.4 x 10^6 spores m^{-2}. Spores are currently produced on solid substrates with some modification of the procedure (84).

A potential mycoherbicide for control of Bathrust burr (Spiny cocklebur or spiny clotbur=*Xanthium spinosum)* has been developed in New South Walves, Australia, with a strain of *Colletotrichum orbiculare* that is indigenous in the area where the weed is an economic problem (3). Small scale field tests with the pathogen indicate that it kills large weed plants in relatively dry conditions when heavily inoculated. The pathogen does not kill common cocklebur *(Xanthium pennsylvanicum)* when spray inoculated, but if infected into young stems it causes death. The fungus is known as a pathogen of species in the family cucurbitaceae (71).

Another, host pathogen interaction that has been examined in detail for mycoherbicide potential is *Cercospora rodmanii* on Waterhyacinth. *(Eichhornia crassipes)* (17). Mycelium of this leaf spot inducing pathogen was grown for 3 weeks on potato dextrose broth containing 5% yeast extract, blended and applied at the rate of 1.1 gm wt weight m^{-2}. After inoculation of the plants grown in buckets of nutrient supplemented water disease stress caused a significant reduction in the net rate of leaf production. However, success of this bicontrol strategy was limited by a low rate of plant kill and the ability of the host to compensate for disease loss by rapid leaf production. Charudattan *et al.* (17) concluded that *C. rodmanii* could be an effective control for water hyacinth when used with multiple applications of inoculum early in the growing season or under conditions favouring low to moderate host growth rates, including combining with sublethal rates of chemical herbicides or insect biocontrol agents.

Loss of virulence in culture is a pervasive problem is plant pathology. Undoubtedly, some reported loss of virulence actually involve loss of viability or improper handling of pathogens and plant material. Reversible loss of virulence has been attributed to changes in characters determined by cytoplasmic inheritance. Another problem with potential mycoherbicides is a host range that is too broad. For example *Pyricularia grisea* on crabgrass may affect rice, and *Gloeocercospora sorghi,* a pathogen of shatter cane. Considerable progress has been made during the past decade in the use of fungi as mycoherbicides, and the pace of this progress appears to be accelerating. Thus far, only naturally occurring strains, indigenous to limited geographic sites, have been tried successfully.

8. CONCLUSION

Considerable progress has been made during the past decade in use of fungi as mycoherbicides, and the pace of this progress appears to be accelerating. Thus far, only naturally occurring strains, indigenous to limited geographic sites, have been tried successfully. Many naturally occurring strains remain to be examined for their potential, and the need for biological pesticides is increasing. There are many more site specific weed problems arising as a result of the wide spread use of broad spectrum, chemical herbicides. Resistant species become weedy as competition from other weeds is removed. These resistant species are often closely related to the crop, and thus require a higher order of selectivity for control than existing products provide. An increasing role of biologicals can be expected due to pesticide resistance problems and the need for better integrated pest management strategies that utilize biologicals.

Progress in development of natural strains of fungi as mycoherbicides can be anticipated; there are many of explore and develop. However, all are likely to have innate biological limitations that affect their utility. Currently this is the case with those already commercialized or nearing commercialization, there have been a great number of naturally occurring strains researched for possible use as mycoherbicides, but only a small proportion have been developed to commercial products. Considerable advances are expected from use of existing biotechnology in this area. Goals to improve virulence, host specificity, and environmental capability with genetic engineering techniques hold much promise. The expectations of society to have non-chemical alternatives for pest control appears to be broadly based and deeply rooted. Thus it seems reasonable to expect greater public and private efforts in biotechnology to enhance fungi for control of weeds.

9. REFERENCES

1. Amor, R. L. and Miles, B. A. 1974. Taxonomy and distribution of *Rubus fruticosus* L. agg. (Rosaceae) naturalised in victoria. Muelleria, **3** : 37-62.

2. Andres, L. A., Davis, C. J., Harris, P. and Wapshere, A. J. 1976. Biological control of weeds. In Theory and Practice of Biological control, Eds. C. B. Huffkaker, and P. S. Messenger. pp. 481-499. Academic Press, London.

3. Auld, B.A., Mc Rae, C. F. and Say, M. M. 1988. Possible Control of *Xanthium spinosum* by a fungus. Agriculture, Ecosystems and Environment, **21** : 219-223.

4. Batra, S.W.T. 1981. *Puccinia xanthii* forma specialis *Ambrosiatrifidae* . A microcyclic rust for the biological control of giant rage weed, *Ambrosia trifida.* Mycopathologia, **73** : 61-64.

5. Billings, F.H. and Glen, P.A. 1911. Results of the artificial use of site fungus disease in Kansas. USDA Agricultural Bureau Entomological Bulletin 107 Ed. L. O. Howard, USDA : Washington, DC.

6. Bowers, R. C. 1986. Commercialization of Collego an industrialist view. Weed Science, **34** (Suppl. 1) : 24-25.

7. Brooker, N. L. 1988. Production and regeneration of *Alternaria crassa* protoplasts, M S. thesis, University of Arkansas, Fayetteville.

8. Bruzzese, E. and Hasan, S. 1986. The collection and selection of isolates of *Phragmidium violaceum* pathogenic to species of European blackberry naturalised in Australia. Aun. Appl. Biol. **108** : 527-533.

9. Burnett, H.C., Tucker, D.P.H., Patterson, M. E. and Ridings, W. H. 1973. Biological control of milkweed vine with a race of *Phytophthora citrophthora*. Pro. Fla. State Horticult. Soc. **86** : 111-115.

10. Burnett, H.C., Tucker, D.P.H. and Ridings, W.H. 1974. *Phytophthora* root and stem rot of milkweed vine Plant Dis. Report, **58**:355-357.

11. Calloway, M. B., Pathak, S. C. and Well, H. D. 1985. Studies on alternate hosts of the rust *Puccinia canaliculata,* a potential biological control agent for nutsedges. Plant Disease, **69** : 924-926.

12. Charudattan, R. 1985 The use of natural and genetically altered strains of pathogens for weed control. In: Biological Control in Agricultural IPM Systems, Eds. M. A. Hoy and D. C. Herzoge, pp. 347-372. Academic Press, Inc. New York.

13. Charudattan, R. 1986. Integrated control of water hyacinth *(Eichhornia crassipes)* with a pathogen, insects and herbicides. Weed Science, **34** (Suppl 1) : 26-30.

14. Charudattan, R. 1988. Inundative control of weeds with indigenous plant pathogens. In : Fungi in biological control systems. Ed. M.N. Burge, Manchester University Press, pp. 86-110.

15. Charudattan, R., McKimey, D.E. and Hepting, K. T. 1981. Production, Storage, Germination and infectivity of uredospores of *Uredo eichnorniae* and *Uredo pontederia.* Phytopathology, **71** : 1203-1207.

16. Charudattan, R. and De Loach, C. J. Jr. 1988. Management of pathogens and insects for weed control in agroecosystems. In : Weed management in agroecosystems : ecological approaches. Eds. M.A. Altieri and M. Liebman, CRC Press, Boca Raton, Florida, pp. 245264.

17. Charudattan, R., Linda, S. B., Kluepfel, M. and Osman, Y. A. 1985.. Biological efficacy of *Cercospora rodmanii* of waterhyacinth. Phytopathology, **75** : 1263-1269.

18. Chisaka, H. 1975. Trends in weed control for low land rice in Japan. Inform., **22** : 11-16.

19. Conway, K. E. and Cullen, R. E. 1978. The effect of *Cercospora rodmanii* a potential biological control of waterhyacinth on the Fish, *Gambusia affinis.* Mycopathologia, **66** : 113-116.

20. Cullen, J. M. 1978. Evaluating the success of the program for the biological control of *Chondrilla juncea.* Proc. IV Int. Symp. Biol. Control Weeds. Gainseville, Fla., 1976, 117-121.

21. Cullen, J. M., Kable, P. F. and Catt, M. 1973. Epidemic spread of a rust imported for biological control. Nature, London, **244** : 462-464.

22. Dale, J. 1983. Application of the principles of microbial genetics to biotechnology. "In Principles of Biotechnology" Ed. A. Wiseman, pp. 56-93. Chapman & Hall, New York.

23. Daniel, J. T., Templeton, G. E., Smith, R. J. Jr. and Fox, W. T. 1973. Biological control of northern jointvetch in rice with an endemic fungal disease. Weed Science, **21** : 303-307.

24. Ennis, W. B. Jr. 1967. Crop losses due to weeds. Proc. FAO Symposium, pp. 127-145.

25. Griffith, C. A. 1970. Annual Report. Samuel Roberts Noble Foundation, Inc, Ardmore, Oklahoma.

26. Halstead, B. D. 1894. Weeds and their most common fungi. M. J. Experimental Station Report, 379.

27. Harris, P. 1971. Current approaches to biological control of weeds. Tech. Comm. CIBC, **4** : 67-76.

28. Harris, P. 1986. Biological control of weeds. In : Biological plant and health protection Ed. J. M. Franz, G. Fischer Verlag, Germany, pp. 123-138.

29. Hasan, S. 1970. The possible control of skeleton weed, *Chondrilla juncea* L. using *Puccinia chondrillina* Bubak & Syd. Proceedings, of the IVth International Symposium on the Biological Control of weeds Ed. T. E. Freeman), pp 11-14. University of Florida, Gainesville, FL.

30. Hasan, S. 1972. Behaviour of the powdery mildews of *Chondrilla juncea* L. in the Mediterranean. Actas III Congr. Un. Fitopat. Medit. Oeiras, Postngal, 1972. pp. 171-176.

31. Hasan, S. 1974. First introduction of a rust fungus in Australia for the biological control of skeleton weeds. Phytopathology, **84** : 253-254.

32. Hasan, S. 1980. Plant pathogens and biological control of weeds. Rev. Plant Pathology, **59** : 349-356.

33. Hasan, S. 1981. Present status and prospects of the programme in Europe for the microbiological control of Australian weeds. Proc. V Int. Symp. Biol. Control weeds, Brisbane, Australia, 1980., pp. 333-340.

34. Hasan, S. 1988. Biocontrol of weeds with microbes. In : Bicontrol of plant disease. Eds. K. G. Mukerji and K. L. Garg, CRC Press, Boca Raton, Florida, pp. 129-151.

35. Hasan, S. and Cullen, J. M.1984. Studies in the Mediterranean region on biological control of Australian weeds. Proc. EWRS 3rd Symp. on Weed Problems in the Mediterranean Area., Oeiras, Portugal, 1984. 365-371.

36. Heath, M. C. 1987. Host Vs. non host resistance. In: Molecular Strategies for Crop protection. Eds. C.J. Arntzen and C. Rayan, pp. 25-34. Alan, R. Liss Inc. New York.

37. Holm, L., Weldom, L. W. and Blackburn, R. D. 1969. Acquatic weeds. Science, **166** : 699-709.

38. Huffaker, C. B. 1970. Fundamentals of Biological weed control. In: Biological control of Insect Pests and Weeds, Ed. P. De Bach, pp. 631-649. London, Champman & Hall.

39. Inoue, Y., Goto, K. and Ohata, K. 1957. Over wintering and mode of infection of leaf blight bacteria of rice plant. Bull. Div. Plant Breed. and Cult., Tokai-Kinki Natl. Agric. Exp. Stn. **4** : 74-82.

40. Jutsuw, A. R. 1988. Commercial application of biological control : Status and prospects. Philosophical Transactions of the Royal Society of London, Series B, 318, pp. 357-373.

41. King, L. J. 1966. Harmful aspects of weeds. In: Weeds of The World Grampian Press, London, pp. 91-114.

42. Laslie, J. F. 1983. Some genetic techniques for *Gibberella zeae*. Phytopathology, **73** : 1005-1008.

43. Lethbridge, G. 1989. An industrial view of microbial inoculants for crop plants. In: Microbial Inoculation of crop plants, Society for General Microbiology special publication 25 Eds. R. C. Campbell and R. M. Macdonald, IRL Press : Oxford.

44. Matsunaka, S. 1970. Weed control in rice. Proc. FAO Int. Conf. on Weed Control., pp. 7-23.

45. Mitchell, J. K. 1987. Host and environmental factors used to model Johnsongrass control with the bioherbicides *Colletotrichum graminicola* and *Gloeocereospora sorghi.* Phytopathology, 77, 1771 (abstract).

46. Mortensen, K. 1988. The potential of an endemic fungus, *Colletotrichum gloeosporioides* for biological control of round leaves mallow *(Malva pusilla)* and Velvet leaf *(Abutilon theophrasti)*. Weed Science, **36** : 473-478.

47. Morin Louise, Bruce, A. Auld, John, F. Brown and Abdul Cholil, M. 1994. Pathogenic fungi occurring on the *Noogoor a burr.* Complex in Australia. Agriculture Can. P. O. Box 440, Regina, SK S4P 3A2, CAN. Proceedings of the Linnean Society of New South Walves, **114 (1-4)** : 133-147.

48. Noda, K. 1976. Integrated weed control of rice. Eds. J. D. Fryer, and Shooichi Matsunaka, pp. 17-46.

49. Ochreens, E. 1977. Biological control of the blackberry through the introduction of rust, *Phragmidium violaceum* in Chick. Food and Agricultural Organization Plant Protection Bulletin, **8** : 25-26.

50. Panaccione, D. G., Mc Kiernan, M. and Hanay, R. M. 1988. *Colletotrichum graninicola* transformed with homologus and heterologus benomyl resistance gene & retains expected pathogenicity to corn. Molecular Plant Microbe Interactions, **1** :113-120.

51. Peberdy, J. F. 1979. Fungal protoplasts: isolation, reversion and fusion. Annual Review of Mirobiology, **33** : 21-39.

52. Peberdy, J. F. 1983. Genetic recombination in fungi following protoplast fusion and transformation In Fungal Differentiation : A Contemporary Synthesis, Vol. 4, Ed. J. E. Smith, pp. 559-581. Marcel Dekker Inc. New York.

53. Phatak, S. C., Summer, D. R., Wells, H.D., Bell, D. K. and Glaze, N. C. 1983. Biological control of yellow nutsedge with the indigenous rust fungus *Puccinia caualiculata.* Science, **219** : 1446-47.

54. Politis, D. J., Watson, A. K. and Bruckast, W. L. 1984. Susceptibility of musk thistle and related composites to *Puccinia carduorum.* Phytopathology, **74** : 687-691.

55. Ridings, W. H., Mitchell, D. J., Schonlties, C. L. El. and Gholl. N. E. 1978. Biological control of milkweed wine in Florida citrus groves with a pathotype of *Phytophthora citrophthora.* Proc. IV Int. Symp. Biol. Control Weeds Gainesville, Fla., 1976. pp. 224-240.

56. Ridings, W. H. 1966. Biological control of strangler vine in citrus-a researcher's view. Weed Science, **34** (Suppl 1.) : 31-32.

57. Rodriguez, R. J. and Yoder, O.C. 1978. Selectable genes for transformation of the fungul plant pathogen *Glomerella cigulata* f. sp. *phaseoli (Colletotrichum lidemuthianum).* Gene, **54** : 73-81.

58. Scheepens P.C. and Van Zon, H.C.J. 1982. Microbial Herbicides. In microbial and viral pesticides. Ed. E. Kubstack, pp. 623-624. Marcel Dekker, New York.

59. Scheepens, P.C. and Van Zon, H.C.J. 1982. Microbial herbicides. In Microbial and Viral Pesticides, Ed. E. Kubstak, New York, Marcel Dekker, pp. 624-641.

60. Schroeder, D. 1983. Biological control of weeds. In : Recent Advances in Weed Research, Ed. E.W. Fletcher, Slough. Commonwealth Agric. Bureau, U. K. pp. 41-78.

61. Schroeder, D. 1992. Biological control of weeds: a review of principles and trends. Pesq. agropee. bras. Brantia, **27** : 191-212.

62. Shimamura, M. 1970. The status and problems of the chemical weed control in grasslands. Weed Res. Japan, **10**:10-14.

63. Shivas, R. G. 1994. Controlled environment studies on the infection of *Emex australis* by *Phomopsis emicis.* Plant Pathol. Branch. Dep. Agric. South Perth 1651, WA, AUL. Plant Pathology, **43 (3)** : 547-553.

64. Singh, J. and Faull, J. L. 1988. Hyperparasitism and biological control. In : Biocontrol of Plant Pathogens, Vol. 2. Eds. K. G. Mukerji and K. L. Garg, pp. 167-169. CRC Press. Boca Raton, Florida.

65. Smith, R. J., Jr. and Shaw, W. C. 1966. Weeds and their control in rice production. USDA Agricultural Hand Book No. 292.

66. Smith, R. J. Jr. and Shaw, W. C. 1968. Weeds and their control in rice production. U. S. Dept. Agr., 292 U.S. Govt. Printing Office, Washington, D. C. pp. 64.

67. Smith, R. J. Jr. 1982. Integration of microbial herbicides with existing pest management programs. In : Biological Control of Weeds with Plant Pathogens, Eds. R. Charudattan, H. L. Walker, John Wiley, New York : pp. 189-203.

68. Smith, R. J. 1986. Biological control of northern joinvetch *(Aeschynomene virginica)* in rice *(Oryza sativa)* and Soyabeans *(Glycine max)* - researcher's view. Weed. Science, **34 (Suppl. 1)** : 17-23.

69. Snow, F. H. 1890. Experiments for the destruction of Chineh bugs. 21st Reports of the Entomological Society of Ontario. pp, 93-97.

70. Snow, F. H. 1895. Contagious disease of the Chinch bug. 4th Annual Report of the Directorate of the University of Kansas.

71. Sutton, B. C. 1980.. The Coetomycetes : Fungi Imperfecti with Pycnidia. Acervuli and Stromata. Commonwealth Mycological Institute, Kew, Surrey, UK.

72. Te Beest, D. O. 1984. Introduction of tolerance of benomyl in *Colletotrichum gloeosporioides* f. sp.*aeschynomene* by ethyl methane sulfonate. Phytopathology, **74**: 864 (abstract).

73. Te Beest, D.O. and Templeton, G. E. 1985. Mycoherbicides : Progress in the biological control of weeds. Plant Dis. **69** : 6-10.

74. Templeton, G. E., Te beest, D. O. and Smith, R.J. Jr. 1979. Biological weed control with mycoherbicides. Ann. Rev. Phytopathol. **17** : 301-310.

75. Templeton, G. E. 1982. Status of weed control with plant pathogens. In : Biological Control of Weeds with Plant Pathogens, Eds. R. Charudattan, and H. L. Walker, pp. 29-44, New York, John Wiley.

76. Templeton, G. E. 1986. Mycoherbicide research at the University of Arkansas, past, present and future. Weed Science, **34 (Suppl. 1)** : 35-37.

77. Templeton, G. E. 1987. Mycoherbicides-achievements, developments and prospects. In : Proceedings, of the Eighth Australian Weeds Conference, September 21-25 1987. pp. 489-497. Weed Society of New South Wales, Sydney, Australia.

78. Timmons, F. L. 1971. Benefits and consequences of using herbicides in irrigation systems. Proc. FAO Int. Conf. on Weed Control. pp. 387-400.

79. Turgeon, G. and Yoder, O. C. 1985. Genetically engineered fungi for weed control. In : Biotechnology : Applications and Research, Eds. P. N. Cheremisinoff & R. P. Ouellette, pp. 221-230. Technomic Publishing Co., Inc : Lancaster, Pennsylvania.

80. Turner, S. K., Kwaitkowski, A. and Fay, P. K. 1982. Stimulation of *Puccinia obtegens* teliospore germination (Abstr.). Phytopathology, **72** : 711.

81. Upadhyay, Rajeev K., Mukerji, K.G. and Rajak, R.L. (Eds.) 1996. IPM System in Agriculture, Vol. I. Principles and Perspectives, Aditya Books Pvt. Ltd., New Delhi pp. 1-23.

82. Upadhyay, Rajeev K., Kenfield, D., Strobel, G. A.S. and Hess, W. M. 1991. *Ascochyta cypericola* sp. nov., causing leaf blight of purple nutsedge *(Cyperus rotundus)*. Can. Jour. of Botany, **69** : 797-802.

83. Walker, H. L. and Riley, J. A. 1982. Evaluation of *Alternaria cassiae* for the bicontrol of sickle pod (*Cassia obtusifolia*). Weed Science, **30** : 651-654.

84. Walker, H. L. and Connick, Jr. W. J. 1983. Sodiumalginate for production and formulation of mycoherbicides. Weed Science, **31** : 333-338.

85. Walker, R. 1987. Foreword. In : Toxicological Aspects of Food, Ed. K. Miller, pp. V-IX. Elsevier applied Science, London.

86. Wapshere, A. J. 1979. Recent progress in the biological control of weeds. European and Mediterranean Plant Protection Organisation, Bulletin, **9 (1)** : 95-105.

87. Wapshpere, A. J., DalFosse, E. S. and Cullen, J. M. 1989. Recent development in biological control of weeds. Crop Protection, **8** : 227-250.

88. Watson, A. K. and Brunetti, K. 1984. *Puccinia carduorum* on *Carduns tenuiflorus* in California. Plant Disease, **68** : 1003-1005.

89. Wilson, C. L. 1965. Consideration of the use of *Persium* on wilt as a silvicide for weed persimmons. Plant Disease Reporter, **49** : 789-791.

90. Wilson, C. L. 1969. Use of Plant Pathogens in Weed control. Annual Review of Plant Pathology, **7** : 411-433.

91. Wymore, L. A., Watson, A. K. and Gotlieb, A. R. 1987. Interaction between *Colletotriclum coccodes* and thidiazuron for control of velvet leaf (*Abutilon theophrasti*). Weed Science, **35**, :377-383.

25 Weeds : Their Ecology and Control

V. N. Saraswat

CONTENTS

Integrated Pest and Disease Management
Rajeev K. Upadhyay, K.G. Mukerji, B.P. Chamola and O.P. Dubey (eds.)
APH Publishing Corporation, New Delhi-110002, India

ABSTRACT

Weeds cause enormous loss in crop yield. The present paper describes the different types of weeds, their classification and distribution in relation to yield losses. Emphasis has been given on the distribution of weeds in kharif and rabi season and those growing in orchards, waste and vacant lands and alongwith plantation crops. Attempts have been made to give a methodological analysis of weed management growing with different types of crops.

1. INTRODUCTION

Weeds cause enormous losses in crop yields both in quantity and quality besides they create health and aesthetic problems. The losses are of subtle nature and are not recognized immediately but if not controlled timely and properly, the yield losses in different cereals, pulse, fibre and oilseed crops have been reported varying from 16 to 58% under different agro-climatic conditions (6). In their first attempt Mehta and Joshi estimated the losses to a tune of Rs. 4200 millions per annum by weeds in India in principal agricultural crops. Later Rao worked out these losses Rs. 50,000 to 60,000 millions when calculated at a conservative estimate of 10-15% losses in yield. Most recently Sahu and Saraswat, worked out these losses to a tune of 9.28 million tonnes in cereals, 0.74 million tonnes in pulses, 0.57 million tonnes in oilseeds and 7.20 million tonnes in fibre and other commercial crops valued at Rs. 2799 crores per annum. Besides the losses in field crops, the weeds also cause considerable losses in plantation, horticultural and vegetable crops and interferes in production of fauna and flora in aquatic system and impade water flow in irrigation canals. They also cause problems in highways, rail tracks and industrial sites.

Not much research work was done in India on different aspects of crop-weed association, losses in crop yield due to weeds, weed biology and ecology and weed management in different crops and cropping systems upto the first quarter of this century except the extensive identification of flora of India inclusive of Pakistan, Ceylon and Burma was made by several workers which is well documented. The 'Hand Book of Indian Flora' by H. Drury published in three volumes was a good attempt in this direction which was followed by 'Flora of British India' by J.D. Hooker in seven volumes, which covers flora of India, Pakistan, Bangladesh, Cylon and Burma. In the later publication 'Bengal Plants' by David Prain covering Bihar, Orissa, West Bengal and Bangladesh is also well documented. Among the later publications 'Botany of Bihar and Orissa' by H.H. Haines is a good attempt. Sporadic attempts have also been made later to survey weed flora of individual crops or area and their association with specific crop till recently. The research on different aspects of weed biology and ecology and their control measures was intensified in India in the second half of this century at different Agriculture Universities and ICAR Institutes. However, it was geared up in a coordinated way with the inception of 'All India Coordinated Research Programme on Weed Control' since 1978.

1.1 Weed as defined

'Weed is a plant growing where it is not desired' in a simple language the weed was first defined by Jethro Tull (15). The definition is simple as one may note, it may be a crop plant but is not desired where it is growing. Even a absolute weed viz. *Chicorium intybus* may not be weed when it is grown for specific purpose. Likewise, though *Cynodon dactylon* is amongs the world's most difficult weeds but when put in lawns, it is not a weed. *Imperata cylindrica* is a tough and hard to control weed of uncropped and plantation crops but when used as a soil binder it may not be called a weed. While a gram plant may be weed if it is growing in wheat field where it is not desired. Weeds in general are characterized as a plant having better competitive ability and grow profusely under marginal conditions and escapes adverse conditions by shortening their life cycles. Contrary, the crop species fails under such situations. Mortimer (22) defined weeds as 'natural hazards to the activities of man'. The European Weed Research Society used the word weed as 'any plant or vegetation interfering with the objectives of people'.

Gupta (9) described the weeds as 'those plants growing in places and at times when man wants either some other plants to grow or no plants to grow at all'. So long as a plant is growing at a place and time without interfering with the man's affairs, it is not looked as a weed. In other words, while all weeds are unwanted plants, all unwanted plants may not be weeds. Alstrom (1) has defined agriculture weeds as that part of the non-crop vegetation that reduces the quality or quantity of the existing crop to such an extent that any net benefit of the weed plants themselves can not compensate for their unfavourable effects on crops. The various researchers described the weeds depending upon the situations viz. weed is plant growing out of place - Prof. Beal; weeds are those species of plants which grow unwanted or are not useful often prolific and persistent interfere with agricultural operations, increase labour cost and reduces crop yields - Robinson; a weed is unsightly, useless or injurious plant growing where it is not desired and where something else should grow - Webster; weed is a plant whose potentialities for harm is greater than its potentialities for good-Peter; weeds refer to undesirable, injurious, unsightly and troublesome plant which interfere with cultivated crops and affects human affairs - C. Thakur; weed is an honest and independent competitor for food material in the struggle of existence - Cambell; weed is a plant that grows so luxurilantly or plentifully that it chokes out all other plants that possess more valuable nutritive properties - Brenehley. Most recently Quayum (25) has described the 'weeds as those prolific generation of plant which contributes much on ecological balance and having greater economic potentialities but affects adversely human affairs and crops'.

From a farmer's point of view, a weed is that species of plant kingdom which grows profusely in a place where he want to grow only specific plant and that competes for nutrient, moisture and light and reduces the quality and quantity of the produce. Further, the man has to make special efforts to eliminate such unwanted vegetation which consumes larger portion of his labour, time and adds to cost of cultivation; in aquatic bodies it competes with the crop, interferes in growth of fauna and flora and impedes the flow of irrigation water.

1.2 History of weeds

The history of weed is as old as the human civilization when the man started cultivating crops and then he soon realized that crop yields were more when grown in weed free environment. Thus, monoculture, contrary to natural way was identified and concept of weeds as unwanted plant was developed. A lot of exercise was made to check the spread of undesirable plants through seed cleaning, preparatory cultivation and various interculture practices and efforts are continuing to eliminate undesirable plants from crops which the man wants to grow. Hay identified the stages in the evaluation of weed control practices viz. (i) 10,000 B.C. - removing weeds by hand, (ii) 6,000 B.C. - use of preventive hand tools to till the lands and destroy weeds, (iii) 1,000 B.C. - animal power drawn implements like harrows, plough etc., (iv) 1920 A.D. - mechanical power drawn implements and improved animal drawn implements like cultivator, blades, harrows, finger weeders, rotatory hoes, rod weeder etc., (v) 1930 A.D. - biological weed control, (vi) 1947 - chemical weed control with commercial development of organic herbicides like 2,4-D and MCPA.

Moretimer (22) has indicated three modes viz. pre-adaptation, evaluation and alien immigration by which the weed species incorporates in the weed flora of agro-ecosystems. The pre-adapted weed is defined as those species that are resident with the plant community and within dispersal distance of the crop and becomes predominant with the change in the management practices or cropping system. This does not necessitate mean that there is change in the genetic composition of that species rather it is the practice adopted which favoured that species to grow better. For example, with the introduction of dwarf wheat, the problem of grassy weeds have developed contrary to the traditional problem of broad leaved weed in wheat. Further, continuous use of herbicides for controlling these weeds has resulted in the emergence of broad leaved weeds in this crop. In other instance, where herbicides are used for controlling weeds in pulse crops using pendimethalin which controls most of the grassy and broad leaved weeds but leaves *Commelina communis* Walk which later becomes a dominant weed in the surrounding area. In shift cultivation system, the natural flora of the area is completely changed when arable crops are grown after natural vegetation is removed and the weeds associated with that crop dominates in just a few years and natural vegetation disappears.

The evolution of weed may occur at three levels viz. (i) genetic changes in the existing species due to unpredictable disturbance or in the habitat, (ii) the race formation in which locally adopted races of a weed species occur; and (iii) crop mimicry, where the crop and weeds are so intimately associated with man that it has lead to the evolution of mimetic weeds. The alien immigration is the import of weed seed through crop seed or any other commodities which settles in the new environment and later becomes problem weeds. The most burning examples are recent introduction of *Parthenium hysterophorus* in India from USA through the import of wheat grain for consumption around 1956. *Phalaris minor* and *A. ludoviciana* are some of the other burning examples. Similarly, *Eichhornia crassipes* which has come in India through sea transport long back around in 1890 from South America has created havock in aquatic system in India.

1.3 Characteristics of weeds

The characteristics of the weeds have been narratted by many authors depending upon the situations, however, Moretimer (22) refers a weed possessing ' a general all purpose genotype' that ensures a strategy of species persistence in wide range of unpredictable habitat conditions, which further refers that under very marginal ecological conditions, the species thrive well, have high reproductive capacity, high competition ability, well developed power of dispersal and short life cycle. Baker and Stebbin enumerated a number of characters which an ideal weed may possess, they are (i) seed germination could take place in many environments (ii) degraded seed germination due to internal dormancy and considerable and considerable longevity of seed (iii) rapid growth through vegetative phase to flowering and maturity (iv) seed production in wide range of environmental conditions (v) continuous seed production for as long as condition for growth permits (vi) self compatible but not completely self pollinated (vii) possession of traits for short and long distance dispersal (viii) when cross pollinated, unspecialized pollinator visits or wind pollinated (ix) if a clonal species has vigorous vegetative growth and regeneration from fragments (x) if clonal species, has brittleness or leaf part assuring survival of main plant and (xi) shows strong specific competitions by special mechanism *i.e.* allelopathic chemicals. The authors have further indicated that these are the characters a true weed should posses but luckily it is not found in any weed and also characters varies with the habitat and ecological conditions surrounding it. It dominates under the unchecked conditions and in more favourable environment but man's efforts should continue to check such conditions which favours the weed and provide conditions conducive to a level, it does not effect the interests of the man.

2. CLASSIFICATION OF WEEDS

The weeds differ in their morphological characters, habitat and environment, life cycle, crop-weed association, origin etc. and many of these weeds have similarity to respond to common weed management measures therefore, it becomes necessary to group these weed species for developing better control measures. Though, diverse approaches have been made by various workers to group these weeds for different purposes. Joshi (16) has classified weeds based on morphological characters, life cycle, habitat, botanical system, seasons, ecological affinities, soil type, origin, special weeds etc. Exhaustives classification was done for aquatic weeds by Gupta, giving emphasis on their habitat, biological type and botany. Rao (26) has classified the weeds simply based on their life period. Most recently Alstrom (1), while describing the weeds in cereal crops of hot climate have grouped weeds as (i) annual grasses, (ii) perennial creeping grasses, (iii) perennial sedges and (iv) parasitic dicotyledons. However, the weeds of Indian agriculture may be classified based on botanical classification i.e. based on phylogeny and morphological characters (i) classification based on life cycle of weeds (ii) classification based on ecological conditions (iii) classification based on crop-weed association and (iv) as per origin of the weed species.

2.1Botanical classification

The botanical classification is based on phylogeny and morphology and early attempts in this direction were made by George Benthiam and Joseph Dalton Hooker set up a system of classification in their 'Genera Plantarum'. Though this system is not strictly phylogentic but has been found suitable for practical purposes. The classification is based on cotyledones viz. monocotyledones or dicotylendones which has been further divided as series, order, family, genus, species, sub-species, variety etc. The another system was proposed by Engler and Pranti. This system is illustrated work with phylogenetic arrangements and with modern keys, providing means for quick identification, a number of Indian flora are available which could be referred. Some of the important flora are 'Flora of British India' by J.D. Hooker in 7 volumes; Grasses of Burma, Cylon, India and Pakistan by N.L. Bor; 'Bengal Plants' by D. Prain; 'Flora of Jammu & Plants of neighbourhood' by B.M. Sharma and P. Kachroo etc. In recent past. Thakur (37), Holm *et al*. (12) and Moody *et al*. (21) have published the major weeds alongwith illustrations and photographs, including biology, ecology and crop-weed association. The major weeds in this classification has been mainly grouped as grasses, sedges, broadleaf weeds, algae and ferns.

2.2Based on life cycle

On the basis of life span of the weeds, they may be classified into three major groups viz. annuals, biannuals and perennials, some of the weeds even completes their life cycle within a fortnight, such short lived weeds are called ephemerals, *Phyllanthus niruri* Lin an ideal example of such weeds. They, however, escapes control measures or escapes adverse conditions.

2.2.1. Annual weeds

The weeds in this group completes their life cycle within a year or in a season only. They multiply through seeds and are usually season bound as rabi, kharif or zaid season weeds. They grow fast alongwith the crop and complete their life cycle before the crop is harvested. Under moisture stress conditions, some of these weeds shorten their life cycle and sheds seeds while the crop foils. However, their control measures through herbicides is easy, less costly, and effective.

2.2.2. Biennial weeds

Weeds in this group, covers two seasons from seed to flowering, maturity and death. The vegetative phase covers first year and second year covers flowering, seed formation, maturity and seed dispersal. The weeds of this group does not find place if they are removed properly in the first year. Most of the weeds in this group find shelter on field bunds, canals or road sides and encroach crop field. Some of the good example of such weeds are *Circium arvense* Scop. *Convolvulus arvensis* Linn., *Xanthium strumarium* Linn. etc.

2.2.3. Perennial weeds

Weeds in this group survives for more than two years, they propagate by seeds, rhizomes, stolon, bulbs etc. They are shallow rooted or deep rooted and prolific in nature and are difficult to control through herbicides, or mechanical methods. The weeds in this group may be from grass family, sedges or from broad leaf group. The *Cynodon dactylon* Pers, *Imperata cylindrica* Beauv., and *Agropyrons repens* (grass family), while *Cyperus rotundus* (sedges) and *Solanum elagnifolium* or *Oxalis latifolia* (broad leaved weeds) are deep rooted perennials these weeds are also called perennial weeds.

2.2.4. Based on ecological conditions

The weeds are also classified based on habitat such as weeds of crop land, pasture lands, orchards or plantation crops or depending upon moisture conditions such as aquatic or terrestrial or parasitic. These weeds have further been divided by many workers depending upon their mode of adaptation. Warming classified plants based on their occurrence on soil and other substracts into eight major groups. However, Hanson and Churchill (11) put them in three broad categories viz. (i) Hydrophytes which includes amphibians, floatings and submerged plants, (ii) Mesophytes which includes sun and shade plants and (iii) Xerophytes which includes plants of dry habitat. Saraswat (27) classified the weeds based on moisture conditions as upland, midland and lowland weeds of jute fields.

2.3 Based on crop-weed association

Alstrom (1) has categorielly described the weeds of hot climate agriculture with special reference to India with their association with crop. Hore and Tripathi (13) while compiling the weed survey done in West Bengal have indicated that *Echinochloa colonum* and different species of *Cyperus rotundus* are associated with lowland paddy in almost all districts of the state. Gupta (9) has described the crop-weed association of three types viz. (i) season bound, these weeds are associated with the crops of that season viz. winter or summer crops, (ii) crop bound weeds, they are those weeds which are usually parasitic on the host crop and (iii) crop associated weeds viz. *Allium* sp. with onion and garlic and *Phalaris minor* is associated with cereal crops in winter season, since the seeds of these weed species rapidly mixes with the crop easily.

2.4 Based on origin of the weeds

Phalaris minor is a burning example to introduced weed in India during large scale import of wheat in 1960. There are such many other examples through out the world. Gupta (9) has described many such examples indicating that many of the weed species have introduced from their native land or from some other countries, where they had earlier migrated. Of them *Eichhornia crassipes, Parthenium hysterophorus, Argemone maxicana, Croton bonplandianum* and *Lantana camara* are some of the highly troublesome weeds which have introduced in India. Further *Ageratum conyzoides* has escaped from new world tropics and became one of the commonest in warmer parts of the world (22).

2.5 Other classifications are based on season as kharif or rabi or summer season weeds; ecological affinities i.e. weeds of drylands, paddy lands, garden lands etc.; area infested viz. weeds of crop lands, pasture lands, waste lands, play grounds, road sides etc.; soil types viz. weeds of black cotton soils, red soil, saline or alkaline soils or special weeds viz. poisonous weeds, pasture land weeds etc. (16).

3. YIELD LOSSES DUE TO WEEDS

In India, the total losses due to different parts have been reported to be around Rs. 5000 crores during 1973-74, of which 33% contributed by weeds alone (Financial Express, 7th April, 1975). The amount lost due to pests, other than weeds were : disease 26%, insect 20%, rodents 6% and storage

Table 1. Average yields (q/ha) under different weed management practices in major cereal crops (30)

Crops	Weed free	Chemical control	Manual weeding (once or twice)	Unweeded	Integrated chemical + manual and/or mechanical
Weat	37.60	34.73 (7.5%)*	35.40 (5.85%)	27.81 (26.0%)	37.42
Rice (T)	45.02	43.76 (2.8%)	44.13 (2.0%)	37.43 (16.9%)	45.15
Rice (D)	17.81	12.20 (31.5%)	15.07 (15.4%)	9.40 (47.2%)	17.40
Maize	41.28	40.43 (2.6%)	40.65 (1.5%)	24.46 (40.7%)	41.22
Sorghum	18.50	16.60 (10.3%)	17.20 (7.0%)	10.20 (44.8%)	18.61
Pearlmillet	11.22	11.10 (1.1%)	11.18 (0.4%)	4.88 (56.5%)	11.33
Total	197.76	179.76 (9.1%)	188.91 (4.5%)	127.51 35.5%)	171.13

* (Figures in parenthesis are percent losses in yield)

6-8%. In Colombia, Longe, 1967 report the yield losses varying from 16.6% to 54.4% in different cereal, pulse and fibre crops, while losses were also observed varying from 12.8% to 24.4% in herbicide treatments over manual weeding. At ICRISAT (International Crop Research Institute for Semi Arid Tropics), Hyderabad, the per cent yield reduction due to weed competition in sorghum upto 70%, pearlmillet 60% and chickpea 40.1% were observed. The data of the numerous field experiments conducted during 10 years i.e. 1978 to 1987, on different aspects of weed management both under AICRP-Weed Control and else were computed to estimate losses in yield in cereal, pulse, oilseed, fibre and other commercial crops. The results indicated that yield losses were maximum in unweeded crops which varied from 16.9% in transplanted rice to 56.5% in pearlmillet, which was very less in manual weeded and chemically weeded crops, when compared with completely weed free crop (Table-1). The losses in other crops like oilseeds were upto 71.2%, pulse 38.8% and commercial crops 50.4% in unweeded crops (30).

The extensive studies made in India on crop-weed competition in different crops, to find out critical period of crop-weed competition have indicated that a weed free period of 15 to 60 days in upland drilled rice; 30 to 60 days in transplanted rice; 15-45 days for fingermillets, soybean, groundnut and sorghum, 15 to 30 days in blackgram; 15 to 45 days in greengram; 20 to 40 days in rainfed cotton; 15 to 45 days in jute and mesta and 30 to 60 days in sugarcane after sowing or planting were necessary for avoiding losses caused by weeds. For rabi pulses like chickpea and lentil initially 30 to 60 days, while for pea 30 to 45 days were more critical for weed competition. In dwarf wheat, it was between 30 to 45 days and in tall wheat if falls between 35 to 50 days after sowing.

4. DISTRIBUTION OF WEEDS

The climate factors viz. temperature, light, rainfall pattern, humidity etc. have the major influence in the composition and occurrence of different weed species. However, the other important factors are edaphic and biotic, which are responsible for growth and distribution of different species in particular area. There is wide diversity in the climatic conditions owing to the great size and position of the country. There are stricking contrants of meterological conditions in the country i.e. one part of the country lies in tropics and other within it. In the north-west lies the great Thar desert with the average rainfall of less than 120 mm; in north-east are the 'Khasi' hills with an average rainfall of 10600 mm at Cherapunji, Dras in Kashmir has recorded a minimum temperature of -10° C, while Ganganagar in Rajasthan has several times recorded a temperature over 49° C. The hill stations such as Shimla may be shrouded in clouds for several days with humidity of 100%, while in December they may be over run by air of nearly 0% humidity. The geographical position and climate have an important bearing on the vegetation and animal life. The large extent of area and great variety in physical features and climate combined with natural fertility of soil enabled to produce variety of vegetation in exuberance except in grasslands and savannahs.

Due to its wide diversity in climate, edaphic and system of cultivation, the weed flora also widely differs from one region to another due to different seasons, cropping patterns adopted and association with the crops. Though there is wide variation in the weed flora of the major field crops, however, an attempt has been made to focus on dominant weeds of certain broad agro-situations besides mentioning the major weeds associated with the crops in different zones. Some of the specific weeds which are problematic have been separately discussed.

4.1 Weed flora of kharif season

Rice : The rice crop is cultivated under varied situations such as upland, midland, lowlands and in deep water conditions and in different systems. The cultivation is spread all over Indo-gangetic region, coastal belt of Orissa and Deccan Plateau. The weed flora differs under different situations. In Punjab, where climate is sub-tropical, sub-humid, rainfall varies from 770 to 1150 mm in sub-mountainous districts, to 520-760 mm in central and southern districts of which 70-75% rain is received during kharif season. The plains and part of Indo-gangetic alluvium are composed of silt and clays. In foot hill zones, the sand, stones, clays and boulders of the middle miocene are found. The major weeds occur in this region are *Echinochloa crusgalli* (L) Beauv., *E. colonum* Linn, *Cyperus iria* Linn, *Cyperus difformis* Linn, *Fimbristylis tenera* (Steud) and *Eragrostis japonica* Trin, *Ischaemum rugosum* Salisb., *Paspalum disticum* Linn, *Setaria glauca* (L) Beauv., *Leptochloa penicea* Retz., etc. The other important weeds reported are *Eclipta alba* (L) Hassk, *Ammania senegelensis* Linn, *Caesulia axillaris* Roxb., *Phyllanthus niruri* Linn, *Physalis minima* Linn, *Cyperus compressus* Linn, *Scirpus roylei* (Ness) Parker, etc.

The Kashmir valley is teetonic in origin and enjoys a temperate climate except in area like Uri with moderate cool winters and warm summers. Rice and maize forms the principal cereal crops. The rice crop is transplanted in May as monocrop and the major weeds observed in post transplanted crop are *Ammania auriculata* Willd., *A. bacifera* Linn, *Marsilea quadrifoliata* Linn, *Rotala densiflora* (Retz), *R. indica* (Willd) Koehne, *Monochoria vaginalis* (L) Presl, *Depatrium junceum* (Roxb), *Butomus umbellatus* Linn, *Cara foliosa* D. Don, *C. glonosus* All., *Cyperus iria* Linn, *Fimbristylis dichotoma* (L) Vahl, *Scirpus juncoides* Roxb., *Eriocaulon sieboldianum* Siab & Zucc, *Veronica* sp. etc. In whole of the Indo-gangetic plains i.e. Uttar Pradesh, Bihar, West Bengal and Brahmaputra valley are alluvial plains, the rainfall varies from 750 mm to 2000 mm, rice is grown in different types of situations. Most of the rains in these areas are received during 'monsoon'. The major weeds in upland direct sown rice occur are *Cynodon dactylon* (L) Pers, *Dactyloctanium aegypticum* (L) Beauv, *Digitaria sanguinalis* (L) Scop., *Echinochloa colonum* (L) Link, *Eragrostis japonica* Trin, *Eleusine indica* (L) Gaertn, *Paspalum distichum* L., *Panicum maximum* Jacq, *Setaria glauca* (L) Beauv, *Cyperus rotundus* Linn, *C. iria* Linn, *C. difformis* Linn, *Fimbristylis miliaceae* (L) Vahl, *F. aestivallis* Vahl, *Scirpus grossues* Linn, *S. juncoides* Roxb, *S. triquester* Linn, *Ammania multiflora* Roxb., *Alternanthera sessillis* (L) DC, *Commelina benghalensis* Linn, *C. nodiflora* Linn, *Celosia argentea* Linn, *Cleome viscoa* Linn, Digera arvensis Forsk, *Euphorbia hirta* Linn, *Lindernia crustacea* (L) F.V. Muell, *Physalis minima* L., *Solanum nigrum* Linn and *Corchorus acutangulus* Linn. In transplanted paddy, *Echinochloa crusgalli* (B) Link, *E. colonum* (L) Beauv., *Panicum repens* Linn, *Cyperus stoloniferus* Retz., *C. iria* Linn, *C. esculantus* Linn, *Scirpus* sp., *Ammania beccifera* Linn and *Ludwigia parviflora* Roxb. etc.

In lowland paddy, various forms of aquatic and semi-aquatic weeds from rooted submerged to free floating have been observed, they are *Eichhornia crassipes* (Mart) Solms, *Ipomea aquatica* (L) Forssk, *Eleocharis spirallis* Brown, *Spillanthes paniculata* Linn, *Sagittaria sggitiyolia* Linn, *Jussiaea repens* Linn, *Najas* sp., *Utricularia vulgaris, Rotala indica* (Willd) Koehne, *Marsilea quadrifolia* Linn, *Cenetella* sp., *Salvinia rotundifolia* and *Ludwigia perennis* (L) D.C. besides many already listed in transplanted paddy. In coastal paddy areas, there is also localised dominance of higher and lower algae like *Chara* sp., *Nitella* sp., *Pithophora* sp., *Cladophora* sp. and *Phytoplankton (Anabaena* sp.) etc.

In Karnataka, Tamil Nadu, Andhra Pradesh, Orissa and Kerala where soils are red loam, light red soils, black or river alluvial or red earths which includes the course textured sandy loam and medium fine textured loams, the alluvium mostly in delta regions of Krishna and Godavari districts and laterite and lateritic soils, where paddy is cultivated. The annual rainfall varies from 830-1200 mm, average 1200 mm, 500-1700 mm, 1500 - 4000 mm respectively. The major weeds in paddy fields occur are *Echinochloa colonum* (L) Link, *E. crusgalli* (L) Beauv, *Paspalum scorbiculatum* Linn, *Penicum repens* Linn, *Cyperus rotundus* Linn, *C. iria* Linn, *C. difformis* Linn, *Fimbristylis miliacea* Vahl, *Scirpus* sp., *Ammania baccifera* Linn, *Marsilea quadrifolia* Linn, *Eclipta alba* (L) Hassk, *Cyanotis cuculata* Roem et. Schult, *Sphenochlea zeylanica* Roxb., *Ludwigia parviflora* Roxb. etc. In lowland and flooded conditions in Orissa, weeds like *Monochoria vaginalis* (Presl), *Hydrilla* sp., *Chara* sp., *Salvinia natans* Linn, *Eichhornia crassipes* (Mart) Solms are also observed.

Maize and sorghum : The predominant weeds observed in maize in Punjab, Kashmir valley, Himachal Pradesh, Uttar Pradesh, Bihar and Gujarat are *Dactyloctenium aegyptica* (L) Beauv., *Digitaria marginata* Forsk., *Acrachne racemosa* (Heyne) Ohulld, *Eragrostis tenella* Beauv., *Panicum colonum* Linn, *Trianthema monogyna* Linn., *Cyprus rotundus* Linn., *C. esculentus* Linn. etc.

The other important weeds are *Leptochloa panicea, Eleusine indica* (L) Gaerth., *Digitaria sanguinalis* (L) Scop., *Sorghum halepense* (L) Pers., *Digera arvensis* Linn, *Euphorbia hirta* Linn, *E. microphylla* Heyne ex Roth, *Celosia argentea* Linn, *Phyllanthus niruri* (Muller), *Cleome viscosa* Linn, *Commelina benghalensis* Linn, *Leucas aspera* (Spreng), *Corchorus tridens* Linn, *Malvastrum coromondelanum, Eclipta alba* (L) Hassk, *Brachiaria reptens* (L) Gard et Hubb, *Setaria glauca* (L) Beauv., *Cyperus compressus* Linn, *Tribulus terreshis* Linn, *Vicoa indica* (Dec.) etc. The frequency and distribution of these weeds varied from one region to another and in some places of the species mentioned above are completely absent. In pure sorghum, the dominant weeds observed in Maharashtra, Madhya Pradesh, Karnataka and Andhra Pradesh are *Cynodon dactylon* (L) Pers., *Digitaria marginata* (Forssk), *Acanthosperum hispedium* D.C., *Brachiaria mutica, Setaria glauca* (L) Beauv, *Oplismanus burmani, Panicum maximum* Linn, *Saccharum spontaneum* Linn, *Cyperus compressus* Linn, *Digitaria sangunalis* (L) Scop., *Celosia argentea* Linn, *Achyranthes corymbosa* Linn, *Phyllanthus niruri* (Muller), *Echinochloa colonum* (L) Link, *Dinebra retraflexa* (Vahl) Panz., *Centella asiatica* (L) Urb., *Vicoa indica* (Dec.), *Merremiya emerginata* , (Burm) Hall-F, *Leucas aspera* (Spreng), *Ocimum sanctum* Linn, *Physalis minima, Striga asiatica* Lour etc.

Pearlmillet : The crops are grown on light sandy to sandy loam soils in Rajasthan and Gujarat and poor red or shallow black soils in northern western Deccan Plateau which including Maharashtra, part of Karnataka, Andhra Pradesh and Tamil Nadu. The major weeds occur are *Dactyloctanium aegypticum* (L) Beauv., *Cynodon dactylon* (L) Pers, *Digitaria sanguinalis, Echinochloa colonum* (L) Link, *Sonchus arvensis* Linn, *Dinebra* sp., *Çyperus rotundus* Linn, *Trienthema portulacastrum* Linn, *Euphorbia microphylla* Heyne ex Roth, *Amaranthus viridis* Linn, *Flaveria australasica, Celosia argentea* Linn, *Alysicarpus rugosus* (L) D.C. etc. in Maharashtra, Gujarat and Tamil Nadu. While *Dactyloctanium aegyptium* (L) Beauv, *Cencharus setigerus* Heyne. *C. biflorus* Roxb, *Convolvulus arvensis* Linn, *Corchorus tridens* Linn, *Crotolaria medicaginea* (Lam), *Cucumis callosus* Roxb, *Cyperus rotundus* Linn, *Digera arvensis* Forsk. *Ipomea pestigridis* Linn etc. in Rajasthan.

Sugarcane : In sugarcane the predominant weeds observed are *Sorghum halepense* Wall, *Eleusine aegyptica* Roxb., *E. indica* (L) Gaertn., *Eragrostis tenella* (L) Beauv., *Setaria glauca* Linn, *Cyperus rotundus* Linn, *Vicoa indica* (Dcd.), *Conyza striata* (Wall), *Ipomea pestigridis* Linn, *Trianthema monogyna* Linn, *Euphorbia hirta* Linn in northern states like Punjab, Uttar Pradesh, Bihar and West Bengal. While the weeds like *Cynodon dactylon* (L) Pers., *Dinebra retraflexa* (Vahl) Panz., *Digitaria marginata* (W) Link, *Panicum repense* Linn, *Merrimiya emerginata* (Burn) Hall. F., *Alloteropsis cimicina, Corchorus acutangulus* (Lam), *Convolvulus arvensis* Linn, *Sonchus arvensis* Roxb, *Caesulia axillaris* Roxb., *Commelina benghalensis* Linn, *Tridax procumbens* Linn, *Acalypha indica* Linn, *Cyperus rotundus* Linn, *Ageratum conyzoides* Linn, *Leucas aspera* Spreng, *Ludwigia parviflora* Roxb, *Sida rhomboidea* Linn, *S. acuta* Burm, *Euphorbia microphylla* Heme ex Roth, *Eclipta alba* (Hassk), *Cleome chelidoni* (Linn) etc. are found in dominance in one region or the other in states like Karnataka, Tamil Nadu, Maharashtra and Gujarat.

Grounanut and Soybean : The major weeds in rainfed groundnut crop are *Eleusine indica* Gaerth, *Dactylocanium* (L) Beauv., *Digitaria sanguinalis* (L) Scop, *D. marginata* (L) Link, *Eragrostis pilosa* (L) Beauv, *E. tenella* (L) Beauv., *E. uniloides* Roxb., *Brachiaria reptans* (L) Gaerth ex Hubb., *B. remosa* (L), *Setaria glauca* Linn, *Paspalum* sp., *Cyperus rotundus* Linn, *Bulbostylis barbata* (Rottb.), *Euphorbia hirta* Linn, *Digera ciliaris* Forsk., *Portulaca oleracea* Linn, etc. In soybean, *Echinochloa colonum* (L) Link, *E. crusgalli* (L) Beauv., *Oplismanus burmanii, Setaria glauca* Linn, *Digitaria adscendense* (L) Link, *Eleusine indica* (L) Gaerth, *Cynodon dactylon* (L) Pers, *Brachiaria mutica, Cyperus rotundus* Linn, *C. iria* Linn, *Digera arvensis* Forsk, *Celosia argentea* Linn, *Solanum nigrum* Linn, *Eclipta alba* Hassk, *Phaseolus trilobus* Alt. etc. are found to be the predominant weeds.

Cotton : In light textured sandy loam alluvial soils of Punjab and western Uttar Pradesh, the major weeds observed in cotton field are *Eleusine aegyptiaca* Desf. *Digitaria sanguinalis* (L) Scop, *Eragrostis tenella* , *E. pilosa* (L) P. Beauv., *Acrachene racemosa* (Heyne) Ohwi, *Brachiaria reptans* (L) Gaerth, *Panicum colonum* Linn, *Cyperus rotundus* Linn, *Trianthema monogyna* Linn, *Commelina benghalensis* Linn, *Conyza striata* (Wall), *Physalis minima* Linn, *Ipomea pestigridis* Linn, *Convolvulus arvensis* Linn, *Tribulus terestris* Linn, *Vicoa indica* (D.C.), *Portulacee oleracea* Linn, *Corchorus tridens* Linn, *Xanthium strumarium* Linn, *Diodia teres* (Wall), *Cucumis callosus* (Rotti) Cogn. etc. In medium to deep black cotton soils in Tamil Nadu, Marathawada, Gujarat and parts of Karnataka the weeds observed are *Denebra retreflexa* (Vahl) Panz, *Cynodon dactylon* (L) Pers, *Echinochloa colonum* (L) Link, *Dactyloctanium aegypticum* (L) Beauv, *Ischaemum pilosum* Linn, *Convolvulus arvensis* Linn, *Phyllanthus niruri* Linn, *Trianthema portulacastrum* Linn, *Amaranthus viridis* Linn, *Boerhaavia diffusa* Linn, *Leucas aspera* Spreng, *Celosia argentea* Linn, *Oscimum bacilicum* Linn, *Hibiscus labatus* (Munro) Kuntze, *Tribulus terrestris* Linn, *Vernonia cinera* Less etc.

Kharif pulses : The major kharif pulses grown in India are pigeonpea, greengram, blackgram, cowpea, mothbean and horsegram. They are sown with the onset of monsoon in June-July under varied climatic conditions, usually on light textureș uplands. The major weeds observed in northern states are

Digitaria sanguinalis (L) Scop, *Dichanthium annulatum* (Forsk) Stapf., *Eragrostis tenella* (L) Beauv, *Eleusine indica* Gaerth, *Setaria glauca* (L) Beauv, *Cyperus rotundus* Linn, *Leucas aspera* Spreng, *Achyranthes aspera* Linn, *Scoparia dulcis* Linn, *Mazus rogusus* Lour, *Convolvulus arvensis* Linn, *Cleome viscosa* Linn, *Digera arvensis* Forsk, *Celosia argentea* Linn etc. while in Karnataka, Tamil Nadu, Andhra Pradesh, Maharashtra and Gujarat, the weeds observed in these crops are *Cleome viscosa* Linn, *Portulaca quadrifida* Linn, *Fimbristylis aestivalis* (Vahl), *Adenostemna rugosus* (Wight), *Lindernia parviflora* (Roxb) Haines, *Panicum repens* Linn, *P. maximum* Linn, *Monochoria vaginalis* Presl., *Acalypha indica* Linn, *Commelina benghalensis* Linn, *Celosia argentea* Linn, *Cyperus rotundus* Linn, *Digitaria sanguinalis* (L) Scop, *Vernonia cinerea* Less, *Tribulus terrestis* Linn, *Digera arvensis* Forsk, *Commelina benghalensis* Linn, *Cucumis trigonus* Roxb, etc.

Jute and Mesta : The jute and mesta is grown mainly on alluvial and lateratic soils, under different moisture conditions, jute prefers relatively high temperature, humidity and rainfall and deep rich alluvial soils while mesta is grown on marginal and comparatively dry climate (rainfall 500-750 mm). The weed flora widely differs in different situations. However, the most predominant weed species observed are *Eleusine indica* (L) Gaerth, *Dactyloctanium aegyptiacum* (L) Beauv, *Echinochloa colonum* (L) Link, *Brachiaria reptens* (L) Gaerth, *B. ramosa* (L) Stapc, *Leptochloa chinesis* (L) Nees, *Paspalum commersonii* (Lam), *Eragrostis uniloides* Roxb, *Setaria glauca* (L) Beauv, *Axonopus compressus* Linn, *Cyperus rotundus* Linn, *C. iria* Linn, *C. elatus* Rottb., *Fimbristylis aestivalis* Vahl, *F. dichotoma* Vahl, *Croton bractiferus, Portulaca oleracea* Linn, *Phyllanthus niruri* (Muller), *Physalis minima* Linn, *Triumffetta rotundifolia* Lam, *Euphorbia hirta* Linn, *E. nerifolia* Linn. *E. microphylla* Roth, *Launea pinnatifida* Linn, *Desmodium trifolium* D.e., *Sco-paria dulcis* Linn etc. The frequency and distribution of these weeds varies under different situations and stages of growth.

4.2 Weed flora of rabi season

Wheat : The survey of weed flora of wheat fields revealed that *Phalaris minor* (L) Retz and *Avena ludoviciana* Dur were the major weeds through out the northern states. However, where herbicides are used for their control, weeds like *Lathyrus aphaca* (Linn), *Melilotus indica* (L) All, *Vicia sativa* Linn and *Anagalis arvensis* Linn are becoming major problem. The problem of new grassy weeds like *Polypogon monspeliensis* (L), *Poa annua* (Linn) Desf. and *Lolium temulentum* (Linn) are also increasing in some parts of these states. Infestation of *Circium arvense* (L) Scop in patches was also observed. *Silene conoidea* Linn was quite serious in light textured soils. The other important weeds in irrigated wheat observed were *Chenopodium album* Linn, *C. murale* Linn, *Melilotus alba* Desr, *Spergula arvensis* Linn, *Rumex dentatus* Linn, *Vicia faba* Linn, *Fumaria parviflora* Lam, *Convolvulus arvensis* Linn, *Euphorbia simplex, Asphodelus tenuifolius* Cav., *Medicago denticulata* Linn, *Lepidium sativa, Trigonella polycerata, Carthamus oxycantha* (L) M.Bieb, *Argemone maxicana* Linn, *Vicia sativa* Linn, *V. hirsuta* (L) S.F. Gray, *Polygonum plebejum* R.Br., *Solanum nigrum* Linn, *Launea pinnatifida* Linn, *L. asplenifolia* (Willd) Hook F., *Cynodon dactylon* (L) Pers. etc. In unirrigated wheat in Madhya Pradesh, the major weeds observed were *Heliotropium eichwaldi* Steud et D.C., *Chenopodium album* Linn, *Convolvulus arvensis* Linn, *Asphodelus tenuifolius* Cav., *Digera alternifolia, Phaseolus trilobs* Ait., *Indigofera linifolia* Retz, *Leaunea aspenlifolia* (Willd) Hook, *Medicago denticulata* Linn, *Anagalis arvensis* Linn, *Nicotiana plumbaginifolia* Viv., *Gnaphallium purpurium* Linn etc.

In gram and berseem the major weeds observed are *Asphodelus tenuifolius* Cav., *Chenopodium album* Linn, *Trigonella polycerata* Linn, *Fumaria parviflora* Lam, *Vicia sativa* Linn, *V. faba* Linn, *Launea aspalenifolia* (Willd) Hook.F., *Lathyrus aphaca* Linn, *Argemone maxicana* Linn, *Silene conoidea* Linn, *Hibiscus micranthus, Sonchus arvensis* D.C., *Chrozophora plicata* Var. Rottleri, *Melilotus alba* Desr etc. While in berseem crop, the most predominant grassy weeds were *Poa annua* Linn, *Polypogon monsplensis* (L) Desf. and *Phalaris minor* (L) Retz. and broad leaved weeds were *Cichorium intybus* Linn and *Lepidium sativa* . The other important weeds were *Melilotus indica* (L)

All, *Anagalis arvensis* Linn, *Arenaria serphyllifolia* Linn, *Medicago denticulata* Linn, *Vicia sativa* Linn, *Sonchus arvensis* D.C., *Euphorasia simplex* D.Don, *E. prostrata* Alt., *Trifolium resupinatum* Linn, *Gnaphalium purpium* Linn, *Sphaeranthus indicus* Roxb., *Volutarella divericata* Benth etc.

4.3 Weed flora of orchard and plantation crops

Orchards : The dominant weeds in orchards in hills were *Leucas aspera* Spreng, *Galinsoga parviflora* Cav., *Myractus* sp., *Epilobium* sp. etc. In other orchard crops in plains the major weeds observed were *Cynodon dactylon* (L) Pers, *Dichanthium annulatum* , *Lagascea mollis* Linn, *Cyperus rotundus* Linn, *Corchorus tridens* Linn, *Phyllanthus niruri* Muller, *Lantana camera* Linn, *Amaranthus viridis* Linn and *Ageratum conyzoides* Linn.

4.4 Weed flora of waste and vacant lands

The major weeds of waste and vacant lands were *Cassia tora* Linn, *Dichanthium annulatum* Linn, *Dactyloctenium aegyptiacum* (L) Beauv., *Taphrosia purpurea* (L) Pers, *Aristida* sp., *Tribulus terrestris* Linn, *Cynodon dactylon* (L) Pers, *Apluda varia* Hack, *Alysicarpus monilifer* (DeC), *Boerhaavia diffusa* Linn, *Achyranthes aspera* Linn, *Ageratum conyzoides* Linn, *Parthenium hysterophorus* Linn, *Croton bonplandianum* Baill. etc. The dominant shrubs observed were *Calotropis procera* (Alt.) B.R., *Zizyphus rotundifolia* Roth., *Acacia* sp. and *Lantana camera* Linn etc.

4.5 Other dominant weeds

In Tamil Nadu and Karnataka the recent spread of problematic and hard to control weed i.e. *Solanum elaegnifolium* Cav. is contemplated to exaggerate the existing weed problem considerably in crops like sorghum, sunflower, blackgram etc. The incidence of *Ageratum conyzoides* Linn in recent years reported to be a problem of maize in the hilly terrain of Himachal Pradesh and has proved to be hazardous to live stock. An uncommon weed species of transplanted rice viz. *Mollugo pentaphylla* Linn was observed to have high population as 650 per sq.m. constituting about 90% of the total population, in the western Uttar Pradesh, though *Cyperus iria* Linn is considered to be the most dominant weed of the region as a whole. Based on the observations in Madhya Pradesh, *Parthenium hysterophorus* Linn earlier considered to be a problem of waste land was observed to be a common weed of kharif crops like rice (upland), millets, jowar, cotton and soybean while that of rabi crops like irrigated wheat, berseem, vegetable and orchard crops, particularly in the vicinity of the urban areas.

The weed like *Saccharum spontanium* Linn has infested larger areas in Madhya Pradesh, Uttar Pradesh, Rajasthan and Maharashtra. It encroaches cultivated area from bunds of the fields and occupies rapidly the neglected areas through its air borne seeds, during early winter period. *Pennisetum* sp. (*P. polystachyon* and *P. pedicellatum*) and *Chromolaena odorata (Eupatorium odoratum* Linn) are the serious weeds in the plantation crops of central and northern Kerala and citrus orchards in northern and eastern region. *Lantana camera* Linn introduced as ornamental hedge plant has now escaped to pasture, gardens, forests and even cultivated lands both in plains in the south and east and upto hills in Himchal Pradesh. It has no specific soil adaptation and grows in varied climatic conditions. *Imperata cylindrica* (L) Beauv is also a problem weed in cultivated and neglected field and plantation crops in eastern states and more particularly in West Bengal, Assam, Meghalaya and northern parts of Bihar. *Pluchea lanceolata* Oliv. and *Alhagi cemelorum* Fisch. are the weeds of light textured soils. Due to extensive root system, they resist drought conditions for fairly long period. They infest winter crops particularly grain crops and have infested larger areas around Delhi, Haryana, Rajasthan, western Uttar Pradesh and Madhya Pradesh.

Orobanche spp. (*O. aegyptiaca* Pers, *O. cernua* Loef. and *O. indica* Spreng) a parasitic weed is problem in crops like tobacco, brinjal, tomato, rapeseed and mustard in Tamil Nadu and Andhra

Pradesh. The another parasitic weed *Striga* spp. (*S. densiflora* (Benth), *S. lutea* (Lour), *S. hirsuta* (Benth), *S. orobanchoides* (Benth) etc). effects crops like sorghum, pearlmillet, sugarcane, maize, ragi and kedaon in many states like Andhra Pradesh, Maharashtra, Rajasthan, Uttar Pradesh, Tamil Nadu and Punjab. *Striga* infestation is greater in light shallow and red soils. The other parasitic weeds *Cuscuta* spp. (*C. chinenses* (Linn), *C. hyalina* (Roth) and *C. reflexa* (Roxb) etc.) causes heavy losses to lucerne. It also effects linseed, blackgram, greengram, redgram and sugarcane. In Orissa, it has invaded niger crop.

5. WEED MANAGEMENT TECHNOLOGY

The old age traditional method of weed control viz. manual and mechanical have continued to be the mainstay in India irrespective of field, orchard, plantation or vegetable crops till recently. However, the herbicides have been found to be very effective tools for controlling these weeds in diversified farming systems all over the world. In India, the use of herbicides got momentum only after 1978, when the farmers of Punjab and Haryana began using herbicides like metoxuron, methabenzthiazuron and trillate to control problem weed like wild Canary grass (*Phalaris minor* L) in wheat and butachlor and 2,4-D to control *Echinochloa colonum* L. and *E. crusgalli* L in rice. The national market of herbicide was around 15 tonnes in 1970 to 1425 tonnes in 1977-78, 4862 metric tonnes in 1990-91 and is anticipated to 7620 metric tonnes in 1994-95 (Table-2). The extensive research work done on different aspects of weed management in different crops and cropping systems

Table-2 : Consumption of technical grade pesticides in India 1965-90 (Quantity in Metric Tonnes)

Years	Insecticides	Fungicides	Herbicides	Source
1965	10,428	1480	-	
1970	21,741	1774	15	Pesticide management &
1975-76	46,605	9183	890	pesticide- B.V. David
1976-77	41,250	8553	975	
1977-78	46,775	9683	1425	
1986-87	29,670	17125	3670	Pesticide Association of India & Fundamental of weed management in hot climate peasant agriculture - S. Alstrom
1988-89	67,670	16365	4160	Pesticide information XVII (i) :
1989-90	56,424	17530	4730	April-June 1991-92
1990-91	47,074	17770	4862	
1994-95*	51,755	22895	7620	Anticipated

is summarized below.

The different herbicides evaluated for different crops under AICRP-WC and elsewhere through out India over these years are given in Table-3 while weed management practices developed for various crops are summarized below.

Rice

Rice in India is cultivated in different cultures viz. direct seeded upland rainfed, direct seeded lowland rainfed, direct seeded paddled and transplanted rice. It is also sown as pregerminated direct sown paddled rice. The weed problem is very severe in upland direct seeded rainfed rice. In absence of flooded water the weeds provide severe competition to the crop and the yield losses are reported to the extent of 100%, if weeds are not removed timely and properly. Usually, two to three manual weedings at an interval of 2-3 weeks are given after sowing. In broadcast sown crop, spike tooth harrow is used to reduce the weed infestation in early stages of growth. The use of higher seed rate, sowing the crop in rows, good land preparation, stale seed bed technique, application of nitrogen fertilizers just after weedings are suggested to avoid severe competition from weeds in this crop.

Since manual weeding operation is slow, time consuming, labour intensive, expensive, difficult to perform due to adverse climatic conditions and fields are unfit for the operation during rainy season besides grassy weeds are undistinguishable with the crop during early stages of growth the efforts were made to develop weed management technology using herbicide in combination with manual or mechanical weedings.

Upland rice : A number of herbicides have been evaluated for upland rice which includes premergence application of nitrofen, butachlor, thiobencarb, pendimethalin, oxadiazon, anilophos, oxyfluorfen and postemergence foliar application herbicides like 2,4-D and propanil.

Preemergence application of nitrogen, 2-3 kg/ha or a mixture of 2,4-D or MCPA with propenil as postemergence application followed by flooding 3-5 days after sowing gave excellent control of all type of weeds in drilled rice. Oxadiazon 1.0 kg/ha or butachlor or thiobencarb 1.5 kg/ha as preemergence are also recommended. From the work done under AICRP-WC, a combination of oxadiazon, nitrogen, molinate or thiobencarb at 1.0 kg/ha with postemergence application of propanil at 2-3 kg/ha, 20 to 25 days after sowing is suggested. However, flooding the field 2-3 days after

Table-3 : Herbicide for different field, vegetable, orchard and plantation crops

Rice (upland)	:	Thiobencarb, oxadiazon, butachlor, benthiocarb, pendimethalin, oxyfluorfeb, propanil, butachlor, 2-4DEE, nitofen, molinate and piperophos.
Rice (lowland/transplanted	:	Butachlor, piperophos 2,4-D Na salt, pendimethalin, oxadiazon, benthiocarb, propanil, nitrofen, butachlor + 2,4-DEE and anilophos + 2,4-DEE
Wheat	:	Isoproturon, metoxuron, trillate, diuron, chlorotoluron, 2,4-D Na salt, methabenzthiazuron, fluroxypyr, isoproturon + 2,4-D Na salt, methabenzthiazuron, fluroxypyr, isoproturon + 2,4-DEE, tralkoxydim and tralkoxydim + 2,4-DEE
Maize	:	Simazine, cynazine, atrazine, pendimethalin and metolachlor.
Perarlmillet/sorghum	:	Atrazine, simazine and cynazine.
Groundnut	:	Fluchloralin, nitrofen, fluazifop-p-butyl, imazethapyr, pendimethalin, alachlor, oxyfluorfen, butachlor, phenoxypropethyl and terbutryn.
Soybean		Metolachlor, fluchloralin, metribuzin, dimethazon, sethoxydim, haloxyfop-methyl, butachlor, oxyfluorfen and fluazifop-p-butyl.
Sesamum		Alachlor, fluchloralin, butachlor, thiobencarb and nitrofen.
Sunflower		Pendimethalin, fluchloralin, oxadiazon, metolachlor and haloxyfop-methyl.
Castor		Pendimethalin, metolachlor and fluchloralin.
Linseed		Isoproturon, fluchloralin, methabenzthiazuron, diclofop-methyl and alachlor.
Rapeseed-mustard		Isoproturon, metoxuron, alachlor, pendimethalin, haloxyfop-methyl, bifenox and oxadiazon.
Lentil and gram		Oxyfluorfen, fluchloralin, pendimethalin, isoproturon, metribuzin, linuron, trifluralin and terbutryn.
Cowpea, chickpea and pea		Bentazon, fluchloralin, prometryn, metribuzin, metolachlor and linuron.

Cotton	Fluchloralin, oxyfluorfen, oxadiazon and pendimethalin.
Jute and Mesta	Fluchloralin, MSMA, tetrapion, dalapon, dalapon + MSMA and fluazifop-p-butyl.
Sugarcane	Atrazine, metribuzin, atrazine + metribuzin, diuron, butachlor, butachlor + 2,4-DEE.
Coffee and Tea	Dalapon, paraquat, glyphosate, dalapon + 2,4-D, glyphosate + diuron, haloxyfop-methyl, glyphosate + uphar surfactant 0.1%.
Potato	Oxyfluorfen, metolachlor, atrazine, fluchloralin, alachlor, methabenzthiazuron, linuron and metribuzin.

application of propanil is essential for effective control. The work done at UAS, Bangalore has indicated that in drilled paddy preemergence application of anilophos (Aniloguard 30 EC), 0.4 kg/ha, 2,4-DEE 0.6 kg/ha + butachlor 1.0 kg/ha or pendimethalin 2.0 kg/ha gave good control of weeds. At Faizabad, preemergence application of butachlor or thiobencarb 1.5 kg/ha or pendimethalin 1.0 kg/ha or postemergence application of 2,4-D Na salt at 0.5 kg/ha reduced weed biomass considerably as compared to unweeded check and the yield in these treatments were comparable with manual weeded treatment. At Jorhat, butachlor 1.0 kg/ha as preemergence followed by operating dryland weeder (25 DAS) or anilophos 0.4 kg/ha following mechanical weeding with dryland weeder or thiobencarb 1.0 kg/ha as preemergence and followed by mechanical weeding with dryland weeder provided weed free conditions. The better weed control efficiency was also observed when the weeding tools were operated twice at 20 and 40 days after sowing. Dryland weeder, peg type and twine wheel hoe were used for this purpose.

Transplanted/direct seeded puddled rice : The puddling operation not only reduces the infiltration of water to the lower horizon but also provides better conditions for the rice plant and suppresses the resprouting of perennial weeds. It also suppresses the germination of most of the grassy weeds except *E. crusgalli* and *E. colonum* . However, in puddled conditions, the sedges other than *C. rotundus* and broad leaved weeds become the dominant weeds. Usually one or at the most two manual weedings are done by pulling the weeds in standing water. A pull and push type rotary weeder is also used to suppress initial growth of weeds. In some parts of India viz. Orissa, Madhya Pradesh and West Bengal, the farmers control the weeds by cross ploughing their fields after flooding the fields when rice is 30-40 days old.

The use of herbicide is very less in traditional rice belt of eastern India, but, has become popular in recent years in Punjab, Haryana, Jammu & Kashmir and eastern parts of Uttar Pradesh (non traditional rice belts). The herbicides are also being used in this crop now in southern states like Tamil Nadu, Andhra Pradesh and to some extent in Karanataka. The earlier methods recommended for weed management were postemergence application of 2,4-D or MCPA, 30 days after transplanting to control sedges and broad leaved weeds. Later granular formulation after transplanting as preemergence to weeds were recommended to control grassy and also broad leaved weeds. However, it was suggested that the applications be made on moist soil.

In recent, years, butachlor, 1.25 kg/ha; thiobencarb 1.0 kg/ha, oxadiazon 0.75 to 1.0 kg./ha, pendimethalin 1.0 to 1.5 kg/ha, oxyfluorfen 0.125 kg/ha, anilophos 0.4 kg/ha, anilophos 0.3 kg/ha + 2,4-D, 0.3 kg/ha, molinate 1.5 kg/ha as preemergence to weeds 7 days after transplanting as spray or mixed with urea or granular formulation or pre or postemergence application of 2,4-D (Weltable powder), 0.8 kg/ha are suggested for effective control of weeds in this crop. These herbicides could also be used at above rates in direct seeded puddled rice (2).

Maize and millets

Maize (*Zea mays* L.). The traditional practices in this crop is to go for manual weedings. Usually two manual weedings are done, the first weeding is done at 30-35 days after sowing followed by another after 15 days depending upon the availability of labour, soil and climate conditions. In maize, which is sown in wide spaced rows, different types of bullock drawn cultivators and hoes are used later on a light hand weeding is given to remove weeds from maize rows. Usually 2 to 3 intercultivations are done followed by one manual weeding. In northern parts of the country, earthing with the help of Spade or khurpi is also done during second weeding operation. In Punjab, for maize simazine or atrazine at 1.0 kg/ha as preemergence followed by earthing up between 30-40 days is suggested, the herbicides controls most of the grassy sedges and broad leaved weeds, however, this system has replaced the earlier recommendation of 2,4-D or MCPA which has selective control of certain species of weeds. The other promising herbicides which could be used in this crop are alachlor, pendimethalin or metolachlor at 1.0 to 1.5 kg/ha as preemergence. Linuron, 1.0 kg/ha also been reported for control of weeds in this crop.

At Hyderabad, a combination of preemergence application of atrazine 1.25 kg/ha followed by 2,4-D Na salt 1.0 kg/ha at 30 days after sowing or directed application of glyphosate provided weed free conditions in this crop and was comparable with two hand weedings done at 20 and 40 days after sowing. The other promising herbicides are application of oxyfluorfen, 0.20 kg/ha or pendimethalin at 1.5 kg/ha as preemergence application. Postemergence application of cynazine or cyprazine each at 0.75 to 2.0 kg/ha or thiobencarb, alachlor, prometryn and isoproturon at 1.0 kg/ha are also suggested for control of these weeds in this crop. A mixture of dicamba and 2,4-D has also been recommended for control of problem weeds like *Trianthema monogyna, Portulaca oleracea* and *Ipomoea* sp.

Millets : Sorghum (*Sorghum bicolor* (L.) Moench) and pearlmillet (*Pennisetum glaucum* (L) R. Bic. emend Stuntz. *americasium* (L.) Lacke) are the major millet crops grown in wide range of climatic and soil conditions. The general practice of weed removal is by hand weedings using khurpi Usually two weedings are done within 45-50 days after sowing. 2,4-D at 1.0 kg/ha could be applied as postemergence application for control of broad leaved weeds. A mixture 2,4-D and dicamba or atrazine has proved quite effective in controlling problem weeds like *Tribulus terrestris* in pearlmillet. At HAU, Hisar, in pearlmillet atrazine 0.375 kg/ha at 7-14 days after sowing is suggested. The work done at various centres of AICRP-WC indicated that in sorghum, preemergence application of metribuzin -.25 kg/ha, terbutryn or linuron 0.25 kg/ha or thiobencarb, pendimethalin or alachlor 1.0 kg/ha or oxyfluorfen 0.15 kg/ha (for rabi sorghum only) is recommended. For pearlmilled (Bajra), preemergence application of simazine or cynazine 0.50 kg/ha and 0.25 kg/ha, respectively is suggested.

Wheat

The general practice of weed control in wheat is through cultural practices by engaging labour for manual weeding. Usually one or at the most two manual weedings are done within 45 to 50 days after sowing. However, the chemical control of broad leaved weeds in wheat using 2,4-D is well known and is adopted all over the world. In recent past, with the introduction of dwarf wheat, the problem of grassy weeds like Canary grass (*Phalaris minor*), wild oats (*Avena ludoviciana*) and most recently grassy weeds like *Polypogon monspelnsis, Poa annua* and *Lolium temulentum* has increased in some parts of northern states. A number of herbicides has been suggested for controlling these weeds. The most effective herbicides for control of wild oats are triallate and dichlo-fop-methyl, while for control of *Phalaris minor* and other broad leaved weeds, the application of isoproturon 1.0 kg/ha and 2,4-D 0.250 kg/ha as postemergence 30-35 days after sowing (after first irrigation) as spray is suggested. The other promising herbicides for wheat suggested are methabenzthioxuron 1.5 kg/ha, metozuron 1.5 kg/ha, chloroturon, pendimethalin 1.0 kg/ha, fluroxypyr 1.50 kg/ha and tralkoxydim 0.250-075 kg/ha which could be used in different solutions. For control of *Carthamus oxyacantha* clopyralid 200 gm/

ha as postemergence may be used (2).

At Ludhiana, tralkoxydim 0.25 kg/ha as postemergence application 28 days after sowing gave effective control of both *P. minor* and *A. ludoviciana*. A combination of fluroxypyr and isoproturon 0.2 kg/ha and 0.75 kg/ha gave better control of grassy and broad leaved weeds and was comparable with isoproturon 1.0 kg/ha + 2,4-D Na salt 0.6 kg/ha. At Jabalpur, fluroxypyr 0.2 kg/ha provided effective control of dicot weeds except *Medicago denticulata* while clopyralid was also effective against most of the dicot weeds but *Chenopodium album* and *Spergula arvensis* were not controlled. Both the herbicides were ineffective against grassy and sedges weeds, significantly lower weed biomass was recorded under fluroxypyr 150 gm/ha + isoproturon 1.0 kg/ha as postemergence.

Studies on the effect of wheat crop densities on the growth and development of *Phalaris minor* indicated that the higher seed yield of 120 kg/ha induced an appreciable smoothering effect of this weed, though the grain yield obtained was at par with the recommended seed rate of 80 kg/ha. The grain yield with cross sown wheat crop was significantly higher than that of a normal row seeded crop. Skip-row sowing (skipping 1 row after every 5 rows 4-5-15-15-15-15-45 cm) in combination with one each of hoeing and light hand weeding produced grain yields comparable to that of isoproturon treated crop. Closer row spacing or cross row sowing in combination with half the recommended rate of isoproturon, i.e. at 0.5 kg/ha applied two weeks after sowing controlled *Phalaris minor* and *Avena* sp. more effectively as compared to their independent effects. Besides, the tall genotypes of wheat effected reduction in dry matter of important weeds like *Phalaris minor, Chenopodium album* and *Melilotus alba* by 51,45 and 50%, respectively. Inducing germination of weeds by presowing irrigation and destroying weeds either by cultivator or use of paraquat, gave significantly higher yield over unweeded crop, however, these yields were lower than isoproturon treated fields.

Pulse crops

The pulse crops in India are grown under different agro-climatic and edaphic conditions. They are usually grown on marginal conditions as rainfed, therefore, the weed intensity is not as intense as in other cereal or vegetable crops. Further due to their profuse canopy, these crops suppresses germination and growth of weeds specially when they are established. The important pulse crops grown are redgram (*Cajanas cajan* L. Millsp), blackgram (*Vigna mungo* L. Hepper), greengram (*V. radiata* L. Wilezek), cowpea (*V. anguiculata* L. Walp), horsegram (*Myerotyloma uniflorum* (Lam) Verdc., mothbean (*V. aconitifolia* Jacq. Marechal) are grown during kharif season (June-September) and chickpea (*Cicer arietinum* L), pea (*Pisum sativum* L), lentil (*Lens culinaris*), french bean (*Phaseolus vulgaris* L.) and lathyrus (*Lathyrus sativa* L.) during rabi season (Oct. to March). Greengram and blackgram are also grown during zaid season (February-April).

The analysis made of the reasons of the low productivity of pulse crops in India indicated that weeds are the major constraints in realizing the optimum yields in pulse crops. The weed problem in kharif pulse crops has been more predominant due to heavy infestation of grassy weeds besides sedges specially *Cyperus rotundus* Linn and a few broadleaved weeds. The hot humid climate and frequent rains favours profuse growth of weeds and crops suffers, if the weeds are not controlled in time. The problem of weeds in rabi pulses is of lesser magnitude and can be managed easily.

The research work done on critical period of crop-weed competition in different pulse crops has indicated that increasing length of crop-weed competition resulted in decrease in grain yield of pulse crops. The initial 20-40 days period was critical for short duration kharif and summer pulses and 15-60 days for medium and late maturing crops like pigeonpea, whereas in rabi pulses, the critical period of crop-weed competition was found to be between 30-60 days after sowing. The field experiments conducted under AICRP-WC at different centres indicated that presence of weeds in different crops cause heavy losses in grain yield which varied from crops to crops. In redgram, it is 28.2%, blackgram 82.2%, lentil 75.2%, chickpea 76.30% and in pea 82.2% (30). The critical period of crop weed competition observed in different crops have been summarized in Table - 4.

Cultural practices : A good amount of weed infestation in pulse crops can be managed by preparatory cultivations, adjustment in time of sowing, time and method of fertilizer application, hoeing and interculture practices. Usually one or two manual weedings are done for different pulse crops depending upon the duration of the crop, intensity of weeds and weed suppression ability of crop. In long duration crops like pigeonpea one more weeding or mechanical weeding using different types of hoes becomes essential. In short duration pigeonpea, single weeding at 30 days or two weedings at 20 and 40 days provided effective control of weeds and resulted in at part seed yields obtained in weed free treatment at Faizabad. In lentil and chickpea, a single weeding at 25-30 days after sowing gave good control at Pusa.

The experiments conducted at Jorhat to study the effects of placement of herbicides and the

Table-4 : Critical period of crop-weed competition in different crops (33)

S.No.	Name of the pulse crops	Critical period of crop-weed competition (Days after sowing)
1.	Pigeonpea	15-60
2.	Greengram	15-45
3.	Blackgram	15-45
4.	Cowpea	15-35
5.	Chickpea	30-60
6.	Pea	30-45
7.	Lentil	30-60

performance of various weeding tools on the efficiency of weed control in pea fields, indicated that more numbers of broadleaved weeds were observed where fertilizers were broadcasted whereas narrow leaved weeds were in profuse number in line placement of fertilizers. Lower population of weeds were observed under grubbing 25 days after sowing. However, maximum efficiency was recorded in hand weeding done at 20 and 40 days after sowing (2).

Chemical control : In redgram, preemergence application of metribuzin 0.25 kg/ha or alachlor or nitrofen 1.0 kg/ha or oxadiazon 1.5 kg/ha or preplant soil incorporation of fluchloralin 0.5 kg/ha is suggested for effective control of weeds (6). Most recently pendimethalin or nitrofen 1.0 kg/ha have also been suggested for control of weeds in this crop. In blackgram, preplant soil incorporation of fluchloralin 1.5 kg/ha (Jorhat); preemergence application of metolachlor 1.0 kg/ha (Bhubaneshwar and Ludhiana) and pendimethalin (Pantnagar) have been found quite effective in controlling weeds (Table-5).

In greengram, preplant soil incorporation of fluchloralin or trifluralin 0.5 to 1.0 kg/ha or application of thiobencarb 1.0 kg/ha or pendimethalin 0.75 kg/ha or oxyfluorfen 0.15 kg/ha or terbutryin 0.50 kg/ha were found quite effective (2). In a separate study, preplant incorporation of fluchloralin 1.0-1.5 kg/ha or preemergence application of pendimethalin 1.5-2.0 kg/ha effectively controlled *Trianthema portulacastrum* and *Echinochloa colonum* in summer mungbean (17). In cowpea, preemergence application of prometryn 1.0 kg/ha gave higher yield at Jabalpur (6). Screening

Table-5 : Mean grain yield of blackgram as influenced by different herbicides treatments (32)

Herbicides	Rate (kg/ha)	Time of application	Jorhat	Bhubaneshwar	Pantnagar
Fluchloralin	1.5	PPI	1082	-	-
Metolachlor	1.0	Pre	-	544	-
Metolachlor	2.5	Pre	960	-	-
Fluazifop-p-butyl	0.25	Post	720	400	-
Butachlor	0.50	Pre	-	517	-
Pendimethalin	0.50	Pre	-	528	-
Thiobencarb	0.50	Pre	-	333	-
Pendimethalin	2.0	Pre	-	-	1067
Fluchloralin	1.0	PPI	-	-	883
Weed free	-	-	-	-	1267
Weedy check	-	-	402	100	217

trials conducted at various centres to find out suitable herbicides for chickpea indicated that premergence application of oxadiazon (Jabalpur), linuron (Pantnagar) and terbutryn (Pusa), each at 1.0 kg/ha provided effective control of weeds in chickpea and grain yields were comparable with weed free check (Table-6).

Table-6 : Mean grain yield of chickpea (kg/ha) as affected by different herbicide treatments (1984-86) (32)

Herbicides	Rate (kg/ha)	Time of application	Jorhat	Bhubaneshwar	Pantnagar
Fluchloralin	1.0	PPI	1745	194	1447
Oxadiazon	1.0	Pre	1664	777	-
Methabenzthiazuron	1.0	Pre	1578	410	-
Terbutryn	1.0	Pre	1926	694	-
Thiobencarb	1.0	Pre	1754	-	-
Pendimethalin	1.0	Pre	1711	288	1254
Isoproturon	1.0	Pre	-	632	-
Metribuzin	1.0	Pre	-	-	1904
Linuron	1.0	Pre	-	-	2100
Weed free	-	-	2107	-	1918
Weedy check	-	-	1179	225	455

The research work done at Pantnagar indicated that in pea, the yield reduction due to weed infestation was to an extent of 70%. Preemergence application of linuron 1.5 kg/ha or metribuzin 0.5 kg/ha or pendimethalin 1.0 kg/ha or preplant soil incorporation of flichburlin 1.0 kh/ha effectively controlled weeds in peas. Sharma and Vats (35) also observed that pendimethalin was the safest herbicide and very effective against broadleaved weeds. Kumar and Kolar (18) found that application of terbutryn 0.6-1.2 kg/ha, methabenzthiazuron 1.0-1.5 kg/ha and pendimethalin 1.0-1.5 kg/ha were very effective against broadleaved weeds except *Medicago denticulata* .

It has been observed that fluchloralin controls quite effectively all types of annual grasses and broadleaved weeds. At higher doses it causes stunted plant growth, reduced root growth and dry matter accumulation in crop seedlings at early stages. In chickpea and lentil, it delayed germination. Pendimethalin is another effective herbicide for control of annual grasses and other non-grassy weeds in rabi pulses, however, *Avena* sp. and *Commelina* sp. are not controlled by pendimethalin. Metribuzin, terbutryn and cyanazine, if applied under high soil moisture condition cause toxicity to peas, lentil and chickpea during early growth stage.

Integrated weed management : Integrated approach of weed management revealed the scope for using herbicides at reduced rates by combining it with interculture operations and/or by agronomic manipulations for suppressing later flush of weeds or by growing intercrops in between the rows. In redgram, application of fluchloralin followed by one or two hand weedings at 30-40 and 50 DAS gave seed yields comparable to the weed free treatments at Faizabad (7). The major weeds of pigeonpea were effectively controlled by preemergence application of terbutryn 1.0 kg/ha or cyanazine 0.5 kg/ha or oxadiazon 0.5 kg/ha or preplant soil incorporation of fluchloralin at 1.0 kg/ha followed by one manual weeding performed at 45 DAS (5). In mungbean, the study revealed that postemergence application of 1.0 kg/ha bentazone in combination with hand weeding 30 DAS produced grain yield comparable with that of weed free conditions.

In irrigated blackgram at Coimbatore, highest grain yield (11.70 q/ha) was obtained with isoproturon 0.5 kg/ha fb one hand weeding at 40 DAS and was comparable with pendimethalin 0.75 kg/ha + 1 HW (11.10 q/ha). Application of pendimethalin 1.0 kg/ha fb one late manual weeding was found to increase the crude protein content (30.1%) and per day production (3.34% kg/ha/day) (7). In chickpea, preemergence application of pendimethalin 0.90 kg/ha fb one hand weeding at 6 weeks after sowing (at Parbhani) and preplant soil incorporation of 0.5 kg/ha fluchloralin with 1 hand weeding done at 21 days after sowing (at Sriniketan) appeared most effective in controlling the weeds.

Oilseed crops

The oilseeds in India are grown in over 21 million hectares of lands under wide range of agro-ecological conditions, cropping systems and management practices. The average productivity is very low as compared to other countries mainly due to low inputs in the form of selection of land, land preparation, application of low levels of fertilizers, seed quality, irrigation, use of pesticides and untimely and inadequate weed management practices. The major oilseed crops grown during kharif season are groundnut (*Arachis hypogaea* L.), sesamum (*Seamum indicum* L.), sunflower (*Helianthus annus* L.), caster (*Ricinus communis* L.) and niger (*Guizotia abyssininica* L.) while sunflower (*Carthamus tinetorius* L.), linseed (*Linum usitatissium* L.) and rapeseed-mustard (*Brassica* sp.) are grown during rabi season. Sunflower is also grown during rabi season in many parts of the country. The kharif crops are grown as rainfed while rabi crops are grown with residual moisture of the previous crops or with limited irrigation facilities. The research work done on weed management in India is summarized below :-

The extensive weed survey conducted during 1978-91 in different states of the Indian Union has resulted in identifying the major weeds associated with the oilseed crops. In kharif season *Echinochloa colonum* L., *Digitaria adscendens* (L.) Link, *Eleusine indica* Gaertn, *Brachiaria eruciformis* L. etc. were most dominant grassy weeds and *Celosia argentea* L., *Ageratum conyzoides* L., *Commelina benghalensis* L., *Amaranthus polygomous* L. and *A. viridis* were the dominant broadleaved weeds. In rabi season, *Phalaris minor* Retz. and *Avena ludoviciana* Dur. amongst grasses and *Argemone maxicana* L. *Chenopodium album* L., *Chichorium intybus* L. and *Anagallis arvensis* L. among broadleaved weeds have been identified besides annual and perennial grasses viz. *Cynodon dactylon* (L) Pers, *Panicum* spp., *Echinochloa colonum* (L) Link and sedges like *Cyperus rotundus* L. cause serious problem in both rabi and kharif oilseed crops. Amongst the parasitic weeds, *Cuscuta* spp. has been reported infesting niger and linseed, while *Orobanche* spp. to mustard crops.

Oilseed crops take relatively longer time to establish in their early growth stage, except a few crops like groundnut and soybean. If the weeds are managed to a certain stage of crop, the yield of oilseeds can be increased. Crop-weed competition studies conducted under AICRP-WC have indicated that initially 30-60 days weed free for groundnut, 15-40 days for mustard, sesamum, niger and sunflower, 20-45 days for linseed, 30-60 days for soybean and 30-60 days for castor are necessary and the crops during this period should be kept weed free to avoid yield losses which vary from 18% in castor to 76% in soybean (6).

Cultural practices : A major portion of weeds can be managed by preparatory cultivation, adjustment in time of sowing, planting methods, selection of varieties, time, rate and method of fertilizer application, intercultivation with wheel hoes etc. In general one or two manual weedings are done during critical period of crop-weed competition. In groundnut, two manual weeding at 20 and 45 days after sowing provides satisfactory weed control. The interculturing with bullock drawn hoes fitted with sharp blades at 20 and 40 days after sowing also provided satisfactory control of weeds in this crop. Weeding with wheel hoe fitted with double blade (Philippines model) is found effective in sunflower. Intercultivations with bullock drawn cultivator or manual push and pull type wheel hoe fitted with harrows during critical period of crop-weed competition effectively controlled the weeds in castor. In niger, manual weeding between 30-35 days is the effective method for weed control. In rapeseed-mustard under poor fertility and light textured soils, only one hoeing in between the rows at 30 days after sowing eliminated maximum weeds. Experiments conducted at Jorhat revealed that the Dryland Weeder or grubbing operation could be adopted for weed control in mustard crop. Singh (36) reported that the highest accumulation of dry matter of weed under broadcast method of fertilizer application and lowest was recorded under band placement. In linseed, two manual weedings, first between 15-30 days after sowing and second at 40-50 days after sowing have been found quite effective in controlling weeds.

Chemical methods : Several herbicides for weed control in groundnut have been tried at various centres. Preemergence application of pendimethalin, 1.0 kg/ha (at Ranchi), alachlor 1.0 1.5 kg/ha (at Shillong and Pusa), preplant soil incorporation of fluchloralin, 1.0 kg/ha (at Pusa) and postemergence application of fluazifop-p-butyl at 0.125 (at Coimbatore) were found quite effective in controlling the weeds and provided higher yield of kharif groundnut (Table-7). Other promising herbicides reported from elsewhere are oxadiazon, 0.5 kg/ha, imazethapyr 0.10 kg/ha and oxyfluorfen, 0.2 kg/ha have been

Table -7 : Effect of herbicide treatments in weed control and pod yield of kharif groundnut at various centres

Herbicides	Rate (kg/ha) and time of application	(Yield (kg/ha)				
		Shillong	Bangalore	Coimbatore	Pusa	Kanke
Pendimethalin	1.0 PE	-	926	3447	1985	1950
Fluchloralin	1.0 PE/PPI	-	876	1962	2138	-
Alachlor	1.0-1.5 PE	1426	-	-	2095	1600
Butachlor	1.0-1.5 PE	1391	-	-	1902	-
Fluazifop-p-butyl	0.125 Postem.	-	1045	-	-	-
Weedy check		320	531	1826	1162	1260

PE= Preemergence, PPI = Preplant incorporation, Postem. = Postemergence reported for controlling weeds in this crop.

In soybean, preplant soil incorporation of fluchloralin, 1.0 kg/ha or preemergence application of alachlor, metribuzin, pendimethalin or oxyfluorfen at 0.5-1.0 kg/ha were found quite effective in

Table-8 : Effect of chemical weed control treatments on yield of soybean at various AICRP-WC centres (1978-91)

Herbicides	Rate (kg/ha) and time of a application	Grain yield (kg/ha) at different centres							
		Jabalpur	Pantnagar	Coimbatore	Pusa	Jorhat	Shillong	Kanke	Palampur
Fluchloralin	1.0-1.5 PE	2900	2130	1592	1582	1756	-	3256	1389
Fluazifop-p-butyl	0.25 Postem.	2551	2367	-	-	1578	-	-	-
Metolachlor	2.00 PE	-	-	933	-	1710	-	-	1408
Alachlor	1.00 PE	-	-	-	-	-	-	3273	1552
Oxadiazon	0.7-1.50 PE	2647	-	-	1666	-	2200	2527	-
Pendimethalin	1.0 PE	3025	-	1632	-	-	1550	3164	-
Weedy check		1593	639	672	987	1411	324	1597	643

PE = Preemergence, Postem. = Postemergence

controlling the weeds. The effect of various herbicides on grain yield of soybean at various centres have been summarized in Table-8.

In sunflower, application of oxadiazon, 1.5 kg/ha, alachlor, 1.0-1.5 kg/ha, metribuzin, 0.35-1.0 kg/ha, methabenzthiazuron, 1.0 kg/ha or preplant incorporation of fluchloralin 1.0-1.5 kg/ha or trifluralin, 1.0 kg/ha have been found very effective. Maximum seed yield, protein content and oil content were obtained with presowing application of fluchloralin, 0.75 kg/ha (14). In sesamum, application of fluchloralin, 1.0 kg/ha as preplant soil incorporation or preemergence application of alachlor or nitrogen, 1.0 kg/ha were found effective both at Kharagpur and Jabalpur. In niger, preemergence application of pendimethalin, 1.0 kg/ha or alachlor, 1.0 kg/ha or butachlor, 2.0 kg/ha can be used against grassy and broadleaved weeds of this crop.

Preliminary work on chemical control in sunflower has revealed that weeds in this crop are effectively controlled by oxadiazon, 0.75 kg/ha or fluchloralin, 1.0 kg/ha or metolachlor, 0.5 kg/ha as preemergence without any phytotoxic effect on the crop. In a separate study, preemergence application of alachlor or pendimethalin, 0.5-1.0 kg/ha were quite effective against grassy weeds. For control of broadleaved weeds and grassy weeds, isoproturon at 0.75-1.0 kg/ha or oxadiazon, 1.0-1.5 kg/ha can be used. Postemergent application of haloxyfop-methyl, 0.75 kg/ha have been found quite effective against grassy weeds. In linseed, preemergence application of fluchloralin or pendimethalin, 1.0 kg/ha or oxyfluorfen, 0.20 kg/ha or oxadiazon, 0.75 or isoproturon, 1.0 kg/ha have been found quite effective.

Integrated weed management : It has been observed that the effect of combination of mechanical, cultural and chemical weed control is more effective in controlling the weeds than their isolated application in various oilseed crops. In irrigated groundnut, application of fluchloralin at 0.50 kg/ha and one interculture with star type weeder controlled the weeds very effectively. Similarly, preemergence application of metolachlor (1.0 kg/ha), methabenzthiazuron (0.25 kg/ha,), oxyfluorfen

(0.108 kg/ha), oxadiazon (0.625 kg/ha) and controlled irrigation alongwith one manual or mechanical weeding between 30-40 DAS has been found to be promising. Panwar, *et. al.* (23) observed that preplant soil incorporation of fluchloralin or preemergence application of pendimethalin, both at 1.0 kg/ha followed by one hoeing resulted in highest pod yield of groundnut. In soybean at Pantnagar, preemergence application of metribuzin at 1.0 kg/ha and oxyfluorfen at 0.10 kg/ha each followed by one hand weeding at 45 DAS were found very effective for higher yields. Singh and Sharma, observed the lowest dry weight of weeds with narrow row spacing (30 cm) and 2 hand weedings at 20 and 40 DAS. At Jabalpur, preplant incorporation of fluchloralin at 1.0 kg/ha + 1 hand weeding showed significantly increased the number of capsule per plant and produced higher yield (628 kg/ha) of sesamum over weedy check. Bansode and Shelke (3) reported that preemergence application of alachlor 0.75-1.5 kg/ha fb hand weeding/hoeing three weeks after sowing gave the highest yield of sesamum (689 kg/ha).

Studies in Gobhi-sarson *(Brassica napus)* at Ludhiana indicated that preemergence application of isoproturon 0.50 kg/ha fb one hand weeding gave effective weed control and yielded at par with two hand weedings. *Avena ludoviciana, Melilotus* spp., *Fumaria parviflora* and *Anagallis arvensis* were not completely controlled even with 0.75 kg/ha of isoproturon.

Weed management in oilseed based intercropping systems : Research on weed management in intercropping system is very meagre. However, in recent years a number of evaluation trials have been conducted at different centres of AICRP-WC. In Groundnut + sunflower intercropping system at Hyderabad, preemergence application of metolachlor or pendimethalin both at 1.5 kg/ha or imazethapyr 0.50 kg/ha or postemergence application of fluazifop-p-butyl at 0.25 kg/ha were found effective for controlling grassy and broadleaved weeds. Girijesh and Patil, reported that application of fluazifop-p-butyl at 0.187 kg/ha fb hand weeding three times provided the highest yield and N uptake to both the crops. In paddy + groundnut intercropping system, sowing of alternate rows or paired rows of groundnut alongwith preemergence application of pendimethalin (1.0-1.5 kg/ha) followed by hoeing gave yields comparable to two manual weedings at Kharagpur. In Maize + groundnut inercropping at Kanke, hand weeding thrice at 20, 30 and 40 DAS was the effective method of weed control. Among herbicides, preplant soil incorporation of fluchloralin at 1.0 kg/ha was found promising. Intercropping of soybean with pigeonpea is an effective cropping system for reducing weed competition. Weed management research conducted at Kanke revealed that application of fluchloralin at 1.5 kg/ha as preplant soil incorporation or 1.5 kg/ha alachlor as preemergence were effective treatments for controlling the weeds. However, maximum yields of pigeonpea, soybean and pigeonpea equivalent was obtained under hand weeding at 20, 30 and 40 DAS.

In lentil + linseed intercropping system at Kanpur, use of prometryn (0.8 kg/ha), fluchloralin (0.67 kg/ha), pendimethalin (1.0 kg/ha) or isoproturon (0.75 kg/ha) was found safe and effective in controlling weeds. However, *Convolvulus arvensis* L. remained unaffected with either of the herbicides used. In another study involving Groundnut + pearlmillet, application of prometryn (0.8 kg/ha) established its significance in preventing the germination of most of the annual grasses and broadleaved weeds particularly *Trianthema monogyna* L. and enhanced the yield of component crops. At Palampur, in Wheat + mustard or Wheat + raya cropping system, weeds could be effectively controlled by using pendimethalin at 1.5 kg/ha as preemergence or isoproturon 1.0 kg/ha as postemergence. Mehta, *et. al.* (20) observed that postemergence application of isoproturon at 0.75-1.0 kg/ha was quite effective against broadleaved weeds and *Phalaris minor* and selective to sugarcane + Indian mustard intercropping system.

Weed management in oilseed based sequential cropping systems : In view of better utilization of available resources more efficiently and also to reduce the costs of application of herbicides, experiments were initiated, to study the effect of herbicides in fixed cropping systems. In such studies, the residual effect of herbicides applied on the weeds in the first crop were evaluated on the weed control in second crop. In groundnut-maize cropping system, application of fluchloralin 0.5-1.0 kg/ha as preplant incorporation to maize followed by one manual or mechanical weeding at 30-40 DAS gave yields comparable to two manual weedings to each crop. Among various herbicides tested in soybean-wheat cropping sequence, preemergence application of pendimethalin 1.5 kg/ha, oxadiazon 0.5 kg/ha, alachlor 1.5 kg/ha, oxyfluorfen 0.2 kg/ha and butachlor 2.0 kg/ha have been found very effective in controlling the weeds. Studies conducted at Anand in cotton-pearlmillet-mustard cropping system indicated that integration of preemergence fluchloralin at 0.675 kg/ha with one interculturing between cotton rows was found at par with traditional method of four interculturings and weedings.

Control of parasitic weeds : For control of parasitic weed *Cuscuta chinensis* in niger and linseed, application of pronamide as postemergence at 20 DAS has been found promising. Other herbicides viz. fluchloralin or pendimethalin each at 1.0 kg/ha as preplant incorporation or preemergence can also be used to reduce *Cuscuta* infestation. Control of *Orobanche* spp. in mustard crop have been studied at various centres of AICRP-WC. Application of Copper sulphate 10% solution, growing of resistant variety of mustard viz. Durgamati and use of herbicides viz. pendimethalin or fluchloralin at 1.0-1.5 kg/ha as preemergence or preplant incorporation are found promising.

Sugar crops

Sugarcane is a long duration crop and therefore it is infested by a variety of weed species both of kharif and rabi season. The crop needs repeated cultural practices for controlling these weeds. Since crop take about a month to sprout the cane, the blind hoeing is done initially and later a number of cultivation both mechanical and manual are performed in between the rows of sugarcane. The crop then earthed up 2-3 month after planting which also controls the later flushes of weeds. In Haryana, the local practice is to control weeds using Kasola (bullock drawn hoe) repeatedly. However, in February and March planted crop, atrazine 2.0 kg/ha, 4-7 days after planting is applied followed by one hoeing with Kasola at 40 days after sowing. In case of late plant crop, herbicide combinations do not work and therefore hand weeding remains only the solution.

A combination of herbicides viz. simazine or atrazine 1.0-2.0 kg/ha as preemergence followed by intercultivations with different types of bullock drawn hoes in between the rows and supplemented with manual weeding and earthing up is suggested. For later flushes of weeds, application of MSMA or DSMA at 4-5 kg/ha or dalapon 6.0 kg/ha at directed application is recommended.

Fibre crops

Cotton : The weed problem in cotton is more severe than other crops since it is planted in wide spaced rows ranging from 60 to 100 cm apart. However, in early planted crop i.e. April and May, the intensity of weeds is very low. The usual practice of weeding in cotton is by blind hoeing with different types of traditional and improved bullock drawn weeders. It is supplemented with hand weeding once or twice. About 80% of weeds are removed by interculture operations. The intensity of weeds is higher in rainfed late sown crop. The interculture operation also some times becomes difficult due to vagaries of climate and soil conditions. Binswanger and Shetty (4) surveyed the cotton fields and observed that interculture is more common than traditional method of manual weeding. They indicated that integrated approach was more conducive to traditional method of weed control.

In recent years, preemergence application of diuron 0.5-1.5 kg/ha, alachlor 1.0-3.0 kg/ha, dinitramine 1.0-2.0 kg/ha, crotoran 1.2-3.0 kg/ha, nitrofen 3.0-6.0 kg/ha or postemergence directed application of MSMA 4.0-6.0 kg/ha or diuron 1.0-2.0 kg/ha have been found quite effective in controlling weeds in this crop.

In Haryana, the general practice recommended for cotton is to apply pendimethalin 1.5 kg/ha as preemergence, which controls most of the grassy and broadleaved weeds including *Trianthema monogyna* and to some extent *Digera arvensis*. The herbicide application provides initially 4-6 weeks weed free condition. This is followed by 2-3 intercultivations with Kasola, a bullock drawn hoe, which controls later flushes of weeds, conserve moisture and improves soil aeration. In medium black soils of Maharashtra, pre-sown or preemergence application of fluchloralin, 2.0-3.0 kg/ha was found to be economical. In deep black cotton soils, preplant application of fluchloralin 1.25 kg/ha followed by diuron 0.75 kg/ha as postemergence 4-6 weeks after sowing registered significantly higher seed cotton yield than other treatments (24). In Andhra Pradesh, alachlor 1.0 kg/ha as preemergence followed by intercultivations are suggested. The results of the experiments conducted at Coimbatore indicated that pendimethalin 1.0 kg/ha or butachlor 1.5 kg/ha could be used for controlling weeds in this crop.

A number of experiment conducted all over India under AICRP-WC and elsewhere indicated that a combination of preemergence application of herbicides like fluchloralin, pendimethalin, diuron, butachlor, alachlor, oxadiazon, oxyfluorfen, nitorfen or a combination of diuron + pendimethalin or postemergence directed application or fluazifop-p-butyl and followed by 2-3 intercultivations with different types of bullock or tractor drawn cultivator/hoes provided satisfactory control of weeds in this crop. Most recently at HAU, Hisar, Mon 13200 (Code No.) 1.0 kg/ha preplant incorporation gave season long control of all weeds except *Cyperus rotundus*. Trifluralin 1.0 kg/ha was also found equally effective.

Jute and mesta : Jute (*Corchorus olitorius* L. and *C. capsularis* L.) is grown on wide range of soil and climate conditions. It prefers deep, fertile alluvial soils and well distributed plaintiful rainfall for its growth, which also encourages profuse growth of weeds. If weeds are not checked in time, the crop as such is to be plough down. The crop is sown with premonsoon showers and all interculture operations i.e. weeding, thinning, hoeing and application of fertilizers are done between premonsoon showers in April or early May and with the on set of monsoon in early June. Usually the manual weeding is followed by 3-4 interculture operations with three tine wheel hoe. In broadcast sown crop, a larger portion of initial weeds are remove by blind hoeing with bida (harrow), which removes 70-80% weed population and also reduces the excess plant population. Once the crop, attains a height of 45-60 cms, the crop itself suppresses further germination and growth of weeds with its dense canopy. The extensive research work done at Jute Agricultural Research Institute, Barrackpore, during 1971 to 1981 on weed survey, screening of herbicides and standardizing rates, time and mode of application of herbicides have identified a number of herbicides for this crop.

Preplant or preemergence application of fluchloralin 0.90-1.0 kg/ha, or preplant application of tertrapion (2,2,3,3, tetrafluropropionate) 2-4 kg/ha depending upon soil type followed by 3-4 intercultivation with hand pushed wheel hoe in line sown crop or using bida in broadcast sown crop, at an interval of 10-15 days after sowing or postemergence application of dalapon, 6.0 kg/ha (partly directed) or MSMA or DSMA, 4.0-5.0 k/ha as directed application or a combination of MSMA + dalapon at 2.0-3.0 kg/ha and 3.0-4.0 kg/ha, respectively in row sown crop depending upon the composition of weed flora. In case, sedges including nutsedge is a problem, the application of MSMA or DSMA have been observed more effective while in case of grassy weeds including *Cynodon dactylon*, application of dalapon has been found very effective (27,29,30).

In case of mesta (*Hibiscus subdariffa* L. var. *altissima* and *H. cannabinus* L.) which is grown on comparatively marginal soils and in drier regions, the intensity of weeds is not so intense. The herbicides recommended are alachlor 1.0-2.0 kg/ha as premergence followed by intercultivation with wheel hoe in row planted crop or bida in broadcast sown crop. The preemergence application of dalapon, MSMA or DSMA could also be used as in jute in row sown crop followed by 2-3 intercultivation for effective control of weeds.

Vegetable crops

The major vegetable crops are cruciferous like cabbage, cauliflower, knolkhol and raddish; curcurbitaceous crops like watermelon, muskmelon, bottle gourd, ridge gourd, bitter gourd, squash, pumpkin; leguminous crops like dolichos beans and frenchbeans and beet root and amaranth. Out of many herbicides tested only alachlor, butachlor, fluchloralin, oxyfluorfen and fluazifop-p-butyl were found effective for these crops. For most of the curcurbitaceous and leguminous crops, alachlor or butachlor, 1.0-2.0 kg/ha as preemergence application; for cabbage, cauliflower and knolkhol, fluchloralin 1.0 kg/ha or oxyfluorfen 0.20 kg/ha as premergence application could be used for controlling the weeds. In general, herbicides in these crops are more economical than repeated manual weedings (8).

In potato, a number of herbicides viz. oxyfluorfen 0.2 kg/ha, metolachlor 1.0 kg/ha, atrazine 0.5 kg/ha, fluchloralin 1.0 kg/ha, alachlor or methabenzthiazuron or linuron or metribuzin 1.0 kg/ha as preemergence have been recommended for effective control of weeds.

Plantation crops

Rubber: Kerala is the lead rubber producing state and the state has an area of 3.67 lakhs hectares under rubber plantation. The best practice of weed management is growing a cover crop viz. *Peuraria phaseoloides* before trapping stages. However, if there is no cover crop, the weeds can be controlled by applying paraquat 0.4 kg/ha three times a year at monthly intervals starting from June-July. Where ever perennial weeds are predominant, it could be controlled by single application of glyphosate 0.8 kg/ha. A higher incidence of *Synedrella nodiflora* is observed after the use of glyphosate probably due to the secondary germination of this annual weed after the existing weeds were controlled by the herbicides. In such situations a combination of glyphosate with diuron could be used. *Mimosa pudica* could be effectively controlled by repeated application of paraquat or with a combination of paraquat and diuron as postemergence application. *Chromolaena odorata* , a major problem weed in rubber gardens could completely be controlled by repeated application of paraquat or with a combination of paraquat and diuron or glyphosate 0.8 kg/ha or by manual weeding.

Pennisetum pedicellatum and *P. polystachyon* are two perennial weeds called Siam weed are serious problem in early stages of plantation crops such as oil palm, cocoa, rubber, cashewnut and coconut. Experiments conducted at KAU, Trichur indicated that at one or two months stage of the weed, the most effective treatment is glyphosate 0.8 kg/ha or a combination of paraquat 0.4 kg/ha + diuron 0.1 kg/ha. However, at 3 months stage only glyphosate 1.2 kg/ha resulted in satisfactory control of this weed.

Coffee : It is cultivated in hilly tracts of the Kerala ad Karnataka states. Weed competition is very severe during the early part of the crop growth. The experiments conducted at Cardamom Research Station, Pampadumpara showed that the diverse weed flora could be controlled by application of dalapon 5.0 kg/ha + 2,4-D 1.0 kg/ha postemergence (tank mixed) twice at monthly interval or glyphosate along 1.2 kg/ha or glyphosate 0.8 kg/ha + diuron 1.0 kg/ha. In cardamom, postemergence directed application of paraquat 0.4 kg/ha resulted in control of weeds upto 2 months. Glyphosate 0.8 kg/ha effectively control all grassy weeds. A combination of atrazine or diuron 1.0 kg/ha with glyphosate gave prolonged weed control. In tea plantation (young), atrazine 2.0 kg/ha as preemergence or glyphosate 1.0 kg/ha as postemergence directed application gives satisfactory weed control in established tea plantation, (Fig. 1 A B) Glyphosate 1.0 kg/ha or haloxyfop-methyl 0.5 kg/ha as postemergence directed application is recommended.

6. REFERENCES

1. Alstrom, S. 1990. Fundamentals of weed management in hot climate peasant agriculture. Deptt. of Crop Production Sc., Swedish Univ. Agric. Scies., Uppasala, Sweden.

2. Annual Report. 1990. All India Coordinated Research Programme on Weed Control, NRCWS, Jabalpur.

3. Bansode, B.V. and Shelke, D.C. 1991. Integrated weed management in sesamum. J. Maharashtra Agric. Univ., **16(2)** : 273.

4. Binswanger, H.P. and Shetty, S.V.R. 1977. Economic aspects of weed control in semi-arid and tropical areas of India. Proc. Ind. Soc. Weed Sci. Conf., Hyderabad.

5. Brar, L.S., Chander, V. and Kundal, H.C. 1990. Weed control studies in pigeonpea. J. Res. PAU., **27(2)** : 213-217.

6. Consolidated Report (Ist Phase). 1978-84. All India Coordinated Research Programme on Weed Control, CRRI, Cuttack.

7. Consolidated Report (IInd Phase). 1982-87. All India Coordinated Research Programme on Weed Control, NRCWS, Jabalpur.

8. Final Technical Report. 1982-87. All India Coordinated Research Programme on Weed Control. Centre IIHR, Bangalore.

9. Gupta, O.P. 1984. Scientific weed management. Today and Tomorrow's Printer & Publishers, New Delhi.

10. Hance, R.J. and Holly, K. 1989. Weed Control Hand Book - Principles. British Crop Prod. Coun., Blackwell Sci. Publ., Oxford.

11. Hanson, H.C. and Churchill, E.D. 1965. The plant community. Affiliated East West Press Pvt. Ltd., New Delhi.

12. Holm, L.G., Plucknett, D.L., Pancho, J.V. and Herberger, J.P. 1977. The world's worst weeds : Distribution and biology. Univ. Press., Hawaii, Honolulu, Hawaii.

13. Hore, D.K. and Tripathi, R.S. (Eds). 1985. Ecological survey of weeds of West Bengal. Deptt. Agric. Engg., Kharagpur.

14. Jay Kumar, R., Sekar, M.P., Kempuchetty, N. and Subramanian, S. 1988. Effect of integrated weed management on yield and quality of sunflower. Madras Agri. J. **75(3-4)** : 85-88.

15. Jethro, Tull. 1931. Horse hoeing husbandry, Barkshire, MDCC. 33.

16. Joshi, N.C. 1987. Manual of weed control. Researchco Publication, Delhi.

17. Kumar, S. and Kairon, M.S. 1988. Weed Control in summer mungbean (*Vigna radiata* Wilozek). India. J. Weed Sci., **20(1)** : 64-67.

18. Kumar, K. and Kolar. 1989. Effect of chemical control and *Rhizobium* inoculation on the yield of lentil. J. Res. PAU, **26(1)** : 19-24.

19. Macneish, R.S. 1964. The origin of new world civilization. Sci. Amer. **211** : 29-37.

20. Mehta, S.P., Kanwar, R.S. and Bhatia, R.S. 1989. Herbicidal control of winter weeds in Autumn cane + Indian mustard intercropping system. Ind. J. Weed Sci., **21 (3-4)** : 37-41.

21. Moody, K., Munro, C.E., Lubigen R.T. and Paller, E.C. Jr. 1984. Major weeds of Philippines. UPLB College, Laguna, Philippines.

22. Moretimer, A.M. 1989. The biology of weeds 1. In: Weed Control Hand Book - Principles. R.J. Hense, and Holly, K. British Crop Protection Coun., Black Sci. Pub., Oxford.

23. Panwar, R.S., Malik, R.K. and Bhan, V.M. 1989. Cultural weed control in groundnut. Ind. J. Agro. **33(4)** : 458-549.

24. Patel, R.G., Patel, R.B., Patel, H.C. and Patel, T.D. 1985. Weed competition in summer groundnut. Proc. Ind. Soc. Weed Sc. Conf. GAU, Anand.

25. Quayum, A. 1992. Personal communication.

26. Rao, V.S. 1983. Principles of weed science. Oxford & IBH Publications Co, New Delhi.

27. Saraswat, V.N. 1976. Sodium 2, 2, 3, 3, tetrafluoropropionate, a new herbicide for jute fields. PANS, **22(3)** : 391-8.

28. Saraswat, V.N. 1980. Ecology of weeds of jute fields in India. Tropical Pest Management, **26(1)** : 20-28.

29. Saraswat, V.N. 1985. A decade of weed control research in jute. Ind. Jr. Weed Sci. **17(4)** : 58-60.

30. Sarawat, V.N. and Sahoo, K.M. 1988. Magnitude of losses in yields of major crops due to weed competition in India. Pesticide Information, **14(2)** : 2-9.

31. Saraswat, V.N. 1993. Major weeds of Indian Agriculture - their distribution and ecology. Proc. Int. Sym. IPM Sust. Agric., HAU, Hisar.

32. Saraswat, V.N. and Mishra, J.S. 1993. Weed management in pulse crops. Proc. Int. Sym. IPM Sust. Agric., HAU, Hisar.

33. Saraswat, V.N. and Mishra, J.S. 1993. Status of weed management research in oilseed crops in India, Ibid.

34. Setty, S.V.R. 1979. Weed control in sorghum in tropics. In : Weed control in tropical crops. Ed. K. Moody. Weed Sci. Soc. Philippines, Inc., Philippine Coun. Agric. Resources Research.

35. Sharma, A.R. and Vats, O.P. 1988. Growth, yield and nutrient uptake of peas as affected by different herbicides. J. Res. PAU, **28(4)** : 563-568.

36. Singh, S.S. 1992. Effect of fertilizer application and weed control on the yield of mustard (*Brassica juncea*). Ind. J. Agro. **37(1)** : 196-8.

37. Thakur, C. 1984. Weed Schience, Metropolitan, New Delhi.

Fig. 1 a) A tea garden badly infested with *Eupatorium adonophorum*

b) A weed free established tea garden

26 Biological Control of Fusarial Diseases

K. Satyaprasad, I.K. Kunwar and P. Ramarao

CONTENTS

Integrated Pest and Disease Management
Rajeev K. Upadhyay, K.G. Mukerji, B.P. Chamola and O.P. Dubey (eds.)
APH Publishing Corporation, New Delhi-110002, India

ABSTRACT

Diseases caused by *Fusarium* spp. account for enormous losses of major crops throughout the world. Basically Fusaria are vascular wilt pathogens besides inciting root and stalk rots. Fusarial pathogens are soil-borne while some are seed borne. Management of soil-borne diseases is complicated and a difficult proposition. As Fusaria are capable of producing perennating structures like chlamydospores, they sustain in soil for a long period. Fusaria are also capable of surviving saprophytically in the dead plant parts either infected during parasitic phase or colonized as a competitive saprophyte. Competitive saprophytic ability of these pathogens enable them to over come the interhost periods and also to build-up the inoculum in soil. Role of soil microorganisms exerting disease control was realised long back in the form of *Fusarium* -suppressive soils in nature. *Fusarium* -suppressive soils provided an impetus to the idea of biological control as it is a modern non-polluting and comprehensive approach to the management of soil-borne diseases. Suppressive soils harboured several plant growth promoting pseudomonads which produce siderophores. Suppressiveness is inducible and durable for years. Solar heating has great potential in disease suppression. Solarization works selectively on the pathogenic forms and encourages non-pathogenic and PGPR organisms. Recently, non-pathogenic fusaria were implicated in the control of fusarial diseases in addition to the several antagonistic bacteria and fungi. Mycorrhizal fungi are also proved useful in the management of soil-borne diseases. Biological control measures are going to form an important component in the integrated management of soil-borne diseases.

1. INTRODUCTION

To maintain the current caloric intake and to feed over 7.5 billion expected world population by 2010 A.D., the food production has to be increased at least by 40% (47). Of the several constraints affecting food production, plant pathogens form an important obstacle for the control of which we are heavily dependent on the use of chemical pesticides and fungicides. In view of the growing world energy crisis, escalating prices of pesticides and fungicides, abuse of pesticides resulting in hazards to human and animal health and environment, alternatives were searched to overcome the above disadvantages. Hence biological control has become one of the priority components of integrated pest management besides host resistance and cultural control.

Management of pathogenic fungi is no doubt a difficult proposition but to control the root infecting soil-borne pathogens is the most difficult task. The complexity of the dynamic soil medium, the variety of pathogenic fungi and the host-pathogen-soil interaction with the environment have made the problem of soil-borne disease management much more complicated. Pathogens enter into a definite association with the host during their interaction such as neutralism, competition, mutualism, proto cooperation, commensalism, amensalism, parasitism and predation. Inspite of the presence of a multitude of potential pathogenic forms in nature, they are under natural check due to biological equilibrium. This forms the basis for biological control due to which the diseases are not usually catastrophic.

The occurrence of plant disease is an indication of disturbance in some aspect of biological equilibrium while the restoration of the disease inhibiting balance in nature is the process (phenomenon) called biological control.

Among soil-borne diseases root rots, basal stem rots and vascular wilts are of major concern for economic losses to both cereals and legumes. Fusarial pathogens incite all the three types of diseases of economic consequence on cereals and legumes. *Fusarium* spp. are ubiquitous plant pathogens present in almost all agricultural soils.

In view of the numerous reports and reviews on various aspects of biological control (1,2,11,12,13,14,29,75,80,81,103) we intend to emphasize on various fusarial diseases and their integrated biological management employing non-pathogenic isolates, soil solarization,.plant growth promoting rhizobacteria (PGPR) and VA mycorrhizal fungi in enhancing the efficiency of the biocontrol organisms.

Natural disease suppression in specific soils has been recognised for many years but only during the last two decades the biocontrol of plant pests and pathogens gained momentum in elucidating the nature of suppression, mechanisms of mycoparasitism and antibiosis, in view of the recent progress made and the discovery of numerous biocontrol agents, management of soil-borne diseases through biological means has become a promising practical reality.

Mycoparasites are effective only at high doses of inoculum and colonize effectively in sterile, fumigated or soilless mix in green house but not under field conditions. The inability of the biocontrol agents to establish and occupy the new ecological niche is the prime reason for their failure at field level to displace the pathogens.

A comprehensive approach and understanding is lacking on the ecology of plant pathogens and biocontrol agents in soil. Population dynamics of the resident microflora in soil and their role in affecting the pathogen as well as the biocontrol agent is the main factor responsible for the success of biocontrol of soil-borne diseases. The role of saprophytic organisms in the exclusion of the niche occupied by the soil-borne plant pathogen is all the more important component in biological control of fusarial diseases. In general Fusaria are soil-borne pathogens while some are seed borne and seed transmitted, also the losses due to fusarial diseases may range from about 5% to almost 100% in the case of severe infection.

1.1 Mode of infection

The wilt pathogens are unique in their ability to become established systemically within the vascular tissues of the host plants through roots. Infection may be through direct penetration of the young and injured root tissues or may be facilitated by the injury or wounds caused by various agents. Fusaria enter the roots without the aid of specialized penetrating structures such as appressoria. After reaching the xylem of the susceptible plant, the pathogen proliferates and gets distributed by means of hyphal growth and conidia. The detached conidia are swept upward in transpiration stream and become systemic.

1.2 Survival of the pathogen

Fusarial pathogens survive in soil in the absence of suitable host and under unfavourable conditions in the form of mycelial fragments, conidia and dormant propagules like chlamydospores. They may also survive in the soil to tide over inter-host periods, in competition with other soil microorganisms by competitive saprophytic colonization of available dead organic substrates and saprophytically surviving on these in addition to saprophytic survival in the infected plant material infected during its parasitic phase (3,42). McRae and Shaw (71) reported prolonged survival of *Fusarium udum* upto 8 years. Nene *et al.* (76) were able to detect *F. udum* in stubble fragments from vertisol up to $2^1/_2$ years and from Affisol up to 3 years.

Saprophytic survival (S50 period) of *F. oxysporum* f.sp. *ciceri* in black soil was slightly more (19 months) than its median survival period (17 months) in red sandy loam soil. Guar root rot pathogen, *F. solani* could survive in the colonized plant debris up to nearly two years (S50 period - 23 months) in red sandy loam soil and more than two years (25 months) in black soil. Inorganic soil amendment with zinc and manganese completely nullified the competitive saprophytic colonization by *F. oxysporum* f.sp. *ciceri*. Nitrogen amendments with Na NO_3, NH_4Cl, urea, chitin and cowdung were highly fungistatic to *F. oxysporum* f. sp. *ciceri* and *F. solani* (94,95,96,97).

Various species of *Fusarium* colonize dead plant tissues and survive saprophytically in soil (31,43,126). Schlub *et al.* (101) observed an increase in inoculum density of *F. oxysporum* on addition of dead plant residues. Karunakar (53) reported the successful saprophytic colonization of fusarial pathogens involved in root and stalk rot of sorghum.

Fusarium moniliforme var. *subolutinans* survived saprophytically in sorghum stem substrates for 68 weeks (S50 period) while *F. moniliforme* survived up to 45 weeks. Crop sequence, use of fertilizer and other micronutrients and organic amendments influence the behaviour and survival of the pathogen in soil which forms basic means for biological control.

3. BIOLOGICAL CONTROL STRATEGIES

3.1 Suppressive soils

The development of the disease is suppressed in suppressive soils although virulent pathogens and susceptible hosts are present. Suppression is an umbrella term that encompasses various phenomenon like soil fungistasis, competition and antagonism. Induction of soil suppressiveness to fusarial diseases may provide a long term plant protection.

In nature evidence of *Fusarium* suppressive soils dates back to a century (9). It is now a well known fact that there are soils that are suppressive to plant pathogens. The most notable among the suppressive soils are take-all decline (13,93,107) and *Fusarium* - suppressive soils (5,122,129).

The metz fine sandy loam soils of the Salinas valley California are suppressive to *Fusarium* wilt diseases (112) and soil suppressiveness was transferred to conducive soil by introducing fluorescent pseudomonads from the suppressive soil (36,99). Fluorescent pseudomonads isolated from a take-all suppressive soil, have controlled both take-all in barley and Fusarium wilt in flax (59), due to production of fluorescent siderophores by the bacteria. Siderophores helps in the suppression of several formae specials of *F. oxysporum* (99). The addition of *Pseudomonas putida* to a conducive soil rendered the soil suppressive to Fusarium wilt pathogens of flax, cucumber and radish. The activity of siderophore-producing pseudomonads leading to Fe competition was suggested as a possible mechanism for the Fusarium wilt suppression. The catechol hydroximate siderophores of fluorescent pseudomonds have a higher stability constant than the hydroximate siderophores of *F. oxysporum* and this enables the pseudomonads to compete successfully for available Fe (99).

Root colonizing bacteria play an important role in making the soils suppressive (104). Lockwood (64) postulated that rapid deprivation of plant root exudates by soil microflora may result in disease suppression. Enhanced suppression was obtained with the supply of nutrients (glucose) in *Fusarium* suppressive soils as evidenced by the changes in respiratory rates due to increased microbial activity (7) and dehydrogenase activity alongwith other factors in soil amended with composited municipal sludge (67).

Chlamydospore germination of *F. oxysporum* f.sp. *cucumerinum* was affected by the highly fluorescent siderophore producing pseudomonad isolates present in suppressive soils. However, some non-fluorescent bacteria also suppressed the disease (112). Simeoni *et al.* (108) reported significant reduction in chlamydospore germination of *F. oxysporum* f.sp. *cucumerinum* in the presence of siderophore-producing *P. putida*.

The noteworthy point in the biological control of fusarial diseases by suppressive soils is the slowness with which such changes occur. Though the pathogen suppressive soils are rare, those identified are excellent examples of the full potential of biological control of soil-borne pathogens (127). It will be worthwhile to search for such natural biocontrol mechanisms like disease suppressive soils in India, against these fusarial pathogens of legumes and cereals.

3.2 Soil solarization

Soil solarization is a method of hydrothermal disinfestation accomplished by covering moist soils with transparent plastic tarps during summer. The process is simple, non-hazardous and does not involve toxic materials (88,89). With the appropriate use of solarization approach to kill the target organisms, and with the knowledge gained in the change in microbial ecology of solarized soil and technology of application, this technique can effectively be employed for controlling soil-borne disease in tropical countries including India (8). Katan and his co-workers were the first plant pathologists to extensively use this approach to control soil-borne diseases (38,45,54,55). Control of soil-borne diseases by solarization is a successful strategy for reducing crop losses.

Direct thermal effects are probably the major phenomenon implicated in the control of soil-borne pathogens during solar heating. Moisture plays a crucial role in reducing the population of pathogens, that is why soils which were irrigated before solarization had higher LD values.

Microbial activities are affected by soil solarization and may result in increased antagonistic activity inducing soil suppressiveness (44,115). Increased plant growth responses following soil solarization usually indicate successful application of the treatment (55, 89, 113, 114).

In beans solarization significantly increased the stand, reduced *Fusarium* root rot, and increased yield by 24% (116). Induced suppressiveness against *F. oxysporum* f.sp. *vasinfectum* in cotton was observed in solarized soil (41) due to fluorescent pseudomonads as the effective agents of suppression against major and minor pathogens in soil. Long term control of disease is possible by soil solarization. In cotton, disease control of *F. oxysporum* f.sp. *vasinfectum* was reported even after third year of solar heating (38,55). Gamliel and Katan (40) observed an increase in growth promoting fluorescent pseudomonads and a decrease in population densities of *P. pinophylum* , which suppresses plant growth. A possible beneficial effect of fluorescent pseudomonads is the suppression of deleterious and other harmful microorganisms through aggressive colonization of roots and antibiotic production (58,59,100). Gamliel and Katan (40) reported reduction in the populations of bacteria and fungi (including thermotolerant fungi) by solarization. But actinomycetes were least affected. Hardy and Sivasithamparam (46) presented direct evidence to show that solarized soils become suppressive to carnation wilt caused by *F. oxysporum* f.sp. *dianthi*. Suppression of chlamydospore formation by *F. oxysporum* f.sp. *lycopersici* was evident in solarized soil (44).

3.3 Non-pathogenic fusaria

Fusarium suppressive soils are known to have harboured an abundance of fusarial species that are non-pathogenic to crops grown in those soils. Saprophytic isolates of *F. oxysporum* and *F. solani* were implicated in the induction of suppressiveness to fusarial wilts in certain alluvial soils of France and soils of Salines valley of California. Biological control of Fusarium wilt has become possible by introducing non-pathogenic forms of *Fusarium* spp. in soil and on to the root surfaces (6,17,77,79,87,102,130). The mechanism of disease suppression by nonpathogenic forms is attributed to three different phenomenon: (a) The suppression may be due to nutrient competition between pathogenic and saprophytic species of *Fusarium* (6). (b) The reduced incidence of Fusarium wilt of celery induced by non-pathogenic fusaria, may be because of competition for infection sites on root surface (102). (c) Enhanced resistance induced by non-pathogenic *Fusarium* spp.

Lauvet *et al.* (65) concluded that suppression was due to competition between pathogenic and saprophytic clones of *Fusarium* spp. for ecological possession of soil niche. Prior inoculation of non-pathogenic *F.oxysporum* offered protection against *Fusarium* wilt of sweet potato (78). Paulitz *et al.* (87) established suppressiveness to *Fusarium* wilt of cucumber by the addition of non-pathogenic isolates of *F.oxysporum* is raw soil. Park *et al.* (85) recorded increase in population densities of fluorescent pseudomonads in the rhizosophere of cucumber in the presence of non-pathogenic isolates of *F. oxysporum* in alkaline soils (pH 8.1). They attributed the enhanced activity of fluorescent pseudomonads to the increased root exudation as a consequence of non-pathogenic fusaria.

The saprophytic strains of *F.oxysporum* (C5 and C14) have significantly reduced the germination of chlamydospores of *F. oxysporum* f.sp. *cucumerinum* in the rhizosphere of cucumber. Exposure of wounded and unwounded roots to strain 14 enhanced systemic resistance. Therefore, three-phase mechanism was noticed in the biological control of *Fusarium* wilt involving competition in the rhizosphere and infection sites, as well as induction of enhanced resistance in the host (68).

3.4 Plant growth promoting rhizobacteria (PGPR)

Colonization of plant roots or rhizosphere by bacteria has been widely studied (86). Considerable attention has been given to the control of plant pathogenic fungi with antagonistic rhizobacteria (103,104,118). Colonizing ability of the introduced plant growth promoting rhizobacteria (PGPR) on an elongating root system is the crucial factor in the ultimate success of the bacteria either in the increase of plant growth or in curtailing the major or minor pathogens. Rhizosphere competence and root colonization form the driving forces that determine the associations, interactions and functions of rhizobacteria in plant growth. Rhizobacteria are selected mainly for their plant growth promoting abilities ~d secondly as biocontrol agents.

Plant gr:wth promoting pseudomonads may function through direct antagonism towards major pathogens, sup ressing deleterious microorganisms and minor pathogens, through antibiotic production, nutrient comp etition for iron by siderophore production or more directly through plant growth promotion (30,37,41,58,100,104,119). Seed treatment with *Pseudomonas fluorescens* strain WCS 374 suppressed the fusarial wilt of radish caused by *F.oxysporum* f.sp. *conolutinans* (61). They also recorded effective control of root rot of tomato caused by *F. oxysporum* f.sp. *radices-lvcopersici* by simultaneous application of bacterial strain C7 and non-pathogenic *Fusaruim* FO 47. *Fusarium* wilt of carnation caused by *F. oxysporum* f.sp. *dianthi* was reduced by *Pseudomonas* sp. strain WCS 358r. Suppression of wilt by WCS 358r appeared to depend on only siderophore mediated competition (35). *Streotomyces* strains are also reported to show antagonistic properties against *F. oraminearum* and *F. tricinctum* (20).

Suppression of *Penicillium pinophilum* and *F.oxysporum* f.sp. *vasinfectum* by fluorescent pseudomonads was more pronounced in the roots than in the rhizosphere (41). Fluorescent pseudomonads were not only the competent colonizers of the rhizosphere (37,59,100,104) but also seem to be highly competent colonizers of the root tissue (40,124).

Gamliel and Katan (41) reported that inoculation of plant growth promoting strains of *Pseudomonas putida*, *P. fluorescens* and *P. alcaligenes* reduced incidence of *Sclerotium rolfsii* in bean, *Fusarium* in cotton and tomato and also reduced the colonization of root infecting pathogens of cotton. Kumar and Dube (60) reported 52% reduction in chickpea wilt caused by *F. oxysporum* f.sp. *ciceri* through seed bacterization with a siderophore producing fluorescent pseudomonad RBT 13.

Synergistic effect in disease suppression was noticed when PGPR are applied with non-pathogenic fusaria. Park *et al.* (85) found that fluorescent pseudomonads and non-pathogenic isolates of *F. oxysporum* effectively suppressed the Fusarium wilt of cucumber, when added to soil simultaneously but were ineffective in inducing suppression when added separately. Lemanceau & Alabouvette (62) observed significant control in Fusarium wilt of flax (caused by *F. oxysporum* f.sp. *lini*) when fluorescent strain of *Pseudomonas* was applied in association with non-pathogenic strain of *F. oxysporum* (FO 47).

Fluorescent strains of pseudomonad species enhanced protection given by non-pathogenic *F. oxysporum* strain FO 47 to tomatoes against *F. oxysporum* f.sp. *lycopercisi* (39).

3.5 Plant growth promoting fungi (PGPF)

In addition to fluorescent pseudomonads, *Trichoderma harzianum* , a common biocontrol agent, is known to be as a significant plant growth promoter independent of any plant disease (15,27). Soils amended with *T.harzianum* and *T. Koningi* produced early germination and flowering, increased number of blooms per plant and enhanced shoot and root biomass (27,128).

3.6 Mycorrhizae

Vesicular-arbuscular mycorrhizal fungi are the major symbionts of plants from tropics. These are not only effective biofertilizers but also help in the control of soil-borne diseases. Several mycorrhizologists reviewed the effect of VAM fungi on root diseases (2,32,98,106) and microbial interactions in mycorrhizosphere (10).

The ectomycorrhizal fungus *Laccaria bicolor* (Syn. *Laccaria laccata*) provided partial or temporary protection of primary roots of Douglas-fir from *F. oxysporum* infection in field as well as controlled environment apparently due to enhanced host resistance by the accumulation of condensed tannins within cortical cells (110,120). Protection of Douglas-fir roots from *F.oxysporum* by *L. bicolor* was variable and the resistance was expressed to a non-aggressive strain at lower temperature of 17° but not aggressive strains (117). The susceptibility and root protection depends on multiple factors such as the stage and rate of primary root development, specific tissues inoculated with *F. oxysporum*, temperature and aggressiveness of the pathogenic strain. The intercellular restriction of the pathogen in protected roots was associated with the stimulation of flavanolic wall infusions induced by *L. bicolor* (117). This fungus also offered protection to the roots of *P. sylvestris* from damage by *Cylindrocarpon destructans, F. moniliforme* and *Rhizoctonia solani* (23, 24, 25).

Protection of primary roots of *Pinus resinosa* from *F. oxysporum* f.sp. *pini* was obtained by *Paxillus involutus* before mycorrhizal formation (34). The increased survival ability of *P. resinosa* to *Fusarium* was attributed to the fungistatic activity of root exudates in the rhizosphere due to the ectomycorrhizal fungus *P. involutus.*

Investigations on VA Mycorrhiza-root pathogen interaction during the last decade provided valuable information in inducing suppression of soil-borne plant pathogens (50). Several species of *Glomus* are involved in the reduction of root diseases incited by *Fusarium* spp. VA mycorrhizal fungi in association with diazotrophic bacteria enhanced the microbial activity in the mycorrhizosphere. Jalali (49) reported the positive effect of VAM fungi on the control of root diseases caused by fungal pathogens in chickpea, pigeonpea, tobacco, groundnut and tomato. Caron *et al.* (21,22) reported the control of crown and root rot of tomato caused by *F. oxysporum* f.sp. *radicis-lycopersici* by using *Glomus intraradicis* and was not affected by the phosphorus nutrition. Mycorrhizal root with *G. intraradicis* exhibited reduction in root necrosis due to *Fusarium* irrespective of the sequence of inoculation. VAM endophyte (*G.fasciculatum*) has significantly reduced the wilt incidence in chickpea caused by *F. oxysporum* f.sp. *ciceri* (51). No influence of mycorrhizal inoculations with three species of *Glomus* was recorded in pigeonpea wilt to *F. udum* (91) while *Glomus mosseae* controlled the fusarial rot caused by *F. solani* in soybean (132).

The mycorrhizal protection to root diseases was attributed to competition, by the formation of mechanical barrier to penetration, morphological alterations resulting in thickened cells walls and stronger vascular bundles, favourable rihzosphere dynamics and liberation of inhibitory compounds. It is necessary to screen several species or isolates to obtain an effective isolate and to use several mycorrhizal fungi together against various cultivars of the host.

3.7 Bacterial antagonists

The widespread commercial success in the control of the ubiquitous crown gall disease by an avirulent strain of *Agrobacterium radiobacter* K 84 (56), has triggered the attention on the use of bacteria as biocontrol agents against plant pathogenic fungi. Antagonism is the phenomenon involved in the interaction. Antagonism can be described as killing, injury or inhibition of the growth of one microorganism by another. The possible modes of bacterial action in the inhibition of microorganisms are (i) direct penetration (ii) antibiotic production (iii) competition for nutrient (iv) stimulation of host defence and (v) antibiosis.

Bacillus species are the promising biocontrol agents as they produce endospores that are tolerant to heat and desiccation. Seed bacterization with *B. subtilis* has successfully controlled the diseases of groundnut and onion. Mitchel and Alexander (73) obtained biological control of Fusarium wilts and Pythium sp. by the use of cell wall degrading lytic rhizosphere bacteria. Fusarium wilt of carnation was effectively controlled by the chitionlytic arthobacter species isolated from rhizosphere. *Fusarium* basal rot of carnation was controlled when cuttings were dipped into a suspension of *B.subtilis* or of a psandomonads (109). Change and Kommedahl (26) obtained the control of maize seedling blight caused by F. *roseum* f. sp. *cerealis,* by coating kernels with *B. subtilis.* Successful biocontrol of Fusarium blight of maize seedings is attributed to the antibiotics produced by *B. subtilis.* Seed bacterization by *B. subtilis* was widely used in the erstwhile Soviet Union to control the pigeonpea wilt. The suspension was prepared in molasses and groundnut cake, which may presumably increased the antibiotic production around seed coat and rhizosphere (55, 196, 199, 200). Singh and Singh (111) found lysis of mycelium of *F. udum* in soil amended with organic matter, which was associated with increased *B. subtilis* population in soil. Satyaprasad and Ramarao (96) reported reduced lysis of conidia and decreased chlamydospore production by *F. oxysporum* f. sp. *ciceri* , in chitin and cowdung amended soils which stimulated the microflora affecting the survival of the pathogen in soil. *Micromonospora globosa* is another antoagonist belonging to actinomycetes which causes lysis of *F. udum* .

Sneh *et al.* (112) evaluated over 700 bacteria and actinomycetes isolated from *Fusarium* suppressive soil for their ability to induce suppressiveness to *F. oxysporum* f. sp. *cucumerinum.* Among non-fluorescent bacteria only *Enterobacter cloacae* inhibited chlamydospore germination of *F. oxysporum* f. sp. *cucumerium.* They pointed out a direct correlation between siderophore producing bacteria and the inhibition of chlamydospore germination in soil.

Introduced bacterial populations in the rhizosphere and root zones were able to disperse vertically at least up to 50 cm. deep and 30 cm. laterally from tap root rather than horizontally from the application site (63). They have stated that *B. megaterium* B 153-2-2 was a potential candidate for the management or microflorae in the rhizosphere. Bacterization of carnation roots with WCS 417r controlled the Fusarium wilt by inducing photoalexin accumulation (125).

3.8 Fungal antagonists

Various fungi are involved in the biocontrol of soil-borne diseases as they exhibit widespread mycoparasitism and antagonism towards pathogenic forms. The subject of mycoparasitism has been reviewed extensively (1, 13, 28, 30, 66, 80, 81, 82,). Fungal biocontrol agents like species of

Trichoderma,. Gliocladium, Penicillium and *Aspergillus* are used in the control of fusarial diseases. Papavizas (81) discussed in detail on various aspects of *Trichoderma* and *Gliocladium* in relation to their biology, ecology and biocontrol potential, Ramarao *et al.* (90) have cited some examples of successful biological control of soil-borne diseases caused by *F. oxysporum, Sclerotium. rolfsii, Rhiozoctonia solani* by species of *Trichoderma* .

Kim and Jee (57) controlled wilt of çucumber caused by *F. oxysporum* f. sp. *cucumerinum* by introducing *T. harzianum* and *P. flourescens.* Carnations wilt caused by *F. oxvsporum* f. sp. *dianthi* was controlled by a potential biocontrol organism like *T. harzianum* (72). Several species of *Trichoderma* proved promising against *F. moniliforme* on sorghum. F. *oxysporum* f. sp. *licopersici* on tomato, *F. oxysporum* f. sp. *chrysanthemi* on chrysanthemum (33, 70).

Antibiosis, competition and mycoparasitism are the phenomenon involved in the *Trichoderma* -plant pathogen interactions.

Gliocladium catenulatum was found to be antagonistic against *Fusarium* spp. (48). Papavizas (81) opined that *Gliocladium* exerts similar antagonistic effect as that of *Trichoderma.*

Tomato crown rot caused by *F. oxysporum* f. sp. *radicislycopersici* was reduced by *Penicillium funiculosum* along with other antagonists in the field (69). *Penicillium cyclopium* exhibited antagonism against *F. avenaceum, F. culmorum. F. oxysporum* f. sp. *melonis, F. oxysporum* f. sp. *lycopersici* and *F. oxysporum* f. sp. *niveum* (19, 52). Reyes (92) observed suppression of cabbage yellows incited by *F. oxysporum* f. sp. *conglutinans* by *F. oxalicum* and *T, viride.* Muskmelon wilt caused by *F. solani* and watermelon wilt caused by *F. oxysporum* were effectively controlled by different strains of *Aspergillus niger* (74, 105). Chickpea wilt pathogen *F. oxysporum* f. sp. *ciceri* was curtailed by *A. niger* in pot culture experiments. Bora (18) also observed antagonism by *A. niger* on damping off disease of eggplant caused by F. solani and other fungi. Linseed wilt caused by *F. oxysporum* f. sp. *lini* was effectively reduced by *A. flavus* (4).

In tomatoes, Yeshia *et al.* (131) controlled pre-and post-emergence damping off cussed by *F. oxysporum* f. sp. *lycopersici* and other pathogens by *T. viride, Streptomyces griseus* and *B. subtilis. Fusarium udum* was suppressed in pigeonpea fields by *A. niger, A. flavus, Penicillium* sp. and *Micromonospora globosa* (123).

Bashar and Rai (16) reported antagonistic potential of *T. harzianum, T.viride* against wilt disease of chickpea caused by *F. oxysporum* f. sp. *ciceri.*

Antagonistic fungi are exposed to the fungicides applied to the soil so the strains which are resistant to the fungicides may perform better in such soil. Benomyl tolerant isolates of *T. harzianum* with enhanced biocontrol capabilities were introduced into the soil (83).

4. CONCLUSIONS

Plant health is dependant on a critical biological equilibrium between the harmful and beneficial biotic agents and its prolonged maintenance. Biological control is a modern, comprehensive, non-polluting approach to the management of diseases (121). Use of antagonists in combination with growth promoting and nutrient mobilising organisms like PGPR and VAM is very much desired, instead of employing individual antagonists.

Rhizosphere competence, rapid colonization of substrates and long term survival in soil are the prime criteria used in the selection of biocontrol agents. Solarization of soils particularly in tropical countries like India should be explored more as it is non-hazardous, non-polluting, cost effective and simple method for the suppression of soil-borne diseases.

The ability of biocontrol agent, PGPR, to control the disease and improve plant growth in a given soil depends on the genetic potential of their populations on the biotic and abiotic environmental conditions prevailing in the rhizosphere and other agricultural practices. Search for the new and efficient biocontrol agents in the soil and rhizosphere should be a continuous process. Improvement of genetic potentialities of antagonists and other beneficial organisms is the best strategy to achieve success in biocontrol programmes. Baker (12) noted three methods that have the most potential for genetic manipulation of fungi without functional sexual stages, these are: (i) conventional mutagenesis (ii) protoplast fusion and (iii) transformation.

Biocontrol efficiency of the antagonists can be improved by the modern biotechnological methods and biocontrol measures should be exploited along with other disease management practices.

ACKNOWLEDGEMENTS

One of the authors (IKK) gratefully acknowledges the financial assistance of C. S. I. R.

5. REFERENCES

1. Adams, P. B. 1990. The potential of mycoparasites for biological control of plant diseases. Annu. Rev. Phytopathol. **248**: 59-72.

2. Agarwal, G. P. 1991. Biological plant protection - recent developments, sectional presidential address (Part II) at the 78th Science Congress, ISCA, 20pp.

3. Agnihotrudu, V. 1954. Soil conditions and wilt diseases in plants, rhizosphere microflora in relation to fungal wilts. Ph. D. Thesis, Madars University.

4. Akhtar, C. M., Amin, C. M. and Salim, C. M. 1982. Investigations on the biological control of *Rhizoctonia solani* and *Fusarium oxysporum* f. sp. *lini* causing root rot and wilt diseases in cotton and linseed crops respectively. Faislabad, Pakistan Ayub Agric. Res. Inst. 63p.

5. Alabouvette, C., Rouxel, F. and Louvet, J. 1979. Characteristics of Fusarium wilt suppressive soils and prospects for their utilization in biological control. In: Soil-bone plant pathogens, Eds. B. Schippers and W. Gams, Acad. press, London, pp. 165-182.

6. Alabouvette, C., Couteaudier V., and Louvet, J. 1984. Recherches sur la resistance des sols aux maladies. IX Dynamique des populations du *Fusarium* spp. et de *Fusarium oxysporum* f. sp. *melonis* dins un sol resistant et dans un sol sensible aux fusarioses vasculaires. Agronomic Paris, **4**: 729-733.

7. Alabouvette, C., Couteaudeir, Y. and Louvet, J. 1985. Soils suppressive to Fusarium wilt: Mechanism Ecology and Management of Soil-borne Plant pathogens, Eds. C. A. Parker *et al* Amer. Phytophathol. Soc., St. Paul MN. 358pp.

8. Arora, D. K. and Pandey, A. K. 1989. Soil solarization for control of soil-borne diseases. Theory and application. In: Perspectives in phytopathology (R. S. Singh Festschrift Volume). Eds. V. P. Agnihotri, N. singh, H. S. Chaube, U. S. Singh and T. S. Dwivedi, pp. 429 -438.

9. Atkinson, G. F. 1982. Some diseases of cotton. Ala. Agr. Exp. Stn. Bull. **41**:65p.

10. Bagyaraj, D. J. 1984. Biological interaction with VAM fungi. In: VA Mycorrhizae, Eds. C. L. Powell and D. J. Bagyarai, CRC Press, Boca Raton, FL. 234 pp.

11. Baker, R. 1986. Biological control: an overview. Can. J. Plant pathol., **8**: 218 -221.

12. Baker, R. 1986. Some perspectives on the application of molecular approaches to biocontrol problems. In: Biotechnology of fungi for improving plant growth, Eds. J. M. Wipps and R. D. Lumsden, Cambridge Univ. Press. p. 220-233.

13. Baker, K. F. and Cook, R. J. 1974. Biological control of plant pathogens, W. H. Freeman and Co. San Fransico. 433 pp.

14. Baker, R. and Scher, F. M. 1987. Enhancing the activity of biological control agents, p. 1-17 In: Innovative approaches to plant diseases control, Ed. I. Chet, John Wiley & Sons. Inc. 372 pp.

15. Baker, R., Elad, Y. and Chet, I. 1984. The controlled experiment in the scientific methods with special emphasis on biological control. Phytophathology, **74**: 1019-1021.

16. Bashar, M. A. and Rai, B. 1991. Antagonistic potential of root region microflora of chickpea against *Fusarium oxysporum* f. sp. *ciceri*. Proc. Int. Bot. Conf., Dhaka (Bangladesh) p.10-12 June. P. 8.

17. Biles, C. L. and Martyn, R. D. 1989. Local and systems resistance induced in watermelon by formae speciales of *Fusarium oxysporum*. Phytophathology, **79**: 856-860.

18. Bora, T. 1977. *In vitro* and *in vivo* investigations on the effect of some antagonistic fungi against the damping off diseases of egg plant. J. Turk. Phytophath. **6**:7-13.

19. Bora, T., Yildiz, M., Akinci, C. and Nemli, T. 1982. Investigations of fungistasis with respect to wilt diseases in important cultivated soils of the western aeglan region. J. Turk Phytopath. **6**: 7-13.

20. Borghi, A. L.., Degado, C. J. and Bracalenti, B. J. C. De. 1990. *Streptomyces* populations from soils under cereals and legumes. II. Antagonistic properties under toxigenic *Fusarium* strains. Cienca del Suelo, **8**: 177-180.

21. Caron, M., Fortin, J. A. and Richard, C. 1986. Effect of *Glomus intraradices* infection by *Fusarium oxysporum* f. sp. *radicis-pyloperpisa* in tomatoes a 12 week period. Can J. Bot., **64**: 552.

22. Caron, M., Fortin, J. A. and Richard, C. 1986. Effect of phosphorus concentration and *Glomus intraradices* on fusarium crown and root rot of tomatoes. Phytopathology, **76**: 942.

23. Chakravarthy, P. and Unestam, T. 1986. Role of mycorrhizal fungi in protecting against damping-off of *Pinus sylvestris* L. seedlings. In : Proc. 1st Eur. Symp. Mycorrhizae, 1985, Eds. V. Gianiazzi-Pearson and S. Gianinazzi, pp. 811-814.

24. Chakravarty, P. and Unestam, T. 1987. Mycorrhizal fungi prevent disease in stressed pine seedlings. J. Phtopathol. **118**: 335-340.

25. Chakravarty, P. and Unestam, T. 1987. Differential influence of ectomyccrrhizae on plant growth and disease resistance in *Pinus sylvestris* seedlings. J. Phytopathol. **120**: 104-120.

26. Chang, I. P. and Kommedhal, T. 1968. Biological control of seedling blight of corn by coating kernels with antagonistic microorganisms. Phytopathology, **58**: 1395-1401.

27. Chang, Y. C., Chang, Y.C., Baker, R., Kleifeld, O and Chet, I. 1986. Increased growth of plants in the presence of the biological control agent *Trichoderma harzianum*. Plant Dis., **70**: 145-148.

28. Chet, I. 1987. *Trichoderma* -application, mode of action and potential as a biocontrol agent of soil-borne plant pathogenic fungi, In: Innovative approaches to plant disease control. Ed. I. Chet. John Wiley & sons, 372pp.

29. Cook, R. J. 1985. Biological control of plant of pathogens: Theory to application. Phytopathology, **75**:25-29.

30. Cook, R. J. and Baker, K. F. 1983. The nature and practice of biological control of plant pathogens. An. Phytopathol. Soc. St., Paul. Mn. 539pp.

31. Cook, R. J. and Bruehi, G. W. 1968. Relative significance of parasitism versus saprophytism in colonization of wheat straw by *Fuasrium roseum culmorum* in the field. Phytopathology, **58**: 306.

32. Dehene, H. W. 1982. Interaction between vesicular arbuscular mycorrhizal fungi and plant pathogens. Phytopathology, **72**:1115.

33. D' ercole, N. 1983. Impiego di *Trichoderma* in lotta biologica. Informat Fitopatol, **33**: 37-40.

34. Duchesne, L. C., Perterson, R. L. and Ellis, B. E. 1988. Interaction between the ectomycorrhizal fungus *Paxillus involutus* and *Pinus resinosa* induces resistance to *Fusarium oxysporum*. Can. J. Bot., **66**: 558-562.

35. Duff, B. J., Meiier, J. W., Bakker, P. A. H. M. and Schippers, B. 1991. Suppression of Fusarium wilt of carnation by *Pseudomonas* in soil: Mode of action. Bull. SROP, **14**: 152-157.

36. Dupler, M. and Baker, R. 1984. Survival of *Pseudomonas putida*, a biological control agent in soil. Phytopathology, **74**: 195-200.

37. Elad, Y. and Baker, R. 1985. The role of competition for iron and carbon in suppression of chlamydospore germination of *Fusarium* spp. by *Pseudomonas* spp. Phytopathology, **75**: 1053-1059.

38. Elad, Y., Katan, J. and Chet, I. 1980. Physical, biological and chemical control integrated for soil-borne disease in potatoes. Phytopathology, **70**: 418-422.

39. Fuchs, J. and Defago, G. 1991. Protection of tomatoes against *F. oxysporum* f. sp. *lycopersici* by combining a non-pathogenic *Fusarium* with different bacteria in untreated soil. Bull SROP, **14**: 51-56.

40. Gamliel, A. and Katan, J. 1991. Involvement of fluorescent pseudomonads and other microorganisms in increased growth response of plants in solarized soils. Phytopathology, **81**: 494-502.

41. Gameliel, A. and Katan, J. 1993. Suppression of major and minor pathogens by fluorescent pseudomonads in solarized and nonsolarized soils. Phytopathology,**83**: 68-75.

42. Garrett, S. D. 1970. Pathogenic root-infecting fungi, Cambridge Univ. press 293pp.

43. Garrett, S. D. 1975. Cellulolysis rate and competitive saprophytic colonization of wheat straw by foot-rot fungi. Soil Biol. Biochem, **7**: 323.

44. Greenberger, A., Yogev, A. and Katan, J. 1987. Induced suppressiveness in solarized soils. Phytopathology, **77**: 1663-1667.

45. Grinstein, A., Katan, J., Abdul-Razik, A., Zeydan, O. and Elad, Y. 1979. Control of *Sclerotium rolfsii* and weeds in peanuts by solar heating of soil. Plant Dis. Reptr. **63**: 1056-1059.

46. Hardy, G. E. and Sivasithamparam, K. 1985. Soil solarization: Effects of Fusarium wilt of carnation and Verticillium wilt of egg plant. In: Ecology and Management of Soil-borne plant pathogens. Eds. C. A. Parker, A. D. Rovira, K. J. Moore, P. T. W. Wong and J. F. Kollmoregen, A PS. St. Paul. MN.

47. Hess, C. E. 1992. New challenges in Agriculture. Phytopathology, **82**: 40-42.

48. Huang, H. C. 1978. *Gliocladium catenulatum*. Hyperparasite of *Sclerotiorum* and *Fusarium* species. Can. J. Bot. **56**: 2243-2246.

49. Jalali, B. L. 1991. Mycorrhizal systems in Management of Plant diseases. Mycorrhiza News, **3**:3.

50. Jalali, B. L. and Tharaia, M. L. 1981. Suppression of Fusarium wilt of chickpea in VA-Mycorrhiza inoculated soils. Int. Chickpea Newsletter, **4**: 21-23.

51. Jalali, B. L. and Jalali, I. 1991. Mycorrhiza in plant disease control. In: Hand book of applied mycology I. Soil and plants, Eds. D. K. Arora, K. G. Mukerji and G. R. Knudsen, pp. 131-154. Marcel Dekkjer Inc. New York.

52. Kamyshika, O. P., Tupenevich, S. M., Chumakov, A. E. and Shekunova, E. G. 1976. Mikroflora v eksperiamental nomesevooborote i ee antagonistich eskays activnost K pochverrnym fitopatogenam. Mikol. Fitoputol. **10**: 32-36.

53. Karunkar, R. I. 1989. Studies on root and stalk rot of sorghum with special reference of soil survival of stalk rot pathogens. Ph. D. Thesis, Osm. Univ. Hyd.

54. Katan, J. 1980. Solar pasteurization of soils for disease control: Status and prospects. Plant Dis. **64**: 450-454.

55. Katan, J. 1981. Solar heating (Solarization) of soil for control fo soil-borne pests. Annu. Rev. Phytopathol., **19**: 211-236.

56. Kerr, A. 1980. Biological control of crown gall through production of agrocin 84. Plant Dis. **64**: 24-30.

57. Kim, H. K. and Jee, H. J. 1988. Influence of rhizosphere antagonists on suppression of cucumber wilt, increased cucumber growth and density fluctuation of *Fusarium oxysporum* f. sp. *cucumerinum*. Korean J. Pl. Pathol . **4**: 10-18.

58. Kloepper, J. W. and Schroth, M. N. 1981. Relationship of *in vitro* antibiosis of PGPR to plant growth and displacement of root microflora. Phytopathology, **71**: 1020-1024.

59. Kloepper, J. W. Schroth, M. N. and Miller, T. D. 1980. Effects of rhizosphere colonization by PGPR on potato plant development and yield. Phytopathology, **70**: 1078-1082.

60. Kumar, B. S. D. and Dube, H.C. 1992. Seed bacterization with fluorescent *Pseudomonas* for enhanced plant growth, yield and disease control. Soil Biol Biochem. **24**: 539-542.

61. Leeman, M., Scheffer, R. S., Pelt, J. A., Van, Bakker, P. A. H. M. and Schippers, B. 1991. Control of Fusarium wilt of radish by *Pseudomonas flourescens*. Bull. SROP, **14**: 34-38.

62. Lemanceace, P. and Alabouvette, C. 1991. Biological control of *Fusarium* diseases by the association of fluorescent *Pseudomonas* and non-pathogenic *Fusarium*. Bull. SROP, **14**: 45-50.

63. Liu, Z. L. and Sinclair, J. B. 1992. Population dynamics of *Bacillus megaterium* strain 153-2-2 in the rhizosphere of soyabean. Phytopathology, **82**: 1297-1301.

64. Lockwood, J. L. 1986. Soil-borne plant pathogens: concepts and connections. Phytopathology, **76**: 20-27.

65. Louvet, J., Alabouvette, C., and Rouxel, F. 1981. Fusarium wilt. In: Fusarium diseases, biology and taxonomy. Eds. P. E. Nelson, T. A. Toussan and R. J. Cook, Penn. State Univ. Press, Univ. Park., pp. 261-275.

66. Lumsden, R. D. 1981. Hyperparasitism for control of plant pathogens. In : RC Hand Book of Pest Management in Agriculture, Ed. D. Pimmentel, Boca Raton, FL, CRC press, pp. 475-484.

67. Lumsden, R. D., Millner, P. D. and Lewis, J. A. 1986. Suppression of lettuce drop caused by *Sclerotinia minor* with composted sewage sludge. Plant Dis. **70**: 197-201.

68. Mandeel, G. and Baker, R. 1991. Mechanisms involved in biological control of Fusarium wilt of cucumber with strains of nonpathogenic *Fusarium oxysporum*. Phytopathology, **81**: 462-469.

69. Marois, J. J., Mitchell, D. J. and Sonoda, R. M. 1981. Biological control of Fusarium crown rot of tomato under field conditions. Phytopathology, **71**: 1257-1260.

70. Marois, J. J. and Locke, J. C. 1985. Population dynamics of *T. viride* in steamed plant growth medium. Phytopathology, **75**: 115-118.

71. Mc Rae, A. T. K. and Shah, F. J. F. 1933. Part I. Influence of superphosphate, cattle manure and green manure on the wilt of *Caianus indicus* caused by *F. vasinfectum.* Imp. Council. Agri. Res. Sci. Monograph, **7**: 1-68.

72. Mirkova, E. 1982. *In vitro* study of antagonistic activity of *Trichoderma* spp. towards some soil pathogens. Gardinarska i Lozarska Nauka, **19**: 73-81.

73. Mitchel, R. and Alexander, M. A. 1963. Addition of fungal cell wall components to soil for biological control. Phytopathology, **53**: 308-312.

74. Naik, M. K. 1989. Wilt of watermelon caused by *Fusarium oxysporum* Sch. Ph. D. Thesis IARI, New Delhi, p. 184.

75. Naik, M. K. and Sen, B. 1992. Biocontrol of plant diseases caused by *Fusarium* species. In : Recent Developments in Biocontrol of Plant Diseases, Eds., K.G. Mukerji, J.P. Tewari, D.K. Arora and Geeta Saxena, Aditya Books Pvt. Ltd., New Delhi. pp. 37-52.

76. Nene, Y. L., Kannaiyan, J., Hawre, M. P. and Reddy, M. V. 1979. Review of work done at ICRISAT on soil-borne diseases of pigeonpea and chickpea. In : Proc. of consultants group discussion on the resistance of soil-borne diseases of legumes, pp. 3-35.

77. Ogawa, K. and Komada, H. 1984. Biological control of Fusarium wilt of sweet potato with cross-protection by non-pathogenic *Fusarium oxysporum.* In: Ecology and Management of soil-borne plant pathogens. Eds. C.A. Parker, A. D. Rovira, K. J. Moore, P.T.W. Wong and J. F. Kollmargen. Am. Phytopathihol. Sic. St. Paul. MN, pp. 121-124.

78. Ogawa, K. and Komada, H. 1982. Biological control of Fusarium wilt of sweet potato with cross-protection by prior inoculation with non-pathogenic *Fusarium oxysporum.* JARQ, **19** : 20-25.

79. Ouchi, S. 1983. Induction of resistance or susceptibility. Annu. Rev. Phytopathol. **21** : 289-315.

80. Papavizas, G. C. 1981. Biological control in crop production *totowa. NJ.* Allahneld. Osmun. 461pp.

81. Papavizas, G. C. 1985. *Trichoderma* and *Gliocladium :* Biology, ecology and potential for biocontrol. Annu. Rev. Phytopathol. **23** : 23-54.

82. Papavizas, G. C. and Lumsden, R. D. 1980. Biological control of soil-borne fungal propagules. Annu. Rev. Phytopathol. **18** : 389-413.

83. Papavizas, G. C., Lewis, J. A. and Abd-El Moity, T. H. 1982. Evaluation of new biotypes if *Trichoderma harzianum* for tolerance to benomyl and enhanced biocontrol capabilities. Phytopathology, **72**: 126-132.

84. Parke, J. L. 1991. Root colonization by indigenous and introduced microorganisms. In : The Rhizosphere and plant growth, Eds. D. L. Keister and P. B. Gregan, pp. 33-42. Kluwer Acad. Publ Dordrecht. Neth.

85. Park. C. S., Paulitz, T. C. and Barker, R. 1988. Biocontrol of Fusarium wilt of cucumber resulting from interactions between *Pseudomonas putida* and nonpathogenic isolates of *Fusarium oxysporum.* Phytopathology, **78** :190-94.

86. Parke, J. L., Liddel, C. M. and Clayton, M. K. 1990. Relationship between soil mass adhering to pea tap roots and recovery of *Pseudomonas flourescens* from the rhizosphere. Soil Biol. Biochem, **22** : 495-499.

87. Paulitz, T. C., Park, C S. and Baker, R. 1987. Biological control of Fusarium wilt of cucumber with nonpathogenic isolates of *F. oxysporum*. Can. J. Microbiol. **3** : 349-353.

88. Pullman, G. S., De Vey, J. E. and Garber, R. H. 1981. soil Solarization and thermal death : Logarithic relationship between time and temperature for four soil-borne plant pathogens. Phytopathology, **71** : 959-963.

89. Pullman, G. S., De Vey, J. E., Garber, R. H. and Wienhold, A. R. 1981. Soil solarization: Effects on Verticivium wilt of cotton and soil-borne populations of *Verticivium dahliae, Pythium* spp. *Rhizoctonia solani* and *Thielaviopsis basicola*. Phytopathology, **71** : 954-959.

90. Ramarao, P., Manoharachary, C. and Vaidehi, B. K. 1987. Rhizosphere microflora and root-borne fungal pathogens. ISCA symposium, pp. 239-258.

91. Reddy, M. V., Rao, J. N. and Krishna, K. R. 1989. Influence of VA Mycorrhiza on Fusarium wilt of pigeonpea. Int. Pigeonpea Newsletter. **9** : 23.

92. Reyes, A. A. 1984. Suppressive activity of different fungi against cabbage yellows in the soil and *Fusarium oxysporum* f. sp. *conglutinans in vitro*. Phytoprotection, **65** : 27-33.

93. Rovira, A. D. and Wildermuth, G. B . 1981. The nature and mechanisms of suppression. In : Biology and control of Take-all. Eds. M.I.C. Asher and P. J. Shipton, Acad. Press. London, p. 385-415.

94. Satyaprasad, K. 1980. Studies on soil survival of *Fusarium solani* causing root rot of guar and *F. oxysporum* f. sp. *ciceri* causing wilt of chickpea. Ph. D. Thesis, Osm. Uni. Hyd.

95. Satyaprasad., K. and Ramarao, P. 1983. Saprophytic colonization of *Fusarium oxysporum* f. sp. *ciceri* in soil. Folia Microbiol. **28** : 173-178.

96. Satyaprasad., K. and Ramarao, P. 1983. Influence of various amendments on the lysis of conidia and chlamydospore production by *F. oxysporum* f. sp. *ciceri*. Indian Phytopath., **36** : 67-71.

97. Satyaprasad., K. and Ramarao, P. 1984. Competitive saprophytic colonization by *Fusarium solani*, Indian Phytopath., **37** : 75-78.

98. Schenck, N. C. 1987. Vesicular-arbuscular mycorrhizal fungi and the control of fungal root diseases. In: Innovative approaches to plant disease control. Ed. I. Chet, John wiley & sons. New York. 372 pp.

99. Scher, F. M. and Baker, R. 1982. Effect of *Pseudomonas putida* and a synthetic iron character on inducing of soil suppressiveness to Fusarium wilt pathogens. Phytopathology, **72** : 1572-1573.

100. Schippers, B., Bakker, A. W. and Bakker, P.H.M. 1987. Interactions of deleterious and beneficial rhizosphere microorganisms and the effect of cropping practice. Annu. Rev. Phytopathol., **25** : 339-358.

101. Schulub, R. L., Lackwood, J. L. and Komada, H. 1981. Colonization of soyabean seeds and plant tissue by *Fusarium* species in soil. Phytopathology, **71** : 693.

102. Schneider, R. W. 1984. Effect of non pathogenic strains of *F. oxysporum* on celery root, infection by *F. oxysporum* f. sp. *apii* and a novel use of the line weaver bark double reciprocal plot technique. Phytopathology, **74** : 646-653.

103. Schroth, M. N. and Hancock, J. G. 1981. Selected topics in biological control. Annu. Rev. Microbiol. **35** : 453-476.

104. Schroth, M. N. and Hancock, J. G. 1982. Disease suppressive soils and root colonizing bacteria. Science, **216** : 1376-1381.

105. Sharma, J. 1989. Spectrum and application of the antagonist in the biological control of Fusarium wilt of cucurbits. Ph.D. Thesis IARI, New Delhi.

106. Sharma, M. and Mukerji, K. G. 1992. Mycorrhiza-Tool for Biocontrol. In: Recent developments in biocontrol of plant diseases, Eds. K. G. Mukerji, J. P. Tewari, D. K. Arora and Geeta Saxena, pp. 52-80.

107. Shipton, P J. 1975. Take-all decline during cereal monoculture. In : Biology and control of soil borne plant pathogens, Ed. G. W. Bruel. The Am. Phytopath. Soc. St. Paul, MN, p. 137-144.

108. Simeoni, L. A., Lindsay, W. L. and Baker, R. 1987. Critical iron level associated with biological control of Fusarium wilt. Phytopathology, **77** : 1057-1061.

109. Sinclairm, H.V. 1989. *Bacillus subtilis* as a biocontrol agent for plant diseases. In: Perspectives in plant pathology. Today & Tomorrow Publ. New Delhi. pp. 367-374.

110. Sinclair, W. A., Cowles, D.P. and Hee, S. M. 1975. *Fusarium* root rot of Douglas-fir seedlings: Suppression by soil fumigation, fertility, management and inoculation with spores of the fungal symbiont *Laccaria laccata.* For Sci., **21** : 390-399.

111. Singh, N. and Singh, R. S. 1981. Lysis of mycelium of *F. oxysporum* f. sp. *udum* in soil amended with organic matters. Plant and Soil, **59** : 9-15.

112. Sneh, B., Dupler, M., Elad, Y. and Baker, R. 1984. Chlamydospore germination of *F. oxyoporum.* f. sp. *cucumerinum* and affected by fluorescent and lytic bacteria from *Fusarium* suppressive soil. Phytopathology, **74** : 1115-1124.

113. Stapleton, J. J. and De Vay, J. E. 1982. Effect of soil solarization on population of selected soil-borne microorganisms and growth of deciduous fruit tree seedlings. Phytopathology, **72** : 723.

114. Stapleton, J. J. and De Vay, J. E. 1983. Response of phytoparasitic and free living nematodes to soil solarization and 1,3-dichloropropane in California. Phytopathology, **73** : 1429-1436.

115. Stapleton, J. J. and De Vay, J. E. 1984. Thermal components of soil solarization as related to changes in soil and root microflora and increased plant growth response. Phytopathology, **74** : 255-259.

116. Strausbaugh and Forster, R. L. 1992. The influence of fall soil solarization on soil-borne diseases of potatoes and beans. Phytopathology, **82** : 1172.

117. Strobel, N. E. and Sinclair, W. A. 1991. Influence of temperature and pathogen aggressiveness on biological control of *Fusarium* root rot by *Laccaria bicolor* in Douglas-fir. Phytopatholqgy, **81** : 415-425.

118. Suslow, T. V. 1982. Role of root-colonizing bacteria in plant growth: In: Phytopathogenic Prokaryotes Vo. I. Academic Press. pp. 187-223.

119. Suslow, T. V. and Schroth, M. N. 1982. Rhizobacteria of sugar beets: effects of seed application and root colonization on yield. Phytopathology, **72** : 199-206.

120. Sylviya, D. M. and Sinclair, W. A. 1983. Phenolic compounds and resistance to fungal pathogens induced in primary roots of Douglas - fir seedlings by the ectomycorrhizal fungus *Laccaria laccata.* Phytopathology, **73** : 390-397.

121. Tauber, M. J. and Baker, R. 1988. Every other alternative but biological control. Bio-Science, **38** : 660.

122. Toussoun, T. A. 1975. Fusarium-suppressive soils. In : Biology and control of soil-borne plant pathogens. Ed. G. W. Bruehi, Am. Phytopathol. Soc. St. Paul. MN 216, pp. 145-151.

123. Upadhyay, R. S. and Rai, B. 1983. Fungistatic activity of different Indian soils against *Fusarium udum.* Plant and Soil, **63** : 181-188.

124. Van Peer, R. and Schippers, B. 1989. Plant growth response to bacterization with selected *Pseudomonas* spp. strains and rhizosphere microbial development in hydroponic cultures. Can J. Microbiol. **35** : 456-463.

125. Van Peer, R., Niemann, G. J. and Schippers, B. 1991. Induced resistance and phytoalexin accumulation in biological control of Fusarium wilt of carnation by *Pseudomonas* sp. strain WCS 417. Phytopathology, **81** : 728-734.

126. Warren, H. L. and Kommedahl, T. 1973. Root infecting species of *Fusarium* in soil and in roots, rhizosphere and residues of Oats. Phytopathology, **63** : 1401.

127. Weller, D. M. 1988. Biological control of soil-borne plant pathogens in the rhizosphere with bacteria. Annu. Rev. Phytopathol. **26** : 379-407.

128. Windham, M. T., Elad, Y. and Baker, R. 1986. A mechanism for increased plant growth induced by *Trichoderma* spp. Phytopathology, **76** : 518-521.

129. Wong, P.T.W. and Baker, R. 1984. Suppression of wheat take all and *Ophiobolus* patch by fluorescent pseudomonads from a Fusarium suppressive soil. Soil Biol. Biochem. **16** : 397-403.

130. Wymore, L. A. and Baker, R. 1982. Factors affecting cross-protection in control of Fusarium wilt of tomato. Plant Dis., **66** : 908-910.

131. Yeshia, A. H., El-Hassan, S. A. and Ismail, F. K. 1981. Studies on damping off disease of tomato seedlings and its biological control. Mesopot, S. Agric. **11** : 115-124.

132. Zambolim, L. and Schenck, N. C. 1983. Reduction of the effect of pathogenic root infecting fungi on soyabean by the mycorrhizal fungus *Glomus mosseae*. Phytopathology, **73** : 1402.

27 Management of Diseases of Edible Mushrooms in India with Particular Reference to White Button Mushroom (*Agaricus bisporus*)

H.S. Sohi

CONTENTS

Integrated Pest and Disease Management
Rajeev K. Upadhyay, K.G. Mukerji, B.P. Chamola and O.P. Dubey (eds.)
APH Publishing Corporation, New Delhi-110002, India

ABSTRACT

The problem of protein malnutrition is becoming very acute in the developing countries due to rise in population. It is therefore very essential to exploit all the alternative sources of protein production since availability of pulses is going down. Recent technological break-throughs in cereal production have resulted in the availability of large quantities of surplus agricultural and industrial wastes (agrowastes). Their disposal needs priority attention to avoid pollution. Edible fungi can easily use these wastes for their growth producing nutritious food. Eight edible mushrooms (out of about 2000) are being commercially cultivated with a production of about 38 lakh tonnes (in 1988-89). Amongst these, white button contributes about 37% followed by oyster (24%) and shiitake (10%). The cultivation of white button mushroom has become popular in India during recent years. Being temperate in nature, it is more suited for cultivation in the hilly areas throughout the year. One crop is taken in the northern plains during winter. Most of the growers are not able to exploit its optimum yield potential as a result of the use of unpasteurized compost. Cropping rooms lack proper facilities for environmental control. This is mostly responsible for higher incidence of diseases and pests resulting in lower and inconsistent yields. The productivity and quality of produce is adversely affected by a number of biotic and abiotic factors. The adverse effects of most common biotic causes including diseases, competitor/indicator and weed moulds and their management has been discussed for guidance of Indian mushroom growers.

1. INTRODUCTION

The population of the world is increasing at an alarming rate (1.7% annually). It will be over six billion by the year 2000. About 700 million human beings are living today in absolute poverty. Unfortunately their number is more in under-developed an developing countries due to steep rise in population. In addition, there is acute malnutrition as a consequence of lesser protein intake. Average daily consumption per capita (in g) in South East Asia is only about half of that in U.S. and even less than the world average. The per capita daily consumption of total animal proteins is almost the lowest in India, only next to Indonesia (23). Energy intake is only 1998 calories (per capita per day) against an optimal requirement of 2800 (62). The availability of pulses (main protein source) has gone down drastically during the last few decades (from 75 to 37g). With a land area of 2.4% it has to support a population of 15%. Inspite of three times increase in grain production during 1951 to 1988-89, the per capita availability increased only marginally (from 395g in 1951 to 463g in 1985) as the grain production increased by 2.32% per annum against the population rise to 2.24%. The rate of agricultural growth has not been enough to make a perceptible impact on the problems of poverty and unemployment. Nearly 90% of our population is mostly dependent on agriculture and lives in villages. Agricultural development has to serve as a means of producing more food as also an instrument of generating more jobs and rural income. It is, therefore essential to improve quality of life in villages by creating more employment in rural areas to check migration of rural poor to urban areas. The constantly rising cost of production and reduction in size of holdings is making agriculture less profitable. The present population of 844 million in 1991 will grow to about one billion by the turn of the century. The country would need about 225 mt of food grains by then in addition to 35 mt of pulses. Providing balanced diet to our people is the enshrined goal of our national policy. Indian diet is primarily cereal based, quite rich in calories but highly deficient in protein. The total national availability of proteins from all natural sources was short of requirement by 6.15 mt for a population of 680 m in 1981 (66). This will become more acute in due course. Under the prevailing circumstances, it is essential to exploit all the alternative sources of protein production.

2. USE OF AGROWASTES IN MUSHROOM PRODUCTION

Edible mushroom have been recommended by the FAO as food contributing to the protein nutrition of the developing countries depending largely on cereals (22). Recent technological breakthroughs in the production of cereal crops have resulted in availability of large quantities of crop residues (agrowastes). According to Goldstein annual production of agrowastes was of the order of 50.10^9 tonnes in 1981. About half of this quantity remains unused. The production of cereal straw in 1978 was estimated to be 2516.5 mt. Of this Indian share was 302 mt. (9, 10, 12). In India alone about 120 million tonnes of straw (out of 269 mt.) is available for recycling. Agricultural wastes rich in cellulose, hemicellulose, and lignin will continue to accumulate as methods of crop production become more efficient. Proper disposal of different agrowastes needs priority attention to avoid pollution and build up of pests and pathogens alongwith saprophytic moulds. In the underdeveloped and developing countries, there is need to convert the surplus agricultural/industrial wastes in food protein. Mushroom culture is currently the only economic way of up-grading lignocellulosic wastes. These organisms can easily use these wastes for their growth producing nutritious food. With a protein content of about 3.7, and high percentage of all the 9 essential amino acids, low in calories will traces of sugar and rich in water soluble vitamins, mushroom constitute a valuable source of nutritive and protective foods (22, 46). Mushroom production has shown rapid progress in India in the last decade but the tempo is slow as compared to other countries. The per unit productivity is comparatively very low resulting in higher cost of production. Major share of production still is contributed by small farms doing seasonal cultivation under natural conditions, using unpasteurized compost, in ordinary buildings lacking facilities for environmental control. Yields are low and inconsistent with frequent crop failures. Among different mushrooms, white button *(Agaricus bisporus)* still enjoys the largest consumer preference but its share in world production (of 38 lakh tonnes in 1988-89) in the last decade has decreased from 75% to 56%. Production of all the edible mushrooms showed an average annual increase of 24.5% during 1986 to 1989-89 (11). The production of all the 8 cultivated mushrooms showed an increase ranging from 16% of white button of 437% of oyster *(Pleurotus* spp). In India, out of the three commercially cultivated mushrooms, white button contributes about 85 to 90% followed by oyster. The farmer needs highly sophisticated and capital intensive technology because of its temperate nature and involves the use of pasteurized compost. Oyster and paddy straw mushrooms are more suited for cultivation in tropical and sub-tropical regions of the country involving low cost technology. These need to be popularised and are equally nutritious. Their cost of production shall be comparatively lower. It is heartening to note that a few mushroom farms of white button with most uptodate pasteurization and cropping facilities have come up during the last 2-3 years. Sizeable quantity will be produced at these farms. A number of other units are in planning stage since the government has announced some incentives for export of mushrooms.

Good deal of information is available in literature on various aspects of cultivation. Since majority of the Indian growers, especially the seasonal ones, will continue to use unpasteurized compost alongwith growing in ordinary buildings lacking proper environment for cropping, the chances of pest build up are more. It was, therefore, thought worth while to compile information on disease management in *A. bisporus* based on Indian work for guidance of growers.

3. DISEASES AND COMPETITOR/INDICATOR MOULDS

Mushrooms, like any other living organisms are subject to attack by diseases. These are grown in specially prepared compost (substrate) which favours the selective growth of the mushroom mycelium. However, if compositing is not carried out properly, weed moulds develop, some of which can reduce yield by as much as 40% or even more depending upon their incidence and stage of infection. Since the presence of weed moulds in compost indicates that compositing has not been done properly, these are referred to as indicator moulds. Weed moulds compete with mushrooms for compost nutrients and are also called competitor moulds. Most of the pest and disease problems are eliminated with the use of pasteurized compost under hygienic conditions thereby emphasizing the need for use of properly pasterurized compost and casing soil. The productivity and quality of produce is adversely affected by a number of biotic and abiotic factors. The most common biotic causes include parasitic and antagonistic fungi, bacteria, viruses and insect pests. The main abiotic factors are temperature, RH, CO_2 concentration, excess of moisture in compost and casing and the presence of toxic chemicals in the substrate or atmosphere. Disease problems of white button mushroom have been intensively studied abroad and suitable control measures are known in most cases (14, 16, 27, 28, 56, 60, 71). Three avenues of disease control are available :

i) Ecological-manipulation of environmental factors such as temperature, humidity and ventilation.

ii) Biological-use of antagonistic organisms through incorporation of biocontrol agents and organic amendments.

iii) Chemical.

A close relationship exists between management practices and some endemic disease problems like dry bubble, brown blotch and truffle.

General distribution of various pathogenic fungi and moulds is as follows :

A. Those occurring mainly in compost :

Olive green mould *(Chaetomium globosum & C. olivaceum);* Ink caps *(Coprinus* spp.); Green moulds *(Aspergillus); Penicillium* and *Trichoderma* spp.); Black moulds *(Mucor & Rhizopus* spp.)and others *(Myricoccum praecox, Sporotrichum* sp., *Sepedonium* sp., *Oedocephalum* sp., *Fusarium solani, F.* spp., *Cephalosporium roseum, Gliocladium* sp. and *Papulaspora byssina*).

B. Fungi occurring in compost and on casing soil :

White plaster mould *(Scopulariopsis fimicola)*; Brown plaster mould *(P. byssina)*; lip stick mould *(Geotrichum candidum, Sporondonema purpurescens;* False truffle *(Diehliomyces microsporus)* and green moulds.

C. Fungi occurring on and in the casing soil and on growing fruit bodies:

Cinnamon mould *(Peziza ostracoderma);* Wet bubble *(Mycogone perniciosa);* Dry bubble *(Verticillium fungicola* and its vars); Cobweb *(Dactylium dendroides & Cladobotryum variospermum);* Pink mould *(T. roseum);* Green moulds and others *(Fusarium* spp.).

D. Fungi attacking the fruit bodies only :

Fusarial spot *(F.* spp.); Green mould and Wet rots *(Verticillium & Cladobotryum* spp.).

E. Bacterial and viral infections.

The following diseases have been recorded/worked upon from India (34, 60):

3.1. Dry bubble *(Verticillium fungicola, f. fungicola* syn. *V. malthousei)*

It is also named as brown spot, fungus spot or Verticillium disease. A serious and wide spread disease, perhaps the most destructive and has been reported from almost all the mushroom growing countries including England (59, 68), U.S.A. (31), Canada & Switzerland (73) and India (52). It can cause epiphytotics under favourable weather conditions infecting most of the sporophores resulting in serious crop losses alongwith deterioration in quality. It was predominant in England from the late 1960's to 1980. It is quite common in Chail, Taradevi (H.P.) and Sonepat (Haryana) areas especially during the rainy season and in late flushes. Whitish mycelial growth initially is noticed on the casing soil which has a tendency to turn greyish yellow. It produces small deformed undifferentiated masses of tissues (bubbles) upto 2 cm in diameter, covered with a greyish conidial layer. Late infection leads to imperfectly formed partially differentiated caps with distorted shaggy swollen stipes showing downward peeling of stem tissue and tilted caps. In fully developed sporophores it produces localized light brown depressed spots. Adjacent spots coalesce and form irregular brown blotches like those of bacterial blotch but lighter in colour and some what sunken at the centre. Diseased caps shrink in blotched areas, turn leathery, dry and show cracking. Infected mushrooms are covered with a fine grey white mycelial growth with small pimple like out growth. Severely infected fruit bodies are malformed, onion shaped without a distinct stipe, ultimately becoming irregular swollen masses of dry leathery tissue. The fungus produces numerous one-called, thin walled, oblong to cylindrical, hyaline conidia, 3.5-15.9 X 1.5-5µ on lateral or terminal, verticillately branched conidiophores (200-800 X 1.5-5µ). Conidia accumulate in clusters surrounded by a sticky mucilage. The fungus is soil brone. Spores can survive in moist soil for one year (15). It also perpetuates through resting mycelium from dried bulbils and in spent compost. Primary infection mostly comes through contaminated casing soil and is also carried by workers, flies and mites from spent compost. A number of wild growing fleshy fungi are also susceptible and provide inoculum. Secondary spread is brought about by water splashes, flies and pickers. It spreads fast due to sticky spores. High humidity, lack of proper air circulation resulting in condensation of moisture on fruit body surface, delayed removal of diseased mushrooms from beds and temperature above 60°F (16°C) favour its development (29). All the commercial strains of *A. bisporus* are susceptible, whereas *A. bitorguis* is less sensitive. Use of sterilized casing soil, proper disposal of spent compost, control of insects/nematodes and better hygiene is essential to avoid primary infection (60).

Spraying with Dithane 2.78 at 0.2 to 0.5% (52), Parzate at 1 lb. in 100 gallons of water once before first break (73), Prochlorazmanganese complex or sporgon at 1.5g ai. per m^2 nine days after casing (47) has been found effective. Benzimidazole fungicides were very effective initially but the fungus developed resistance (72). Panacide M and chlorathalonil are effective against this and other diseases (31, 68). Topsin M and Daconil were found effective (6, 75). Two other varieties of *V. fungicola*, namely *aleophilum* and *Psalliotae* are reported to attack but do not cause distortion of fruit body. The former is more common on *A. bitorguis* and attacks. *A. bisporus* when the cropping temperature is higher. The latter variety has sickle shaped conidia.

The incidence of dry bubble has been greatly minimised at most of the farms with the use of pasteurized compost and casing soil, regulation of RH an temperature during cropping with good hygiene and proper disposal of spent compost.

3.2. Wet bubble *(Mycogone perniciosa)*

It is also known as bubble, white mould or Mycogone disease. A common and wide spread disease. It also attacks many wilt agaricus in Kandaghat and Chail forest areas of Himachal Pradesh (62). Reported from England (40), U.S.A. (2, 73), France (8) and India (60). It adversely affects yield and quality. Small fluffy white mycelial patches appear on the surface of casing soil. Infected mushrooms have malformed swollen stalks and smaller caps and the surface is covered with white mycelium. Early infection results in the development of distorted aggregations of mushroom tissue (Sclerodermoid masses 5cm or more) which are initially white and fluffy but turn brown and necrotic on ageing (59). Under humid conditions, amber to dark brown drops of liquid (ooze) accumulate on the surface of undifferentiated diseased tissues, whereas in dry weather these show dry rot. Diseased fruit bodies show wet rot with foetid odour. Brown discoloration alongwith whitish mycelial growth is noticed at the base of stipe of infected buttons. The fungus produces two type of spores, (a) conidia hyaline, one to two celled, thin walled, oblong, 12-20' 3-4µ, borne terminally on erect, septate, hyaline, verticillately branched conidiophores measuring 150-450 x 4-5μ, tapering at tips to 1-2μ, (b) chlamydospores or aleuriospores borne terminally either on mycelium on lower branches of conidiophores, 2-called, upper cell thick walled covered with minute warts or spines, spherical, hyaline to light brown, 14-20 x 16-23μ, lower cell thin walled, hyaline and cup or funnel shaped. These spores can survive in casing soil upto three years and help in perpetuation of the fungus (59). The contaminated casing soil results in early infection leading to serious yield loss or even crop failure. The spores are carried by sciarid flies, pickers, on tools and water splashes. All the commercial strains are susceptible.

Proper disinfection of casing soil (at 140°F for 2 hr or 122°F for 4 hr or 106°F for 6 hr) and avoidance of primary infection from spent compost through control of insects and good hygiene at all stages of crop growth give good control. Diseased mushrooms may be picked and destroyed before watering. Fumigation of casing soil with formaldehyde (1-2% at half a gallon per 10 lb of soil) or spraying with 0.8% on the casing surface immediately after putting has been found effective (25, 33). Spraying with sporgon (at 1.5g a i. per m^2) 3-9 days after casing (47) and drenching with Benomyl (at 7oz. in 25-50 galls. of water per 1000 sq. ft.) at first watering has been recommended (2). Cooking of spent compost after cropping is essential to kill infection.

3.3. Brown plaster *(Papulaspora byssina)*

It has been reported from U.S.A., U.K., Germany, Brazil, China, Rumania, Belgium, Australia and India (13, 36, 43). Infection results in yield loss and delayed cropping (71). Hine and coworkers reported an yield loss of 45% from Taiwan in 1964. Yield losses have also been recorded from Denmark (1), Italy (7) and New South Wales (8). It caused 90-92% reduction in yield of white button at Chail and Delhi, Complete crop failure of oyster mushroom has been reported from Kasauli (43).

Infection is very common at almost all the farms using unpasteurized compost. Incidence is more when the cultivation is done in polythene bags with wet compost, compactly filled in larger quantity. Incidence is even noticed at farms undertaking mushroom cultivation for the first time.

It is first noticed as whitish mycelial growth on the exposed surface of compost and casing soil in trays as well as on sides in bags due to moisture condensation. It is not found in compost. This develops further into large dense patches gradually changing colour through shades of tan, light brown to cinnamon brown; ultimately becoming rust coloured. The mycelium is brownish, septate; later produces small clusters of brown coloured many celled, spherical bulbils, 60-130 X 30 -190μ. These are interwoven with a net work of hyphae and are set free later with the death of the mycelium. Primary infection comes through air borne bulbils (36), or containers, compost and casing soil and workers. It's development is favoured by wet, soggy and wrongly prepared compost higher temperature during spawn run (28-30°C) and lower temperature during cropping favour its spread (69). Bulbils can survive inadequate composting and ineffective pasteurization (39, 69). Fig. (A) These are killed at 140°F (21). It can be controlled by proper pasteurization of compost and casing soil, maintenance of proper moisture in compost (70% during spawn run and 80% during cropping, required temperature during different stages of crop growth and proper hygiene to avoid primary infection. Spraying with Benomyl and Bavistin (0.05%) is effective against mycelial infection (31). Infected patches can be treated with formaldehyde (2%) spray (2). Use of wet compost be avoided and should be filled loosely especially in bags.

3.4. Mildew *(Dactylium dendroides* - the conidial stage of *Hypomyces rosellus)*

It is also known as soft decay and cobweb disease. It generally attacks later flushes but assumes serious proportions if the conditions are favourable for its development. Cosmopolitan in distribution. Reported from India by Seth and Dar (51). Irregular to circular whitish loose mycelial patches are noticed on the casing surface Fig. (B). These eventually turn brown or rosy red. Fruit bodies are attacked at all stages of growth and are covered with a dense cotton wool type of mycelial growth. This later turns reddish. Infected buttons die with a soft wet rot. Sterile hyphae form a tuft and are prostrate, branched, septate and hyaline with opposite erect branches which divide above into three pointed branchlets. Conidiophores are erect, simple or branched in many whorls. Conidia oblong, elongate, apiculate at base, 2-3 septate, slightly constructed at the septa, hyaline to yellowish, 20-30 x 10-12μ. Sexual stage has been frequently observed on decaying dried fruit bodies of wild mushrooms in Himachal Pradesh. Casing soil is the primary source of infection. It spreads further through air borne spores by insects, tools, workers and water splashes. Sudden drop in temperature and excess of moisture over casing surface favour disease build up. Ascospores also spread infection and help in perpetuation.

Disinfection of cropping rooms and trays before use with formaldehyde (15%) minimise chances of infection. Remove diseased mushrooms before watering. Benzimidazole sprays are very effective. Protective dusting with Zeieb and spot application of PCNB 20% or Calcium hypochlorite (75% powder) is also recommended (65). Use of properly sterilized casing is essential. Avoid excessive watering and sudden temperature drops (2).

3.5. White plaster mould *(Scopulariopsis fimicola)*

The infection is quite similar to brown plaster initially but remains white throughout. It was reported from India by Bhardwaj *et al.* (4). It inhibits spawn run and indicates over composting and high pH of compost. Mycelium white, septate, fasciculate; conidiophores short, commonly borne along the fasciculate hyphae, mostly branched and produce at apex a cluster of sporogenous cells, which elongate slightly before producing succeeding conidia. These are hyaline or sub-hyaline, 1-celled, globose-ovate, base truncate, produced in basipetal chains or clusters. Infection is favoured by wet and wrongly prepared compost and high temperature during spawn run. It spreads faster because of profuse sporulation. Sprays of Dithane Z-78 or Bavistin (0.05%) or formalin (2%) are effective.

3.6. Green mould *(Trichoderma* spp.)

It is also known *Trichoderma* spore or blotch or mildew. A number of species are reported to be involved. Green colonies with wide whitish border appear on casing soil and compost. It also attacks grain in spawn bottles. Produces dark brown or greenish lesions/spots/discolouration on pileus as well as stipe. It inhibits spawn run and fruiting and indicates uncomposted organic matter in compost with excess of carbohydrates. *T. konigii* grows as cottony weft of mycelium over the casing, also covers fruit bodies with a cottony web producing wet rot. Produces purple brown spots (not depressed) with a dry cracked surface. Conidia oblong to elliptic, smooth and hyaline. The loose sage green clusters of spore heads develop on affected tissues. Diseased caps turn brown on one side. Soil borne, therefore exclusion measures are essential. *T. viride* is associated with decaying organic mater in the casing soil. It produces large, dry brown, sunken lesions on cap and stem. Mushrooms are lopsided. Mycelium abundant with bright green spores. Chlamydospores are also produced. Inhibits the development of mushroom mycelium.

T. hamatum spores are dark green *T. harzianum* produces less spores which are pale green and smooth. Conidial areas are powdery. The infection generally comes through compost and air and is more serious with addition of unsterilized supplements. Dead mushroom stumps on beds and high humidity during cropping favour infection. All the species are highly sensitive to benomyl or bavistin (25-30 ppm) sprays.

3.7. False truffle *(Diehliomyces microsporus)*

It has been described as a parasite as well as a competitor mould. The infection was first noticed in USA in 1929 which was described in 1930 (7), followed by its occurrence in UK (70) and India in 1965 (63).

It is generally noticed in later flushes during April-May in the hilly areas and February - March in northern plains. Incidence is more at farms lacking proper temperature control alongwith faulty aeration during cropping. It is favoured by temperature beyond 22°C (66°F) with optimum growth at 28°C (83°F). Infection is noticed as small wefts of white to cream coloured mycelium in compost and casing soil, usually the layer where compost and casing soil meet and on the casing soil. Gradually the mycelial growth becomes thicker and develops into solid rounded to irregular, wrinkled, cream coloured lobed masses resembling small brains or peeled walnuts. These ascocarp vary in size (1 mm to 2.5 cm), become pink then reddish on maturity, disintegrate into a powdery mass emitting a characteristic chlorine like odour. The spawn in the affected areas degenerates into a soggy mass and disappears altogether without giving any yield. Ascocarp produce numerous, reddish brown, oval sub-spherical, short or long stalked asci. Ascorpores are sulphur coloured, spherical and with one distinct oil drop. These can survive for long periods in soil and spent compost and are the main source of primary infection. Lembert (40) reported that spores can withstand a temperature of 82°C for 6 hours. The mycelium in/or on wooden trays and in beds is not completely killed during cooking out and can cause infection in the subsequent crops. It can be prevented by treatment of trays with sodium pentachlorophenate. Gandy *et al.* (32) recommended the use of copper sulphate (1 lb per tonne of compost) towards the end of Phase-I but this has been contradicted by others. Stoller (64) reported that fumigation of empty houses with ammonia at a concentration of 0.15g per litre of air prevents carry over to the next crop. Methyl bromide treatment has also been tried (24). Maintenance of temperature of 72°F during spawn run and below 60°F during cropping alongwith proper aeration is essential to avoid infection. Strict hygiene is necessary against primary infection. Infected trays/bags may be removed and destroyed to avoid further spread. Prepare compost on a concrete platform to avoid infection from soil. Benomyl/bavistin (0.05%) sprays are effective against mycelial infection. Disease is best managed through temperature control.

8. Bacterial blotch *(Pseudomonas tolaasii)*

It is also called brown blotch and bacterial spot. Cosmopolitan in distribution and serious (35). Yellow to brown spots appear on the cap which can turn dark brown if the infection is severe. Sometimes slightly sunken pits filled with slimy contents are noticed. Brown streaks are also produced on stipes. In dry weather, the infected pin heads show radial cracking. It affects only the surface layers. The infected tissues become water soaked and greyish. Diseased mushrooms are usually greasy to touch. Temperature beyond 20°C (68°F), high relative humidity (more than 85%) and fluctuation in temperature during cropping resulting in condensation of moisture on fruit body surface favour its development. Pin heads attain normal shape, if the infection is light otherwise it affects yield and quality adversely. Infected mushrooms are not saleable and show faster determination in quality during transit and marketing. It assumes epidemic proportions very fast. Bacterium is omnipresent in soil and the primary infection comes through casing soil or compost (not pasteurized properly) and through contaminated tools/containers used in cultivation. It spreads further through water splashes, flies, nematodes and pickers. *A. bisgorus* and *A. bitorguis* are susceptible. The disease is best controlled by manipulation of crop environment. Proper circulation of air after watering is essential to dry the pinheads. Early and careful trashing is required alongwith good hygiene. Casing should be sterilized properly alongwith control of saprophagus nematodes. Avoid fluctuation of temperature during cropping. Spraying with chlorinated water containing 500 ppm of available chlorine (125 ml hypochlorite in 100 l. water) from the beginning of first flush once or twice a week gives good control.

Dreching of beds with stabilized chlorine dioxide has shown promise in controlling infection alongwith improvement in shelf life (42). Use of bacterial antagonistis of the pathogen has been recommended (45). Mummy diseases *(Pseudomonas* sp.) caused by an interacellular bacterium causes serious losses in certain stains of *A. bisporus,* whereas it has never been observed in *A. bitorguis.* Infection is generally observed in a few spots/patches on the beds. Mycelial growth as well as pin head development is slowed down in diseased patches. Diseased mushrooms grow in a distorted manner with croaked stipes surrounded at the base with fluffy mycelium. These soon become necrotic showing brownish discolouration. Infected mycelium does not die but produces successive flushes of diseased mushrooms. Affected patches may be treated with formaldehyde (2%) to avoid further spread.

3.9. Virus diseases

These are named as La France, Die-back, Watery stipe and X-disease and are one of the most serious and fast multiplying diseases of cultivated mushrooms. It also attacks. *A. campestris,* whereas *A. bitorguis* is resistant. These are reported from USA (5, 7), England (30), Netherlands (18), Australia (44) and India (67). Disease symptoms are very variable hence difficult to recognize. These include : sparse growth of mycelium from tissue isolates, suppression of development, rapid necrosis, or dwarfing and early maturity of sporophores (49). Drum stick formation (long stalks and small caps), premature opening of veil; bending, splitting and swelling of the stalk, bare patches on beds where mycelium does not grow into the casing soil or only slightly, producing mushrooms of poor quality and loose attachment on casing soil are the main symptoms (18, 19, 30). The spawn run before casing appears to be normal. Infected caps open early and spread infection through spores. Early infection before casing significantly reduces yield and can even lead to complete crop failure Virus particles of six different shapes and sizes have been observed in diseased fruit bodies. Five of these are polyhedreal or spherical (19, 25, 29, 35) and 50 nm dia) and sixth one bacilliform (19 x 50nm). Particles of 19 x 50, 25 and 35 nm are most commonly observed (19, 30). More than one type of particles have been observed in the same fruit body. Infection mostly comes through diseased cultures and spores from infected mushrooms. Diseased spores show quick germination. Virus infected mycelium can also survive in wood (of trays) and can transmit the disease to fresh mycelium of succeeding crop through anastomosis. Strict farm hygiene is the best means of prevention of spread. Pick only closed mushrooms to avoid dispersal of infected spores. Disinfect all trays, growing rooms, implements and machinery after each operation including the work areas and growing rooms. Sterlize trays after cropping for 12 hr at 160°F. Cook out spent compost (150°F for 24 hr or 160°F for 12 hr or fumigated with methyl bromide after taking crop (2, 24). However, higher spawning rate with shorter cropping period can help to reduce the severity of future attacks. Entry of diseased spores into spawn laboratory and cropping rooms can be restricted by providing filters.

4. WEED MOULDS

4.1. Ink caps ***(Coprinus atramantarius, C. logopus, C. comatus, C. fimetarius):*** These appear during spawn run and cropping and show profuse unrestricted whitish mycelial growth followed by sudden appearance of fruit bodies in large numbers. It indicates under composting and presence of ammonia in compost. Becomes serious with wet compost, inhibits spawn run and fruiting. Young fruit bodies of the mould may be picked to avoid its further spread.

4.2. Brown mould ***(Peziza ostracoderma):*** Numerous cup shaped, brown to dark brown fruit bodies appear on compost surface. High humidity and low ventilation in growing room appear to encourage its growth. Brown mould is also caused by *Oedocephalum* sp. as unrestricted growth in casing. It indicates residues of ammonia in compost.

4.3. Lip-stick mould ***(Geotrichum candidum)*****:** Produces red colonies with white border in compost or casing and inhibits spawn run. Infection comes through air.

4.4. Yellow mould ***(Sepedonium* sp.):** Produces small yellow colonies in compost and affects production. (Fig. C).

4.5. Mat disease ***(Chrysosporium luteum)*** *:* Small white colonies becoming yellow appear underneath casing.

Most of these weed moulds can be controlled through good hygiene, proper compost preparation followed by pasteurization and providing desired growing conditions. Spent compost needs to be disposed carefully away from the cropping site.

5. REFERENCES

1. Anon. 1936. Plant disease in Denmark in 1936. Tidaskr. Plant-eayl. **2** : 189-242.

2. Atkins, P,. and Atkins, F. C. 1971. Major diseases of the cultivated white mushroom. MGA. 1-23.

3. Beach, W. S. 1937. The *Verticillium* disease or Brown spot. Pennsylvania Agricultural Experiment Station Bulletin 351.

4. Bhardwaj, S. C., Seth, P. K. and Guleria, D. S. 1989. A note on the occurrence of white plaster disease of cultivated mushroom *(A. bisporus)* caused by *Scopulariopsis fimicola*-a new fungus from India. Curr. Sci. **58** : 320-321.

5. Bisset, P. G., Colhoun, J. and Gandy, D. G. 1982. Germination of *Diehlimyces microsporus* ascospores and determination of their TDP. Trans. Brit. Mycol. Soc. **78** : 540-542.

6. Bollen, G. J. and Zaayen, A. van 1975. Resistance to benzimidazole fungicide in pathogenic strains of *V. fungicola.* Nether. Jour. Plant Pathology, **81 (5)** : 157-167.

7. Borzini, G. 1942. On the fungi most injurious to our cultivation of edible mushrooms *(Psalliota compestris* and *P. arvensis).* Bull. staz. Pat. Veg. Roma NS, **3-4**: 131-172.

8. Brown, H. P. 1937. Mushroom bed invaders. Their identity and means of control. Agric. Gaz. N.S.W., **8** : 436-439.

9. Chang, S. T. 1980a. Food Policy, **5** : 64-65.

10. Chang, S. T. 1980b. Bioscience, **30** : 399-401.

11. Chang, S. T. 1991. Mushroom J. **504** : 15-17.

12. Chang, S. T. and Mikles, P. G. 1989. Recent trends in world production of cultivated edible mushrooms. The Mushroom J. **504** : 15-18.

13. Charles, V. K. and Lambert, B. 1933. Plaster mould occurring in mushroom green houses. Jour. Agric. Res. **12** : 1089-1093.

14. Conte, Ginseppe. 1985. Mushroom Information II. No. 1. p. 26.

15. Cross, M. J. and Jacobs, L. 1969. *Verticillium* and *Mycogone* diseases of the mushroom. MGA. Bull. **238**: 440-447.

16. Delmas, J. 1978. Protection against pests and diseases. In: The biology and cultivation of edible mushrooms, pp. 279-291.

17. Diehl, W. W. and Lambert, E. B. 1930. A new truffle in mushroom beds of cultivated mushrooms. Mycologia, **22** : 223-226.

18. Dieleman-van Zaalen, A. 1970. Means by which virus disease in cultivated mushrooms is spread. MGA. Bull. **244** : 158-178.

19. Dieleman-van Zaayen, A. 1979. Mushroom viruses. In : Viruses and Plasmids in Fungi. Ed. Paul a. Lemke, pp. 239-324.

20. Dieleman-Van Zaayen, A. and Temmink, J. H. M. 1968. Nether. Jour. Plant Pathology, **74** : 48-51.

21. Edwards, R. L. 1949. Science in the mushroom industry. The Journal of the Ministry of Agriculture, **9** : 434-442.

22. FAO. 1970. Amino acid contents of foods and biological data on proteins. Nutritional studies No. 24.

23. FAO. 1978. FAO Production Year Book. Vol. 32.

24. Flegg. P. B. 1968. Methyl bromide and mushrooms. MGA. Bull. **228** : 582.

25. Fletcher, J. T. and Ganney, G. W. 1968. Experiments on the biology and control of *Mycogone perniciosa.* Mushroom Science, **7** : 221-237.

26. Fletcher, J. T., Hims, M. J. and Hall, R. J. 1983. The control of bubble diseases and cobweb diseases of mushrooms with prochloraz. Plant Pathol. **32** : 123-131.

27. Fletcher, J. T., White, P. F. and Gaze, R. H. 1986. Mushroom Pests and Diseases. Intercept Ltd. England.

28. Gandy, D. G. 1954. Your pests and diseases. MGA, Bull. **58** : 327-328.

29. Gandy, D. G. 1972. Observations on the development of *V. malthousei* in mushroom crops and the role of cultural practices in its control. Mushroom Science, **8** : 171-108.

30. Gandy, D. G. and Hollings, M. 1961. Die-Back of mushrooms; a disease associated with a virus. GCRI. Annual Report 1961, 103-108.

31. Gandy, D. G. and Spencer, D. M. 1976. The use of chlorathalonil for the control of benzimidazole tolerant strains of *V. fungicola.* Scientia Horticulaurae, **5 (1)** : 13-21.

32. Gandy, D. G., Duncan, C. W. and Edwards, R. L. 1953. The use of copper sulphate for control of Truffle. Mushroom Science, **2** : 167-174.

33. Ganney, G. W. and Fletcher, J. T. 1968. Summary of experiments in the biology and control of the mushroom diseases *Mycogone perniciosa.* MGA, Bull. **224** : 413-418.

34. Garcha, H. S. 1984. A manual of mushroom growing. PAU. Publication, Ludhiana, 71 pp.

35. Gaze, R. H. and Fletcher, J. T. 1975. ADAS survey of mushroom diseases and fungicide usage 1974-75. Mushroom J. **24** : 468-470.

36. Hotson, J. W. 1912. Cultural studies of fungi producing bulbils and similar propagative bodies. Proc. Am. Acad. Arts and Science, **48** : 227-306.

37. Huang, C. H. 1979. Studies and experiments on mushroom culture in Taiwan. Mushroom Science, **10** (Part II) : 575-587.

38. Kneebone, R. L. 1958. Control of mushroom diseases by sanitation and cultural practices. MGA, Bull. **112** : 109-112.

39. Kneebone, R. L. and Merek, E. L. 1961. Brief outline of the controls of mushroom pathogens, indicator moulds and weed moulds or competitors. III Mushroom Industry short course, MRC. The Penn. State University, U.S.A. Miscel. Publ. 24 pp.

40. Lambert, E. B. 1930. Studies on the relation of temperature to the growth parasitism, thermal death points and control of *Mycogone perniciosa.* Phytopathology, **20** : 75-83.

41. Lambert, E. B. 1948. The control of Bubble. MGA, Bull. **12** : 128-130.

42. Markowitz, Norman I. and Bodine, David. 1989. Studies on the use of stabilized chlorine dioxide for improving mushroom quality and shelf life. Mushroom News, April 1989 : p. 16.

43. Munjal, R.L. and Seth, P. K. 1974. Brown plaster mould-a new disease in white button mushroom. Indian Hort. **18 (4)** : 13.

44. Nair, N. G. 1984. Diseases of cultivated mushrooms. Agfact H8. AB, 30. pp. 11.

45. Nair, N. G. and Fahy, P. C. 1972. Bacterial antagonistic to *Pseudomonas tolaasii* and the control of brown blotch of the cultivated mushroom, *A. bisporus.* J. app. Bact. **35** : 439-442.

46. Rai, R. D. and Sohi, H. S. 1988. How protein rich are mushrooms. Indian Hort. **33 (2)** : 2-3.

47. Russell, P. 1984. Sporgon on mushrooms. Mushroom Jour. No. **141** : 299-300.

48. Sarazin, A. 1950. La Mole Malldie du champignon de couches. Mushroom Science, **1** : 93-95.

49. Schisler, L. C., Sinden, J. W. and Siegel, E. M. 1967. Etiology, symptomatology and epidemiology of a virus disease of cultivated mushrooms. Phytopathology, **57** : 5, 519.

50. Seth, P. K. 1977. Pathogens and competitors of *A. bisporus* and their control. Indian J. Mushr. **3 (1)** : 31-38.

51. Seth, P. K. and Dar, G. M. 1989. Studies on *Cladobotryum dendroides* (Boll : Merat) W. Gams et Hoozem; causing cobweb disease of *Agaricus bisporus* (Lange) Sing. Mushroom Science, **12** : 711-723.

52. Seth, P. K., Kumar, S. and Shandilya, T. R. 1973. Combating dry bubble of mushrooms. Indian Hort. **18 (2)** : 17-18.

53. Sinden, J. W. 1936. Thermal death points of *Dactylium.* Rep. Pa. Agr. Exp. Sta. 1936-37, 3841.

54. Sinden, J. W. 1955. Disease control and sanitation programme for mushroom nurseries, MGA, Bull. **71** : 832-841.

55. Sinden, J. W. 1971. The diseases of cultivated mushroom. Mushroom classics. Am. Rev. Phytopath.

56. Sinden, J. W. 1985. Ecological control of pathogens and weed moulds in mushroom culture. Mushroom News. May : 10-31.

57. Sinden, J. W. and Hauser, E. 1950. Report on two new diseases. Mushroom Science, **1** : 96-100.

58. Sinden, J. W. and Yoder, J. B. 1949. Effect of copper ar certain dithiocarbamates on the control of *Verticillium* spot and bubble of mushrooms. Phytopathology, **39** : 1949.

59. Smith, F. E.W. 1934. Three diseases of cultivated mushrooms. Trans. Brit. Mycol. Soc. **10** : 81-97.

60. Sohi, H. S. 1986a. Diseases of edible mushrooms and their control NCMRT. Chambaghat, Solan. Ext. Bull. No. 2 : 12 pp.

61. Sohi, H. S. 1986b. Diseases of white button *(Agaricus bisporus)* mushroom in India and their control. Indian J. Mycol. Pl. Pathol. **18** : 10-18.

62. Sohi, H. S. 1992. Presidential address, IPS. Bhavnagar, Gujarat.

63. Sohi, H. S., Seth, P. K. and Kumar, S. 1965. Occurrence of truffle diseases *(Pseudobalsamia microspora* Diehl and Lambert) in mushroom beds. Curr. Sci. **31** : 488-489.

64. Stoller, B. B. 1943. The use of ammonia in controlling the truffle disease of the mushroom. Phytopathology, 33.

65. Stoller, B. B., West, R. E. and Bailey, J. F. 1956. Controlling the mildew disease of cultivated mushroom. MGA., Bull. **82** : 316-322.

66. Swaminathan, M. S., Austin, A., Kaul., A. K. and Naik, M. S. 1981. Proc. Intern. Atomic Energy Agency, Vienna, 91-96.

67. Tiwari, R. P. and Singh, S. J. 1984. Mushroom virus disease in India. The Mushroom Jour. **142** : 354-355.

68. Ware, W. M. 1933. A disease of cultivated mushroom caused by *Verticillium malthouisei* sp. nov. Ann. Bot. **47** : 763-782.

69. Ware, W. M. 1934. Plaster mould in mushroom beds. Gdnr's Chson. **93** : 463-465.

70. Williams, P. H. 1936. A disease of mushrooms new to Great Britain. The Gardener's chronicle, **2591** : 147.

71. Wuest, P. J. 1965. Diseases and their control. Mushroom Industry short course. The Penn. State University, June 28 - 30, 1-8.

72. Wuest, P. J., Cole (Jr.) H. and Sanders, P. L. 1974. Tolerance of *Verticillium malthousei* to benomyl. Phytopathology, **64** : 331-334.

73. Yoder, J. B., Sinden, J. W. and Houser, E. 1950. Experience with zinc ethylene bis dithiocarbamate as a fungicide in mushroom cultivation. Mushroom Science, **1** : 100-108.

74. Zaayen-van, A. 1976. Immunity of strains of *Agaricus bitorguis* to mushroom virus disease. Neth. J. Path. **82** : 121-131.

75. Zaayen-van, A. and Rutijens, A. J. 1978. The application of daconil for the control of bubble. Champignon culture, **22 (4)** : 121-123.

Fig A. *Papulaspora byssina* Brown plaster mould

Fig. B. *Dactylium dendroids* Mildew of mushrooms

Fig. C. *Sepedonium sp.* yellow mould of mushrooms

28 Plant Disease Management with Potassic Fertilizers

Vinayak M. Hegde, Bineeta Sen, Anil Kumar Patibanda and Krishanu Mukherjee

CONTENTS

Integrated Pest and Disease Management
Rajeev K. Upadhyay, K.G. Mukerji, B.P. Chamola and O.P. Dubey (eds.)
APH Publishing Corporation, New Delhi-110002, India

ABSTRACT

Work done on role of potassic fertilizers in the management of plant diseases, indicates that potassium reduces plant diseases in most cases. Some of the important diseases of several economic crops are known to be reduced by proper K nutrition. Mechanisms of disease reduction due to K is not vert clear. However, from the available literature, it can be concluded that K reduces the diseases by adversely affecting the pathogen, imparting structural and biochemical resistance to host plants and favouring pathogen antagonistic microorganisms in soil. Proper K nutrition increases the thickness of sclerenchymatous layer in rice, maize and wheat. Silica content increases in rice varieties supplied with potassium. Rice and muskmelon plants well supplied with K show higher phenol content and lower disease incidence. Potassium application increases population of total fungi, bacteria and actinomycetes in soil.

1. INTRODUCTION

Plant disease epidemics develop as a result of timely combination of susceptible hosts, virulent pathogen and favourable environmental conditions over a thirty long period of time, Nutrition influences all parts of this disease pyramid (52). Among nongenic conditions which may affect the susceptibility of plants to diseases, fertilizers are very important especially for soil borne plant diseases (106). So, nutrition manipulation of plants by application of inorganic fertilizers provides ample scope to modify the disease resistance. Knowledge of host nutrition in relation to disease development helps in modifying cultural weapons against plant pathogens and development of suitable disease management strategies.

Beneficial effect of potassic fertilizers in reducing plant disease was recognised by the end of last century (68, 99, 141). High levels of potassium is known to reduce the severity of more than 100 fungal diseases , more than 20 bacterial diseases and 10 diseases of each caused by viruses and nematodes (15). A large volume of literature is available on disease control effects provided by potassic fertilizers some of which have been review in recent years (72, 73, 74, 128, 129, 155). According to International Potash Institute's (IPI) research review on " Potassium and plant health" potassium has a beneficial effect in seven out of 10 cases reported. However no single generalisation on the effect of potassium in diseases can be made (128). Effect of K on some of the crop plant diseases have been listed in Tables 1, 2 and 3.

Potassium nutrition not only reduces the disease incidence, but also increases efficacy of pesticides. Application of N, P, and K or K alone increased efficacy of captafol seed treatment in reducing root rot of mungbean (*Vigna radiata*) caused by *Macrophomina phaseolina* (78).

2. MECHANISMS OF DISEASES REDUCTION WITH POTASSIUM

Apart from academic interest, knowledge about the mechanism of disease reduction by potassium helps in efficient utilization of fertilizers, especially in modern agriculture where application of inorganic fertilizers has become an integral part of crop production. Although, potassium is known to reduce many diseases the mechanism involved therein is not well understood in most cases.

Beneficial effect of potassium in reducing certain plant diseases may be due to :-

(i) It's direct effects on the pathogen,

Table 1. Plant diseases ameliorated by application of potassic fertilizers

Host	Diseases	Pathogen	Reference
Apple	Scab	*Venturia inequalis*	84, 101
	Dry eye rot	*Botrytis cinerea*	96
Apricot	Brown rot	*Sclerotinia frusticola*	187
Avocado	Root rot	*Phytophthora cinnamomi*	20
Banana	Fusarium wilt	*Fusarium oxysporum* f. sp. *cubense*	142
Barley	Net blotch	*Helminthosporium teres*	161
	Powdery mildew	*Erysiphe graminis*	181
	Common root rot	*H. sativum*	59, 177
	BYD	Barley yellow	
		Dwarf virus	83
Beans	Anthracnose	*Colletotrichum lindemuthianum*	47
	Mosaic	TMV	123
Beets	Damping off	*Pythium ultimum*	193
Betel vine	Wilt	*Phytophthora palmivora*	134
Cabbage	Gray mold	*Botrytis cinerea*	131
	Yellows	*Fusarium oxysporum* f. sp. *conglutinans*	50, 188
	Soft rot	*Erwinia carotovora*	184
Carnation	Wilt	*Fusarium spp.*	58, 97
Celery	Yellows	*F. oxysporum* f. sp. *apii*	152
Chilli (*Capsicum frutescens*)	Mosaic	Virus	112
Chilli (Pepper)	Wilt	*F. oxysporum* f. sp., *redolens*	150
Chrysanthemum	Root rot	*Phoma chrysanthemicola*	105
Clover	Wilt and Root rot	*Fusarium* spp.	33
Cotton	Root rot	*Phymatotrichum omnivorum*	103, 183
	Wilt	*Verticillium albo-atrum*	65, 176
	Wilt	*Fusarium oxysporum* f. sp. *vasinfectum*	32, 45, 145, 148, 156, 178, 194
	Seedling blight	*Rhizoctonia solani*	138
	Leaf blight	*Cercospora gossypina*	107
	Leaf blight		

Table 1. (Contd.)

Host	Diseases	Pathogen	Reference
	Angular leaf spot	*Alternaria solani*	107
	nematode disease	*Xanthomonas compestris* pv. *malvacearum*	143
Coconut	Leaf spot	*Rotylenchus reniformis*	121
		Pestalotia palmarum	118
Coffee	Rust	*Dreschlera incurvata*	53
	Brown spot	*Hemileia vastatrix*	14
Cocoa	Black pod	*Cercospora coffeecola*	79
Cumin	Wilt	*Phytophthora palmivora*	6
Flax	Wilt	*Fusarium oxysporum* f. sp. *cumini*	28
	Rust	*Fusarium lini*	41
Grape	Fruit rot	*Melampsora lini*	157
	Downy mildew	*Botrytis cinerea*	94
Groundnut	Rust	*Plasmopara viticola*	37
Jute	Anthracnose	*Puccinia arachidis*	113
	Root rot	*Colletotrichum corchorum*	132
Lentil	Root rot	*Rhizoctonia solani*	55
		Rhizoctonia solani +	135
Lucerne	Root rot	*Sclerotium solfsii*	
Maize	Northern leaf blight	*Phytophthora megasperma*	12
	Stalk rot	*Exserohilum turcicum*	50, 114, 173
		Fusarium spp.	2, 120
		Gibberellea zeae	50, 75, 76
	Stem rot	*Diplodia zeae*	98
	Stewart wilt	*Fusarium culmorum*	159
Muskmelon	Wilt	*Erwinia stewartii*	104
		Fusrium oxysporum	31, 137, 163, 165, 192
Onion	Basal rot		
Oil palm	Wilt	*Fusarium oxysporum*	16
Pea	Root rot	*F. oxysporum* f. sp. *elaeidis*	119

Table 1. (Contd.)

Host	Diseases	Pathogen	Reference
		Aphanomyces euteiches	
		Pythium vexans	
	Wilt	*Rhizoctonia solani*	124, 164
	Pod rot	*F. oxysporum* f. sp. *pisi*	153
Pine	Rust	*Rhizoctonia solani*	51
	Root rot	*Peridermium* spp.	77
Pineapple	Root rot	*Macrophomina phaseolina*	144
Potato	Late blight	*Phytophthora cinnamomi*	13
		Phytophthora infestans	11, 27, 166
	Canker		
	Mosaic	*Rhizoctonia solani*	85
	Leaf roll	Potato mosaic virus	85
Pumpkin	Fruit rot	Potato leaf roll virus	133
Rice	Blast	*Sclerotinia sclerotiorum*	3, 4, 102
	Brown leaf	*Pyricularia oryzae*	109, 116
		Helminthosporium oryzae	1, 7, 70, 117
	Sheath blight		
		Rhizoctonia solani	82, 90, 111
	Sheath rot		
	Stem rot	*Sarocladium oryzae*	8
		Helminthosporium sigmoideum	82
	Bacterial blight	*Sclerotium oryzae*	86
	Stem rot	*Xanthomonas campestris* p.v. *oryzae*	108, 125
		Helminthosporium sigmoideum	*82*
	Bacterial blight	*Sclerotium oryzae*	86
Snapdragon	Wilt	*Xanthomonas campestris* p. v. *oryzae*	108, 125
Soybean	Pod rot	*Verticillium dahliae*	48
	Purple seed stain	*Diaporthe sojae*	88
	Seed rot	*Cercospora kikuchii*	26
Squash	Foot rot	*Phomopsis* sp.	160
		Fusarium solani f. sp. *cucurbitae*	62

Table 1. (Contd.)

Host	Diseases	Pathogen	Reference
Sugarcane	Eye spot	*Helminthosporium sacchari*	100
Sunhemp	Root rots	*Rhizoctonia solani, Pythium butleri, Sclerotium rolfsii*	122
Sunflower	*Sclerotinia* disease	*Sclerotinia sclerotiorum*	4
Tobacco	Leaf blight	*Alternaria* spp. *Cercospora* spp. *Sclerotinia* spp.	95
	Mosaic	TMV	10, 162, 186
Tomato	Wilt	*F. oxysporum* f. sp. *lycopersici*	50, 103, 188
	Leaf spot	*Alternaria solani*	19, 94
Turfgrass	Fusarium pactch	*Fusarium nivale*	61
Wheat	Leaf blotch	*Septoria tritici*	56, 171
	Glume blotch	*S. nodorum*	39
	Take-all	*Gaumannomyces graminis*	38, 67, 140, 170
	Powdery mildew	*Erysiphe graminis*	24, 172
	Stem rust	*Puccinia graminis tritici*	5, 94
	Leaf rust	*Puccinia recondita*	182
	Stripe rust	*P. striiformis*	146
	Bunt	*Tilletia* spp.	182
	Common root rot	*Helminthosporium sativum*	60

(ii) it may alter the soil microflora which inturn may influence the survival of plant pathogens,

(iii) it may impart structural and biochemical resistance to plant tissues.

2.1. Direct effect of potassium on pathogens

Wensley and McKeen reported reduction in population of muskmelon wilt fungus *F. oxysporum* f. sp. *melonis* in calcium and potassium amended soil (192). Mycelial formation of *F. nivale* was reduced in rye grass due to application of potassium in soil (115). Potassium is also known to have inhibiting effect on spore germination and fusaric acid production by *F. oxysporum* f. sp. *melonis* (73). Higher levels of K application is also known to reduce the population of *F. oxysporum* f. sp. melionis in near rhizosphere soil of muskmelon (136). Hegde and Sen found 50% reduction in the population of muskmelon wilt pathogen 10 weeks after sowing in soil amended with 110 kg K_2O/ ha (unpublished). Habibullah and Prasad (64) reported the inhibitory effect of potassium on growth and production of pectinolytic and cellulolytic enzymes of *F. oxyspourm* f. sp. *melonis* (64). Survival

Table 2. Plant diseases aggravated by application of potassic fertilizers

Host	Diseases	Pathogen	Reference
Apple	Fire blight	*Erwinia amylovora*	30
Apple	Gloesporium fruit rot	*Pezicula melicorticis*	149, 154
Aster	Wilt	*Phialophora asteris*	25
Bean	Root rot	*Rhizoctonia solani*	18
Cabbage	Club root	*Plasmodiophora brassicae*	49, 103, 190
Cabbage	Downy mildew	*Peronospora parasitica*	35
Cucurbits	Root knot	*Meloidogyne incognita*	80
Cucumber	Bacterial wilt	*Erwinia tracheiphila*	191
Cedar red	Blight	*Phomopsis juniperovora*	127
Citrus	Gummosis	*Phytophthora parasitica*	29
Conifers	Root rot	*Armillaria mellea*	158
Cowpea	Damping off	*Rhizoctonia solani*	92
Cotton	Seedling rot	*Rhizoctonia solani*	22
Melon	Gummy stem blight	*Micosphaerela melonis*	168
Oat	BYD	Barley yellow dwarf virus	174
Onion	Purple blotch	*Alternaria porri*	17
Oil palm	"Boyomi"	*Fusarium bulbigenum*	69
Pea	Leaf roll	Pea leaf roll virus	174
Peanut	Pod rot	*Rhizoctonia solani*	66
Pine	Rust	*Cronartium fusiforme*	144
Potato	Late blight	*Phytophthora infestans*	130
Potato	Scab	*Streptomyces scabies*	46
Tomato	Root knot	*Meloidogyne incognita*	81
Watermelon	Wilt	*Fusarium oxysporum* f. sp. *niveum*	191
Wheat	Root rot	*Fusarium* spp.	54

and intensity of rice sheath blight pathogen *Rhizoctonia solani* and rice stem rot pathogen *Sclerotium oryzae* in rhizosphere and non-rhizosphere soil was reduced by K and NPK fertilizers (86, 91). High levels of N and K decreased the mycelial growth of *Helminthosporium sativum* , suppressed both germination and germtube elongation and reduced the phytotoxic effect of the fungus culture filtrate (151).

Table 3. Plant disease not influenced by application of potassic fertilizers

Host	Diseases	Pathogen	Reference
Broad bean	Charcoal spot	*Botrytis fabae*	44
Barley cowpea	Powdery mildew	*Erysiphe graminis*	71, 110
Bean	Rust	*Uromyces phaseoli*	139
Cotton	Wilt	*F. oxysporum* f. sp. *vasiinfectum*	9
Corn	Root rot	*Gibberella zeae*	126
Cotton	Root rot	*Meloidogyne incognita*	121
Maize	Stalk rot	*Diplodia zeae*	98
Oat	BYD	Barley yellow dwarf virus	36
Pearlmillet	Downy mildew	*Sclerospora graminicola*	43
Rice	Blast	*Pyricularia oryzae*	1
Sorghum	Sugary disease	*Sphacelia sorghi*	63
Soybean	Pod rot	*Diaporthe sojae*	167
Wheat	Common root rot	*Helminthosporium sativum*	56
Wheat	Leaf rust	*Puccinia recondita*	40

2.2. Effect of potassium on host plant in relation to disease development

Use of host plant resistance is a practicable method to control majority of plant diseases. Resistance depends upon both genetical and environmental factors. Among the environmental factors nutrition plays an important role in determining plant health. As potassium is involved in various metabolic functions in plants, it provides ample opportunity to modify host-resistance or susceptibility.

A sufficient amount of potassium tends to harden plant structures including stronger cuticle, epidermis, sclerenchyma tissue, and stimulates lignification and silicification of tissues (129). Potassium supply is known to stimulate natural cork formation in carrot under proper mineral nutrition (147). Reduced stem diameters and poorly lignified structural elements make K-deficient maize plants susceptible to lodging (179). An increase in thickness of sclerenchymatous tissue layer by improved K-nutrition is reported inter alia for maize (23), wheat (179) and rice (185). Adequate Kenhances lignification and resists the penetration of cell wall by the pathogen (57). An increase in K supply increased the silica content in culms of two rice varieties kannagi IR-21 and increased the thickness of sclerenchymatous layer (160). In K deficient plants, stomata remain open for longer time which may facilitate infection by pathogens (185).

The major function of potassium in plants is activation of a number of enzymes (60) involved in photosynthesis and metabolism of carbohydrates and protein. It also assists in membrane permeability, stomatal regulation and water utilization (169). Accumulation of inhibitory amino acids, phytoalexins, phenols and auxins around infection sites depends on the level of K and other minerals present (94). Resistance to *Phytophthora* in potato is associated with K-induced accumulation of fungistatic levels of arginine in leaves (11). As inorganic N accumulates under low K conditions, phenols that have fungicidal properties are rapidly broken down. This decreased phenol content of potato, rice and wheat tissues is associated with increased susceptibility to *Verticillium* wilt, bacterial blight, rusts and mildews (94). Potassium nutrition increased both total and ortho dihydroxphenols in both healthy and inoculated susceptible cultivators of muskmelon, with a concomitant reduction in wilt-incidence (137). Leaves of K deficient niger (*Guizotia abyssinica*) plants showed enhanced activities of phenyl phosphatase, ribonuclease and protease as well as increased concentration of sugars and phenols (21). Resistance of some plants has been correlated with higher levels of K than observed in tissues of susceptible cultivar. A wilt resistant flax takes up and contains more K^+ than a susceptible variety (41). Hegde and Sen (unpublished) also found higher K^+ content in muskmelon plants supplied with potassium (110 kg K_2O/ha), with concomitant reduction in wilt incidence. Increased susceptibility of potato tubers to *Phytophthora infestans* from plants supplied with N and P, but deficient in K is associated with effects of N/K ratio on N components regulated by K (166).

Nutritionally induced changes in host composition are manifested in plant exudates. High levels of K in leachates from a susceptible juniper (red cedar) but not from a resistant juniper (Chinese juniper) stimulate germination of *Phomopsis juniperovora* spores (127). Cotton cultivars exuding more carbohydrate and K and less Mg and Ca have less seedling diseases (183).

Potassium plays an important role in water relations of plants. Susceptibility of winter wheat plant to *Gaumannomyces graminis tritici* colonisation may be reduced by chemical potential of the water in plant (34). Since osmotic potential in wheat plants changes readily with chlorine (Cl^-) salts, fertilization with chlorine salts like KCl or NaCl provides an opportunity to mange plants water potential (34). In their experiment on effect of K on water relations of muskmelon wilt. Hegde and Sen (unpubished) found that, higher levels of potassium, increases the relative water content and total leaf water potential of *F. oxysporum* f. sp. *melonis* inoculated plants of muskmelon.

2.3. Effect of potassium on rhizosphere microflora

An increase in soil microbial activity may alter the pathogens ability to cause disease by competing for nutrients, space and oxygen or by increasing the changes of parasitisation of soil borne pathogens or by stimulating the antibiotic producing soil microorganisms. Foliar spray with muriate of potash, increased the fungal, bacterial and actinomycete population in the rhizosphere of muskmelon and reduced the population of wilt pathogen *F. oxysporum* f. sp. *melonis* (89). Soils amended with NPK fertilizers reduced fungal but increased actinomycetes and bacterial population. Species of *Penicillium, Trichoderma* and members of *Mucorales* were the dominant fungi present in treated soil (42). Increased doses of potassium was also found to increase the population of total fungi, bacteria and actinomycetes in rhizosphere soil of muskmelon plants.

3. CONCLUSION

Lot of literature is available on the effect of potassium on plant diseases. There is no doubt from the literature that severity of most of the diseases can be greatly reduced by applying proper dose of potassium, cultural, chemical, genetical and biological control methods of many plant pathogens can be enhanced by integrating with K-fertilizer (31).

Although, a few reports exists, on the mechanism of tolerance / resistance due to potassium, systematic investigations are lacking, especially the role of K in the physiology of healthy plants need to be compared in context to disease resistance.

4. REFERENCE

1. Abdel-Hak, T. M., Sirry, A. R., Kamel, S. M. and Ashour, W. A. 1973. Effect of different fertilizers on the incidence of blast and brown spot diseases of rice. In: U. A. R. Agric. Res. Rev. Cairo, **51 (3)**: 45-62.

2. Abdel-Rahim, M. F., Shata, H. M., El-Fahl, A. M., E-Assiuty, E. M. and Gouda, A. A. 1984. Cultural practices in relation to control of maize stalk rot complex. Agril. Res. Rev. **62 (2)**: 63-74.

3. Abia, J. A. and Smith, B. N. 1980. Mineral nutritional status of pumpkin and infection by *Sclerotinia sclerotiorum.* Plant Physiol. **65** (Suppl.6): 106.

4. Abia, J. A. and Smith, B. N. 1984. Mineral nutrient deficienceies and susceptibility of cotton to *Verticillum albo-atrum* and *Fusarium oxysporum* f. sp. *vasinfectum.* Phytopathol. **57** : 457.

5. Agarwal, S. C., Singh, S. P. and Gupta, K. P. 1968. Role of potash application in reducing black rusts of wheat. Indian J. Agron. **13** : 99-102.

6. Ahenkorah, Y., B. J., Appiah, M. R., Akrofi, G. S. and Yirenskyi, J. E. K. 1987. Twenty year's result from a shade and fertilizer trial on Amazon. Cocoa (*Theobroma cacao*) in Ghana. Expt. Agric. **23 (1)**: 31-39.

7. Akai, S. 1982. Application of potash and occurrence of *Helminthosporium* leaf spot on rice. Potash Rev. Subject 23, 27th issue.

8. Alagarasamy, G. and Bhaskaran. 1986. Effect of foliar spray of macro- and micro nutrients on sheath rot (Sh R) incidence and rice grain yield. Internatl. Rice Res. Newslett. **11 (4)**: 27-28.

9. Albert W. B., Keller, W. B. and Armstrong, G. M. 1941. Mineral nutrition and wilt resistance in a susceptible and wilt resistance cotton variety. South Carolina Agric . Exp. Station Annual Report, **54** : 62-65.

10. Allington, W. B. and Laird, E. F. 1954. The infection of *Nicotiana glutinosa* with tobacco mosaic virus as affected by potassium nutrition. Phytopathology, **44** : 297-299.

11. Alten, F. and Orth, H. 1941. Untersuchungen uber den Aminosauregehalt und die Anfalligkeit der kartofel gegen die kraut-and Knollen faule (*Phytophthora infestans* de By.) Phtopathologishche, **13** : 243-271.

12. Alva, A. K., Lanyon, L. E. and Leath, K. T. 1985. Influence of P and K fertilization on *Phytophthora* root rot or excess soil water injury of alfalfa cultivars. Communications in soil Sci. Pl. Analy. **16 (2)**: 299-243.

13. Anderson, E. J. 1951. The *Phytophthora cinnamomi* problem in pine apple fields of Hawaii. Phytopathology (Abst.), **41** : 1-2.

14. Anonymous, 1983. Potash and folair fungicides team up to increase yields. Better crops. 22-23.

15. Anonymous, 1988. Potash : Its need and use in modern Agriculture. Potash and Phosphate Inst. Canada.

16. Ashour, W. A., Elewa, I. S., Ali, A. A. and Dabash, T. 1980. The role of some systemic and non systemic fungicides and fertilization on the enzyme activity and control of *F. oxysporum* f. sp. *cepae* the cause of basal rot disease of onion. Agric. Res. Rev. **58 (2)**: 145-160.

17. Awad, M. A. Z., El-Shenawy, A. F., Orman and Shatla, M. N. 1978. Cultural practices in relation to purple blotch disease of onion. Sci. Hortic. **9** : 237-243.

18. Bateman, D. F. and Lumsden, R. D. 1965. Relation of calcium content and pectic substance in bean hypocotyles of different ages to susceptibility to an isolate of *Rhizoctonia solani.* Phytopathology, **54** : 887.

19. Bedi, P. S. and Dhiman, J. S. 1983. Influence of nitrogen, phosphorus and potassium on the development of early blight of tomato. Indian Phytopathology, **36 (3)**: 546-548.

20. Bingham, F. T., Zentmyer, G. A. and Martin, J. P. 1958. Host nutrition in relation to *Phytopathora* root rot of avocado seedlings. Phytopathol. **48** : 144-148.

21. Bist, S. S., Kalpana Chaturvedi and Alka Tripathi. 1989. Effect of varying levels of potassium supply on some biochemical constitutes and hydrolytic enzymes in niger (*Guizotia abyssinica*). J. Pot. Res . **5 (1)**: 35-41.

22. Blair, W. C. and Curl, E. A. 1974. Influence of fertilizer regimes on the rhizosphere microflora and *Rhizoctonia* diseases of cotton seedlings, Proc. Amer. Phytopath. Soc. **1** : 28-29.

23. Boeiriu, J. 1970. The role of potassium in increasing maize production on acids soils. An. Inst. Cercet. Cereale plane Teh Fundulea Acad stunte Agric Silvice Ser. B. **36** : 307-314.

24. Boquet, D. J. and Johnson, C. G. 1987. Fertilizer effects fields, grain composition an foliar disease of double crop soft red winter wheat. Agron. J. **79** : 135-141.

25. Burge, M. N. and Issac, I. 1977. Predisposition of Aster to *Phialophora* wilt Ann. Appl. Biol. **86** :353-358.

26. Camper, H. M., and Lutz, J. A. 1977. Plowsole placement of fertilizer for soybeans and response to tillage of plowsole. Agron. J. **69** : 701-704.

27. Carnegie, S. F. and Colhoun, J. 1983. Effects of plant nutrition on susceptibility of potato leaves to *Phytophthora infestans.* Phytopathologische Zeritschrift, **108**, (3): 342-350.

28. Champawat, R. S. and Pathak, V. N. 1988. Role of nitrogen, phosphorus and potassium fertilizers and organic amendments in cumin (*Cuminum cyminum*) wilt incited by *Fusarium oxysporum* f. sp. cumini. Indian J. agric. Sci.

29. Champman, H. D. and Brown, S. M. 1942. Some fungal infections of citrus in relation to nutrition. Soil. Sci. **54** : 303-312.

30. Chase, F. E., Corke, C. T. and Robinson. J. B. 1968. Nitrifying bacteria in soil, p. 593-611. In: Ecology of soil bacteria, Eds. T.R.G. Grug and D. Parkinson, Toronto Univ. Press, Toronto.

31. Chattopadhyay, C. and Sen, B. 1993. Potassium in integrated management of muskmelons wilt caused by *Fusarium oxysporum melonis* Schlecht.

32. Chester, K. S. 1947. The nature and prevention of plant disease, 2nd ed. Blackston, Philadelphia.

33. Chi, C. C. and Hanson, E. W. 1962. Nutrition in relation to root and top growth of *Trifolium pratense* and severity of *Fusarium* root rot. Phytopathol. **52** : 728.

34. Christensen, N. W., Taylor, R. G., Jackson, T. L. and Mitchell, B. L. 1981. Chloride (Cl^-) effects on water potentials and yields of winter wheat infected with take-all root rot. Agron. J. **73** : 1053-1058.

35. Chupp, G. 1930. The effect of potash and phosphorus on tip burn and mildew of cabbage. Phytopathology, **20** : 307-318.

36. Comeau, A. and Barnett, G. 1979. Effect of barley yellow dwarf virus on N, P, K fertilizer efficiency and harvest index of oats. Can. J. Plant. Sci. **59** : 43-54.

37. Condei, G., Lepadatu, V., Uglea, C. and Gliick, P. 1980. Influence de la fertilisation sur la qualtie' de al production et al resistance de al uigne. Compte rendu du coll. francorounmain sur les relations "Sol-Vigne" Bordeaux/France, 209-223.

38. Cooke, G. W. 1979. Contribution to the discussion on the effects of fertilizers on the infection of roots of crops by fungi. In: Fertilizer use and plant health. Proc. Colloq. Int. Potash Inst. **12** : 183-185.

39. Cunfer, B. M., Touchton, J. T. and Johnson, J. W. 1980. Effects of phosphorus and potassium fertilization on a *Septoria* glume blotch of wheat. Phytopathology, **70** : 1196-1199.

40. Das, N. R. and Sen, S. 1971. Incidence of brown rust in transplanted wheat as affected by N, P and K fertilizers and crop rotations, Ind. J. Agric. Sci. **41 (10)**: 872-875.

41. Dastur, R. H. and Bhat, J. G. 1964. Relation of potassium to Fusarium wilt of flax, Nature (London). **201** : 1243-1244.

42. Deb, R. 1990. Effect of chemical fertilizers on soil microorganisms. Acta Botanica Indica, **18 (2)**: 177-183.

43. Deshmukh, S. S., Mayee, C. D. and Kulkarni, B. S. 1978. Reduction of downy mildew of pearl millet with fertilizer management. Phytopathology, **68** : 1350-1353.

44. Deverall, B. J. and Wood, R. K. S. 1961. Infection of bean plants (*Vicia faba* L..) with *Botrytis cinerea* and *B. fabae.* Ann. Appl. Biol. **49** : 461-472.

45. Dick, J. B. and Tisdalle, H. B. 1983. Fertilizers in relation to incidence of wilt as affecting a resistant and a susceptible variety. Phytopathol. **28 (9)**: 666-667.

46. Doyel, J. J. and Maclean, A. A. 1960. Relationships between Ca: K ratio, pH and prevalence of potato scab. Can J. Plant. Sci. **40** : 616-619.

47. Drobney, H. G., Haffmann, G. M. and Amberger, A. 1984. Einfluss der mineralsoffversorgung auf die Predisposition von *Phaseolus vulgaris* gegan *Collabotrichum lindemuthianum* (Sacc. et. Magn.) Br. et. Cv. 1. Wirkung Verschiedener Nahrstoffe und. des. Blattatters auf die symptomanuspragung. Z. Pflanzenernaech. Blderk. **147(2)**: 242-254.

48. Dutta, B. K. and Isaac, I. 1979. Effect of inorganic amendments (N, P and K) to soil on the rhizosphere microflora of *Antirhinum* plants infected with *Verticilium dahliae* Kleb. Plant and Soil, **52** : 561-569.

49. Eddins, A. H. 1952. Diseases, deficiencies and injuries of cabbage and other crucifers in Florida. Bull. Fla. Agric. Exp. Stn. **492** : 5-63.

50. Ellett, C. W. 1973. Soil fertility and disease development. Better crops. Plant food, **57 (3)** : 6-8.

51. El-Wakil, A. A., El-Deeb, A. A., Shalaby, H. A. and Hilal, A. A. 1984. Pods rots of peanut (*Arachis hypogaea* L.) and its control. Agril. Res. Rev. **62 (2)**: 201-209.

52. Engelhard, A. W. 1989. Soil-borne plant pathogens: Management of diseases with macro and micro elements. APS Press, St. Paul. Minnesota. USA. pp. 237.

53. Fagan, H. J. 1985. Effect of nitrogen and potassium on severity of *Dreachslera* leaf spot and growth of coconut seedlings in sand culture. Oleagineux, **40 (5)**: 245-256.

54. Fehrmann, H. and Duben, J. 1980. Vorkommen and pathogenitat von *Fusarium* -Arien an Winterwezien in Budesepublic Deutchland. IV. Abhangigkeit des *Fusarium* -Befulls von Standart eigenschaften und anderen factoren. Z. pflanzenkr. Pflanzensutz. **87** : 281-289.

55. Finlow, R. S. 1918. Rhizoctonia in juite: The inhibiting effect of potash manuring. Agric. J. India, **13** : 65-72.

56. Fixen, P. E., Buchenau, G. W., Gelderman, R. H., Schemacher, T. E., Germwing, J. R., Cholick, F. A. and Farber, B. G. 1986. Influence of soil and applied chloride on several wheat parameters. Agron. J. **78(4)**: 736-740.

57. Friend, J. 1976. Lignification in infected tissue. In: Biochemical aspects of plant parasite relationship, Acad. Press, London.

58. Gasiorkiewicz, E. W. 1960. Influence of nitrogen and potassium levels on the development of *Fusarium* systemic wilt of carnations. Phytopathol. **50** : 636.

59. Goose, R. J., Johson, B. E. and Holmes, B. M. 1987. Effect of potassium chloride fertilization on two barley cultivars differing in common root rot reaction. Can J. Pl. Sci. **67 (2)**: 395-401.

60. Goose, R. L. 1986. Effects of KCl fertilization on small grains in North Dakota. Chloride and crop production. PPI Special Bulletin No. 2, 52-61.

61. Goss, R. L. and Gould, C. J. 1966. Plant nutrients help control turf grass diseases, Better crops plant food, **50 (3)**: 10-13.

62. Gries, G. A. 1946. Physiology of *Fusarium* foot rot of squash, Connecticut Agric. Exp. Sta. Bull. 500.

63. Govil, J. N. and Krishnan, K. S. 1979. Role of potash in controlling diseases in millets. Indian Pot. J. **4 (3)**: 16-20.

64. Habibullha, V. M. and Prasad, N. N. 1976. Effect of potassium on growth and production of pectinolytic and cellulolytic enzymes of *F. oxysporum* f. sp. *melonis*. Indain J. expt. Biol. **14** : 733-734.

65. Hafez, A. A. R., Stout, P. R. and DeVavy, J. E. 1975. Potassium uptake by cotton in relation to Verticillium wilt Agron. J. **67**: 359-361.

66. Hallock, D. L. and Garren, K. H. 1968. Pod breakdown, yield and grade of Virginia type peanuts as affected by Ca, Mg and K sulfates. Agron. J. **60** : 253-257.

67. Halsey, M. and Powelson, R. 1981. The influence of NH_4Cl with KCl on antagonism between fluorescent pseudomonades (FP) and *Gaeumannomyces graminis* var. tritici (GGT) on the wheat rhizosplane. Phytopathology, **71** : 105.

68. Hecke, L. 1898. Untersuchungen uber *Phytophthora infestans* de Bary. als Ursache der Kartoffelkrasnkehjis. J. Landwirtch. **46**: 71-74, 97-142.

69. Heim, R. 1949. Introduction al' etude au Boyomi. Rev. Mycol. **14** (Sippl.2): 41-49.

70. Hiremath, P. C. and Hedge R. K. 1975. Effect of fertilizer application on the grain infection paddy by Drechslera oryzae. Curr. Res. **4** : 9, 153-154.

71. Hofengart, H. 1953. Die varanderug der Mehltauanfalligketi von Sommegerste bei verschiedener Mineral salzemuhrung. A. Acker-Und-falbau, **96** : 75-150.

72. Huber, D. M. 1978. Disturbed mineral nutrition, p. 163-181. In: Plant Diseases: An advanced treatise, Eds. J. G. Horsfall and E. B. Cowling, vol. 3 Acad. Press, New-York.

73. Huber, D. M. 1980. The role of mineral nutrition in defence. p. 381-406. In: Plant Disease: An advanced treartise, Eds. J. G. Horsfall and E. B. Cowling, vol. 5 Acad. Press, New York.

74. Huber, D. M. and Amy, D. C.. 1985. Interactions of potassium with plant disease. In: Potassium in Agriculture, Ed. R. D. Munson, Publ. by ASA-CSSA-SSA. p. 467-488.

75. Huber , D. M. and Nelson, D. W. 1974. Control of corn stalk rot through stabilisation of ammonium nitrogen with nitrapyrin, Down to Earth, **30** : 1-3.

76. Huber, D. M., Warren, H. L. and Tsai, C. Y. 1977. Nitrification inhibitors-new tools for food production. Bio Science, **27** : 523-529.

77. Hutchinson, W. G. 1935. Resistance of *Pinus sylvesteris* to gall-forming *Peridermium*. Phytopathology. **25** : 819-843.

78. Indra Hooda and Grover, R. K. 1989. Influence of host nutrition on the efficacy of fungicide in controlling *Macrophomina phaseolina* on mungbean. Plant Dis. Res. **4 (2)**: 170-173.

79. Isla, L. H., Rodrquez, C. A. and Morales Bomez, A. Y. 1984. Influencia de la fertilisation con NPK on cafeto al sol sobre la incidencia de *Cercospora coffeicola* Breaky, CKE. Centro Agricola, **11 (1)**: 79-86.

80. Ismail, W. 1980. The influence of potassium on the relationship between *Meloidogyne incognita* and cucurbits. Indian J. Bot. **3** : 24-30.

81. Ismail, W. and Saxena, S. K. 1976. Effect of different levels of potassium on the growth and nitrogen,, phosphorus, and potassium status of tomato *Lycopersicon lycopersicum, L. peruvianum* and development of root knot. Indian J. nematol. **6** : 23-31.

82. Ismunadji, M. 1976. Rice diseases and physiological disorders related to potassium deficient. In : Fertilizer use and plant health. Proc. Colloq. Int. Potash. Inst. **12** : 47-60.

83. Jackson, T. L., Foote, W. H. and Dickson, E. A. 1962. Effect of fertilizer treatments and planting dates on yield and quality of barley. Oreg. Expt. St. Tech. Bull. 65.

84. Jagdish Kumar and Gupta, G. K. 1986. Influence of NPK fertilizers on the susceptibility of apple plants to scab pathogen (*Venturia inequalis*). Indian Phytopath. **39** : 480-481.

85. Janssen, J. J. 1930. Influence of fertilization on the health of the potato. Am. potato *J.* **7** : 79-80.

86. Jayaraj, J., Ramanujum, K. and Ramabadran, R. 1991. Effect of K application on the survival and population of *Sclerotium oryzae* in soil. J. Pot. Res. **7 (2)** : 126-131.

87. Jayasekhar, M. and Prasad, N. N. 1986. Effect of levels of potassium and nitrogen on pectinolytic and cellulolytic enzymes and *Sarcoladium oryzae* inoculation in rice. J. Pot. Res. **2 (4)**: 144-147.

88. Jeffers, D. L., Schmitthenner, A. F. and Knoetz, M. E,. 1982. Potassium fertilization effects on *Phomopsis* seed infection, seed quality, and yield of soybeans, Agron. J. 886-890.

89. Kannaiyan, S. and Prasad, N. N. 1974. Influence of foliar nutrient sprays on the population of *Fusarium oxyspourm* f. sp. *melonis* and other soil microflora in the rhizosphere of muskmelon. Indian Phtopath. **27 (4)**: 527-531.

90. Kannaiyan, B. and Prasad, N. N. 1981. Effect of NPK nutrition on the survival and saprophytic activity of *Rhizoctonisa solani.* Indian Potash J. **6 (1)**: 19-25.

91. Kannaiyan, S. and Prasad, N. N. 1983. Effect of nitrogen, phosphorus and potassium'on the survival and saprophytic activity of *Rhizoctonia solani.* Madras agric. J . **70** : 506-510.

92. Kataria, H. R. Yadav, J. P. S. and Grover R. K. 1981. Interactions of chemical fertilizers with seed dressing fungicides in controlling *Rhizoctonia solani.* A. Pflanzenkr. Pflanzenschutz, **88** : 641-650.

93. Kesavan, R. and Prasad, N. N. 1975. Effect of certain carbon and nitrogen source on *in vitro* production of fusaric acid by muskmelon wilt pathogen. Indian phytopath. **33** : 29-32.

94. Kiraly, Z. 1976. Plant disease resistance as influenced by biochemical effects of nutrients in fertilizers. In Fertilizer use and plant health. Proc. Colloq. Int. Potash Inst. **12** : 33-46.

95. Klien, E. K. 1957. Uber den Einfluss der Mineral salzernahrung auf den Gehalt des Blattes an frien tminosauren und Monosaccharidesn und seine Bedeutung fur die empfauglichkeit der Pflanze gegenuber parasitaren Pilzen. Soils. Fer. **20** : 303.

96. Kolbe, W. 1977. Untersuchungen uber den Einfluss mineralischer und organischer Diingung auf den Krankheits befall in Otstbab bei einheitlichen Pflanzenschutzmass nahmen. 1961-1976. Pflanzenschutz-Nachrichten. Bayer, **30 (92)** : 138-152.

97. Kovacikova, and Altmanova, O. 1989. Experience with combined application of fertilizers and fungicides to control Fusarium wilt of carnations. Sbornik UVT12 Zahradnictvi, **16 (1)**: 39-45.

98. Kruger, W. 1976. The influence of fertilizer on fungal diseases of maize. In : **Fertilizer use and Plant** Health. Proc. Colloq. Int. Potash. Inst. **12** : 145-156.

99. Laurent, E. 1899. Researches experimental sur less maladies des plantes. Ann. Del **Institu, Pasteur, 13** :1.

100. Lee, H. A. and Martin. J. P. 1986. Effect of fertilizer constituents on the eye spot disease of sugarcane. Ind. Engg. Chem. **20** : 220-224.

101. Lefter, G. and Pascu, I. 1970. Influenta factorilor climaticici technici ampra sensibilitsatii la rapan a soiuluis golden delicious. Protectia cultur. horti. viticole, **6** : 95-100.

102. Luke, K., Hess, W. M. and Smith, B. N. 1979. Investigation of mineral nutrient deficiencies upon infection of cucurbita-pepo by *Sclerotinia sclerotiorum.* Plant Physiol. 63 (suppl.) **5** : 163.

103. McNew, G. L. 1953. The effects of soil fertility. p. 100-114. In : Plant diseases-the year book of agriculture-U. S. Department of Agriculture. U. S. Government Printing Office, Washington, D. C.

104. McNew, G. L. and Spencer , E. L. 1939. Effect of nitrogen supply of sweet corn on the wilt bacterium, Phytopathol. **29** : 1051-1067.

105. Menzies, S. A. and Colhoun, J. 1976. Mechanism of control of *Phoma* root rot of chrysanthemum by the use of fertilizers. Trans. Br. Mycol. Soc. **67** : 463-467.

106. Meyer, J. A. 1970. Soil-borne diseases and fertilizer responses. Potash symposium. Annual Meeting Internatl. Potash Inst. 449-460.

107. Miller, J. W. 1969. The effect of soil moisture and plant nutrition on the *Cercospora, Alternaria* leaf. blight complex of cotton in Missouri, Phytopathol. **59** : 767-769.

108. Mohanty, S. K. Reddy, P. R. and Sridhar, R. 1983. Effect of major nutrients on the susceptibility of rice plants to bacterial leaf light. Z. Pflanzenkr. Pflanzenschtz. **90** : 50-54.

109. Moscol Riofrio, L. and Morales Abanto, A. 1975. Potassium and the resistance of rice tissues to the attack of *Pyricularia oryze,* Cav. Departmento de cienvias Biologicas. Universidad, Nacional Tecnica Puira, Peru, Peru-Turrialba (Costao-Rica), **25 (3)**: 316-322.

110. Nagy, G. 1985. Correlations between nutrient status and juvenile powdery mildew susceptibility of barley variety. Favourite in nutrient solution experiments. Hungarian Agril. Review., **4** : 785.

111. Nandi, Sujat, and Chakrabarthi, N. K. 1986. Effect of NPK nutrition on susceptibility of rice of sheath blight disease. Indian J. Mycol J. Mycol. and Plant Pathol. **6 (91)**: 55-58.

112. Narasimhan, V. and Algiangalingam, M. N. 1986. Potassium in the management of chilli mosaic disease. J. Potassium Res. **2** : 59-64.

113. Narayana Reddy, P. and Khare, M. N. 1986. Influence of nitrogen, phosphorus and potassium nutrition on groundnut rust disease. Andhra agric J. **33** : 273-274.

114. Nelson, D. W. 1963. The relationship between soil fertility and the incidence of *Diplodia* stalk rot and northern leaf blight in *Zea mays* . M. Sc. Thesis, Univ. of Illinois, Urbana.

115. Nissinen, O. 1970. Effects of different minerals on the resistance of English rye grass to *Fusarium nivale.* (Fr.) ces. Preliminary results of laboratory experiments. Peat. and Pl. News., **3** : **1, 3-11.**

116. Noguchi, Y. Sugavara, T. 1966. Potassium and Japanica rice. Internatl. Potash Inst. Berne, Switzerland.

117. Okamota, H. 1958. Relation between rice blast (*Pyricularia oryzae*) and sesame leaf spot (*Ophioblus miyabeanus*) and potassium. Jap. Potassium symp. Tokyo. 76-89.

118. Ollgnier, M., Ochs, R., Pomier, M. and DeTaffin, G. 1983. Action due chlora sur le cocotier hybride PB-121en (ote d' Invoire et en Indonesi). Development, tolerance a la secheresse, production. Oleagineux, **38 (5)**: 309-321.

119. Ollagnier, M. and Renard, J. L. 1976. The influence of fertilizers on the resistance of oil palms to *Fusarium* " Fertilizer and Plant health" Proc, 12th coll. internall. Potash. Inst. Bern. 157-166.

120. Orlovious, K. 1986. Kalidiingung in Maisanbau, Mais. **14 (4)**: 12-15.

121. Oteifa, B. A. and Elgindi, A.Y. 1976. Potassium nutrition of cotton, *Gossypium barbednse,* in relation to nematode infection by *Meloidogyne incognita* and *Rotylenchulus reniformis.* In : Fertilizer use and Plant Health. Proc. Colloq. Int. Potash Inst. **12** : 301-306.

122. Pal, A. K. and Choudhry, K. C. B. 1980. Effect of phosphorus and potassium on disease development in sunhemp. Indian J. Agric. Sci. **50 (12)**: 952-954.

123. Panzer, J. D. 1979. The effect of mineral salts onlocallession formation by *Phaseolus vulgaris* inoculated with tobacco mosaic virus. Phytopathol. **47** : 748-751.

124. Papavizas, G. C. and Davey, C. B. 1962. Control of *Aphanomyces* root rot or peas in the greenhouse with methionine and norbeucine. Plant Disease Reporter, **46(9)**: 646-650.

125. Parashar, R. D. and Sindhan, G. S. 1986. Effect of nitrogen and potassium nutrition of rice plants on bacterial blight, carbohydrates, and phenolic contents. Ind. J. Mycol. and plant Pathol. **16** : 185-188.

126. Parker, D. T. and Burrows, W. G. 1958. Root and stalk rot in corn as affected by fertilizer and tillage treatment . Agron. J. **51** : 4140-417.

127. Pero, R. W. and Howard, F. L. 1970. Activity of junifer diffusate on spores of *Phomopsis juniperovora.* Phytopathology, **60** : 491-495.

128. Perrenoud, S. 1977. Potassium and Plant Heath. IPI Research Topics No.3, IPI, Bern. Switzerland.

129. Perrenoud, S. 1977. Potassium and Plant Health, IPI Research Topics No.3, IPI, Bern. Switzerland.

130. Phuken. S. N. and Baruah, C. K. 1990. Peroxides activity and potassium nutrition of potato plants in relation to infection by *Phytopathora infestans* Indian J. Mycol. Plant Pathol. **19 (2)**: 239-241.

131. Polegaev, V. L., Chzhao, A. E. and Avilov, S. Y. 1979. The effect of combined fertilizers on resistance of cabbage to gray mould and on it keeping ability. Izv. Timiryazebvsk. Skh. Akad. **2** :193-195.

132. Purukayasta, R. P. and Ray, C. 1977. Effect of foliar application of plant hormones and mineral nutrition of the host on the development of anthracnose disease of jute. Z. pflazenkr. Pflanzenschutz. **84** : 193-201.

133. Quelhas, dos Santos, J. 1979. Potassium and the potato leaf roll virus. Potash Review Subject. 23, 59th suite.

134. Ribindran, R. Marimuthu, T., Sivakumar, M. and Natarjan, S. 1987. Role of graded doses of potash on incidence of Phytophtora wilt of betelivne. J. Potassium Res. **3** : 155-159.

135. Rajendra Prasad, B. and Bascuhoudary, K. C. 1984. Influence of potash and phosphorus on mortality of lentil due to *Rhizoctonia solani* and *Sclerotium rolfsii.* Agric. Sci. Dig. **4** : 59-61.

136. Ramasamy, K. and Prasad, N. N. 1974. Influence of potassium nutrition on wilt incidence in muskmelon. Potash News. Letter, **9** : 3-6.

137. Ramasamy, K. and Prasad, N. N. 1975. Effect of potassium nutrition on phenol metabolism of melon wilt. Madras Agric. J. **63 (2)**: 313-317.

138. Ramasami, R. and Shanmugam, N. 1976. Effect of nutrients on the incidence of *Rhizoctonia* seedling disease of cotton. Indian Phtopathol. **29** : 465-466.

139. Rawal, R. D., Sohi, H. S. and Sokhi, S. S. 1974. Effect of different levels of N, P and K on cowpea rust caused by *Uromyces phaseoli* var. *vingniae*. Indian Phytopathol. **27** : 405-407.

140. Reis, E. M., Cook, R. J. and McNeal, B. L. 1982. Effect of mineral nutrition on take-all or wheat. Phytopathology, **72 (2)**: 224-229.

141. Remy, T. 1898. Mitleilungen aus dem Instit fur Garungsgewerb, Berlin.

142. Rishbeth, J. 1957. Fusarium wilt of banas in Jamaica II. Some aspects of host-parasite relationship. Ann. Bot. **21**: 215-245.

143. Rolfs, F. M. 1915. Angular leaf spt of cotton, South Carolina Agric. Expt. Stn. Bull. 184.

144. Rowan, S. J. 1977. Fertilizer induced changes in susceptibility to Fusarium rust vary among families of slash and lobolly pine. Phytopathology, **67**: 1280-1284.

145. Russell, E. J. 1956. Soil conditions and plant growth. 8th ed. Longmans., Green and Co., London.

146. Russell, G. E. 1978. Some effects of applied sodium and potassium Chloride on yellow rust in winter wheat. Ann. Appl. Biol. **90** : 163-169.

147. Sabourova, P. V. 1951. Physiological-anatomical basis of increasing carrot resistance to *Sclerotinia* under the influence of mineral nutrition. Trudy. Vsesoyunznogo, Instituta zashchity Rastenii. **3** : 38-48.

148. Sadasivan, T. S. 1965. Effect of mineral nutrients on soil microorganisms and plant disease. p. 460-470. In : Ecology of soil borne plant pathogens: Prelude to biological control. Eds. K. F. Baker and W. C. Synder, . University of California Press, Berkeley.

149. Sansavini, S. and Montevecchi, P. 1986. Influenza della concimazione potassica and azotata sulmelo "Imperator" Banca Popolare di Verona. Atto del Convegno "La Fertilizzazione delle Piante da Frutto". 171-180.

150. Sarhan, A. R. T. and Hegazi, A. M. A. 1988. Effect of N, P and K on sensitivity of pepper plants to *Fusarium* wilt toxin. Acta Phytopathologica et. Entomologica Hungarica. **23** : 57-60.

151. Sarhan, A. R. T. and Jala, T. K. 1988. Effect of N and K nutrition on leaf disease of barley II. Growth of the pathogen and response of plants to toxic effect of culture filtrate. Arab. J. Plant Protection, **6 (1)**: 18-26.

152. Schneider, R. W. 1985. Suppression of fusarium yellows of celery with potassium, chloride an nitrate, Phytopathol. **75 (1)**: 40-48.

153. Schroeder, W. T., and Walker, J. C. 1942. Influence of controlled environment and nutrition on the resistance of gardenpea to Fusarium wilt. J. Agic. Res. **65** : 221-248.

154. Schultz, F. A. 1976. The effect of potassium on catabolisim of rot infected apple fruti callus. In : Fertilizer use an plant health. Proc. Colloq. Int. Potash. Inst. **12** : 61-68.

155. Sen, B. 1991. Potassium and disease resistance in fruit and vegetable crops. Potash Review, Switzerland, **4** : 1-7.

156. Sharoubeem, H. H., Naim, M. S. and Habib, A. A. 1967. Potassium nitrogen and phophorus in relation to the incidence of cotton wilt caused by *Fusarium oxysporum* f. *vasinfectum*. J. Microbiol. U.A.R. **2** (1) : 1-8.

157. Sharvelle, E. G. 1936. The nature of resistance of flax to *Melampsor lini.* J. agric. Res. **53** : 81-127.

158. Shields, W. J. Jr. and Hobbs, S. D. 1979. Soil nutrient levels and pH associated with *Armillariella mellea* on conifers in northern Idaho. Can. J. For. Res. **9** : 45-48.

159. Siebold, M. 1974. Einfluss der Kalidiingung auf die Stengelfaiile bei Kormermais. Gesunde Pflanze. **26** : 65-68.

160. Sij, J. W., Turner, F. T. and Whitney, N. G. 1985. Suppression of anthracnose and phomopsis seed rot on soyabean with potassium fertilizer and benomyl. Agron. J. **77 (4)** : 639-642.

161. Singh, S. 1963. Effect of mineral nutrition on the incidence of net-blotch of barley. Indian J. agric. Sci. **33 (1)** : 1-7.

162. Spencer, E. J. 1937. Influence of host. nutrition on system-development of tobacco mosaic. Plant Physiol. **12** : 825-832.

163. Spizel, Y. and Netzer, D. 1984. Effect of nitrogen from at various levels of potassium, on the *Meloidogyne* Fusarium wilt complex in muskmelon. Plant and Soil, **81 (1)**: 85-92.

164. Srihuttagum, M. and Sivasithamparam, K. 1991. The influence of fertilizers on root rot of field peas caused by *Fusarium oxysporum, Pythium vexans* and *Rhizoctonia solani* inoculated singly or in combination. Plant and soil, **132** : 21-27.

165. Stoddard, D. L. 1942. Fusarium wilt of cantaloupe and studies on the relation of potassium and nitrogen supply to susceptibility. Trans. penin. Hort. Soc. **31** : 91-93.

166. Struchtemeyer, R. A. 1966. Potassium suppressor of diseases and other disorders in potatoes. Better crops plant food, **50 (3)**: 14-16.

167. Svec. L. V., Andrews, A. K. and Crittenden, H. W. 1976. Soybean (*Glycine max* (L.) Merr.) yield and disease incidence with potassium fertilization. Common. Soil Sci. Plant. anal. **7** : 7270741,

168. Takada, K. 1980. Studies on breedings of melon, *Cucumis melon* resistant to gummy stem blight, *Mycosphaerella melonis.* Evaluation of inoculation methods and growth stage and nutritional status of test plants used for screening breeding materials. Yasai Snikenjo Hokoku, A 0 **(7)**: 23-24.

169. Tandan, H. L. S. and Shekon, G. S. 1988. Potassium research and agricultural production in India. FDCO Pub. New Delhi, India. pp. 144.

170. Taylor, R. G., Jackson, T. L., Powelson, R. L. and Chirstiansen, N. W. 1983. Chloride, nitrogen from, lime and planting date effects on take-all root rot of winter wheat. Plant Dis. **67** : 1116-1120.

171. Temiz, K. 1976. Interaction of fertilizers with Septoria leaf blotch of wheat. In : Fertilizer use and Plant health. Proc. Colloq. Int. Potash Inst. **12** : 129-132.

172. Thier, A. L., Sammons. D. L., Grybauskas, A. P. and Tomerlin, J. R. 1986. The effect of chloride on the incidence and severity of powdery mildew in soft red winter wheat. Chloride and crop production PPI special Bulletux. No. 2, 62-72.

173. Thomas, W. T., Baker, D. E. and Carig, W. F. 1966. Potassium and corn leaf blight. Better crops, **50 (3)**: 38-40+13.

174. Thomson, A. D. and Ferguson, J. D. 1976. Effect of varying the nutrient supply on response of pea plants to leaf roll. N. Z. J. Agric. Res. **19** : 529-533.

175. Thomson, A. D. and Ferguson, J. D. 1980. Response of oat, ryegrass, and barley infected with barley yellow dwarf virus to variation in nutrient supply. N. J. Agric. Res. **23** : 367-372.

176. Thomson, W. R. 1983. P and K produce higher yields plus better quality cotton. Better crops. 22-23 summer.

177. Timm, C. A., Goos, R. J., Johnson, B. E., Sobolik, F. J. and Stack, R. W. 1986. Effect of potassium fertilizers of malt barley infected with common root rot. Agronomy J. **78** : 192-200.

178. Tisdale, H. B. and Dick, J. B. 1942. Cotton wilt in Alabama as affected by potash supplements and as related to varietal behaviour and other important agronomic problem. J. Amer. Soc. Agron. **34** : 405-426.

179. Trolldenier, G. 1965. Einflub der Dungung auf die standfe stigkeitvon Weizen. Kali-Bariefe 7, Fachgeb. **11** : 1-4.

180. Trolldenier, G. 1971. Recent aspects of the influence of potassium on stomatal opening and closure, In: Potassium in Biochemistry and physiology. Int. Pot. Inst. p. 130-133.

181. Trolldenier, G. 1983. Wechsel winkungen zwischen Mehtaubefatl, Kaliumernahrung und fungizideisatz auf den Entrag jweier, Sommergerstensorten. Phytopathol. Z. **104** : 337-344.

182. Trunov, G. A. 1957. The role of mineral nutrition in wheat disease resistance (In Russian). Jr. Khark. Skh. Inst. im. V. V. Dekuchaera. B **50** : 107-116.

183. Tsai, H. Y. and Bird, L. S. 1975. Microbiology of host pathogen interactions for resistance to seedling disease and multi adversity resistance in cotton. p. 39-45. In : Proc. of the Bettwide cotton production research conf. Memphis. 6-8 January. Ed. J.M. Brown. National cotton council of America, Menphis.

184. Upti, I. E. 1970. Effektivnost udorenii podkormok kapusty s poivnovi vodi na torfyanobolotnnykh pochvakh Agrokniniya B: 78-83.

185. Vathilingam, G. and Balasubramanitian, M. 1976. Effect of different leaves of K on two rice varieties with reference to feeding of brown plant hopper, sclerenchyma thickness and silica control. 5th Internol. Pot. Inst. competition for young Research Workers, Berg.

186. Veram, G. S. and Verma, J. P. 1961. Potassium nutrition of Turkish tobacco (*Nicotiana tabacum* L.) (N. T. L.) plants in relation to multiplication of tobacco mosaic virus. Indian J. Microbiol. **1** : 97.

187. Wade, G. G. 1956. Aphanomyces root rot of peas-the effect of a potassium fertilizer on the severity of the disease in a potassium deficient soil. J. Agust. Inst. Agric. Sci. **21** : 260-263.

188. Walker, J. C. 1969. Plant Pathology 3rd ed. Mc Graw-Hill Book Co., New York.

189. Walker, J. and Foster, R. E. 1946. Plant nutrition in relation to disease development III. Fusarium wilt of tomato. Am. J. Bot. **33** : 259-264.

190. Walker, J. C. and Hooker, W. J. 1945. Plant nutrition in relation to disease development. I : Cabbage Yellows. Am. J. Bot. **32** : 314-320.

191. Wei, C. T. and Scheffer, R. P. 1951. Relation of host nutrition to the development of bacterial wilt of cucumber and fusarial wilt of watermelon. Phytopathology, **41** : 38.

192. Wensley, R. N. and McKeen, C. D. 1962. Relations between plant nutrition, fungal population and incidence of Fusarium wilt of muskmelon. Can. J. Microbiol. **11** : 581-594.

193. Yale, J. W. Jr. and Vaughon, E. K. 1962. Effect of mineral fertilizers on dumping off of table beets. Phytopathology, **52** : 1285-1287.

194. Young. V. H., Janssen, G. and Were, J. O. 1932. Cotton wilt studies. Arkanass Agric. Exp. Stn. Bull. 272.

29 Botanical Toxicants as Potential Component in the Strategy of Integrated Pest Management

S.N. Tewari

CONTENTS

Integrated Pest and Disease Management
Rajeev K. Upadhyay, K.G. Mukerji, B.P. Chamola and O.P. Dubey (eds.)
APH Publishing Corporation, New Delhi-110002, India

ABSTRACT

Plant protection through synthetic chemicals contributed significantly in reducing losses due to diseases/insect pests and thereby increased agricultural production. However, pesticidal share to environmental pollution through air, water, soil and by induction of resistance/resurgence, amongst pathogens/insect pests, currently caused wide spread concern, and hence the need to look for an alternative safer means for plant disease and pest control is felt greater now than ever before.

Use of botanicals for prevention of diseases/insect pests without the mention of their harmful effect has been recorded in ancient Indian literature. Since then there has been several recorded and unrecorded mention of their uses against a large number of microbial organisms. This article describes major specific preparation of the plants grouped under aqueous, ethanolic, essential oil extracts and active ingredients reported to possess antimicrobial activity so as to get a quick idea of their feasibility for utilization as a component in the strategy of Integrated Pest Management under each group. From these groups 33 plants showing promising potentials against two fungal pathogens of rice and two of groundnut viz., *Pyricularia grisea, Rhizoctonia solani, Aspergillus niger* and *A. flavus* inciting blast, sheath blight, collar rot and yellow root diseases respectively have been short listed for further clarity. Success of *Aegle marmelos* and *Ocimum sanctum* plant potentials for the control of a serious disease of rice, blast, achieved effectively in the field has also been described.

Nevertheless, a systematic evaluation of pesticidal properties of plants by utilizing modern techniques such as GLC, UV, IR, NMR and Mass Spectro-scopic analysis, may not only help unravel precisely the chemical(s) and its involvement in disease/pest control mechanism but also indicate the possibilities of exploiting those plants having similar chemical compound against specific microbial agent.

The natural abundance of botanicals, their genetic multiplicity globally and little pollution problems are the key factors for their attractive prospect of application in the strategy of Integrated Pest Management.

1. INTRODUCTION

Every year about 20000 species of field and storage pest destroy approximately one-third of world's food production. Maximum loss occurs in staple crops such as rice and maize in tropics of Asian countries (36). Chemical pesticides in general and broad spectrum pesticides in particular gave convincing means of controlling the plants pests, since 1950. Plant protection coverage in India alone increased from 2.4 to 13.7 million hectare (Target) and consumption of technical grade pesticides from 2350 M.T. to 75000 M. T. between first plan period to seventh plan period (69).

2. SYNTHETIC FUNGICIDE AND OUR CONCERN

A regular use of these also brought new problems with them. More alarming amongst them are pollution of air, water and soil; residual toxicity; development of resistance/resurgence in pathogen/insect pest thereby the need to apply them more with their escalating prices and harmful effect on nontarget organisms. Consequently there has been alarming development of harmful environment for human beings.

Current views are therefore to examine and introduce ecologically sound and environmentally safer alternative means of plant protection. Appropriate cropping system, better seeds with built in resistance through conventional means or development of transgenic resistance, suitable cultural and agronomical practices coupled with integration of judicious plant protection measures are the possible environmentally safer strategies that would help sustain agricultural production and productivity in future.

3. BOTANICALS A PROMISING SUBSTITUTE TO SYNTHETIC CHEMICALS

Contrary to the problems associated with the use of synthetic chemicals, botanicals are environmentally non-pollutive renewable, inexhaustible, indigenously available, easily accessible, largely non-phytotoxic, systemic ephimeral, thus readily biodegradable, relatively cost effective and hence constitute as a suitable plant protectant in the strategy of integrated pest management.

4. INSTANCES OF BOTANICAL USES IN PAST

Green plants are known to provide a vast reservoir of potential drugs and not more than 10% of these have so far been examined in detail for their biological activity against human diseases (64) much less against plant diseases. The earliest mention of poisonous plants, or those with pest control properties is found in Indian Rig Veda, the classic book of Hinduism, compiled during second millenium B. C. (36). Democritus tried plant extracts for controlling plant disease as early as in 470 B.C. (76). Decoction of tobacco and alder buds with lime and sulphur, against mildew diseases of trees was recommended by Forsyth (23). Bradely (7) used pepper dust to protect trees against blight to their blossoms.

5. CURRENT STATUS OF BOTANICAL PRODUCT EXPLOITATION

During the last two decades many reviews have appeared on botanical products exhibiting antimicrobial activity (18, 31, 36, 39, 44, 45, 53, 61, 66, 73, 86, 87, 89, 92, 93, 97). In this compilation, Grainge *et. al.* (29) stated that out of 1600 plant species tested only 94 and 11 were reported to possess antifungal and anti-bacterial properties respectively. Seventy one genera of plants belonging to 47 families were screened against fungal and bacterial pathogens of rice (86). The present paper discusses some of the important finding on a few major types of plant product preparations in relation to their specific antimicrobial activity.

The types of plant product preparations that are discussed in the ensuing article are as follows:

6. BOTANICAL PREPARATIONS

6.1. Aqueous extract and their antimicrobial potentials

6.1.1. Preparation method

The Plant materials are extract in cold water, boiling water or by autoclaving. Aqueous extract preparations displaying antimicrobial potential are considered to be the easiest approach that can be exploited in the strategy of plant protections in areas where the test plants are abundant. Most of these preparations can not be kept for a longer period for use, following quick decomposition of macerated cellular structure in water, hence these should be utilized before inactivation of their antimicrobial properties. Nevertheless a major factor, that might affect reproducibility of the result in aqueous extract preparations are lack of uniform extraction and dilution procedure, influence of agro-climatic variability on qualitative/quantitative fungitoxic character of botanicals, season influence and stage at which the plant parts are examined.

6.1.2. Antimicrobial activity

Webb and Agnihorti (103) reported that the aqueous extracts of the pericarp of *Acer saccharum* inhibited the mycelial growth of several pathogenic soil fungi including *Pythium* and *Rhizoctonia* spp. The extracts also controlled the damping off of *Pinus resimosa* Pelletier and Comeau (68) isolated water soluble inhibitors from young leaves of oats which they felt if purified and identified would be valuable for the control of cereal spot diseases. An aqueous extract from fresh bulb tissue of *Urginea altissima* showed considerable antifungal activity against *Phytophthora capsici* (58). Out of 93 plants tested, *Impatiens balsamina, Pseudocalymna alliaecum* garlic and *Tagets erecta* extracts were found active as either therapeutant and/or protectant against *Helminthosporium oryzae* on rice (44). Aqueous extract of *Catharanthus roseus* stem though did not inhibit the spore germination of *H. oryzae*, but it delayed the germination (45).

Extract of garlic greatly reduced the growth of *Fusarium oxysporum* f. sp. *ciceri* and *Sclerotinia sclerotiorum in vitro*. Extract treated gram seeds produced wilt free seedlings when sown in soil infested with these pathogens (77). When leaf extracts of *Polyalthia longifolia* was tested against *Alternaria alte.nata, Colletotrichum gloesporioides* and *Curvularia lunata* it inhibited the growth of these fungi (3).

Grainge *et al.* (29) compiled 1600 plant species possessing pest control properties out of which 94 plants have been reported to exhibit antifungal activity against *H. oryzae* and *Pyricularia grisea (P. oryzae)*, the causal agents of brown spot and blast of rice respectively. These were the extracts from *Aerva lantana, Allium sativum, Amaranthus spinosus, Antidesma pentadrum, Blumea balsamifera, Brassica integrifolia, Catharanthus roseus, Chrysophyllum oliviforme, Clerodendrum indicum, Chorchorus olitorius, Datura stramonium, Derris elliptica, Erythroxylum coca, Euphorbia hirta, E. pulcherrima, Ipomoea batatas, I. balsamina, Inora coccinea, Jatro pa podagrica, Leucosyke capiteollata, Mentha arvensis, Mirabilis jalapa, Nerium indicum, Plumbago anriculatas, P. indica, Portulaca oleracae, Psidium guajava, Raphanus sativus, Samnea saman, Stachytarpheta jamaicensis, Sterculia foetida, Symphytum officinale, T. erecta, Vitex negundo, Zebrina pendula* and *Zinngiber officinale.*

Seventy one plant species collected from the coastal regions of Orissa, belonging to fortyseven families were tested against six major fungal and bacterial pathogens of rice (86). Out of these, steamed aqueous extract from *Aegle marmelos, A. sativum, Amorphophallus bulbifer, Annona squamosa, D. stramonium, Ficus glomerata, Lawsonia, inermis, Momordica charantia, Nyctanthes arbor-tritis, Ocimum sanctum, Piper betle, Quisqualis indica, Thevetia neriifolia, Zizyphus jujuba* have proved fungitoxic to *P. grisea (P. oryzae), Drechslera oryza (Chochliobolus miyabeanus)* and *R. solani* conidial germination and/mycelial growth. While assaying biotoxicity of *Sapindus emarginatus* in steamed aqueous extract against *P. grisea (P. oryzae), H. oryzae (C. miyabeanus), R. solani, Xanthomonas compestris* pv. *oryzae, X. campestris* pv. *oryzicola* and *Erwinia chrysanthemi*, it was observed that both conidial germination and radial growth of *P. grisea* and *H. oryzae* were inhibited. Growth of *E. chrysanthemi*, a foot rot pathogen of rice was also arrested in steamed aqueous extract of this plant (94).

Mishra *et al.* (56) studied the effect of aqueous extract of *Alangium salvifolium, Cinnamomum zeylanicum, Crysentia cujeta, Lagerstreemia parviflora, P. gaujava, Strychnos nux-vomica, S. potatorum* and *Thuja orientalis* against *P. grisea (P. oryzae), Fusarium moniliforme* and *Curvularia lunata.* It was found that *P. guajava, T. orientalis* and *L. parviflora* aqueous extracts displayed least spore germination followed by *S. potatorum* fruit extracts, against *P. grisea, F. moniliforme* showed minimum spore germination (10%) in *P. guajava* followed by *S. potatorum* fruit extracts. Interestingly, *T. orientalis, L. parviflora* extract treatments deformed *P. grisea* and *C. lunata* germ tubes which in turn thought to render the pathogen incapable for infection. P. *guajava, S. patotorum, T. orientalis* and *L. parviflora* exhibited greater fungitoxicity in aqueous extract compared to ethanolic extract.

6.2. Organic solvent extracts and their antimicrobial potentials

6.2.1 Preparation method

Extraction of fresh or dried botanical products in various organic solvents such as petroleum ether, hexane, benzene, chloroform, methanol acetone, ether, ethyl acetate etc. for a varying length of time have been widely adopted by various workers (3, 5, 6, 9, 24, 34, 45, 63, 78, 86, 88, 89, 90, 94, 108).

Selected portions of the plants, green or shade dried powder either macerated are extracted as such in a suitable organic solvent, for varying length of time, until no more extractable fraction is recovered. The ratio of the suitable solvent needed for complete extraction differs from plants to plants and hence can be standardized with respect to specific plant or parts thereof. Solvents containing extracted materials are then collected avoiding residues either through centrifugation process or decanting them carefully and then evaporated under reduced pressure. Organic solvent extracts of plant products involves only a few steps in its preparations and generally possess longer shelf-life than aqueous extract preparations, hence, these can be considered more suitable for commercial production. Preparations of neem products in alcohol in patented and marketed as Margosan-O in USA (36).

6.2.2. Antimicrobial activity

Nicolls (63) reported that a substance toxic to many fungi was found in 27 species and 4 hybrids of passiflora and in some genera of the related Turneraceae and Flacourtiaceae families. He stated that the active principle moved from living tissues into agar and water and could be extracted from diffusates by organic solvents.

Zayed *et al.* (108) tested 27 plants including tomato, onion peel, apple, guava and sugarcane and reported ether extracts to be the best followed by alcohol and then aqueous extracts against 40 different soil organisms. They found gram +ve bacteria to be more sensitive than gram -ve bacteria, except in sugarcane extract. Mahadevan and Mohan (45) found alcoholic extract of *Catharanthus roseus* roots inhibitory to linear mycelial growth, conidial germination and germ tube elongation of *Helminthosporium oryzae* even upto 10-fold dilution. The extract however did not inhibit the growth of *X. oryzae.* Methanol or petroleum ether extracts of entire *Pinus leucodermis* Ant. needles did not inhibit the growth of 5 aquatic hyphomycetes while methanol extract of needle powder was found inhibitory (5). According to Chaudhury and Sen (9) benzene extract from *Piper nigrum* showed considerable fungitoxic activity most prominent on *Sclerotium rolfsii,* moderate on *R. solani* and least on *Sclerotinia sclerotiorum.* The extract was more inhibitory to mycelial growth than sclerotial germination.

When leaf extracts of *Polyalthia longifolia* in different solvents were tested against *Alternaria alternata, C. gloeosporioides* and *C. lunata,* methanol extract inhibited the growth of fungi but chloroform and ethyl acetate extracts were inactive (3). The growth of *Shizophyllum commune* and *Pycnoporus sanguineus* were inhibited *in vitro* by the methanol extract of 6 hard wood tree species (37).

Extracts from *Hedera helix* (English ivy) in alcohol, *T. erecta* and *Vitex negundo* in ether and alcohol, *Argemone mexicana* in kerosene and petroleum ether and *Nerium indicum* in alcohol, kerosene and petroleum ether have been found to exhibit fungitoxicity against *Venturia inae-qualis, P. oryzae, A. tenuis, Helminthosporium* sp. and *H. oryzae* respectively (29).

Extract from dried leaf powder of *Aegle marmelos* in ethanol, hexane, benzene, chloroform and acetone reduced the radial growth of *P. grisea (P. oryzae).* Maximum inhibitory effect was seen in benzene soluble fraction (86, 87). Mishra *et al.* (56) extracted certain plant species in ethanol and found that *P. grisea* spore germination was inhibited maximum in *T. orientalis* followed by *L. parviflora.* Absolute inhibition of mycelial growth of *P. grisea* was registered with *C. zeylanicum, P. guajava, A. salvifloium, S. potatorum, T. orientalis* and *S. nux-vomica* at 1% treatment; *R. solani* in *C. zeylanicum, A. salvifolium* (at 0.1 and 1%) and in *P. guajava, S.potatorum, S. nux-vomica* and *L. parviflora* at 1% *F. moniliforme* in *c. zeylanicum, A. salvifolium, S. potatorum, T. orientalis* and *S. nux-vomica* at 1% and *A. niger* in *L. parviflora* at 1%.

S. emarginatus fruits extracted in ethanol inhibited spore germination of *P. grisea, C. miyabeanus* and *R. sonali* (94). Ethanolic extract of *P. longifolia* inhibited spore germination and mycelial growth of *P. grisea, C. lunata, F. moniliforme, A. niger* and *R. solani* (55).

Ethanolic leaf extracts of four plants, *Piper betle, O. sanctum, Nyctanthes arbor-tristis* and *Citrus lemon* were observed effective in reducing radial growth of *P. grisea, C. miyabeanus* and *R. solani. P. betle* was found to be the best followed by *O. sanctum* in reducing the growth of the pathogens completely *in vitro* and checking the spread of blast, brown spot and sheath blight diseases of rice *in vivo.* Though, the leaf extracts from the other two plant species tested (*N. arbor-tristis* and *C. lemon)* reduced the radial growth of the pathogens *in vitro* at a higher concentration, they failed to check their spread effectively in the glasshouse. Thus, *P. betle* and *O. sanctum* could be used as source of a pesticide of plant origin to combat the above three pathogens of rice in the field has been proposed (91).

Antifungal activity of ethanolic, acetone and benzene extracts for *Aegle marmelos* leaves was studied against rice blast incitant *P. grisea* both *in vitro* and *in vivo* (90). It was found that germ tube inhibitions of *P. grisea* increased with increasing concentration of ethanolic extract from 10 ppm and above. At 100 ppm cell wall ruptured cytoplasmic content granulated or germ tube grew short swollen and deformed. At 500, 1000 and 10000 ppm concentrations were observed spore deformities intense.

Ethanolic extract of *P. longifolia* registered fungitoxic activity against five rice fungal pathogens including *P. grisea (P. oryzae)* (55). It was found that crude ethanolic extract showed persistence of toxicity for a longer period than other extracts tested, and hence suggested that this extract could be exploited further to control rice blast but only after further extensive trial both *in vitro* and *in vivo.* Ethanolic extracts of *Calotropis procera, Azadirachta indica* and *D. stramonium* leaves have been found to inhibit both conidial germination and mycelial growth of *F. moniliforme* and *C. lunata* (54).

Extraction of plant products through water or organic solvents may involve primary disruptions of cellular material leading to release of materials rich in sugars, enzymes, nitrogenous compounds, organic acids, salts, amino acids, nucleotides and nucleic acids. These may be toxic to plant growth (105). However, some plant products are non-phytotoxic also (20, 91, 92). Hence, selection of a suitable plant product, based on effective dose response relation for plant protection, like any other synthetic agro-chemicals, would be one of the determinant factors for its utilisation.

6.3. Essential oil and their antimicrobial potentials

6.3.1 Preparation method

Essential oils from the green portions of the plants are generally extracted in laboratory through hydrodistillation process using Clevenger's apparatus, and the oil thus extracted is dried over anhydrous Na_2SO_4, and utilised for further studies.

6.3.2 Antimicrobial activity

Several species of storage fungi cause loss both in quality and quantity in storage grains (11) and other food materials. They produce mycotoxins, which results in physiological malfunctioning in man and animal, when consumed (99, 107). Chemical fumigants have been widely used to prevent them but their pollutive, carcinogenic, teratogenic or sometimes lethal effects due to consumption of treated food grains have mostly been overlooked. Volatile constituents from plants may play a vital role in protecting both storage and field pathogens if they are systematically evaluated.

Uchida (98) and Okazaki and Homma (65) extracted the essential oil from *Thujopsis dolabrata* and observed toxicity against *Merulius lacrymans, Polyporus lacrymanus* and species of *Epidermophyton* and *Trichophyton.*

The antifungal activity of essential oil from *Allium sativum* and *A. cepa* has been described by Skinner (79) and Maruzzella and Balter (48) against *Alternaria alternata, A. niger, Botrytis allii, Claviceps purpurea, Microsporum gypseum* and other storage fungi. Germicidal properties of *Citrus reticulata* oil was first recorded by Murdoch and Allen (60). Further Subba Rao *et al.* (84) reported that the essential oil from this plant was found toxic to *Aspergillus* spp. and *Torula* spp. Korta and Starzyk (42) evaluated *Angelica archenglica* oil against several fungi and found effective against *Candida albicans, C. krusei, C. parakrusei, C. pelliculosa, C. silosis, C. stellatoidea* and *C. tropicallis. Carum carvi* volatiles proved toxic against *A. niger* (80).

Eucalyptus genii essential oil was found inhibitory to *A. niger* (32). Do Vgich (15) reported the toxicity of essential oil from *Coriandrum sativum, Foeniculum vulgare, Hyssopus officinalis* and *Pastinaca sativa* against *A. tenuis, A. oryzae* and *Penicillium chrysogenum.* Rao and Joseph (70) screened *Anethum graveolans, Apium graveolens, Cicer arietinum, C. aurantium, C. amada, C. zedoaria, Leucas aspera, O. canum* and *O. thesiforum* oil and found toxic against *Cephalosporium sacchari, Ceratocystis paradoxa, C. lunata, F. moniloforme, H. sacchari, Physalospora tucumanensis, Sclerotium rolfsii, Candida albicans, Epidermophyton floccosum, Trychophyton mentagrophytes* and *T. rubrum*, Garg (26) reported *Bosewellia serrata* oil toxicity against *C. lunata, C. pallescens, F. culmorum, F. oxysporum* and *P. aphanidermatum.* Zutshi *et al.* (109) and Geda *et al.* (28) extracted *Caesulia axillaris* volatile fraction and found fungicidal to *H. oryzae.*

Bothricloa pertusa oil inhibited *A. robi, A. flavus, A. fumigatus, A. niger, A. nidulans, C. herbarum, Rhizopus nigricans* and *Trichothecium roseum* (38). Tripathi (96) isolated 0.4% oil from rhizomes of *Alpinia galanga* and found effective against *A. alternata, Absidia ramosa, A. fumigatus, A. versicolor, F. solani, F. moniloforme* and other fungi. He, however, found that the antifungal properties of the oil was destroyed by autoclaving and not by heating alone upto 100°C and storage upto 91 days. Garg and Oswal (27)) found the volatile fraction of *Chloroxylon swietenia* fungitoxic against *A. oryzae, A. terreus* and *C. albicans.* Kala *et al.* (40) extracted oil from *Cymbopogan citratus* and reported prevention of production of hepatocarcinogenic aflatoxin by *A. parasitiacus.* Pathak and Dixit (67) reported inhibition of *A. niger, Botryodiplodia theobrome, M. gypseum, R.nodosus* and several other fungal pathogens when treated with the essential oil obtained from *Glossocardia bosvallia.*

Essential oil from *Inula cuspidata* showed toxicity against *A. flavus, A. fumigatus, A. glaucus, A. nidulans* and *A. terreus* (10). Mall *et al.* (47) reported wood oil toxicity from *Juniperus verginians* against *Pseudotropicallis, E. floccosum M. gypseum* and *T. rubrum.* Dixit (12) isolated eugenol from the oil of *O. gratissimum* and stated that the use of this could prevent bakery bread from fungal deterioration during storage.

Dube *et. al.* (17) reported that the essential oil from *O. basilicum* inhibited 22 species of fungi including *A. flavus* and *A. parasiticus,* responsible for mycotoxin production. Mishra and Tewari (54)

extracted the essential oil from *P. longifolia* and reported its toxicity against *P. grisea (P. oryzae), R. solani, F. moniliforme, A. niger* and *C. lunata.*

Influence of *O. sanctum* essential oil was observed by Mishra and Tewari (53) against *P. grisea (P. oryzae), R. solani* and *F. moniliforme.* Growth of all the pathogens were checked when treated with this oil. Dube *et al.* (17) reported *Zanthoxylum alatum* essential oil inhibiting 24 fungi including *A. flavus* and *A. parasiticus.*

6.4. Active ingredients and their antimicrobial potential

6.4.1 Isolation method of active principle

Isolation of biologically active antimicrobial principle differs widely depending upon the specific chemicals and the microbial agents against which the active ingredients are tried. However, broadly the isolation procedure of active principle involves extractions of plant material (green or dried powder) in suitable solvent(s) and passed through column containing appropriate type of acid, basic or neutral, resin and them chromatographed through polarity gradient based solvent system in to several band-wise fractions. These fractions are then separated further through either thin layer or paper chromatographic procedure and biologically active distinct spot(s) are subjected for chemical and spectral (UV, IR, NMR, GC-MS or MS) analysis for chemical characterizations and structural elucidation.

6.4.2 Antimicrobial activity

Purified active ingredients to serve as antimicrobial agents have been reported from several plants. However, it is sometimes believed that crude form of plant products are more potent and feasible in controlling diseases and pests in plants. Preparation of crude extracts to protect the crops from diseases/pests are much easier, hence feasible for the farmers to adopt. Yet the investigations on the pure active moiety from plant products must proceed to unfold mechanisms of antimicrobial action more precisely (19, 25, 33, 41, 57, 72, 101, 104).

Plants are known to contain a wide range of antifungal compound with diverse skeletal patterns. Compounds such as flavonoids, terpenoids, alkaloids, cynogenic glycosides, coumarins, pterocarpens have been extensively reported (19, 33, 85, 101, 104) thus constituting vast reservoir of potential plant chemotherapeutant, of which many have been implicated in natural resistance. A widely quoted example is 'Wyerone' isolated from broad bean tissue (21, 22, 82, 102). Similarly a large contigent of microbial organisms causing variety of diseases are reported to be sensitive to specific chemicals isolated from certain plant species. Several workers have tried to assess the antimicrobial potential of these chemicals against more important pathogenic orgnaisms.

Triterpenic heteroside (anagalloside) extracted from the stem and leaves of *Anagallis arvensis* var. *phoenica* inhibited a large number of fungi *in vitro.* Staron et *al.* (83) found it to be slightly inhibitory to *C. lindemuthianum* and *Pythium ultimum* and was too toxic for practical use.

Allen (2) isolated steroid-glycolalkaloides from the peel of Irish potato tubers which were inhibitory to *Helminthosporium (C. carborum).* In an extensive survey of plants for antifungal properties, Gupta and Banerjee (30) found only *Curcuma zedoaria and Brassica* sp. contained active inhibitors against *Trichophyton rubrum* and *A. niger* while others were inactive against *T. rubrum.* According to Webb and Agnihotri (103) the inhibitor gallotannin isolated from *Acer saccharum* pericarp inhibited *P. irregulare* sporangia. Limonene and flavedo extracts from orange fruits caused considerably greater inhibition of growth of *Guignardia citricarpa* in culture than did albedo extract (8). Zayed *et al.* (108) isolated antibacterial substance, tomatine from tomato, quercitin from onion peel, apple and guava leaves and phloretin from apple and other rosaceae plants, that were found to be more sensitive to gram +ve bacteria than gram -ve. Wyk *et al.* (106) isolated P-Hydroxybenzoyl-Heryanin an antifungal substance from the root bark of *Protea cynaroides* which inhibited the growth *P. cinnamomi.* The active compound isolated from *Pistacia matica* was inhibitory only to gram +ve bacteria and considered to be a phenolic (46).

Capric acid, a 10 carbon straight chain fatty acid isolated from seeds of *Ulmus americanus* showed antifungal activity against *Ceratocystis ulmi* and several other fungi (14). Lycorine isolated from the fresh bulb tissue of *Urginea altissima* showed considerable antifungal activity against *P. capsici* (58). The gallic acid isolated from rose flowers was reported to inhibit the growth of *C. pallescens, Cephalosporium sacchari* and *F. nivale* (13). Gupta *et al.* (30) isolated three antifungal agents from dried rhizomes of *C. zedoaria,* of which the most abundant was ethyl P-methoxy cinnamate. *F. oxysporum, H. oryzae (C. miyabeanus)* and *Sclerotium rolfsii* were inhibited at 25 μg/ml. The growth of *A. tenuis, C. lunata, F. nivale* and *H. gramineum* was totally inhibited by the leaf extracts from *Clematis gouriana* and the fungitoxic principle was found to be protoanemonin (50).

Deltozid and deltonin isolated from rhizome of deltoid yam were inhibitory to conidia of *F. solani* and zoospores of *P. infestans* (100). A non-phytotoxic and non-systemic but highly effective antifungal property against a number of fungi was observed in *Ranunculus scleratus* (51). Subsequently this antifungal compound from *R. scleratus* was identified as protoanemonin (52), reported earlier from *C. gouriana.*

A broad spectrum fungitoxic principle identified as Lawsone as isolated from the leaves of *L. inermis* (97). The minimum effective dose against *H. oryzae* was 1000 ppm and it was non-phytotoxic. Hartman *et al.* (35) in their search on bacterial and fungal inhibiting compounds in higher plants observed that isopimaric and dehydroabuietic acid inhibited gram +ve bacteria but not gram -ve with one exception. The extracts of odoriferous organic compounds were more inhibitory than 500 ppm of aureofungin to growth of *A. tenuis* and *C. lunata,* the incitants of leaf spot disease of betelvine ((*Piper betle* L.) and that from *Trachyspermum ammi* was found to be the best inhibitor against them (78).

Jasmonic acid isolated from the leaves of *Oryza officinalis* inhibited 100% spore germination of *P. grisea (p. oryzae)* at 250 ppm, 86% at 200 ppm and 10% at 100 ppm (62). The appressorium formation in germinated spores were also inhibited to 30% at 100 ppm. Carvine the active principle in *Nigella sativa* strongly inhibits *R. solani* (71).

Rowan *et al.* (74) isolated the steroidal alkaloides, 5-solasodan-3-one and solasodone antifungal stress metabolites from *Solanum aviculare.* The constitutive leaf alkaloides solasodine and solamargin also showed antifungal activity against *C. gloeosporioides.* Fresh rhizomes of *Zingiber officinale* were toxic to *A. solani* and the active principle was 3, 7, dimethyl 2, 6-octadieneal (Citral). Citral showed a broad spectrum of activity at 1:400, inhibiting all the fungi tested but served as specific toxicant at 1:1600 inhibiting only *Abridia romosa* and *F. nivale.* Citral was 8 times more effective than Ziram and Blitox-50 and 4 times more than Dithane M-45 and Dithane Z-78 (78).

Reuveni *et al.* (72) isolated cineol, linalool, methyl chavicol and eugenol from *O. basilicum* oil. He found that linalool and methyl chavicol gave 100% inhibition of *R. stolonifer* while eugenol gave 100% inhibition of *F. oxysporum* f. sp. *vasinfectum.* A number of pathogenic (6) and non-pathogenic (5) fungi were inhibited by triticenes extracted from aqueous wheat homogenate (81).

Kobayashi *et al.* (41) reported pisiferic acid derivatives as an antifungal constituent of *Chrymaecyparis pisifera* var. *plumosa* toxic to *P. grisea (P. oryzae).* They also suggested that the size of the substituents at C_{10} and C_{12} and their presence and electro-negativity constituents of oxygen atoms were structural determinants for activity. Mori *et al.* (59) stated that two of the fatty acids derived from *Miscanthus sinensis* showed antifungal activity against rice blast pathogen, *P. grisea(P. oryzae).* These fatty acids were also found in infected plant leaves following treatments with probenazole. They found that free monocarboxylic fatty acid with C_{10} - C_{13} carbon length was the basic structural factor for inducing antifungal activity against rice blast pathogen.

Dubey and Tripathi (16) isolated eugenol of *Piper betle* and found it fungistatic against *A. flavus* and 13 other fungi. *Costus speciosus,* a medicinal plant produced a compound, methyl-3(4-hydroxyphenyl)-2 (e)-propenoate and was found inhibitory to *Curvularia* spp. and other fungi (4). They recorded the inhibitory effect of this compound comparable with standard fungicide benlate.

Gärcia *et al.* (25) isolated 3β, 7α-dihydroxyergosta-5, 24(28)-diene from the lipid fractions of *Entandrophragma utile* and found fungicidal against *P. grisea (P. oryzae).* This compound has also been isolated from the sponge *Halidona oculata* whose activity against blast pathogen is not indicated.

O. gratissimum, O. viride, O. canum (O. americanum), O. basilicum, Cymbopogan winterianus and *C. martinii* predominantly containing eugenol, geraniol, thymol, linaloof, methyl chavicol, citronellal and geraniol respectively were screened against *R. solani* Out of these thymol, linalool and geramiol checked the growth of this pathogen (95).

β-(Isoxazolin-5-on-2-yl)-alanine (βIA), a heterocyclic non-protein amino acid from root extracts and root exudates of pea seedlings have been found to be a potent growth inhibitor of several eukaryotic organisms including yeast (75). They reported that biotic effect on bakers yeast was reversed by L-methionine, L-cysteine and L-homocysteine. Mycelial growth and sclerotial production of *R. solani* grown on agar containing (βIA) was also inhibited. They further affirmed that *Bradyrhizobium japonicum* gave a positive chemostatic response to βIA. Miles *et al.* (49) isolated 3-formyl-2', 4',',6', trihydroxy dihydrochalcone from *P. acutangulam* leaves and twigs extracted in dichloromethane and reported antifungal properties against *R. solani.* Achenbach and Hemrich (1) isolated 12 alkaloids from the dichloromethane soluble parts of methanol extracts fractionated from petroleum extract of *Oxymitra velutina* twigs. He reported that out of these aporphinoids including lysicamine showed antifungal properties against *R. solani.* The antifungal activity of saponins isolated from the defatted flour of *Madhuca butyracea (Diploknema butracea)* against *Fusarium* sp. has been found (43).

7. PROMISING BOTANICALS

Out of several plant products screened, certain commonly available are listed below (Table 1):

Table 1 : Table showing potenaials of certain commonly available promising botanicals against two major fungal pathogens each of rice and groundnut

Botanical source	Preparation	Pathogens	References
Acer saccharum	Aqueous extract	*Rhizoctonia* spp.	103
Aegle marmelos	Aueous extract, steamed aqueous extract, ethanolic extract, hexane extract benzene extract, acetone extract	*Pyricularia oryzae* *Rhizoctonia solani*	86,87,90
Allium sativum	Aqueous extract, steamed aqueous extract, ethanolic extract, essential oil	*P. oryzae* *R. solani* *A. niger*	28, 48, 86

Botanical source	Preparation	Pathogens	References
Amaranthus spinosus	Aqueous extract	*P. oryzae*	29
Blumea balsamifera	Aqueous extract	*P. oryzea* *R. solani*	29, 86
Datura stramonium	Aqueous extract Ethanolic extract	*P. oryzae* *R. solani*	29, 54, 86, 90
Derris elliptica	Aqueous extract	*P. oryzae*	29
Euphorbia hirta	Aqueous extract Ethanolic extract	*P. oryzae* *R.. solani*	29, 86
E. pulcherrima	Aqueous extract	*P. oryzae*	29
Ipomoea batatas	Aqueous extract	*P. oryzae*	29
I. cornea	Aqueous extract Ethanolic extract	*P. oryzae* *R. solani*	86
Mirabillis jalapa	Aqueous extract Ethanolic extract	*P. oryzae* *R. solani*	29, 86
Psidium guajava	Aqueous extract Ethanolic extract	*P. oryzae* *R. solani*	29, 56
Vitex negundo	Aqueous extract Ethanolic extract	*P. oryzae* *R. solani*	29, 86
Amorphophallus bulbifer	Steamed aqueous extract Ethanolic extract	*P. oryzae* *R. solani*	86
Annona squamosa	Steamed aqueous extract Ethanolic extract	*P. oryzae* *R. solani*	86
Momordica charantia	Steamed aqueous extract Ethanolic extract	*P. oryzae* *R. solani*	86
Piper betle	Steamed aqueous extract Ethanolic extract	*P. oryzae* *R. solani*	16, 86
Zizyphus jujuba	Steamed aqueous extract Ethanolic extract	*P. ozyzae* *R. solani*	86
Sapindus emarginatus	Steamed aqueous extract	*P. oryzae* *R. solani*	94
Thuja orientalis	Aqueous extract Ethanolic extract	*P. oryzae*	56
Lagerstroemia parviflora	Aqueons extract Ethanolic extract	*P. oryzae* *A. niger*	56
Cinnamomum zeylanicum	Ethanolic extract	*R. solani*	56
Polyalthia longifolia	Ethanolic extract Essential oil	*P. grisea* *A. niger* *R. solani*	55

Botanical source	Preparation	Pathogens	References
Eucalyptus genii	Essential oil	*A. niger*	32
Zanthoxylum alatum	Essential oil	*A. flavus*	17
Oryza officinalis	Jasmonic acid	*P. oryzae*	62
Nigella sativa	Carvine	*R. solani*	71
Ocimum basilicum *O. canum* *O. gratissimum* *O. sanctum* *O. viride*	Steamed aqueous extract, Ethanolic extract, Essential oil, Eugenol, Geranial, Thymol, Linalool	*P. oryzae* *R. solani* *A. flavus*	17, 30, 86, 95

8. EVIDENCE OF SUCCESS FOR RICE BLAST CONTROL WITH *AEGLE MARMELOS* AND *OCIMUM SANCTUM* EXTRACTS

8.1. *In vitro* fungitoxic performance evaluation of *Aegle marmelos* and *Ocimum sanctum*

The three preparations viz., aqueous/steamed aqueous, ethanolic and essential oil extracts from the renewable sources, leaves of *Aegle marmelos* Corr. and *Ocimum sanctum* L. has been tested against more than one of these pathogens inciting blast and sheath blight of rice and collar rot and yellow root diseases of groundnut in that order. Bioassy tests in laboratory and field against these have been studied (90). Properties of extracts from *A. marmelos* and *O. sanctum* displayed independently varying degree of fungitoxic pattern on conidial germination as well as on mycelial growth. At the lowest concentration, *P. grisea* conidia displayed production of thinner and coiled germ tube compared to the normal in control. Subsequently on increasing the dose conidia responded showing drastic reduction in the germ tube length followed by formation of bulbous (round) structure at the germ tube tip or on top of it coupled with sometimes twisted germ tube. Septal dissolution granulation or aggregation of cytoplasmic contents or bursting of cell wall have also been observed in *in vitro* tests at higher concentrations with the extracts of the above two plants *(A. marmelos* and *O. sanctum)* consequently resulting into the leaching out of cellular contents thus rendering the conidia deformed and making it incapable of inciting disease.

Similarly increase in fungitoxic pattern on mycelial growth was observed in ascending order corresponding to the increase in ethanolic extract dosages of either plants against *P. grisea, R. solani, A. niger* and *A. flavus* pathogens. The pattern observed between 0.01-12% concentration treatments were as follows. At 0.01% concentration it reduced sporulation, no sporulation or delayed/no sclerotial production. Strikingly fluffy (luxurious) mycelial growth or concentric ring type restricted discontinuous radial growth was the typical fungitoxic performance observed with rice blast pathogen *P. grisea* especially when the rate of radial growth was drastically reduced in treated petriplates compared to control.

8.2. Effectivity of the product in greenhouse

Considering fungitoxic performance of both *A. marmelos* and *O. sanctum* preparation *in vitro,* these were tested for rice blast control in greenhouse. Out of these, the ethanolic extract and essential oil of *A. marmelos* and the essential oil of *O. sanctum* followed by ethanolic extract or steamed aqueous extract gave maximum disease reduction (Table 2).

8.3. Effect of *A. marmelos* and *O. sanctum* preparations in control of rice blast in field

Encouraged with the result obtained in the greenhouse test the ethanolic extract preparation of both *A. marmelos* and *O. sanctum* were evaluated in the field for the control of rice blast during three

Table 2 : Effect of A. marmelos and O. sanctum preparations on the reduction of foliar blast of rice in greenhouse

Treatment	Disease score (0-9 scale)	
	A. marmelos	*O. sanctum*
Aqueous extract *(A. marmelos)* & steamed aqueous extract *(O. sanctum)* (0.1%)	0.4	2.4
Ethanolic extract (0.1%)	0.1	2.2
Essential oil (0.1%)	0.1	0.8
Carbendazim	0.6	0.6
Control	8.0	8.0
C.D.	0.4	

test seasons. The disease in the field was reduced in treatments with both the extracts (1.5 to 9.1%) and was found comparable with ediphenphos (5.9%). Control produced 46.6-73.9% foliar blast in field.

It was further decided to conduct studies on ethanolic extracts of both *A. marmelos* and *O. sanctum* subsequently coded as NPP-I (1985-1987) and NPP-II (1986-87) respectively under various agro-climatic conditions through Directorate of Rice Research Hyderabad during wet season.

8.4. Test locations

The test locations of these natural products during the reported years were Aduthurai, Coimbatore, Almora, Cuttack, Lonavala, Ponnampet, Nawgaon, Khudawani, Malan, Kanke, Rewa, Pattambi, Igatpur, Nellore, Hazaribagh and Wangbal which were hot spots for the occurrence of rice blast.

8.5. Effectiveness of the product

Largely, significant reduction of the foliar and neck blast of rice and consequent increase in the yield by the application of either of the products was recorded at various agro-climatic regions of the country. The performance of these products were comparable with various synthetic fungicides such as Topsin-M, Kitazin, Fujione, Carbendazim, Fongoren, Ediphenphos and Berfran in reducing the foliar and neck blast *(P. grisea)* and increasing the yield of paddy at various centres.

Thus it is evidently clear from the above citations that choice for integrating botanical based products such as aqueous extract, extract in organic solvents, essential oil or active ingredient and others in plant protection strategy is vast. These can be appropriately selected, formulated and introduced in a phased manner in order to reduce further compounding of problems arising out of the use of synthetic chemicals.

9. PRACTICAL PROBLEMS IN USING BOTANICAL TOXICANTS IN THE STRATEGY OF INTEGRATED PEST MANAGEMENT

9.1 Lack of organised approach

Owing largely to un-organised approach, so far, in exploitation and utilization of botanical products as a promising strategy in plant disease and pest management, non-availability of the product at an appropriate time of its application becomes a limiting factor for cultivators.

9.2 Proneness to rapid decomposition

Since botanical based products when applied in the field have the tendency for their rapid decomposition, hence for better efficacy of these a precise dose response relationship in response to duration of their residual persistence in field must be worked out to determine quantity frequency and the interval of treatments in field.

9.3 Qualitative and quantitative variability

Qualitative and quantitative diversity in plant and plant products, due to agro-ecological and seasonal variations may also influence the performance of any such product. Therefore these factors must be accounted, when use of plant products are to be advocated.

9.4 Patenting policies complexities

Patenting policies of the plant products if compared with synthetic produces, the process of commercialization of the former products will needlessly be delayed, hence this must be simplified.

9.5 Lack of network approach on information retrieval and its exchange

In order to get a modest advantage of plant protection through the use of botanical products as chemotherapeutants, scattered efforts achieved in this field must be consolidated and integrated through periodical exchange of informations and coordinated network approaches.

10. PROSPECTS OF BOTANICALS IN PEST MANAGEMENT STRATEGY

Pesticides have played an undisputable role in protecting agricultural crops, thus convincingly boosted up production to cater food requirement of growing population world over. However, environmental pollution associated with their use attracted greater public attention during the recent past. Its resultant impact has been seen as gradual disappearance of certain hazardous synthetic pesticide products from the markets.

11. FUTURE APPROACH

Diverse plant genetic resources and their wide spectrum utility are the gift of the nature, largely remaining untapped for the betterment of mankind and the environment. Having gone through several hazardous effect experiences of the synthetic pesticides in the past it is genuine now to exploit the natural plant resources and integrate their use as fungicide/insecticide in the strategy of plant disease/ pest management.

(a) The approach must therefore constitute to prioritize plants having greater potential against the important diseases/pests of important agricultural crops, after a systematic survey and biological activity analysis both *in vitro* and *in vivo*.

(b) Availability of the plant (s) or products thereof that have proven potential against specific pathogens/pests must be insured at the time of need. This can be achieved by either mapping out the naturally occurring plant population or encouraging plantation of the desired plant species at appropriate places.

(c) Plant products must suitably be formulated, their shelf-life, thermal and photolytic activity evaluated, appropriately packed before their recommendation for final use.

(d) Exchange of informations through conferences, symposia, seminars, publications of findings in scientific journal etc. regarding successful products needs to be accelerated.

(e) Coordinated network approach should be made globally to derive maximum advantages of the findings under varied agro-ecological conditions.

(f) Legislative control should be liberal for such products and industrialists should be encouraged for production and sale of plant product based pesticides.

(g) Merits of the plant products should receive wide publicity based on authentic informations.

12. CONCLUSIONS

In view of the data base generated as cited above and several others, the plants constitute a strong potential against several pathogens, hence renaissance of natural pesticide research is being experienced now with a greater thrust. In tropical and sub-tropical countries where profound botanical wealth with precious genetic diversity in spread over more than temperate regions, however, needs a systematic and scientific investigation, with special reference to their potential against devastating disease/pest of crops in field and harvest in storage. Experiences of tribals in several parts of hilly areas on the utility of herbs must also be understood and evaluated scientifically.

Over a dozen companies in industrialized countries and several in India and Burma are now working on commercial neem products. Giessen University was the first to obtain a neem extract using an alcohol solvent in 1982. Margosan-0 was produced in the United States and has since been registered for marketing as a commercial neem pesticide by the American Environmental Protection Agency (36). *P. betle* and *O. sanctum* extracts have proved to prevent spread of a serious disease of rice causing blast in field (86, 91). These and many other plant products, thus holds a good promise for their suitable integration into plant protection strategies in future, must, though, in a phased manner.

Now it has been consciously believed and appreciated that botanicals are renewable one of the environmentally non-pollutive inexhaustible, largely non-phytotoxic and a cheaper indigenous ephimeral resource thus constitute a prospective appropriate component in the strategy of safer integrated pest management.

13. REFERENCES

1. Achenbach, H. and Hemrich, H. 1991. Alkaloids, flavonoids and phenylpropanoids of the West African plant, *Oxymitra vehetina.* Phytochemistry, **30(4)** : 1265-1267.

2. Allen, E. H. 1970. The nature of antifungal substances in the peel of Irish potato tubers. Phytopathol. Z. **69** : 151-152.

3. Annapurna, Y., Mitra, S., Iyengar, D. S., Rao, S. N. and Bhalerao, U. T. 1983. Antimicrobial activity of leaf extracts of *Polyalthia longifolia.* Phytopathol. Z. **106** : 183-185.

4. Bandra, B.M.R., Fernando, I.H.S., Hewage, C. M., Karunarate, V., Adikaram, N. K. B. and Wijesundara, D.S.A. 1989. Antifungal activity of some medicinal plants of Sri Lanka. J. Natn. Sci. Coun. Sri Lanka, **17 (1)** : 1-13.

5. Barlocher, F., Oertli, J. J. and Guggenheim, R. 1979. Accelerated loss of antifungal inhibitors from *Pinus leucodermis* needles. Trans. Brit. Mycol. Soc. **72** : 277-281.

6. Bendall, G. M. 1975. The allelopathic activity of California thistle *(Cirsium arvense* (L) Scop) in Tasmania. Weed Res. **15** : 77-81.

7. Bradley, R. 1983. New Improvements of Planting and Gardening - Both Philosophical and Practical. London, Butterworth, Hitch, Pemberton, Knapton and Brown, p. 221.

8. Brodrick, H. T. 1971. Toxicity of limonene and citrus peel extracts to *Guignardia citricarpa* Kiely. Phytophylactica, **3** : 69-72.

9. Chaudhury, T. and Sen, C. 1982. Effects of some plant extracts on three sclerotia forming fungal pathogens. Zeitschrift Fur. Pflanxcakrankheiter und Pflanzenschutz, **89** : 852-585.

10. Chauhan, P. and Zaxena, V. K. 1985. Antifungal activity of essential oil of leaves of *Inula caspidata.* Indian J. Pharm. Sci. **47(4)** : 160-161.

11. Christensen, C. M. and Lopez, F.L.C. 1963. Pathology of stored seeds. Proc. Int. Seed Test. Ass. **28** : 701-711.

12. Dixit, K. 1986. Evaluation of some higher plants and their products for preservation of some food-stuffs against fungal deterioration during storage. Ph.D. thesis, Univ. of Gorakhpur, Gorakhpur, India.

13. Dixit, S. N., Tripathi, S. C. and Upadhyaya, R. R. 1976. The antifungal substance of rose flowers *(Rosa indica).* Economic Botany, **30** : 371-374.

14. Doskotch, R. W., Keely, S. L. Jr. and Schreiber, L. R. 1975. Isolation and identification of an antifungal agent from seeds of American elem. Phytopathol. **65** : 634-635.

15. Do Vgich, N. A. 1971. Antimicrobial effect of essential oils. Mikerobiol. Zh. **33** : 253-259.

16. Dubey, P. and Tripathi, S. C. 1987. Studies on antifungal, physico-chemical and phytotoxic properties of the essential oil of a *Piper betle.* Journal of Plant Disease and Protection, **94(3)** : 235-241.

17. Dube, S., Upadhayay, P. D. and Tripathi, S. C. 1989. Antifungal, physico-chemical and insect-repelling activity of the essential oil of *Ocimum basilicum.* Canadian Journal of Botany, **67** : 2085-2087.

18. Egawa, H., Tatsuyama, K., Kobata, K. I. and Fumoto, J. 1973. Antifungal activity of Ericaceae plants. B. Facul. Agr. Shimane u. **7** : 31-33.

19. Fawcett, C. H. and Spencer, D. M. 1969. Natural antifungal compounds. Fungicides, Vol. 2, Ed. D. C. Torgeson, Academic Press, New York and London, p. 637.

20. Fawcett, C. H. and Spencer, D. M. 1970. Plant chemotherapy with natural products. Ann. Rev. Phytopath. **8** : 403-418.

21. Fawcett, C. H., Spencer, D. M., Wain, R. L., Jones, Sir E.R.H., Laquan, M., Page, C. B. and Thaller, V. 1965. An antifungal acetylenic keto-ester from a plant of the Papilionaceae family. Chem. Commun. **1** : 422.

22. Fawcett, C. H., Spencer, D. M., Wain, R. L., Fallis, A. G., Jones, Sir E. R. H., Lequan, M., Page, C. B., Thaller, V., Shubrook, D. C. and Whitham, P. M. 1968. Natural acetylenes. Part XXVIII. An antifungal acetylenic furanoid keto-ester (Wyerone) from shoots of broad bean *(Viaia faba* L.) Fam. Papilionaceae. J. Chem. Soc. 2455.

23. Forsyth, W. 1802. Treatise on the Culture and Management of Fruit Trees. Longman, Rees, Cadel, Davis and Debrett, London, p. 250.

24. Friedman, O. T. and Horowitz, M. 1971. Biological active substance in Subterranean parts of purple nutsedge. Weed Sci. **19** : 389-401.

25. Garcia, J., Morin, C., Nyassee, B., Sondenga, M.B.L., Tchouankeu, J. C. and Tsamo, E. 1991. Complete structural assignments of an ergesterol derivative from *Entandrophragma utile.* J. Nat. Prod. **54 (1)** : 136-142.

26. Garg, S. C. 1974. Antifungal activity of some essential oils. Indian J. Pharm. **36** : 46-47.

27. Garg, S. C. and Oswal, V. B. 1982. *In vitro* antifungal activity of the essential oil of *Chloroxylon sweitenia* Roxb Coram. Indian Drugs, **19** : 189-190.

28. Geda, A., Saxena, S. P. and Bokadia, M. M. 1978. *In vitro* antimicrobial efficiency of essential oil against human pathogenic bacterial and phytopathogenic fungi. Indian Drugs, **156** : 253-255.

29. Grainage Michael, Saleem Ahmed, Wallace, C., Mitchell and Hylin John, W. 1985. Plant species reportedly possessing pest control properties - AN-EWC/UH Data BAS, pp. 249.

30. Gupta, S. and Banerjee, A. B. 1970. A rapid method of screening antifungal antibiotic producing plants. Indian J. Exp. Biol. **8** : 148-149.

31. Gupta, S. K., Banerjee, A. B. and Achari, B. 1976. Isolation of ethyl p-methoxycinnamate, the major antifungal principle of Curcuma zedoaria. Lloydia, **39** : 218-219.

32. Haerdtl, H. 1962. Fungicidal properties of oils and solid filters. Deuts and Pothek Ztd. **102** : 447-448.

33. Harborne, J. B. 1972. Phytochemical Ecology. Academic Press, London, pp. 272.

34. Harborne, J. B. 1973. Phytochemical Methods. Champman & Hall, London, pp. 278.

35. Hartmann, E., Renz, B. and Jung, J. A. 1981. Investigations of bacterial and fungal inhibiting compounds in higher plants. Isolation, identification and activity spectrum of two resin acids from Sprucebark. Phytopathol. Z. **101** : 31-32.

36. Hoddy, Elizabeth. 1991. Nature's Bitter Boon : The Neem Tree-a substitute for pesticides. Development and Cooperation, No. 5, pp. 29-30.

37. Hong, L. T. and Abdul Razak, M. A. 1983. Antifungal properties of methanol extractives from some tropical hard woods. Malaysian Forester, **46** : 138-139.

38. Jain, P. K., Charia, A. K., Sharma, S. K. and Bokadia, M. M. 1979. Study of some essential oils for their antifungal activities. Indian Drugs, **16** : 122-123.

39. Janardhan, K. K., Ganguly, D., Baruah, J. N. and Rao, P. R. 1963. Fungitoxicity of extracts from tannin-bearing plants. Curr. Sci. **32** : 226-227.

40. Kala, P. K., Tripathi, R. K., Gupta, K. C. and Singh, A. K. 1984. Effect of some essential oils on growth and aflatoxin production by *Aspergillus parasiticas* in stored grains. Pesticides, **18** : 43-46.

41. Kobayashi, K., Nishino, G., Tomita, H. and Fukushima, M. 1987. Antifungal activity of pisiferic acid derivatives against the rice blast fungus. Phytochemistry, **26(12)** : 3175-3179.

42. Korta, J. and Starzyk, J. 1963. Investigation on antibiotic properties of essential oils of certain species of Umbelliferae. Acta. Botanica Croncoviensia, **6** : 149-160.

43. Lalitha, T. and Venkataraman, L. V. 1991. Antifungal activity and mode of action of saponins from *Madhuca butyracea* Macb. Indian J. Exptl. Biol. **29(5)** : 558-562.

44. Lapis, D. B. and Dumancas, E. E. 1978. Fungicidal activity of crude plant extracts on *Helminthosporium oryzae*. Philippine Phytopathol. **14** : 23-27.

45. Mahadevan, A. and Mohan, N. 1978. Prohibition in *Catharanthus roseus*. Curr. Sci. **7** : 57-58.

46. Malekzadeh, F. 1974. An antimicrobial compound in two Pistacia species. Mycopathologia and Mycologia Applicata, **54** : 73-77.

47. Mall, H. V., Asthana, A., Dubey, N. K. and Dixit, S. N. 1985. Toxicity of *Cedar* wood oil against some dermatophytes. Indian Drugs, **22** : 296-298.

48. Maruzzella, J. C. and Balter, J. 1959. The action of essential oils on phytopathogenic fungi. Pl. Dis. Reptr. **43(11)** : 1143-1147.

49. Miles, D. H., Rosa De Medeiros, J. M., Chittawong, V., Bedin, P. A., Swithenbank, C. and Lidert, Z. 1991. 3'-Formyl 2', 4', 6' trihydroxydihydrochalone from *Psidium acutangulum.* Phytochemistry, **30(4)** : 1131-1132.

50. Misra, S. B. and Dixit, S. N. 1977. Fungicidal properties of *Clematis gouriana.* Indian Phytopath. **30** : 577-579.

51. Misra, S. B. and Dixit, S. N. 1978a. Antifungal properties of leaf extracts of *Ranunculus scleratus* L. Experentia, **34** : 1442-1443.

52. Misra, S. B. and Dixit, S. N. 1978b. Antifungal activity of *Ranunculus scleratus* Linn. Sci. & Cult. **44** : 369-370.

53. Mishra, Debjani and Tewari, S. N. 1989. Influence of *Ocimum sanctum* essential oil on the two major fungal pathogens of *Sanctum* essential oil on the two major fungal pathogens of rice. National Symp. on recent trends in plant disease control. April 13-14, North-Eastern Hill Univ., Medziphema, Nagaland, p. 27.

54. Mishra, Manasi and Tewari, S. N. 1990. Ethanolic extract toxicity of three botanicals against five fungal pathogens of rice. Nat. Acad. Sci. Lett. **13** : 409-412.

55. Mishra, Manasi and Tewari, S. N. 1992. Toxicity of *Polyalthia longifolia* against fungal pathogens of rice. Indian Phytopath. **45** : 59-61.

56. Mishra, Manasi, Mallik, S. S. and Tewari, S. N. 1992. Alleopathic effect of certain botanicals against six fungal pathogens of rice. Proceedings of First National Symposium "Allelopathy in agro-ecosystems". Eds. P. Tauro and S. S. Narwal, Indian Society of Allelopathy, Haryana Agricultural University, Hisar, India, pp. 191-193.

57. Mitra, S. R., Chaoudhury, A. and Choudhury, A.N. 1984. Protection of antifungal compounds of higher plants-A review of recent researches. Pl. Physiol. & Biochem. **11** : 53-77.

58. Miyakado, M., Kato, R., Ohno, N. and Koshimiz, U. K. 1975. Alkaloids of *Urginea altissimac* and their antimicrobial activity against *Phytophthora capsici.* Phytochemistry, **14** : 2717.

59. Mori, A., Anoki, N., Shinozuka, K., Nishino, C. and Fukushima, M. 1987. Antifungal activity of fatty acids against *P. oryzae* related to antifungal constituents of *Miscanthus sinensis.* Agril & Biol. Chem. **51(12)** : 3403-3405.

60. Murdoch, D. I. and Allen, W. E. 1960. Germicidal effect of orange peel oil and d-limonene in water and orange juice. I. Fungicidal, properties against yeast. Food Technol. **14** : 441-445.

61. Murthy, M. B. K. and Amonkar, S. V. 1974. Effect of natural insecticide from garlic (*Allium sativum* L.) and its synthetic form (diallyldisulphide) on plant pathogenic fungi. Indian J. Expt. Biol. **12** : 208-209.

62. Neto, G. C., Kono, Y., Hyakutake, Watanabe, M., Suzuki, Y. and Sukurai, A. 1991. Isolation and identifications of (-)-jasmonic acid from wild rice *Oryza officinalis* as an antifungal substance. Agriculture and Biological Chemistry, **55** : 3097-3098.

63. Nicolls, J. M. 1970. Antifungal activity in *Passiflora* species. Ann. Bot. **34** : 229-237.

64. Nitya Nand. 1977. Research on indigenous drugs. Sci. & Cult. **43** : 17-21.

65. Okazaki, K. and Homma, A. 1953. Antibacterial activity of higher plants. XXVI. Antimicrobial effect of essential oil. I. Oil of *Thujopsis dolabrata* and its components. J. Pharm. Soc. Japan, **74** : 174-176.

66. Pariya, S. and Chakravarti, D. K. 1977. Antifungal activity of some Indian medicinal plant extracts on phytopathogenic fungi. Phytopathologia Mediterranea, **16** : 33-34.

67. Pathak, A. and Dixit, V. K. 1984. Antifungal and antibacterial studies on essential oil of *Glossocardia bosvallia.* Indian Perfumer, **28(2)** : 91-94.

68. Pelletier, G. and Comeau, A. 1972. Growth and sporulation of *Septoria avenae* Frank as inhibited by factors present in leaves of *Avena sativa* L. Proc. Canad. Phytopathol. Soc. **39** : 48.

69. Rajak, R. L. 1992. Pest management and pesticides : Indian Scenario (Plant Protection Past Forty Years - An Overview), Ed. B. Vasanthraz David. Namrutha Publications, Madras, 384 pp.

70. Rao, B.G.V.N. and Joseph, P. L. 1971. Activity of some essential oils towards phytopathogenic fungi. Riechest Aromes Koer, **21** : 405-410.

71. Rathe, P. S., Mishra, S. H. and Kaushal, R. 1982. Antimicrobial activity of essential oil, fixed oil and unsaponifiable matter of *Nigella sativa* Linn. Indian J. Pharm. Sci. **44** : 8-10.

72. Reuveni, R., Fleischer, A. and Putiersky, E. 1984. Fungistatic activity of essential oils from *Ocimum bacilicum* chemotypes. Phytopathol. Z. **110** : 20-22.

73. Ribeiro, A. S. and Cardoso, C.O.N. 1977. Effects of rice leaf extracts in *Pyricularia oryzae* CAV. Summa. Phytopathologica, **3** : 201-204.

74. Rowan, D. D., MacDonald, P. E. and Skipp, R. A. 1983. Antifungal stress metabolites from *Solanum aviculare.* Phytochem. **22** : 2102-2104.

75. Schenk, S. U., Lambein, F. and Werner, D. 1991. Broad antifungal activity of β-isoxazolinonyl-alanine, a non-protein amino acid from roots of pea *(Pisum sativum* L.) seedlings. Biology and Fertility of Soils, **11(3)** : 203-209.

76. Sharville, E. G. 1960. The Nature and Use of Modern Fungicides. Burgess Publishing Co., Minnesota, U.S.A., p. 3.

77. Singh, U. P., Pathak, K. K., Khare, M. N. and Singh, R. B. 1979. Effect of leaf extract of garlic on *Fusarium oxysporum* f. sp. *ciceri, Sclerotinia sclerotiorum* and on gram seeds. Mycologia, **71** : 556-564.

78. Singh, B. S., Johari, J. K. and Balasubramanyam, U. R. 1982. Effect of some Odoriferous organic compounds on the growth of *Alternaria tenuis* Auct. and *Curvularia lunata* (Wakker) Boedijun causing leaf spot diseases of betelvine *(Piper betle* L.). New Botanist, **9** : 13-14.

79. Skinner, F. A. 1955. Antibiotics. In : Modern methods of plant analysis. Vol. III. Eds. K. Peach, M.N. Tracey. Springer-Verlag, Berlin, p. 626-725.

80. Slavenas, I. 1963. Physiological role of essential oils. Obschch Zakonomer Rosta. Razk. Rast. Dokl. Nauk. Knof. Vilnyus, pp. 313-338.

81. Spendley, P. J. and Ride, J. P. 1984. Fungitoxic effects of two alko-2,4-dienals and triticene, isolated from wheat. Trans. Brit. Mycol. Soc. **82** : 283-284.

82. Spencer, D. M., Topps, J. H. and Wain, R. L. 1957. An antifungal substance from the tissues of *Vicia faba.* Nature, London, **179** : 651.

83. Staron, T., Allard, C., Beitz, Suzanne and Billet, Dinise. 1969. Purification and antifungal properties of a triterpenic heteroside extracted from the stem and leaves of *Angallis arvensis* var. *Phoenicea phytiat.* Phytopharm. **18** : 161-166.

84. Subba Rao, M. S., Sumithiri, T. C. and Surnarayana Rao, P. 1967. Antimicrobial action of citrus oils. J. Food Sci. **32** : 225-227.

85. Swain, T. 1977. Secondary compounds as protective agents. Ann. Rev. Plant Physiol. **28** : 479-501.

86. Tewari, S. N. 1986a. Evaluation of some of the indigenous plants for antifungal and antibacterial properties against major fungal and bacterial pathogens of rice. Ph.D. thesis, Utkal University, Orissa, India.

87. Tewari, S. N. 1986b. A new technique for bioassay of natural plant product. Curr. Sci. **22** : 1137-1139.

88. Tewari, S. N. 1992. Botanicals in allelopathy march-A current view and the thrust for future (Abstr.). Proceedings of First National Symposium on Allelopathy in Agro-ecosystems. Eds. P. Taure and S. S. Narwal, Indian Society of Alleolopathy, Haryana Agricultural University, Hisar, India, p. 19.

89. Tewari, S. N. and Dath, A. P. 1984. Effect of leaf extract media of some plants on the growth of three fungal pathogens of rice. Indian Phytopath. **37** : 458-461.

90. Tewari, S. N. and Mishra, Manasi. 1990. *Aegle marmelos* Corr., a botanical source of fungicide. Extended Summary. Proc. Int. Symp. on Rice research-New frontiers. November 15-18, 1990. Directorate of Rice Research, Hyderabad, p. 250-251.

91. Tewari, S. N. and Nayak, Mandakini, 1990. Activity of four plant leaf extracts against three fungal pathogens of rice. Trop. Agric (Trinidad), **4** : 373-375.

92. Tewari, S. N., Shukla, H. S., Biswal, K. M. and Nayak, Mandakini 1988. Fungitoxic properties of some leaf extracts. Nat. Acad. Sci. Lett. **11** : 369-373.

93. Tewari, S. N., Prakash, A. and Mishra, Manasi. 1989. Toxicity impact of *Thevetia nerifolia* on certain plant pathogenic fungi and stored grain pest (Abstr.). National Symposium on Research Trends in Plant Disease Control. April 12-14, North Eastern Hill Univ., Medziphema, Nagaland, 120 pp.

94. Tewari, S. N., Prakash, A. and Mishra, Manasi. 1900. Biological insecticidal property of *Sapindus emarginatus* Vahl. to rice pathogens and insect pests in storage. J. Appl. Zool. Res. **1** : 52-56.

95. Thakur, R. N., Singh, P. and Khosla, M. K. 1989. Studies on antifungal activities of some aromatic oil. Indian Perfumer, **33(4)** : 257-260.

96. Tripath, N. N. 1980. Fungitoxicity in some higher plants. Ph.D. thesis, Univ. of Gorakhpur, Gorakhpur, India.

97. Tripathi, R. D., Srivastava, H. S. and Dixit, S. N. 1978. A fungitoxic principle from the leaves of *Lawsonia inermis* Lam. Experientia, **34** : 51-52.

98. Uchida, S. 1928. In essential oils of leaves of 'Hiba' wood and its relation to the resistance of the wood to destruction by fungi. J. Soc. Chem. India, **31** : 501-508.

99. Uraguchi, K. and Yamazaki, M. 1978. Toxicology, Biochemistry and Pathology of Mycotoxins. John Wiley & Sons, New York, p 288.

100. Vasyukova, N. K., Paseshni, V. A., Chenko, Darydova, M. A. and Chalenko, G. I. 1977. Study on the fungitoxic properties of steroid saponins from the rhizome of deltoid yam. Izychenie fungi toksicheskikh svoistv steroidnykh saponinov iz kornev shch dioskorei del. toridwoi. Prikl. Biokhimiya. Mikrobiologiya, **13** : 172-176.

101. Wain, R. L. 1969. Naturally occurring fungicides. Proc. Symp. Potentials in crop protection, New York, p. 26.

102. Wain, R. L., Spencer, D. M. and Fawcett, C. H. 1961. Antifungal compounds in seedlings of *Vicia faba*. S.C.I. Monograph No. 15, p. 109.

103. Webb, D. P. and Agnihotri, V. P. 1970. Presence of a fungal inhibitor in the percarps of *Acer saccharum* fruits. Can. J. Bot. **48** : 2109-2116.

104. Whittaker, R. H. 1970. The biochemical ecology of higher plants. In : Chemical ecology. Eds. E. Sondhermer and J. B. Simeone. Academic Press, New York, 336 pp.

105. Woods, F. W. 1960. Biological antagonisms due to phytotoxic root exudates. Bot. Rev. **26** : 546-569.

106. Wyk, P.S., Van and Koeppen, B. H. 1974. p-Hydroxybenzoyl calleryanin, an antifungal substances in the root bark of *Protea cynaroides.* South African J. Sci. **70** : 121-122.

107. Wyllie, T. D. and Morehouse, L. G. (eds.) 1978. Mycotoxic Fungi, Mycotoxins and Mycotoxicoses. Marcel Dekker Inc., New York and Basel.

108. Zayed, M. N., Saber, M.S.M., Abd-El-Malik, Y. and Monib, M. 1971. Antibacterial substances in dry residue of certain higher Egyptian plants. Zent. Bl. Bakt. Parasiktde Abt. **126** : 615-629.

109. Zutshi, S. K., Joshi, S. K. and Bokadia, M. M. 1975. Antimicrobial activity of some Indian essential oils. Indian J. Pharm. **37** : 129-130.

30 Present Status of Cereal Downy Mildews and their Management

Vasanthi U. Kumar and H. Shekar Shetty*

CONTENTS

Integrated Pest and Disease Management
Rajeev K. Upadhyay, K.G. Mukerji, B.P. Chamola and O.P. Dubey (eds.)
APH Publishing Corporation, New Delhi-110002, India

* Corresponding Author

ABSTRACT

Cereal downy mildews are epidemiologically explosive and dangerous group of pathogens causing diseases characterised by yellowing, downy sporulation on the abaxial leaf surfaces, malformed inflorescence and stunted growth. The causal agents are Oomycetous, biotrophic fungi belonging to the order Peronosporales. The importance of this disease is mainly due to its effect on grain production. The disease can appear in epidemics causing loss of millions of acres of crop area. There are a number of strategies described for managing the downy mildews of cereals and each method seems to have some inherent limitation for application. This review analyses the downy mildew diseases of cereals and reasons out the best possible way to combat the disease. Work in the authors' laboratory has shown the emergence of biotechnological methods like Systemic Acquired Resistance and Somaclonal Variations for selection of resistant host cell lines via cell, tissue and organ culture as attractive methods for downy mildew disease management. This review, while describing the significance of these method, also enlightens the readers on various aspects of the pathogen like its biology, biochemistry and pathogenicity at cellular and molecular levels.

1. INTRODUCTION

'Mildews' are a group of plant diseases characterized by delicate white or highly coloured growth of the infecting fungi from plant organs. Riley (134) differentiated two types of mildews based on their appearance and classified them as 'downy mildews' and 'powdery mildews'. Downy mildews are of common occurrence affecting cereals, pulses, vegetables, oilseeds and plantation crops and have been the cause for a number of epidemics. Severe loss in productivity attracted wider research on these plant diseases.

Downy mildews are caused by fungi belonging to the order Peronosporales and the family Peronosporaceae. They are known to affect graminaceous crops and some species cause considerable yield loss in food crops like maize (*Zea mays* L.), sorghum [*Sorghum bicolor* (L.) Moench], and pearl millet [*Pennisetum glaucum* (L.) R. Br.]. In many of the underdeveloped countries, particularly in Africa and Asia, large number of people depend on these cereal crops for food and it also serves as a fodder for millions of cattle in tropical and sub-tropical countries. Lesser yield of these food grains due to downy mildew disease has resulted in hunger and malnutrition in the population of underdeveloped countries. In India, maize, sorghum and pearl millet are grown in an area of 16,18 and 11.3 million hectares with an annual production of 14, 16 and 18 million tonnes respectively. Since they occupy the third, fourth, and fifth positions of the Indian cereal crops, downy mildew disease management has been a study of importance in this country.

There are several reviews on the subject of downy mildews and the readers are encouraged to refer to many of these reviews (46, 83, 135, 138, 151, 152, 155, 161, 171, 186, 215). The foregoing discussion on the subject is not an intention to add up to the available list, but, essentially summarizes conclusions that have emerged from over two decades of research on downy mildew disease in this laboratory. The review also highlightens several new informations on the cereal host-downy mildew pathogen system and disease management strategies. It is hoped that this will stimulate scientists to work on the downy mildews, for they are a group of fungi with a lot of variability providing an ideal system for studies on host specificity.

2. DISTRIBUTION AND ECONOMIC IMPORTANCE

Most graminaceous downy mildews are 'old world in origin'. *Sclerospora graminicola, Sclerophthora macrospora, S cryophila* and *S. farlowii* were reported from Western hemisphere. Ever since the report of *Peronosclersopora sorghi* in 1964 (134) from Texas, it has spread to 9 others states in the USA. *P. sorghi* and *S. graminicola* have their origins in Africa (26) and from there it was disseminated to Asia by producing millions of oospores that were carried by seed and plant refuse (56, 216). The other possibility is that these pathogens infecting grasses in India were transferred to sorghum and pearl millet when these were introduced from Africa. Species of *Peronosclerospora* probably originated in tropical regions of the Far East, particular in South East Asia (150) *S. graminicola* has a temperate origin because it is circumpolar on species of *Setaria* and that it has become adapted to plants in tropical habitats, particularly pearl millet (150). The only record on the geographical distribution of graminaceous downy mildews in Asia is that of Frederiksen and Renfro (51). Hence, a thorough understanding of the distribution of downy mildew pathogens is necessary for there are unconfirmed reports of downy mildews on maize from countries which do not have any record of downy mildew disease. Sorghum, maize and other wild perennial grass hosts are widely distributed in the tropical and sub-tropical regions of South and North America. It is assumed that the downy mildew due to *P. sorghi* is spreading to these regions as well (35,99). Certainly, the pathogen is finding its way to these regions from Asia and Australia. Less studied downy mildew of pearl millet is the disease caused by *Plasmospora penniseti* (80). This disease was located in a remote area of Ethiopian highlands where pearl millet was introduced in 1966. Though claimed to be of Indian origin, the pathogen has not been isolated in India or elsewhere. The reported geographical distribution of the graminaceous downy mildews are given in Table 1.

Table 1. The reported geographical distribution of the graminaceous downy mildews

Region/ Continent	Species[b]	Countries
	P. dichanthicola	India
Asia	*P. heteropogoni*[a]	India
	P. maydis[a]	Indonesia
	P. miscanthi	Philippines, Taiwan
	P. philippinensis[a]	India, Indonesia, Nepal, Philippines, Thailand
	P. sachhari[a]	India, Nepal, Philippines, Taiwan, Thailand
	P. sorghi[a] (sorghum pathotype)	India
	P. sorghi[a] (maize pathotype)	Thailand
	P. spontanea	Philippines, Thailand
	P. westonii	India
	S. graminicola[a]	China, India, Pakistan
	Sc. macrospora	Several countries
	Sc. rayssiae var *zeae*[a]	India, Nepal, Pakistan, Thailand
Middle East	*P. sorghi*[a] (sorghum pathotype)	Israel
	S. graminicola[a]	Israel

Host infection at the seedling stage, by the downy mildew pathogens result in loss of grain and different degrees of loss in fodder. This is mainly due to colonization of meristematic host tissues by the pathogen, leading to the production of sterile inflorescence or prevention of inflorescence development of death of plants even when they are young. Though reduction in grain yield and fodder quality are reported by local lesion type of disease development, economic loss is not as high in systemic infection. Further, introduction of high yielding cultivars of exotic crop varieties is inhibited due to their susceptibility to downy mildew diseases. The available reports in the literature clearly bring out the detrimental effects on grain and forage yield of a variety of host plants by the downy mildews. Major losses due to downy mildews in pearl millet, sorghum and maize in India were discussed in the report of Safeeulla (136). Maximum grain loss up to 74% in highly susceptible cultivars of sorghum and maize was reported by Anahosur (2). The all India Coordinated Millet Improvement Programme estimated an average annual loss of 30% in high yielding pearl millet varieties amounging to >270 million US$ due to this disease (160). According to economists, successful research on the single most important disease of pearl millet, the downy mildew, can lead to an extra production worth 118 million US$ a year. King and Webster (84) recorded the economic losses of pearl millet due to downy mildew in West Africa. Similarly, Ramakrishnan's (128) review highlighted the loss from downy mildews in foxtail and fingermillets in Asia. The estimates of economic loss due to downy mildews reported by various authors has been described in Table 2.

Table 2. Selected information on the economic importance of downy mildews in tropical and sub-tropical cereal crops

Pathogen	Crop	Country	Statement on severity and/or economic loss	Reference
Sclerospora graminicola	Pearl millet	Nigeria	Average annual yield losses estimated to be 10% losses of more than 50% have been seen in some fields.	84
		India	Annual losses about 30% in high yielding varieties.	
			Downy mildew epidemics take a heavy toll annually	136
			Downy mildew control in a susceptible hybrid resulted in yields of 170-354% of unprotected check plots.	217
	Foxtail millet	China, India	Losses reported to go upto 20% Damage varied from little to loss of about 50%	128
Peronosclero-spora sorghi	Maize & sorghum	USA	Production losses based on estimated diseases incidence were valued at $2.5mn.	50
	Sorghum	Argentina	Damage on sorghum estimated as 15-20% in the major grain sorghum producing areas.	51
	Maize	Mexico	Maize downy mildew caused extensive losses in certain years in northeast Mexico. In	
	Sorghum	Insdia	Karnataka and Tamil Nadu incidence from 30-70% reported	

			with estimates losses of about 100,000 metric tonnes.	51
	Maize	Thailand	In 1974, 100,000 ha infected by downy mildew, and the reduction in yield varied from 10-100%.	
	Forage sorghum	Israel	The disease was severe that hybrid forage sorghum is hardly grown today. As much as 50%	
	Sweet corn	Israel	infection could found.	78
Peronosclero-spora heteropogoni	Maize	India	Estimates, provided with some reservations about their accuracy, indicate that in the state of Rajasthan losses may average up to 1.5%.	
Peronosclero-spora philippinensis	Maize	Nepal	It caused serious damage to local cultivars and was especially damaging to late sown maize.	
		Philippines	Yield losses in farmers fields may range from 15-40% and in some instances as high as 80-95%. The overall loss in 1974-75 was estimated as 8%, valued at $22.6 mn	
		India	Losses may be as high as 60%.	51
Peronosclero-spora maydis	Maize	Indonesia	High annual losses occur at about 40%.	
Peronosclero-spora sachhari	Maize	Taiwan	In 1964, two-thirds of the total acreage of hybrid maize was affected and infection was as high as 90-95%.	31
		India	Incidence was 30% or more in hybrids and as much as 64.2% in composites in the Tarai area of Uttar Pradesh.	
Peronosclero-spora rayssiae var. zeae	Maize	India	Grain yield reduction range from 20-90% in certain states in Northern India.	
Sclerophthora macrospora	Rice	(Generally)	Downy mildew does not cause serious damage to the rice crop.	
	Maize	USA	Crazy top of maize is generally of minor economic importance and it is only in very localised areas in fields where appreciable damage is sustained.	
	Finger millet	India	Severe outbreak... was recorded in 1948, the damage being so severe in some fields as to render the crop not worth harvesting	128

3. SYMPTOMS

Majority of the graminaceous downy mildews produce either 'local' or 'systemic' symptoms (Table 3). While local lesions are rarely induced by *Sclerospora graminicola* (78), sorghum pathotype of *Peronosclerospora sorghi* produces both types of symptoms. The symptoms of downy mildew can appear at any growth stage of the host plant starting from seedling to flowering. Generally, the systemic symptoms start from second leaf stage onwards as mild yellowing confining mostly to basal part of the leaf with a distinct margin between the diseased and non-diseased areas. Subsequent leaves show increased areas of disease and the progression of the symptom results in complete yellowing of the leaf. The infected leaf with chloritic yellowing produce abundant sporangia on the abaxial surface. During optimum conditions, sporulation in highly susceptible cultivars can be seen on the adaxial leaf surfaces also. Bhatia and Thakur (20) studied the various symptoms in NHB-3 cultivar of pearl millet infected by *S. graminocola* under field conditions. In maize and sorghum infected by *P. sorghi*, distinction between the diseased and non-diseased areas by a distinct margin has been referred as "half-leaf" symptom (51). Here, mild yellowing on the first few leaves progresses rapidly to produce distinct longitudinal stripes interpersed with light green and mildly yellowed are in the subsequent leaves.

Finger millet [*Eleusine coracana* (L.) Gaertn.] infected by *Sclerophthora macrospora* (Sacc.) Thiram., Shaw and Naras produce glabrous leaves. Pale yellow, chlorotic, translucent spots appear extending from leaf sheath to the tip. However, the basal part of the newly formed leaves appear uniformly yellow without characteristic spots and some of the leaves may regain the green colour. Downy growth of sporangia in the infected leaf is not a characteristic symptom of finger millet downy mildew.

Downy mildew infected leaves are more susceptible than healthy leaves to many other fungal pathogens such as anthracnose and leaf blight. In pearl millet, *Fusarium* species and *Drechslera setariae* cause irregular necrotic lesions on systemically diseased leaf which can be severe during periods of intermittent rains.

Teratological changes of the earheads due to systemic colonization of the meristematic region by the fungus are of many types. Colonization results in lack of inflorescence production or production of barren or malformed inflorescence. In pearl millet, *S. graminicola* causes malformed inflorescence (Fig. 1-A) with various degrees of inflorescence malformation into leafy structures. Sometimes, fertile, grains are found in between proliferations. Sorghum plants systemically infected by *P. sorghi* produce barren inflorescence or partially barren inflorescence. On maize, the pathogen incites malformed tassel (Fig. 1-B) and shoots. Infection by *S. macrospora* induces malformed inflorescence in many crops as 'crazy top'.

The stage at which the pathogen invades the growing point of host plant determines the stage at which disease appears on the host and the degree of damage to the ear-bearing organ. After the complete colonization of the host tissues, the pathogen starts producing oospores abundantly between the vascular bundles and leaf lesions. The leaf lamina then breaks down and shredding begins, dispersing the oospores. Leaf shredding is common in sorghum and maize infected by *P. sorghi*, Setaria italica infected by *S. graminicola* and sugarcane colonized by *P. sacchari* and *P. miscanthi*.

Variation from the general symptoms that occur include severe stunting of infected plants with yellow mottling of the diseased leaves in pearl millet and wilting of the emerging leaf whorl in maize. Stem pith discoloration and formation of brace roots have been observed in maize systematically colonized by *P. sorghi* in the America (99) and brown discoloration of maize nodal tissue has been observed in plants systemically colonized by *P. sorghi* in South India (41).

Table 3. The downy mildews of the Gramineae, their hosts and type of symptoms produced [a]

Pathogen	Crops hosts reported ser-vises disease	Symptoms		Common fisease name	Ooospores in major crop host	Other reported hosts or host genera[b]
		Systemic	Local			
Sclerospora (Schroet.) de Bary		+	Rarely	Green-ear	+	Several *Setaria* spp.,
S. graminicola (Sacc.) Schoroet	Pearl millet Foxtail millet			Graminicola DM	+	*Euchelaena, Echinochloa, Zea mays, (Saccharum, Sorghum)*
Peronosclerospora (Ito) Shirai & K. Hara						
P. heteropogoni Siradhana, Dange, Rathore & Singh	Maize	+	-	—	—[c]	*Heteropogon contortus,*
P. maydis (Racib). C. G. Shaw	Maize	+	-	Java DM	_	*(Euchlaena mexicana, Tripsacum)*
P. philippinensis (Weston) C. G. Shaw	Maizw	+	-	Philippine DM	_	*Saccharum, Euchlaena, Sorghum, Tripsacum, (Miscanthus, Avena)*
P. sacchari (T. Miyake) Shirai & K. Hara	Maize Sugarcane	+	+	Sugarcane DM	_	*Euchlaena, Miscanthus, Sorghum, Tripsacum (Andropogon, Botriochloa, Schizachyrium)*
P. sorghi (Weston & Uppal) C. G. Shaw	Sorghum	+	+	Sorghum DM	+	*Euchlaena, Panicum (Andropogon)*
P. spontanea (Weston) C. G. Shaw	Maize	+	-	Spontaneum DM	-	*Saccharum, Euchlaena, Miscanthus, Sorghum*
P. dicanthicola (Thirum & Naras.) C. G. Shaw	_	+	-	_	-	*Dichanthium annulatum*
P. noblei (Weston) C. G. Shaw						
P. miscanthi (T. Miyake apud	_	+	-	Leaf splitting DM	-[c]	*Sorghum plumosum, Sorghum leiocladum*
P. westonii (Srin. Naaras. & Thirum). C. G. Shaw	_	+	-	-	-c	*Saccharum, Sorghum, (Zea mays)*
		_	-	-		*Iseilema laxum*
Sclerophthora Thirum., Shaw & Naras.						

S. macrospora (Sacc.) Thirum, Shaw & Naras.	Maize Rice Wheat	+	-	Crazy top Yellow wilt	+	140 species in many genera
S. rayssiae var. *rayssiae* Kenneth Koltin & Wohl	Barley	-	+	Blotch DM	+	–
S. rayssiae var. *zeae* Payak & Renfro	Maize	-	+	Brown stripe DM	+	*Digitaria sanguinalis*
S. cryophila W. Jones	–	-	+	–	-[c]	*Dactylis, Apluda, Dicanthium, Digitaria, Heteropogon*
S. farlowii (Griffiths) Kenneth	–	-	+	–	-[c]	*Chloris elegans, Cynodon*
S. lolii Kenneth	–	-	+	–	-[c]	*Lolium*
Basidiophora Roze & Cornu	–	-	+	–	-[c]	*Eragrostis aspera, E. plumosa, E. tremula*
B. butleri (Weston) Thirum & Naras.						
Bremia Regel						
B. graminicola Naoumov	–	-	+	–	-	*Arthraxon ciliaris*
B. graminicola var.*indica* Patel	–	-	+	–	-	*Arthraxon lancifolius*
Plasmopara Schroet						
P. oplismeni Veionnot-Bourgin	–	-	+	–	-	*Oplimenus*
P. penniseti Kenneth & Karantz	–	-	+	-	-	*Penninsetum americanum*
Unknown genetic affinity (no asexual stage konwn)						
(*Sclerospora*)[d] iseilematis Thirum	–	+	-	–	-	*Iseilema laxum*
(*Sclerospora*)[d] *northi* Weston	–	+	-	–	-	*Erianthus maximum,*
(*Sclerospora*)[d] *secalina* Naoumov	–	-	+	-		*Saccharum officinarum Secale, cereale*

a Based primarily on Kenneth (78)
b Hosts in parentheses indicate infection by inoculation. For full treatment of species
c Oospores are reported from hosts other than major crop host.
d Species temporarily assigned to *Sclerospora* pending discovery of their asexual states.

Sclerophthora rayssiae var.*zea* induces solely local symptoms on maize characterized by parallel longitudinal stripes variables in length and 3-7 mm wide, with well defined margins. The stripes are initially chlorotic or yellow and later turn reddish to purple. The development of many lesions close together results in severe striping and blotching. No leaf shredding and malformation of inflorescences have been observed. Asexual and sexual spores are produced in the infected tissues (121).

Local lesions produced by *P. sorghi* (sorghum pathotype) are generally rectangular, 1-4 mm x 5-15 mm, limited by lateral veins and yellowish. Individual leions often coalesce to form longer ones. Heavy sporulation occur on these lesions which later turn reddish brown without forming oospores. In overall appearance and grain production, the plants with local lesions were similar to the healthy plants. Teratological changes of the earheads due to systemic colonization of the meristem region are of many types.

Table 4 : Present binomials and synonyms of the graminaceous downy mildews referred to in the literature by more than one binomial combination[a]

Present binomial[b]	Synonyms
Sclerospora graminicola	Protomyces graminicola Sacc., Perenospora graminicola Sacc., Perenospora setariae Pass., Ustilago (?) urbani Magnus
Peronosclerospora dichanthiicola	Sclerospora dicanthhiicola Thirum. & Naras.
Peronosclerospora heteropogoni	Sclerospora sorghi Weston & Uppal
Peronosclerospora maydis	Perenospora maydis Racib., Sclerospora javanica Palm, Sclerospora maydis (Racib.) Butler
Peronosclerospora miscanthi	Sclerospora miscanthi T. Miyake apud. Sacc.
Peronosclerospora noblei	Sclerospora noblei Weston
Peronosclerospora philippinensis	Sclerospora indica I Butler, Sclerospora philippinensis Weston
Peronosclerospora sacchari	Sclerospora sacchari Miyake
Peronosclerospora sorghi	Sclerospora graminicola var.andropogonis-sorghi Kulk., Sclerospora sorghi-vulgaris (Kulk.) Mundkur, Sclerospora sorghi (Weston & Uppal).
Peronosclerospora spontanea	Sclerospora spontanea Weston
Peronosclerospora westonii	Sclerospora westonii Srin., Naras. & Thirum.
Sclerophthora butleri	Sclerophthora butleri Weston
Sclerophthora farlowii	Sclerophthora farlowii (Griffiths
Sclerophthora macrospora	Sclerophthora macrospora Sacc. Phytophthora macrospora (Sacc.) Ito & Tanaka

a Cited from Williams, 1984

b Authorities provided in previous Table

4. CAUSAL AGENT

4.1 Pathogen

The graminaceous downy mildews are obligate parasites including 21 species belonging to 6 genera-*Sclerospora* (1 sp.), *Peronosclerospora* (10 spp.), *Scleropthora* (5 spp.), *Plasmopara* (2 spp.), *Bremia* (2 spp.) and *Basidiophora* (1 sp.). The first three genera are known to affect only graminaceous crops. Shaw's review (150) on the phylogenetic relationships among the members of Peronosporales, distinguished *Sclerospora* and *Peronosclerospora* as 2 different genera based on the differences in the germination of annual spores (149, 150). Present binomials and synonyms of downy mildews are listed in Table 4.

4.2 Morphological, cultural, biochemical and physiological characteristics

The graminaceous downy mildews are characterized by mycelium, asexual and sexual phases. The mycelium has been observed in all parts of affected plants. It is intercellular coenocytic, profusely branched and found mostly in the laminar portion of diseased leaves. The fungi derive nutrition from hosts cells by intercellular haustoria. They may be simple or branched, knob shaped or finger shaped (214) as in the case of *S. graminicola. Sclerophthora macrospora* produce pointed and filamentous haustoria which are borne as short lateral branches. From the mycelium both asexual and sexual structures develop.

After the mycelia invade the leaf tissue, they develop profusely. Clusters of hyphal protrusions crowd beneath the stomata of infected leaves (Fig. 1-C) and give rise to sporangiophores on which sporangia are produced externally. Thus, asexual phase is characterized by downy, whitish growth on the surface of infected leaves in all the species of downy mildews with the exception of *S. macrospora.* However, it is highly ephemeral and not possible to preserve as herbarium material. Sporangiophores are short, stout and dichotomously branched. In *S. graminicola,* sporangia are produced on sterigmata located at the tip of sporangiophore branches. Fully developed sporangia are hyaline, thin walled, ellipsoidal or broadly elliptical and papillate with dimensions of 15-22 x 12-21 mm (135).

In *S. macrospora,* a large sporangium is formed first at the apex that germinates indirectly. Subsequently, sporangia in a cluster for 4-5 are developed in lateral position. Mature sporangia are large, lemon shaped with a prominent papilla and pedicel measuring 60-100x43-64m. In nature, sporangia are produced in abundance during midnight and early hours of morning and highly influenced by the environmental factors like temperature and relative humidity.

Peronosclerospora sorghi produce condia that germinate directly. Conidia are suborbicular 15-28.9 x 15-26.9µ, most frequently 21-24.9 x 19-22.9µ hyaline, thin-walled without any papilla of dehiscence. The host genotype and prevailing environmental conditions are known to influence the morphology of downy mildew pathogens. Varietal influence of maize on morphology of asexual propagules and sporulation of *P. sorghi* was studied by Muralidhara-Rao *et al.* (106). The result indicated that morphology of spores varied with host genotypes. Their studies also observed increasing sporulation with increasing susceptibility.

In *S. graminicola* and *S. macrospora* sporangia germinate indirectly by producing zoospores. The number of zoospores per sporangium vary from 1-12 as in the case of *Sclerospora graminicola* (128, 151) to 24-48 as observed in *S. macrospora* (135). They are released through the papillar region and are ellipsoidal to reniform in shape, hyaline and unequally biflagellate. They pass through the three distinct phases viz., motility, encystment and germination.

Sexual phases of downy mildews often incite a marked distortion of host parts. It is initiated as antheridia and oogonia and culminates in the formation of oospores. The oospore phase is persistent, conspicuous and they are usually spherica, brownish-yellow and have three wall layers. They are produced in rows in the interveinal areas of the infected mature host leaf. As a rule, in *Sclerospora* spp. the oogonial wall is fused with the oospore wall which is a major identifying feature of the genus. Report of Michelmore *et al.* (105), apart from illustrating the heterothallic nature also suggest some evidence of the existence of homothallism in *S. graminicola*. Though there are reports on the germination of oospores in *S. graminicola* (117, 176), these results have generally not been reproducible.

Reports are not available on the growth of downy mildew fungi on any synthetic medium due to their obligate nature. However, they have been grown and maintained in dual cultures on their host callus tissue for indefinite periods. Inherent attempts were made to culture *S. cryophila* (135) and *S. rayssiae* var. *zeae* (206). Establishment of downy mildew pathogens in host cells have been reported for *S. graminicola* (8, 17, 18, 56) and *Peronosclerospora sorghi* (61, 76). Infected callus produced abundant mycelium, sexual and asexual spores. *P. sorghi* produced only oogonia in the callus. Although oospores were formed in infected calluses, they were not formed in the aerial mycelia. In *S. graminicola* sporangia produced from the mycelium grown on callus tissues were bigger in size than that produced on diseased pearl millet plants. Its mycelium was infective to both pearl millet and finger millet calluses. Biologically, it appears significant because the growth of this obligate parasite is supported by a non-host species and genus indicating the presence of host derived substances in the medium. The tissue culture technique to determine viability of internally seed-borne mycelium of *S. graminicola* in pearl millet seeds has also been shown (125).

Information on the biochemical and physiological characteristics of cereal downy mildews is limited due to their obligate nature and difficulties in growing them under laboratory conditions. Until recently, only amino acid content of sporangia of *S. graminicola* was available (135). With the advent of molecular techniques, isolated protoplasts of *S. graminicola* have been used as elegant source material for studying pathological and biochemical aspects of the pathogen (203). The results indicated the sporangia of *S. graminicola* to contain mainly β-1,3 glucanase with chitinase and protease in trace amounts. (Fig. 4)

4.3 Host range, pathogenic variability and physiological specialization

The details of different downy mildew pathogens, their principal hosts, the generally accepted common names of the diseases they cause along with symptoms that they induce are presented in Table-3.

Bhat (17) reported 14 graminaceous species belonging to 8 genera as hosts of *S. graminicola*. The genus *Peronosclerospora* is found scattered throughout the orders of higher plants. Eleven hosts belonging to the group Pnicoideae of the family Gramineae have been reported for *P. sorghi*. Kenneth (78) drew attention to the relatively narrow host range of *Sclerospora* and *Peronossclerospora*. *S. macrospora* is known to have a wider host range of 141 species belonging to Gramineae apart from the 8 grass hosts found distributed in and around finger millet fields. This variability in pathogenicity among downy mildew pathogens is of major concern for plant pathologists, taxonomists, geneticists, plant breeders and for quarantine scientists. In fungal plant pathogenes, variations can occur at any stage of the growth leading to changes in the virulence pattern, morphology, its host range, inoculum potential, infectivity and virulence. Therefore, constant monitoring of pathogen populations for shifts in virulence with changes in host cultivar and environmental factors is necessary. Under monocultural conditions, spectacular shift in virulence of a particular pathogen population makes high yielding cultivars of a crop vulnerable for infection resulting in disease epidemics.

In downy mildew fungi, pathogenic vulnerability occurs and within species at host species level and within host-species at cultivar level. According to the definition in the Commonwealth Mycological

Institute Publication, 'A Guide to the use of terms in Plant pathology' the term 'Pathotype' is used only to characterize sub-division of a downy mildew species specialized to particular host species and the term 'physiological race' is used to characterize specialization to different cultivars of one host species. Intraspecific variations in plant pathogenic fungi become apparent from studies of pathogenic specialization towards different host species and cultivars within one host species (47).

Pathogenicity differs among populations of *P. sorghi* from different geographic areas (51). The form of *P. sorghi* found in Rajasthan is pathogenic to maize and tangle head but not to sorghum. In contrast, *P. sorghi* in Karnataka attacks maize and sorghum but not tangle head. Maize downy mildew is reported to be caused by *P. sorghi* in Thailand and it is pathogenetically distinct at host species level from *P. sorghi* that causes downy mildew of sorghum and maize in South India and America. Schmitt and Freytag (141) demonstrated that the isolate of *P. sorghi* from Thailand differed from the from Texas in its inability to infect sorghum and its greater virulence to maize differentials.

Craig and Frederiksen (37) reported the existence of pathotype 1 and 2 in *P. sorghi* based on their differential pathogenecity in seedlings of sorghum lines. Subsequent report (38) recorded 3 pathotypes depending on the basis of their ability to sporulate on sorghum genotypes-Pathotype 1 that does not sporulate on TX 430 and CS 3541 lines, Pathotype 2 that sporulates only on CS 3541, and Pathotype 3 that sporulates on both TX 430 and CS 3541 lines. Three *P. sorghi* that exhibit differential pathogeneticity to cultivars with different host resistance genes are known in Texas (168). Micales *et al.* (104) used isoenzyme analysis to identify and differentiate *P. sorghi* from *P. sacchari* and *P. maydis.*

In *S. graminicola*, the fungus exists in different forms pathogenically distinct at host species level. The use of correct terms such as 'pearl millet pathotype', 'Setaria pathotype', and 'Panicum pathotype' immediately after the pathogen name *S. graminicola* enables us to understand the particular pathogenic form under reference. *S. graminicola* from setaria or panicum failed to infect pearl millet and vice versa indicating existence of different pathytypes. Tasugi (199) reported the inability of *S. graminicola* from *Setaria italica* to infect *S. viridis* and vice versa.

Variations occur at host-cultivar level, as far as the infective ability of the pathogenic population is concerned. Pathogenic variability in *S. graminicola* was first reported by Bhat (17). Girard (60) reported that some pearl millet varieties resistant to certain pathotypes were found susceptible to others. At Mysore, the hybrids HB3, NHB3, BJ104 and BK 560 are immune to *S. graminicola* populations occurring on the land race cultivars in fields but they are highly susceptible to *S. graminicola* occurring in the Departmental plots in Mysore University and also in many parts of India. Bhat (17) noticed that the pearl millet hybrid BH3 was resistant at Mysore and susceptible in many other locations in India. Similarly, some of the pearl millet cultivars resistant at certain places were found susceptible elsewhere (60). King *et al.* (85) and Werder and Ball (212) showed that some genotypes such as MBH 110, NHB 3 and 81 B differ in their downy mildew reactions to pathogen populations suggesting, further evidence for the differences in the virulence of *S. graminicola* in Western Africa and India. Virulence differences were also found in greenhouse tests in United Kingdom in which oospore collections from Africa and India were compared under the same environment to ascertain whether variability in cultivar response was genetically or environmentally determined. Generally, the collections from sub Sahel regions of West Africa were more pathogenic than those from Senegal, Zambia and India (11, 12, 13, 14).

In India, the populations of *S. graminicola* vary in their infection potential on a set of host cultivars (154, 177). Four pathotypes were identified based on their infection potential and differential interaction with a set of host genotypes. Several genotypes have been identified for differentiating the pathotypes of *S. graminicola.* Pathotype 1 which is the cause for many epidemics in India, occurs on

Tift 23A-based hybrids. Pathotype 2 occurs on traditional local cultivars and is not capable of infecting many of the hybrids. Pathotype 3 causes low levels of infection on all other hybrids except MBH 10 in Maharashtra. Pathotype 4 occurs mainly on 7042 S. It infects Tift 23 A-based hybrids, but not the inbred line 852 B. Morphological, biometrical, and cytological variatins in the asexual structures of these four pathotypes have been studied (154, 163). Serological studies correlated with these results and specific antibodies were recognized for identification and quantification of the pathogen *in situ.* Serological analysis of the pathotypes indicated the pathotypes 1 and 4 to be more related, pathotypes 1 and 3 to be partially related and pathotypes 1 and 2 to be serologically distinct (94).

4.4 Histopathology and pathogenesis

Infection process by downy mildews begins with the germination of oospore, zoospore on sporangium leading to the formation of appressorium at the junctionof epedermal cells, directly over the epidermal cells or over stomata (74, 222).

Entry of *P. sorghi* on both susceptible and resistant sorghum leaves is through stomata followed by vesicle formation in sub-stomatal cavities (74, 222). However, subsequent mycelial development was distinctly different in resistant and susceptible cultivars (222). The resistnat cultivars showed hypersensitive type of reaction resulting in restricted fungal development. Infection of maize by sorghum pathotype of P. sorghi also followed a similar pattern (145). Studies with *P. sorghi* and *S. macrospora* reported penetration between epidermal cells in the mesocotyl region of young seedlings by the infective propagules (215). Similar susceptibility of mesocotyi region of young seedlings but not unfolded leaves by zoospores of *S. macrospora* has been reported in *Lolium perenne* as well (215). Yeh and Frederiksen (222) observed greater downy mildew development when the colcoptiles of sorghum seedlings were exposed to conidia of *P. sorghi* than when the radicles were exposed. Shetty and Safeeulla (153) studied the factors affecting infection by *P. sorghi* on sorghum and observed the age of the plant is important in the development of downy mildew.

In pearl millet downy mildew system earlier reports enlighten the behaviour of the pathogen during infection and further establishment in susceptible pearl millet cultivars (19, 135, 192). Subramanya's work (192) has shown root, coleoptile and leaves as excellent sites for infection by zoospores. Mauch-Mani *et al.* (102) compared the infection process in seedlings and in suspension cultured cells and observed only a few instances of penetration of the roots of both susceptible and resistant lines by the pathogen. Though they recorded the entry of root hairs by germ tube, further growth was not observed. However, the latest reports (146, 147) recorded mycelium in the roots of both resistant and susceptible seedlings with increased establishment in susceptible cultivar (Fig. 1-D). They noticed accumulation of zoospores at the site of infection. In highly resistant lines, membranous sheath around the fungal structures were observed. In most of the highly resistant lines, post-haustorial resistance was seen, whereby, the pathogen was arrested following the formation of atleast one haustorium.

Hypersensitive response, as characterized by a local necrosis of the tissues immediately surrounding the infection site, has been demonstrated in pearl millet against *S. graminicola.* Nagarathna (110) established a relationship between the rate of hypersensitive response of pearl millet seedlings and their resistance to *S. graminicola*. Histological observations at the necrotic region revealed a restricted growth of the pathogen in seedlings of both highly resistant and resistant genotypes. Induction of hypersensitive reaction in pearl millet seedlings were observed by treatment with arachidonic acid (54). In induced quick hypersensitive reaction, fast induction of lipoxygenase (LOX) and more number of cell deaths at the vicinity of treatment, in highly resistant genotypes, both in calluses and seedlings was also established.

Studies on pearl millet downy mildew system from this laboratory reported involvement of pathogen specific, cell surface protein in host recognition phenomenon (94). Distribution of this protein in the pathogen propagules in relation to its host infectivity was studied using immunofluorescent technique (204). Immunofluorescence studies with cell wall, isolated protoplasts and emerging and germinating zoospores indicated higher concentration of host recognition protein in the protoplast as evidenced by high fluorescence, that the rest of the test samples. Significantly, the test protein appeared concentrated at the sporangial papillae, the site of zoospore release. During the development of zoospores, there was sequential transfer of host recognition protein to zoospores and finally to germ tubes. Concentration of this protein more at the germ tube tips was attributed to its involvement in host recognition and infection.

In general, infection of host plants from oospores of the downy mildew pathogen occurs in coleorhizae, radicles, lower portion of the coleoptiles of seedlings and in roots and underground portion of stem bases in older plants (215). Shetty and Safeeulla (153) and Pratt (126) have demonstrated the infection process in the host root by the germinating oospores of *P. sorghi.*

Infection of host by downy mildews leads to the establishment of intercellular mycelium towards the miristematic region. Colonization of the developing leaves or inflorescence at the growing point results in development of systemic symptoms. In cereal downy mildews, the host plants are more susceptible to systemic colonization by the pathogen at a younger stage than at the older age.

Shetty and Safeeulla (153) studied the inoculum density and disease development in sorghum plants inoculated with *P. sorghi.* They showed higher infection percentage and earlier symptom expression when more sporangia were used for inoculation. Ramesh (130) studied the influence of increasing sporangial load of *S. graminicola* on downy mildew disease expression in pearl millet. Due to the lack of correlation between the inoculum density and the infection percentage, the results described by these studies were not fully convincing. A critical evaluation of the pathogenicity of *S. graminicola* in relation to its inoculum density was undertaken by Kumr *et al.* (92). They used an Enzyme-Linked Immunosorbent Assay (ELISA) to detect *in vivo S. graminicola* in infected pearl millet tissues and quantified the concentration of zoospores of *S. graminicola* required to cause maximum infection (Fig. 2).

4.5 Host-Pathogen interaction

The first event in a parasitic interaction is the host infection by a pathogen (53). In downy mildew fungi, due to their biotrophic nature, this stage is important for determining their virulence and host compatibility. Due to a lack of sufficient methodology, this even was invariably not taken into account in the studies of host-pathogen relationship of downy mildew disease. Recently, an ELISA was described for quantification of the downy mildew fungus *S. graminicola* in host tissues (93). This ELISA procedure was shown to forecast host resistance since incompatibility of the host was detected by a reduced fungal biomass *in yivo* (93). These observations were further confirmed in histopathological studies by Sharadha *et al.* (147). Another important feature that determined pearl millet cultivar specificity by pathotypes of *S. graminicola* was the phenomenon of recognition. Kumar *et al.* (94) while serologically relating pathotypes of *S. graminicola* identified a specific cell surface protein of 90kDa that differentiated the pathotypes. The serological similarity was traced to host specificity in that, the protein of the virulent strain was recognized by the compatible host as seen by the binding reaction. This obviously suggested the probable operation of recognition for disease development (94). Efforts to further evidence this observation resulted in the localization of the protein (203, 204) in the papillary region of sporangia, protoplasts, zoospores and their infective tips all these being the infective stages of the pathogen.

Increasing informations have also been accumulating on the biochemical basis of host pathogen interaction in downy mildew diseases of cereals in particular reference to pearl millet downy mildew. Recent reports on the host's defense related enzymes substantiate their role in the resistance mechanism of pearl millet against pathogen infection. The lipoxygenase (LOX) enzyme, oxidizing unsaturated fatty acids that have a cis 1,4 pentadione structure has been reported to be involved in the initiation of lipid peroxidation during the early development of HR in incompatible, host pathogen combinations. Phenylalanine ammonia lyase (PAL), the first enzyme of phenyl propanoid metabolism which plays an important role in the biosynthesis of lignin, suberin, phytoalexins and phenolic compounds, is associated with the defense mechanism of host plants against pathogens. β-1,3-glucanase and chitinase that act on β-1,3-glucans and chitin of fungal cell walls were suggested to inhibit the growth of pathogenic fungi and release elicitors from fungal cell walls what can stimulate phytoalexin accumulation in the host plant. In pearl millet, the activities of these enzymes were tested in the seeds of different cultivars with different degrees of downy mildew reaction. Results indicated a good relation between the activities of LOX, PAL, β-1,3-glucanase and chitinase in the seeds and their downy mildew reaction in the field (86, 87, 113, 114, 116). Maximum activities were recorded in the seeds of highly resistant genotypes and minimum in highly susceptibles one (Fig. 3).

The enzyme activities of LOX, PAL and chitinase were also studied in resistant and susceptible genotypes of pearl millet after inoculation with *S. graminicola*. In the resistant genotypes, the enzyme activities increased significantly after inoculation, whereas, it decreased in the susceptible ones when compared with respective uninoculated controls. The change in activities of these enzymes over a period of time after inoculation showed a considerable difference between resistant and susceptible genotypes (Fig. 4). Studies on the activities of these enzymes in different parts of pearl millet seedlings revealed that in the resistant genotypes, enzyme activities significantly increased 24 h after inoculation in the shoot portion whereas in the root, the activities decreased. This post-infectional increase in the enzyme activities established the role of these enzymes in the initiation of defence reaction in pearl millet.

β-1,3-glucanase was recently purified from a resistant cultivar of pearl millet. This protein, a basic enzymes, was found to play a role in the host resistance reaction (87) based on *in vivo* studies that revealed antifungal activity of the purified enzyme. Its activity could also be used to determine the genetics of host resistance. A role for peroxidases and esterases in the resistance to downy mildew was also suggested by Gupta *et al.* (63) and Arora *et al.* (6). Their studies were based on the changes in the isoenzyme patterns of these enzymes in pearl millet cultivars infected with downy mildew pathogens. Studies of Sreedhara *et al.* (188) suggested a positive role for the peroxidase enzyme in the activities of resistant response in pearl millet against downy mildew infection. They demonstrated organ-specific activity, inoculum density-dependent induction and temporal behaviour of the enzymes.

Resistance in pearl millet hosts shows continuous variation (15, 44, 162, 169) due to genetic variability between pathogen populations and this hinders the studies on the understanding of pearl millet resistance. Research on genetics of downy mildew resistance in pearl millet have found additive dominance and epistatic effects to be important components in the inheritance of resistance. Nature of gene action governing the incidence of disease and phenotypic stability of some pearl millet crosses with regard to downy mildew was reported by Bharadwaj and Webster (16) and Pethani *et al.* (122). They found a variable expression of downy mildew in crosses of several populations of pearl millet. A system of biparental and sibmating in F_2 and F_3 families of best hybrids was reported by Murthy and Ahmed (108) for the development of inbreds with disease resistance. Appadurai *et al.* (5) in their attempt to infer the mode of inheritance of pearl millet downy mildew disease indicated, major gene action for disease resistance The inheritance of resistance of downy mildew was studied by Gill *et al.* (58, 59) in the F_1 and F_2 hybrids of crosses involving resistant and susceptible parents respectively. The

F_1 data indicated a complex nature of inheritance pattern involving genetic interactions and the F_2 and test crosses showed the presence of duplicate dominant factors. Singh *et al.* (169) indicated the quantitative nature of inheritance to downy mildew and observed significant additive and non-additive genetic variance. They also found susceptibility to be dominant over resistance. Studies of Basavaraju *et al.* (15) and Shinde *et al.* (162) under artificial epiphytotic conditions revealed the contribution of additive and non-additive gene effects in the inheritance of resistance. The dominance effects were higher than additive while epistatic effects were more than dominance. The combing effects of some of the downy mildew resistant lines in pearl millet was studied by Pokhriyal *et al.* (123, 124). The hybrids developed disease resistance with high stability over a wide set of environments.

With the development of DNA markers and construction of extensive genetic maps, analysis of recombination frequencies at meiosis using entire genomes was carried out in pearl millet by Busso *et al.* (27) using reciprocal three-way crosses. Their analysis revealed no differences in recombinant distance recovered from either male or female gametes. However, a few individual linkages favouring increased recombination through the male were observed. Similar investigation has also been described by Liu *et al.* (Personal communication) as well.

The first molecular map of pearl millet nuclear genome was published by Liu *et al.* (97) using an F_2 population from the cross LGD - 1B-10 XICMP 85410. The map is short which is unusual among grass genomes that spans 303 CM. The individual linkage groups (LG) vary from 90 CM for LGI to only 30 CM for LG6. They reported highly polymorphic nature of pearl millet genome based on RFLP (Restriction Fragment Length Polymorphism) and genetic mapping with millet genomic probes. Recently, they added a few DNA probes of known function from wheat. A number of RFLP probes have been reduced to sequence-tagged-site (STS) markers (Mohey *et al.* unpublished). Jones *et al.* (75) mapped quantitative trait loci (QTL) for resistance to pathogen populations of *S. graminicola* from India. Nigeria, Niger and Senegal using a susceptible and resistant pearl millet cross. Their study showed independent inheritance of resistance to pathogen population demonstrating the existence of differing virulence and pathotype-specific resistance as major mechanisms of downy mildew resistance. Howrah et al. (70) reported two RFLP based skeleton map for a population derived from a cross between H71/833-2 and ICMP 451 of pearl millet to tag genes controlling seedling heat tolerance. Their results indicated the suitability of the skeleton map for identifying QTLs for traits having high heritablities.

Studies on inheritance to resistance to downy mildew in sorghum revealed that susceptibility was dominant to resistance and was controlled by two complimentary factors (127). Rana *et al.* (131) found that sorghum downy mildew resistance is quantitatively inherited. Heterosis for resistance was observed in hybrids between susceptible and resistant parents. Additional doses of resistant parents in backcrosses increased the level of resistance proportionately. High degrees of determination of F_2 by F_1 and F_3 by F_2 indicated the highly heritable nature of sorghum downy mildew resistance (131). In a recent study Weltzein and King (211) subjected one highly susceptible population of pearl millet to two cycles of recurrent selection for downy mildew resistance using a modified greenhouse screening method. A rapid decline of genotype variance was observed in advanced cycles of selection, indicating that a number of genes control downy mildew resistance in this population. Their study also suggested the use of sib-progenies in recurrent selection for increased downy mildew resistance.

In sorghum downy mildew, resistance segregates by one or two dominant genes (39, 168). Oh *et al.* (119) and Xu *et al.* (219) mapped the gene conferring resistance to pathotype 1 of *P. sorghi* in cultivars SC 325 and BT X 623 respectively. A RAPD technique to locate markers for a downy mildew resistance gene in sorghum was described by Gowda *et al.* (62). They have also described a RFLP for mapping population to find markers linked to a downy mildew resistance gene in one of the parents, accession BT X 623. Their screening of 674 random primers for polymorphism revealed two amplified fragments linked to a downy mildew resistance gene in sorghum line SC 414.

Studies of Handoo *et al.* (64) revealed that resistance to brown stripe downy mildew *s. rayssiae* var. zeae (Payak and Renfro) in maize was controlled by several genes some of which may have major effects. Singh and Asnani (170) reported the importance of additive, dominance and epistatic effects for the inheritance of brown stripe downy mildew in maize. (Yao *et al.* (220) developed DNA probes from *P. maydis* genomic library for the detection and distribution downy mildew pathogen in maize leaves and seeds. Detection and idenfication of *P. sacchari* in maize by DNA hybridization was reported by Yao *et al.* (221).

Use of DNA finger printing to detect genetic variation among the population of *S. graminicola* has also been demonstrated recently (140, 185). Sastry *et al.* (140) studied genetic variability in six host-genotype-specific pathotypes of pearl millet downy mildew pathogen at the molecular level using mini and macrosatellites. The dendrogram constructed on the basis of finger prints, classified the six pathotypes into five distinct groups and according to them, corresponded with the characterization of the pathogens on differential host reaction (120). Earlier, Yao *et al.* (220) described a 1.5 kb DNA pebble of 1.5kb selected from *Peronosclerospora maydis* library for the identification of *Peronosclerospora* spp.

5. DISEASE CYCLE

The infective propagules of graminaceous downy mildews are:

(a) Oospores, which are thick-walled, long-lived resting spores produced by sexual reproduction to enable the pathogen to survive the long, hot, dry, crop-free periods characteristic of many regions where the diseases are of major importance.

(b) Sporangia produced by *Peronosclerospora* sp. are fragile and ephemeral that germinate by germ tubes.

(c) Zoospores, which are ephemeral, fragile, initially motile, asexually produced from sporangia by *S. graminicola* and *S. macrospora.*

Primary infection and recurrence of the downy mildew diseases in fields is generally due to oospores (in soil, plant residues and on seed surface), and mycelium in the embryo. When the infection is severe, the seedlings die within 30 days without producing oospores. But plants that express disease symptoms about 20 days after sowing, produce millions of sporangia. Secondary infection and spread occur in the field by water-borne and air-borne sporangia and zoospores. The annual infective propagules can travel up to a distance of 2 kms (181). Later in the season, numerous oospores are produced in the leaves which get mixed with soil and initiate the disease in the next season. Seeds from healthy looking plants can also carry oospores and mycelium. Infection of unfertilized florets during the heading stage lead to the establishment of mycelium in seeds (191).

In finger millet, the oospores of the pathogen present in the debris germinate and produce zoosporangium. The zoospores infect the seedlings under water logged conditions. After the establishment of primary infection, secondary infection takes place through zoospores produced by sporangia.

6. EPIDEMIOLOGY

Epidemiology, a branch of science dealing with the spread of disease leads to disease management as a technology. In epidemics of cereal downy mildews, the pathogen population starts from a low level of 'initial inoculum' and increases through successive cycles during growing seasons of the host. Thus, downy mildews can be grouped into 'compound interest diseases' (209). Graminaceous downy mildews initiate infection within a field or a crop from seed-borne or soil-borne oospore and/ or from asexual spores arising from an adjacent crop or weed host. Secondary infection occurs by asexual spores and seasonal increase of the pathogen population is based on the asexual phase. The relative importance of oospores, asexual spores and alternative hosts in the epidemiology of downy mildews varies from one downy mildew to another.

Abundant oospores are produced by *P. sorghi, P. heteropogoni* and *S. graminicola as* their hosts are annuals grown traditionally in dry environment. On the other hand, *P. maydis* and *P. philippinensis* do not produce oospores as these pathogens are prevalent in humid tropical environments (215). Oospore producing downy mildew pathogens transmit the pathogen through oospores carried on the seed and plant debris (79, 138, 172) and they occur widely in different continents. On the other hand, downy mildews that do not produce oospores in cereal crop hosts such as *P. philippinensis, P. maydis, P. heteropongoni* and Thailand maize infecting *P. sorghi* have limited geographical distribution In *P. maydis* which is confined to Indonesia, the pathogen perpetuation is only by the asexual spores produced on the maize crop which is continuously grown in the region (143) since it does not produce oospores.

Oospore production is related to its necessity and to the environment in which the host is growing. They disperse to long distances by associating with seeds lots as free oospores or in leaf debris mixed with seeds or in glumes or pericarps. The oospores of downy mildews are viable for varying periods from a few months up to 10 years (24, 117) due to the thick wall that protect them from desiccation.

The secondary spread of downy mildew is by sporangia and hence the factors affecting sporulation are very important to understand the epidemiology of downy mildews. Sporulation generally takes place during the early hours of the morning as that helps the spores to be blown to new areas and to initiate infection before sunrise to avoid desiccation. The process of sporulation is based on the supply of photosynthate in the infected leaves, air temperature, relative humidity, wind, light, and the presence of dew on plants (21, 22, 23, 73, 77, 98, 117, 135, 139, 152, 196, 213). Although there are conflicting reports regarding the role of asexual propagules in initiation and spread of the disease, the recent reports (144, 181, 191, 190) clearly demonstrate that asexual propagules play an important role in the epidemiology and secondary spread of the disease in pearl millet. Similarly, the study of Ramalingam and Rajasab (129) confirmed the major role of sporangia in sorghum downy mildew epiphytotics.

Apart from the sexual and asexual spores serving as sources of inoculum, the alternate hosts of downy mildew pathogens also serve as reservoirs of both sexual and asexual inocula as observed in the case of *P. sorghi* (99), *P. heteropogoni* (43) *P. philippinesis* (193). The cereal crop plants are highly susceptible to downy mildews from coleoptile stage to 3-4 weeks of emergence. Then onwards, susceptibility periods decreases and adult plant resistant sets in.

There is extensive data on the epidemiology of downy mildews of cereals in relations to *P. sorghi, P. heteropogoni, P. phillippinensis, P. sacchari and P. maydis* (21). Variations in the infectivity on different hosts and effects of environment on the sporangial and oospore phases of the pathogens have been described.

There is a lot of concern about the movement of downy mildew pathogens as they are considered as epidemiologically explosive and dangerous groups of pathogens. Reports available on the seed-borne nature and transmission aspects of downy mildew pathogens are conflicting. Generally, they occur in seed as dormant mycelium in the embryonic tissue and other seed parts as oospore. Seedborne nature and transmission of *P. sorghi* in maize seed have also been reported (9, 106, 107, 135, 137). *P. sacchari* is transmitted if infected maize seed is planted soon after harvest without drying (33). Chang (31) reported that drying of seeds of below 20% seed moisture and prolonged storage inactivates the pathogen and prevents transmission. He also believed that the pathogens was inadvertently introduced into Taiwan in 1909 during the import of diseased sugar cane sets from Australia.

Many reports are available on the association of downy mildew pathogen with the seed, serving as a source of primary infection, in addition to the oospores present in soil (7, 28, 135, 152, 195, 205).

Seed-borne nature and seed transmission of *S. graminicola* in pearl millet have also been reported (156, 157, 159). There are also reports available on the seed-borne nature of *P. philippinensis* (49) and *P. maydis* (143) as well.

7. DISEASE MANAGEMENT

The critical period during which cereal crop plants have to be protected from downy mildews is from planting to the completion of panicle formation which is 25-30 days after planting. In pearl millet, the crop needs protection till 35-40 days due to the formation of tillers. Hence, the control strategy should take into account the sources of primary and secondary inoculum. The available strategies of downy mildew control in cereals include use of host resistance, cultural practices, chemical, biological and biotechnological approaches.

7.1 Cultural practices

Disease management by crop rotation is based primarily on the reduction of primary inoculum Reduction in sorghum disease management was achieved by crop rotation in Texas (126, 207) by growing sorghum cultivar in oospore infested soils in which non-host species had been grown for 15 days. However, its applicability for all graminaceous downy mildews in highly reduced mainly due to (i) the longevity of oospores; (ii) their differential dormancy; (iii) differential response of pathogen species to new host; and (iv) sources of asexual inoculum from neighboring fields.

'Deep ploughing' as recommended by Tuleen *et al.* (207) to reduce oospore inoculum from oil surface and 'over seeding' (50) to maintain higher plant population has not be found to be effective method of controlling cereal downy mildews due to con-practicability.

'Roguing' of infected plants has also been suggested by many workers to manage to downy mildew diseases in cereal crops (99, 194, 198, 200). This method helps by preventing the accumulation of asexual propagules including oospores. However, the success of this method depends on the stage at which the infected plants are removed.

'Date of planting' as a method of downy mildew control, is that which allows the seedlings and young-plant growth to occur during a time that is not favourable for sporangial development, formation and infection or during a time when the inoculum source of the crop has been harvested (198). Though, large scale planting of the cereal crop with the first rains considerably reduces downy mildew incidence, delayed plantings generally expose young plants to higher amount of inoculum (10, 30, 184, 198). The available reports on the effect of fertilizers on the downy mildew incidence (98, 193, 198, 200) indicate that downy mildews cannot be efficietntly controlled in cereal crops by the applications of fertilizers.

7.2 Use of host resistance

For intensive cultivation of a crop, improved varieties or hybrids are most essential to boost up production. Breeding for disease resistance is the most practical, effective and economical method of disease management. Resistant varieties check the epidemics of the pathogen and help in maintaining the biological balance in the environment. In a breeding programme, the first requirement is the development of reliable screening techniques that identify resistant cultivars.

Methods of resistance screening : Among the different techniques available for identification of resistance of downy mildew diseases, some are effective while others are not.

Field screening : This technique depends on the sick plot development utilizing mainly oospores in the soil as primary inoculum (117). Secondary inoculum is supplied by the infector row plants planted 3 weeks before the candidate lines are sown. Disease incidence is recorded 30 days after sowing and again at 50% flowering stage. The advantages of this system are the uniform distribution of sporangia throughout the field and inoculation in a manner similar to that normally occurring under field conditions. However, its success depends on night time environmental factors such as high relative humidity, moderate temperature and dew formation especially during early plant growth. This technique has been used effectively to identify resistance to maize pathotype (215) of *P. sorghi* sorghum pathotype of *P. Sorghi* (4) and *S. graminicola* (30, 217)

Greenhouse screening : This technique basically involves inoculation of plants at the coleoptile stage with sporangial suspension, incubation for 12 h at 20°C and high relative humidity, and then evaluation for downy mildew incidence after about 15 days. This technique is suitable for mass screening and is now extensively practiced to supplement field screening methodology (82). However, it has the disadvantage of determining downy mildew reaction only at the seedling stage.

Multilocational testing is an essential aspect of any downy mildew screening programme is it allows expression of disease under different environments and exposure to potentially different pathotype.

Laboratory screening : In this procedure, plants are assessed for their reactions to downy mildew 15 days after inoculation. The method is more labour intensive, not suited for screening large number of materials and do not give a true picture of disease reaction of a cultivar in field conditions. This technique includes inoculation of seedling/seeds by a suspension of zoospores/sporangia and transplantation into pots.

In vitro screening techniques: In the routine screening experiments for downy mildew resistance programme, the data obtained are highly variable on account of environmental factors (82) and/or pathogen or host variability. In recent years, attention has been directed towards *in vitro* screening of host cultivars. This screening method helps in screening of a large number of cultivars at one time and the system avoids maintenance of downy mildew sick plots of interference by environmental factors affecting disease development.

Histological markers : The degree of disease resistance by different pearl millet cultivars were correlated with their cellular reactions, percentage of host cells colonized and number of haustorial mother cells, haustoria and papillae formed (147). These responses can be used as a rapid screening technique for determining resistance of pearl millet cultivars to downy mildew.

Biochemical markers : Hypersensitive responses and induction of defense related enzymes like LOX, PAL, and ß-1,3-glucanase in pearl millet infected with *S. graminicola* demonstrated the existence of relationship between different biochemical events and downy mildew resistance. (86, 113, 114). Hypersensitive responses occurred faster in resistant cultivars and the time required for the appearance of necrosis increased with decreasing host resistance (112). The activities of constitutive and inducible LOX, PAL, and ß-1,3-glucanase eyzymes were higher in highly resistant cutivars and they declined with the reduction in the host resistance (86, 114, 116). Since the activity of each of these enzymes and field reaction of pearl millet cultivars to downy mildew showed a good correlation, these enzymes were useful as markers for screening resistance reaction of unknown cultivars of pearl millet.

Use of cell culture and ELISA for screening host resistance : An indirect enzyme linked immunosorbent assay (ELISA) based on differentiating susceptible and resistant cultivar by the binding reaction of a 90 kDa cell surface protein of a virulent pathotype of *S. graminicola* (94) was developed recently (167). Differential binding reactions of the pathogen protein as indicated by the intensity of the colour reaction in ELISA, formed the basis for screening resistant and susceptible cultivars of pearl millet.

Stability and durability of resistance: In the use of host resistance, most important point to be taken into consideration are the resistance stability and its durability. The degree of stability and durability were dependent upon several interacting factors (215) that include pathogenic variability, the ability of the pathogen to produce and maintain virulent genotypes, genetics of host resistance and epidemiology of the downy mildew disease in the region where the resistant cultivars are to be grown. Durability of host resistance can be predicted by following multilocational trials or cultivating varieties over a large area for many years. However, basic research to study the principles underlying the mechanism of resistance is essential to identify and utilize resistant genes for crop improvement programmes.

Sources of resistance : In the development of resistant cultivars as a strategy to manage downy mildews, basic information on the sources of host resistance is necessary (161). The sources included selection from a population (57, 101, 174, 175, 215), use of germplasm from wild and related species in breeding programme (25, 65), new sources of genetic variability by hybridization (66) pooling the resistant genes in cultivars that have not be subjected to selection pressure for resistance during development providing opportunities to exploit variability for resistance to downy mildew (45, 215, 218), recovery resistance (178, 179), isolation of resistant lines from highly susceptible cultivars by selecting for resistance from variability within the lines (173, 183) and use of genetic variability observed among the progeny of plants regenerated from somatic cells cultured *in vitro* somaclonal variations (115).

Present status of cereal hosts in relation to downy mildews : Continuous effort are being made by different research organizations to expand the sources of resistance in cereals, to downy mildews.

In sorghum, 10, accessions (IS 18512, IS 18552, IS 18713, IS 18714, IS 190018, IS 19096, IS 19105, IS 19239, IS 20049 and IS 20205) were identified as resistant to downy mildew by ICRISAT. IS 18716, IS 19019, IS 19506 and IS 19971 were shown to develop less than 5% disease. The All India Coordinated Maize Improvement Programme has developed inbred lines resistant to downy mildews. Among the hybrids, Ganga 11, Y-142 M, PAC - 90E02 and K4-5992 were considered highly resistant. In pearl millet, latest released hybrids, HHB 50, ICMH423, PUSA 23 and VBH4 had tolerance levels of resistance against downy mildew. But GHB 181, ICMH 451 and H77/833-2 were found susceptible to the disease with a disease record of 14.3, 11.6 and 14.6 respectively.

The ICRISAT has developed a top cross hybrid of pearl millet, ICMH 88088 (ICMH 312) by crossing male sterile line 81A and pollinator population BSECTCP 2 (ICMR 132). The resistance nature of this hybrid was mainly contributed by the genetically heterogeneous nature of the pollinator. The experimental hybrids ICMA841 and ICMB 841 developed from 5141 A and 5141 B were resistant to downy mildew. ICMBA 841 also had higher grain yield potential than other CMS lines currently available in India (71).

7.3 Chemical control

As there is no durable resistance for downy mildew in improved host cultivars, the other alternative method of manage the disease is by chemical control. Until the discovery of metalaxyl, a systemic fungicide in 1977, contact fungicides were used to reduce downy mildew disease incidence in pearl millet, sorghum and maize (100, 103, 187). Thakur and Kanwar (202) reduced the downy mildew incidence by more than 70% by treating pearl millet seeds with agrosan GN and Thiram. Venugopal and Safeeulla (210) effectively controlled downy mildew of pearl millet, sorghum and maize using CGA-48988 as seed treatment at concentration of 2 g kg-1 of seed. However, there are reports on the failure of most of the contact fungicides to control many of the downy mildews under high inoculum pressure and on farm scale because they are not systemic in action (49, 51). Necessity for the protection of host plants from downy mildew fungi for 25-40 days after sowing and planting led to the discovery of chemicals with systemic fungicides, metalaxyl fosetyl-Al, akomin, cymoxanil and oxadixyl are the major fungicides found effective against Peronosporales in particular and oomycetes in general (142).

Metalaxyl in many formulations are available for soil and seed treatment and for foliar spray and many of these are used against control of downy mildews of cereals with promising results (1, 3, 32, 42, 95, 109, 118, 210, 217). The efficacy of metalaxyl formulations and its systemic remissive property in pearl millet against downy mildew have been tested by a number of workers (132, 166, 182, 217). Jain *et al.* (72) confirmed the effective control of pearl millet downy mildew using two formulations of metalaxyl. Their study also indicated an increase of 49.53% in yield with a seed treatment of 2 g (a.i)/kg of seed followed by two foliar sprays at 1000 ppm. l concentrations. However, in pearl millet hybrids with high tillerings, seed treatment with metalaxyl does not control downy mildew in the later developing tillers due to external inoculum source (217). Control of *P. sorghi* on maize using preparations of metalaxyl has been reported (29, 42, 148, 197). In maize and sorghum, due to the highly effective systemic action, complete control of downy mildew was possible with effective seed dressing. There are reports on successful control of *P. philippinensis* and *P. sacchari* on maize (48, 171) using metalaxyl. Singh and Shetty (180) reviewed the work done on the control of pearl millet downy mildew using metalaxyl. According to them, all the formulations of metalaxyl tested as seed treatment/or foliar spray controlled downy mildew with increased grain yield.

Recommended dosages of metalaxyl for sorghum, maize and pearl millet are:

Apron 35 SD-seed treatment 2 g (a.i)/kg of seed

Ridomil MZ 72 WP 4g/1 - (foliar spray) 20 days after sowing for pearl millet

Ridomil M2 72 WP 2g/1- (foliar spray) 30-40 days after sowing for maize and sorghum.

Other fungicides used against downy mildews includes akomin against pearl millet downy mildew under greenhouse conditions (166) and flurazole against sorghim downy mildew (40). However, the most effective fungicide against downy mildews is metalaxyl. The effectiveness is achieved by its systemic nature, curative effect and larger residual activity. Metalaxyl has been found to be fungistatice against pearl millet downy mildews and fungitoxic against many other oomycetous pathogens. Different host cultivars vary in their response to this fungicide (163, 173, 208).

7.4 Biological control

Even though chemicals are effective in controlling downy mildews in cereals, the ever increasing resistance of pathogens to fungicides and concern on environmental pollution, limits their usage. Hence there is a need for an integrated complex of control measures In recent years, biological control has emerged as a promising strategy in crop disease management. Biological control of downy mildews as theorized by Kenneth *et al.* (181) based on the reports of a number of fungi and bacteria associated with the oospores. These microbes are antagonistic towards downy mildews. Foliar sprays of culture filtrates of these organisms on different growth steps of plants reduced downy mildew disease (164, 165). Culture filtrates sprayed before, simultaneous with, or after inoculation of seedling with zoospore suspension suggested its possible use for prophylactic and curative treatments. Field trials of antagonists, either individually or in combination with metalaxyl formulations decreased downy mildew incidence in pearl millet (173).

Apart from the reports on microbes as efficient biocontrol agents, informations are available on use of many plant products that are known to have antifungal activity, against several plant pathogens. Control of *P. sacchari* on maize using a combination of seed treatment with chloroneb and spraying with *Azadirachta* oil was reported by Lal *et al.* (96). Efficacy of natural plants products against pearl millet downy mildew has also been studied (163). Their study indicated that all the extracts tried were fungicidal in action on *Sclerospora graminicola* and pre-soaking with these extracts hardened the susceptible hybrid seeds thus reducing downy mildew incidence upon artificial inoculation with the pathogen.

7.5 Biotechnological approaches

Of the different biotechnological methods used for the management of plant diseases, the following two systems have been found to be more efficient against pearl millet downy mildew diseases.

7.5.1 Systemic acquired resistance-a biotechnological, non-pesticide strategy for downy mildew management in cereals: Enhancement of resistance in susceptible plants against pathogens in response to an extrinsic stimulus without a known alteration of the genome is called 'induced resistance' (88, 189). The agents involved induce a host response which in turn protects either treated part of the plant (Localized Acquired Resistance-LAR) or parts of the plant distinct from treated area (Systemic Acquired Resistance-SAR or Induced Systemic Resistance-ISR)

SAR is not based on the elimination of the pathogen but on the activation of enhancement of plant's defense mechanism. Such utilization of a plant's defense mechanism by application of technologies to induce resistance has recently emerged as an alternative method of plant protection. This system of resistance of protection of cereals against downy mildews is apparent because it is highly effective, inexpensive, natural, safe, durable and ecofriendly. It would also complement resistance breeding and fungicide use in protection with approaches that would replace or supplement conventional methods. The efficiency of SAR has been demonstrated in over 25 crops including cereals against a broad spectrum of bacterial, fungal and viral pathogens. Specifically, operation of SAR has been described in the downy mildew diseases of cucumber (120), tobacco (34) and pearl millet (92). Extensive studies in the last decade has convincingly established the effectiveness of SAR in protecting crop plants under field conditions as well (88).

Table 5 : Downy mildew disease incidence due to reinoculation *Sclerospora graminicola* in pearl millet plants protected by prior inoculation with *Sclerospora graminicola*

S.No.	Treatment	Part of the plant reinoculated	Downy mildew disease incidence				Mean of % protection over control
			Expt. 1	Expt. 2	Expt. 3	Mean	
1.	Plants in which resistance was induced by prior inoculation followed by challenge inoculation with zoospore suspension of *S. graminicola*	Primary tillers	19.50	21.93	19.77	20.44[a]	69.82[a]
		Nodal tillers	16.80	20.30	17.95	18.35[a]	71.32[a]
		Inflorescence	4.30	6.06	4.70	5.02[b]	84.91[b]
2.	Plants that were not induced (control)	Primary tillers	87.08	87.06	86.87	87.01[c]	—
		Nodal tiller	84.55	84.68	84.78	84.67[c]	—
		Inflorescence	87.10	87.96	87.98	87.68[c]	—

Studies in this laboratory have shown for the first time, that pearl millet downy mildew can be managed by immunizing them with sublethal concentrations of the virulent pathotype of *S. graminicola* (68, 89, 92). Resistance induced at the seedling stage remained operative against subsequent inoculation by *S. graminicola* through out the various growth stages (Table 5). This suggested the systemic nature of induced resistance (92). Subsequent studies indicated the efficiency of a number of biotic agents like other pathotypes of *S. graminicola*, other downy mildew pathogen (68) and a biotic agents like salicyle acid, acetyl salicytic acid, acotyl salicytic acid, aruchidi onic acid (54, 68, 90, 158) in inducing systemic resistance in otherwise susceptible cultivars of pearl millet. Different cultivars of pearl millet with different basic levels of susceptibility to *S. graminicola* were found to develop systemic resistance to the same level. The level of SAR that was achieved was independent of the cultivar and not related to the constitutive resistance (Table 6.)

Table 6 : Downy mildew indicence *in pearl millet cultivars due to prior and challenge inoculations with *S. graminicola*

Treatment	Cultivars					
	23 B	HB 3	7042 S	J104	852 B	5141 B
Prior inoculation only	13.07[a]	14.76[a]	17.69[a]	11.84[a]	9.72[a]	12.43[a]
Challenge inoculation only	90.69[b]	89.92[b]	90.85[b]	60.10[b]	47.00[b]	65.73[b]
Prior & Challenge inoculation	35.12[c] (61.27)	34.00[c] (62.18)	36.05[c] (60.32)	23.05[c] (59.98)	19.75[c] (57.97)	24.70[c] (62.42)
Routine inoculation experiment	89.60[b]	91.10[b]	91.80[b]	60.05[b]	45.40[b]	64.20[b]
s,e,m,	±0.64	±0.72	±0.68	±0.58	±0.65	±0.59

* Average of three independent experiments.

b means in the same column followed by different letters differ significantly ($P \leq 0.05$) according to Newman and Keul's test.

Figure in parenthesis indicate percentage protection over unprotected control.

An interesting observation made during induction of resistance in pearl millet was the enhancement in growth parameters like tillering, rooting, early onset of flowering of induced resistant plants (68). Seeds of induced resistant plants were bigger in size and 1000 grain weight showed their increased biomass as well (68, 91).

Physiological and biochemical studies suggested the induction of resistance to higher activities of enzymes like β-1,3-glucanase and peroxidase (68, 69). Increased activities of these enzymes probably explain our earlier observation of significant reduction in the fungal mycelium of protected plants. The total enzyme activities were higher in protected plants than in the control plants from 24h after prior inoculation and enhanced further after challenge inoculation, indicating the active response offered. Results also indicated tissue-specific expression of these enzymes and their differential regulation by the pathogen.

7.5.2 Cell, tissue and organ culture: It is one of the tools that can be exploited to provide genetically modified material which can be integrated into the established programmes. Somaclonal variation is gaining importance as a potential alternative to plant improvement as it serves as a novel source for agriculturally useful traits including disease resistance. Somaclonal variation refers to the genetic variation observed among the progeny of plants regenerated from somatic cells cultured *in vitro*. It reduces the time required to produce new varieties or new breeding lines.

Development of downy mildew resistant pearl millet lines has been described from this laboratory using cell and tissue culture techniques (55, 110, 111, 115). Regenerants were developed from different explants of various pearl millet cultivars with varying susceptibility to downy mildew disease. The R_1 and R_2 somaclonal families derived from R_0 seeds were screened in the downy mildew sick plot along with the respective non-tissue-culture-derived control plants for their resistance reaction to downy mildew. Maximum number of resistant regenerants was obtained from moderately resistant lines.

Genetic analysis for somaclonal variations for downy mildew resistance was carried out in the somaclonal families of 23B and the results indicated that two R_1 lines SC6 and SC 7, obtained from the seed cultures, of 23 B, segregated for susceptibility and resistance in a 3:1 ratio. Mutations in SC4 and SC5, derived from inflorescence cultures segregated for resistance and susceptibility close to 7:9 ratio. Resistant somaclones against downy mildew *(Sclerospora sacchari)* of sugarcane have been identified and incorporated into breeding programmes (67).

8. CONCLUSIONS

Despite considerable studies, downy mildews continues to be the major limiting factor in improving cereal production. The fact that downy mildews still ravage cereal crops suggests that the knowledge gained on this disease is insufficient. Plant disease is an exception rather than the rule and there is a continuous development of newer pathogens together with newer hosts developed for resistance. However, the vast literature available on this subject has failed to recognize this phenomenon. More fundamental research is needed to understand the biology of the organism, since, recent reports have shown the occurrence of races in the pathogen. It is a possibility that more races occur and their non-identification has resulted in periodic observations of resistance breakdown in cultivars bred for downy mildew resistance. Lack of this information is also affecting control measures since, studies have not accounted for the development of resistance to fungicides in the unknown races. Reports have identified genes determining host resistance to downy mildew diseases. This strategy epitomizes effective biological disease control, since resistance will be heritable and inexpensive.

Biocontrol as a method of plant protection has also been envisaged and introduction of genetically engineered microbes detrimental to the pathogen is being attempted. However, the success of the story is still ambiguous since there have been no methods to authenticate the ability of these organisms of survive in natural soils.

9. FUTURE PROSPECTS

The ultimate interest in the study of any plant disease is towards increasing plant productivity to cater to the ever increasing population. In this regard, all studies eventually lead to a possible and or an effective control measure to prevent grain loss due to pathogens. The present day stress on environmental pollution has restricted the use of chemicals as plant protectants necessitating consolidation of newer methods. In this regard, Systemic Acquired Resistance emerges as an important method of plant disease management. The method of control is especially attractive as it takes care of pathogen variability that includes the unidentified races that may exist.

Another important feature that should be considered in disease management programmes of downy mildews is the accurate diagnostic procedures for potential 'on site' use to identify the pathogen and its resistance hosts. ELISA and DNA finger printing have been suggested as the methods for easy diagnosis of downy mildews and sources for host resistance.

Genetic interactions of downy mildew fungi is a very important study for future consideration. Understanding of this phenomenon can lead to procedures that prevent pathogen development. Basic studies in this regard should concentrate on differentiating heritable and non-heritable variations. It will be an attractive feasibility to carry out laboratory studies using protoplasts of the pathogens for identifying such variations.

10. REFERENCES

1. Anahosur, K. H. 1986. Effect of metalaxyl seed treatment on seedling emergence and downy mildew incidence in sorghum. Indian Phytopathology, **39** : 75-77.

2. Anahosur, K. H. 1992. Sorghum Disease in India : Knowledge and Research Needs. In : Sorghum and Millets Diseases : A second World Review. Eds. W.A.J. de Milliano, R. A. Frederiksen and G. D. Bengston, ICRISAT, Andhra Pradesh, India, pp. 45-56.

3. Anahosur, K. H. and Patil, S. H. 1980. Chemical control of sorghum downy mildew in India. Plant Disease, **64** : 1004-1006.

4. Anahosur, K. H. and Hegde, R. K. 1979. Assessment of the techniques used in screening sorghum genotypes to downy mildew. Mysore Journal of Agricultural Sciences, **13** : 449-451.

5. Appadurai, R., Parambaramani, C. and Natarajan, U.S. 1975. Note on the inheritance of susceptibility of pearl millet to downy mildew. Indian Journal of Agricultural Science, **45** : 179-180.

6. Arora, Y. K., Mehta, N., Thakur, D.P. and Wagle, D. S. 1986. Enzyme changes associated with host-parasite interactions between pearl millet and downy mildew fungus. Phytopathology Zeitschrift, **116**:97-105.

7. Arya, H. C. and Sharma, R. 1962. On the perpetuation and recurrence of the green ear disease of bajra *(Pennisetum typhoides)* caused by *Sclerospora graminicola* (Sacc.) Schroet. Indian Phytopathology, **15**: 166-172.

8. Arya, H. C. and Tiwari, M. M. 1969. Growth of *Sclerospora graminicola* on callus tissues of *Pennisetun typhoides* and in culture. Indian Phytopathology, **22** : 446-452.

9. Bain, D. C. and Alford, W.W. 1969. Evidence that downy mildew *(Sclerospora sorghi)* of sorghum is seed-borne. Plant Disease Reporter, **53** : 802-803.

10. Balasubramanian, K. A. 1974. Role of date of seeding, soil moisture, temperature and pH in the incidence of downy mildew of sorghum. Plant and Soil, **41** : 233-241.

11. Ball, S. L. 1983. Pathogenic variability of downy mildew *(S. graminicola)* on perrl millet. I. Host cultivar reactions to infection by different pathogen isolates. Annals of Applied Biology, **102**:257-264.

12. Ball, S. L. and Pike, D. J. 1983. Pathogenic variability of downy mildew *(S. graminicola)* on pearl millet. II. Statistical techniques for analysis of data. Annals of Applied Biology, **102**:265-273.

13. Ball, S. L. and Pike, D. J. 1984. Intercontinental variation of *S. graminicola.* Annals of Applied Biology, **104** : 41-51.

14. Ball, S. L., Pike, D. J. and Burridge, C. Y. 1986. Characterisation of populations of *S. graminicola.* Annals of Applied Biology, **108** : 519-526.

15. Basavaraju, R., Safeeulla, K. M. and Murthy, B. R. 1980. The role of gene effects and heterosis for resistance to downy mildew in pearl millet. Indian Journal of Genetics and Plant Breeding, **40** : 537-548.

16. Bharadwaj, B. D. and Webster, D. J. 1971. Studies of F_1 hybrids between photoperiod sensitive and non-sensitive *Pennisetum* millets. Crop Science, **11** : 291.

17. Bhat, S. S. 1973. Investigations on the biology and control of *S. graminicola* on bajra. Ph.D. Thesis, University of Mysore, India, 165p.

18. Bhat, S. S., Safeeulla, K. M. and Shaw, C. G. 1980. Growth of *S. graminicola* in host tissue cultures. Transactions of British Mycological Society, **75** : 303-309.

20. Bhatia, J. N. and Thakur, D.P. 1992. Symptom variation in pearl millet due to downy mildew under sick plot. Environment and Ecology, **10** : 888-891.

21. Bonde, M. R. 1982. Epidemiology of downy mildew diseases of maize, sorghum and pearl millet. Tropical Pest Management, **28** : 49-60.

22. Bonde, M. R. and Melching, J. S. 1979. Effects of dew period, temperature on sporulation, germination of conidia and systemic infection of maize by *Peronosclerospora sorghi.* Phytopathology, **69** : 1084-1086.

23. Bonde, M. R. Peterson, G. L. and Duck, N. B. 1985. Effects of temperature on sporulation, conidial germination and infection of maize by *Peronosclerospora sorghi* from different geographical areas. Phytopathology, **75** : 122-126.

24. Borchhardt, A. 1927. Operations of the phytopathological section of the agricultural experiment station in the Eastern Steppe region in the year 1925. Diepropertropvok (Ekaterinoslaff) : 13-37., Russian Abstract in Botanical Centrallbl, N. S. **11** : 440-441, 1928.

25. Bruken, J. N. 1977. A systemic study in *Pennisetum* sect *Pennisetum* (Gramineae). American Journal of Botany, **64** : 161-176.

26. Brunken, J. N., de Wet, J. M. M. and Horland, J. R. 1977. The morphology and domestication of pearl millet. Economic Botany, **31** : 163-174.

27. Busso, C. S., Liu, C. J., Hash, C. T., Witcombe, J. R., Devos, K. M., de Wet, J. M. J. and Gale, M. D. 1995. Anslysis of recombination rate in female and male gametogenesis in pearl millet *(Pennisetum glaucum)* using RFLP markers. Theoretical and Applied Genetics, **90** : 242-246.

28. Butler, E. J., 1918. Fungi and Disease in Plants. Thacker, Spink and Co., Calcutta, 547 p.

29. Castellani, E. and Mohamed, M. I. 1981. *Peronospora* disease of maize in Somalia. Rivista Agricoltura Subtropicale Tropicale, **75** : 223-230.

30. Chahal, S. S., Gill, K. S. and Phul, P. S. 1978. Relationship between date of sowing and incidence of downy mildew in pearl millet *Pennisetum typhoides* in the Punjab State. Crop Improvement, **5** : 165-166.

31. Chang, S. C. 1970. A review of studies on downy mildew of maize in Taiwan. Indian Phytopathology, **23** : 270-274.

32. Chang, S. C. 1978. Control of corn downy mildew by slurry seed treatment with Ridomil (Cuba-Geigy). Report of the Corn Research Centre Taiwan Dias No. **12** : 5-8.

33. Chang, S. C. and Twu, M. I. 1965. Studies on the Seed Transmission of Corn Downy mildew I. Report of the Corn Research Centre, Taiwan, **2** : 1-4.

34. Cohen, Y. 1994. β-ammunobutrynic acid induces systemic resistance against *Peronospora tabacina*. Physiological and Molecular Plant Pathology, **44** : 273-288.

35. Craig, J. 1980. Sorghum downy mildew research at Texas A. & M. University. In : Sorghum diseases-A World Review, Eds. R. J Williams, R. A. Frederiksen, L. K. Mughogho and G. D. Bengston), ICRISAT, Andhra Pradesh, India, pp. 195-199.

36. Craig, J. 1987. Tiered temperature system for producing and storing conidia of *Peronosclerospora sorghi*. Plant Disease, **71** : 356-358.

37. Craig, J. and Frederksen, R. A. 1980. Pathotypes of *Peronosclerospora sorghi*. Plant Diseases, **64** : 778-779.

38. Craig, J. and Frederiksen, R. A. 1983. Differential sporulation of pathotypes of *Peronosclerospora sorghi* on inoculated sorghum. Plant Disease, **67** : 278-289.

39. Craig, J. and Schert, K. F. 1985. Inheritance of resistance in sorghum to three pathotypes of *Peronosclerospora sorghi*. Phytopathology, **75** : 1077-1078.

40. Craig, J., Frederiksen, R. J., Odvody, G. N. and Szerszen, J. 1987. Effects of herbicide antidotes to sorghum downy mildew. Phytopathology, **77** : 1530-1532.

41. Dang, J. K., Thakur, D. P. and Grover, R. K. 1983. Control of pearl millet downy mildew caused by *S. graminicola* with systemic fungicide in an artificially contaminated plot. Annals of Applied Biology, **102** : 99-106.

42. Dange, S. R. S. and Williams, R. J. 1980. The ICRISAT sorghum downy mildew programme. In : Sorghum Diseases-A World Review, Eds. R. J. Williams, R. A. Frederiksen, L. K. Mughogho and G. D. Bengston, ICRISAT, Andhra Pradesh, India pp. 209-212.

43. Dange, S. R. S., Jain, K. L., Siradhana, B. S. and Rathore, R. S. 1974. Perpetuation of sorghum downy mildew *(Sclerospora sorghi)* of maize of *Heteropogon contortus* in Rajasthan, India. Plant Disease Reporter, **58** : 265-286.

44. Dass, S., Kapoor, R. L., Paroda, R. S. and Jatasra, D. S. 1984. Gene effects for downy mildew *(S. graminicola)* resistance in pearl millet. Indian Journal of Genetics, **44** : 280-285.

45. De Leon, C. 1979. Sources of resistance and gene management for control of downy mildew in maize. Unpublished position paper presented at the International Conference on the Graminaceous Downy Mildew Diseases, Bellagio, Italy (Available from the Rockfeller Foundation Publication Office, 1133 Avenue of the Americas, NY 100036, USA).

46. de Milliano, W.A.J., Frederiksen, R. A. and Bengston, G. D. (eds.)1992. Sorghum and Millets Diseases: A Second World Review, ICRISAT, Andhra Pradesh, India, 370 p.

47. de Wit, P. J. G. M. 1992. Molecular characterisation of gene for gene systems in plant fungus interactions and the application of a virulence genes in control of plant pathogens. Annual Review of Phytopathology, **30** : 391-418.

48. Dey, S. K., Khehra, A. S. and Dhillon, B. S. 1983. Control of Philippine downy mildew of maize through seed treatments with metalaxyl. Pesticides (Bombay), **17** : 26

49. Exconde, O. R. 1975. Corn in the Philippines : Its production and research activities with emphasis on downy mildew. Tropical Agricultural Research, **8** : 21-30.

50. Frederiksen, R. A. 1980. Sorghum downy mildew in the United States. Overview and outlook. Plant Disease, **64** : 903-908.

51. Frederiksen, R. A. and Renfro, B. L. 1977. Global status of maize downy mildew. Annual Review of Phytopathology, **15** : 249-275.

52. Futrell, M. C. and Webster, O. J. 1965. New sources of resistance to the downy mildew disease of sorghum. Plant Disease Reporter, **50** : 641-644.

53. Gaumann, E. 1950. Prnciples of Plant infection. Crosby Lockwood & Sons Ltd., London, 543 p.

54. Geetha, S. 1996. Elicitation of defense mechanism against downy mildew caused by *S. graminicola* in pearl millet. Ph.D. Thesis, University of Mysore, Mysore, India, 136 p.

55. Geetha Upadhayaya. 1994. Tissue culture studies with reference to host-pathosystem in pearl millet downy mildew and crop improvement. PhD. Thesis, University of Mysore, Mysore, India, 183 p.

56. Geetha Upadhyaya, Shivanna, M. S., Prakash, H. S. and Shetty, H. S. 1992. A novel approach to the establishment of dual cultures of pearl millet and *S. graminicola*. Plant Cell Tissue and Organ Culture. Vol. **31** : 203-306.

57. Gianinazzi, S. 1984. Genetic and molecular aspects of resistance induced by infections or chemicals. In: Plant-microbe interactions-Molecular and Genetic Prospectives, Vol. I. Eds. T. Kosnge and E. W. Nester, Macmillan Publishing Company, New York, London, pp. 321-342.

58. Gill, K. S., Phul, P. S., Singh, N. B. and Chahal, S. S. 1975. Inheritance of resistance to downy mildew in pearl millet-A Preliminary Report. Crop Improvement, **2**: 1-2.

59. Gill, K. S., Phul, P. S., Chahal, S. S. and Singh, N. B. 1978. Inheritance of resistance to downy mildew disease in pearl millet. Cereal Research Communications, **6** : 71-74.

60. Girad, J. C. 1974. Three years of study on millet and sorghum diseases at CNRA of Bombay. Senegal (1972-74) 1-5.

61. Gowda, P.B.S. and Bhat, S. S. 1992. Reaction of sorghum lines to downy mildew *Peronosclerospora sorghi*. Mycological Research, **96** : 889-890.

62. Gowda, P.B.S. Xu, G. U., Frederik, A. and Magill, C. W. 1995. DNA markers for downy mildew resistance genes in sorghum. Genome, **38** : 823-826.

63. Gupta, D. N., Roy, M. K., Mehta, S. L., Singhal, G. S. and Murthy, B. R. 1980. Peroxidase and esterase isoenzymes in downy mildew susceptible and resistant cultivars of pearl millet. In : Trends in Genetical Research on Pennisetum. Eds. V. P. Gupta and J. L. Minocha, ICAR, New Delhi, India, pp. 173-178.

64. Handoo, M. J., Renfro, B. L. and Payak, M. M. 1970. On the inheritance of resistance to *Sclerophthora rayssiae* var. *zeae* in maize. Indian Phytopathology, **23** : 231-249.

65. Hanna, W. W. and Dujardin, M. 1985. Interspecific transfer of apomixis in *Pennisetum*. Presented at the XV International Grassland Congress, 24-31 August, Kyoto, Japan.

66. Harinarayana, G. and Rai, K. N. 1989. Use of pearl millet germ plasm and its impact on crop improvement in India. In : Collaboration on genetic research : Summary. Proceedings of a joint ICRISAT/NBPGR (ICAR) workshop on germ plasm exploration and evaluation in India, ICRISAT, India, pp. 89-92.

67. Heinz, D. J, Krishnamurthi, M., Nickell; L. G. and Marketzmi, A. 1977. Cell Tissue and Organ Culture. Eds. J. Reiners and Y.P.S. Bajaj, Berlin and Springer pp. 3-17.

68. Hindumathy, C. K. 1996. Systemic Acquired Resistance in Pearl Millet against Downy Mildew. Ph.D. Thesis, University of Mysore, Mysore, India, 220 p.

69. Hindumathy, C. K., Kumar, V. U. and Shetty, H. S. 1995. Effect of systemic acquired resistance on pearl millet seed and the role of pathogenesis related proteins. Proceedings of National Seminar on Seed Pathology, 23-25 March, Mysore, India.

70. Howarth, C. J., Cavan, G. P., Layton, R. W. H., Harsh, C. T. and Witcombe, J. R. 1994. Mapping QTLs for heat tolerance in pearl millet. In : Use of Molecular markers in sorghum and pearl millet breeding for developing countries. Eds. J. R. Witcone, and R. R. Duncan, London, U. K., : Overseas Development Administration, pp. 80-85.

71. ICRISAT 1994. Pearl millet male-sterile line ICMA 841 and its mountaineer ICMB 841. Plant Material Description, **58** : 4 p.

72. Jain, A. K., Tripathi, S. K. and Gupta, J. C. 1993. Metalaxyl on downy mildew of pearl millet. Current Research, **22** : 133-134.

73. Jones, B. L. 1970. A simple technique of inoculating sorghum with *Sclerospora sorghi* using conidia as inoclum. Plant Disease Reporter, **54** : 603-604.

74. Jones, B. L. 1971. The mode of *Sclerospora sorghi* infection of *Sorghum bicolor* leaves. Phytopathology, **61** :L 406-408.

75. Jones, E. S., Liu, C. J., Gale, M. D., Hash, C. T. and Witcombe, J. R. 1995. Mapping quantitative trait loci for downy mildew resistance in pearl millet. Theoretical and Applied Genetics, **91** : 448-456.

76. Kaveriappa, K. M. 1973. Investigations on the downy mildew of sorghum. Ph.D. Thesis, University of Mysore, India, 171 p.

77. Kenneth, R. 1970. Downy mildew of gramineae in Israel. Indian Phylopathology, **23** : 371-377.

78. Kenneth, R. 1981. Downy mildews of graminaceous crops. In : The downy mildews, Ed. D. M. Spencer. Academics Press, London and New York, pp. 367-394.

79. Kenneth, R. and Klein, Z. 1970. Epidemiological studies of sorghum downy mildew *(S. graminicola* Weston and Uppal) on sorghum and corn in Israel. Israel Journal of Agricultural Research, **20** : 183.

80. Kenneth, R. and Kranz, J. 1973. *Plasmospara penniseti* sp. nov., a downy mildew of pearl millet in Ethiopia. Transactions of British Mycological Society, **60** : 590-593.

81. Kenneth, R., Cohen, E. and Shahor, G. 1975. A species of *Phyctochytrium* attacking nematodes and oospores of downy mildew fungi. Phytoparasitica, **3** : 1.

82. King, S. B. 1990. Are our screening procedure satisfactory. Paper presented at the group discussion of downy mildew of pearl millet. November 19-20, ICRISAT Centre, India.

83. King, S. B. 1992. World Review of Pearl Millet Disease : Knowledge of Future Research Needs. In : Sorghum and Millets Diseases : A Second World Review. Eds. W. A. J. de Milliano, R. A. Frederiksen and G. D. Bengston. ICRISAT, Andhra Pradesh, India, pp. 95-108.

84. King, S. B. and Webster, O. J. 1970. Downy mildew of sorghum in Nigeria. India Phylopathology, **23** : 342-349.

85. King, S. B., Werder, J. and Singh, S. D. 1989. Pathogenic variation of *S. graminicola* on pearl millet. Pathology, **79** : 1390 (Abstract).

86. Kini, R. K., Kumar, V. U. and Shetty, H. S. 1996. Activities of properties of β-1,3-glucanase from downy mildew infected seedlings of pearl millet and its role in disease management. In : Proceedings of National Seminar on Integrated Crop Disease Management for Sustainable Agriculture, February 26-27. Aurangabad, India.

87. Kini, R. K., Kumar, V. U. and Shetty, H. S. 1995. Involvement of β-1,3-glucanase and chitinase of downy mildew resistance in pearl millet. In : Proceedings of 64th Annual Meeting of Society of Biological Chemists (India), October 6-8, Lucknow, India.

88. Kuc, J. 1995. Induced Systemis Resistance-An Overview. In : Induced Resistance to Disease in Plants. Eds. R. Hammerschmidt and J. Kuc. Kluwer Academic Publishers, Netherlands, pp. 169-173.

89. Kumar, V. U., Hindumathy, C. K. and Shetty, H. S. 1994. Protection of pearl millet against the downy mildew pathogen *S. graminicola* by preimmunisation. In : National Symposium on Recent Developments in Biocontrol of Plant Pathogens 11-12 November, Madras.

90. Kumar, V. U., Hindumathy, C. K. and Shetty, H. S. 1996. Systemic Acquired Resistance-An ecofriendly strategy for downy mildew management in pearl millet. Proceedings of National Seminar on Integrated Crop Disease Management for Sustainable Agriculture, February 26-27, Aurangabad, India.

91. Kumar, V. U., Hindumathy, C. K. and Shetty, H. S. 1995. Induction of Systemic Resistance in Pearl Millet against Downy mildew. Technical Bulletin. Ed. S. B. Mathur, University of Mysore, Mysore, 8 p.

92. Kumar, V. U., Meera, M. S., Hindumathy, C. K. and Shetty, H. S. 1993. Induced systemic resistance protects pearl millet plants against downy mildew disease due to *S. graminicola*. Crop Protection, **12** : 458-162.

93. Kumar, V. U., Meera, M. S., Hindumathy, C. K. and Shetty, H. S. 1993. Quantification of *S. graminicola* in tissues of *Sclerospora graminicola* using an enzyme linked immunosorbent assay. Plant Pathology, **42** : 251-255.

94. Kumar, V. U., Shishupala, S., Shetty, H. S. and Umesh Kumar, S. 1993. Seological evidence for the occurrence of races of *S. graminicola* and identification of a race-specific surface protein involved in host recognition. Canadian Journal of Botany, **71** : 1467-1471.

95. Lal, S., Bargava, S. K. and Upadhaya, R. N.M. 1979. Control of sugarcane downy mildew of maize with metalaxyl. Plant Disease Reporter, **63** : 986-989.

96. Lal, S., Nath, K. and Saxena, S. C. 1980. Use of pesticides and natural products in control of *P. sacchari* in maize. Tropical Pest Management, **26** : 286-292.

97. Liu, C. J., Witcombe, J. R., Pittaway, T. S., Nash, M., Hash, C. T., Busso, C. S. and Gale, M. D. 1994. An RFLP-based genetic map of pearl millet *(Pennisetum glaucum)* Theoretical and Applied Genetics, **89** : 487.

98. Malaguti, G. 1976. Downy mildew disease of corn in Venezuela. Kasetsart Journal, **10** : 160-173.

99. Malaguti, G. 1980. Sorghum downy mildew in the Americas. In : 'Sorghum disease - A world review'. Eds. R. J. Williams, L. K. Frederiksen, L. R. Mughogho and G. Bengston. ICRISAT 189-194. Andhra Pradesh, India, pp. 189-194.

100. Martin, H. 1973. The Scientific Principles of Crop Protection. 6th Edition Ed. E. Arnold, London.

101. Mathews, P. 1981. In : The downy mildews. Ed. D. M. Spencer, Academic Press London, U. K. pp. 256-282.

102. Mauch-Mani, B., Schwinn, F. J. and Guggenheim, R. C. 1989. Early infection stages of the downy mildew fungi *Sclerospora graminicola* and *P. sorghi* in plants and cell culture. Mycological Research, **92** : 445-452.

103. McCallan, S. E. A. 1967. Fungicides and advanced treatise. Vol. I, Ed. D. C. Torgeson. Academic Press, New York and London.

104. Micales, J. A., Bonde, M. R. and Peterseon, G. L. 1988. Isoenzyme analysis and antipeptidase activities within the genes *Peronosclerospora*. Pathology, **78** : 1396-1402.

105. Michelmore, R. W., Pawar, M. N. and Williams, R. J. 1982. Heterothallism in *S. graminicola*. Pathology, **72** : 1368-1372.

106. Muralidhara-Rao, B., Prakash, H. S.,, Shetty, H. S. and Safeeulla, K. M. 1984. Production of *P. sorghi* oospores in maize seeds and further studies on the seed-borne nature of the fungus. Indian Pathology, **37** : 278-283.

107. Muralidhara-Rao, B., Shetty, H. S. and Safeeulla, K. M. 1984. Downy mildew in maize seeds : Technique to detect seed-borne inoculum of *P. sorghi* in maize. Seed Science and Technology, **13** : 593-600.

108. Murthy, B. R. and Ahmed, Z. 1974. Components of genetic variation in biparental progenies of a diallelset in dwarf *Pennisetums*. Crop Improvement, **1** : 82.

109. Muthuswamy, M. and Narayanaswamy, P. 1980. Chemotherapeutic action of new systemic fungicide, Ridomil on downy mildew disease of pearl millet. Madras Agricultural Journal, **67** : 487-490.

110. Nagarathna, K. C. 1993. Biotechnological aproach to develop downy mildew disease resistance in pearl millet. Ph.D. Thesis, University of Mysore, Mysore, India, 202 p.

111. Nagarathna, K. C., Prakash, H S. and Shetty, H. S. 1991. Genotype effects on the callus formation from different explants of pearl millet B. lines. Advances in Plant Sciences, **4** : 82-86.

112. Nagarathna, K. C., Shetty, S. A. and Shetty. H. S. 1993. Hypersensitive reaction in *P. sorghi* and associated changes in defense related enzymes in response to infection by *S. graminicola*. VI International Congress of Plant Pathology, July 28-August 6, Montreal, Quebec, Canada.

113. Nagarathna, K. C., Shetty, S. A. and Shetty, H. S. 1993. Phenylalanine ammonia lyase activity in pearl millet seedlings and its relation to downy mildew disease resistance. Journal of Experimental Botany, **44** : 1291-1296.

114. Nagarathna, K. C., Shetty, S. A., Bhat, S. G. and Shetty, H. S. 1992. The possible involvement of lipoxygenase in downy mildew resistance in pearl millet. Journal of Experimental Botany, **43** : 1283-1287.

115. Nagarathna, K. C., Shetty, , S. A., Harinarayana, G. and Shetty, H. S. 1993. Selection for downy mildew resistance from the regenerates of pear millet. Plant Science, **90** : 53-61.

116. Nagarathna, K. C., Shetty, S. A., Prakash, H. S., Bhat, S. G. and Shetty, H. S. 1992. Lipoxygenase activity in fresh, stored and germinating pearl millet seeds resistant and susceptible to downy mildew disease. Canadian Journal of Botany, **70** : 2028-2030.

117. Nene, Y. L. and Singh, S. D. 1976. Downy mildew and ergot of pearl millet. Pest Article News Summary, **22** : 366-385.

118. Odovody, G. N. and Frederiksen, R. A. 1984. Use of systemic fungicides metalaxyl and fosetyl-al for control of sorghum downy mildew in corn and sorghum in South Texas. II : Foliar Application. Plant Disease, **68** : 608-609.

119. Oh, B. J., Gowda, P.B.S., Magill, C. W. and Frderiksen, R. A. 1993. Tagging acremonium wilt, downy mildew, and head smut resistance genes in sorghum using RFLP and RAPD markers. Sorghum Newsletter, **35** : 34.

120. Okuno, T., Nakayama, M., Okaima, N. and Furasawa, J. 1991. Systemic resistance to downy mildew and appearance of acid soluble proteins in cucumber leaves treated with boitic and abiotic inducers. Annual Pathology Society of Japan, **57** : 203-211.

121. Payak, M. M. and Renfro, B. L. 1967. A new downy mildew disease in maize. Pathology, **57** : 394-397.

122. Pethani, K. V., Kapoor, R. L. and Chandra, S. 1980. Gene action and phenotypic stability for incidence of downy mildew disease in pearl millet. Indian Journal of Agricultural Research, **14** : 217-223.

123. Pokhriyal, S. C., Patil, S. R., Dass, R. and Singh, B. 1974. Combining ability of new male sterile lines in pearl millets. Indian Journal of Genetics, **34** : 208-213.

124. Pokhriyal, S. C., Unnikrishnan, K. V., Singh, B., Dass, R. and Patil, R. R. 1976. Combining ability of downy mildew resistant lines in pearl millet. Indian Journal of Genetics and Plant Breeding, **36** : 403-409.

125. Prabhu, M. S. C., Safeeulla, K. M. and Shetty, H. S. 1983. Tissue culture technique to demonstrate the viability of downy mildew mycelium in pearl millet seeds. Current Science, **52** : 1027-1029.

126. Pratt, R. G. 1978. Germination of oospores of *Sclerospora sorghi* in the presence of growing roots of host and non-host plants. Phylopathology, **68** : 1606-1613.

127. Puttarudrappa, A., Kulkarni, B. G. and Kajjari, N. B. 1972. Inheritance of resistance to downy mildew (*S. graminicola*) in sorghum. Indian Phytopathology, **25** : 471-473.

128. Ramakrishnan, T. S. 1963. Diseases of millets. Indian Council of Agricultural Research Publication, New Delhi, 152 p.

129. Ramalingam, A. and Rajasab, A. H. 1981. Epidemiology of sorghum downy mildew VI. Relative importance of oospores and conidia in epidemics of systemic infection. Proceedings. Indian National Sciences Academy, **47** : 725-730.

130. Ramesh, C. R. 1981. Studies on host and parasite systems of the green ear disease of bajra. Ph.D. Thesis, University of Mysore, Mysore, India, 224 p.

131. Rana, B. S., Parameswarappa, R., Anahousur, K. H., Rao, V.J. M., Vasudeva Rao, M. J. and Rao, N. G. P. 1978. Breeding for multiple insect/disease resistance. In : Proceedings of All India Sorghum Workshop. 17-19 April, Dharwad, India.

132. Reddy, M. V. B., Shetty, H.S. and Reddy, M. S. 1989. Mobility and efficacy of metalaxyl against downy mildew of pearl millet growing in different types of soil. International Journal of Tropical Plant Diseases, **7** : 97-104.

133. Reyes, L., Rosenow, D. T., Berry, R. W. and Futrell, M. C. 1964. Downy mildew and head smut mildews. Ed. D. M. Spencer. Academic Press, London, pp. 107-117).

134. Riley, C. V. 1986. Rural New Yorker 45 : 72-87. See : Viennot-Bourgin, G. (1981). History and Importance of downy mildews. Ed. D. M. Spencer, Academic Press, London, New York, San Franscisco, pp. 1-15.

135. Safeeulla, K. M. 1976. Biology and control of the downy mildews of pearl millet, sorghum and finger millet. Wesley Press, Mysore, India, 304 p.

136. Safeeulla, K. M. 1977. Genetic vulnerability : The basis of recent epidemics of downy mildews. Annals of New York Academy of Sciences, **287** : 73-85.

137. Safeeulla, K. M. and Shetty, H. S. 1977. Seed transmission of sorghum downy mildew on corn. Seeds and Farms, **3** : 21-25.

138. Safeeulla, K. M. and Shetty, H. S. 1980. Sorghum downy mildew in Asia : Assessment of present knowledge and future research needs . In : Sorghum diseases : A World Review. Eds. R. J. Williams, R. A. Frederiksen, L. K. Mughogho and G. D. Bengston, ICRIST, India, pp. 173-183.

139. Safeeulla, K. M, Shaw, C. G. and Thirumalachar, M. J. 1963. Sporangial germination of *S. graminicola* and artificial infection of *Pennisetum glaucum*. Plant Disease Reporter, **47** : 679-681.

140. Sastry, J. G., Ramakrishna, W., Sivaramakrishnan, S., Thakur, R. P., Gupta, V. S. and Ranjekar, P. K. 1995. DNA fingerprinting detects genetic variability in the pearl millet downy mildew pathogen *(S. graminicola)*. Theoretical and Applied Genetics, **91** : 856-861.

141. Schmitt, C. G. and Freytag, R. E. 1977. Response of selected resistant maize line to three species of *Sclerospora*. Plant Disease Reporter, **61** : 478-481.

142. Schwinn, F. J., Staub, T. and Urech, P. A. 1977. A new type of fungicide against diseases caused by oomycetes. Meded. Fac, Landabouwwet Rijksuniv. Genetics, **42** : 1181-1188.

143. Semangoen, H. 19970. Studies on downy mildews of maize in Indonesia with special reference to the perennation of the fungus. Indian Phytopathology, **23** : 307-320.

144. Semsi, S. T. and Ball, S. F. L. 1989. Infection of the pearl millet *(Pennisetum americanum)* inflorescence by the downy mildew fungus *(S. graminicola)*. Plant Pathology, **38** : 571-576.

145. Shabani, S. 1978.. Reaction of *Zea mays*. L. to inoculation with *Sclerospora sorghi* Weston and Uppal. M. S. Thesis, Texas A. and M. University, College Station, Texas, USA.

146. Sharadha, M. S. 1996. Histological illustration of the interactions between *Pennisetum glaucum* and *S. graminicola*. Ph.D. Thesis, University of Mysore, Mysore, India, 245 p.

147. Sharadha, M. S., Shetty, S. A. and Shetty, H. S. 1995. Infection process of *S. graminicola* on *Pennisetum glaucum* lines resistant and susceptible to downy mildew. Mycological Research, **99** : 317-322.

148. Sharma, S. C., Kehhra, A. S., Bains, S. S. and Malhi, N. S. 1981. Efficacy of fungitoxicant sprays and seed treatment against Philippine downy mildew of maize. Indian Phytopathology, **34** : 498-499.

149. Shaw, C. G. 1978. *Peronosclerospora* spp. and other downy mildews of the gramineae. Mycologia, **70** : 594-604.

150. Shaw, C. G. 1981. Taxonomy and evolution . In : The downy mildews. Ed. D. M. Spencer. Academic Press, London, U. K., pp. 17-29.

151. Shetty, H. S. 1987. Biology and epidemiology of downy mildew of pearl millet. In : Proceedings of International Perarl millet Workship, 7 - 11 April 1986, ICRISAT, Andhra Pradesh, India, pp. 147-160.

152. Shetty, H. S. 1990. Basic research on downy mildews of cereals and disease Management. In : Basic research for crop disease management. Ed. P. Vidhyasejkaran. Daya Publishing House, Delhi, pp. 316-32.

153. Shetty, H. S. and Safeeulla, K. M. 1980. Studies on the oospore germination of *P. soreghi*. Kavaka, **8** : 47-51.

154. Shetty, H. S. and Ahmad, R. 1981.. Physiological specialisation in *S. graminicola*. Indian Phytopathology, **34** : 307-309.

155. Shetty, H. S., Nagarathna, K. C. and Shetty, S. A. 1993. Tissue culture for management of downy mildew in pearl millet. In : Scientific engineering and tissue culture for crop pest and disease management. Ed. P. Vidhyasekaran. Daya Publishing House, Delhi.

156. Shetty, H. S., Neergaard, P. and Mathur, S. B. 1977. Demonstration of seed transmission of downy mildew or green ear disease, *S. graminicola* in pearl millet, *Pennisetum typhoides*. Proceedings of Indian national Science Academy, **43** : 201-205.

157. Shetty, H. S., Mathur, S. B. and Neergaard, P. 1980. *S. graminicola* in pearl millet seeds and its transmission. Transactions of British Mycological Society, **74** : 127-134.

158. Shetty, H. S., Kumar, V. U. and Hindumathy, C. K. 1996. Systemic Acquired Resistance for Pearl Millet Downy Mildew Disease Management. Paper presented to the Agricultural Science Section of 83rd session of the Indian Science Congress 3-8 January, at Punjabi University, Patiala, Chandigarh.

159. Shetty, H. S., Khanzada, A. K., Mathur, S. B. and Neergaard, P. 1978. Procedures for detecting seed-borne inoculum of *S. graminicola* in pearl millet *(Pennisetum typhoides)*. Seed Science and Technology, **6** : 935-941.

160. Shetty, S. A., Shetty, H.S. and Mathur, S. B. 1995. Downy mildew of pearl millet. Technical Bulletin, University of Mysore, Mysore, 8 p.

161. Shetty, S. A., Shishupala, S., Nagarathna, K. C., Kumar, V. U. and Shetty, H. S. 1995. Integrated approach for the management of cereal downy mildews. Review which contributed for the book series in Integrated Pest Management Systems in Agriculture, IARI, Delhi.

162. Shinde, R. B., Patil, F. B. and Sangve, R. A. 1984. Resistance to downy mildew in pearl millet. Journal of Maharashtra Agricultural University, **9** : 337-338.

163. Shishupala, S. 1994. Variability in pearl millet downy mildew pathogen and the disease management. Ph.D. Thesis, University of Mysore, Mysore, India, 349 p.

164. Shishupala, S. and Shetty, H. S. 1989. Use of antagonists to control the oospore infection in pearl millet downy mildew. Journal of Biological Control, **3** : 117-118.

165. Shishupala, S. and Shetty, H. S. 1990. Bioefficacy of the antagonists culture filtrates against pearl millet downy mildew. Plant Disease Research, **5** : 52-54.

166. Shishupala, S., Lokesha, S. and Shetty, H. S. 1990. Efficacy of metalaxyl and Oxadixyl formulations against pearl millet downy mildew. Advances in Plant Sciences, **30** : 28-33

167. Shishupala, S., Kumar, V. U. Shetty, H. S. and Umesh Kumar, S. 1996. Screening pearl millet cultivars by ELISA for resistance to downy mildew disease. Plant Pathology.

168. Sifuents, E. S. and Frederiksen, R. A. 1988. Inheritance of resistance to pathotypes 1, 2 and 3 of *P. sorghi* in sorghum. Plant Disease, **72** : 332-333.

169. Singh, F., Singh, R. K., Singh, R. M. and Singh, R. B. 1978. Genetic analysis of downy mildew *(S. graminicola)* resistance in pearl millet *[Pennisetum typhoides* (Burm.) S. & H.]. Zeitschrift fur Pflanzenzuchtung, **81** : 54-59.

170. Singh, I. S. and Asnani, V. L. 1975. Gene effects for resistance to brown stripe downy mildew in maize. Indian Journal of Genetics, **35** : 123-127.

171. Singh, I. P. and Lal, S. 1985. Effective spray schedule of metalaxyl formulations for the therapy of sugarcane downy mildew of maize. Tropical Pest Management, **31** : 327-328.

172. Singh, I. P., Renfro, B. L. and Payak, M. M. 1971. Studies on the epidemiology and control of brown stripe downy mildew of maize *(Sclerophthora rayssiae* var. *zeae)*. Indian Phytopathology, **23** : 194-207.

173. Singh, S. D. 1983. Variable cultivar response to metalaxyl treatment in pearl millet. Plant Disease, **67** : 1013-1015.

174. Singh, S. D. 1990. Sources of resistance to downy mildew and rust in pearl millet. Plant Disease, **74** : 871-874.

175. Singh, S. D. 1992. New Sources of Resistance, ICRISAT Annual Report, India, pp., 154.

176. Singh, S. D. 1995. Downy mildew of pearl millet. Plant Disease, **79** : 545-550.

177. Singh, S. D. and Singh, G. 1987. Resistance to downy mildew in pearl millet hybrid NHB-3. Indian Phytopathology, **40** : 178-180.

178. Singh, S. D. and de Milliano, W. A. J. 1989. First report of recovery of sorghum from downy mildew in Zimbabwe. Plant Disease, **73** : 1020 (Abstract).

179. Singh, S. D. and King, S. B. 1988. Recovery resistance to downy mildew in pearl millet. Plant Disease, **72** : 425-428.

180. Singh, S. D. and Shetty, H. S. 1990. efficacy of systemic fungicide metalaxyl for the control of downy mildew *(S. graminicola)* of pearl millet *(Pennisetum americanum)* Indian Journal of Agricultural Sciences, **60** : 575-581.

181. Singh, S. D. and Williams, R. J. 1980. The role of sporangia in the epidemiology of pearl millet downy mildew. Phytopathology, **70** : 1187-1190.

182. Singh, S. D., Gopinath, R., Luther, K. D. M., Reddy, P. M. and Pawar, M. N. 1984. Systemic remissive property of metalaxyl against downy mildew in pearl millet. Plant Disease, **68** : 668-670.

183. Singh, S. D., Williams, R. J. and Reddy, P. M. 1988. Isolation of downy mildew resistant lines from a highly susceptible cultivar of pearl millet. Indian Phytopathology, **41** : 450-456.

184. Siradhana, B. S., Dange, S. R. S., Jain, K. L. and Rathore, R. S. 1975. Effect of sowing dates on the natural incidence of sorghum downy mildew of maize in Udaipur, Rajasthan. Indian Phytopathology, **28** : 140-141.

185. Sivaramakrishnan, S., Harikrishnan, R., Thakur, R. P. and Rao, V. P. 1995. Use of RAPD markers for the detection of genetic variability in pearl millet downy mildew pathogen. In : Proceedings of 2nd Asia-Pacific Conference of Agricultural and Biotechnology, Madras. Eds. R. P. Sharma, V. L. Chopra and M. S. Swaminathan. Oxford IBH Publishng Company, New Delhi.

186. Spencer, D. M. 1981. The downy mildews. Academic Press, London, U. K., 636 p.

187. Spencer, E. Y. 1977. History of fungicides. In : Antifungal Compounds. Eds. M. R. Siegel and H. D. Sisler. Marcel Dekker Inc. New York and Basel, pp. 600.

188. Sreedhara, H. S., Nandhini, B. A., Shetty, S. A. and Shetty, H. S. 1995. Peroxidase activities in the pathogenesis of *S. graminicola* in pearl millet seedlings. International Journal of Tropical Plant Diseases, **13** : 19-32.

189. Steiner, U. and Schonbeck, F. 1975. Induced disease resistance in monocots. In : Induced resistance to disease in plants. Eds. R. Hammerschmidt and J. Kuc. Kluwer Academic Publishers, Netherlands, .

190. Subramanya, S. 1984. Studies on the epidemiology and host-parasite relationship in downy mildew of bajra *[Pennisetum typhoides* (Burm). Stapf & Hubb.] Ph.D. Thesis, University of Mysore, Mysore, India, 342 p.

191. Subramanya, S., Safeeulla, K. M. and Shetty, H. S. 1982. Importance of sporangia in the epidemiology of downy mildew of pearl millet. Proceedings of Indian National Science Academy, **48** : 824-833.

192. Subramanya, S., Shetty, H. S. and Safeeulla, K. M. 1983. Pearl millet downy mildew : Biology of systemic infection by zoospores. Proceedings of Indian National Science Academy, **B49** : 385 - 394.

193. Sun, M. H. 1970. Sugarcane downy mildew of maize. Indian Phytopathology, **23** : 262-269.

194. Sun, M. H., Chang, S. C. and Tseng, C. M. 1976. Research advances in sugarcane downy mildew of corn in Taiwan. Kasetsart Journal, **10** : 89-93.

195. Suryanarayana, D. 1962. Occurrence of an unknown fungal mycelium inside the sound grains produced on partly formed green areas of bajra plants. Science and Culture, **28** : 536.

196. Suranarayana, D. 1965. Studies on the downy mildew diseases of millets in India. Indian Phytopathological Sooiety Bulletin, **3** : 72-78.

197. Takeda, A. S., Nakamura, K., Gimenes-Fernandes, N., Kronka, S. D. N., Gomes, G. and Inove, L. T. 1982. Effect of chemical seed treatment on control of sorghum downy mildew on maize and sorghum. Cientifica (Jaboticabal), **10** : 129-133.

198. Tantera, D. M. 1975. Cultural practices to decrease losses due to corn downy mildew diseases. Tropical Agricultural Research Series, **8** : 165-173.

199. Tsugi, H. 1933. Studies on Japanese Peronosporales. II. Studies on the physiology of conidiophores, conidia and oospores of *S. graminicola* (Sacc.) Schroet. on the Japanese millet [*Setaria intalica* (L.) Beauv.]. Journal of Imperial Agricultural Experiment Station Nishigahara, Tokyo, **2** : 225-262.

200. Thakur, D. P. 1980. Utilisation of downy mildew resistant germplasms and other resources for increased productivity of pearl millet under semi-arid regions of India. Annals of Arid-Zone, **19** : 265-270.

201. Thakur, R. P. 1993. Genetic diversity for virulence and aggressiveness in a field population of the pathogen. Cereals Program. Annual Report 1992. ICRISAT, Andhra Pradesh, India, pp. 63-64.

202. Thakur, D. P. and Kanwar, Z. S. 1977. Reducing downy mildew *(Sclerospora graminicola)* of pearl millet *(Pennisetum glaucum)* by fungicidal and soil treatments. Pesticides (Bombay), **11** : 53-54.

203. Tejaswini, P. H., Kumar, V. U. and Shetty, H. S. 1995. Physiological and pathological studies on the biotrophic fungus *Sclerospora graminicola* using isolated protoplasts. In : National symposium on "Biodiversity, Biology and Biotechnology of Fungi", 20-23 November, Madras.

204. Tejasiwini, P.H., Kumar, V.U. and Shetty, H. S. 1995. Localisation of host recognition protein in infective propagules of *S. graminicola* In : Proceedings 64th Annual meeting Society of Biological Chemists (India). 6-8 October, Lucknow.

205. Tiwari, M. M. and Arya, H. C. 1966. Studies on green ear disease of bajra *(Pennisetum glaucum* Stapf and Hubb). caused by *S. graminicola* (Sacc.) Schroet. Indian Phytopathology, **19** : 125 Abst.

206. Tokura, R. 1965. Studies on the downy mildew of rice plant with special reference to artificial culture and sporulation of the causal fungus, *S. macrospora* Bulletin of Kyoto Gakugei University, **27** : 93-144 (In Japanese).

207. Tuleen, D.M., Frederiksen, R. A., Vudhivanich, P. 1980. Cultural practices and the incidence of sorghum downy mildew in grain sorghum. Phytopathology, **70** :L 905-908.

208. Unnikrishnan, K. V., Govila, O.P., Siddiqui, M. R. and Dass, R. 1987. Chemical control of downy mildew in pearl millet 'BJ sonu104 and BK 560'. Seed Research, **15** : 57-59.

209. Vander Plank, J. E. 1963. 'Plant Diseases' : Epidemics and Control. Academic Press, New York and London, 349 p.

210. Venugopal, M. N. and Safeeulla, K. M. 1978. Chemical control of downy mildews of pearl millet, sorghum and maize. Indian Journal of Agricultural Sciences, **48** : 537-539.

211. Weltzein, R. E. and King, S. B. 1995. Recurrent selection for downy mildew resistance in pearl millet. Plant Breeding, **114** : 308-312.

212. Werder, J. and Ball, S. L. 1992. Variability of pearl millet downy mildew *(S. graminicola)* and its implications for breeding for stable and durable resistance. Tropical Pest Management, **38** : 43-47.

213. Weston, W. H. Jr. 1923. Production and dispersal of conidia in the Philippines *Sclerospora* of maize. Journal of Agricultural Research, **23** : 239-277.

214. Weston, W. H. and Weber, G. F. 1928. Downy mildew *[S. graminicola* (Sacc.) Schroet.] on the everglade millet *Chatochloa magne* (Griseb) Scribn. Phytopathology, **16** : 17.

215. Williams, R. J. 1984. Downy mildew of tropical cereals. In : Advances in Plant Pathology. Vol. 2. Eds. D. S. Ingram and P. H. Williams. Academic Press pp. 1-103.

216. Williams, R. J. and Singh, S. D. 1979. West African pearl millet lines show stable resistance to *S. graminicola.* Newsletter, International Working Group on Graminaceous Downy Mildews, **1** : 3.

217. Williams, R. J. and Singh, S. D. 1981. Control of pearl millet downy mildew by seed treatment with metalaxyl. Annals of Applied Biology, **97** : 263-268.

218. Williams, R. J.M., Dange, S. R. S., Mughogho, L. K. and Rao, K. N. 1982. Identification of QL-3 sorghum, a source of resistance to *Peronosclerospon sorghi* Plant Disease, **66** : 807-809.

219. Xu, G. W., Magill, C. W., Schertz, K. F. and Hart, G. E. 1994. An RFLP map of *Sorghum biocolor* (L.) Moench. Theoretical and Applied Genetics, 8.

220. Yao, C. L., Frederksen, R. A. and Magill, C. W. 1990. Seed transmission of sorghum downy mildew : Detection by DNA hybridisation,. Seed Science and Technology, **18** : 201-207.

221. Yao, C. L., Magill, C. W., Frederksen, R. A., Bonde, M. R., Wang, Y. and Wu, P.S. 1991. Detection and identification of *P. sacchari* in maize by DNA hybridization.

222. Yeh. Y. and Frederiksen, R. A. 1980. Sorghum downy mildew : Biology systemic infection by conidia and a resistant response in sorghum. Pathology, **70** : 372-376.

Fig. 1 A. Pearl millet treated showing the green ear symptom caused by *Sclerospora graminicola*. Healthy earheads are also seen is the photograph.

B. Maize plant infected with sorghum downy mildew pathogen *Peronosclerospora sorghi*, showing malpremed tapped.

C. Vertual section of injected peast millet leas showing accumulation of fungal cells (arrow) in the substomantal Chamber.

D. Pearl millet not colonized by the mycelium of *Sclerospora* showing compatible with propusely developed haciptoria.

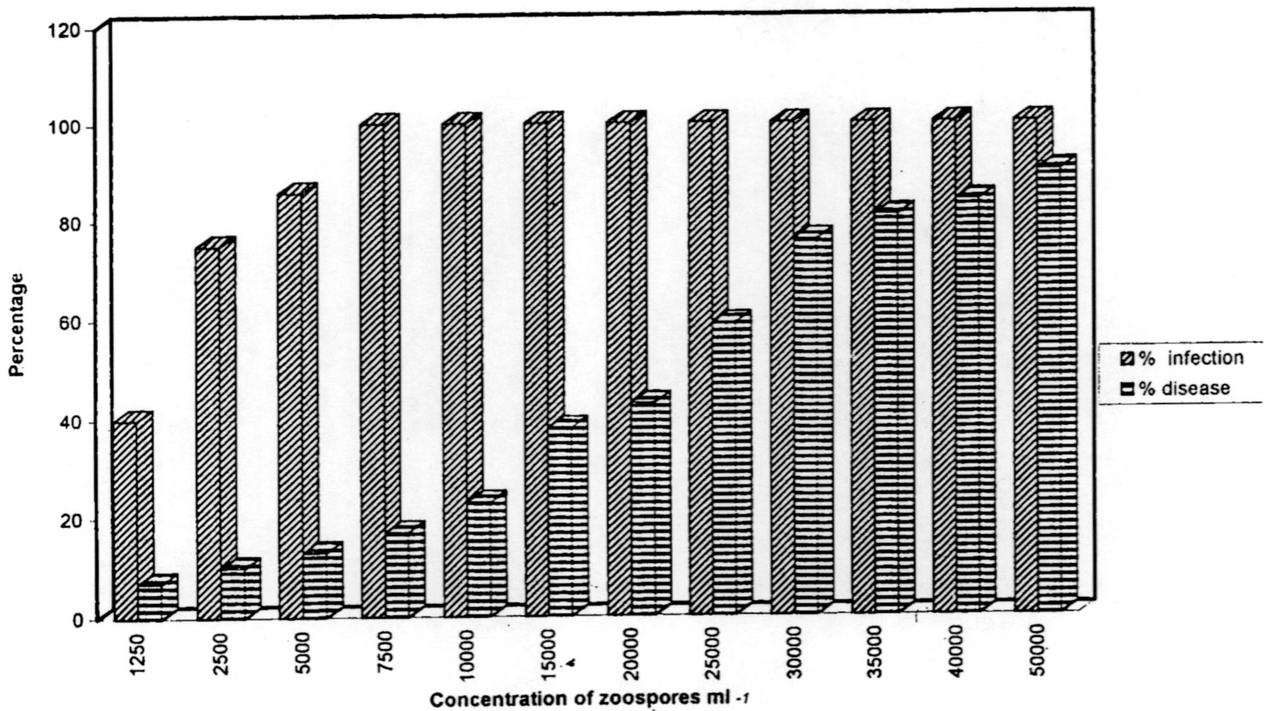

Fig. 2 : Percentage of infection and downy mildew disease incidence in HB3 cultivar of pearl millet treated to different concentrations of *S. graminicola*

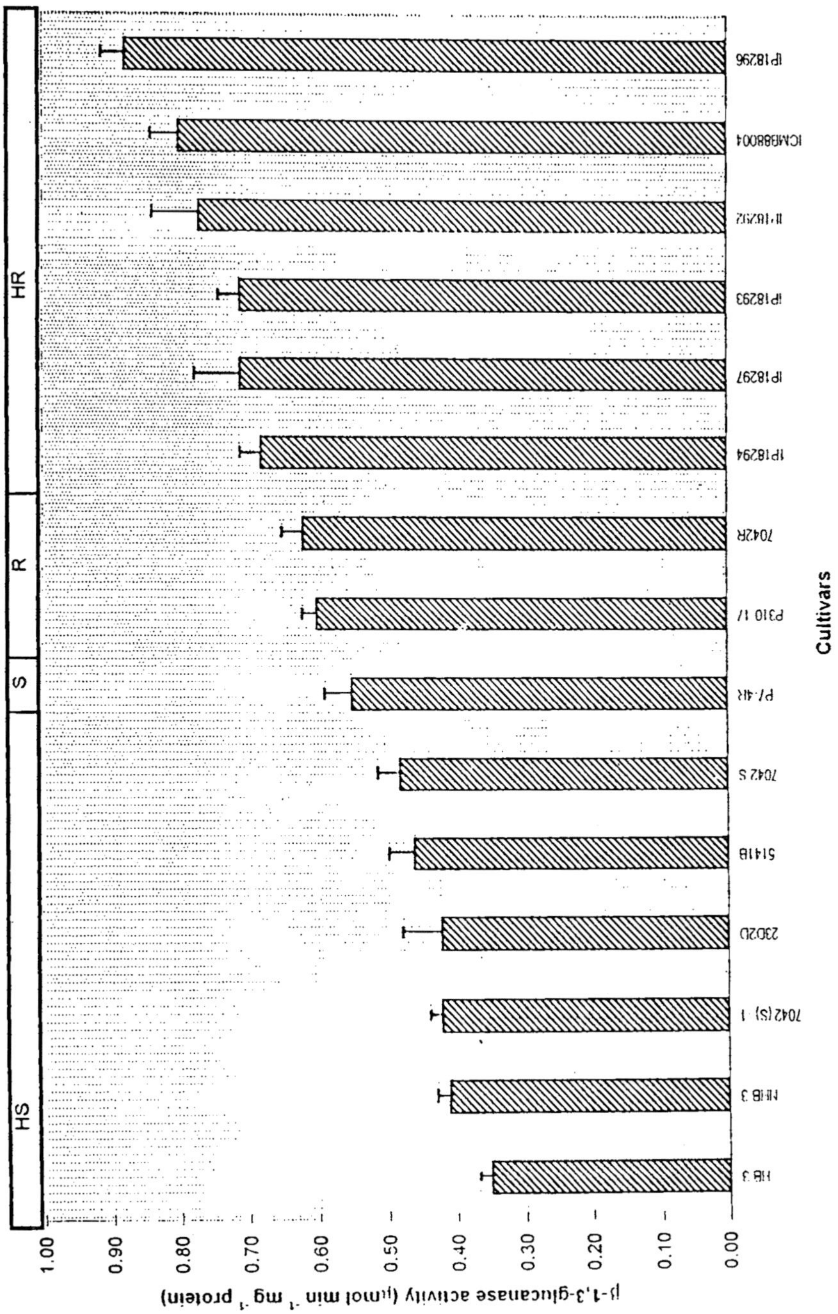

Fig. 3 : β-1, 3-glucanase activity in seeds of different cultivars of pearl millet. Bars indicate ± SE

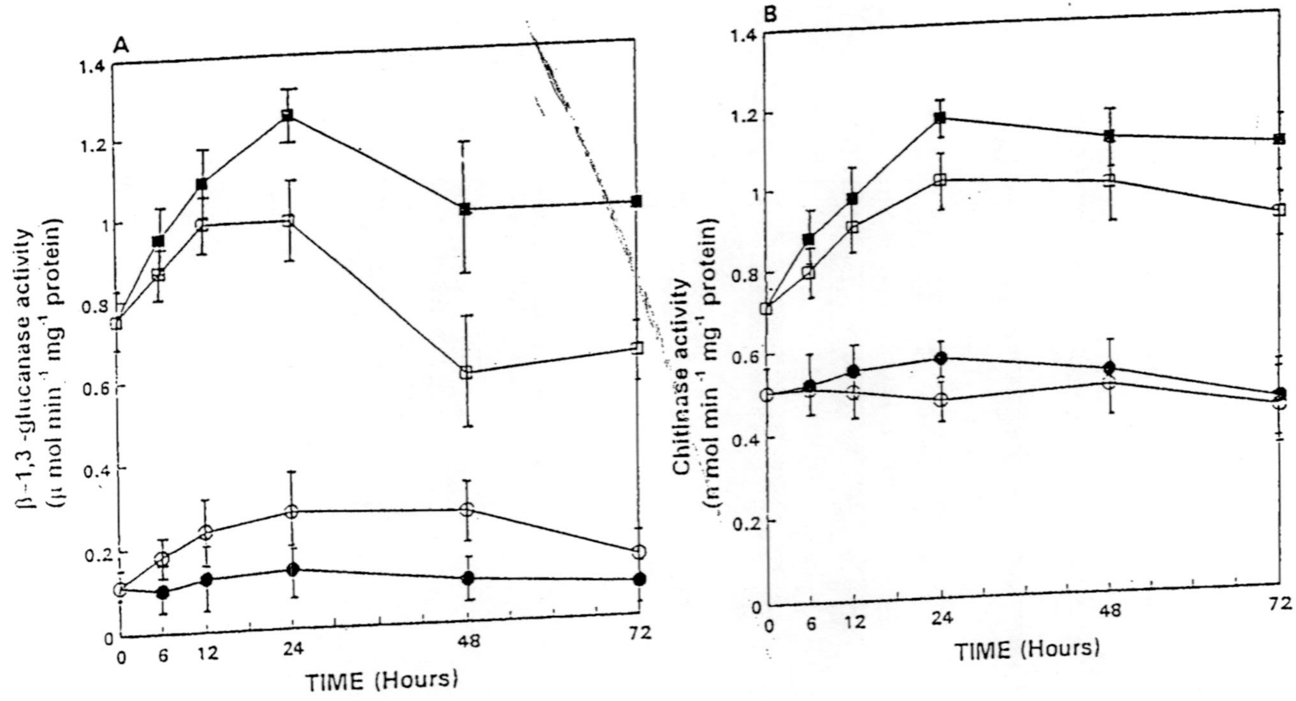

Fig. 4 : Changes in β-1, 3-glucanase (A) and chitinase (B) activity in seedlings of pearl millet cultivars resistant and susceptible to downy mildew as a response to inoculation with *Sclerospora graminicola.*

Resistant inoculated =   **Resistant control =**

Susceptible inoculated =   **Susceptible control =**

Bars indicaate ± SE

INDEX

A

B

INDEX

C

F

I

K

L

M

R